CAMBRIDGE

Brighter Thinking

A Level Further Mathematics for OCR A

Pure Core 2 Student Book (Year 2)

Vesna Kadelburg, Ben Woolley, Stephen Ward, Paul Fannon

CAMBRIDGE
UNIVERSITY PRESS

University Printing House, Cambridge CB2 8BS, United Kingdom

One Liberty Plaza, 20th Floor, New York, NY 10006, USA

477 Williamstown Road, Port Melbourne, VIC 3207, Australia

314–321, 3rd Floor, Plot 3, Splendor Forum, Jasola District Centre, New Delhi – 110025, India

103 Penang Road, #05-06/07, Visioncrest Commercial, Singapore 238467

Cambridge University Press is part of the University of Cambridge.

It furthers the University's mission by disseminating knowledge in the pursuit of education, learning and research at the highest international levels of excellence.

www.cambridge.org
Information on this title: www.cambridge.org/9781316644393 (Paperback)
www.cambridge.org/9781316644249 (Paperback with Cambridge Elevate edition)

First published 2018

20 19 18 17 16 15 14 13 12 11 10 9 8 7 6 5 4 3

Printed in Italy by Rotolito S.p.A.

A catalogue record for this publication is available from the British Library

ISBN 978-1-316-64439-3 Paperback
ISBN 978-1-316-64424-9 Paperback with Cambridge Elevate edition

Additional resources for this publication at www.cambridge.org/education

Contents

Introduction

You have probably been told that mathematics is very useful, yet it can often seem like a lot of techniques that just have to be learnt to answer examination questions. You are now getting to the point where you will start to see where some of these techniques can be applied in solving real problems. However, as well as seeing how maths can be useful, we hope that anyone working through this book will realise that it can also be incredibly frustrating, surprising and ultimately beautiful.

The book is woven around three key themes from the new curriculum:

Proof

Maths is valued because it trains you to think logically and communicate precisely. At a high level, maths is far less concerned about answers and more about the clear communication of ideas. It is not about being neat – although that might help! It is about creating a coherent argument that other people can easily follow but find difficult to refute. Have you ever tried looking at your own work? If you cannot follow it yourself it is unlikely anybody else will be able to understand it. In maths we communicate using a variety of means – feel free to use combinations of diagrams, words and algebra to aid your argument. And once you have attempted a proof, try presenting it to your peers. Look critically (but positively) at some other people's attempts. It is only through having your own attempts evaluated and trying to find flaws in other proofs that you will develop sophisticated mathematical thinking. This is why we have included common errors in our 'work it out' boxes – just in case your friends don't make any mistakes!

Problem solving

Maths is valued because it trains you to look at situations in unusual, creative ways, to persevere and to evaluate solutions along the way. We have been heavily influenced by a great mathematician and maths educator, George Polya, who believed that students were not just born with problem solving skills – these skills were developed by seeing problems being solved and reflecting on the solutions before trying similar problems. You may not realise it but good mathematicians spend most of their time being stuck. You need to spend some time on problems you can't do, trying out different possibilities. If after a while you have not cracked it then look at the solution and try a similar problem. Don't be disheartened if you cannot get it immediately – in fact, the longer you spend puzzling over a problem the more you will learn from the solution. You may, for example, never need to integrate a rational function in future, but we firmly believe that the problem solving skills you will develop by trying it can be applied to many other situations.

Modelling

Maths is valued because it helps us solve real-world problems. However, maths describes ideal situations and the real world is messy! Modelling is about deciding on the important features needed to describe the essence of a situation and turning that into a mathematical form, then using it to make predictions, compare to reality and possibly improve the model. In many situations the technical maths is actually the easy part – especially with modern technology. Deciding which features of reality to include or ignore and anticipating the consequences of these decisions is the hard part. Yet some fairly drastic assumptions – such as pretending a car is a single point or that people's votes are independent – can result in models that are surprisingly accurate.

More than anything else, this book is about making links. Links between the different chapters, the topics covered and the themes just discussed, links to other subjects and links to the real world. We hope that you will grow to see maths as one great complex but beautiful web of interlinking ideas.

Maths is about so much more than examinations, but we hope that if you take on board these ideas (and do plenty of practice!) you will find maths examinations a much more approachable and possibly even enjoyable experience. However, always remember that the results of what you write down in a few hours by yourself in silence under exam conditions is not the only measure you should consider when judging your mathematical ability – it is only one variable in a much more complicated mathematical model!

How to use this book

Throughout this book you will notice particular features that are designed to aid your learning. This section provides a brief overview of these features.

In this chapter you will learn how to:
- use de Moivre's theorem to derive trigonometric identities
- find sums of some trigonometric series.

Before you start…

Chapter 2, Section 1	You should be able to use de Moivre's theorem to raise a complex number to a power.	1	Find $\left(2\left(\cos\frac{\pi}{7}+i\sin\frac{\pi}{7}\right)\right)^{5}$ in modulus-argument form.
Chapter 2, Section 2	You should be able to use the exponential form of a complex number.	2	a Write $4e^{\frac{i\pi}{3}}$ in exact Cartesian form. b Write down the complex conjugate of $2+e^{3i}$.

Learning objectives
A short summary of the content that you will learn in each chapter.

Before you start
Points you should know from your previous learning and questions to check that you're ready to start the chapter.

WORKED EXAMPLE
The left-hand side shows you how to set out your working. The right-hand side explains the more difficult steps and helps you understand why a particular method was chosen.

Key point
A summary of the most important methods, facts and formulae.

PROOF
Step-by-step walkthroughs of standard proofs and methods of proof.

Explore
Ideas for activities and investigations to extend your understanding of the topic.

WORK IT OUT
Can you identify the correct solution and find the mistakes in the two incorrect solutions?

Tip
Useful guidance, including on ways of calculating or checking and use of technology.

Each chapter ends with a **Checklist of learning and understanding** and a **Mixed practice** exercise, which includes **past paper questions** marked with the icon.

In between chapters, you will find extra sections that bring together topics in a more synoptic way.

Focus on …
Unique sections relating to the preceding chapters that develop your skills in proof, problem solving and modelling.

CROSS-TOPIC REVIEW EXERCISE
Questions covering topics from across the preceding chapters, testing your ability to apply what you have learnt.

You will find **practice questions** towards the end of the book, as well as a **glossary** of key terms (picked out in colour within the chapters), and **answers** to all questions. Full **worked solutions** can be found on the Cambridge Elevate digital platform, along with a **digital version** of this Student Book.

Maths is all about making links, which is why throughout this book you will find signposts emphasising connections between different topics, applications and suggestions for further research.

⏮ Rewind

Reminders of where to find useful information from earlier in your study.

📷 Focus on ...

Links to problem solving, modelling or proof exercises that relate to the topic currently being studied.

⏭ Fast forward

Links to topics that you may cover in greater detail later in your study.

ⓘ Did you know?

Interesting or historical information and links with other subjects to improve your awareness about how mathematics contributes to society.

Colour-coding of exercises

The questions in the exercises are designed to provide careful progression, ranging from basic fluency to practice questions. They are uniquely colour-coded, as shown here.

2 For each equation from question 1, write the roots in exact Cartesian form.

5 Let $z = 2e^{\frac{i\pi}{12}}$ and $w = 4e^{\frac{i\pi}{3}}$. Show that $z^2 + w = 2(1+i)(1+\sqrt{3})$.

7 Solve the equation $z^3 - \sqrt{2}(4-4i) = 0$, giving your answers in Cartesian form.

8 Multiply out and simplify $(a+b\omega)(a-b\omega^2)$, where $\omega = e^{\frac{i\pi}{3}}$.

14 In the derivation of $\cosh^{-1} x$ you found that two possible expressions were $\ln(x+\sqrt{x^2-1})$ and $\ln(x-\sqrt{x^2-1})$. Show that their sum is zero and hence explain why the expression chosen in Proof 3 is non-negative.

22 Point A represents the complex number $3+i$ on an Argand diagram. Point A is rotated through $\frac{\pi}{3}$ radians anticlockwise about the origin to point B. Point B is then translated by $\begin{pmatrix} -2 \\ 1 \end{pmatrix}$ to obtain point C.

Black – drill questions. Some of these come in several parts, each with subparts i and ii. You only need attempt subpart i at first; subpart ii is essentially the same question, which you can use for further practice if you got part i wrong, for homework, or when you revisit the exercise during revision.

Green – practice questions at a basic level.

Blue – practice questions at an intermediate level.

Red – practice questions at an advanced level.

Yellow – designed to encourage reflection and discussion.

Purple – challenging questions that apply the concept of the current chapter across other areas of maths.

1 Series and induction

In this chapter you will learn how to:

- use the principle of mathematical induction to prove results about sequences, series and differentiation
- use given results for the sums of integers, squares and cubes to find expressions for sums of other series
- use a technique called the method of differences to find an expression for the sum of n terms of a series
- use the expression for the sum of the first n terms to determine whether an infinite series converges and find its limit.

Before you start…

Pure Core Student Book 1, Chapter 6	You should be able to use mathematical induction to prove results about matrices, divisibility and inequalities.	1 Prove that $\begin{pmatrix} 1 & 3 \\ 0 & 1 \end{pmatrix}^n = \begin{pmatrix} 1 & 3n \\ 0 & 1 \end{pmatrix}$.
GCSE	You should be able to use the nth term formula to generate terms of a sequence.	2 A sequence is defined by $u_n = n^2 + 3n - 1$. Find the first three terms.
GCSE	You should be able to simplify expressions by factorising.	3 Simplify $n(n+1)(2n+3) + n(n+1)(n-3)$.
A Level Mathematics Student Book 2, Chapter 4	You should be able to use sigma notation to write a series.	4 Find $\displaystyle\sum_{k=1}^{5} 2^k$.
A Level Mathematics Student Book 2, Chapter 10	You should know how to differentiate using the product rule and chain rule.	5 Given that $y = (2x+1)e^{3x}$, find $\dfrac{dy}{dx}$.
A Level Mathematics Student Book 2, Chapter 5	You should know how to write an expression in partial fractions.	6 Write $\dfrac{1}{r(r+1)}$ in partial fractions.

Introduction

In Pure Core Student Book 1, Chapter 6, you learnt about the method of proof by induction, which you can use to prove that observed patterns continue forever. The sorts of patterns you looked at included powers of matrices (for example, prove that $\begin{pmatrix} 1 & 0 \\ 1 & 1 \end{pmatrix}^n = \begin{pmatrix} 1 & 0 \\ n & 1 \end{pmatrix}$

for all $n \in \mathbb{N}$), divisibility (for example, prove that $7^n - 3^n$ is divisible by 4 for all $n \in \mathbb{N}$) and inequalities (for example, prove that $2^n > 2n$ for $n \geqslant 3$).

In this chapter you will revisit these ideas, including examples where you need to conjecture (guess) the pattern first, and then see how to extend them to other contexts, such as sums of series and differentiation.

Finding expressions for sums of series is one of the most difficult problems in mathematics as there is no general method that always works. For example, you know how to find the sum of the first n terms of a geometric series: $5^1 + 5^2 + \ldots + 5^n = \dfrac{5(5^n - 1)}{4}$. But what about a series such as $1^5 + 2^5 + \ldots + n^5$?

The method of mathematical induction is useful for proving that a conjectured formula for the sum of a series is correct, but it offers no help in finding what the formula might be. Sometimes you can guess the formula by looking at some examples, but most of the time the general expression is far from obvious. In this chapter you will meet the method of differences, which allows you to find the formula in some cases. You will also learn how to derive formulae for sums of more complicated series by combining results you have already derived.

▶▶) Fast forward

In Chapter 2 you will use induction to prove de Moivre's theorem, a result about powers of complex numbers.

▶▶) Fast forward

In Chapter 8 you will learn about another type of series called the Maclaurin series.

Section 1: Review of proof by induction

You can use induction to prove statements about a sequence or a pattern, where the statement holds for every natural number n. The proof involves two steps:

1   Prove that the statement is true for some starting value of n (usually, but not always, $n = 1$).
2   Assuming that the statement is true for some k, prove that it is also true for $k + 1$.

Then the principle of mathematical induction states the statement is true for all values of n.

Sometimes you need to conjecture the pattern for yourself before using induction to prove it.

◀◀) Rewind

You will need to use the product rule for differentiation. This was covered in A Level Mathematics Student Book 2, Chapter 10.

WORKED EXAMPLE 1.1

Let $y = x e^x$.

a   Find $\dfrac{\mathrm{d}y}{\mathrm{d}x}, \dfrac{\mathrm{d}^2 y}{\mathrm{d}x^2}$ and $\dfrac{\mathrm{d}^3 y}{\mathrm{d}x^3}$.

b   Conjecture an expression for $\dfrac{\mathrm{d}^n y}{\mathrm{d}x^n}$ and prove it by induction.

Continues on next page ...

a $\dfrac{dy}{dx} = e^x + xe^x = (1+x)e^x$

> Use the product rule to differentiate.

$\dfrac{d^2 y}{dx^2} = e^x + (1+x)e^x = (2+x)e^x$

> The factorised form makes it easier to spot the pattern.

$\dfrac{d^3 y}{dx^3} = e^x + (2+x)e^x = (3+x)e^x$

b Conjecture:

$\dfrac{d^n y}{dx^n} = (n+x)e^x$

Proof:

When $n = 1$:

> Start by showing that the statement is true when $n = 1$.

$\dfrac{dy}{dx} = (1+x)e^x$

So the statement is true for $n = 1$.

Assume it is true for $n = k$:

$\dfrac{d^k y}{dx^k} = (x+k)e^x$

> Write down the statement with $n = k$.

When $n = k+1$,

> Think about what you are trying to prove. Remember that you cannot use this result!
>
> You are working towards: $\dfrac{d^{k+1} y}{dx^{k+1}} = (x+k+1)e^x$

$\dfrac{d^{k+1} y}{dx^{k+1}} = \dfrac{d}{dx}\left(\dfrac{d^k y}{dx^k} \right)$

> Relate $\dfrac{d^{k+1} y}{dx^{k+1}}$ to $\dfrac{d^k y}{dx^k}$.

$= \dfrac{d}{dx}\left((x+k)e^x \right)$

> Use the result you have assumed for $n = k$.

$= e^x + (x+k)e^x$

> Differentiate using the product rule.

$= e^k (1+x+k)$

$= e^k (x+(k+1))$

> This the the required result for $n = k+1$.

Hence the result is also true for $n = k+1$.

The result is true for $n = 1$, and if true for $n = k$ it is also true for $n = k+1$. Therefore, the result is true for all $n \in \mathbb{Z}^+$ by the principle of mathematical induction.

> Remember to write the conclusion.

Sequences are often given by a term-to-term rule, but you might want to know a formula for the nth term. You might be able to guess the formula by looking at the numbers and then you can use induction to prove that it works for all n.

For example, the term-to-term rule $u_{n+1} = 3u_n + 2, u_1 = 2$ describes a sequence whose first four terms are 2, 8, 26, 80. You might notice that these are all one less than a power of 3, so the formula for the nth term could be $u_n = 3^n - 1$. You can prove by induction that this formula indeed works for all n.

WORKED EXAMPLE 1.2

A sequence is given by $u_1 = 2$ and $u_{n+1} = 3u_n + 2$ for $n \geqslant 1$. Prove that the nth term of the sequence is $u_n = 3^n - 1$.

When $n = 1$:

$u_1 = 2 = 3^1 - 1$ Show that the result is true for $n = 1$.
So the formula works when $n = 1$.

Assume that the formula works when $n = k$:

$u_k = 3^k - 1$ Assume that the formula works for some k.

When $n = k + 1$,   Think about what you are trying to prove.
$u_{k+1} = 3u_k + 2$ You are working towards: $u_{k+1} = 3^{k+1} - 1$

$\quad\quad = 3(3^k - 1) + 2$

$\quad\quad = 3^{k+1} - 1$ Use the result you assumed for $n = k$.
So, the formula also works when $n = k + 1$.

The formula works when $n = 1$, and if it works for some $n = k$ then it also works for $n = k + 1$.   Remember to write the conclusion.
Hence, by the principle of mathematical induction, the formula works for all $n \in \mathbb{N}$.

Sometimes each term in the sequence depends on more than one previous term. For example, the term-to-term rule

$$u_{n+2} = 5u_{n+1} - 6u_n \text{ with } u_1 = 5 \text{ and } u_2 = 13$$

produces this sequence:

$$u_1 = 5$$
$$u_2 = 13$$
$$u_3 = 5 \times 13 - 6 \times 5 = 35$$
$$u_4 = 5 \times 35 - 6 \times 13 = 97$$

and so on. You can still use proof by induction, but you need to show that the formula works for two starting values of n.

WORKED EXAMPLE 1.3

A sequence is given by the recurrence relation $u_1 = 5$ and $u_2 = 13$, $u_{n+2} = 5u_{n+1} - 6u_n$ for $n \geqslant 2$. Prove that the formula for the nth term of the sequence is $u_n = 2^n + 3^n$.

When $n = 1$:	Check that the formula works for $n = 1$ and $n = 2$.
$\qquad$ RHS $= 2^1 + 3^1$	
$\qquad\quad = 5$	
$\qquad\quad = u_1$	
So, the formula works for $n = 1$.	
When $n = 2$:	
$\qquad$ RHS $= 2^2 + 3^2$	
$\qquad\quad = 13$	
$\qquad\quad = u_2$	
So, the formula works for $n = 2$.	Assume that the formula works for $n = k$ **and** $n = k+1$, and prove that it works for $n = k+2$.
Assume the formula works for $n = k$ and $n = k + 1$:	
$\quad u_k = 2^k + 3^k$	Think about the formula with $n = k$ and $n = k+1$.
$\quad u_{k+1} = 2^{k+1} + 3^{k+1}$	
When $n = k + 2$,	Think about what you are trying to prove. You are working towards: $u_{k+2} = 2^{k+2} + 3^{k+2}$
$\quad u_{k+2} = 5u_{k+1} - 6u_k$	Express u_{k+2} in terms of u_k and u_{k+1}.
$\qquad = 5\left(2^{k+1} + 3^{k+1}\right) - 6\left(2^k + 3^k\right)$	Use the results for $n = k$ and $n = k+1$.
$\qquad = 5 \times 2^{k+1} + 5 \times 3^{k+1} - 6 \times 2^k - 6 \times 3^k$	
$\qquad = \left(5 \times 2 \times 2^k - 6 \times 2^k\right) + \left(5 \times 3 \times 3^k - 6 \times 3^k\right)$	Look at what you are working towards; group the powers of 2 and the powers of 3.
$\qquad = 4 \times 2^k + 9 \times 3^k$	
$\qquad = 2^2 \times 2^k + 3^2 \times 3^k$	
$\qquad = 2^{k+2} + 3^{k+2}$	
So, the formula also works for $n = k + 2$.	
The formula works for $n = 1$ and $n = 2$, and if it works for $n = k$ and $n = k + 1$ then it also works for $n = k + 2$. Therefore, the formula works for all $n \in \mathbb{Z}^+$ by the principle of mathematical induction.	Write a conclusion.

EXERCISE 1A

1 Given that $u_{n+1} = 5u_n - 8, u_1 = 3$, prove by induction that $u_n = 5^{n-1} + 2$.

2 A sequence has first term 1 and subsequent terms defined by $u_{n+1} = 3u_n + 1$. Prove by induction that

$$u_n = \frac{3^n - 1}{2}.$$

3 Given that $u_{n+1} = 5u_n + 4, u_1 = 4$,

   a find the first four terms of the sequence

   b conjecture a formula for the nth term and prove it by induction.

4 Let $\mathbf{A} = \begin{pmatrix} 1 & 0 \\ 1 & 1 \end{pmatrix}$.

   a Find $\mathbf{A}^2, \mathbf{A}^3$ and $\mathbf{A}^4$.

   b Conjecture an expression for $\mathbf{A}^n$ and prove it by induction.

5 Let $\mathrm{f}(n) = 5^n - 1$.

   a Find $\mathrm{f}(n)$ for $n = 1, 2, 3, 4$.

   b Which natural number do all $\mathrm{f}(n)$ seem to be multiples of?

   c Use mathematical induction to prove your conjecture from part **b**.

> **Focus on …**
>
> Powers of matrices have many interesting applications. To explore one of them see Focus on … Modelling 1.

6 A sequence is given by the term-to-term rule $u_{n+2} = 5u_{n+1} - 6u_n$ with $u_1 = 1$ and $u_2 = 5$.

Prove that the general term of the sequence is $u_n = 3^n - 2^n$.

7 Given that $u_1 = 3, u_2 = 36, u_{n+2} = 6u_{n+1} - 9u_n$, prove by induction that $u_n = (3n - 2)3^n$.

8 Let $\mathbf{A} = \begin{pmatrix} 1 & 1 \\ 1 & 1 \end{pmatrix}$.

   a Find $\mathbf{A}^2, \mathbf{A}^3$ and $\mathbf{A}^4$.   **b** Conjecture an expression for $\mathbf{A}^n$ and prove it by induction.

9 **a** Suggest which natural number is a factor of all the numbers of the form $9^n - 4^n$.

   b Prove your claim by induction.

10 Given that $y = \dfrac{1}{1-x}$, use induction to prove that $\dfrac{d^n y}{dx^n} = \dfrac{n!}{(1-x)^{n+1}}$.

11 Given that $\mathrm{f}(x) = \dfrac{1}{1-3x}$, prove by induction that $\mathrm{f}^{(n)}(x) = \dfrac{3^n n!}{(1-3x)^{n+1}}$.

12 Use mathematical induction to show that $\dfrac{d^n}{dx^n}(xe^{2x}) = (2^n x + n2^{n-1})e^{2x}$.

13 Prove by induction that $\dfrac{d^n}{dx^n}(x^2 e^x) = (x^2 + 2nx + n(n-1))e^x$ for $n \geqslant 2$.

14 Given that $y = x\sin x$, use mathematical induction to prove that $\dfrac{d^{2n} y}{dx^{2n}} = (-1)^n(x\sin x - 2n\cos x)$.

15 The Fibonacci sequence is defined by $u_1 = u_2 = 1, u_n = u_{n-1} + u_{n-2}$ for $n \geqslant 3$. Show that the nth term of the

Fibonacci sequence is given by $u_n = \dfrac{1}{\sqrt{5}}\left(\left(\dfrac{1+\sqrt{5}}{2}\right)^n + \left(\dfrac{1-\sqrt{5}}{2}\right)^n\right)$.

Explore

Leonardo Fibonacci (c. 1170 - c. 1250) was an extremely influential mathematician in the Middle Ages, largely responsible for spreading the number system you use today. He also gave his name to the famous Fibonacci sequence. The formula in question 15 shows the link between the Fibonacci sequence and the golden ratio, a quantity $\frac{1+\sqrt{5}}{2}$ which appears in many surprising places in mathematics.

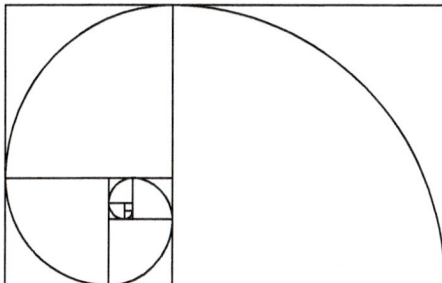

Section 2: Induction and series

A **series** is a sum of the terms of a sequence. If you add the terms up to a certain point you get a **finite series**, such as $\frac{1}{2}+\frac{1}{3}+\frac{1}{4}+\frac{1}{5}$. You can also try to form an **infinite series**, for example $\frac{1}{2}+\frac{1}{4}+\frac{1}{8}+\frac{1}{16}+....$ Some infinite series, such as the geometric series given here, have a finite sum. In this section you will only look at finite series; you will meet some infinite series in Section 4.

Rewind

You met sequences, series and sigma notation in A Level Mathematics Student Book 2, Chapter 4.

You can use **sigma notation** as a shorter way of writing a series. For example,

$$\frac{1}{2}+\frac{1}{3}+\frac{1}{4}+\frac{1}{5}=\sum_{k=2}^{5}\frac{1}{k}$$

$$\frac{1}{2}+\frac{1}{4}+\frac{1}{8}+\frac{1}{16}+...=\sum_{k=1}^{\infty}\frac{1}{2^k}$$

In the A Level Mathematics course you learnt how to find the general formula for the sum of the first n terms of an arithmetic and a geometric series:

$$\sum_{k=1}^{n}\left(a+(k-1)d\right)=\frac{n}{2}\left(2a+(n-1)d\right)$$

$$\sum_{k=1}^{n}ar^{k-1}=\frac{a\left(r^n-1\right)}{r-1}$$

You also learnt about finite and infinite binomial series, for example:

$$1+4x+6x^2+4x^3+x^4 = \sum_{k=0}^{4} {}^4C_k x^k = (1+x)^4$$

$$1-2x+3x^2-4x^3+\ldots = \sum_{k=0}^{\infty} \frac{(-2)(-3)\ldots(-2-(k-1))}{k!} x^k = (1+x)^{-2}$$

In Chapter 8 of this book you will learn about Maclaurin series, which you can use to write a function as an infinite series, for example:

$$\sum_{k=0}^{\infty} \frac{(-1)^k}{(2k+1)!} x^{2k+1} = \sin x$$

These are some examples of finite and infinite series where it is possible to find an exact expression for the sum. In general, finding an expression for the sum of the first n terms of a series can be surprisingly difficult, if not impossible. For example, it is not possible to express a formula for the sum $\sum_{k=1}^{n} \frac{1}{k^2} = \frac{1}{1^2} + \frac{1}{2^2} + \ldots + \frac{1}{n^2}$ in terms of standard functions.

In cases where you manage to guess the formula for the sum of a series, you can then try to prove it by induction. The inductive step relies on making the connection between the sum of the first k terms and the sum of the first $k + 1$ terms; this is done simply by adding the next term of the series.

> **◄◄ Rewind**
>
> Binomial series were covered in A Level Mathematics Student Book 1, Chapter 9, and A Level Mathematics Student Book 2, Chapter 6.

> **🔍 Explore**
>
> Although it is not possible to find a general expression for $\sum_{k=1}^{n} \frac{1}{k^2}$, it is possible to find its exact sum to infinity, $\sum_{k=1}^{\infty} \frac{1}{k^2}$. Find out what it is: the result may surprise you!

> **🔑 Key point 1.1**
>
> If $S_k = u_1 + u_2 + \ldots + u_k$ then
> $$S_{k+1} = S_k + u_{k+1}$$

WORKED EXAMPLE 1.4

Prove by induction that

$$\sum_{r=1}^{n} r(r+2) = \frac{n(n+1)(2n+7)}{6} \text{ for all } n \in \mathbb{Z}^+.$$

For $n = 1$:   Show that the statement is true for the starting value (in this case, $n = 1$).

    LHS = $1 \times 3 = 3$

    RHS = $\dfrac{1(1+1)(2 \times 1+7)}{6} = \dfrac{1 \times 2 \times 9}{6} = 3$

    So, the result is true for $n = 1$.

Continues on next page …

Assume that the result is true for $n = k$:

$$\sum_{r=1}^{k} r(r+2) = \frac{k(k+1)(2k+7)}{6}$$

State the assumption for $n = k$.

Let $n = k + 1$:

$$\sum_{r=1}^{k+1} r(r+2) = \sum_{r=1}^{k} r(r+2) + (k+1)(k+3)$$

Consider S_{k+1} and relate it to S_k by using $S_{k+1} = S_k + u_{k+1}$.

$$= \frac{k(k+1)(2k+7)}{6} + (k+1)(k+3)$$

Substitute in the result for $n = k$ (assumed to be true).

$$= (k+1)\left(\frac{2k^2 + 7k}{6} + \frac{6k+18}{6} \right)$$

$$= \frac{(k+1)(2k^2 + 13k + 18)}{6}$$

$$= \frac{(k+1)(k+2)(2k+9)}{6}$$

Combine this into one fraction and simplify. It is always a good idea to take out any common factors.

$$= \frac{(k+1)((k+1)+1)(2(k+1)+7)}{6}$$

Show that this is in the required form by separating out $k+1$ in each place it occurs.

So, the result is also true for $n = k + 1$.

The result is true for $n = 1$, and if it is true for $n = k$ it is also true for $n = k + 1$. Therefore, the result is true for all $n \in \mathbb{Z}^+$, by induction.

Make sure you write a conclusion.

EXERCISE 1B

1 Prove by induction that, for all $n \in \mathbb{Z}^+$:

$$\sum_{r=1}^{n} 2 \times 3^{r-1} = 3^n - 1$$

2 Prove by induction that, for all integers $n > 1$:

$$\sum_{r=1}^{n} r^2 = \frac{n(n+1)(2n+1)}{6}$$

3 Using mathematical induction prove that, for all positive integers:

$$\sum_{r=1}^{n} r^3 = \frac{n^2(n+1)^2}{4}$$

4 Prove by induction that, for all integers $n > 1$:

$$\sum_{r=1}^{n} \frac{1}{r(r+1)} = \frac{n}{n+1}$$

5 Use mathematical induction to show that, for all integers $n > 1$:

$$\sum_{r=1}^{n} r 2^r = 2\left[(n-1)2^n + 1 \right]$$

6 Prove by induction that, for all $n \in \mathbb{Z}^+$:

$$\frac{1}{1\times3}+\frac{1}{3\times5}+\frac{1}{5\times7}+\ldots+\frac{1}{(2n-1)(2n+1)}=\frac{n}{2n+1}$$

7 Using mathematical induction prove that, for all integers $n > 1$:

$$\sum_{r=1}^{n} r(r!) = (n+1)! - 1$$

8 Prove by induction that, for all positive integers:

$$1^2 - 2^2 + 3^2 - 4^2 + \ldots + (-1)^{n+1}n^2 = (-1)^{n+1}\frac{n(n+1)}{2}$$

9 Prove using mathematical induction that, for all $n \in \mathbb{Z}^+$:

$$(n+1)+(n+2)+(n+3)+\ldots+(2n)=\frac{1}{2}n(3n+1)$$

10 Prove by induction that, for all integers $n > 1$:

$$\sum_{k=1}^{n} k2^k = (n-1)2^{n+1} + 2$$

Section 3: Using standard series

You will now look at finding expressions for the sums of series such as $\sum_{k=1}^{n}(2n^3 - 5n)$ by combining some standard results.

You can use the following formulae without proof (unless the question explicitly asks you to prove them).

> ### 🔑 Key point 1.2
>
> Formulae for the sums of integers, squares and cubes:
>
> - $\displaystyle\sum_{r=1}^{n} r = \frac{1}{2}n(n+1)$
>
> - $\displaystyle\sum_{r=1}^{n} r^2 = \frac{1}{6}n(n+1)(2n+1)$
>
> - $\displaystyle\sum_{r=1}^{n} r^3 = \frac{1}{4}n^2(n+1)^2$
>
> **The second and third formulae will be given in your formula book.**

> ### ⏮ Rewind
>
> You proved the second and third formulae in Exercise 1B, Questions 2 and 3.

The first formula in Key Point 1.2 is just a special case of an arithmetic series.

Explore

The formula for the sum of the first n integers, $\frac{1}{2}n(n+1)$, is the formula for the nth term in the sequence of triangular numbers: 1, 3, 6, 10, 15, 21....

Explore pictorial representations of the other two formulae from Key point 1.2.

Before you use these results, notice how you can split up sums and take out constants. For example:

$$\sum_{r=1}^{n}(3r+2)=(3\times1+2)+(3\times2+2)+(3\times3+2)+(3\times4+2)+\cdots+(3n+2)$$

$$=3(1+2+3+4+\cdots+n)+\underbrace{2+2+2+2+\cdots+2}_{n\text{ times}}$$

$$=3\sum_{r=1}^{n}r+\sum_{r=1}^{n}2$$

where $\displaystyle\sum_{r=1}^{n}2=2n$.

🔑 **Key point 1.3**

You can manipulate series in several ways.

- $\sum(u_r+v_r)=\sum u_r+\sum v_r$
- $\sum cu_r=c\sum u_r$
- $\displaystyle\sum_{r=1}^{n}c=nc$

where c is a constant.

💡 **Tip**

Remember that a constant, c, summed n times is nc and not just c. For example, $\displaystyle\sum_{r=1}^{n}2=2n$ and not 2.

WORKED EXAMPLE 1.5

a Use the formula for $\sum_{r=1}^{n} r$ to show that $\sum_{r=1}^{n}(4r+3)=n(2n+5)$.

b Hence find $\sum_{r=8}^{20}(4r+3)$.

a $\sum_{r=1}^{n}(4r+3)=\sum_{r=1}^{n}4r+\sum_{r=1}^{n}3$

> You need to rearrange the expression into a form to which you can apply the standard formulae. Start by splitting up the sum.

$=4\sum_{r=1}^{n}r+\sum_{r=1}^{n}3$

> Then take 4 out of the first sum as a factor.

$=4\times\frac{1}{2}n(n+1)+3n$

> $\sum_{r=1}^{n}r=\frac{1}{2}n(n+1)$ and $\sum_{r=1}^{n}3=3n$

$=n[2(n+1)+3]$
$=n(2n+5)$

> Notice that it's always a good idea to factorise first. In this case only n factorises, but in more complicated examples this will avoid having to expand and then factorise a higher order polynomial later.

b $\sum_{r=8}^{20}(4r+3)=\sum_{r=1}^{20}(4r+3)-\sum_{r=1}^{7}(4r+3)$

> You can only use the formula in part **a** if the sum starts from $r=1$. Therefore, work out the sum of the first 20 terms and subtract the sum of the first 7 terms.

$=20(2\times20+5)-7(2\times7+5)$
$=900-133$
$=767$

> Now use the formula $n(2n+5)$ with $n=20$ and $n=7$.

WORKED EXAMPLE 1.6

a Use the formulae for $\sum_{r=1}^{n}r$, $\sum_{r=1}^{n}r^2$ and $\sum_{r=1}^{n}r^3$ to show that $\sum_{r=1}^{n}r(2r-5)(r+1)=\frac{1}{2}n(n+1)(n+2)(n-3)$.

b Hence find an expression for $\sum_{r=1}^{2n}r(2r-5)(r+1)$, simplifying your answer fully.

a $\sum_{r=1}^{n}r(2r-5)(r+1)=\sum_{r=1}^{n}r(2r^2-3r-5)$

> Expand the brackets.

$=\sum_{r=1}^{n}(2r^3-3r^2-5r)$

$=\sum_{r=1}^{n}2r^3-\sum_{r=1}^{n}3r^2-\sum_{r=1}^{n}5r$

> Split up the series into separate sums.

Continues on next page ...

$$= 2\sum_{r=1}^{n} r^3 - 3\sum_{r=1}^{n} r^2 - 5\sum_{r=1}^{n} r$$

Take out constants.

$$= 2\left[\frac{1}{4}n^2(n+1)^2\right] - 3\left[\frac{1}{6}n(n+1)(2n+1)\right] - 5\left[\frac{1}{2}n(n+1)\right]$$

Substitute in the standard formulae.

$$= \frac{1}{2}n^2(n+1)^2 - \frac{1}{2}n(n+1)(2n+1) - 5\left[\frac{1}{2}n(n+1)\right]$$

Simplify the first two terms.

$$= \frac{1}{2}n(n+1)[n(n+1) - (2n+1) - 5]$$

Now factorise as many terms as possible. Note that this is much easier than expanding everything first.

$$= \frac{1}{2}n(n+1)[n^2 + n - 2n - 1 - 5]$$

$$= \frac{1}{2}n(n+1)[n^2 - n - 6]$$

$$= \frac{1}{2}n(n+1)(n+2)(n-3)$$

b $$\sum_{r=1}^{2n} r(2r-5)(r+1) = \frac{1}{2}2n(2n+1)(2n+2)(2n-3)$$

Substitute $2n$ for n in the formula found in part **a**.

$$= n(2n+1)(2n+2)(2n-3)$$
$$= 2n(2n+1)(n+1)(2n-3)$$

Simplify and factorise, taking out a 2 from the second bracket.

WORK IT OUT 1.1

Given that $\sum_{r=1}^{n}(r^2 - 2r) = \frac{n}{6}(n+1)(2n-5)$, find an expression for $\sum_{r=n+1}^{2n}(r^2 - 2r)$.

Which is the correct solution? Identify the errors made in the incorrect solutions.

Solution 1	$\sum_{r=n+1}^{2n}(r^2 - 2r) = \frac{2n}{6}(2n+1)(2(2n)-5)$
	$= \frac{n}{3}(2n+1)(4n-5)$
Solution 2	$\sum_{r=n+1}^{2n}(r^2 - 2r) = \sum_{r=1}^{2n}(r^2 - 2r) - \sum_{r=1}^{n}(r^2 - 2r)$
	$= \frac{2n}{6}(2n+1)(2(2n)-5) - \frac{n}{6}(n+1)(2n-5)$
	$= \frac{n}{6}[2(2n+1)(2(2n)-5) - (n+1)(2n-5)]$
	$= \frac{n}{6}[2(8n^2 - 6n - 5) - (2n^2 - 3n - 5)]$
	$= \frac{n}{6}(14n^2 - 9n - 5)$
	$= \frac{n}{6}(14n+5)(n-1)$

Continues on next page ...

Solution 3

$$\sum_{r=n+1}^{2n} (r^2 - 2r) = \sum_{r=1}^{2n}(r^2-2r) - \sum_{r=1}^{n+1}(r^2-2r)$$

$$= \frac{2n}{6}(2n+1)(2(2n)-5) - \frac{n}{6}((n+1)+1)(2(n+1)-5)$$

$$= \frac{n}{6}[2(2n+1)(2(2n)-5) - (n+2)(2n-3)]$$

$$= \frac{n}{6}[2(8n^2-6n-5) - (2n^2+n-6)]$$

$$= \frac{n}{6}(14n^2-13n-4)$$

EXERCISE 1C

In this exercise, you can assume the formulae for $\sum_{r=1}^{n} r$, $\sum_{r=1}^{n} r^2$ and $\sum_{r=1}^{n} r^3$.

1 Evaluate each expression.

    a **i** $\sum_{r=1}^{30} r^2$ **ii** $\sum_{r=1}^{20} r^3$ **b** **i** $\sum_{r=32}^{50} r^3$ **ii** $\sum_{r=25}^{100} r$

2 Find a formula for each series, giving your answer in its simplest form.

    a **i** $\sum_{r=1}^{4n} r$ **ii** $\sum_{r=1}^{3n} r^2$ **b** **i** $\sum_{r=1}^{n-1} r^2$ **ii** $\sum_{r=1}^{n+1} r^3$

3 Show that $\sum_{r=1}^{n} r(3r-5) = n(n+1)(n-2)$.

4 Show that $\sum_{r=1}^{n} 3r(r-1) = n(n^2-1)$.

5 **a** Find an expression for $\sum_{r=1}^{n} (6r+7)$.

    b Hence find the least value of n such that $\sum_{r=1}^{n} (6r+7) > 2400$.

6 **a** Show that $\sum_{r=1}^{n} (r+1)(r+5) = \frac{n}{6}(n+7)(2n+7)$.

    b Hence evaluate $\sum_{r=16}^{40} (r+1)(r+5)$.

7 Show that $\sum_{r=1}^{n} r^2(r-1) = \frac{n}{12}(n^2-1)(kn+2)$, where k is an integer to be found.

8 **a** Show that $\sum_{r=1}^{n} r(r^2-3) = \frac{n}{4}(n+1)(n-2)(n+3)$.

    b Hence find a formula for $\sum_{r=1}^{2n} r(r^2-3)$, fully simplifying your answer.

9 **a** Show that $\displaystyle\sum_{r=1}^{n} r(r+1) = \frac{n}{3}(n+1)(n+2)$.

b Hence find, in the form $\ln 3^k$, the exact value of $2\ln 3 + 3\ln 3^2 + 4\ln 3^3 + \cdots + 20\ln 3^{19}$.

10 Show that the sum of the squares of the first n odd numbers is given by $S = \dfrac{n}{3}(an^2 - 1)$, where a is an integer to be found.

Section 4: The method of differences

Whenever you are investigating a series, start by writing out a few terms to see if any patterns develop.

For example, for the series $\displaystyle\sum_{r=1}^{n} u_r = \sum_{r=1}^{n} [r(r+1) - r(r-1)]$:

$$u_1 = 1(2) - 1(0)$$
$$u_2 = 2(3) - 2(1)$$
$$u_3 = 3(4) - 3(2)$$
$$u_4 = 4(5) - 4(3)$$
$$\vdots$$

You can see that each term shares a common element with the next; in the first term, this element is positive and in the next it is negative. Therefore, when you complete the sum, these common elements will cancel out.

This cancellation continues right through to the nth term.

$$1(2) - 1(0)$$
$$+2(3) - 2(1)$$
$$+3(4) - 3(2)$$
$$+4(5) - 4(3)$$
$$\vdots$$
$$+(n-1)n - (n-1)(n-2)$$
$$+n(n+1) - n(n-1)$$

$$\therefore \sum_{r=1}^{n} r(r+1) - r(r-1) = n(n+1) - 1(0) = n(n+1)$$

In fact, because

$$r(r+1) - r(r-1) = r^2 + r - r^2 + r = 2r$$

you have just shown that

$$\sum_{r=1}^{n} 2r = n(n+1) \Rightarrow \sum_{r=1}^{n} r = \frac{n}{2}(n+1)$$

which is the result for the sum of the first n integers that you used in Section 3.

This process for finding a formula for the sum of the first n terms of a sequence is called the **method of differences**.

Method of differences

If the general term of a series, u_r, can be written in the form $u_r = f(r+1) - f(r)$, then:

$$\sum_{r=1}^{n} u_r = f(n+1) - f(1)$$

💡 **Tip**

The series won't always take exactly this form, so write out several terms to see how the cancellation occurs.

WORKED EXAMPLE 1.7

a Show that $(2r+1)^3 - (2r-1)^3 \equiv 24r^2 + 2$.

b Hence show that $\sum_{r=1}^{n} r^2 = \dfrac{1}{6}n(n+1)(2n+1)$.

a $(2r+1)^3 - (2r-1)^3$

$= (2r)^3 + 3(2r)^2 1 + 3(2r)1^2 + 1^3$

$\quad - [(2r)^3 + 3(2r)^2(-1) + 3(2r)(-1)^2 + (-1)^3]$

$= 8r^3 + 12r^2 + 6r + 1 - [8r^3 - 12r^2 + 6r - 1]$

$= 24r^2 + 2$

> Use the binomial expansion to expand the cubed brackets.

> Simplify to give the result required.

b $\displaystyle\sum_{r=1}^{n}(24r^2 + 2) = \sum_{r=1}^{n}[(2r+1)^3 - (2r-1)^2]$

> Sum both sides of the result from **a**.

$\text{RHS} = (\cancel{3})^3 - (1)^3$

$\quad + (\cancel{5})^3 - (\cancel{3})^3$

$\quad + (7)^3 - (\cancel{5})^3$

$\quad \vdots$

$\quad + (\cancel{2n-1})^3 - (2n-3)^2$

$\quad + (2n+1)^3 - (\cancel{2n-1})^2$

> The RHS is a difference, so you expect cancellation.

> Write out the first few terms ($r = 1, 2, 3, \ldots$) and the last couple of terms ($r = n-1, n$).

$= (2n+1)^3 - 1^3$

> Everything cancels except the terms shown.

$\text{LHS} = 24\displaystyle\sum_{r=1}^{n} r^2 + 2n$

> For the LHS, remember that $\displaystyle\sum_{r=1}^{n} 2 = 2n$.

$\therefore 24\displaystyle\sum_{r=1}^{n} r^2 + 2n = (2n+1)^3 - 1^3$

> Make the expressions for the LHS and RHS equal.

Continues on next page ...

$$24\sum_{r=1}^{n} r^2 + 2n = (2n)^3 + 3(2n)^2 1 + 3(2n)1^2 + 1^3 - 1^3$$

You now need to make $\sum_{r=1}^{n} r^2$ the subject.

Start by expanding the RHS and then simplify.

$$24\sum_{r=1}^{n} r^2 + 2n = 8n^3 + 12n^2 + 6n$$

$$24\sum_{r=1}^{n} r^2 = 8n^3 + 12n^2 + 4n$$

$$6\sum_{r=1}^{n} r^2 = 2n^3 + 3n^2 + n$$

$$6\sum_{r=1}^{n} r^2 = n(2n^2 + 3n + 1)$$

Factorise the RHS.

$$6\sum_{r=1}^{n} r^2 = n(n+1)(2n+1)$$

$$\sum_{r=1}^{n} r^2 = \frac{1}{6}n(n+1)(2n+1)$$

Finally, divide by 6.

Sometimes the cancellation occurs two terms apart.

WORKED EXAMPLE 1.8

a Write $\dfrac{2}{(k+1)(k+3)}$ in partial fractions.

b Find an expression for $\sum_{k=1}^{n} \dfrac{2}{(k+1)(k+3)}$.

a $\dfrac{2}{(k+1)(k+3)} = \dfrac{A}{k+1} + \dfrac{B}{k+3}$

Each factor in the denominator corresponds to one partial fraction.

$\Rightarrow 2 = A(k+3) + B(k+1)$

Multiply through by the common denominator.

$k = -1: \ 2 = A(2) + B(0) \Rightarrow A = 1$
$k = -3: \ 2 = A(0) + B(-2) \Rightarrow B = -1$

Use suitable values of k to make each bracket equal to zero.

$\therefore \dfrac{2}{(k+1)(k+3)} = \dfrac{1}{k+1} - \dfrac{1}{k+3}$

Continues on next page …

Rewind

You learnt about partial fractions in A Level Mathematics Student Book 2, Chapter 5.

b $\displaystyle\sum_{k=1}^{n} \frac{2}{(k+1)(k+3)} = \sum_{k=1}^{n} \frac{1}{k+1} - \frac{1}{k+3}$ .... Sum both sides of the result in **a**.

$$= \frac{1}{2} - \frac{1}{4}$$

$$+ \frac{1}{3} - \frac{1}{5}$$ .... Writing out several terms $(k = 1, 2, 3, \ldots)$ shows that the cancellations in the series occur two terms apart.

$$+ \frac{1}{4} - \frac{1}{6}$$

$$+ \frac{1}{5} - \frac{1}{7}$$

$$\vdots$$

$$+ \frac{1}{n-1} - \frac{1}{n+1}$$

$$+ \frac{1}{n} - \frac{1}{n+2}$$ .... Continue the pattern of cancellation for the last few terms $(k = n-2, n-1, n)$.

$$+ \frac{1}{n+1} - \frac{1}{n+3}$$

$$= \frac{1}{2} + \frac{1}{3} - \frac{1}{n+2} - \frac{1}{n+3}$$ .... This leaves part of the first two terms and part of the last two. You could put this all over a common denominator and combine into one fraction but, as the question doesn't specifically require this, there is no need to do anything else.

As you keep adding more and more terms of a series, the sum could keep increasing without a limit, or it could approach a finite value. In the latter case you say that the series **converges** (is convergent) and you could try to find its sum to infinity. To do this, you can find an expression for the sum of the first n terms and consider what this expression tends to when n gets very large.

WORKED EXAMPLE 1.9

Use the result from Worked example 1.8 to evaluate $\displaystyle\sum_{k=1}^{\infty} \frac{2}{(k+1)(k+3)}$.

From Worked example 1.8:

$$\sum_{k=1}^{n} \frac{2}{(k+1)(k+3)} = \frac{1}{2} + \frac{1}{3} - \frac{1}{n+2} - \frac{1}{n+3}$$

Let $n \to \infty$ in this result. Then:

$$\frac{1}{n+2} \to 0$$

$$\frac{1}{n+3} \to 0$$ .... As the denominator tends to ∞, these fractions tend to zero.

$$\therefore \sum_{k=1}^{\infty} \frac{2}{(k+1)(k+3)} = \frac{1}{2} + \frac{1}{3} = \frac{5}{6}$$

EXERCISE 1D

1 **a** Show that $(r+1)^2 - r^2 \equiv 2r+1$.

    b Hence show that $\displaystyle\sum_{r=1}^{n}(2r+1) = n(n+2)$.

2 **a** Show that $r^2(r+1)^2 - (r-1)^2 r^2 \equiv 4r^3$.

    b Hence show that $\displaystyle\sum_{r=1}^{n} r^3 = \frac{1}{4}n^2(n+1)^2$.

3 **a** Show that $\dfrac{1}{k+1} - \dfrac{1}{k+2} \equiv \dfrac{1}{(k+1)(k+2)}$.

    b **i** Hence show that $\displaystyle\sum_{k=1}^{n}\frac{2}{(k+1)(k+2)} = \frac{n}{n+2}$.

       ii Find $\displaystyle\sum_{k=11}^{24}\frac{2}{(k+1)(k+2)}$.

4 **a** Express $\dfrac{2}{(2r-1)(2r+1)}$ in partial fractions.

    b Use the method of differences to show that $\displaystyle\sum_{r=1}^{n}\frac{2}{(2r-1)(2r+1)} = \frac{2n}{2n+1}$.

    c Find $\displaystyle\sum_{r=1}^{\infty}\frac{2}{(2r-1)(2r+1)}$.

5 **a** Show that $(r+1)! - (r-1)! \equiv (r^2+r-1)(r-1)!$

    b Use the method of differences to show that $\displaystyle\sum_{r=1}^{n}(r^2+r-1)(r-1)! = (n+2)n! - 2$.

6 **a** Show that $\dfrac{1}{k} - \dfrac{1}{(k+2)} \equiv \dfrac{2}{k(k+2)}$.

    b Hence, find $\displaystyle\sum_{k=1}^{\infty}\frac{2}{k(k+2)}$.

7 **a** Show that $\dfrac{1}{2k+1} - \dfrac{1}{2k+3} \equiv \dfrac{2}{(2k+1)(2k+3)}$.

    b Find an expression for $\displaystyle\sum_{k=1}^{n}\frac{2}{(2k+1)(2k+3)}$.

    c Hence show that $\dfrac{1}{3\times5} + \dfrac{1}{5\times7} + \dfrac{1}{7\times9} + \ldots = \dfrac{1}{6}$.

8 Use the method of differences to find $\displaystyle\sum_{r=1}^{n}\frac{1}{(r+1)(r+2)(r+3)}$.

9 **a** Show that $\dfrac{1}{2k} - \dfrac{1}{k+1} + \dfrac{1}{2(k+2)} \equiv \dfrac{1}{k(k+1)(k+2)}$.

    b Use the method of differences to show that

$$\sum_{k=1}^{2n}\frac{1}{k(k+1)(k+2)} = \frac{n(an+b)}{c(n+1)(2n+1)}$$

    where a, b and c are constants to be found.

    c Find $\dfrac{1}{11\times12\times13} + \dfrac{1}{12\times13\times14} + \ldots + \dfrac{1}{20\times21\times22}$

10   **a**   Use the method of differences to find $\displaystyle\sum_{k=1}^{n}\ln\left(1+\frac{1}{k}\right)$.

       b   Hence, prove that the series $\displaystyle\sum_{k=1}^{\infty}\ln\left(1+\frac{1}{k}\right)$ diverges.

Checklist of learning and understanding

- You can use proof by induction to prove results concerning matrices, divisibility, inequalities, sequences, series, powers and differentiation.
- When working with series, use $S_{k+1}=S_k+u_{k+1}$.
- The formulae for the sums of integers, squares and cubes are:
  - $\displaystyle\sum_{r=1}^{n}r=\frac{1}{2}n(n+1)$
  - $\displaystyle\sum_{r=1}^{n}r^2=\frac{1}{6}n(n+1)(2n+1)$
  - $\displaystyle\sum_{r=1}^{n}r^3=\frac{1}{4}n^2(n+1)^2$
- You can manipulate series in these ways:
  - $\displaystyle\sum(u_r+v_r)=\sum u_r+\sum v_r$
  - $\displaystyle\sum cu_r=c\sum u_r$
  - $\displaystyle\sum_{r=1}^{n}c=nc$

  where c is a constant.

- **Method of differences**

  If the general term of a series, u_r, can be written in the form $u_r=\mathrm{f}(r+1)-\mathrm{f}(r)$, then:

  $$\sum_{r=1}^{n}u_r=\mathrm{f}(n+1)-\mathrm{f}(1)$$

- You might need to split an expression into partial fractions before using the method of differences.
- You can check whether an infinite series converges, and find its sum to infinity, by finding an expression for the first n terms of the series and considering what happens to it as n gets very large.

Mixed practice 1

1 Let $f(n) = 3^{2n} + 7$.

   a Evaluate $f(n)$ for $n = 1, 2, 3$ and 4. Hence suggest a natural number which is a factor of each $f(n)$.

   b Prove by induction that your conjecture from part **a** is correct for all $n \in \mathbb{N}$.

2 Prove by induction that, for $n \geqslant 1$

   $$\sum_{r=1}^{n} r(r+1) = \frac{n}{3}(n+1)(n+2).$$

   © OCR AS Level Mathematics, Unit 4725 Further Pure Mathematics 1, June 2010

3 Let $f(x) = e^{3x}$.

   a Find $f'(x), f''(x)$ and $f'''(x)$. Suggest an expression for $f^{(n)}(x)$.

   b Use mathematical induction to prove your conjecture from part **a**.

4 Find an expression for $\displaystyle\sum_{r=n+1}^{2n} (2r+1)$.

5 Use the formulae for $\displaystyle\sum_{r=1}^{n} r$ and $\displaystyle\sum_{r=1}^{n} r^2$ to show that $\displaystyle\sum_{r=1}^{n} (r+2)(r-1) = \frac{n}{3}(n+4)(n-1)$.

6 Use the formulae for $\displaystyle\sum_{r=1}^{n} r^2$ and $\displaystyle\sum_{r=1}^{n} r^3$ to find the value of $\displaystyle\sum_{r=5}^{40} r^2(2r-3)$.

7 **a** Show that $\dfrac{1}{k+4} - \dfrac{1}{k+5} \equiv \dfrac{1}{k^2+9k+20}$.

   b Hence show that $\displaystyle\sum_{k=1}^{n} \frac{1}{k^2+9k+20} = \frac{an}{b(n+5)}$, where a and b are integers to be found.

8 Find $\displaystyle\sum_{r=1}^{2n}\left(3r^2 - \frac{1}{2}\right)$, expressing your answer in a fully factorised form.

   © OCR AS Level Mathematics, Unit 4725 Further Pure Mathematics 1, June 2011

9 Use induction to prove that, for all integers $n \geqslant 1$,

   $$\frac{1}{2!} + \frac{2}{3!} + \frac{3}{4!} + \frac{4}{5!} + \cdots + \frac{n}{(n+1)!} = \frac{(n+1)!-1}{(n+1)!}$$

10 Prove by induction that, for all $n \in \mathbb{N}$,

   $$\sum_{r=1}^{n} \frac{r}{2^r} = 2 - \left(\frac{1}{2}\right)^n (n+2)$$

11 **a** Show that $\cos\left(x + \dfrac{\pi}{2}\right) = -\sin x$.

   b Prove that $\dfrac{d^n}{dx^n}(\cos x) = \cos\left(x + \dfrac{n\pi}{2}\right)$.

12 **a** Show that $\displaystyle\sum_{r=2}^{n} r(r-1)(r+1) = \frac{n}{4}(n^2-1)(n+2)$.

   b Hence find the sum of $(11 \times 12 \times 13) + (12 \times 13 \times 14) + (13 \times 14 \times 15) + \cdots + (38 \times 39 \times 40)$.

13   **a**   Show that $\dfrac{1}{r^2} - \dfrac{1}{(r+1)^2} \equiv \dfrac{2r+1}{r^2(r+1)^2}$.

     b   Hence show that $\displaystyle\sum_{r=1}^{n} \dfrac{2r+1}{r^2(r+1)^2} = \dfrac{n(n+2)}{(n+1)^2}$.

14   **a**   Show that $\dfrac{1}{r!} - \dfrac{1}{(r+1)!} \equiv \dfrac{r}{(r+1)!}$.

     b   Find $\dfrac{1}{2!} + \dfrac{2}{3!} + \dfrac{3}{4!} + \ldots + \dfrac{n}{(n+1)!}$.

     c   Hence find $\displaystyle\sum_{r=1}^{\infty} \dfrac{r}{(r+1)!}$.

15   The sequence $u_1, u_2, u_3, \ldots$ is defined by $u_n = 5^n + 2^{n-1}$.

     i   Find u_1, u_2 and u_3.

     ii   Hence suggest a positive integer, other than 1, which divides exactly into every term of the sequence.

     iii   By considering $u_{n+1} + u_n$, prove by induction that your suggestion in part **ii** is correct.

© **OCR AS Level Mathematics, Unit 4725/01 Further Pure Mathematics 1, June 2014**

16   **i**   Show that $\dfrac{1}{3r-1} - \dfrac{1}{3r+2} \equiv \dfrac{3}{(3r-1)(3r+2)}$.

     ii   Hence show that $\displaystyle\sum_{r=1}^{2n} \dfrac{1}{(3r-1)(3r+2)} = \dfrac{n}{2(3n+1)}$.

© **OCR AS Level Mathematics, Unit 4725/01 Further Pure Mathematics 1, June 2013**

17   Use mathematical induction to prove that $\dfrac{1}{\sqrt{1}} + \dfrac{1}{\sqrt{2}} + \ldots + \dfrac{1}{\sqrt{n}} > \sqrt{n}$ for all $n > 1$.

18   Prove by induction that $\dfrac{1}{\sqrt{1}} + \dfrac{1}{\sqrt{2}} + \ldots + \dfrac{1}{\sqrt{n}} > 2(\sqrt{n+1} - 1)$ for all $n \geqslant 1$.

19   Find the smallest positive integer N for which $3^N < N!$. Prove that $3^n < n!$ for all $n \geqslant N$.

20   Show that $(1+x)^n \geqslant 1 + nx$ for $n \in \mathbb{N}$ and $x \in \mathbb{R}$.

21   Prove by induction that, for any positive integer n:

$2 \times 6 \times 10 \times \ldots \times (4n-2) = \dfrac{(2n)!}{n!}$.

22   **i**   Show that $\dfrac{r}{r+1} - \dfrac{r-1}{r} \equiv \dfrac{1}{r(r+1)}$.

     ii   Hence find an expression, in terms of n, for $\dfrac{1}{2} + \dfrac{1}{6} + \dfrac{1}{12} + \ldots + \dfrac{1}{n(n+1)}$.

     iii   Hence find $\displaystyle\sum_{r=n+1}^{\infty} \dfrac{1}{r(r+1)}$.

© **OCR AS Level Mathematics, Unit 4725 Further Pure Mathematics 1, January 2012**

23 **a** Express $\dfrac{3}{(r-1)(r+2)}$ in partial fractions.

b Hence show that $\displaystyle\sum_{r=2}^{3n+1}\dfrac{1}{(r-1)(r+2)}=\dfrac{n(an^2+bn+c)}{6(3n+1)(3n+2)(3n+3)}$ where a, b and c are constants to be found.

24 Use the method of differences to show that $\displaystyle\sum_{k=1}^{n}\dfrac{3k+4}{k(k+1)(k+2)}=\dfrac{n(an+b)}{c(n+1)(n+2)}$,

where a, b and c are coprime integers to be found.

2 Powers and roots of complex numbers

In this chapter you will learn how to:

- raise complex numbers to integer powers (de Moivre's theorem)
- work with complex exponents
- find roots of complex numbers
- use roots of unity
- find quadratic factors of polynomials
- use a relationship between complex number multiplication and geometric transformations.

Before you start…

Pure Core Student Book 1, Chapter 4	You should know how to find the modulus and argument of a complex number.	1 Find the modulus and argument of $-3 + 4i$.
Pure Core Student Book 1, Chapter 4	You should be able to represent complex numbers on an Argand diagram.	2 Write down the complex numbers corresponding to the points A and B.
Pure Core Student Book 1, Chapter 2, Section 2, Chapter 4	You should know how to work with complex numbers in Cartesian form.	3 Given that $z = 3 - 2i$ and $w = 2 + i$, evaluate: **a** $z - w$ **b** $\dfrac{z}{w}$.
Pure Core Student Book 1, Chapter 4	You should be able to multiply and divide complex numbers in modulus–argument form.	4 Given that $z = 10\left(\cos\dfrac{3\pi}{4} + i\sin\dfrac{3\pi}{4}\right)$ and $w = 2\left(\cos\dfrac{2\pi}{3} + i\sin\dfrac{2\pi}{3}\right)$, find: **a** zw **b** $\dfrac{z}{w}$. Give the arguments in the range $(-\pi, \pi)$.

Continues on next page …

Pure Core Student Book 1, Chapter 4	You should be able to work with complex conjugates.	5 Write down the complex conjugate of: **a** $5i - 3$ **b** $3\left(\cos\dfrac{\pi}{4} + i\sin\dfrac{\pi}{4}\right)$.
Pure Core Student Book 1, Chapter 4	You should know how to relate operations with complex numbers to transformations on an Argand diagram.	6 Let $a = 2 + i$ and z be any complex number. Describe a geometrical transformation that maps: **a** z to z^* **b** z to $z + a$.

Extending arithmetic with complex numbers

You already know how to perform basic operations with complex numbers, both in Cartesian and in modulus–argument form. Modulus–argument form is particularly well suited to multiplication and division. In this chapter you will see how you can utilise this to find powers and roots of complex numbers. This chapter also includes a definition of complex powers which can make calculations even simpler.

You will also meet roots of unity, which are the solutions of the equation $z^n = 1$. They have some useful algebraic and geometric properties. Some of the applications include finding exact values of trigonometric functions.

Because you can represent complex numbers as points on an Argand diagram, operations with complex numbers have a geometric interpretation. You can use this to solve some problems that at first sight have nothing to do with complex numbers. This is just one example of the use of complex numbers to solve real-life problems.

◄◄ Rewind

You met complex numbers in Pure Core Student Book 1, Chapter 4.

►►| Fast forward

You will learn more about links between complex numbers and trigonometry in Chapter 3.

Section 1: De Moivre's theorem

In Pure Core Student Book 1, Chapter 4, you learnt that you can write complex numbers in Cartesian form, $x + iy$, or in modulus–argument form, $r(\cos\theta + i\sin\theta)$, or $r \operatorname{cis}\theta$. You also learnt the rules for multiplying complex numbers in modulus–argument form:

$$|zw| = |z||w| \text{ and } \arg(zw) = \arg z + \arg w$$

You can apply this result to find powers of complex numbers. If a complex number has modulus r and argument θ, then multiplying $z \times z$ gives that z^2 has modulus r^2 and argument 2θ. Repeating this process, you can see that

$$|z^n| = |z|^n \text{ and } \arg(z^n) = n \arg z$$

In other words, when you raise a complex number to a power, you raise the modulus to the same power and multiply the argument by the power.

Key point 2.1

De Moivre's theorem

For a complex number, z, with modulus r and argument θ:

$$z^n = \left(r(\cos\theta + i\sin\theta)\right)^n = r^n\left(\cos n\theta + i\sin n\theta\right)$$

for every integer power n.

For positive integer powers, you can prove this result by induction.

Rewind

For Proof 1 you will also need the compound angle formulae from A Level Mathematics Student Book 2, Chapter 8.

Tip

De Moivre's theorem can also be written as $(r\operatorname{cis}\theta)^n = r^n\operatorname{cis}(n\theta)$ or $[r, \theta]^n = [r^n, n\theta]$.

Focus on ...

See Focus on ... Proof 1 for a discussion of how to extend de Moivre's theorem to all rational n.

PROOF 1

When $n = 1$:
$$\left(r(\cos\theta + i\sin\theta)\right)^1 = r(\cos\theta + i\sin\theta)$$
so the result is true for $n = 1$.

Check that the result is true for $n = 1$.

Assuming that the result is true for some k:
$$\left(r(\cos\theta + i\sin\theta)\right)^k = r^k\left(\cos k\theta + i\sin k\theta\right)$$

Assume that the result is true for some k and write down what that means. (Remember, you will need to use this later.)

Then for $n = k + 1$:
$$\left(r(\cos\theta + i\sin\theta)\right)^{k+1} = r^k\left(\cos k\theta + i\sin k\theta\right) \times r(\cos\theta + i\sin\theta)$$

Make a link between $n = k$ and $n = k + 1$. In this case use $z^{k+1} = z^k z$.

$$= r^{k+1}\left(\cos k\theta \cos\theta + i\cos k\theta \sin\theta + i\sin k\theta \cos\theta - \sin k\theta \sin\theta\right)$$

Expand the brackets, remembering that $i^2 = -1$.

$$= r^{k+1}\left((\cos k\theta \cos\theta - \sin k\theta \sin\theta) + i(\cos k\theta \sin\theta + \sin k\theta \cos\theta)\right)$$

Group real and imaginary parts.

$$= r^{k+1}\left(\cos(k+1)\theta + i\sin(k+1)\theta\right)$$

Recognise the compound angle formulae:

$\cos(A + B) = \cos A \cos B - \sin A \sin B$ and
$\sin(A + B) = \sin A \cos B + \cos A \sin B$.

Hence the result is true for $n = k + 1$.

This is the result you are trying to prove, but with n replaced by $k + 1$.

The result is true for $n = 1$, and if it is true for some k then it is also true for $k + 1$. Therefore, it is true for all $n \geqslant 1$ by induction.

Remember to write the full conclusion.

You can use **de Moivre's theorem** to evaluate powers of complex numbers.

WORKED EXAMPLE 2.1

Evaluate, without a calculator, $\dfrac{(1+i)^{22}}{4i}$.

$|1+i| = \sqrt{1^2 + 1^2} = \sqrt{2}$

First find the modulus and argument of each number.

$\arg(1+i) = \arctan\left(\dfrac{1}{1}\right) = \dfrac{\pi}{4}$

$\therefore 1+i = \sqrt{2}\left(\cos\dfrac{\pi}{4} + i\sin\dfrac{\pi}{4}\right)$

$|4i| = 4,\ \arg(4i) = \dfrac{\pi}{2}$

$\therefore 4i = 4\left(\cos\dfrac{\pi}{2} + i\sin\dfrac{\pi}{2}\right)$

By *de Moivre's theorem*:

$\left(\sqrt{2}\left(\cos\dfrac{\pi}{4} + i\sin\dfrac{\pi}{4}\right)\right)^{22} = \left(\sqrt{2}\right)^{22}\left(\cos\dfrac{22\pi}{4} + i\sin\dfrac{22\pi}{4}\right)$

$\qquad = 2^{11}\left(\cos\dfrac{11\pi}{2} + i\sin\dfrac{11\pi}{2}\right)$

$\qquad = 2048\left(\cos\dfrac{\pi}{2} + i\sin\dfrac{\pi}{2}\right)$

The argument needs to be between $-\pi$ and π: $\dfrac{11\pi}{2} - 5\pi = \dfrac{\pi}{2}$.

Dividing the moduli and subtracting the arguments:

$\dfrac{(1+i)^{22}}{4i} = \dfrac{2048\left(\cos\dfrac{\pi}{2} + i\sin\dfrac{\pi}{2}\right)}{4\left(\cos\dfrac{\pi}{2} + i\sin\dfrac{\pi}{2}\right)}$

$\qquad = \dfrac{2048}{4}(\cos 0 + i\sin 0)$

$\qquad = 512$

💡 **Tip**

If the power of $(1 + i)$ had been smaller, you might have been able to use the binomial expansion with the fact that $i^2 = -1$, $i^3 = -i$ and $i^4 = 1$. For example:

$$(1 + i)^6 = 1 + 6i + 15i^2 + 20i^3 + 15i^4 + 6i^5 + i^6$$
$$= 1 + 6i - 15 - 20i + 15 + 6i - 1$$
$$= -8i$$

The usefulness of complex numbers is that the calculation does not get any longer or more difficult with larger powers.

You can also prove that de Moivre's theorem works for negative integer powers.

WORKED EXAMPLE 2.2

Let $z = r(\cos\theta + i\sin\theta)$.

a Find the modulus and argument of $\dfrac{1}{z}$.

b Hence prove de Moivre's theorem for negative integer powers.

a Multiplying top and bottom by the complex conjugate:

$$\frac{1}{z} = \frac{1}{r(\cos\theta + i\sin\theta)} \times \frac{(\cos\theta - i\sin\theta)}{(\cos\theta - i\sin\theta)}$$

$$= \frac{\cos\theta - i\sin\theta}{r(\cos^2\theta + \sin^2\theta)}$$

$$= \frac{1}{r}(\cos\theta - i\sin\theta) \quad \cdots\cdots\cdots\cdots\cdots$$ Use $\cos^2\theta + \sin^2\theta \equiv 1$.

$$= \frac{1}{r}(\cos(-\theta) + i\sin(-\theta)) \quad \cdots\cdots\cdots$$ To find the modulus and argument, you need to write the number in this form.

Remember that $\cos(-\theta) \equiv \cos\theta$ and $\sin(-\theta) \equiv -\sin\theta$.

Hence $\left|\dfrac{1}{z}\right| = \dfrac{1}{r}$ and $\arg\left(\dfrac{1}{z}\right) = -\theta$. $\cdots\cdots$ This means that you can write

$$\frac{1}{z} = \frac{1}{r}(\cos(-\theta) + i\sin(-\theta)).$$

b Using de Moivre's theorem for positive powers:

Since you have already proved de Moivre's theorem for positive powers, you can use

$$\left(\frac{1}{z}\right)^n = \left(\frac{1}{r}(\cos(-\theta) + i\sin(-\theta))\right)^n \quad \cdots\cdots$$ $z^{-n} = (z^{-1})^n = \left(\dfrac{1}{z}\right)^n$ with the modulus and

$$= \left(\frac{1}{r}\right)^n (\cos(-n\theta) + i\sin(-n\theta))$$

argument of $\dfrac{1}{z}$ found in part **a**.

Hence $z^{-n} = r^{-n}(\cos(-n\theta) + i\sin(-n\theta))$, as required.

WORKED EXAMPLE 2.3

Find the modulus and argument of $\dfrac{1}{\left(1-i\sqrt{3}\right)^{7}}$.

Modulus and argument of $z = 1 - i\sqrt{3}$:

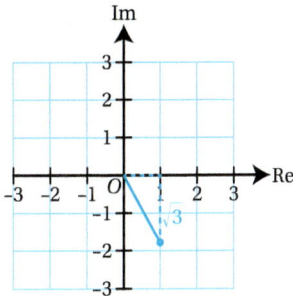

The best way to find the modulus and argument is to sketch a diagram.

$\sqrt{1^2 + \sqrt{3}^2} = 2$

$\tan^{-1}\left(-\dfrac{\sqrt{3}}{1}\right) = -\dfrac{\pi}{3}$

So $z = 2\operatorname{cis}\left(-\dfrac{\pi}{3}\right)$

Applying de Moivre's theorem for negative powers:

$z^{-7} = 2^{-7}\operatorname{cis}\left(-7 \times -\dfrac{\pi}{3}\right)$

$\qquad = \dfrac{1}{128}\operatorname{cis}\dfrac{7\pi}{3}$

This is in the form $r\operatorname{cis}\theta$ so you can just read off the modulus and the argument.

The modulus is $\dfrac{1}{128}$ and the argument is

$\dfrac{7\pi}{3} - 2\pi = \dfrac{\pi}{3}$.

The argument needs to be between 0 and 2π, so you need to take away 2π.

EXERCISE 2A

1 Evaluate each expression, giving your answer in the form $r(\cos\theta + i\sin\theta)$.

a i $\left(2\cos\dfrac{\pi}{5} + 2i\sin\dfrac{\pi}{5}\right)^{6}$
      ii $\left(3\left(\cos\left(-\dfrac{\pi}{3}\right) + i\sin\left(-\dfrac{\pi}{3}\right)\right)\right)^{4}$

b i $\left(\operatorname{cis}\dfrac{\pi}{6}\right)^{2}\left(\operatorname{cis}\dfrac{\pi}{4}\right)^{3}$
      ii $\left(\operatorname{cis}\dfrac{\pi}{8}\right)^{4}\left(\operatorname{cis}\dfrac{\pi}{3}\right)^{2}$

c **i** $\dfrac{\left(\cos\dfrac{2\pi}{3}+i\sin\dfrac{2\pi}{3}\right)^{6}}{\left(\cos\dfrac{\pi}{6}+i\sin\dfrac{\pi}{6}\right)^{3}}$ **ii** $\dfrac{\left(\cos\dfrac{\pi}{4}+i\sin\dfrac{\pi}{4}\right)^{2}}{\left(\cos\dfrac{\pi}{3}+i\sin\dfrac{\pi}{3}\right)^{6}}$

2 Given that $z=\cos\dfrac{\pi}{6}+i\sin\dfrac{\pi}{6}$:

   a write z^2, z^3 and z^4 in the form $r(\cos\theta+i\sin\theta)$

   b represent z, z^2, z^3 and z^4 on the same Argand diagram.

3 **a** Given that $z=\left[1,\dfrac{2\pi}{3}\right]$:

   i write z^2, z^3 and z^4 in modulus–argument form

   ii represent z, z^2, z^3 and z^4 on the same Argand diagram.

   b For which natural numbers n is $z^n=z$?

In questions 4 and 5 you must show detailed reasoning.

4 **a** Find the modulus and argument of $1+i\sqrt{3}$.

   b Hence find $\left(1+i\sqrt{3}\right)^{5}$ in modulus–argument form.

   c Hence find $\left(1+i\sqrt{3}\right)^{5}$ in Cartesian form.

5 **a** Write $-\sqrt{2}+i\sqrt{2}$ in the form $r\operatorname{cis}\theta$.

   b Hence find $\left(-\sqrt{2}+i\sqrt{2}\right)^{6}$ in simplified Cartesian form.

6 Find the smallest positive integer value of n for which $\left(\cos\dfrac{5\pi}{12}+i\sin\dfrac{5\pi}{12}\right)^{n}$ is real.

7 Find the smallest positive integer value of k such that $\left(\operatorname{cis}\dfrac{3\pi}{28}\right)^{k}$ is purely imaginary.

Section 2: Complex exponents

The rules for multiplying complex numbers in modulus–argument form look just like the rules of indices:

Compare

$$r_1\operatorname{cis}\theta_1\times r_2\operatorname{cis}\theta_2=r_1r_2\operatorname{cis}(\theta_1+\theta_2)$$

with

$$k_1\,e^{x_1}\times k_2\,e^{x_2}=k_1k_2\,e^{x_1+x_2}.$$

You can extend the definition of powers to imaginary numbers so that all the rules of indices still apply.

🔑 Key point 2.2

Euler's formula:

$$e^{i\theta} \equiv \cos\theta + i\sin\theta$$

It is important to realise that Euler's formula is a definition, and so it makes no sense to ask why it is true or how to prove it. You can, however, note that it seems to be a sensible definition, since it ensures that imaginary powers follow the same rules as real powers.

You can write a complex number with modulus r and argument θ in **exponential form** as $re^{i\theta}$. You now have four different ways of writing complex numbers with a given modulus and argument.

🔑 Key point 2.3

$$re^{i\theta} = r(\cos\theta + i\sin\theta) = r\operatorname{cis}\theta = [r, \theta]$$

When working with complex numbers in exponential form you can use all the normal rules of indices.

⏩ **Fast forward**

With this definition of imaginary powers, all the usual properties of the exponential function still hold. For example, it turns out that the Maclaurin series, which you will meet in Chapter 8, can be extended to include imaginary powers.

ℹ️ **Did you know?**

Substituting $\theta = \pi$ into Euler's formula and rearranging gives $e^{i\pi} + 1 = 0$. This equation, called Euler's identity, connects five important numbers from different areas of mathematics. It is often cited as 'the most beautiful' equation in mathematics.

WORKED EXAMPLE 2.4

Given that $z = 2e^{\frac{i\pi}{12}}$ and $w = \dfrac{1}{2}e^{\frac{i\pi}{4}}$, find $z^5 w^3$ in the form $x + iy$.

$z^5 w^3 = \left(2e^{\frac{i\pi}{12}}\right)^5 \left(\dfrac{1}{2}e^{\frac{i\pi}{4}}\right)^3$

You can do all the calculations in exponential form and then convert to Cartesian form at the end.

$= 2^5 e^{\frac{5i\pi}{12}} \times \dfrac{1}{2^3} e^{\frac{3i\pi}{4}}$

Use rules of indices for the powers:

$$\dfrac{5}{12} + \dfrac{3}{4} = \dfrac{7}{6}$$

$= 4 e^{\frac{7i\pi}{6}}$

$= 4\left(\cos\left(\dfrac{7\pi}{6}\right) + i\sin\left(\dfrac{7\pi}{6}\right)\right)$

Now write in terms of trigonometric functions and evaluate.

$= 4\left(-\dfrac{\sqrt{3}}{2} - \dfrac{1}{2}i\right)$

$= -2\sqrt{3} - 2i$

You can combine Euler's formula with rules of indices to raise any real number to any complex power.

WORKED EXAMPLE 2.5

Find, correct to three significant figures, the value of:

a e^{2+3i} **b** 3^{2+3i}.

a $e^{2+3i} = e^2 e^{3i}$

Use rules of indices to separate the real and imaginary parts of the power.

$= e^2(\cos 3 + i\sin 3)$

Use Euler's formula for the imaginary power.

$= -7.32 + 1.04i$

Expand the brackets and give the answer to 3 s.f. Remember that the arguments of the trigonometric functions are in radians.

b $3^{2+3i} = (e^{\ln 3})^{2+3i}$

You only know how to raise e to a complex power, so express 3 as a power of e.

$= e^{2\ln 3} e^{(3\ln 3)i}$

Use rules of indices and then Euler's formula.
Note that $e^{2\ln 3} = e^{\ln 9} = 9$ and $3\ln 3 = \ln 27$.

$= 9(\cos(\ln 27) + i\sin(\ln 27))$

$= -8.89 - 1.38i$

The complex conjugate of a number is easy to find when written in exponential form. This is best seen on an Argand diagram, where taking the complex conjugate is represented by a reflection in the real axis. In this case it is best to take the argument between $-\pi$ and π.

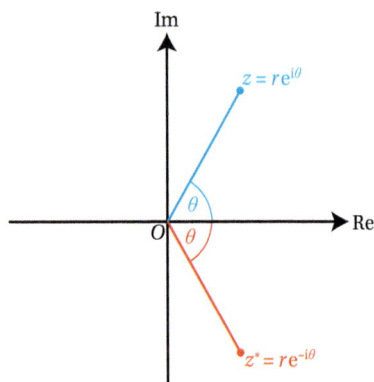

▶▶) Fast forward

You will use the exponential form of complex numbers when solving second order differential equations in Chapter 10.

🔑 Key point 2.4

The complex conjugate of $z = re^{i\theta}$ is $z^* = re^{-i\theta}$.

💡 Tip

Note that if $r = 1$, you also have $z^* = \dfrac{1}{z} = e^{-i\theta}$.

In Pure Core Student Book 1, Chapter 5, you used complex conjugates when solving polynomial equations.

WORKED EXAMPLE 2.6

A cubic equation has real coefficients and two of its roots are 1 and $2e^{\frac{i\pi}{3}}$. Find the equation in the form $x^3 + ax^2 + bx + c = 0$.

The roots are: $1, 2e^{\frac{i\pi}{3}}$ and $2e^{-\frac{i\pi}{3}}$.	Complex roots occur in conjugate pairs, so you can write down the third root.
$-a = 1 + 2e^{\frac{i\pi}{3}} + 2e^{-\frac{i\pi}{3}}$	Use the formulae for sums and products of roots to find the coefficients of the equation: $-a = x_1 + x_2 + x_3$
$= 1 + 4\cos\dfrac{\pi}{3}$ $= 3$ $\therefore a = -3$	Remember that $z + z^* = 2\text{Re}(z)$, and that $\text{Re}\left(e^{\frac{i\pi}{3}}\right) = \cos\dfrac{\pi}{3} = \dfrac{1}{2}$
$b = \left(1 \times 2e^{\frac{i\pi}{3}}\right) + \left(1 \times 2e^{-\frac{i\pi}{3}}\right) + \left(2e^{\frac{i\pi}{3}} \times 2e^{-\frac{i\pi}{3}}\right)$	$b = x_1 x_2 + x_2 x_3 + x_3 x_1$
$= 4\cos\left(\dfrac{\pi}{3}\right) + 4$ $= 6$	
$-c = 1 \times 2e^{\frac{i\pi}{3}} \times 2e^{-\frac{i\pi}{3}} = 4$	$-c = x_1 x_2 x_3$
$\therefore c = -4$	

Hence the equation is $x^3 - 3x^2 + 6x - 4 = 0$.

EXERCISE 2B

You can use your calculator to perform operations with complex numbers in Cartesian, modulus–argument and exponential forms, as well as to convert from one form to another. Do the questions in this exercise without a calculator first, then use a calculator to check your answers.

1 Write each complex number in Cartesian form without using trigonometric functions.

a **i** $3e^{i\frac{\pi}{6}}$ **ii** $4e^{\frac{i\pi}{4}}$

b **i** $4e^{i\pi}$ **ii** $5e^{2\pi i}$

c **i** $e^{\frac{2\pi i}{3}}$ **ii** $2e^{\frac{3\pi}{2}i}$

2 Write each complex number in the form $re^{i\theta}$.

a **i** $5 + 5i$ **ii** $2\sqrt{3} - 2i$

b **i** $-\dfrac{1}{2} + \dfrac{1}{2}i$ **ii** $2 + 3i$

c **i** $-4i$ **ii** -5

3 Write the answer to each calculation in the form $r e^{i\theta}$.

a **i** $4e^{i\frac{\pi}{6}} \times 5e^{i\frac{\pi}{4}}$

ii $\dfrac{5e^{i\frac{3\pi}{4}}}{10e^{i\frac{\pi}{4}}}$

b **i** $\dfrac{\left(2e^{i\frac{\pi}{4}}\right)^3}{\left(5e^{i\frac{\pi}{3}}\right)^2}$

ii $\dfrac{2e^{i\frac{\pi}{3}}}{\left(e^{i\frac{\pi}{6}}\right)^5}$

4 Represent each complex number on an Argand diagram.

a **i** $e^{i\frac{\pi}{3}}$

ii $e^{i\frac{3\pi}{4}}$

b **i** $5e^{i\frac{\pi}{2}}$

ii $2e^{-i\frac{\pi}{3}}$

5 Let $z = 2e^{\frac{i\pi}{12}}$ and $w = 4e^{\frac{i\pi}{3}}$. Show that $z^2 + w = 2(1+i)(1+\sqrt{3})$.

6 **In this question you must show detailed reasoning.**

Let $z = 2e^{\frac{i\pi}{3}}$ and $w = 3e^{-\frac{i\pi}{6}}$. Write each complex number in the form $x + iy$.

a $\dfrac{z}{w}$

b $z^5 w^3$.

7 Write e^{4+3i} in the form $x + iy$, where x and y are real, giving your answer correct to three significant figures.

8 Write $e^{2-\frac{i\pi}{3}}$ in exact Cartesian form.

9 The equation $x^3 + ax^2 + bx + c = 0$ has real coefficients, and two of its roots are 2 and $e^{\frac{i\pi}{3}}$. Find the values of a, b and c.

10 A quartic equation has real coefficients and two of its roots are $e^{\frac{i\pi}{6}}$ and $2e^{-\frac{i\pi}{3}}$. Find the equation in the form $x^4 + ax^3 + bx^2 + cx + d = 0$.

11 Find 5^i in the form $x + yi$.

12 Find 3^{2-i} in the form $x + yi$.

Section 3: Roots of complex numbers

Now that you can use de Moivre's theorem to find powers of complex numbers, it makes sense to ask whether you can also find roots.

In Pure Core Student Book 1, you learnt how to find the two square roots of a complex number by writing $z = x + iy$ and comparing real and imaginary parts. You also know that a polynomial equation of degree n has n complex roots. Just as a complex number has two square roots, it will have three cube roots, four fourth roots, and so on.

You can't always use the algebraic method to find all those roots. De Moivre's theorem gives an alternative method.

> ◄◄ **Rewind**
>
> See Pure Core Student Book 1, Chapter 4, for an example of finding square roots of a complex number.

WORKED EXAMPLE 2.7

Solve the equation $z^3 = 4\sqrt{3} + 4i$.

Let $z = r(\cos\theta + i\sin\theta)$.

Use the modulus–argument form since raising to a power is easier in this form than in Cartesian form.

Then the equation is equivalent to
$$r^3(\cos 3\theta + i\sin 3\theta) = 4\sqrt{3} + 4i$$

Use de Moivre's theorem and then compare the modulus and the argument of both sides.

Find the modulus and argument of the RHS.

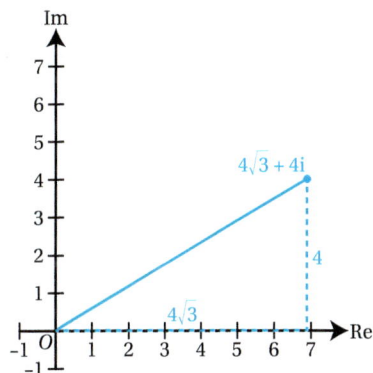

$$\left|4\sqrt{3} + 4i\right| = \sqrt{\left(4\sqrt{3}\right)^2 + 4^2} = 8$$

$$\arg\left(4\sqrt{3} + 4i\right) = \arctan\left(\frac{4}{4\sqrt{3}}\right) = \frac{\pi}{6}$$

Therefore,

$$r^3(\cos 3\theta + i\sin 3\theta) = 8\left(\cos\frac{\pi}{6} + i\sin\frac{\pi}{6}\right)$$

Comparing the moduli:

Remember that, by definition, r is a positive real number.

$$r^3 = 8 \Rightarrow r = 2$$

Comparing the arguments:

If $0 < \theta < 2\pi$ then $0 < 3\theta < 6\pi$.

$$3\theta = \frac{\pi}{6}, \frac{13\pi}{6} \text{ or } \frac{25\pi}{6}$$

$$\theta = \frac{\pi}{18}, \frac{13\pi}{18}, \frac{25\pi}{18}$$

Since adding 2π to the argument returns to the same complex number, there are 3 possible values for 3θ between 0 and 6π.

The roots are:

$$z_1 = 2\left(\cos\frac{\pi}{18} + i\sin\frac{\pi}{18}\right)$$

Write down all three roots in modulus–argument form.

$$z_2 = 2\left(\cos\frac{13\pi}{18} + i\sin\frac{13\pi}{18}\right)$$

$$z_3 = 2\left(\cos\frac{25\pi}{18} + i\sin\frac{25\pi}{18}\right)$$

If you plot the three roots from Worked example 2.7 on an Argand diagram, you will notice an interesting pattern. They all have the same modulus so they lie on a circle of radius 2. The arguments differ by $\dfrac{2\pi}{3}$

so they are equally spaced around the circle. Therefore, the three points form an equilateral triangle.

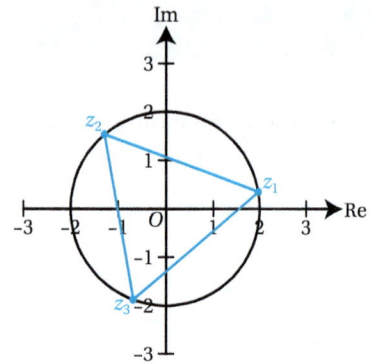

🔑 Key point 2.5

To solve $z^n = w$:

- write w in modulus–argument form
- use de Moivre's theorem to write $z^n = r^n(\cos n\theta + i\sin n\theta)$
- compare moduli, remembering that they are always real
- compare arguments, remembering that adding 2π to the argument does not change the number
- write n different roots in modulus–argument form.

All n roots of the equation $z^n = w$ will have the same modulus, and their arguments will differ by $\dfrac{2\pi}{n}$. This means that the Argand diagram will always show the pattern you noticed in Worked example 2.7.

⏩ Fast forward

You will learn more about the connection between powers and rotations in Section 6.

🔑 Key point 2.6

The roots of $z^n = w$ form a regular polygon with vertices on a circle centred at the origin.

WORKED EXAMPLE 2.8

Draw an Argand diagram showing the roots of the equation $z^6 = 729$.

One root is

$z_1 = \sqrt[6]{729} = 3$ ············· There are six roots, forming a regular hexagon. You only need to find one of them and then complete the diagram.

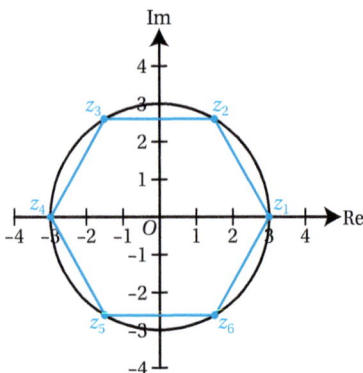

You can also find the equation whose roots form a given regular polygon.

WORKED EXAMPLE 2.9

The diagram shows a regular pentagon inscribed in a circle on an Argand diagram. One of the vertices lies on the positive imaginary axis.

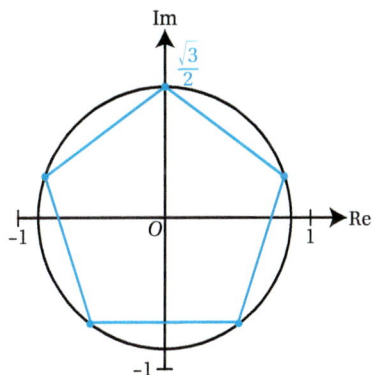

The five vertices of the pentagon correspond to the solutions of an equation of the form $z^n = w$, where w is a complex number. Find the values of n and w.

There are five roots, so $n = 5$.	Any equation of the form $z^n = w$ has n complex roots.
From the diagram: $z = \dfrac{\sqrt{3}}{2} i$ is a root.	Use the fact that one root is given in the question.
Hence	Remember that $i^5 = i$.
$w = \left(\dfrac{\sqrt{3}}{2} i \right)^5 = \dfrac{9\sqrt{3}}{32} i$	

EXERCISE 2C

In this exercise you must show detailed reasoning.

1 Find all three cube roots of each number, giving your answers in the form $r\,e^{i\theta}$.

    a **i** 27   **ii** 100

    b **i** 8i   **ii** i

    c **i** 1+i   **ii** $2 - \sqrt{3}i$

2 Find the fourth roots of each number. Give your answers in the form $r \operatorname{cis}\theta$ and show them on an Argand diagram.

    a   **i** -16 **ii** $81\mathrm{i}$

    b   **i** $8\sqrt{2}+8\sqrt{2}\mathrm{i}$ **ii** $-\dfrac{1}{2}+\dfrac{\sqrt{3}}{2}\mathrm{i}$

3 Solve the equation $z^3 = -8$ for $z \in \mathbb{C}$. Give your answers in the form $x + y\mathrm{i}$.

4   **a**   Find the modulus and the argument of $8\sqrt{3} - 8\mathrm{i}$.

    b   Solve the equation $z^4 = 8\sqrt{3} - 8\mathrm{i}$, giving your answers in the form $r\left(\cos\left(\dfrac{p}{q}\pi\right) + \mathrm{i}\sin\left(\dfrac{p}{q}\pi\right)\right)$, where p and q are integers.

5 Solve the equation $z^3 - \sqrt{2}(4 - 4\mathrm{i}) = 0$, giving your answers in Cartesian form.

6 Find all complex roots of the equation $z^4 + 81\mathrm{i} = 0$, giving your answers in the form $r\left(\cos\left(\dfrac{p}{q}\pi\right) + \mathrm{i}\sin\left(\dfrac{p}{q}\pi\right)\right)$, where p and q are integers.

7   **a**   Write $4 + 4\sqrt{3}\mathrm{i}$ in the form $r\mathrm{e}^{\mathrm{i}\theta}$.

    b   Hence solve the equation $z^4 = 4 + 4\sqrt{3}\mathrm{i}$, giving your answers in the form $r\mathrm{e}^{\mathrm{i}\theta}$.

    c   Show your solution on an Argand diagram.

8 The diagram shows a square with one vertex at $(2, 2)$. The complex numbers corresponding to the vertices of the square are solutions of an equation of the form $z^n = w$, where $n \in \mathbb{N}$ and $w \in \mathbb{R}$.

Find the values of n and w.

9   **a**   Solve the equation $z^4 = -16$, giving your answers in Cartesian form.

    b   Hence express $z^4 + 16$ as a product of two real quadratic factors.

10   **a**   Find all the roots of the equation $z^3 = -8\mathrm{i}$.

    b   Hence solve the equation $w^3 + 8\mathrm{i}(w - 1)^3 = 0$. Give your answers in exact Cartesian form.

11 Consider the equation $z^3 + \left(4\sqrt{2} - 4\sqrt{2}\mathrm{i}\right) = 0$.

    a   Solve the equation, giving your answers in the form $r(\cos\theta + \mathrm{i}\sin\theta)$.

The roots are represented on an Argand diagram by points A, B and C, labelled anticlockwise with A in the first quadrant. D is the midpoint of AB and the corresponding complex number is d.

    b   Find the modulus and argument of d.

    c   Write d^3 in exact Cartesian form.

12   **a**   Find, in exponential form, the three roots of the equation $z^3 = -1$.

    b   Expand $(x + 2)^3$.

    c   Hence or otherwise solve the equation $z^3 + 6z^2 + 12z + 9 = 0$, giving any complex root in exact Cartesian form.

Section 4: Roots of unity

In Section 3 you learnt a method for finding all complex roots of a number. A special case of this is solving the equation $z^n = 1$. Its roots are called **roots of unity**.

WORKED EXAMPLE 2.10

Find the fifth roots of unity, giving your answers in exponential form.

Let the roots be $z = r\,e^{i\theta}$. Then: $\left(r\,e^{i\theta}\right)^5 = 1$	Write z in exponential form and use de Moivre's theorem.
$\Rightarrow r^5 e^{5i\theta} = 1e^{0i}$	1 has modulus 1 and argument 0.
Comparing the moduli: $r = 1$ Comparing the arguments:	Remember that there should be five roots.

$5\theta = 0, 2\pi, 4\pi, 6\pi, 8\pi$

$$\theta = 0, \frac{2\pi}{5}, \frac{4\pi}{5}, \frac{6\pi}{5}, \frac{8\pi}{5}$$

The fifth roots of unity are:

$$1, e^{\frac{2\pi i}{5}}, e^{\frac{4\pi i}{5}}, e^{\frac{6\pi i}{5}}, e^{\frac{8\pi i}{5}}$$

As in Section 3, the five roots form a regular pentagon on the Argand diagram:

The same procedure works for any power n: there will be n distinct roots, each with modulus 1, and with arguments differing by $\dfrac{2\pi}{n}$. Remembering that one of the roots always equals 1, you can write down the full set of roots.

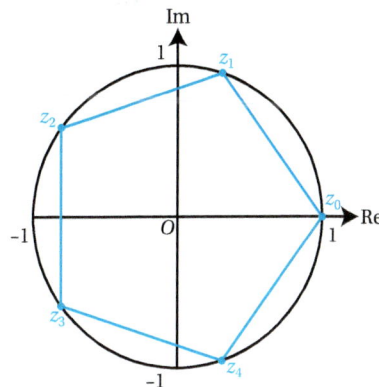

🔑 Key point 2.7

The nth roots of unity are:

$$1, e^{\frac{2\pi i}{n}}, e^{\frac{4\pi i}{n}}, \ldots, e^{\frac{2(n-1)\pi i}{n}}$$

They form a regular n-gon on an Argand diagram.

Notice that all the arguments are multiples of $\dfrac{2\pi}{n}$. But multiplying an argument by a number k corresponds to raising the complex number to the power of k. Hence all the nth roots of unity are powers of $e^{\frac{2\pi i}{n}}$. It is usual to denote the n roots $\omega_0, \omega_1, \ldots, \omega_{n-1}$.

Key point 2.8

You can write the nth root of unity as:

$$\omega_k = \left(e^{\frac{2\pi i}{n}}\right)^k = \omega_1^k, \text{ where } k = 0, 1, \ldots, n-1.$$

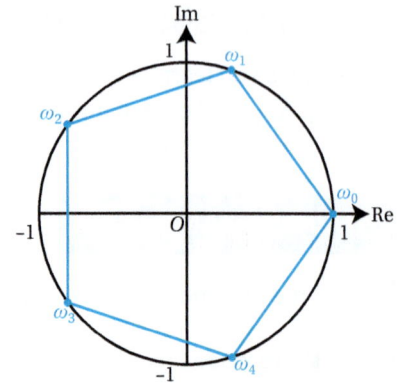

You can use the fact that the roots form a regular polygon to deduce various relationships between them. For example, for $n = 5$, you can use the symmetry of the pentagon to see that $\omega_4 = \omega_1^*$ and $\omega_3 = \omega_2^*$.

One of the most useful results concerns the sum of all n roots. You know from Pure Core Student Book 1, Chapter 4, that adding complex numbers corresponds to adding vectors on an Argand diagram. Since the points corresponding to the n roots of unity are equally spaced around the circle, the sum of the corresponding vectors should be zero.

You can also prove this result algebraically.

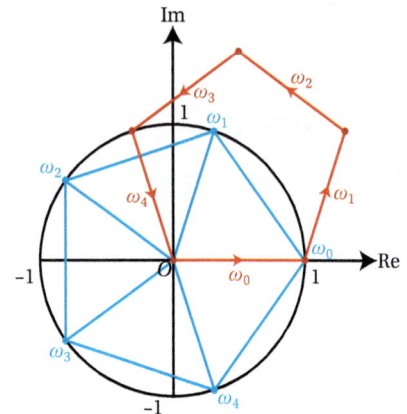

WORKED EXAMPLE 2.11

Let n be a natural number, and let $\omega = e^{\frac{2\pi i}{n}}$. Let $1, \omega_1, \omega_2, \ldots, \omega_{n-1}$ be the nth roots of unity.

a Express ω_k in terms of ω.

b Hence show that $1 + \omega_1 + \ldots + \omega_{n-1} = 0$.

a $\omega_k = \omega^k$

This is the result from Key point 2.8.

b $1 + \omega_1 + \ldots + \omega_{n-1} = 1 + \omega + \omega^2 + \ldots + \omega^{n-1}$

$$= \frac{1 - \omega^n}{1 - \omega}$$

This is a geometric series with first term 1 and common ratio ω.

Note that $\omega \neq 1$ so you can use the formula for the sum of the geometric series.

$$= 0 \text{ (since } \omega^n = 1)$$

ω is an nth root of unity, which means that $\omega^n = 1$.

Key point 2.9

If $1, \omega_1, \omega_2, \ldots, \omega_{n-1}$ are the nth roots of unity, then

$$1 + \omega_1 + \ldots + \omega_{n-1} = 0.$$

You can use the result in Key point 2.9 with a specific value of n to find some special values of trigonometric functions.

WORKED EXAMPLE 2.12

Let $\omega = e^{\frac{2\pi i}{5}}$.

a Show that $\text{Re}(\omega) + \text{Re}(\omega^2) = -\frac{1}{2}$.

b Hence find the exact value of $\cos\frac{2\pi}{5}$.

a

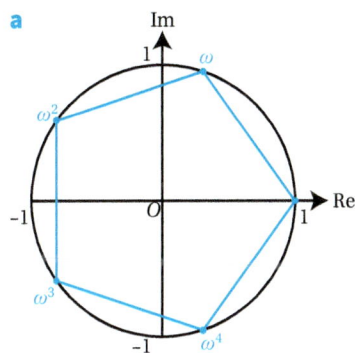

The five points form a regular pentagon.

From the diagram:

$\omega^4 = \omega^*$ and $\omega^3 = (\omega^2)^*$

Hence $\text{Re}(\omega^4) = \text{Re}(\omega)$ and $\text{Re}(\omega^4) = \text{Re}(\omega)$

You are interested in the real parts.

Using the result $1 + \omega + \omega^2 + \omega^3 + \omega^4 = 0$

This is the result from Key point 2.9.

and taking the real part:

$1 + \text{Re}(\omega) + \text{Re}(\omega^2) + \text{Re}(\omega^3) + \text{Re}(\omega^4) = 0$

$\Rightarrow 1 + 2\text{Re}(\omega) + 2\text{Re}(\omega^2) = 0$

Pair up the terms with equal real parts.

$\Rightarrow \text{Re}(\omega) + \text{Re}(\omega^2) = -\frac{1}{2}$

Continues on next page ...

b $Re(\omega) = \cos\dfrac{2\pi}{5}, Re(\omega^2) = \cos\dfrac{4\pi}{5}$

> Use the fact that $\omega = e^{\frac{2\pi i}{5}} = \cos\dfrac{2\pi}{5} + i\sin\dfrac{2\pi}{5}$ and $\omega^2 = \cos\dfrac{4\pi}{5} + i\sin\dfrac{4\pi}{5}$.

But $\cos\dfrac{4\pi}{5} = 2\cos^2\dfrac{2\pi}{5} - 1$, so:

> Use the double angle formula to relate the two values.

$$\cos\dfrac{2\pi}{5} + \cos\dfrac{4\pi}{5} = -\dfrac{1}{2}$$

$$\Rightarrow \cos\dfrac{2\pi}{5} + 2\cos^2\dfrac{2\pi}{5} - 1 = -\dfrac{1}{2}$$

$$\Rightarrow 4\cos^2\dfrac{2\pi}{5} + 2\cos\dfrac{2\pi}{5} - 1 = 0$$

> This is a quadratic equation in $\cos\dfrac{2\pi}{5}$.

$$\Rightarrow \cos\dfrac{2\pi}{5} = \dfrac{-2 + \sqrt{4+16}}{8}$$

> Take the positive root since $\cos\dfrac{2\pi}{5} > 0$.

$$= \dfrac{-1 + \sqrt{5}}{4}$$

EXERCISE 2D

1 Write down, in the form $r(\cos\theta + i\sin\theta)$, all the roots of each equation.

  a i $z^3 = 1$ **ii** $z^2 = 1$

  b i $z^6 = 1$ **ii** $z^4 = 1$

2 For each equation from question 1, write the roots in exact Cartesian form.

3 a Write down, in the form $e^{i\theta}$, the roots of the equation $z^5 = 1$.

  b Represent the roots on an Argand diagram.

4 The diagram shows all the roots of an equation $z^n = 1$.

  a Write down the value of n.

  b Write down the value of $\omega_1 + \omega_2 + \omega_3 + \omega_4$.

  c Which of these statements are correct?

   A $\omega_3 = \omega_1^3$ **B** $\omega_4 = \omega_2^2$

   C $\omega_3^5 = 1$ **D** $\omega_1^3 = -\omega_1^2$

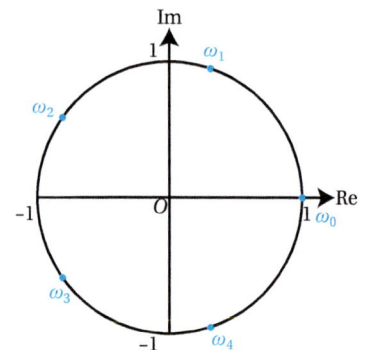

5 Let $\omega = e^{\frac{2\pi i}{7}}$.

  a Express the seventh roots of unity in terms of ω.

  b Is there an integer k such that $\omega^k = -\omega$? Justify your answer.

c Write down the smallest positive integer p such that $\omega^{17} = \omega^p$.

d Write down an integer m such that $\omega^m = (\omega^2)^*$.

6 Let $1, \omega_1, \omega_2, \omega_3, \omega_4, \omega_5$ be the distinct sixth roots of unity.

a Show that $\omega_n = \omega_1^n$ for $n = 2, 3, 4, 5$.

b Hence show that $1 + \omega + \omega^2 + \omega^3 + \omega^4 + \omega^5 = 0$.

7 **In this question you must show detailed reasoning.**

a Find, in exact Cartesian form, all the complex roots of the equation $z^3 = 1$.

b Hence find the exact roots of the equation $(z - 1)^3 = (z + 2)^3$.

8 Multiply out and simplify $(a + b\omega)(a - b\omega^2)$, where $\omega = e^{\frac{i\pi}{3}}$.

9 **In this question you must show detailed reasoning.**

Let $\omega = e^{\frac{2\pi i}{5}}$.

a Write, in terms of ω, the complex roots of the equation $z^5 = 1$.

Consider the equation $(z - 1)^5 = (z + 1)^5$.

b Find, in terms of ω, all roots of the equation.

c Show that the roots can be written as $i\cot\left(\dfrac{k\pi}{5}\right)$ for $k = 1, 2, 3, 4$.

d Show that $(z - 1)^5 = (z + 1)^5$ is equivalent to $5z^4 + 10z^2 + 1 = 0$.

e Hence show that $\cot^2 \dfrac{\pi}{5} + \cot^2 \dfrac{2\pi}{5} = 2$.

10 Let $1, \omega, \omega^2, \omega^3, \omega^4, \omega^5$ be the roots of the equation $z^6 = 1$, where ω is the solution with the smallest positive argument.

a Show these roots on an Argand diagram.

b Write in the form $re^{i\theta}$:

i $\dfrac{1 + \omega}{2}$

ii $\dfrac{\omega^3 + \omega^4}{2}$.

11 **a** Show that $\cos 3\theta = 4\cos^3 \theta - 3\cos\theta$.

Let $\omega = \cos\dfrac{2\pi}{7} + i\sin\dfrac{2\pi}{7}$.

b **i** Show that $1 + \omega + \omega^2 + \omega^3 + \omega^4 + \omega^5 + \omega^6 = 0$.

ii Hence deduce the value of $\cos\dfrac{2\pi}{7} + \cos\dfrac{4\pi}{7} + \cos\dfrac{6\pi}{7}$.

c Show that $\cos\dfrac{2\pi}{7}$ is a root of the equation $8t^3 + 4t^2 - 4t - 1 = 0$.

Section 5: Further factorising

In Pure Core Student Book 1, Chapter 5, you learnt that complex roots of a real polynomial come in conjugate pairs, and how you can use this fact to factorise a polynomial. You used the important result that, for any complex number w,

$$(z-w)(z-w^*) = z^2 - 2z\operatorname{Re}(w) + |w|^2$$

You can now combine this with your knowledge of roots of complex numbers to factorise expressions of the form $z^n + c$.

WORKED EXAMPLE 2.13

a Find all the complex roots of $z^4 = -81$, giving your answers in Cartesian form.

b Hence write $z^4 + 81$ as a product of two real quadratic factors.

a Let $z = r e^{i\theta}$.

> Write z in exponential form to find the roots, then turn answers into Cartesian form.

Then $r^4 e^{i4\theta} = -81 = 81e^{i\pi}$

> The argument of -81 is π.

Comparing the moduli:

$r^4 = 81$, so $r = 3$

Comparing the arguments:

$4\theta = \pi, 3\pi, 5\pi, 7\pi$

> You are looking for four roots, so add 2π three times.

$\theta = \dfrac{\pi}{4}, \dfrac{3\pi}{4}, \dfrac{5\pi}{4}, \dfrac{7\pi}{4}$

The roots are:

> Find the Cartesian form
> $x + iy = (r\cos\theta) + i(r\sin\theta)$

$z_1 = \dfrac{3\sqrt{2}}{2} + \dfrac{3\sqrt{2}}{2}i$

$z_2 = -\dfrac{3\sqrt{2}}{2} + \dfrac{3\sqrt{2}}{2}i$

$z_3 = -\dfrac{3\sqrt{2}}{2} - \dfrac{3\sqrt{2}}{2}i$

$z_4 = \dfrac{3\sqrt{2}}{2} - \dfrac{3\sqrt{2}}{2}i$

b $z^4 + 81 = (z - z_1)(z - z_2)(z - z_3)(z - z_4)$

> The factors of $z^4 + 81$ correspond to the roots of the equation $z^4 + 81 = 0$, which you found in part **a**.

$(z - z_1)(z - z_4) = z^2 - 2\operatorname{Re}(z_1)z + |z_1|^2$

$\qquad\qquad = z^2 - 3\sqrt{2}z + 9$

$(z - z_2)(z - z_3) = z^2 - 2\operatorname{Re}(z_2)z + |z_2|^2$

$\qquad\qquad = z^2 + 3\sqrt{2}z + 9$

> To get real quadratic factors you need to pair up the factors corresponding to the conjugate roots. You can use the shortcut
> $(z - w)(z - w^*) = z^2 - 2z\operatorname{Re}(w) + |w|^2$ and $|z_k| = 3$.

$\therefore z^4 + 81 = \left(z^2 - 3\sqrt{2}z + 9\right)\left(z^2 + 3\sqrt{2}z + 9\right)$

In this exercise you must show detailed reasoning.

1 **a** Find, in exponential form, all the complex roots of the equation $z^4 = -16$.

    b Write your answers from part **a** in exact Cartesian form.

    c Hence express $z^4 + 16$ as a product of two real quadratic factors.

2 By solving the equation $z^8 = 16$, express $z^8 - 16$ as a product of four real quadratic factors.

3 Show that $z^5 - 1 = (z-1)(z^2 - (2\cos\theta)z + 1)(z^2 - (2\cos\phi)z + 1)$, where $\theta, \phi \in (0, \pi)$.

4 Let $\omega = e^{\frac{2i\pi}{5}}$.

    a Write the roots of the equation $z^5 - 1 = 0$ in terms of ω.

    b Hence evaluate $(2-\omega)(2-\omega^2)(2-\omega^3)(2-\omega^4)$.

5 **a** Show that $t^2 + t + 1 = \left(t - e^{\frac{2i\pi}{3}}\right)\left(t - e^{-\frac{2i\pi}{3}}\right)$.

    b Solve the equation $z^4 = e^{\frac{2i\pi}{3}}$.

    c Hence write $z^8 + z^4 + 1$ as a product of four real quadratic factors.

6 Let $\omega = e^{\frac{2i\pi}{7}}$.

    a Write down the non-real roots of the equation $z^7 = 1$ in terms of ω.

    b Show that $\cos\dfrac{2\pi}{7} + \cos\dfrac{4\pi}{7} + \cos\dfrac{6\pi}{7} = -\dfrac{1}{2}$.

    c Hence show that $\cos\dfrac{2\pi}{7}$ is a root of the equation $8t^3 + 4t^2 - 4t - 1 = 0$.

Section 6: Geometry of complex numbers

Multiplication of complex numbers has an interesting geometrical interpretation. On an Argand diagram, let A be the point corresponding to the complex number $z_1 = r_1 \text{cis} \theta_1$, and let B be the point corresponding to the complex number $z_1 \times r_2 \text{cis} \theta_2 = r_1 r_2 \text{cis}(\theta_1 + \theta_2)$.

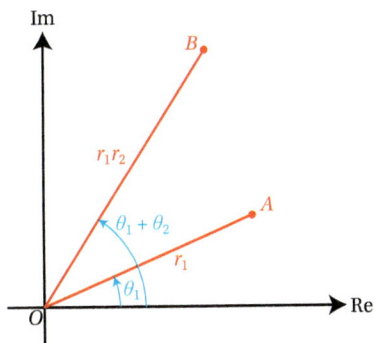

Then $OA = r_1$, $OB = r_1 r_2$, and $\angle AOB = (\theta_1 + \theta_2) - \theta_1 = \theta_2$. Hence the transformation that takes point A to point B is a rotation through angle θ_2 followed by an enlargement with scale factor r_2.

Key point 2.10

Multiplication by $r\operatorname{cis}\theta$ corresponds to a rotation about the origin though angle θ and an enlargement with scale factor r.

⏮ Rewind

You already know, from Pure Core Student Book 1, Chapter 4, that adding a complex number $a + ib$ corresponds to a translation with vector $\begin{pmatrix} a \\ b \end{pmatrix}$, and that taking the complex conjugate corresponds to a reflection in the real axis.

WORKED EXAMPLE 2.14

Points A and B on an Argand diagram represent complex numbers $a = \sqrt{3} + i$ and $b = 2\sqrt{2} + 2i\sqrt{2}$, respectively.

a Find the modulus and argument of a and b.

b Hence describe a combination of two transformations which maps A to B.

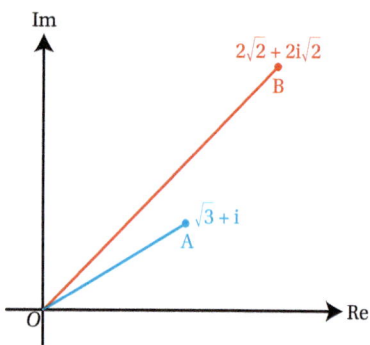

A diagram helps to find the modulus and the argument.

a $|a| = \sqrt{3+1} = 2$, $\arg(a) = \arctan\left(\dfrac{1}{\sqrt{3}}\right) = \dfrac{\pi}{6}$

$|b| = \sqrt{8+8} = 4$, $\arg(b) = \arctan\left(\dfrac{2\sqrt{2}}{2\sqrt{2}}\right) = \dfrac{\pi}{4}$

b Enlargement with scale factor 2

$|b| = 2|a|$

and rotation through $\dfrac{\pi}{4} - \dfrac{\pi}{6} = \dfrac{\pi}{12}$ about the origin.

The angle of rotation is the difference between the arguments.

The result from Key point 2.10 is remarkably powerful in some situations that have nothing to do with complex numbers.

WORKED EXAMPLE 2.15

An equilateral triangle has one vertex at the origin and another at $(1, 2)$. Find one possible set of coordinates of the third vertex.

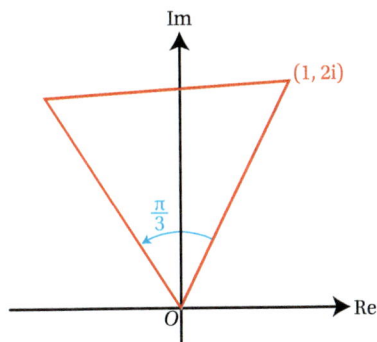

On an Argand diagram the point $(1, 2)$ corresponds to the complex number $1 + 2i$.

You can obtain the third vertex by rotation through $60°$ anticlockwise about the origin and no enlargement. This corresponds to multiplication by the complex number with modulus 1 and argument $\dfrac{\pi}{3}$.

The complex number corresponding to the third vertex is

$$(1+2i)\left(\cos\frac{\pi}{3}+i\sin\frac{\pi}{3}\right)=(1+2i)\left(\frac{1}{2}+\frac{\sqrt{3}}{2}i\right)$$

$$=\left(\frac{1}{2}-\sqrt{3}\right)+\left(\frac{\sqrt{3}}{2}+1\right)i$$

So the coordinates are

$$\left(\frac{1}{2}-\sqrt{3},\frac{\sqrt{3}}{2}+1\right)$$

Tip

There is another equilateral triangle with vertices $(0, 0)$ and $(1, 2)$. You can obtain it by rotating clockwise through $60°$, corresponding to multiplication by $\cos\left(-\frac{\pi}{3}\right)+i\sin\left(-\frac{\pi}{3}\right)$.

Focus on ...

You can use several different approaches to solve the problem from Worked example 2.15. You could use coordinate geometry and trigonometry, or you could use a matrix to carry out the rotation. In Focus on ... Problem solving 1 you will explore different approaches to similar problems.

Rewind

You studied rotation matrices in Pure Core Student Book 1, Chapter 3.

Division by $\text{cis}\,\theta_2$ is the same as multiplication by $\text{cis}(-\theta_2)$:

$$(r_1\text{cis}\,\theta_1) \div (\text{cis}\,\theta_2) = r_1\text{cis}(\theta_1 - \theta_2) = (r_1\text{cis}\,\theta_1) \times (\text{cis}(-\theta_2))$$

Geometrically, this represents a rotation through angle $-\theta_2$.

Key point 2.11

Division by $r\,\text{cis}\,\theta$ corresponds to a rotation about the origin though angle $-\theta$ and an enlargement with scale factor $\frac{1}{r}$.

Tip

If the angle θ is positive then multiplication corresponds to an anticlockwise rotation and division corresponds to a clockwise rotation.

Since raising to a positive integer power is repeated multiplication, in an Argand diagram it corresponds to repeated rotation and enlargement.

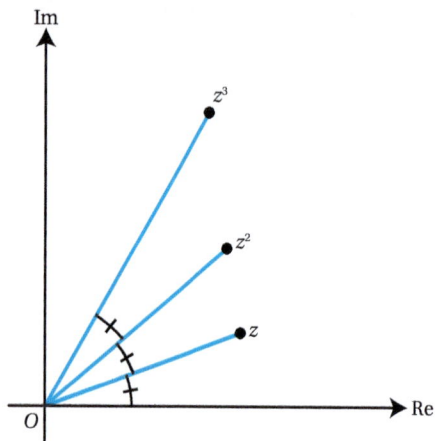

Although you can solve Worked example 2.16 just by doing the algebra, thinking about rotations might help you visualise what's going on.

WORKED EXAMPLE 2.16

Find the smallest positive integer value of n for which $\left(3\text{cis}\left(\dfrac{3\pi}{16}\right)\right)^n$ is pure imaginary.

$\left(\text{cis}\left(\dfrac{3\pi}{16}\right)\right)^n$ has argument $\dfrac{3n\pi}{16}$.

If $\dfrac{3n\pi}{16} = \dfrac{\pi}{2}$, $n = \dfrac{8}{3}$ which is not an integer.

If $\dfrac{3n\pi}{16} = \dfrac{3\pi}{2}$, $n = 8$.

Raising $3\text{cis}\left(\dfrac{3\pi}{16}\right)$ to a power corresponds to repeated rotation through $\dfrac{3\pi}{16}$ (combined with an enlargement with scale factor 3).

The question is therefore: how many rotations through $\dfrac{3\pi}{16}$ are needed to reach either $\dfrac{\pi}{2}$ or $\dfrac{3\pi}{2}$?

EXERCISE 2F

1 Points A and B represent complex numbers $a = 4 + i$ and $b = 5 + 3i$ on an Argand diagram.

    a Find the modulus and argument of a and b.

    b Point A is mapped to point B by a combination of an enlargement and a rotation. Find the scale factor of the enlargement and the angle of rotation.

2 Points P and Q represent complex numbers $p = 3 + 5i$ and $q = -\sqrt{30} + 2i$, respectively.

    a Show that $|p| = |q|$.   **b** Describe a single transformation that maps P to Q.

3 The complex number corresponding to the point A in the diagram is $z_1 = 3 + 2i$. The distance $OB = OA$. Find, in surd form, the complex number corresponding to the point B.

4 The diagram shows a square $OABC$, where A has coordinates $(5, 2)$.

Find the exact coordinates of B and C.

5 Point A in the diagram corresponds to the complex number a. The complex number z equals $\text{cis}\left(\dfrac{\pi}{6}\right)$.

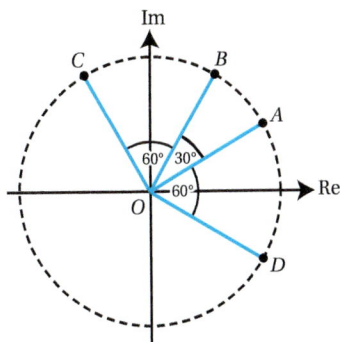

Write the complex numbers b, c and d corresponding to the points B, C and D in terms of a and z.

6 The diagram shows a right angled triangle OAB with angle $AOB = 30°$. The coordinates of A are $(6, 3)$.

   a Find the exact length OB.

   b Using complex numbers, or otherwise, find the coordinates of B.

7 Let $z = 0.6 + 0.8i$.

   a Represent z, z^2 and z^3 on an Argand diagram.

   b Describe fully the transformation mapping z to z^3.

8 The diagram shows line l through the origin with gradient $\sqrt{3}$ and the point A representing the complex number $\sqrt{2} + \sqrt{2}i$.

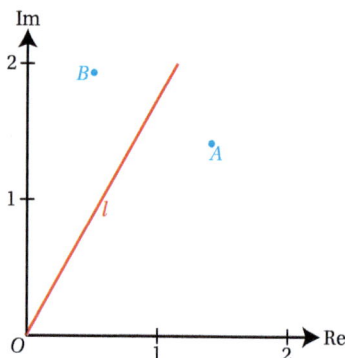

   a The line l is the locus of $z \in \mathbb{C}$ which satisfy $\arg z = \theta$. Find the exact value of θ.

   Point B is the reflection of point A in the line l.

   b Find the size of the angle AOB.

   c Use complex numbers to find the exact coordinates of B.

9 The diagram shows an equilateral triangle with its centre at the origin and one vertex $A(4, -1)$.

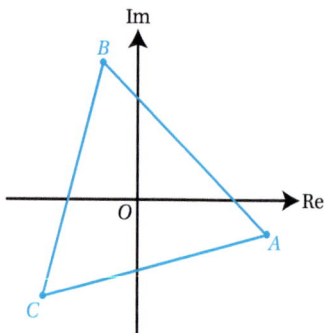

  a Write down the complex number corresponding to the vertex A.

  b Hence find the coordinates of the other two vertices.

10 The diagram shows a regular pentagon with one vertex at $z = 2$.

Write down, in the form $r\operatorname{cis}\theta$, the complex numbers corresponding to the other four vertices.

11 **a** The point representing a complex number z on an Argand diagram is reflected in the real axis and then rotated 90° anticlockwise about the origin. Write down, in terms of z, the complex number representing the resulting image.

  b If the rotation is applied before the reflection, show that the resulting image represents the complex number $-iz^*$.

12 **a** The point representing the complex number p on an Argand diagram is rotated through angle θ about the point representing the complex number a. The resulting point represents complex number q. Explain why $q - a = (p - a)e^{i\theta}$.

  b Find the exact coordinates of the image when the point $P(1, 3)$ is rotated 60° anticlockwise about the point $A(2, -1)$.

13 **a** On an Argand diagram, points A, B and C represent complex numbers a, b and c, respectively. C is the image of B after a rotation through angle θ, anticlockwise, about A.

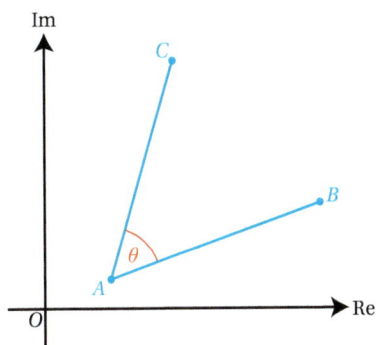

Express the complex number $c - a$ in terms of $b - a$.

  b The point $(4, 1)$ is rotated 45° anticlockwise about the origin. The image is then rotated 30° anticlockwise about the point $(-1, 2)$. Find the coordinates of the final image.

14 The diagram shows two equilateral triangles on an Argand diagram.

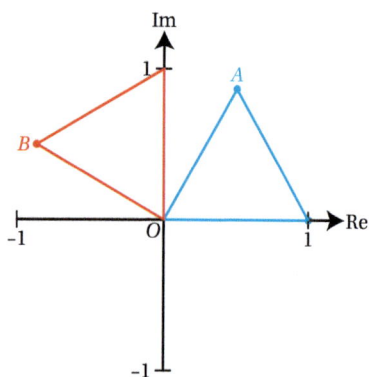

Find the complex number corresponding to the midpoint of AB.
Give your answer in exact Cartesian form.

✎ Checklist of learning and understanding

- De Moivre's theorem: $\left(r(\cos\theta + i\sin\theta) \right)^n = r^n \left(\cos n\theta + i\sin n\theta \right)$ for $n \in \mathbb{Z}$.
- Exponential form of a complex number: $re^{i\theta} = [r, \theta] = r\operatorname{cis}\theta = r(\cos\theta + i\sin\theta)$.
- To solve $z^n = w$:
  - write w in modulus–argument form and write $z^n = r^n \left(\cos n\theta + i\sin n\theta \right)$
  - compare moduli, remembering that they are always real
  - compare arguments, remembering that adding 2π to the argument does not change the number
  - the n roots form a regular polygon on an Argand diagram.
- The nth roots of unity (solutions of $z^n = 1$):

  - You can write them as ω^k, where $\omega = e^{\frac{2\pi i}{n}}$ and $k = 0, 1, 2, \ldots, n-1$.
  - $1 + \omega_1 + \ldots + \omega_{n-1} = 0$.
- You can use roots of the equation $z^n = a$ to factorise the expression $z^n - a$. Real quadratic factors are found by combining each root with its complex conjugate:

  $$\left(z - re^{i\theta}\right)\left(z - re^{-i\theta}\right) = z^2 - 2r\cos\theta\, z + r^2$$

- Multiplication by $re^{i\theta}$ corresponds to an enlargement with scale factor r and a rotation through θ anticlockwise about the origin. Division by $r\operatorname{cis}\theta$ corresponds to rotation around the origin through angle $-\theta$ and an enlargement with scale factor $\frac{1}{r}$.

Mixed practice 2

1 If $z^* = r\mathrm{e}^{\mathrm{i}\theta}$, write $\dfrac{1}{z}$ in exponential form.

2 **a** Find the modulus and argument of $-1+\mathrm{i}\sqrt{3}$.

b Hence find $\left(-1+\mathrm{i}\sqrt{3}\right)^5$ in exact Cartesian form.

3 **a** Write down, in the form $r\mathrm{e}^{\mathrm{i}\theta}$, all the roots of the equation $z^5 = 1$.

b Show the roots on an Argand diagram.

4 Find $\left(\cos\dfrac{\pi}{3}+\mathrm{i}\sin\dfrac{\pi}{3}\right)^4\left(\cos\dfrac{\pi}{4}+\mathrm{i}\sin\dfrac{\pi}{4}\right)^5$ in the form $r\mathrm{e}^{\mathrm{i}\theta}$, where $-\pi < \theta \leqslant \pi$.

5 **a** Find the modulus and argument of $8-8\mathrm{i}$.

b Hence solve the equation $z^4 = 8-8\mathrm{i}$, giving your answers in the form $r\mathrm{e}^{\mathrm{i}\theta}$.

6 **a** Find the modulus and argument of $1+\mathrm{i}$.

b A regular hexagon is inscribed in a circle on an Argand diagram, centred at the origin, and one of its vertices is $1+\mathrm{i}$. Find an equation whose roots are represented by the six vertices of the hexagon.

7 **i** Express $\dfrac{\sqrt{3}+\mathrm{i}}{\sqrt{3}-\mathrm{i}}$ in the form $r\mathrm{e}^{\mathrm{i}\theta}$, where $r>0$ and $0 \leqslant \theta < 2\pi$.

ii Hence find the smallest positive value of n for which $\left(\dfrac{\sqrt{3}+\mathrm{i}}{\sqrt{3}-\mathrm{i}}\right)^n$ is real and positive.

© **OCR A Level Mathematics, Unit 4727 Further Pure Mathematics 3, January 2009**

8 If $\arg\!\left((a+\mathrm{i})^3\right)=\pi$, where a is real and positive, find the exact value of a.

9 Find the exact value of $\dfrac{1}{\left(\sqrt{3}+\mathrm{i}\right)^6}$, clearly showing your working.

10 **a** Express $\dfrac{\sqrt{3}}{2}-\dfrac{1}{2}\mathrm{i}$ in the form $r(\cos\theta+\mathrm{i}\sin\theta)$.

b Hence show that $\left(\dfrac{\sqrt{3}}{2}-\dfrac{1}{2}\mathrm{i}\right)^9 = c\mathrm{i}$ where c is a real number to be found.

c Find one pair of possible values of positive integers m and n such that

$$\left(\dfrac{\sqrt{3}}{2}-\dfrac{1}{2}\mathrm{i}\right)^m = \left(\dfrac{\sqrt{2}}{2}+\dfrac{\sqrt{2}}{2}\mathrm{i}\right)^n.$$

11 Use trigonometric identities to show that

a $\dfrac{1}{\mathrm{cis}\,\theta}=\mathrm{cis}(-\theta)=\mathrm{cis}(2\pi-\theta)$ **b** $\dfrac{\mathrm{cis}\,\theta_1}{\mathrm{cis}\,\theta_2}=\mathrm{cis}(\theta_1-\theta_2)$

12 If ω is a complex third root of unity and x and y are real numbers, prove that:

a $1+\omega+\omega^2 = 0$

b $\left(\omega x+\omega^2 y\right)\left(\omega^2 x+\omega y\right)=x^2-xy+y^2$.

13 If $0<\theta<\dfrac{\pi}{2}$ and $z=\left(\sin\theta+\mathrm{i}(1-\cos\theta)\right)^2$, find in its simplest form $\arg z$.

14 If $z=\cos\theta+\mathrm{i}\sin\theta$, prove that $\dfrac{z^2-1}{z^2+1}=\mathrm{i}\tan\theta$.

15 **a** Express i in the form $r\mathrm{e}^{\mathrm{i}\theta}$.

b Hence state the exact value of i^{i}.

16 The complex numbers 0, 3 and $3e^{\frac{1}{3}\pi i}$ are represented in an Argand diagram by the points O, A and B respectively.

i Sketch the triangle OAB and show that it is equilateral.

ii Hence express $3 - 3e^{\frac{1}{3}\pi i}$ in polar form.

iii Hence find $\left(3 - 3e^{\frac{1}{3}\pi i}\right)^5$, giving your answer in the form $a + b\sqrt{3}i$ where a and b are rational numbers.

© OCR A Level Mathematics, Unit 4727/01 Further Pure Mathematics 3, June 2013

17 Let $\omega = e^{\frac{2i\pi}{5}}$.

a Write ω^2, ω^3 and ω^4 in the form $e^{i\theta}$.

b Explain why $\omega^1 + \omega^2 + \omega^3 + \omega^4 = -1$.

c Show that $\omega + \omega^4 = 2\cos\dfrac{2\pi}{5}$ and $\omega^2 + \omega^3 = 2\cos\dfrac{4\pi}{5}$.

d Form a quadratic equation in $\cos\dfrac{2\pi}{5}$ and hence show that $\cos\dfrac{2\pi}{5} = \dfrac{\sqrt{5}-1}{4}$.

18 Let 1, ω, ω^2 be the solutions of the equation $z^3 = 1$.

a Show that $1 + \omega + \omega^2 = 0$.

b Find the value of

   i $(1+\omega)(1+\omega^2)$ **ii** $\dfrac{1}{1+\omega} + \dfrac{1}{1+\omega^2}$.

c Hence find a cubic equation with integer coefficients and roots 3, $\dfrac{1}{1+\omega}$ and $\dfrac{1}{1+\omega^2}$.

19 Point A has coordinates $(1, 2)$. Triangle OAB has $OB = 2OA$ and angle $AOB = \dfrac{\pi}{6}$.

a Write down the complex number corresponding to the point A.

b Find the two possible pairs of coordinates of B.

20 Let Z and A be points on an Argand diagram representing complex numbers z and a, respectively. The complex number z_1 represents the point obtained by translating Z using the vector $\overrightarrow{OA}$ and then rotating the image through angle θ anticlockwise about the origin. The complex number z_2 corresponds to the point obtained by first rotating Z anticlockwise through angle θ about the origin and then translating Z by vector $\overrightarrow{OA}$.

Show that the distance between the points represented by z_1 and z_2 is independent of z.

21 **i** Show that $(z - e^{i\phi})(z - e^{-i\phi}) \equiv z^2 - (2\cos\phi)z + 1$.

ii Write down the seven roots of the equation $z^7 - 1$ in the form $e^{i\theta}$ and show their positions in an Argand diagram.

iii Hence express $z^7 - 1$ as the product of one real linear factor and three real quadratic factors.

© OCR A Level Mathematics, Unit 4727/01 Further Pure Mathematics 3, June 2007

22 Point A represents the complex number $3 + i$ on an Argand diagram. Point A is rotated through $\dfrac{\pi}{3}$ radians anticlockwise about the origin to point B. Point B is then translated by $\begin{pmatrix} -2 \\ 1 \end{pmatrix}$ to obtain point C.

a Find, in Cartesian form, the complex number corresponding to B.

b Find the distance AC.

23 **a** Points P and Q on an Argand diagram correspond to complex numbers $z_1 = x_1 + iy_1$ and $z_2 = x_2 + iy_2$. Show that $PQ = |z_1 - z_2|$.

b The diagram shows a triangle with one vertex at the origin, one vertex at the point $A(a, 0)$ and one vertex at the point B such that $OB = b$ and $\angle AOB = \theta$.

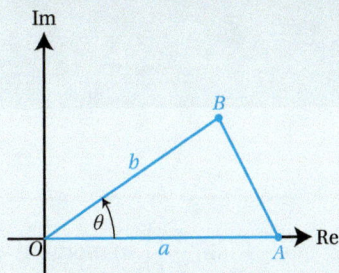

i Write down the complex number corresponding to point A.

ii Write down the complex number corresponding to point B in modulus–argument form.

iii Write down an expression for the length of AB in terms of a, b and θ.

iv Hence prove the cosine rule for the triangle AOB:

$$AB^2 = OA^2 + OB^2 - 2(OA)(OB)\cos AOB.$$

3 Complex numbers and trigonometry

In this chapter you will learn how to:

- use de Moivre's theorem to derive trigonometric identities
- find sums of some trigonometric series.

Before you start…

Chapter 2, Section 1	You should be able to use de Moivre's theorem to raise a complex number to a power.	1 Find $\left(2\left(\cos\frac{\pi}{7}+i\sin\frac{\pi}{7}\right)\right)^5$ in modulus–argument form.
Chapter 2, Section 2	You should be able to use the exponential form of a complex number.	2 **a** Write $4e^{-\frac{i\pi}{3}}$ in exact Cartesian form. **b** Write down the complex conjugate of $2+e^{3i}$.
A Level Mathematics Student Book 1, Chapter 9	You should be able to use the binomial expansion for positive integer powers.	3 Expand and simplify $(x-2y)^5$.
Chapter 2, Section 2, Pure Core Student Book 1, Chapter 4	You should know how to divide complex numbers.	4 Find the real and imaginary part of $\dfrac{1-e^{ix}}{1+e^{ix}}$.
A Level Mathematics Student Book 2, Chapter 4	You should be able to use the formulae for the sum of a geometric series.	5 **a** Find an expression for the sum of the first n terms of the geometric series $e^x+e^{2x}+e^{3x}+...$ **b** For which values of x does the series in part **a** have a sum to infinity?

Using complex numbers to derive trigonometric identities

The modulus–argument form of a complex number provides a link between complex numbers and trigonometry. This is a powerful tool for deriving new trigonometric identities.

These trigonometric identities are one example of the use of complex numbers to establish facts about real numbers and functions. Other such applications include a formula for cubic equations, calculations involving alternating current, and analysing the motion of waves. The fact that complex numbers proved correct results in a concise way was a major factor in convincing mathematicians that they should be accepted.

Section 1: Deriving multiple angle formulae

You can raise a complex number to a power in two different ways. You can either use the Cartesian form and multiply out the brackets, or you can write the complex number in modulus–argument form and use de Moivre's theorem. Equating these two answers allows you to derive formulae for trigonometric ratios of multiple angles.

⏪ **Rewind**

You have already met double angle formulae, such as $\cos 2\theta = 2\cos^2 \theta - 1$, in A Level Mathematics Student Book 2, Chapter 8.

WORKED EXAMPLE 3.1

Derive a formula for $\cos 4\theta$ in terms of $\cos\theta$.

Let $z = \cos\theta + i\sin\theta$.

Then $z^4 = (\cos\theta + i\sin\theta)^4$.

First using the binomial theorem:

Start with an expression for a complex number involving $\cos\theta$, and find z^4 in two different ways.

$$z^4 = \cos^4\theta + 4\cos^3\theta(i\sin\theta) + 6\cos^2\theta(i\sin\theta)^2$$
$$+ 4\cos\theta(i\sin\theta)^3 + (i\sin\theta)^4$$

$$= \cos^4\theta + 4i\cos^3\theta\sin\theta$$
$$- 6\cos^2\theta\sin^2\theta - 4i\cos\theta\sin^3\theta + \sin^4\theta$$

$i^2 = -1,\ i^3 = -i,\ i^4 = 1.$

Now using de Moivre's theorem:

$$z^4 = \cos 4\theta + i\sin 4\theta$$

Equating real parts:

$$\cos 4\theta = \cos^4\theta - 6\cos^2\theta\sin^2\theta + \sin^4\theta$$

The two expressions for z^4 must have equal real parts and equal imaginary parts.

$$= \cos^4\theta - 6\cos^2\theta(1 - \cos^2\theta)$$
$$+ (1 - \cos^2\theta)^2$$

You want the answer in terms of $\cos\theta$ only, so use $\sin^2\theta = 1 - \cos^2\theta$.

$$\therefore \cos 4\theta = 8\cos^4\theta - 8\cos^2\theta + 1$$

Simplify the final expression.

⏩ **Fast forward**

By equating imaginary parts of the two expressions in Worked example 3.1, you can obtain a similar expression for $\sin 4\theta$ (see question 1 in Exercise 3A).

ℹ️ **Did you know?**

These expressions for sines and cosines of multiple angles can also be derived through repeated application of compound angle identities. However, the calculations become increasingly long.

EXERCISE 3A

1 **a** Find the imaginary part of $(\cos\theta + i\sin\theta)^4$.

   b Hence show that $\sin 4\theta = 4\cos\theta(\sin\theta - 2\sin^3\theta)$.

2 Use the binomial expansion to find the real and imaginary parts of $(\cos\theta + i\sin\theta)^3$. Hence find an expression for $\sin 3\theta$ in terms of $\sin\theta$.

3 **a** Expand $(\cos\theta + i\sin\theta)^5$.

   b Hence or otherwise express $\sin 5\theta$ in terms of $\sin\theta$.

4 **a** Show that $\cos 5\theta = 16\cos^5\theta - 20\cos^3\theta + 5\cos\theta$.

  b Hence solve the equation $\cos 5\theta = 5\cos\theta$ for $\theta \in [0, 2\pi]$.

5 **a** Find the values of A, B and C such that $\sin 5\theta = A\sin^5\theta - B\sin^3\theta + C\sin\theta$.

  b Given that $4\sin^5\theta + \sin 5\theta = 0$, find the possible values of $\sin\theta$.

6 **a** Find the real and imaginary parts of $(\cos\theta + i\sin\theta)^4$.

  b Hence express $\tan 4\theta$ in terms of $\tan\theta$.

7 **a** Show that $\tan 6\theta = \dfrac{6\tan\theta - 20\tan^3\theta + 6\tan^5\theta}{1 - 15\tan^2\theta + 15\tan^4\theta - \tan^6\theta}$.

  b Hence solve the equation:

  $\tan^6\theta + 6\tan^5\theta - 15\tan^4\theta - 20\tan^3\theta + 15\tan^2\theta + 6\tan\theta - 1 = 0$

  for $\theta \in \left[0, \dfrac{\pi}{2}\right]$.

8 **a** Use the binomial expansion to find the real and imaginary parts of $(\cos\theta + i\sin\theta)^5$.

  b Hence show that $\dfrac{\sin 5\theta}{\sin\theta} = 16\cos^4\theta - 12\cos^2\theta + 1$.

  c Assuming that θ is small enough that the terms in θ^4 and higher can be ignored, find an approximate expression, in increasing powers of θ, for $\dfrac{\sin 5\theta}{\sin\theta}$.

Section 2: Application to polynomial equations

In Section 1 you learnt how to express $\sin n\theta$ and $\cos n\theta$ as a polynomial in $\sin\theta$ or $\cos\theta$. For example, $\cos 4\theta = 8\cos^4\theta - 8\cos^2\theta + 1$. You can now use the roots of the polynomial and the solutions of $\cos 4\theta = 0$ to find the values of $\cos\theta$.

WORKED EXAMPLE 3.2

  a Find all the values of $\theta \in [0, 2\pi)$ for which $\cos 4\theta = 0$.

  You are given that $\cos 4\theta = 8\cos^4\theta - 8\cos^2\theta + 1$.

  b Write down the roots of the equation $8c^2 - 8c^4 = 1$ in the form $\cos\theta$, where $\theta \in [0, \pi)$.

  c Hence find the exact value of $\cos\dfrac{3\pi}{8}$.

  a $\theta \in [0, 2\pi) \Rightarrow 4\theta \in [0, 8\pi)$

  $4\theta = \dfrac{\pi}{2}, \dfrac{3\pi}{2}, \dfrac{5\pi}{2}, \dfrac{7\pi}{2}, \dfrac{9\pi}{2}, \dfrac{11\pi}{2}, \dfrac{13\pi}{2}, \dfrac{15\pi}{2}$

  $\theta = \dfrac{\pi}{8}, \dfrac{3\pi}{8}, \dfrac{5\pi}{8}, \dfrac{7\pi}{8}, \dfrac{9\pi}{8}, \dfrac{11\pi}{8}, \dfrac{13\pi}{8}, \dfrac{15\pi}{8}$

  b Write $c = \cos\theta$.

  Then

  $8c^2 - 8c^4 = 1$

  $\Leftrightarrow 8c^4 - 8c^2 + 1 = 0$

  $\Leftrightarrow \cos 4\theta = 0$

  > Making the substitution relates the equations from parts **a** and **b**.

Continues on next page ...

Hence

$$c = \cos\theta$$

$$= \cos\frac{\pi}{8}, \cos\frac{3\pi}{8}, \cos\frac{5\pi}{8}, \cos\frac{7\pi}{8}$$

The equation from part **a** has eight solutions but the equation from part **b** should only have four (since it is a degree 4 polynomial). This is because, for example,

$$\cos\frac{\pi}{8} = \cos\frac{15\pi}{8}.$$

c $8c^4 - 8c^2 + 1 = 0$

$$c^2 = \frac{2 \pm \sqrt{2}}{4}$$

$$c = \pm\sqrt{\frac{2 \pm \sqrt{2}}{4}}$$

You can actually solve the equation from part **b** exactly, as it is a quadratic in c^2.

$\cos\dfrac{3\pi}{8}$ is the smallest positive one of the four solutions from part **b**.

$\cos\dfrac{3\pi}{8}$ is one of these four solutions. You can see from the $\cos$ graph that it is the smallest positive one of the four numbers.

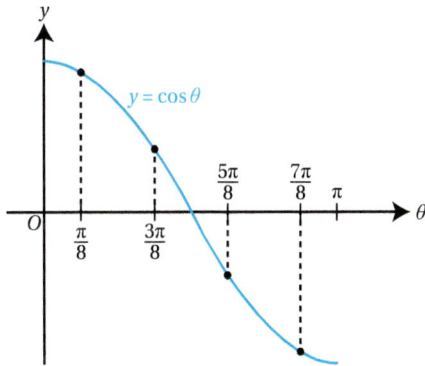

$$\therefore \cos\frac{3\pi}{8} = \sqrt{\frac{2 - \sqrt{2}}{4}}$$

Sometimes you can't solve the polynomial equation, but you can still use the results about sums and products of roots to derive expressions involving combinations of trigonometric ratios.

Rewind

See Pure Core Student Book 1, Chapter 5, for a reminder about roots of polynomials.

WORKED EXAMPLE 3.3

a Show that $\tan 3\theta = \dfrac{3\tan\theta - \tan^3\theta}{1 - 3\tan^2\theta}$.

b Show that the equation $t^3 - 3t^2 - 3t + 1 = 0$ can be written as $\tan 3\theta = k$, where $t = \tan\theta$, and state the value of k.

c Hence find the exact value of $\tan\dfrac{\pi}{12} + \tan\dfrac{5\pi}{12}$.

Continues on next page ...

a Write $c = \cos\theta$ and $s = \sin\theta$.

Then

$$\cos 3\theta + i\sin 3\theta = (\cos\theta + i\sin\theta)^3$$

$$= c^3 + 3ic^2 s - 3cs^2 - is^3$$

| Use de Moivre's theorem. |

Hence

$$\tan 3\theta = \frac{\sin 3\theta}{\cos 3\theta} = \frac{3c^2 s - s^3}{c^3 - 3cs^2}$$

| Separate real and imaginary parts to find sin and cos. |

$$= \frac{3t - t^3}{1 - 3t^2}$$

| Divide top and bottom by c^3 and use $\dfrac{s}{c} = \tan\theta$. |

where $t = \tan\theta$.

b $t^3 - 3t^2 - 3t + 1 = 0$

$\Leftrightarrow 1 - 3t^2 = 3t - t^3$

$\Leftrightarrow \dfrac{3t - t^3}{1 - 3t^2} = 1$

$\Leftrightarrow \tan 3\theta = 1$

(so $k = 1$)

| Rearrange the equation into the form from part **a**. |

c $\tan 3\theta = 1$

$$3\theta = \frac{\pi}{4}, \frac{5\pi}{4}, \frac{9\pi}{4} \dots$$

$$\theta = \frac{\pi}{12}, \frac{5\pi}{12}, \frac{9\pi}{12} \dots$$

| Solve the cubic equation by solving $\tan 3\theta = 1$. |

$$t = \tan\theta = \tan\frac{\pi}{12}, \tan\frac{5\pi}{12} \text{ or } \tan\frac{9\pi}{12}$$

| Although there are infinitely many values of θ, they only give three different values of $\tan\theta$ (since tan is a periodic function). |

Hence

$$\tan\frac{\pi}{12} + \tan\frac{5\pi}{12} + \tan\frac{9\pi}{12} = -\frac{-3}{1} = 3$$

| Use the result about the sum of the roots of a cubic polynomial: $$p + q + r = -\frac{b}{a}$$ |

$$\Rightarrow \tan\frac{\pi}{12} + \tan\frac{5\pi}{12} + (-1) = 3$$

| $\tan\dfrac{9\pi}{12} = \tan\dfrac{3\pi}{4} = -1$ |

$$\Rightarrow \tan\frac{\pi}{12} + \tan\frac{5\pi}{12} = 4$$

EXERCISE 3B

1 **a** Write down an expression for $\cos 2\theta$ in terms of $\cos\theta$.

  b Given that $\cos 2\theta = \dfrac{\sqrt{3}}{2}$, find a quadratic equation in c, where $c = \cos\theta$.

  c Hence find the exact value of $\cos\dfrac{\pi}{12}$.

2 **a** Given that $\tan 2\theta = 1$, show that $t^2 + 2t - 1 = 0$, where $t = \tan\theta$.

   b Solve the equation $\tan 2\theta = 1$ for $\theta \in (0, \pi)$.

   c Hence find the exact value of $\tan\dfrac{5\pi}{8}$.

3 You are given that $\cos 5\theta = 16\cos^5\theta - 20\cos^3\theta + 5\cos\theta$.

   a Find the possible values of $\theta \in [0, \pi]$ for which $16\cos^4\theta - 20\cos^2\theta + 5 = 0$.

   b Hence show that $\cos\dfrac{\pi}{10}\cos\dfrac{3\pi}{10} = \dfrac{\sqrt{5}}{4}$.

4 **a** Show that $\sin 3\theta = 3\sin\theta - 4\sin^3\theta$.

   b Given that $\theta \in [0, 2\pi]$ and that $\sin 3\theta = \dfrac{1}{2}$, find the possible values of θ.

   c Hence show that $\sin\dfrac{\pi}{18}$ is a root of the equation $8x^3 - 6x + 1 = 0$ and find, in a similar form, the other two roots.

5 You are given that $\tan 4\theta + \tan 3\theta = 0$.

   a Show that $\tan 7\theta = 0$.

   b Let $t = \tan\theta$. Express $\tan 4\theta$ and $\tan 3\theta$ in terms of t. Hence show that $t^7 - 21t^5 + 35t^3 - 7t = 0$.

6 You are given that $\sin 7\theta = 7\sin\theta - 56\sin^3\theta + 112\sin^5\theta - 64\sin^7\theta$.

   a Show that the equation $64s^6 - 112s^4 + 56s^2 - 7 = 0$ has roots $\sin\left(\pm\dfrac{\pi}{7}\right)$, $\sin\left(\pm\dfrac{3\pi}{7}\right)$ and $\sin\left(\pm\dfrac{5\pi}{7}\right)$.

   b Hence find the exact value of $\sin\dfrac{\pi}{7}\sin\dfrac{3\pi}{7}\sin\dfrac{5\pi}{7}$.

Section 3: Powers of trigonometric functions

Another important link with trigonometry comes from considering the exponential form of complex numbers:

$$e^{i\theta} = \cos\theta + i\sin\theta$$

and

$$e^{-i\theta} = \cos(-\theta) + i\sin(-\theta)$$
$$= \cos\theta - i\sin\theta$$

By adding and subtracting these two equations you can establish two very useful identities.

Key point 3.1

$$\cos\theta = \frac{e^{i\theta} + e^{-i\theta}}{2}$$

$$\sin\theta = \frac{e^{i\theta} - e^{-i\theta}}{2i}$$

You can further generalise this result.

Key point 3.2

If $z = e^{i\theta}$, then

$$z^n + \frac{1}{z^n} = 2\cos n\theta$$

$$z^n - \frac{1}{z^n} = 2i\sin n\theta$$

PROOF 2

Using de Moivre's theorem for positive and negative integers:

$z^n = \cos n\theta + i\sin n\theta$

$\frac{1}{z^n} = z^{-n} = \cos n\theta - i\sin n\theta$

Adding the two equations:

$z^n + \frac{1}{z^n} = 2\cos n\theta$

Subtracting the two equations:

$z^n - \frac{1}{z^n} = 2i\sin n\theta$

Remember that $\cos(-x) = \cos(x)$ and $\sin(-x) = -\sin x$.

You can use these results to derive another class of trigonometric identities, expressing powers of trigonometric functions in terms of functions of multiple angles. For example: $\cos^2\theta = \frac{1}{2}(\cos 2\theta + 1)$.

WORKED EXAMPLE 3.4

Show that $\sin^5\theta = \frac{1}{16}\sin 5\theta - \frac{5}{16}\sin 3\theta + \frac{5}{8}\sin\theta$.

Let $z = \cos\theta + i\sin\theta$.

Using the binomial expansion:

$\left(z - \frac{1}{z}\right)^5 = z^5 + 5z^4\left(-\frac{1}{z}\right) + 10z^3\left(-\frac{1}{z}\right)^2 + 10z^2\left(-\frac{1}{z}\right)^3 + 5z\left(-\frac{1}{z}\right)^4 + \left(-\frac{1}{z}\right)^5$

$= z^5 - 5z^3 + 10z - \frac{10}{z} + \frac{5}{z^3} - \frac{1}{z^5}$

$= \left(z^5 - \frac{1}{z^5}\right) - 5\left(z^3 - \frac{1}{z^3}\right) + 10\left(z - \frac{1}{z}\right)$

Simplify the fractions, taking care with negative signs.

Group the terms to get expressions of the form $z^n - \frac{1}{z^n}$.

Continues on next page …

So

$$(2i\sin\theta)^5 = 2i\sin5\theta - 10i\sin3\theta + 20i\sin\theta$$

$$32i\sin^5\theta = 2i\sin5\theta - 10i\sin3\theta + 20i\sin\theta$$

$$\therefore \sin^5\theta = \frac{1}{16}\sin5\theta - \frac{5}{16}\sin3\theta + \frac{5}{8}\sin\theta$$

> On both sides of the equation, use the result from Key point 3.2:
>
> $$z^n - \frac{1}{z^n} = 2i\sin n\theta$$

Trigonometric identities such as these are very useful when integrating powers of trigonometric functions.

Rewind

In A Level Mathematics Student Book 2, Chapter 11, you used the identity $\cos^2\theta = \frac{1}{2}(\cos2\theta+1)$ to find $\int\cos^2 x\,dx$.

WORKED EXAMPLE 3.5

a Expand and simplify $\left(z+\dfrac{1}{z}\right)^6$.

b Show that $\cos^6 x = \dfrac{1}{32}\cos6x + \dfrac{3}{16}\cos4x + \dfrac{15}{32}\cos2x + \dfrac{5}{16}$.

c Hence find $\int\cos^6 x\,dx$.

a Using the binomial expansion:

$$\left(z+\frac{1}{z}\right)^6 = z^6 + 6z^5\left(\frac{1}{z}\right) + 15z^4\left(\frac{1}{z}\right)^2 + 20z^3\left(\frac{1}{z}\right)^3 + 15z^2\left(\frac{1}{z}\right)^4 + 6z\left(\frac{1}{z}\right)^5 + \left(\frac{1}{z}\right)^6$$

$$= z^6 + 6z^4 + 15z^2 + 20 + \frac{15}{z^2} + \frac{6}{z^4} + \frac{1}{z^6}$$

b Let $z = \cos x + i\sin x$.

$$\left(z+\frac{1}{z}\right)^6 = \left(z^6 + \frac{1}{z^6}\right) + 6\left(z^4 + \frac{1}{z^4}\right) + 15\left(z^2 + \frac{1}{z^2}\right) + 20$$

$$\Rightarrow (2\cos x)^6 = (2\cos6x) + 6(2\cos4x) + 15(2\cos2x) + 20$$

$$\Rightarrow \cos^6 x = \frac{1}{32}\cos6x + \frac{3}{16}\cos4x + \frac{15}{32}\cos2x + \frac{5}{16}$$

> Group the terms on the right so that you can use the result from Key point 3.2.

> Divide by $2^6 = 64$.

c Using the result from part **a**:

$$\int\cos^6 x\,dx = \int\left(\frac{1}{32}\cos6x + \frac{3}{16}\cos4x + \frac{15}{32}\cos2x + \frac{5}{16}\right)dx$$

$$= \frac{1}{192}\sin6x + \frac{3}{64}\sin4x + \frac{15}{64}\sin2x + \frac{5}{16}x + c$$

> Don't forget to divide by the coefficient of x.

EXERCISE 3C

1 Let $z = \cos\theta + i\sin\theta$. Express each of these as a sum of terms of the form $\cos k\theta$ or $\sin k\theta$.

a i $\left(z + \dfrac{1}{z}\right)^3$ **ii** $\left(z + \dfrac{1}{z}\right)^4$

b i $\left(z - \dfrac{1}{z}\right)^4$ **ii** $\left(z - \dfrac{1}{z}\right)^5$

2 Let $z = \cos\theta + i\sin\theta$.

a Show that $z^n + z^{-n} = 2\cos n\theta$.

b Hence show that $32\cos^5\theta = A\cos 5\theta + B\cos 3\theta + C\cos\theta$ where A, B and C are constants to be found.

3 **a** Use the expansion of $\left(z - \dfrac{1}{z}\right)^6$, where $z = e^{i\theta}$, to show that $32\sin^6\theta = 10 - 15\cos 2\theta + 6\cos 4\theta - \cos 6\theta$.

b Hence find the exact value of $\int_0^{\frac{\pi}{3}} \sin^6\theta\, d\theta$.

4 A complex number is defined by $z = \cos\theta + i\sin\theta$.

a i Show that $\dfrac{1}{z} = \cos\theta - i\sin\theta$.

ii Use de Moivre's theorem to deduce that $z^n - \dfrac{1}{z^n} = 2i\sin n\theta$.

b i Expand $\left(z - \dfrac{1}{z}\right)^5$.

ii Hence find integers a, b and c such that
$16\sin^5\theta = a\sin 5\theta + b\sin 3\theta + c\sin\theta$.

c Find $\int \sin^5 2x\, dx$.

5 Let $z = \cos\theta + i\sin\theta$.

a Show that $z^n - z^{-n} = 2i\sin n\theta$.

b Expand $\left(z + z^{-1}\right)^6$ and $\left(z - z^{-1}\right)^6$.

c Hence show that $\cos^6\theta + \sin^6\theta = \dfrac{1}{8}(3\cos 4\theta + 5)$.

6 **a** Write down expressions for $\sin x$ and $\cos x$ in terms of e^{ix}.

b Hence evaluate $\int_0^\pi \sin^3 x\cos^4 x\, dx$, clearly showing your working.

Section 4: Trigonometric series

In Section 1 you learnt about expressions for sine and cosine of multiple angles. What happens if you add several such expressions together? For example, is it possible to simplify a sum such as $\sin x + \sin 2x + \sin 3x + \sin 4x$?

Explore

Sums like these come up when combining waves (interference). They are also used in Fourier series, which is a way of writing other functions in terms of sines and cosines.

You can simplify certain sums of this type using the exponential form of complex numbers and the formula for the sum of geometric series. This is because $\sin kx$ is the imaginary part of e^{ikx}, and the numbers e^{ix}, e^{2ix}, e^{3ix}, e^{4ix} form a geometric series.

> ⏮ **Rewind**
>
> You met geometric series in A Level Mathematics Student Book 2, Chapter 4.

WORKED EXAMPLE 3.6

a Find an expression for $e^{ix} + e^{2ix} + e^{3ix} + \ldots + e^{nix}$.

b Hence show that $\sin x + \sin 2x + \sin 3x + \ldots + \sin 10x = \dfrac{\sin x + \sin 10x - \sin 11x}{4\sin^2\left(\dfrac{x}{2}\right)}$.

a Geometric series with $a = e^{ix}$, $r = e^{ix}$.

$\therefore e^{ix} + e^{2ix} + e^{3ix} + \ldots + e^{nix} = \dfrac{e^{ix}\left(1 - e^{nix}\right)}{1 - e^{ix}}$

> This is a geometric series with common ratio e^{ix}. Use $S_n = \dfrac{a\left(1 - r^n\right)}{1 - r}$ for the sum of the first n terms.

b $\sin x + \sin 2x + \ldots + \sin 10x = \text{Im}\left(e^{ix} + e^{2ix} + \ldots + e^{10ix}\right)$

$\qquad = \text{Im}\left(\dfrac{e^{ix}\left(1 - e^{10ix}\right)}{1 - e^{ix}}\right)$

> This is the imaginary part of the series from part **a**, with $n = 10$.

$= \text{Im}\left(\dfrac{e^{ix}\left(1 - e^{10ix}\right)}{1 - e^{ix}} \times \dfrac{1 - e^{-ix}}{1 - e^{-ix}}\right)$

$= \text{Im}\left(\dfrac{e^{ix} - 1 - e^{11ix} + e^{10ix}}{1 - e^{ix} - e^{-ix} + 1}\right)$

> Multiply top and bottom by the complex conjugate of the denominator in order to separate real and imaginary parts.

$= \text{Im}\left(\dfrac{e^{ix} - 1 - e^{11ix} + e^{10ix}}{2 - 2\cos x}\right)$

> Use $e^{ix} + e^{-ix} = 2\cos x$ in the denominator.

The imaginary part is:

$\dfrac{\sin x - \sin 11x + \sin 10x}{2 - 2\cos x}$

> Now the denominator is real, so you just need to take the imaginary part of the numerator.

$= \dfrac{\sin x + \sin 10x - \sin 11x}{4\sin^2\left(\dfrac{x}{2}\right)}$

> Use the double angle formula in the denominator: $2\cos x = 2\left(1 - 2\sin^2\left(\dfrac{x}{2}\right)\right)$.

If the modulus of the common ratio is smaller than 1, a geometric series also has a sum to infinity.

WORKED EXAMPLE 3.7

a Show that the geometric series $1+\dfrac{1}{2}e^{i\theta}+\dfrac{1}{4}e^{2i\theta}+\ldots$ converges, and find an expression for its sum to infinity.

b Hence evaluate $\displaystyle\sum_{k=0}^{\infty}\dfrac{1}{2^k}\cos k\theta$.

a The geometric series has $|r|=\left|\dfrac{1}{2}e^{i\theta}\right|=\dfrac{1}{2}<1$, hence it converges.

> The common ratio is $\dfrac{1}{2}e^{i\theta}$.

Using $S_{\infty}=\dfrac{a}{1-r}$:

$$S_{\infty}=\dfrac{1}{1-\frac{1}{2}e^{i\theta}}=\dfrac{2}{2-e^{i\theta}}$$

b $\displaystyle\sum_{k=0}^{\infty}\dfrac{1}{2^k}\cos k\theta=\mathrm{Re}\left(1+\dfrac{1}{2}e^{i\theta}+\dfrac{1}{4}e^{2i\theta}+\ldots\right)$

> The required sum is the real part of the sum from part **a**.

$$=\mathrm{Re}\left(\dfrac{2}{2-e^{i\theta}}\right)$$

$$=\mathrm{Re}\left(\dfrac{2}{2-e^{i\theta}}\times\dfrac{2-e^{-i\theta}}{2-e^{-i\theta}}\right)$$

> Multiply top and bottom by the complex conjugate of the denominator to separate real and imaginary parts.

$$=\mathrm{Re}\left(\dfrac{4-2e^{-i\theta}}{4-2e^{i\theta}-2e^{-i\theta}+1}\right)$$

$$=\mathrm{Re}\left(\dfrac{4-2e^{-i\theta}}{5-4\cos\theta}\right)$$

> Use $e^{i\theta}+e^{-i\theta}=2\cos\theta$.

$$=\dfrac{4-2\cos\theta}{5-4\cos\theta}$$

> Now take the real part of the numerator, using $e^{-i\theta}=\cos\theta-i\sin\theta$.

Another series you know how to sum is the binomial expansion.

WORKED EXAMPLE 3.8

By considering the expansion of $\left(e^{i\theta}+1\right)^5$, or otherwise, show that

$$\sin 5\theta+5\sin 4\theta+10\sin 3\theta+10\sin 2\theta+5\sin\theta=32\cos^5\dfrac{\theta}{2}\sin\dfrac{5\theta}{2}.$$

Using the binomial expansion:

$$\left(e^{i\theta}+1\right)^5=e^{5i\theta}+5e^{4i\theta}+10e^{3i\theta}+10e^{2i\theta}+5e^{i\theta}+1$$

$$\sin 5\theta+5\sin 4\theta+10\sin 3\theta+10\sin 2\theta+5\sin\theta=\mathrm{Im}\left(\left(e^{i\theta}+1\right)^5\right)$$

> The required series is the imaginary part of this.

Continues on next page …

$$e^{i\theta} + 1 = (\cos\theta + 1) + i\sin\theta$$

Use the double angle formulae.

$$= 2\cos^2\frac{\theta}{2} + 2i\sin\frac{\theta}{2}\cos\frac{\theta}{2}$$

$$= 2\cos\frac{\theta}{2}\left(\cos\frac{\theta}{2} + i\sin\frac{\theta}{2}\right)$$

$$= \left(2\cos\frac{\theta}{2}\right)e^{i\frac{\theta}{2}}$$

Hence

$$\left(e^{i\theta} + 1\right)^5 = \left(2\cos\frac{\theta}{2}\right)^5 e^{i\frac{5\theta}{2}}$$

so

$$\text{Im}\left(\left(e^{i\theta} + 1\right)^5\right) = \left(2\cos\frac{\theta}{2}\right)^5 \sin\frac{5\theta}{2}$$

Therefore, the sum of the series equals $32\cos^5\frac{\theta}{2}\sin\frac{5\theta}{2}$.

EXERCISE 3D

1   **a**   Find an expression for the sum to infinity of the geometric series $1 + \frac{1}{3}e^{i\theta} + \frac{1}{9}e^{2i\theta} + \dots$.

    b   Hence evaluate $\displaystyle\sum_{k=0}^{\infty} \frac{1}{3^k}\cos k$.

2   **a**   Show that the geometric series $1 + \frac{1}{2}e^{i\theta} + \frac{1}{4}e^{2i\theta} + \dots$ converges and find an expression for its sum to infinity.

    b   Hence show that $\dfrac{1}{2}\sin\theta + \dfrac{1}{4}\sin 2\theta + \dfrac{1}{8}\sin 3\theta + \dots = \dfrac{2\sin\theta}{5 - 4\cos\theta}$.

3   Use the geometric series $e^{ix} - \frac{1}{2}e^{3ix} + \frac{1}{4}e^{5ix} - \dots$ to evaluate $\sin 1 - \frac{1}{2}\sin 3 + \frac{1}{4}\sin 5 - \dots$.

4   Use the expansion of $\left(e^{i\theta} + 1\right)^4$ to show that $\cos 4\theta + 4\cos 3\theta + 6\cos 2\theta + 4\cos\theta + 1 = 16\cos^4\left(\dfrac{\theta}{2}\right)\cos 2\theta$.

5   By considering $\left(e^{i\theta} - 1\right)^5$ or otherwise, show that $\sin 5\theta - 5\sin 4\theta + 10\sin 3\theta - 10\sin 2\theta + 5\sin\theta = 32\sin^5\left(\dfrac{\theta}{2}\right)\cos\left(\dfrac{5\theta}{2}\right)$.

6   **a**   Find an expression for the sum of the series $e^{i\theta} + e^{3i\theta} + e^{5i\theta} \dots + e^{(2n-1)i\theta}$.

    b   Hence prove that $\cos\theta + \cos 3\theta + \cos 5\theta \dots + \cos(2n-1)\theta = \dfrac{\sin(2n\theta)}{2\sin\theta}$.

    c   Find all the solutions to the equation $\cos\theta + \cos 3\theta + \cos 5\theta = 0$ for $0 < \theta < \pi$.

Checklist of learning and understanding

- By expanding $(\cos\theta + i\sin\theta)^n$ and comparing the real and imaginary parts to $\cos n\theta + i\sin n\theta$ you can derive expressions for $\sin n\theta$ and $\cos n\theta$ in terms of powers of $\sin\theta$ and $\cos\theta$.
  - Considering these expressions as polynomials in $\sin\theta$ or $\cos\theta$ you can find some exact values of trigonometric functions.
- If $z = e^{i\theta}$, then $z^n + \dfrac{1}{z^n} = 2\cos n\theta$ and $z^n - \dfrac{1}{z^n} = 2i\sin n\theta$.
  - In particular, $\cos\theta = \dfrac{e^{i\theta} + e^{-i\theta}}{2}$ and $\sin\theta = \dfrac{e^{i\theta} - e^{-i\theta}}{2i}$.
  - You can use these expressions, together with the binomial expansion, to express powers of $\sin\theta$ and $\cos\theta$ in terms of sin and cos of multiples of θ.
- By considering real and imaginary parts of geometric or binomial series involving $e^{i\theta}$ you can derive expressions for sums of trigonometric series.

Mixed practice 3

1 **a** Expand and simplify $(\cos\theta + i\sin\theta)^4$.

  b Hence find constants A and B such that $\dfrac{\sin 4\theta}{\cos\theta} = A\sin\theta - B\sin^3\theta$.

2 Use de Moivre's theorem to show that $\cos 5\theta = 16\cos^5\theta - 20\cos^3\theta + 5\cos\theta$. Hence find the largest and smallest values of $\cos\theta - 4\cos^3\theta + \dfrac{16}{5}\cos^5\theta$.

3 **a** By considering $\left(z + \dfrac{1}{z}\right)^5$, where $z = \cos\theta + i\sin\theta$, find the values of constants A, B and C such that
  $\cos^5\theta = A\cos 5\theta + B\cos 3\theta + C\cos\theta$.

  b Hence find the exact value of $\int_0^{\frac{\pi}{2}}\cos^5\theta\,d\theta$.

4 Show that $\sin 5\theta = 16\sin^5\theta - 20\sin^3\theta + 5\sin\theta$. Hence show that $\sin\dfrac{13\pi}{30}$ is a root of the equation $32x^5 - 40x^3 + 10x - 1 = 0$.

5 By considering the expansion of $(1+i)^{10}$, show that $\dbinom{10}{1} - \dbinom{10}{3} + \dbinom{10}{5} - \dbinom{10}{7} + \dbinom{10}{9} = 32$.

6 Show that $1 + 4\cos 2\theta + 6\cos 4\theta + 4\cos 6\theta + \cos 8\theta = 16\cos 4\theta\cos^4\theta$.

7 **i** By expressing $\cos\theta$ in terms of $e^{i\theta}$ and $e^{-i\theta}$, show that
  $$\cos^5\theta \equiv \frac{1}{16}(\cos 5\theta + 5\cos 3\theta + 10\cos\theta).$$

  ii Hence solve the equation $\cos 5\theta + 5\cos 3\theta + 9\cos\theta = 0$ for $0 \leqslant \theta \leqslant \pi$.

© OCR A Level Mathematics, Unit 4727/01 Further Pure Mathematics 3, June 2008

8 Let $z = \cos\theta + i\sin\theta$.

  a Show that $2\cos\theta = z + \dfrac{1}{z}$.

  b Show that $2\cos n\theta = z^n + \dfrac{1}{z^n}$.

  c Consider the equation $3z^4 - z^3 + 2z^2 - z + 3 = 0$.

   i Show that the equation can be written as $6\cos 2\theta - 2\cos\theta + 2 = 0$.

   ii Find all four complex roots of the original equation.

9 **a** By considering $(\cos\theta + i\sin\theta)^3$, find expressions for $\cos 3\theta$ and $\sin 3\theta$.

  b Show that $\tan 3\theta = \dfrac{3\tan\theta - \tan^3\theta}{1 - 3\tan^2\theta}$.

  c Hence show that $\tan\dfrac{\pi}{12}$ is a root of the equation $x^3 - 3x^2 - 3x + 1 = 0$.

  d Show that $(x+1)$ is a factor of $x^3 - 3x^2 - 3x + 1$ and hence find the exact solutions of the equation $x^3 - 3x^2 - 3x + 1 = 0$.

  e By considering $\tan\dfrac{\pi}{4}$, explain why $\tan\dfrac{\pi}{12} < 1$.

  f Hence state the exact value of $\tan\dfrac{\pi}{12}$.

10 i Use de Moivre's theorem to prove that

$$\cos 6\theta = 32\cos^6\theta - 48\cos^4\theta + 18\cos^2\theta - 1.$$

ii Hence find the largest positive root of the equation

$$64x^6 - 96x^4 + 36x^2 - 3 = 0,$$

giving your answer in trigonometrical form.

© OCR A Level Mathematics, Unit 4727/01 Further Pure Mathematics 3, June 2007

11 Convergent infinite series C and S are defined by

$$C = 1 + \frac{1}{2}\cos\theta + \frac{1}{4}\cos 2\theta + \frac{1}{8}\cos 3\theta + \dots,$$

$$S = \frac{1}{2}\sin\theta + \frac{1}{4}\sin 2\theta + \frac{1}{8}\sin 3\theta + \dots.$$

i Show that $C + \mathrm{i}S = \dfrac{2}{2 - \mathrm{e}^{\mathrm{i}\theta}}$.

ii Hence show that $C = \dfrac{4 - 2\cos\theta}{5 - 4\cos\theta}$, and find a similar expression for S.

© OCR A Level Mathematics, Unit 4727 Further Pure Mathematics 3, June 2010

4 Lines and planes in space

In this chapter you will learn how to:

- find the equation of a plane in several different forms
- find intersections between lines and planes
- calculate angles between lines and planes
- calculate the distances between objects in three-dimensional space.

Before you start…

Pure Core Student Book 1, Chapter 2	You should be able to find the vector and Cartesian equation of a line in three dimensions.	1	A line passes through the points $(3, -1, 2)$ and $(5, 1, 8)$. **a** Find a vector equation of the line. **b** Write down a Cartesian equation of the line.
Pure Core Student Book 1, Chapter 2	You should be able to find the point of intersection of two lines.	2	Find the point of intersection of the line from question **1** and the line $\mathbf{r} = \begin{pmatrix} 2 \\ 2 \\ 9 \end{pmatrix} + \mu \begin{pmatrix} -1 \\ 1 \\ 2 \end{pmatrix}$.
Pure Core Student Book 1, Chapter 2	You should know how to calculate the scalar product of two vectors and use it to calculate an angle between two lines.	3	Find the acute angle between the two lines from question **2**.
Pure Core Student Book 1, Chapter 2	You should know how to calculate the vector product of two vectors and use it to find a vector perpendicular to two given vectors.	4	Find a vector perpendicular to both of the lines from question **2**.

Introduction

In Pure Core Student Book 1, Chapter 2, you learnt about equations of lines in three dimensions, and how to find intersections and angles between lines. You know that two lines might be skew (not intersecting but not parallel). In this chapter you will learn how to find the distance between two skew lines, as well as between two parallel lines.

You will also learn how to describe planes (flat surfaces) in three-dimensional space.

Section 1: Equation of a plane

You are already used to describing positions of points in the x-y plane using unit vectors parallel to the x- and y-axes: for example, the position vector of the point $P(3, 2)$ is $\mathbf{r}_p = 3\mathbf{i} + 2\mathbf{j}$.

However, you can also use two directions other than those of $\mathbf{i}$ and $\mathbf{j}$. In the second diagram, the same point P is reached from the origin by moving 2 units in the direction of vector $\mathbf{d}_1$ and 2 units in the direction of vector $\mathbf{d}_2$. Hence its position vector is $\mathbf{r}_p = 2\mathbf{d}_1 + 2\mathbf{d}_2$. In the same way, every point in the x-y plane has a position vector of the form $\lambda\mathbf{d}_1 + \mu\mathbf{d}_2$, where λ and μ are scalars.

Consider now a plane that does not pass through the origin. To reach a point R in the plane starting from the origin, you can go to some other point in the plane first, and then move along two directions which lie in the plane, as shown.

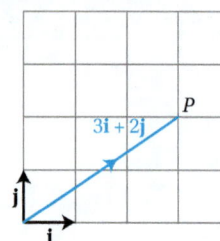

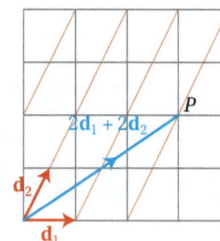

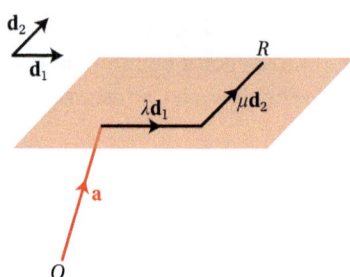

Tip

A **plane** is a flat surface in three-dimensional space. It extends indefinitely in all directions.

Key point 4.1

The vector equation of the plane containing a point with position vector $\mathbf{a}$ and parallel to the directions of vectors $\mathbf{d}_1$ and $\mathbf{d}_2$ is $\mathbf{r} = \mathbf{a} + \lambda\mathbf{d}_1 + \mu\mathbf{d}_2$.

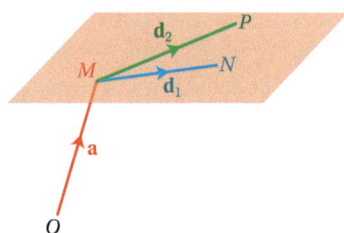

WORKED EXAMPLE 4.1

Find a vector equation of the plane containing points $M(3, 4, -2)$, $N(1, -1, 3)$ and $P(5, 0, 2)$.

$\mathbf{r} = \mathbf{a} + \lambda\mathbf{d}_1 + \mu\mathbf{d}_2$

You need one point and two vectors parallel to the plane. Draw a diagram to see which vectors to use.

$\underline{a} = \begin{pmatrix} 3 \\ 4 \\ -2 \end{pmatrix}$

You can choose any of the three given points to find $\mathbf{a}$, as they all lie in the plane.

Continues on next page ...

$$\underline{d}_1 = \overrightarrow{MN} = \begin{pmatrix} 1 \\ -1 \\ 3 \end{pmatrix} - \begin{pmatrix} 3 \\ 4 \\ -2 \end{pmatrix} = \begin{pmatrix} -2 \\ -5 \\ 5 \end{pmatrix}$$

Vectors $\overrightarrow{MN}$ and $\overrightarrow{MP}$ are parallel to the plane.

$$\underline{d}_2 = \overrightarrow{MP} = \begin{pmatrix} 5 \\ 0 \\ 2 \end{pmatrix} - \begin{pmatrix} 3 \\ 4 \\ -2 \end{pmatrix} = \begin{pmatrix} 2 \\ -4 \\ 4 \end{pmatrix}$$

$$\underline{r} = \begin{pmatrix} 3 \\ 4 \\ -2 \end{pmatrix} + \lambda \begin{pmatrix} -2 \\ -5 \\ 5 \end{pmatrix} + \mu \begin{pmatrix} 2 \\ -4 \\ 4 \end{pmatrix}$$

Use $\mathbf{r} = \mathbf{a} + \lambda \mathbf{d}_1 + \mu \mathbf{d}_2$.

In Worked example 4.1, the plane was determined by three points. Two points do not determine a plane: there is more than one plane containing the line determined by points A and B as shown in this diagram.

You can pick out one of these planes by requiring that it also passes through a third point, point C for example, which is not on the line AB, as illustrated here. This suggests that a plane can also be determined by a line and a point outside of that line.

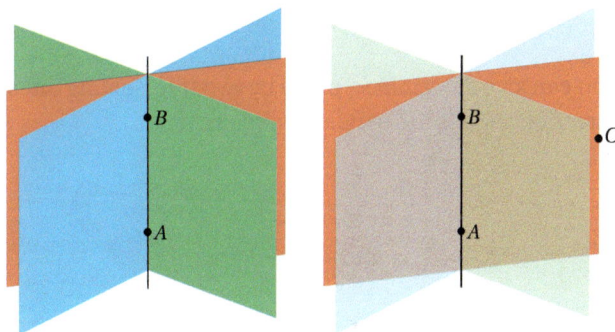

WORKED EXAMPLE 4.2

Find a vector equation of the plane containing the line $\mathbf{r} = \begin{pmatrix} -2 \\ 1 \\ 2 \end{pmatrix} + t \begin{pmatrix} -3 \\ 1 \\ 1 \end{pmatrix}$ and point $A\,(4, -1, 2)$.

$$\underline{a} = \begin{pmatrix} 4 \\ -1 \\ 2 \end{pmatrix}$$

Point A lies in the plane.

$$\underline{d}_1 = \begin{pmatrix} -3 \\ 1 \\ 1 \end{pmatrix}$$

The direction vector of the line is parallel to the plane.

Continues on next page ...

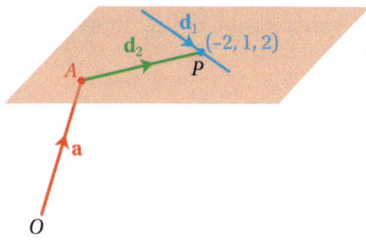

You need another vector parallel to the plane. You can use any vector between two points in the plane. One point in the plane is A. For the second point, you can pick any point on the line, for example $P(-2, 1, 2)$.

$$d_2 = \begin{pmatrix} -2 \\ 1 \\ 2 \end{pmatrix} - \begin{pmatrix} 4 \\ -1 \\ 2 \end{pmatrix} = \begin{pmatrix} -6 \\ 2 \\ 0 \end{pmatrix}$$

$$r = \begin{pmatrix} 4 \\ -1 \\ 2 \end{pmatrix} + \lambda \begin{pmatrix} -3 \\ 1 \\ 1 \end{pmatrix} + \mu \begin{pmatrix} -6 \\ 2 \\ 0 \end{pmatrix}$$

Now use $\mathbf{r} = \mathbf{a} + \lambda \mathbf{d}_1 + \mu \mathbf{d}_2$.

You can also determine a plane by two intersecting lines whose two direction vectors are parallel to the plane.

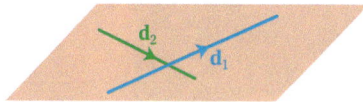

WORKED EXAMPLE 4.3

Find a vector equation of the plane containing the lines $\mathbf{r} = \begin{pmatrix} 3 \\ -1 \\ 2 \end{pmatrix} + \lambda \begin{pmatrix} -1 \\ 1 \\ 2 \end{pmatrix}$ and $\mathbf{r} = \begin{pmatrix} 3 \\ -1 \\ 2 \end{pmatrix} + \mu \begin{pmatrix} 3 \\ 0 \\ 2 \end{pmatrix}$.

$$r = \begin{pmatrix} 3 \\ -1 \\ 2 \end{pmatrix} + \lambda \begin{pmatrix} -1 \\ 1 \\ 2 \end{pmatrix} + \mu \begin{pmatrix} 3 \\ 0 \\ 2 \end{pmatrix}$$

You can tell that the two lines intersect at the point $(3, -1, 2)$, so you can take that as one point in the plane. The two lines' direction vectors give two different directions in the plane.

If you have four points, they don't all necessarily lie in the same plane.

WORKED EXAMPLE 4.4

Determine whether points $A(2, -1, 3)$, $B(4, 1, 1)$, $C(3, 3, 2)$, and $D(-3, 1, 5)$ lie in the same plane.

Plane containing A, B, C:

$\underline{r} = \overrightarrow{OA} + \lambda\overrightarrow{AB} + \mu\overrightarrow{AC}$

$\underline{r} = \begin{pmatrix} 2 \\ -1 \\ 3 \end{pmatrix} + \lambda\begin{pmatrix} 2 \\ 2 \\ -2 \end{pmatrix} + \mu\begin{pmatrix} 1 \\ 4 \\ -1 \end{pmatrix}$

The plan is to find the equation of the plane containing points A, B and C (as in Worked example 4.1) and then check whether the point D lies in that plane.

$\underline{r} = \overrightarrow{OD}$:

$\begin{cases} 2 + 2\lambda + \mu = -3 \\ -1 + 2\lambda + 4\mu = 1 \\ 3 - 2\lambda - \mu = 5 \end{cases}$

For D to lie in the plane, you need values of λ and μ which make **r** equal to the position vector of D.

$\begin{cases} 2\lambda + \mu = -5 \\ 2\lambda + 4\mu = 2 \end{cases}$

$\lambda = -\dfrac{11}{3}, \mu = \dfrac{7}{3}$

You can solve the first two equations, and then check whether the solutions satisfy the third equation.

$3 - 2\times\left(-\dfrac{11}{3}\right) - \dfrac{7}{3} = 8 \neq 5$

D does not lie in the same plane as A, B and C.

There are no values of λ and μ that satisfy all three equations.

You can now summarise all possible ways to determine a plane.

🔑 Key point 4.2

A plane is uniquely determined by:

- three points, not on the same line, or
- a line and a point outside that line, or
- two intersecting lines.

Cartesian equation of a plane

The vector equation of the plane can be a little difficult to work with, as it contains two parameters. It is also difficult to see whether two equations represent the same plane, because the two vectors parallel to the plane are not unique.

Now you will look at the question: is there a way to describe the 'direction' of the plane using just one direction vector?

The diagram shows a plane and a vector **n** perpendicular to it. This vector is perpendicular to every line in the plane, and it is called the **normal vector** of the plane.

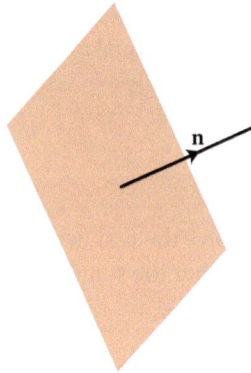

Suppose A is a fixed point in the plane and let P be any other point in the plane. The normal vector is perpendicular to the line AP, so $\overrightarrow{AP} \cdot \mathbf{n} = 0$. This means that $(\mathbf{r} - \mathbf{a}) \cdot \mathbf{n} = 0$, which gives another form of an equation of the plane.

> ### 🔑 Key point 4.3
>
> The **scalar product equation** of the plane is:
>
> $$\mathbf{r} \cdot \mathbf{n} = \mathbf{a} \cdot \mathbf{n}$$
>
> where **n** is the normal to the plane and **a** is the position vector of a point in the plane.

Remember that the position vector of a point is related to its coordinates. This means that you can use the scalar product equation to write the Cartesian equation of a plane.

Write $\mathbf{n} = \begin{pmatrix} n_1 \\ n_2 \\ n_3 \end{pmatrix}$ and $\mathbf{r} = \begin{pmatrix} x \\ y \\ z \end{pmatrix}$. The scalar product $\mathbf{a} \cdot \mathbf{n}$ is a constant, denoted d in Key point 4.4, and the scalar product $\mathbf{r} \cdot \mathbf{n}$ can be expanded to get an expression in terms of x, y and z.

> ### 🔑 Key point 4.4
>
> The Cartesian equation of a plane can be written in the form
> $$n_1 x + n_2 y + n_3 z = d.$$

WORKED EXAMPLE 4.5

Vector $\mathbf{n} = \begin{pmatrix} 2 \\ 4 \\ -1 \end{pmatrix}$ is perpendicular to the plane Π which contains

point $A(3, -5, 1)$.

a Write an equation for Π in the form $\mathbf{r} \cdot \mathbf{n} = p$.

b Find the Cartesian equation of the plane.

a $\underline{r} \cdot \underline{n} = \begin{pmatrix} 3 \\ -5 \\ 1 \end{pmatrix} \cdot \begin{pmatrix} 2 \\ 4 \\ -1 \end{pmatrix} = 6 - 20 - 1$

> The equation of the plane is $\mathbf{r} \cdot \mathbf{n} = \mathbf{a} \cdot \mathbf{n}$.

$\therefore \underline{r} \cdot \underline{n} = -15$

b $\begin{pmatrix} x \\ y \\ z \end{pmatrix} \cdot \begin{pmatrix} 2 \\ 4 \\ -1 \end{pmatrix} = -15$

> The Cartesian equation involves x, y, and z (the coordinates of P), which are the components of the position vector $\mathbf{r}$.

$2x + 4y - z = -15$

Tip

The letter Π (capital π) is often used as the name for a plane.

You can convert from a vector to a Cartesian equation of the plane. This involves using the **vector product** to find the normal. The Cartesian equation is very convenient for checking whether a point lies in the plane: you just need to check that the coordinates of the point satisfy the equation.

◄◄ Rewind

You met the vector product in Pure Core Student Book 1, Chapter 2.

WORKED EXAMPLE 4.6

a Find the Cartesian equation of the plane with vector equation $\mathbf{r} = \begin{pmatrix} 1 \\ -2 \\ 5 \end{pmatrix} + \lambda \begin{pmatrix} 1 \\ 1 \\ 3 \end{pmatrix} + \mu \begin{pmatrix} 2 \\ -3 \\ 5 \end{pmatrix}$.

b Show that the point $(2, 9, 10)$ lies in the plane.

a $\underline{r} \cdot \underline{n} = \underline{a} \cdot \underline{n}$

> To find the Cartesian equation you need the normal vector and one point.

$\underline{a} = \begin{pmatrix} 1 \\ -2 \\ 5 \end{pmatrix}$

> Point $(1, -2, 5)$ lies in the plane.

Continues on next page …

$$\underline{n} = \begin{pmatrix} 1 \\ 1 \\ 3 \end{pmatrix} \times \begin{pmatrix} 2 \\ -3 \\ 5 \end{pmatrix} = \begin{pmatrix} 14 \\ 1 \\ -5 \end{pmatrix}$$

n is perpendicular to all lines in the plane, so it is perpendicular to the direction vectors $\begin{pmatrix} 1 \\ 1 \\ 3 \end{pmatrix}$ and $\begin{pmatrix} 2 \\ -3 \\ 5 \end{pmatrix}$.

The vector product of two vectors is perpendicular to both of them.

$$\underline{r} \cdot \begin{pmatrix} 14 \\ 1 \\ -5 \end{pmatrix} = \begin{pmatrix} 1 \\ -2 \\ 5 \end{pmatrix} \cdot \begin{pmatrix} 14 \\ 1 \\ -5 \end{pmatrix}$$

$$14x + y - 5z = -13$$

To get the Cartesian equation, write **r** as $\begin{pmatrix} x \\ y \\ z \end{pmatrix}$.

b $14(2) + 9 - 5(10) = -13$

Hence the point lies in the plane.

A point lies in the plane if its coordinates satisfy the Cartesian equation.

You can also convert from a Cartesian to a vector equation by finding two vectors that are perpendicular to the normal.

WORKED EXAMPLE 4.7

Find a vector equation of the plane with Cartesian equation $2x - 5y + z = 15$.

Vector equation:

$$\underline{r} = \underline{a} + \lambda \underline{d}_1 + \mu \underline{d}_2$$

In the vector equation of the plane, **a** is the position vector of one point in the plane and $\mathbf{d}_1$ and $\mathbf{d}_2$ are two direction vectors parallel to the plane.

Finding $\underline{a}$:

$$2x - 5y + z = 15$$

When $x = y = 0$:

$$2(0) - 5(0) + z = 15 \Rightarrow z = 15$$

The coordinates of A satisfy the Cartesian equation of the plane. You can choose any three numbers (x, y, z) that satisfy this equation. For example, you can set $x = y = 0$ and then find z.

Hence a possible position vector is $\underline{a} = \begin{pmatrix} 0 \\ 0 \\ 15 \end{pmatrix}$.

Finding $\underline{d}_1$ and $\underline{d}_2$:

Write $\underline{d}_1 = \begin{pmatrix} s \\ t \\ u \end{pmatrix}$. Then $\underline{d}_1 \cdot \underline{n} = 0$ so:

$$2s - 5t + u = 0.$$

The two direction vectors parallel to the plane must be perpendicular to the normal, which is $\mathbf{n} = \begin{pmatrix} 2 \\ -5 \\ 1 \end{pmatrix}$.

Continues on next page ...

When $s = 0$ and $t = 1$:

$$2(0)-5(1)+u=0 \Rightarrow u=5$$

As before, you can choose values for s and t and then find u. To make the calculation simple, take $s=0$ and $t=1$.

(Notice that, in this case, you can't set $s=t=0$ because that would make $u=0$ as well.)

Write $\underline{d}_2 = \begin{pmatrix} 1 \\ 0 \\ u \end{pmatrix}$. Then:

$$2(1)-5(0)+u=0 \Rightarrow u=-2$$

Repeat for $\mathbf{d}_2$, but this time take $s=1$ and $t=0$.

Hence the two direction vectors are

$$\underline{d}_1 = \begin{pmatrix} 0 \\ 1 \\ 5 \end{pmatrix} \text{ and } \underline{d}_2 = \begin{pmatrix} 1 \\ 0 \\ -2 \end{pmatrix}$$

A possible vector equation of the plane is

$$\underline{r} = \begin{pmatrix} 0 \\ 0 \\ 15 \end{pmatrix} + \lambda \begin{pmatrix} 0 \\ 1 \\ 5 \end{pmatrix} + \mu \begin{pmatrix} 1 \\ 0 \\ -2 \end{pmatrix}$$

Put all this together into the vector equation,

$$\mathbf{r} = \mathbf{a} + \lambda \mathbf{d}_1 + \mu \mathbf{d}_2$$

You should remember that the vector equation you found in Worked example 4.7 is not unique. You could have chosen any other point that satisfies the Cartesian equation, and there are infinitely many choices for pairs of direction vectors that are parallel to the normal.

EXERCISE 4A

1 Write down the vector equation of the plane parallel to vectors **a** and **b** and containing point P.

a i $\mathbf{a} = \begin{pmatrix} -1 \\ 5 \\ 2 \end{pmatrix}, \mathbf{b} = \begin{pmatrix} 1 \\ -2 \\ 3 \end{pmatrix}, P(1, 0, 2)$ **ii** $\mathbf{a} = \begin{pmatrix} 0 \\ 4 \\ -1 \end{pmatrix}, \mathbf{b} = \begin{pmatrix} 5 \\ 3 \\ 0 \end{pmatrix}, P(0, 2, 0)$

b i $\mathbf{a} = 3\mathbf{i}+\mathbf{j}-3\mathbf{k}, \mathbf{b} = \mathbf{i}-3\mathbf{j}, \mathbf{p} = \mathbf{j}+\mathbf{k}$ **ii** $\mathbf{a} = 5\mathbf{i}-6\mathbf{j}, \mathbf{b} = -\mathbf{i}+3\mathbf{j}-\mathbf{k}, P(1, -6, 2)$

2 Find a vector equation of the plane containing points A, B and C.

a i $A(3, -1, 3), B(1, 1, 2), C(4, -1, 2)$ **ii** $A(-1, -1, 5), B(4, 1, 2), C(-7, 1, 1)$

b i $A(9, 0, 0), B(-2, 1, 0), C(1, -1, 2)$ **ii** $A(11, -7, 3), B(1, 14, 2), C(-5, 10, 0)$

3 Find a vector equation of the plane containing line l and point P.

a i $l:\mathbf{r} = \begin{pmatrix} -3 \\ 5 \\ 1 \end{pmatrix} + t \begin{pmatrix} 4 \\ 1 \\ 2 \end{pmatrix}, P(-1, 4, 3)$ **ii** $l:\mathbf{r} = \begin{pmatrix} 9 \\ -3 \\ 7 \end{pmatrix} + t \begin{pmatrix} 6 \\ -3 \\ 1 \end{pmatrix}, P(11, 12, 13)$

b i $l:\mathbf{r} = \begin{pmatrix} 4 \\ 4 \\ 1 \end{pmatrix} + t \begin{pmatrix} 0 \\ 0 \\ 1 \end{pmatrix}, P(-3, 1, 0)$ **ii** $l:\mathbf{r} = t \begin{pmatrix} 2 \\ 1 \\ 1 \end{pmatrix}, P(4, 0, 2)$

4 A plane has normal vector **n** and contains point A. Find the equation of the plane in the form $\mathbf{r} \cdot \mathbf{n} = d$, and the Cartesian equation of the plane.

a i $\mathbf{n} = \begin{pmatrix} 3 \\ -5 \\ 2 \end{pmatrix}$, $A(3,3,1)$ **ii** $\mathbf{n} = \begin{pmatrix} 6 \\ -1 \\ 2 \end{pmatrix}$, $A(4,3,-1)$

b i $\mathbf{n} = \begin{pmatrix} 3 \\ -1 \\ 0 \end{pmatrix}$, $A(-3,0,2)$ **ii** $\mathbf{n} = \begin{pmatrix} 4 \\ 0 \\ -5 \end{pmatrix}$, $A(0,0,2)$

5 Find a normal vector to the plane given by the vector equation:

a i $\mathbf{r} = \begin{pmatrix} 5 \\ 0 \\ 1 \end{pmatrix} + \lambda \begin{pmatrix} 1 \\ 2 \\ 3 \end{pmatrix} + \mu \begin{pmatrix} 5 \\ -2 \\ 2 \end{pmatrix}$ **ii** $\mathbf{r} = \begin{pmatrix} 0 \\ 0 \\ 1 \end{pmatrix} + \lambda \begin{pmatrix} -3 \\ 6 \\ 2 \end{pmatrix} + \mu \begin{pmatrix} -1 \\ 1 \\ 2 \end{pmatrix}$

b i $\mathbf{r} = \begin{pmatrix} 7 \\ 3 \\ 5 \end{pmatrix} + \lambda \begin{pmatrix} -5 \\ 1 \\ 2 \end{pmatrix} + \mu \begin{pmatrix} 0 \\ 0 \\ 1 \end{pmatrix}$ **ii** $\mathbf{r} = \begin{pmatrix} 3 \\ 5 \\ 7 \end{pmatrix} + \lambda \begin{pmatrix} 6 \\ -1 \\ 2 \end{pmatrix} + \mu \begin{pmatrix} -1 \\ -1 \\ 3 \end{pmatrix}$

6 Find the equations of the planes from question **5** in the form $\mathbf{r} \cdot \mathbf{n} = p$.

7 Find the Cartesian equations of the planes from question **5**.

8 Find the Cartesian equation of the plane containing points A, B and C.

a i $A(7,1,2)$, $B(-1,4,7)$, $C(5,2,3)$

  ii $A(1,1,2)$, $B(4,-6,2)$, $C(12,12,2)$

b i $A(12,4,10)$, $B(13,4,5)$, $C(15,-4,0)$

  ii $A(1,0,0)$, $B(0,1,0)$, $C(0,0,1)$

9 In each of the following show that point P lies in the given plane Π.

a $P(-4,8,13)$, $\Pi: \mathbf{r} = \begin{pmatrix} 2 \\ 1 \\ 1 \end{pmatrix} + \lambda \begin{pmatrix} 4 \\ 1 \\ 2 \end{pmatrix} + \mu \begin{pmatrix} -1 \\ 4 \\ 7 \end{pmatrix}$

b $P(4,7,5)$, $\Pi: \mathbf{r} \cdot \begin{pmatrix} 4 \\ -1 \\ 2 \end{pmatrix} = 19$

c $P(1, 1, -2)$, $\Pi: 2x - 3y - 7z = 13$

10 A plane contains the point $(3, -2, 5)$. The vector $6\mathbf{i} + \mathbf{j} - 3\mathbf{k}$ is perpendicular to the plane. Find the Cartesian equation of the plane.

11 A plane contains points $A(5, 1, 5)$, $B(-3, 1, 2)$ and $C(0, 1, 5)$.

Find the vector equation of the plane in the form $\mathbf{r} = \mathbf{a} + \lambda \mathbf{d}_1 + \mu \mathbf{d}_2$.

12 A plane contains points $P(3, 0, 2)$, $Q(-1, 1, 2)$ and $R(0, 5, 1)$.

   a Find $\overrightarrow{PQ} \times \overrightarrow{PR}$.

   b Hence find the Cartesian equation of the plane.

13 A plane is determined by the points $A(3, 1, 5)$, $B(-1, 4, 0)$ and $C(0, 0, 3)$.

   a Find the equation of the plane in the form $\mathbf{r} \cdot \mathbf{n} = k$.

   b Determine whether the point $D(1, 1, 4)$ lies in the same plane.

14 **a** Calculate $\begin{pmatrix} 4 \\ 4 \\ 1 \end{pmatrix} \times \begin{pmatrix} 1 \\ -1 \\ 3 \end{pmatrix}$.

   b Hence find the Cartesian equation of the plane with vector equation $\mathbf{r} = \begin{pmatrix} 1 \\ 1 \\ 5 \end{pmatrix} + \lambda \begin{pmatrix} 4 \\ 4 \\ 1 \end{pmatrix} + \mu \begin{pmatrix} 1 \\ -1 \\ 3 \end{pmatrix}$.

15 **a** Calculate $\begin{pmatrix} -1 \\ 0 \\ 2 \end{pmatrix} \times \begin{pmatrix} 0 \\ 1 \\ 3 \end{pmatrix}$.

   b Two lines have equations

$$l_1 : \mathbf{r} = \begin{pmatrix} 7 \\ -3 \\ 2 \end{pmatrix} + t \begin{pmatrix} -1 \\ 0 \\ 2 \end{pmatrix} \text{ and } l_2 : \mathbf{r} = \begin{pmatrix} 1 \\ 1 \\ 26 \end{pmatrix} + s \begin{pmatrix} 0 \\ 1 \\ 3 \end{pmatrix}$$

     i Show that l_1 and l_2 intersect.

     ii Find the coordinates of the point of intersection.

   c Plane Π contains lines l_1 and l_2. Find the Cartesian equation of Π.

16 Determine whether the points $(0, 3, 1)$, $(1, 1, 5)$, $(1, 0, 4)$ and $(3, 8, 5)$ lie in the same plane.

17 A plane has Cartesian equation $x - 3y + 4z = 16$.

   a Write down the normal vector, $\mathbf{n}$, of the plane.

   b Find the values of p and q such that the vectors $\begin{pmatrix} 1 \\ 0 \\ p \end{pmatrix}$ and $\begin{pmatrix} 0 \\ 1 \\ q \end{pmatrix}$ are perpendicular to $\mathbf{n}$.

   c Hence find a vector equation of the plane in the form $\mathbf{r} = \mathbf{a} + \lambda \mathbf{d}_1 + \mu \mathbf{d}_2$.

18 **a** Find the Cartesian equation of the plane with vector equation $\mathbf{r} = \begin{pmatrix} 0 \\ 1 \\ 5 \end{pmatrix} + \lambda \begin{pmatrix} -3 \\ 1 \\ 2 \end{pmatrix} + \mu \begin{pmatrix} 2 \\ 5 \\ 2 \end{pmatrix}$.

   b Another plane has Cartesian equation $x - 3y + z = 7$. Find a vector equation of this plane in the form $\mathbf{r} = \mathbf{a} + \lambda \mathbf{d}_1 + \mu \mathbf{d}_2$.

19 Find, in the form $\mathbf{r} = \mathbf{a} + \lambda \mathbf{d}_1 + \mu \mathbf{d}_2$, a vector equation of the plane $3y - z = 5$.

Section 2: Intersection between a line and a plane

The method introduced in this section requires equations of planes to be in Cartesian form and equations of lines to be in vector form. If in a question they are given in a different form, you will need to convert them first.

The coordinates of the intersection point (if there is one) must satisfy both the equation of the line and the equation of the plane.

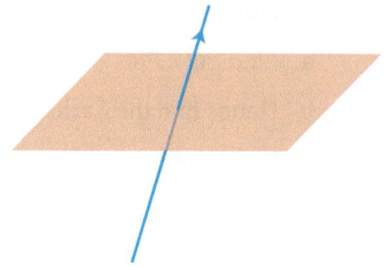

WORKED EXAMPLE 4.8

Find the intersection between the given line and plane, or show that they do not intersect.

a $r = \begin{pmatrix} 4 \\ -3 \\ 1 \end{pmatrix} + \lambda \begin{pmatrix} 3 \\ 0 \\ -2 \end{pmatrix}$ and $2x - y + 2z = 5$

b $\dfrac{x-1}{-1} = \dfrac{y}{-3} = \dfrac{z+4}{2}$ and $x - 3y - 4z = 12$

a $2(4+3\lambda) - (-3) + 2(1-2\lambda) = 5$
$$2\lambda = -8$$
$$\lambda = -4$$

Any point on the line has coordinates $(4+3\lambda, -3, 1-2\lambda)$. The intersection point must also satisfy the equation of the plane, so substitute $x = 4+3\lambda$, $y = -3$, $z = 1-2\lambda$ into the equation of the plane.

$r = \begin{pmatrix} 4 \\ -3 \\ 1 \end{pmatrix} - 4 \begin{pmatrix} 3 \\ 0 \\ -2 \end{pmatrix} = \begin{pmatrix} -8 \\ -3 \\ 9 \end{pmatrix}$

Now use this value of λ to find the coordinates.

The intersection point is $(-8, -3, 9)$.

b $\begin{cases} x - 1 = -\mu \\ y = -3\mu \\ z + 4 = 2\mu \end{cases}$

Change the equation of the line into the vector form and then follow the same procedure as in part **a**.

$\begin{pmatrix} x \\ y \\ z \end{pmatrix} = \begin{pmatrix} 1-\mu \\ -3\mu \\ -4+2\mu \end{pmatrix}$

$(1-\mu) - 3(-3\mu) - 4(-4+2\mu) = 12$
$$17 = 12$$

Substitute the coordinates of a point on the line into the equation of the plane.

Impossible to find μ.

It is impossible to find a value of μ for a point that satisfies both equations. This means that the line and the plane have no common points.

The line and plane do not intersect.

Notice that if a line and a plane do not intersect, this means that they are parallel. For example, in part **b** of Worked example 4.8, the direction vector of the line is perpendicular to the normal of the plane: $(-i - 3j + 2k) \cdot (i - 3j - 4k) = 0$; this shows that the line is parallel to the plane. By contrast, in part **a**, the direction of the line and the normal of the plane are not perpendicular: $(3i - 2k) \cdot (2i - j + 2k) = 2$.

It is possible for a line to lie entirely in a given plane.

WORKED EXAMPLE 4.9

Show that the line $\mathbf{r} = \begin{pmatrix} 3 \\ -1 \\ 1 \end{pmatrix} + t \begin{pmatrix} 3 \\ 1 \\ 0 \end{pmatrix}$ lies in the plane $x - 3y = 6$.

$(3 + 3t) - 3(-1 + t) = 6$ A point on the line has coordinates $(3 + 3t, -1 + t, 1)$.

$6 = 6$ You need to show that every such point satisfies the equation of the plane.

Every t is a solution.   The equation is satisfied for all values of t. This means that every point on the line also lies in the plane.

So the line lies in the plane.

EXERCISE 4B

1   Find the coordinates of the point of intersection of line l and plane Π.

   a **i** $l : \mathbf{r} = \begin{pmatrix} 2 \\ 1 \\ 2 \end{pmatrix} + \lambda \begin{pmatrix} 5 \\ 0 \\ -1 \end{pmatrix}, \Pi : 4x + 2y - z = 29$

       ii $l : \mathbf{r} = \begin{pmatrix} -5 \\ 1 \\ 1 \end{pmatrix} + \lambda \begin{pmatrix} 7 \\ 3 \\ -3 \end{pmatrix}, \Pi : x + y + 5z = 11$

   b **i** $l : \dfrac{x - 2}{5} = \dfrac{y + 1}{2} = \dfrac{z}{6}, \Pi : \mathbf{r} \bullet \begin{pmatrix} 1 \\ -4 \\ 1 \end{pmatrix} = 4$

       ii $l : \dfrac{x - 5}{-1} = \dfrac{y - 3}{2} = \dfrac{z - 5}{1}, \Pi : \mathbf{r} \bullet \begin{pmatrix} 2 \\ -1 \\ 1 \end{pmatrix} = 21$

2   Show that plane Π contains line l.

   a $\Pi : x + 6y + 2z = 7, l : \mathbf{r} = \begin{pmatrix} 5 \\ 0 \\ 1 \end{pmatrix} + t \begin{pmatrix} 2 \\ -1 \\ 2 \end{pmatrix}$

   b $\Pi : 5x + y - 2z = 15, l : \dfrac{x - 4}{1} = \dfrac{y + 1}{1} = \dfrac{z - 2}{3}$

c $\Pi : \mathbf{r} \cdot \begin{pmatrix} 1 \\ 0 \\ -4 \end{pmatrix} = -5, l : \mathbf{r} = \begin{pmatrix} -1 \\ 0 \\ 1 \end{pmatrix} + t \begin{pmatrix} 8 \\ 3 \\ 2 \end{pmatrix}$

d $\Pi : \mathbf{r} \cdot \begin{pmatrix} -2 \\ -2 \\ 5 \end{pmatrix} = \begin{pmatrix} 5 \\ 3 \\ 1 \end{pmatrix} \cdot \begin{pmatrix} -2 \\ -2 \\ 5 \end{pmatrix}, l : \dfrac{x-3}{2} = \dfrac{y}{3} = \dfrac{z+1}{2}$

3 Find the point of intersection of the line $\mathbf{r} = \begin{pmatrix} 3 \\ -1 \\ 2 \end{pmatrix} + \lambda \begin{pmatrix} 2 \\ 1 \\ 1 \end{pmatrix}$ and the plane $x - y + z = 18$.

4 A line has equation $\dfrac{x-2}{-1} = \dfrac{y+1}{2} = \dfrac{z-2}{1}$.

    a Write down the direction vector of the line.

    b Find the coordinates of the point where the line intersects the plane with equation $\mathbf{r} \cdot \begin{pmatrix} 3 \\ 3 \\ 2 \end{pmatrix} = 16$.

5 Find the coordinates of the point of intersection of the line $\dfrac{x-2}{3} = \dfrac{y-1}{2} = z+1$ with the plane $2x - y - 2z = 7$.

6 The plane with equation $12x - 3y + 5z = 60$ intersects the x-, y-, and z-axes at points P, Q and R, respectively.

    a Find the coordinates of P, Q and R.

    b Find the area of the triangle PQR.

Section 3: Angles between lines and planes

The angle between a line l and a plane Π is the smallest possible angle that l makes with any of the lines in Π. In the diagram, this is the angle labelled θ between l and the line AP, where PR is perpendicular to the plane (so that $\overrightarrow{PR}$ is in the direction of the normal). Drawing a two-dimensional diagram of triangle APR makes the angles clearer.

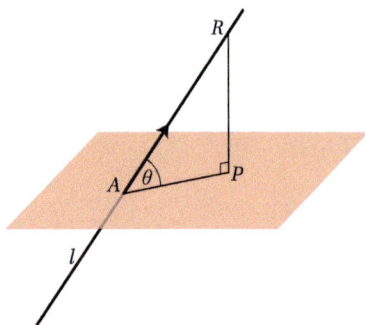

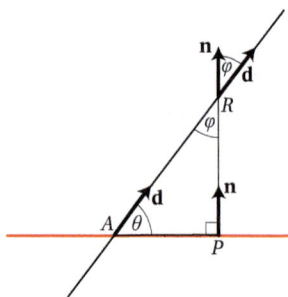

Key point 4.5

The angle between the line with direction vector **d** and the plane with normal **n** is $90° - \phi$, where ϕ is the acute angle between **d** and **n**.

Rewind

You found the angle between two vectors using the scalar product in Pure Core Student Book 1, Chapter 2.

WORKED EXAMPLE 4.10

Find the angle between the line with equation $\mathbf{r} = \begin{pmatrix} 4 \\ 0 \\ 7 \end{pmatrix} + \lambda \begin{pmatrix} 3 \\ 3 \\ 2 \end{pmatrix}$ and the plane with equation $5x - y + z = 7$.

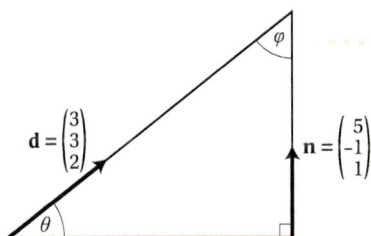

Draw a diagram, labelling vectors and angles as in Key point 4.5.

$$\cos\phi = \frac{\mathbf{d} \cdot \mathbf{n}}{|\mathbf{d}||\mathbf{n}|}$$

$$= \frac{\begin{pmatrix} 3 \\ 3 \\ 2 \end{pmatrix} \cdot \begin{pmatrix} 5 \\ -1 \\ 1 \end{pmatrix}}{\sqrt{9+9+4}\sqrt{25+1+1}}$$

$$= \frac{14}{\sqrt{22}\sqrt{27}}$$

$\therefore \phi = 54.9°$

The angle between the line and the plane is

The angle between a line and a plane is $90° -$ (angle between the line's direction vector and the plane's normal).

$\theta = 90° - \phi = 35.1°$

You can use a similar method to find the angle between two planes. Again, a diagram is helpful so you can see where the relevant angle is. The sum of angles in a quadrilateral is 360°, so the two angles marked θ are equal.

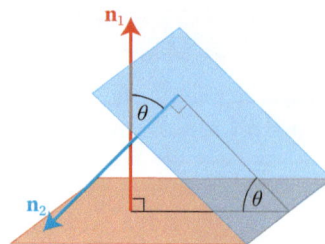

🔑 **Key point 4.6**

The angle between two planes is equal to the angle between their normals.

WORKED EXAMPLE 4.11

Find the acute angle between the planes with equations $4x - y + 5z = 11$ and $x + y - 3z = 3$.

$$\cos\theta = \frac{\mathbf{n_1} \cdot \mathbf{n_2}}{|\mathbf{n_1}||\mathbf{n_2}|}$$

You need to find the angle between the normals.

The components of the normal vector are the coefficients in the Cartesian equation.

$$= \frac{\begin{pmatrix} 4 \\ -1 \\ 5 \end{pmatrix} \cdot \begin{pmatrix} 1 \\ 1 \\ -3 \end{pmatrix}}{\sqrt{16+1+25}\sqrt{1+1+9}}$$

$$= \frac{-12}{\sqrt{42}\sqrt{11}}$$

$$\therefore \theta = 123.9°$$

$$180° - 123.9° = 56.1°$$

You need the acute angle.

The angle between the planes is 56.1°.

EXERCISE 4C

1 Find the acute angle between line l and plane Π, correct to the nearest 0.1°.

a **i** $l : \mathbf{r} = \begin{pmatrix} 4 \\ -1 \\ 2 \end{pmatrix} + \lambda \begin{pmatrix} 1 \\ -1 \\ 3 \end{pmatrix}$, $\Pi : \mathbf{r} \cdot \begin{pmatrix} 4 \\ -1 \\ 2 \end{pmatrix} = 7$

ii $l : \mathbf{r} = \begin{pmatrix} 2 \\ -3 \\ 1 \end{pmatrix} + \lambda \begin{pmatrix} -3 \\ 1 \\ 1 \end{pmatrix}$, $\Pi : \mathbf{r} \cdot \begin{pmatrix} -1 \\ -2 \\ 2 \end{pmatrix} = 1$

b **i** $l: \dfrac{x}{2} = \dfrac{y-1}{5} = \dfrac{z-2}{5}, \Pi: x - y - 3z = 1$

   ii $l: \dfrac{x+1}{-1} = \dfrac{y-3}{3} = \dfrac{z+2}{-3}, \Pi: 2x + y + z = 14$

2 Find the acute angle between each pair of planes.

  a $3x - 7y + z = 4$ and $x + y - 4z = 5$

  b $x - z = 4$ and $y + z = 1$

3 Line l has Cartesian equation $\dfrac{x-3}{2} = \dfrac{y+1}{3} = \dfrac{z-5}{-1}$.

  a Write down the direction vector of l.

  b Find the angle between l and the plane with equation $x - 3y + 5z = 7$.

4 Plane Π_1 has Cartesian equation $3x - y + z = 7$.

  a Write down a normal vector of Π_1.

  Plane Π_2 has equation $x - 5y + 5z = 11$.

  b Find, correct to the nearest degree, the acute angle between Π_1 and Π_2.

5 Line l has equation $\dfrac{x-5}{4} = \dfrac{y+1}{2} = \dfrac{z-2}{3}$.

  a Write down the direction vector of l.

  b Find the acute angle that l makes with the plane $\mathbf{r} \cdot \begin{pmatrix} -1 \\ 4 \\ 3 \end{pmatrix} = 7$.

6 A plane has vector equation $\mathbf{r} = \begin{pmatrix} 3 \\ 7 \\ 1 \end{pmatrix} + \lambda \begin{pmatrix} -1 \\ 3 \\ 1 \end{pmatrix} + \mu \begin{pmatrix} 2 \\ 2 \\ 5 \end{pmatrix}$.

  a Find the normal vector of the plane.

  b Find the angle that the plane makes with the line $\mathbf{r} = \begin{pmatrix} 1 \\ 2 \\ -1 \end{pmatrix} + \lambda \begin{pmatrix} 3 \\ 1 \\ 5 \end{pmatrix}$.

7 Show that the planes with equations $3x + y + 4z = 7$ and $x + 9y - 3z = 8$ are perpendicular to each other.

8 Line l has Cartesian equation $\dfrac{4-x}{5} = z + 1, y = -3$.

  a Find the direction vector of l.

  b Find the acute angle that l makes with the plane $4x - 3z = 0$.

9 Plane Π has equation $5x - 3y - z = 1$.

  a Show that point $P(2,1,6)$ lies in the plane Π.

  b Point Q has coordinates $(7, -1, 2)$. Find the exact value of the sine of the angle between PQ and Π.

  c Find the exact distance PQ.

  d Hence find the exact distance of Q from Π.

> ▶▶ **Fast forward**
>
> In Section 4 you will meet a formula for the distance from a point to a plane.

Section 4: Distances between points, lines and planes

Distance between a point and a plane

Given a plane with equation $\mathbf{r} \cdot \mathbf{n} = p$ and a point M outside of the plane, the shortest distance from M to the plane is equal to the distance MP, where the line MP is perpendicular to the plane. This means that the direction of PM is $\mathbf{n}$. (Point P is called the foot of the perpendicular from the point to the plane.)

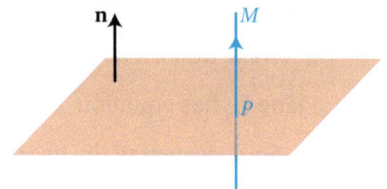

To find the distance MP:

- write down the vector equation of the line with direction $\mathbf{n}$ through point M
- find the intersection, P, between the line and the plane
- calculate the distance MP.

This procedure leads to the formula in Key point 4.7, which you can use unless the question explicitly asks you to carry out the three steps listed here (see questions 14 and 15 in Exercise 4D).

> 🔑 **Key point 4.7**
>
> The shortest distance between the point with position vector $\mathbf{b}$ and the plane with equation $\mathbf{r} \cdot \mathbf{n} = p$ is given by
>
> $$D = \frac{|\mathbf{b} \cdot \mathbf{n} - p|}{|\mathbf{n}|}$$
>
> **This will be given in your formula book.**

WORKED EXAMPLE 4.12

Plane Π has Cartesian equation $3x - y + z = 8$. Find the shortest distance between Π and the point $M(13, -3, 10)$.

Π has equation $\underline{r} \cdot \underline{n} = 8$ with $\underline{n} = \begin{pmatrix} 3 \\ -1 \\ 1 \end{pmatrix}$. ⋯⋯⋯⋯ To use the formula you need to identify the normal vector.

$\underline{b} \cdot \underline{n} = \begin{pmatrix} 13 \\ -3 \\ 10 \end{pmatrix} \cdot \begin{pmatrix} 3 \\ -1 \\ 1 \end{pmatrix} = 52$

and $|\underline{n}| = \sqrt{3^2 + (-1)^2 + 1^2} = \sqrt{11}$

The distance is ⋯⋯⋯⋯ Now use the formula from Key point 4.7.

$D = \dfrac{|\underline{b} \cdot \underline{n} - p|}{|\underline{n}|}$

$= \dfrac{|52 - 8|}{\sqrt{11}} = 4\sqrt{11}$

Distance between a point and a line

Given a line with direction vector **d** and a point M not on the line, the shortest distance between M and the line is the distance from M to the point P on the line such that MP is perpendicular to **d**.

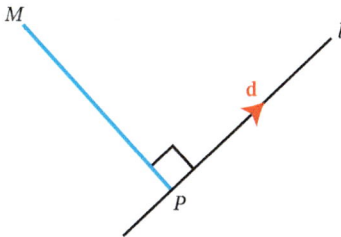

In this course you only need to find this distance in two dimensions. In that case the equation of the line can be written as $ax + by = c$ and you can use the formula in Key point 4.8.

Rewind

In Pure Core Student Book 1, Focus on ... Problem solving 1, you explored various methods for finding the shortest distance between a point and a line in three dimensions. For a reminder, see question 16 in Exercise 4D.

Key point 4.8

The shortest distance between the point with coordinates (x_1, y_1) and the line with equation $ax + by = c$ is given by

$$D = \frac{|ax_1 + by_1 - c|}{\sqrt{a^2 + b^2}}$$

This will be given in your formula book.

WORKED EXAMPLE 4.13

Line l passes through the points $(-2, 5)$ and $(1, 9)$. Find the shortest distance between l and the point $(-4, 7)$.

Gradient of l:

$$m = \frac{1+2}{9-5} = \frac{3}{4}$$

| To use the formula you need to write the equation of l in the form $ax + by = c$. First you need the gradient. |

Equation of l:

$$y - 5 = \frac{3}{4}(x+2)$$

$$\Leftrightarrow 4y - 20 = 3x + 6$$

$$\Leftrightarrow 3x - 4y = -26$$

Use $y - y_1 = m(x - x_1)$.

$$D = \frac{|ax_1 + by_1 - c|}{\sqrt{a^2 + b^2}}$$

Now use the formula from Key point 4.8.

$$= \frac{|3(-4) - 4(7) - (-26)|}{\sqrt{3^2 + 4^2}}$$

$$= \frac{|-14|}{5} = \frac{14}{5}$$

Distance between two skew lines

Consider points M and N moving along two skew lines. The distance between them is the minimum possible when MN is perpendicular to both lines. It may not be immediately obvious that such a position of M and N always exists, but it does. In sketching a diagram of this situation, it is useful to envisage a cuboid, where one line runs along an upper edge, and the other runs along the diagonal of the base, as shown in the diagram. The shortest distance between the two lines is then the height of the cuboid.

You can find the position of the points M and N by using the fact that MN is perpendicular to both lines' direction vector (see question 18 in Exercise 4D). However, in most questions you can use the following formula.

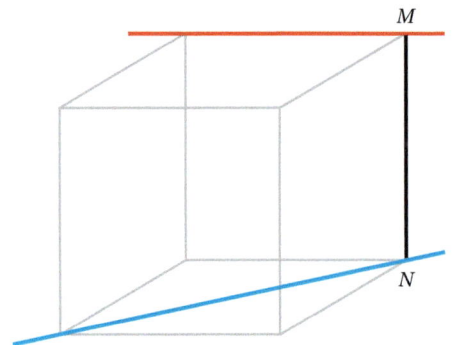

🔑 Key point 4.9

The shortest distance between two skew lines with equations $\mathbf{r} = \mathbf{a} + \lambda \mathbf{d}_1$ and $\mathbf{r} = \mathbf{b} + \mu \mathbf{d}_2$ is given by

$$D = \frac{|(\mathbf{b} - \mathbf{a}) \cdot \mathbf{n}|}{|\mathbf{n}|} \text{ where } \mathbf{n} = \mathbf{d}_1 \times \mathbf{d}_2$$

This will be given in your formula book.

You may need to use the equations of the lines to identify **a** and **b** and then calculate **n**.

WORKED EXAMPLE 4.14

Find the shortest distance between the lines with equations $\dfrac{x-1}{3}=\dfrac{y+1}{4}=\dfrac{3-z}{3}$ and $\dfrac{x+2}{1}=\dfrac{y-1}{-1}=\dfrac{z}{4}$.

The direction vectors are:

The components of the direction vectors are given by the denominators in the Cartesian equation when written in the form $\dfrac{x-a}{p}=\dfrac{y-b}{q}=\dfrac{z-c}{r}$.

For the first line, the last numerator is $3-z$ rather than $z-3$, so the corresponding component of the direction vector is -3.

$$\underline{d}_1=\begin{pmatrix}3\\4\\-3\end{pmatrix},\ \underline{d}_2=\begin{pmatrix}1\\-1\\4\end{pmatrix}$$

$$\underline{n}=\begin{pmatrix}3\\4\\-3\end{pmatrix}\times\begin{pmatrix}1\\-1\\4\end{pmatrix}=\begin{pmatrix}13\\-15\\-7\end{pmatrix}$$

Use the vector product to calculate **n**.

$$D=\frac{|(\underline{b}-\underline{a})\bullet\underline{n}|}{|\underline{n}|}$$

Now use the formula from Key point 4.9 with $\mathbf{a}=\begin{pmatrix}1\\-1\\3\end{pmatrix}$ and $\mathbf{b}=\begin{pmatrix}-2\\1\\0\end{pmatrix}$.

$$=\frac{\left|\begin{pmatrix}-3\\2\\-3\end{pmatrix}\bullet\begin{pmatrix}13\\-15\\-7\end{pmatrix}\right|}{\sqrt{169+225+49}}$$

$$=\frac{48}{\sqrt{443}}=2.28\ (3\text{ s.f.})$$

Distance between two parallel lines

Consider two parallel lines, both with a direction vector **d**. You can measure the distance between them from any point A on the first line: it is the distance AP, where P is the point on the second line such that AP is perpendicular to **d**.

As you can see from the diagram, you can actually find the distance without finding the position vector of P. It equals $|\mathbf{a}-\mathbf{b}|\sin\theta$, where B is any point on the second line and θ is the angle between AB and **d**.

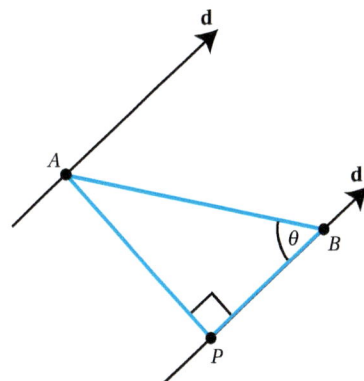

🔑 Key point 4.10

The distance between parallel lines with equations $\mathbf{r} = \mathbf{a} + \lambda\mathbf{d}$ and $\mathbf{r} = \mathbf{b} + \mu\mathbf{d}$ is given by

$$D = |\mathbf{a} - \mathbf{b}|\sin\theta \text{ where } \cos\theta = \frac{(\mathbf{a} - \mathbf{b}) \cdot \mathbf{d}}{|\mathbf{a} - \mathbf{b}||\mathbf{d}|}$$

The formula in Key point 4.10 will not be given in the formula book, so it is a good idea to draw a diagram to make sure you get it right.

⏩ Fast forward

You can also find the distance from point A to the second line by using the fact that AP is perpendicular to the direction vector. See question 17 in Exercise 4D for an example.

WORKED EXAMPLE 4.15

Show that the lines with equations $\dfrac{x-2}{\frac{5}{2}} = \dfrac{y-1}{3} = \dfrac{z-5}{-2}$ and $\dfrac{2x-6}{10} = \dfrac{y+3}{6} = \dfrac{4-z}{4}$ are parallel and find the distance between them.

The first line has direction vector	You need to identify the direction vectors of the two lines.

$$\mathbf{d} = \begin{pmatrix} 2.5 \\ 3 \\ -2 \end{pmatrix}$$

Second line:	For the second line you need to write each term in the form, $\dfrac{x-a}{p}$.

$$\frac{x-3}{5} = \frac{y+3}{6} = \frac{z-4}{-4}$$

has the direction vector

$$\begin{pmatrix} 5 \\ 6 \\ -4 \end{pmatrix} = 2\mathbf{d}$$

The two lines have parallel direction vectors so they are parallel.	The second direction vector is a multiple of the first direction vector.

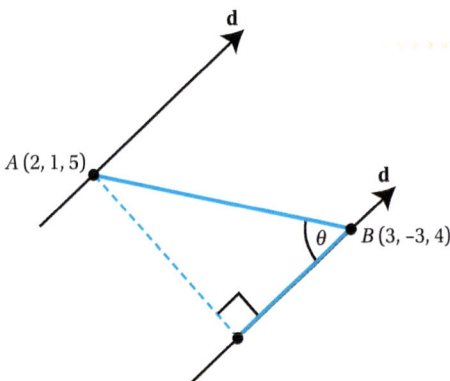

Draw a diagram to identify the required distance.

You can see from the equation that the point $A(2, 1, 5)$ lies on the first line and the point $B(3, -3, 4)$ lies on the second line.

$$\overrightarrow{AB} = \begin{pmatrix} 1 \\ -4 \\ -1 \end{pmatrix}, \underline{d} = \begin{pmatrix} 2.5 \\ 3 \\ -2 \end{pmatrix}$$

The distance is $AB \sin \theta$ where θ is the angle between AB and **d**.

$$\cos\theta = \frac{\overrightarrow{AB} \cdot \underline{d}}{|\overrightarrow{AB}||\underline{d}|}$$

$$= \frac{\frac{5}{2} - 12 + 2}{\sqrt{18}\sqrt{\frac{77}{4}}} = -0.403\ldots$$

$$\Rightarrow \sin\theta = 0.915\ldots$$

$$D = |\overrightarrow{AB}|\sin\theta$$

$$= \sqrt{18} \times 0.915 = 3.88 \,(3\,\text{s.f.})$$

EXERCISE 4D

1 Find the distance from the point M to the plane Π.

  a   **i** $M(1, -1, 3)$, Π: $\mathbf{r} \bullet \begin{pmatrix} 2 \\ 1 \\ 1 \end{pmatrix} = 23$ **ii** $M(3, 5, 10)$, Π: $\mathbf{r} \bullet \begin{pmatrix} 6 \\ -1 \\ 0 \end{pmatrix} = 1$

  b   **i** $M(4, 1, 2)$, Π: $3x - y + 2z = 17$ **ii** $M(0, 3, -5)$, Π: $x - 4z = 8$

2 Find the distance between the point M and the line l.

  a   **i** $M(1, -3)$, l: $3x + 8y = 6$ **ii** $M(0, 4)$, l: $x - 3y = 12$

  b   **i** $M(5, 2)$, l: $y = 4x + 5$ **ii** $M(-6, 1)$, l: $y = -3x + 2$

3 Find the distance between the lines l_1 and l_2.

  a   **i** l_1: $\mathbf{r} = \begin{pmatrix} 1 \\ 2 \\ 3 \end{pmatrix} + t\begin{pmatrix} -1 \\ 1 \\ 2 \end{pmatrix}$ and l_2: $\mathbf{r} = \begin{pmatrix} -4 \\ -4 \\ -11 \end{pmatrix} + s\begin{pmatrix} 5 \\ 1 \\ 2 \end{pmatrix}$ **ii** l_1: $\mathbf{r} = \begin{pmatrix} 4 \\ 0 \\ 2 \end{pmatrix} + t\begin{pmatrix} 2 \\ 0 \\ 1 \end{pmatrix}$ and l_2: $\mathbf{r} = \begin{pmatrix} -1 \\ 2 \\ 3 \end{pmatrix} + s\begin{pmatrix} 1 \\ -2 \\ -2 \end{pmatrix}$

  b   **i** l_1: $\dfrac{x-2}{-3} = \dfrac{y+1}{2} = \dfrac{z-1}{2}$ and l_2: $\dfrac{x+1}{1} = \dfrac{y}{6} = \dfrac{z-1}{3}$ **ii** l_1: $\dfrac{x+1}{2} = \dfrac{y-6}{-2} = \dfrac{z+7}{1}$ and l_2: $\dfrac{x-2}{-1} = \dfrac{y+1}{3} = \dfrac{z}{5}$

4 Show that the lines l_1 and l_2 are parallel and find the distance between them.

  a   **i** l_1: $\mathbf{r} = \begin{pmatrix} 0 \\ 0 \\ 5 \end{pmatrix} + t\begin{pmatrix} 1/2 \\ -1/3 \\ 2 \end{pmatrix}$ and l_2: $\mathbf{r} = \begin{pmatrix} 1 \\ 0 \\ 1 \end{pmatrix} + t\begin{pmatrix} -3 \\ 2 \\ -12 \end{pmatrix}$

      ii l_1: $\mathbf{r} = \begin{pmatrix} 1 \\ 1 \\ 2 \end{pmatrix} + t\begin{pmatrix} 0.6 \\ 0.4 \\ -1 \end{pmatrix}$ and l_2: $\mathbf{r} = \begin{pmatrix} 3 \\ 0 \\ -1 \end{pmatrix} + t\begin{pmatrix} 3 \\ 2 \\ -5 \end{pmatrix}$

b **i** $l_1: \mathbf{r} = \begin{pmatrix} 1 \\ -7 \\ 1 \end{pmatrix} + t \begin{pmatrix} 2 \\ -3 \\ -2/3 \end{pmatrix}$ and $l_2: \dfrac{2x-1}{4} = \dfrac{y-2}{-3} = \dfrac{6-3z}{2}$

  ii $l_1: \mathbf{r} = \begin{pmatrix} 1 \\ 1 \\ 2 \end{pmatrix} + t \begin{pmatrix} 3 \\ 1/2 \\ -2 \end{pmatrix}$ and $l_2: \dfrac{2-3x}{18} = \dfrac{1-y}{1} = \dfrac{2z+3}{8}$

5 Find the shortest distance between the lines with equations $\mathbf{r} = (2\mathbf{i} - \mathbf{k}) + t(4\mathbf{i} - \mathbf{j} + 3\mathbf{k})$ and $\mathbf{r} = (4\mathbf{j}) + t(\mathbf{i} - 5\mathbf{j})$.

6 Find the shortest distance between the point $(3, -1, 5)$ and the plane with equation $3x + y - z = 11$.

7 Find the exact distance of the plane $x - 4y + z = 6$ from the origin.

8 Find the perpendicular distance between the point with coordinates $(-3, 4)$ and the line passing through the points $(1, 3)$ and $(5, -2)$.

9 Show that the lines with equations $\dfrac{x-2}{3} = y+1 = \dfrac{z}{5}$ and $\dfrac{x+1}{3} = \dfrac{y-3}{4} = \dfrac{2-z}{2}$ do not intersect and find the shortest distance between them.

10 Points $A(5, -1)$, $B(1, 2)$ and $C(0, -9)$ form a triangle. Find the height of the triangle corresponding to the side BC and hence find the exact area of the triangle.

11 Show that the lines $\dfrac{x+1}{3} = \dfrac{4-y}{5} = \dfrac{-z}{2}$ and $\dfrac{2x-5}{6} = \dfrac{y+2}{-5} = \dfrac{1-3z}{6}$ are parallel and find the distance between them.

12 Two planes have equations:

$\Pi_1 : x - 3y + z = 6$, and $\Pi_2 : 3x - 9y + 3z = 0$.

a Show that Π_1 and Π_2 are parallel.

b Show that Π_2 passes through the origin.

c Hence find the distance between the planes Π_1 and Π_2.

13 **a** Show that the planes $\Pi_1 : x - z = 4$ and $\Pi_2 : z - x = 8$ are parallel.

b Given that the point $(4, -1, p)$ lies in Π_1 find the value of p.

c Use your answer from part **b** to find the exact distance between the two planes.

14 Plane Π has equation $2x + 2y - z = 11$. Line l is perpendicular to Π and passes through the point $P(-3, -3, 4)$.

a Find the equation of l.

b Find the coordinates of the point Q where l intersects Π.

c Find the shortest distance from P to Π.

15 Plane Π has equation $6x - 2y + z = 16$. Line l is perpendicular to Π and passes through the origin.

a Find the coordinates of the point of intersection of l and Π.

b Find the shortest distance of Π from the origin, giving your answer in exact form.

16 Line l has equation $\mathbf{r} = \begin{pmatrix} 4 \\ 2 \\ -1 \end{pmatrix} + \lambda \begin{pmatrix} 2 \\ -1 \\ 2 \end{pmatrix}$ and point P has coordinates $(7, 2, 3)$. Point C lies on l and PC is

perpendicular to l. Find the coordinates of C.

Hence find the shortest distance of P from l.

17 Two lines are given by Cartesian equations:

$$l_1 : \frac{x-2}{3} = \frac{y+1}{-1} = \frac{z-2}{1}$$

$$l_2 : \frac{x-5}{3} = 1-y = z+4$$

a Show that l_1 and l_2 are parallel.

b Show that the point $A(14, -5, 6)$ lies on l_1.

c Find the coordinates of the point B on l_2 such that AB is perpendicular to the two lines.

d Hence find the distance between l_1 and l_2, giving your answer to 3 significant figures.

18 Two lines have vector equations $l_1 : \mathbf{r} = \begin{pmatrix} 1 \\ 3 \\ 1 \end{pmatrix} + \lambda \begin{pmatrix} 1 \\ -1 \\ 2 \end{pmatrix}$ and $l_2 : \mathbf{r} = \begin{pmatrix} 5 \\ -1 \\ -6 \end{pmatrix} + \mu \begin{pmatrix} 1 \\ 1 \\ 3 \end{pmatrix}$.

The point A on l_1 and the point B on l_2 are such that AB is perpendicular to both lines.

a Show that $\mu - \lambda = 1$.

b Find a second equation linking λ and μ.

c Hence find the shortest distance between l_1 and l_2, giving your answer as an exact value.

🔖 Checklist of learning and understanding

- The vector equation of a plane is $\mathbf{r} = \mathbf{a} + \lambda \mathbf{d}_1 + \mu \mathbf{d}_2$, where $\mathbf{d}_1$ and $\mathbf{d}_2$ are two vectors parallel to the plane and $\mathbf{a}$ is the position vector of one point in the plane.
- The Cartesian equation of a plane has the form $n_1 x + n_2 y + n_3 z = k$. This can also be written in the scalar product form $\mathbf{r} \cdot \mathbf{n} = \mathbf{a} \cdot \mathbf{n}$, where $\mathbf{n} = \begin{pmatrix} n_1 \\ n_2 \\ n_3 \end{pmatrix}$ is the normal vector of the plane, which is perpendicular to every line in the plane. To derive the Cartesian equation from a vector equation, use $\mathbf{n} = \mathbf{d}_1 \times \mathbf{d}_2$.
- The angle between two planes is the angle between their normals.
- The angle between a line and a plane is $90° -$ (angle between line's direction vector and plane's normal).
- To find the intersection between a line and a plane, express x, y, z for the line in terms of λ and substitute into the Cartesian equation of the plane.
- The shortest distance between the point with position vector $\mathbf{b}$ and the plane with equation $\mathbf{r} \cdot \mathbf{n} = p$ is given by $D = \dfrac{|\mathbf{b} \cdot \mathbf{n} - p|}{|\mathbf{n}|}$.
- The shortest distance between the point with coordinates (x_1, y_1) and the line with equation $ax + by = c$ is given by $D = \dfrac{|ax_1 + by_1 - c|}{\sqrt{a^2 + b^2}}$.
- The shortest distance between two skew lines with equations $\mathbf{r} = \mathbf{a} + \lambda \mathbf{d}_1$ and $\mathbf{r} = \mathbf{b} + \mu \mathbf{d}_2$ is given by $D = \dfrac{|(\mathbf{b} - \mathbf{a}) \cdot \mathbf{n}|}{|\mathbf{n}|}$ where $\mathbf{n} = \mathbf{d}_1 \times \mathbf{d}_2$.
- To find the distance between two parallel lines, find the perpendicular distance from a point on one line to the other line. (A diagram will help you see how to find this distance).

Mixed practice 4

1 Find the exact perpendicular distance from the point $(-4, 1)$ to the line with equation $y = 1 - 3x$.

2 Find the shortest distance of the point $(3, -5, 0)$ from the plane $3x - y + 3z = 16$.

3 The vector $\mathbf{n} = 3\mathbf{i} + \mathbf{j} - \mathbf{k}$ is normal to a plane which contains the point $(3, -1, 2)$.

    a Find an equation for the plane.

    b Find a if the point $(a, 2a, a-1)$ lies on the plane.

4   **a** Calculate $\begin{pmatrix} 2 \\ -1 \\ 1 \end{pmatrix} \times \begin{pmatrix} 3 \\ 1 \\ -1 \end{pmatrix}$.

    b Plane Π_1 has normal vector $\begin{pmatrix} 2 \\ -1 \\ 1 \end{pmatrix}$ and contains point $A\,(3, 4, -2)$. Find the Cartesian equation of the plane.

    c Plane Π_2 has equation $3x + y - z = 15$. Show that Π_2 contains point A.

    d A third plane, Π_3, has equation $\mathbf{r} \cdot \begin{pmatrix} 2 \\ 1 \\ 2 \end{pmatrix} = 12$. Find the angle between Π_1 and Π_3 in degrees.

5 The plane Π passes through the points with coordinates $(1, 6, 2)$, $(5, 2, 1)$ and $(1, 0, -2)$.

    i Find a vector equation of Π in the form $\mathbf{r} = \mathbf{a} + \lambda \mathbf{b} + \mu \mathbf{c}$.

    ii Find a cartesian equation of Π.

© OCR A Level Mathematics, Unit 4727/01 Further Pure Mathematics 3, June 2013

6 Two skew lines have equations

$$\frac{x}{2} = \frac{y+3}{1} = \frac{z-6}{3} \quad \text{and} \quad \frac{x-5}{3} = \frac{y+1}{1} = \frac{z-7}{5}.$$

    i Find the direction of the common perpendicular to the lines.

    ii Find the shortest distance between the lines.

© OCR A Level Mathematics, Unit 4727 Further Pure Mathematics 3, January 2009

7 Line l_1 has equation $\mathbf{r} = \begin{pmatrix} 5 \\ 1 \\ 2 \end{pmatrix} + t\begin{pmatrix} -1 \\ 1 \\ 3 \end{pmatrix}$ and line l_2 has equation $\mathbf{r} = \begin{pmatrix} 5 \\ 4 \\ 9 \end{pmatrix} + s\begin{pmatrix} 2 \\ 1 \\ 1 \end{pmatrix}$.

    a Find $\begin{pmatrix} -1 \\ 1 \\ 3 \end{pmatrix} \times \begin{pmatrix} 2 \\ 1 \\ 1 \end{pmatrix}$.

    b Find the coordinates of the point of intersection of the two lines.

    c Write down a vector perpendicular to the plane containing the two lines.

    d Hence find the Cartesian equation of the plane containing the two lines.

8 The plane Π has equation $x - 2y + z = 20$ and the point A has coordinates $(4, -1, 2)$.

    a Write down the vector equation of the line l through A which is perpendicular to Π.

    b Find the coordinates of the point of intersection of line l and plane Π.

    c Hence find the shortest distance from point A to plane Π.

9 Plane Π has equation $x - 4y + 2z = 7$ and point P has coordinates $(9, -7, 6)$.

    a Show that point $R(5, 1, 3)$ lies in the plane Π.

    b Find the vector equation of the line PR.

    c Write down the vector equation of the line through P perpendicular to Π.

    d N is the foot of the perpendicular from P to Π. Find the coordinates of N.

    e Find the exact distance of point P from the plane Π.

10 Find the shortest distance between the skew lines with equations

    $\mathbf{r} = -3\mathbf{i} + 3\mathbf{j} + 18\mathbf{k} + s(2\mathbf{i} - \mathbf{j} - 8\mathbf{k})$ and $\mathbf{r} = 5\mathbf{i} + 2\mathbf{k} + t(\mathbf{i} + \mathbf{j} - \mathbf{k})$.

11 Consider the four points $A(4, -1, 3)$, $B(1, 1, 2)$, $C(3, 0, 1)$ and $D(6, p, q)$.

    a Given that $ABCD$ is a parallelogram find the values of p and q.

    b Find the distance between the lines AB and DC.

    c Hence find the area of $ABCD$.

12 **a** Find the coordinates of the point of intersection of lines

$$l_1 : \frac{x-1}{3} = \frac{y+1}{4} = \frac{3-z}{3} \text{ and } l_2 : \frac{x+12}{2} = \frac{y}{1} = \frac{z+17}{1}.$$

    b Find a vector perpendicular to both lines.

    c Hence find the Cartesian equation of the plane containing l_1 and l_2.

13 Point $A(3, 1, -4)$ lies on line l which is perpendicular to plane $\Pi : 3x - y - z = 1$.

    a Find the Cartesian equation of l.

    b Find the point of intersection of the line l and the plane Π.

    c Point A is reflected in Π. Find the coordinates of the image of A.

    d Point B has coordinates $(1, 1, 1)$. Show that B lies in Π.

    e Find the distance between B and l.

14 A line l has equation $\dfrac{x-6}{-4} = \dfrac{y+7}{8} = \dfrac{z+10}{7}$ and a plane p has equation $3x - 4y - 2z = 8$.

    i Find the point of intersection of l and p.

    ii Find the equation of the plane which contains l and is perpendicular to p, giving your answer in the form $ax + by + cz = d$.

© OCR A Level Mathematics, Unit 4727 Further Pure Mathematics 3, June 2009

15 A tetrahedron $ABCD$ is such that AB is perpendicular to the base BCD. The coordinates of the points A, C and D are $(-1, -7, 2)$, $(5, 0, 3)$ and $(-1, 3, 3)$ respectively, and the equation of the plane BCD is $x + 2y - 2z = -1$.

    i Find, in either order, the coordinates of B and the length of AB.

    ii Find the acute angle between the planes ACD and BCD.

© OCR A Level Mathematics, Unit 4727/01 Further Pure Mathematics 3, January 2008

16 Line l passes through point $A(-1,1,4)$ and has direction vector $\mathbf{d} = \begin{pmatrix} 6 \\ 1 \\ 5 \end{pmatrix}$. Point B has coordinates $(3,3,1)$.

Plane Π has normal vector $\mathbf{n}$, and contains the line l and the point B.

a Write down a vector equation for l.

b Explain why $\overrightarrow{AB}$ and $\mathbf{d}$ are both perpendicular to $\mathbf{n}$.

c Hence find one possible vector $\mathbf{n}$.

d Find the Cartesian equation of plane Π.

17 The plane $3x + 2y - z = 2$ contains the line $x - 3 = \dfrac{2y+2}{5} = \dfrac{z-5}{k}$. Find k.

18 Find the Cartesian equation of the plane containing the lines

$x = \dfrac{3-y}{2} = z - 1$ and $\dfrac{x-2}{3} = \dfrac{y+1}{-3} = \dfrac{z-3}{5}$.

19 Two planes have equations $\Pi_1 : 3x - y + z = 17$ and $\Pi_2 : x + 2y - z = 4$.

a Calculate $\begin{pmatrix} 3 \\ -1 \\ 1 \end{pmatrix} \times \begin{pmatrix} 1 \\ 2 \\ -1 \end{pmatrix}$.

b Show that Π_1 and Π_2 are perpendicular.

c Show that the point $M(1,1,2)$ does not lie in either of the two planes.

d Find a vector equation of the line through M which is parallel to both planes.

20 Four points have coordinates $A(7,0,1), B(8,-1,4), C(9,0,2), D(6,5,3)$.

a Show that $\overrightarrow{AD}$ is perpendicular to both $\overrightarrow{AB}$ and $\overrightarrow{AC}$.

b Write down the equation of the plane Π containing the points A, B and C in the form $\mathbf{r} \cdot \mathbf{n} = k$.

c Find the exact distance of point D from plane Π.

d Point D_1 is the reflection of D in Π. Find the coordinates of D_1.

21 Points $A(8,0,4)$, $B(12,-1,5)$ and $C(10,0,7)$ lie in the plane Π.

a Find $\overrightarrow{AB} \times \overrightarrow{AC}$.

b Find the area of the triangle ABC, correct to 3 significant figures.

c Find the Cartesian equation of Π.

Point D has coordinates $(-7, -28, 11)$.

d Find a vector equation of the line through D perpendicular to the plane.

e Find the intersection of this line with Π, and hence find the perpendicular distance of D from Π.

f Find the volume of the pyramid $ABCD$.

22 A regular tetrahedron has vertices at the points

$$A\left(0,0,\tfrac{2}{3}\sqrt{6}\right),\ B\left(\tfrac{2}{3}\sqrt{3},0,0\right),\ C\left(-\tfrac{1}{3}\sqrt{3},1,0\right),\ D\left(-\tfrac{1}{3}\sqrt{3},-1,0\right).$$

i Obtain the equation of the face ABC in the form

$$x+\sqrt{3}y+\left(\tfrac{1}{2}\sqrt{2}\right)z=\tfrac{2}{3}\sqrt{3}$$

(Answers which only verify the given equation will not receive full credit.)

ii Give a geometrical reason why the equation of the face ABD can be expressed as

$$x-\sqrt{3}y+\left(\tfrac{1}{2}\sqrt{2}\right)z=\tfrac{2}{3}\sqrt{3}$$

iii Hence find the cosine of the angle between two faces of the tetrahedron.

© OCR A Level Mathematics, Unit 4727 Further Pure Mathematics 3, January 2010

23 With respect to the origin O, the position vectors of the points U, V and W are $\mathbf{u}$, $\mathbf{v}$ and $\mathbf{w}$ respectively. The mid points of the sides VW, WU and UV of the triangle UVW are M, N and P respectively.

i Show that $\overrightarrow{UM}=\tfrac{1}{2}(\mathbf{v}+\mathbf{w}-2\mathbf{u})$.

ii Verify that the point G with position vector $\tfrac{1}{3}(\mathbf{u}+\mathbf{v}+\mathbf{w})$ lies on UM, and deduce that the lines UM, VN and WP intersect at G.

iii Write down, in the form $\mathbf{r}=\mathbf{a}+t\mathbf{b}$, an equation of the line through G which is perpendicular to the plane UVW. (It is not necessary to simplify the expression for $\mathbf{b}$.)

iv It is now given that $\mathbf{u}=\begin{pmatrix}1\\0\\0\end{pmatrix}$, $\mathbf{v}=\begin{pmatrix}0\\1\\0\end{pmatrix}$ and $\mathbf{w}=\begin{pmatrix}0\\0\\1\end{pmatrix}$. Find the perpendicular distance from O to the plane UVW.

© OCR A Level Mathematics, Unit 4727 Further Pure Mathematics 3, June 2012

5 Simultaneous equations and planes

In this chapter you will learn how to:

- identify different geometrical configurations of two or three planes
- determine whether a set of simultaneous equations has a unique solution, no solutions or infinitely many solutions
- use simultaneous equations to determine the geometrical configuration of three planes.

Before you start…

Pure Core Student Book 1, Chapter 1	You should be able to find the determinant of a 2×2 and a 3×3 matrix and understand what is meant by a singular matrix.	**1** Find the value of a for which the matrix $$\begin{pmatrix} 1 & 1 & -2 \\ a & 1 & 1 \\ 0 & -2 & 3 \end{pmatrix}$$ is singular.
Pure Core Student Book 1, Chapter 3	You should be able to use matrices to solve simultaneous equations in two and three unknowns.	**2** Express simultaneous equations $$\begin{cases} 2x - 3y = 2 \\ 5x - 8y = 3 \end{cases}$$ in the form $\mathbf{M}\begin{pmatrix} x \\ y \end{pmatrix} = \begin{pmatrix} a \\ b \end{pmatrix}$. Find $\mathbf{M}^{-1}$ and hence find x and y.
Pure Core Student Book 1, Chapter 3	You should be able to determine when simultaneous equations do not have a unique solution.	**3** Find the value of a for which the simultaneous equations $$\begin{cases} x + y - 2z = 2 \\ ax + y + z = b \\ -2y + 3z = 1 \end{cases}$$ do not have a unique solution.
GCSE	You should be able to use elimination to solve simultaneous equations.	**4** Use elimination to solve the simultaneous equations $$\begin{cases} 2x - 3y = 2 \\ 5x - 8y = 3 \end{cases}$$
Chapter 4	You should know how to find the Cartesian equation of a plane, and to identify the normal vector of a plane with a given equation.	**5 a** Find the Cartesian equation of the plane with the normal vector $4\mathbf{i} - \mathbf{j} + 2\mathbf{k}$ which contains the point $(0, -1, 1)$. **b** Write down the normal vector of the plane with equation $3x - 4z = 1$.

In Chapter 4 you learnt about equations of planes in three dimensions. A Cartesian equation of a plane is a linear equation with three unknowns. In Pure Core Student Book 1, Chapter 3, you learnt how to solve three simultaneous equations with three unknowns by using an inverse matrix, and also how to tell when the solution is not unique.

In this chapter you will learn how to distinguish between different situations with non-unique solutions. You will then use simultaneous equations to find the intersection of three planes, and to determine the geometric configuration of the planes in the case when the intersection is not a single point.

> 📷 **Focus on …**
>
> Matrices have applications in many other situations. You can explore one of them in Focus on … Modelling 1.

Section 1: Linear simultaneous equations

In Pure Core Student Book 1, Chapter 3, you learnt how to recast a system of two or three simultaneous equations into a matrix problem, and to use an inverse matrix to solve it. This method only works when the corresponding matrix is non-singular, and it leads to a unique solution of the system (a single pair of (x, y) values or a single triple of (x, y, z) values).

> 💡 **Tip**
>
> A non-singular matrix has an inverse and its determinant is not zero.

If the matrix is singular (has a zero determinant) you need to use a different method, such as elimination, to determine whether there are any solutions.

The three possible cases for two equations with two unknowns can be illustrated by the examples in this table.

Unique solution	Infinitely many solutions	No solutions
$\begin{cases} 6x+12y=30 \quad (1) \\ 3x+8y=19 \quad (2) \end{cases}$	$\begin{cases} 6x+12y=30 \quad (1) \\ 4x+8y=20 \quad (2) \end{cases}$	$\begin{cases} 6x+12y=30 \quad (1) \\ 4x+8y=19 \quad (2) \end{cases}$
$\begin{vmatrix} 6 & 12 \\ 3 & 8 \end{vmatrix}=12\neq 0$	$\begin{vmatrix} 6 & 12 \\ 4 & 8 \end{vmatrix}=0$	$\begin{vmatrix} 6 & 12 \\ 4 & 8 \end{vmatrix}=0$
Non-singular matrix	Singular matrix	
$4\times(1)-6\times(2): 6x=6$	$4\times(1)-6\times(2): 0x=0$	$4\times(1)-6\times(2): 0x=6$
Unique solution: $x=1, y=2$	Infinitely many solutions: any (x, y) with $6x+12y=30$	No solutions
Consistent system		**Inconsistent** system

> 🔑 **Key point 5.1**
>
> For a system of simultaneous equations in matrix form, $\mathbf{Mr}=\mathbf{a}$:
>
> - If $\det \mathbf{M} \neq 0$ the equations have a unique solution.
> - If $\det \mathbf{M} = 0$ there is no unique solution. Use elimination to distinguish between two cases:
>   - Consistent equations: there are infinitely many solutions.
>   - Inconsistent equations: there are no solutions.

WORKED EXAMPLE 5.1

Consider simultaneous equations

$$\begin{cases} 4x+10y=23 \\ 10x+ay=b \end{cases}$$

a Find the value of a for which the equations do not have a unique solution.
b For this value of a, find the value of b for which the equations are consistent.

a $\begin{vmatrix} 4 & 10 \\ 10 & a \end{vmatrix}=0$ — The solution is not unique when the determinant is zero.

$\Leftrightarrow 4a-100=0$

$\Leftrightarrow a=25$

b $\begin{cases} 4x+10y=23 \quad (1) \\ 10x+25y=b \quad (2) \end{cases}$ — Using $a=25$, try to solve the equations by elimination; eliminate y.

$5\times(1)-2\times(2): 0x=115-2b$

For solutions: — For the system to be consistent, the RHS of the last equation must be zero.

$115-2b=0$

$b=\dfrac{115}{2}$

You can apply the same method to a system of three equations with three unknowns.

WORKED EXAMPLE 5.2

For the system of simultaneous equations

$$\begin{cases} x+y-2z=2 \\ ax+y+z=b \\ -2y+3z=1 \end{cases}$$

where a and b are constants:

a find the value of a for which there is not a unique solution
b for the value of a found in part **a**, find the value of b such that the equations are consistent.

a $\begin{pmatrix} 1 & 1 & -2 \\ a & 1 & 1 \\ 0 & -2 & 3 \end{pmatrix}\begin{pmatrix} x \\ y \\ z \end{pmatrix}=\begin{pmatrix} 2 \\ b \\ 1 \end{pmatrix}$ — Recast the problem in matrix form as $\mathbf{M}\begin{pmatrix} x \\ y \\ z \end{pmatrix}=\begin{pmatrix} p \\ q \\ r \end{pmatrix}$.

Continues on next page ...

$$\begin{vmatrix} 1 & 1 & -2 \\ a & 1 & 1 \\ 0 & -2 & 3 \end{vmatrix} = 1\begin{vmatrix} 1 & 1 \\ -2 & 3 \end{vmatrix} - 1\begin{vmatrix} a & 1 \\ 0 & 3 \end{vmatrix} - 2\begin{vmatrix} a & 1 \\ 0 & -2 \end{vmatrix}$$

The determinant tells you whether there is a unique solution.

$$= 5 - 3a - 2(-2a)$$
$$= 5 + a$$

$\therefore a = -5$

No unique solution when det $\mathbf{M} = 0$.

b Using elimination:

$$\begin{cases} x + y - 2z = 2 & (1) \\ -5x + y + z = b & (2) \\ -2y + 3z = 1 & (3) \end{cases}$$

Use elimination to determine when the equations are consistent.

$$\begin{cases} (2) + 5(1): 6y - 9z = 10 + b & (4) \\ (3): \quad -2y + 3z = 1 & (5) \end{cases}$$

Eliminate x from two equations.

$(4) + 3(5): 0 = 13 + b \quad (6)$

Eliminate y from the final equation.

For a consistent solution,
$13 + b = 0$, so $b = -13$

Equation (6) says $0x + 0y + 0z = 13 + b$, which is only possible if $13 + b = 0$.

EXERCISE 5A

1 For each pair of simultaneous equations, determine whether there is a unique solution, no solutions or infinitely many solutions.

a i $\begin{cases} 2x + y = -3 \\ 3x - 6y = 4 \end{cases}$

b i $\begin{cases} 3x + 2y = -3 \\ 9x + 6y = 4 \end{cases}$

c i $\begin{cases} 10x + 5y = 15 \\ 6x + 3y = 9 \end{cases}$

ii $\begin{cases} x - 4y = 5 \\ 2x + y = -1 \end{cases}$

ii $\begin{cases} 5x - 3y = 3 \\ 10x - 6y = 6 \end{cases}$

ii $\begin{cases} 2x - 4y = -3 \\ 3x - 6y = 5 \end{cases}$

2 For each set of simultaneous equations determine whether or not there is a unique solution. Where there is a unique solution, use the inverse matrix to find it. Where there is no unique solution, determine whether the system is consistent or inconsistent.

a i $\begin{cases} 2x + y - z = -3 \\ 3x - 6y + z = 4 \\ 4x + 3y - 2z = 1 \end{cases}$

b i $\begin{cases} 3x + y + z = 8 \\ -7x + 3y + z = 2 \\ x + y + 3z = 0 \end{cases}$

c i $\begin{cases} 3x - y + z = 17 \\ x + 2y - z = 8 \\ 2x - 3y + 2z = 9 \end{cases}$

ii $\begin{cases} 2x + y - 5z = 15 \\ 5x + 6y + z = 7 \\ x - 8y + 5z = 1 \end{cases}$

ii $\begin{cases} 3x - y + z = 17 \\ x + 2y - z = 8 \\ 2x - 3y + 2z = 3 \end{cases}$

ii $\begin{cases} 2x + y - 2z = 0 \\ x - 2y - z = 2 \\ 3x + 4y - 3z = -2 \end{cases}$

3 Show that the system of equations

$$\begin{cases} 2x+3y-2z=2 \\ 3x+2y+z=5 \\ 3x+7y-7z=3 \end{cases}$$

does not have a unique solution.

4 **a** For what values of k does the system of equations

$$\begin{cases} -x+(2k-5)y-2z=2 \\ (1+k)x-y+(k-1)z=5 \\ x+y+2z=1 \end{cases}$$

have no unique solution?

b For each value of k found in part **a**, determine whether the equations have no solutions or infinitely many solutions.

5 **a** Show that the system of equations

$$\begin{cases} kx+3y-z=-2 \\ -3x+(k+4)y+z=-8 \\ x+3y+(k-2)z=4 \end{cases}$$

has no unique solution for $k=1$.

b Find the other values of k for which there is no unique solution.

c For $k=3$, find x, y and z.

d For $k=1$, determine whether the equations are consistent or inconsistent.

6 **a** Show that, for all values of a, the simultaneous equations

$$\begin{cases} 6x+3y+9z=p \\ 2x+ay+3z=q \\ 2x+y+3z=r \end{cases}$$

do not have a unique solution.

b Given that the equations have infinitely many solutions, show that $p=3r$.

7 Find the value of k and the value of c for which the system of equations

$$\begin{cases} 2x+3y-2z=2 \\ 3x+ky+z=5 \\ 3x+7y-7z=c \end{cases}$$

has infinitely many solutions.

Section 2: Intersections of planes

Simultaneous equations in two variables describe lines. If there is a unique solution, it represents the intersection of the lines, and if there is no unique solution this arises because the lines are parallel (no solution) or identical (infinitely many solutions).

Equations in three variables describe planes. Two distinct planes can either intersect (planes intersect in a line) or be parallel.

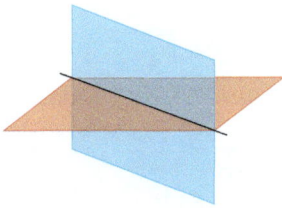

planes intersecting in a line   parallel planes

Key point 5.2

Two planes are parallel if their normal vectors are multiples of each other.

WORKED EXAMPLE 5.3

Five planes have these equations:

$\Pi_1 : 3x - y + z = 17$ $\Pi_2 : x + 2y - z = 8$ $\Pi_3 : -6x + 2y - 2z = 17$

$\Pi_4 : x - 2y + z = 6$ $\Pi_5 : 3x + 6y - 3z = 2$

a Identify pairs of parallel planes.
b Which of the five plane(s) are identical to the plane $\Pi_6 : 2x + 4y - 2z - 16 = 0$?

a Π_1 and Π_3 are parallel.
   Π_2 and Π_5 are parallel.

Parallel planes have normals in the same direction. This means that the LHS of the equations are multiples of each other.

b Π_2 and Π_5 are parallel to Π_6.

First identify the plane(s) with the normals in the same direction.

$\Pi_2 : 2x + 4y - 2z = 16$

$\Pi_5 : 2x + 4y - 2z = \dfrac{4}{3}$

Π_2 is identical to Π_6.

Check whether the RHS is also the same.

With three distinct planes there are several possibilities; altogether, there are five different arrangements, but these fit into three cases:

1 Consistent system: unique solution
The three planes meet at a single point.

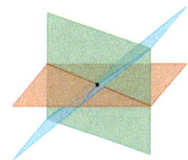

three distinct planes intersecting at a point

2 Consistent system: infinitely many solutions
The three planes intersect in a line (the planes form a sheaf).

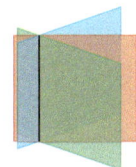

three distinct planes intersecting in a line

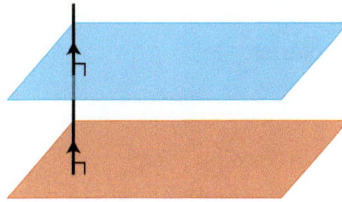

3 **Inconsistent** system: no solutions

There is no point common to all three planes.

a All three planes are parallel.

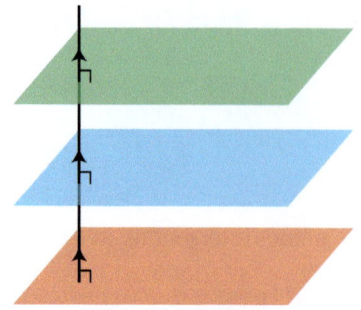

three distinct parallel planes

b Two of the planes are parallel.

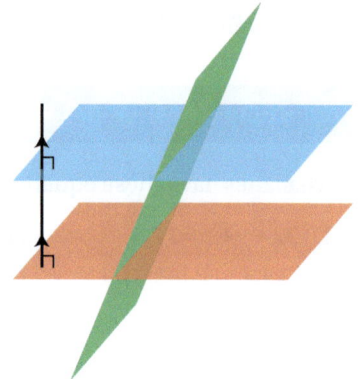

two parallel planes and one non-parallel plane

c The planes enclose a triangular prism, so that each pair of planes intersects in a line, with the three distinct lines running parallel to each other.

three planes forming a triangular prism

You can distinguish between the different cases by using the methods from Section 1.

Key point 5.3

To determine the geometrical arrangement of three planes described by a set of simultaneous equations, use $\mathbf{Mr} = \mathbf{a}$:

- If $\det \mathbf{M} \neq 0$ the three planes meet at a single point.
- If $\det \mathbf{M} = 0$ then:
    - If the equations are consistent the planes meet in a line (form a sheaf).
    - If the equations are inconsistent then either some of the planes are parallel, or the three planes form a prism.

Tip

Remember that two planes are parallel if their normals are multiples of each other. If whole equations are multiples of each other then they describe the same plane.

WORKED EXAMPLE 5.4

The simultaneous equations $\begin{cases} kx - y + 3z = -2 \\ x + 6y - z = 3 \\ 3x - 8y + 7z = 6 \end{cases}$ describe three planes.

a Show that when $k = -4.5$ the planes intersect at a single point and find its coordinates.

b Find the value of k for which there is no unique solution, and describe the configuration of the three planes in that case.

$$\begin{pmatrix} k & -1 & 3 \\ 1 & 6 & -1 \\ 3 & -8 & 7 \end{pmatrix} \begin{pmatrix} x \\ y \\ z \end{pmatrix} = \begin{pmatrix} -2 \\ 3 \\ 6 \end{pmatrix}$$

Recast the problem in matrix form as

$$\mathbf{M}\begin{pmatrix} x \\ y \\ z \end{pmatrix} = \begin{pmatrix} p \\ q \\ r \end{pmatrix}.$$

a Setting $k = -4.5$:

$$\det \begin{pmatrix} -4.5 & -1 & 3 \\ 1 & 6 & -1 \\ 3 & -8 & 7 \end{pmatrix} = -221 \text{ (from calculator)}$$

Show that $\det \mathbf{M} \neq 0$ to establish the existence of a unique solution.

The matrix is non-singular for $k = -4.5$ and therefore there will be a unique solution.

$$\begin{pmatrix} x \\ y \\ z \end{pmatrix} = \begin{pmatrix} -4.5 & -1 & 3 \\ 1 & 6 & -1 \\ 3 & -8 & 7 \end{pmatrix}^{-1} \begin{pmatrix} -2 \\ 3 \\ 6 \end{pmatrix} = \begin{pmatrix} 1 \\ 0.5 \\ 1 \end{pmatrix}$$

Find the solution by using the inverse matrix.

The intersection point is $(1, 0.5, 1)$.

b $\det \begin{pmatrix} k & -1 & 3 \\ 1 & 6 & -1 \\ 3 & -8 & 7 \end{pmatrix} = k \begin{vmatrix} 6 & -1 \\ -8 & 7 \end{vmatrix} + 1 \begin{vmatrix} 1 & -1 \\ 3 & 7 \end{vmatrix} + 3 \begin{vmatrix} 1 & 6 \\ 3 & -8 \end{vmatrix}$

No unique solution when $\det \mathbf{M} = 0$.

$$= 34k + 10 - 78$$
$$= 34(k - 2)$$

There is no unique solution when $k = 2$.

Using elimination:

$$\begin{cases} 2x - y + 3z = -2 \quad (1) \\ x + 6y - z = 3 \quad (2) \\ 3x - 8y + 7z = 6 \quad (3) \end{cases}$$

You now need to use elimination to determine whether the equations are consistent.

$$\begin{cases} (1) - 2(2): -13y + 5z = -8 \quad (4) \\ 3(1) - 2(3): 13y - 5z = -18 \quad (5) \end{cases}$$

Eliminate x from two equations.

$$(4) + (5): 0 = -26 \quad (6)$$

Eliminate y from the final equation.

The equations are inconsistent, so the three planes do not intersect.

The resulting equation has no solutions.

No row of the 3×3 matrix is a multiple of another, so there are no parallel planes.

Check whether any of the planes are parallel.

Therefore the three planes form a triangular prism.

WORKED EXAMPLE 5.5

Planes Π_1, Π_2 and Π_3 are given by the equations:

$\Pi_1 : 3x - y + z = 17$

$\Pi_2 : x + 2y - z = 8$

$\Pi_3 : 2x + py + qz = k$

a Describe the geometric configuration of the three planes in the case when $p = 4$, $q = -2$, $k = 7$.

b In the case $p = -3$ find the value of q and the value of k for which the three planes intersect along a line.

a $\Pi_1 : 3x - y + z = 17$

$\Pi_2 : x + 2y - z = 8$

$\Pi_3 : 2x + 4y - 2z = 7$

$\Pi_2 : 2x + 4y - 2z = 16$

so Π_2 and Π_3 are parallel.

Π_1 is not parallel to them.
There are two parallel planes, with a third plane intersecting them.

It is a good idea to first check whether any of the planes are parallel, as that is easily seen from the equations.

Describe the geometric configuration. A sketch might help you visualise it.

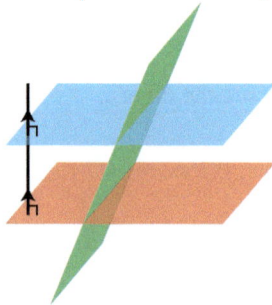

b $\begin{vmatrix} 3 & -1 & 1 \\ 1 & 2 & -1 \\ 2 & -3 & q \end{vmatrix}$

If the three planes intersect along a line then the equations don't have any unique solutions. This means that the determinant is 0.

$= 3\begin{vmatrix} 2 & -1 \\ -3 & q \end{vmatrix} + 1\begin{vmatrix} 1 & -1 \\ 2 & q \end{vmatrix} + 1\begin{vmatrix} 1 & 2 \\ 2 & -3 \end{vmatrix}$

$= 7q - 14$

$= 0$ when $q = 2$.

Using elimination:

$\begin{cases} 3x - y + z = 17 & (1) \\ x + 2y - z = 8 & (2) \\ 2x - 3y + 2z = k & (3) \end{cases}$

You now need to use elimination to determine when the equations are consistent.

$\begin{cases} 3(2)-(1): 7y - 4z = 7 & (4) \\ 3(3)-2(1): -7y + 4z = 3k - 34 & (5) \end{cases}$

Eliminate x from two equations.

$(4)+(5): 0 = 3k - 27$ (6)

Eliminate y from the final equation.

Consistent equations, so
$3k - 27 = 0 \therefore k = 9$

If the planes intersect along a line, the equations must be consistent.

EXERCISE 5B

1 For each set of simultaneous equations, determine whether they are consistent or inconsistent and interpret the geometrical configuration of the planes described by the equations. If there is a unique solution, find it.

a **i** $\begin{cases} 2x+y-z=6 \\ -x+2y+z=3 \\ 3x-5y+2z=3 \end{cases}$ **b** **i** $\begin{cases} x+y-2z=6 \\ -x+2y+z=5 \\ 3x+3y-6z=2 \end{cases}$ **c** **i** $\begin{cases} 9x-3y+3z=5 \\ 4x-2y+z=1 \\ x+5y+2z=6 \end{cases}$ **d** **i** $\begin{cases} 2x+3y+z=2 \\ 5x-6y+z=1 \\ x+3y+2z=3 \end{cases}$

ii $\begin{cases} 4x+2y-z=2 \\ 2x-y+z=5 \\ 3x-3y+2z=8 \end{cases}$ **ii** $\begin{cases} 5x-3y+2z=4 \\ 2x-2y+z=2 \\ x+5y-z=-2 \end{cases}$ **ii** $\begin{cases} 4x+2y-6z=6 \\ -x-2y+3z=-3 \\ 6x+3y-9z=-9 \end{cases}$ **ii** $\begin{cases} 4x+2y-z=-1 \\ -x+3y+z=5 \\ 3x-5y+2z=7 \end{cases}$

2 Consider a set of simultaneous equations $\begin{cases} 3x+y+z=8 \\ -7x+3y+z=2 \\ x+y+3z=0 \end{cases}$

a Show that the equations have a unique solution and find this solution.

b The three equations represent planes. Describe the configuration of the three planes.

3 A system of equations is given by $\begin{cases} x=2 \\ x+y-z=7 \\ 2x+y+z=3 \end{cases}$

a Show that the system has a unique solution.

b The three equations represent planes. Describe the configuration of the three planes.

4 Find the intersection of the planes
$$\begin{cases} x-2y+z=5 \\ 2x+y+z=1 \\ x+2y-z=-2 \end{cases}$$

5 **a** Show that there is no unique solution to the simultaneous equations given by $\begin{cases} 2x-y+z=6 \\ 3x+y+5z=-7 \\ x-3y-3z=8 \end{cases}$

b Show that the equations are inconsistent.

c Interpret this situation geometrically.

6 Describe the configuration of these three planes:

$\Pi_1 : 6x-10y+2z=17$

$\Pi_2 : x+2y-z=8$

$\Pi_3 : 15x-25y+5z=7$

7 **a** Find the value of k such that the planes $\Pi_1 : 3x-y+5z=10$ and $\Pi_2 : 15x-5y+kz=11$ are parallel.

b Now let $k=2$. Describe the configuration of the planes Π_1, Π_2 and $\Pi_3 : 9x-3y-2z=8$.

8 Consider this system of equations:

$$\begin{cases} 2x+y-2z=0 \\ x-2y-z=2 \\ 3x+4y-3z=d \end{cases}$$

a Show that the system does not have a unique solution.

b Find the value of d for which the system is consistent.

c The three equations represent planes. For the value of d found in part **b**, describe the configuration of the three planes.

9 **a** Show that the system of equations $\begin{cases} x+y=0 \\ x-4y-2z=0 \\ \dfrac{1}{2}x+3y+z=0 \end{cases}$ is consistent.

The three equations in part **a** represent three planes.

b Describe the geometrical configuration of the planes.

10 **a** Find the inverse of the matrix $\begin{pmatrix} 1 & -1 & 0 \\ 0 & 1 & 1 \\ 1 & 0 & -1 \end{pmatrix}$.

b Hence find, in terms of d, the coordinates of the point of intersection of the planes $x-y=4$, $y+z=1$ and $x-z=d$.

11 Consider the system of equations $\begin{cases} x-2y-z=-2 \\ 2x+y-3z=9 \\ x+3y-az=3 \end{cases}$

a Find the value of a for which the system does not have a unique solution.

b For the value of a found in part **a**, determine whether the system is consistent, and describe the geometric configuration of the three planes represented by the equations.

12 **a** Find the value of p for which the system of equations $\begin{cases} x-y-z=-2 \\ 2x+3y-7z=a+4 \\ x+2y+pz=a^2 \end{cases}$ does not have a unique solution.

b For the value of p found in part **a**, find the two values of a for which the system is consistent.

c For the value of p from part **a** and the values of a from part **b**, describe the geometric configuration of the three planes represented by the three equations.

13 Find the values of a and b for which the intersection of the planes $\Pi_1 : x+y-2z=7$, $\Pi_2 : x-2y-z=8$ and $\Pi_3 : 2x-y+az=b$ is a line.

✎ Checklist of learning and understanding

- Three different planes could:
  - intersect at a single point
  - intersect along a line (form a sheaf)
  - not intersect because the line of intersection of two of the planes is parallel to the third plane (form a triangular prism)
  - not intersect because two of the planes are parallel or all three planes are parallel.
- Two planes are parallel if their normals are parallel.
- You can represent a system of three linear simultaneous equations in three unknowns in matrix form as

$$\mathbf{Mr} = \mathbf{a} \text{ where } \mathbf{r} = \begin{pmatrix} x \\ y \\ z \end{pmatrix}.$$

  Each row of matrix $\mathbf{M}$ contains the coefficients of x, y and z in the planes described.
- If $\det \mathbf{M} \neq 0$ there is a unique solution representing the point of intersection of the three planes.

$$\mathbf{r} = \mathbf{M}^{-1}\mathbf{a}$$

- If $\det \mathbf{M} = 0$ then the three planes do not intersect at a single point and could be
  - inconsistent (no common intersection: parallel planes or a triangular prism)
  - consistent (a line or a plane as the common intersection).
- When $\det \mathbf{M} = 0$, you can determine the geometrical interpretation using elimination.
  - If you get $0 = 0$ the three planes are identical or intersect along a line.
  - If you get $0 = k \neq 0$ then either at least two of the planes are parallel and distinct, or they form a triangular prism.

Mixed practice 5

1 The planes Π_1, Π_2 and Π_3 are given by the equations:

$\Pi_1 : 3x - y + z = 17$

$\Pi_2 : x + 2y - z = 8$

$\Pi_3 : 3x + y + 2z = 19$

Find the point of intersection of all three planes.

2 By using the determinant of an appropriate matrix, find the values of k for which the simultaneous equations

$\begin{cases} kx + 8y = 1, \\ 2x + ky = 3, \end{cases}$

do not have a unique solution.

© OCR A Level Mathematics, Unit 4725 Further Pure Mathematics 1, June 2011

3 The planes Π_1, Π_2 and Π_3 are given by the equations:

$\Pi_1 : 3x - y + z = 17$

$\Pi_2 : x + 2y - z = 8$

$\Pi_3 : 2x - 3y + 2z = k$

a Show that, when $k = 3$, the three planes form a triangular prism.

b Find the value of k for which the three planes intersect along a line.

4 Consider this system of equations, where a and b are real:

$\begin{cases} ax + 9y + 6z = 6 \\ \quad\;\; ay - z = b \\ \quad x + 6y + z = 4 \end{cases}$

a Given that the system has no unique solution, find all possible values of a.

b When the system has a unique solution, find that solution in terms of a when $b = -1$.

c Find k such that the system is always consistent when $a = kb$.

5 The matrix $\mathbf{M}$ is given by $\mathbf{M} = \begin{pmatrix} a & -a & 1 \\ 3 & a & 1 \\ 4 & 2 & 1 \end{pmatrix}$.

i Find, in terms of a, the determinant of $\mathbf{M}$.

ii Hence find the values of a for which $\mathbf{M}^{-1}$ does not exist.

iii Determine whether the simultaneous equations

$\begin{cases} 6x - 6y + z = 3k, \\ 3x + 6y + z = 0, \\ 4x + 2y + z = k, \end{cases}$

where k is a non-zero constant, have a unique solution, no solution or an infinite number of solutions, justifying your answer.

© OCR A Level Mathematics, Unit 4725 Further Pure Mathematics 1, January 2011

6 The matrix $\mathbf{D}$ is given by $\mathbf{D} = \begin{pmatrix} a & 2 & -1 \\ 2 & a & 1 \\ 1 & 1 & a \end{pmatrix}$.

i Find the determinant of $\mathbf{D}$ in terms of a.

ii Three simultaneous equations are shown below.

$$\begin{cases} ax + 2y - z = 0 \\ 2x + ay + z = a \\ x + y + az = a \end{cases}$$

For each of the following values of a, determine whether or not there is a unique solution. If the solution is not unique, determine whether the equations are consistent or inconsistent.

a $a = 3$

b $a = 2$

c $a = 0$

© OCR A Level Mathematics, Unit 4725 Further Pure Mathematics 1, June 2012

Extending the proof of de Moivre's theorem

De Moivre's theorem states that

$$\left(r(\cos\theta + i\sin\theta)\right)^n = r^n(\cos n\theta + i\sin n\theta)$$

In Chapter 2, Section 1, you saw how to prove this result for integer values of n. However, the result extends, with some careful consideration of conventions, to rational values.

Proving the result for rational numbers

You have to be a little careful when raising a number to a rational power. For example, if you write z as $r(\cos\theta + i\sin\theta)$, then you can also write it as $r(\cos(\theta + 2k\pi)) + i\sin(\theta + 2k\pi)$. If you could apply de Moivre's theorem with a rational power $\dfrac{a}{b}$, where a and b are integers with no common factors, then $z^{\frac{a}{b}}$ would equal

$$r^{\frac{a}{b}}\left(\cos\left(\frac{a}{b}\theta + \frac{2ka\pi}{b}\right) + i\sin\left(\frac{a}{b}\theta + \frac{2ka\pi}{b}\right)\right).$$

This has b different values, corresponding to $k = 0, 1, \ldots, b-1$. Multi-valued expressions are usually considered inconvenient, so you need to apply a convention that, when raising to a rational power, you choose θ to be the smallest positive value and $k = 0$. This is called the principal root.

> **Tip**
>
> Notice that when the power is an integer (for example $b = 1$) then there is no problem: there is only one possible answer.

Question

1. By considering $\left(r(\cos\theta + i\sin\theta)\right)^{\frac{a}{b}} = r^n(\cos m\theta + i\sin m\theta)$, prove that one possible value for m and n is $\dfrac{a}{b}$.

   You can assume de Moivre's theorem for integer powers and that the normal rules for indices hold.

Not proving the result for irrational numbers

It is tempting to think that if de Moivre's theorem can be proved for all rational numbers, then it must hold for irrational numbers. However, this turns out to be difficult to define. For example: consider 1^π.

You could write $1 = \cos 2k\pi + i\sin 2k\pi$. Then, if de Moivre's theorem did extend to irrational numbers, you would have

$$1^\pi = \cos 2k\pi^2 + i\sin 2k\pi^2.$$

Unlike in the rational case there is no period to this expression.

Each different value of k therefore produces a different value for the expression. There are therefore infinitely many values (all lying on the circle with modulus 1), which makes this expression very hard to work with.

Question

2 Use proof by contradiction to prove that if

$$\cos 2m\pi^2 + i\sin 2m\pi^2 = \cos 2p\pi^2 + i\sin 2p\pi^2,$$

then $m = p$.

You can assume that the period of the $\cos\theta + i\sin\theta$ function is 2π and that π is an irrational number.

Using complex numbers to describe rotations

You know two different ways to describe rotations in a plane. The first method is using matrices. You can find the image of a point with position vector $\begin{pmatrix} x \\ y \end{pmatrix}$ after a rotation through an angle θ about the origin by multiplying it by the rotation matrix:

$$\begin{pmatrix} \cos\theta & -\sin\theta \\ \sin\theta & \cos\theta \end{pmatrix}\begin{pmatrix} x \\ y \end{pmatrix} = \begin{pmatrix} x\cos\theta - y\sin\theta \\ x\sin\theta + y\cos\theta \end{pmatrix}$$

> **⏮ Rewind**
>
> You met the rotation matrix in Pure Core Student Book 1, Chapter 3.

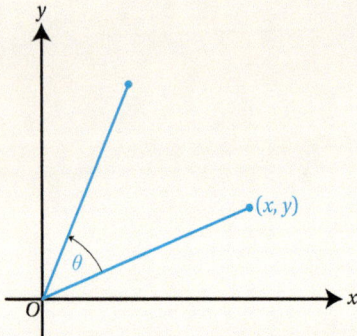

The second method is using complex numbers. When a point corresponding to a complex number $z = x + iy$ is rotated through an angle θ about the origin, you can find the complex number corresponding to the image point by multiplying z by $e^{i\theta}$:

$$(x + iy)(\cos\theta + i\sin\theta) = (x\cos\theta - y\sin\theta) + i(x\sin\theta + y\cos\theta)$$

> **⏮ Rewind**
>
> You met the idea of multiplication by $e^{i\theta}$ representing a rotation in Chapter 2, Section 6.

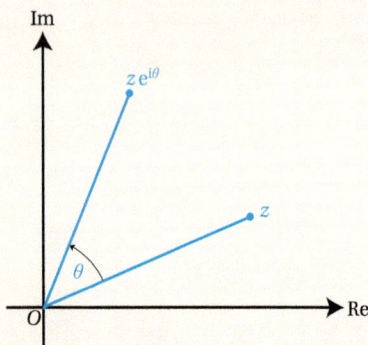

You can see that the two methods give the same coordinates for the image point. The advantage of the complex numbers method is that it results in a single equation, whereas the matrix method results in two equations (one for each component).

Here you will compare the two methods when solving the following problem

Three snails start at the vertices of an equilateral triangle. Each snail moves with the same constant speed towards an adjacent snail: S_1 towards S_2, S_2 towards S_3, and S_3 towards S_1.
Describe the path followed by each snail. After how long (if at all) do the snails meet?

First you need to specify the problem a little more precisely. Set up the coordinate axes so that the origin is at the centre of the equilateral triangle and let the initial position of S_1 be at $(1, 0)$. Let v be the speed of each snail.

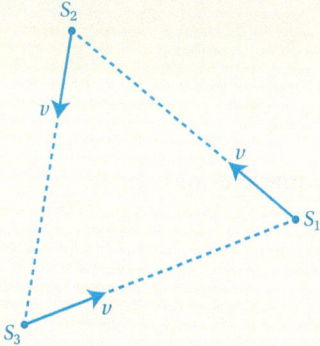

Because of the symmetry of the situation, the three snails will always form an equilateral triangle. You only need to find the path followed by S_1. You can find the position of S_2 by rotating S_1 120° anticlockwise about the origin, and you can find the position of S_3 by rotating S_2 by the same amount.

You will first approach the problem using position vectors and matrices.

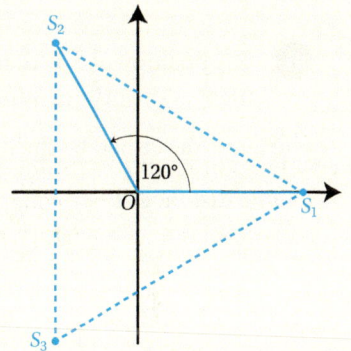

Questions

Let $\mathbf{r} = \begin{pmatrix} x \\ y \end{pmatrix}$ be the position vector of S_1 at time t.

1 Use a rotation matrix to write down the position vector of S_2 in terms of x and y.

2 Explain why $\dfrac{d\mathbf{r}}{dt} = k \begin{pmatrix} -\dfrac{3}{2}x - \dfrac{\sqrt{3}}{2}y \\ \dfrac{\sqrt{3}}{2}x - \dfrac{3}{2}y \end{pmatrix}$ for some constant k. Show that $k = \dfrac{v}{\sqrt{3}\sqrt{x^2 + y^2}}$.

3 Hence show that x and y satisfy the system of differential equations

$$\begin{cases} 2\sqrt{x^2 + y^2}\,\dfrac{dx}{dt} = -v\sqrt{3}x - vy \\ 2\sqrt{x^2 + y^2}\,\dfrac{dy}{dt} = vx - v\sqrt{3}y \end{cases}$$

Although you will learn in Chapter 11 how to solve some systems of differential equations, these equations are non-linear so the methods from that chapter won't work here. You will return to this system of equations later, but for now you will consider a different approach.

Now, let z be the complex number corresponding to the position of S_1 at time t. Then the position of S_2 is given by $ze^{i\frac{2\pi}{3}}$.

Questions

4 Show that $\dfrac{dz}{dt} = \dfrac{v}{\sqrt{3}}\left(e^{i\frac{2\pi}{3}} - 1\right)\dfrac{z}{|z|}$.

5 Write $z = re^{i\theta}$. (Remember that both r and θ vary with time.) Show that

$$\frac{dr}{dt} + ir\frac{d\theta}{dt} = \frac{v}{\sqrt{3}}\left(e^{i\frac{2\pi}{3}} - 1\right).$$

6 By equating real and imaginary parts, obtain this system of differential equations for r and θ:

$$\frac{dr}{dt} = -\frac{\sqrt{3}v}{2}, \frac{d\theta}{dt} = \frac{v}{2r}.$$

7 Given that initially $r = 1$ and $\theta = 0$, show that

$$r = 1 - \frac{\sqrt{3}v}{2}t, \quad \theta = -\frac{1}{\sqrt{3}}\ln\left(1 - \frac{\sqrt{3}v}{2}t\right).$$

8 The position of S_1 is then given by $z = re^{i\theta}$, S_2 by $re^{i\left(\theta+\frac{2\pi}{3}\right)}$ and S_3 by $re^{i\left(\theta+\frac{4\pi}{3}\right)}$. At what time do the snails meet at the origin? What happens to the value of θ as t approaches this value?

The curve described by the equations in question **7** is called a logarithmic spiral. Although each snail travels a finite distance $\left(v\dfrac{2}{\sqrt{3}v} = \dfrac{2}{\sqrt{3}}\right)$, it performs an infinite number of rotations. This diagram shows the paths of all three snails, and their positions when $t = 0$ and $t \approx \dfrac{0.7}{v}$.

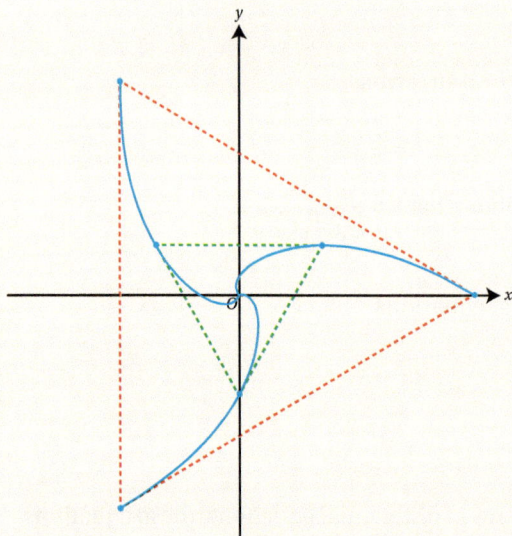

In this problem, using complex numbers resulted in equations you could solve, while this was not the case when using position vectors and matrices.

You should read the rest of this section after you have studied Chapter 9 on polar coordinates.

In question **6** you derived two separate differential equations for r and θ, the modulus and the argument of the complex number representing the position of S_1. This suggests that you might also be able to solve this problem using polar coordinates, which are basically the same as the modulus and argument of a complex number.

Look again at the system of equations from question **3**:

$$\begin{cases} 2\sqrt{x^2 + y^2}\, \dfrac{dx}{dt} = -v\sqrt{3}x - vy \\[2mm] 2\sqrt{x^2 + y^2}\, \dfrac{dy}{dt} = vx - v\sqrt{3}y \end{cases}$$

Let r and θ be the polar coordinates of the point (x, y). Then the 'problem' term in the equations, $\sqrt{x^2 + y^2}$, is simply r. You can rewrite the equations in terms of r and θ.

Questions

9 Show that $\dfrac{dx}{dt} = \dfrac{dr}{dt}\cos\theta - r\dfrac{d\theta}{dt}\sin\theta$ and obtain a similar expression for $\dfrac{dy}{dt}$.

10 Rewrite the system of equations in terms of r and θ. Combine the two equations to show that

$$\frac{dr}{dt} = -\frac{v\sqrt{3}}{2}, \quad \frac{d\theta}{dt} = \frac{v}{2r}$$

11 Obtain an expression for $\dfrac{dr}{d\theta}$ and hence show that $r = e^{-\sqrt{3}\,\theta}$.

The final equation in question **11** is the polar equation of a logarithmic spiral, the path followed by each snail.

Rewind

Remember that $x = r\cos\theta$ and $y = r\sin\theta$.

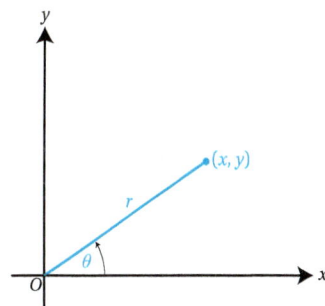

Tip

Notice that these are the same equations you derived in question **6**.

Leslie matrices

Matrices are applied in many different real-life situations. Leslie matrices are a particular application to a biological population structured into different groups such as adults and juveniles.

Imagine a group of rabbits. Each adult on average produces three juvenile rabbits each year. Each year 10% of adult rabbits die and 20% of juvenile rabbits die. Those juveniles who do not die become adults.

The number of adult rabbits in year n is denoted by a_n and the number of juvenile rabbits is denoted by j_n.

Questions

1 Explain why $a_{n+1} = 0.9a_n + 0.8j_n$.

2 Find an expression for j_{n+1} in terms of a_n.

3 The equations found in questions **1** and **2** can be written in a matrix form as

$$\begin{pmatrix} a_{n+1} \\ j_{n+1} \end{pmatrix} = \mathbf{M} \begin{pmatrix} a_n \\ j_n \end{pmatrix}$$

Write down the matrix **M**.

4 An uninhabited island is populated with 200 adult rabbits in year 0. Use technology to find the number of adult rabbits in:

a year 1   **b** year 10.

5 By investigating the sequence formed, find the long-term growth rate of the population.

6 Find the long-term ratio of juveniles to adults.

7 The population of rabbits is infected with a disease which decreases the average number of juvenile rabbits produced per adult rabbit each year to α. Find the smallest value of α so that the population will not become extinct.

8 Describe the assumptions made in creating this model.

9 In an alternative model each adult rabbit produces exactly one juvenile each year and there is no death. If the population starts with one (presumably pregnant) adult, investigate the number of adult rabbits after n years. Can you form a conjecture and prove it by induction?

Explore

The situation in question **9** was described by Fibonacci, leading to his eponymous sequence.

⏮ Rewind

For a reminder of proof by induction see Chapter 1.

CROSS-TOPIC REVIEW EXERCISE 1

1 **a** Show that $\cos z = \dfrac{e^{iz} + e^{-iz}}{2}$.

 b Hence find the value of $\cos 2i$ correct to 3 significant figures.

2 **a** Express $-8i$ in the form $re^{i\theta}$, where $r > 0$ and $-\pi < \theta \leqslant \pi$.

 b Solve the equation $z^6 + 8i = 0$, giving your answers in the form $re^{i\theta}$, where $r > 0$ and $-\pi < \theta \leqslant \pi$.

3 **a** Find the Cartesian equation of the plane Π_1 containing the points $(1, 1, 0)$, $(0, -2, 0)$ and $(0, 1, 2)$.

 The plane Π_2 has Cartesian equation $2x + 3y - 4z = 5$.

 b Find, to 3 significant figures, the acute angle between the planes Π_1 and Π_2.

4 Given that $\displaystyle\sum_{r=1}^{n} (ar^3 + br) = n(n-1)(n+1)(n+2)$, find the values of the constants a and b.

© OCR AS Level Mathematics, Unit 4725 Further Pure Mathematics 1, January 2011

5 Find $\displaystyle\sum_{r=1}^{n} (4r^3 - 3r^2 + r)$, giving your answer in a fully factorised form.

© OCR AS Level Mathematics, Unit 4725/01 Further Pure Mathematics 1, June 2013

6 **i** Show that $(z^n - e^{i\theta})(z^n - e^{-i\theta}) \equiv z^{2n} - (2\cos\theta)z^n + 1$.

 ii Express $z^4 - z^2 + 1$ as the product of four factors of the form $(z - e^{i\alpha})$, where $0 \leqslant \alpha < 2\pi$.

© OCR A Level Mathematics, Unit 4727 Further Pure Mathematics 3, January 2012

7 **i** Solve the equation $z^4 = 2(1 + i\sqrt{3})$, giving the roots exactly in the form $r(\cos\theta + i\sin\theta)$, where $r > 0$ and $0 \leqslant \theta < 2\pi$.

 ii Sketch an Argand diagram to show the lines from the origin to the point representing $2(1 + i\sqrt{3})$ and from the origin to the points which represent the roots of the equation in part **i**.

© OCR A Level Mathematics, Unit 4727 Further Pure Mathematics 3, June 2012

8 The line l_1 passes through the points $(0, 0, 10)$ and $(7, 0, 0)$ and the line l_2 passes through the points $(4, 6, 0)$ and $(3, 3, 1)$. Find the shortest distance between l_1 and l_2.

© OCR A Level Mathematics, Unit 4727 Further Pure Mathematics 3, June 2010

9 By using the determinant of an appropriate matrix, find the values of λ for which the simultaneous equations

$$3x + 2y + 4z = 5,$$
$$\lambda y + z = 1,$$
$$x + \lambda y + \lambda z = 4,$$

do not have a unique solution for x, y and z.

© OCR AS Level Mathematics, Unit 4725/01 Further Pure Mathematics 1, January 2013

10 **a** Given that a and $(a + i)^5$ are both real, with $a > 0$, show that $a^2 = \dfrac{5 \pm 2\sqrt{5}}{5}$.

 b Find, in terms of a, the argument of $(a + i)^5$.

 c Hence show that $\cot\dfrac{\pi}{5} = \sqrt{\dfrac{5 + 2\sqrt{5}}{5}}$.

11 i Show that $\dfrac{1}{\sqrt{r+2}+\sqrt{r}} \equiv \dfrac{\sqrt{r+2}-\sqrt{r}}{2}$.

ii Hence find an expression, in terms of n, for

$$\sum_{r=1}^{n} \dfrac{1}{\sqrt{r+2}+\sqrt{r}}.$$

iii State, giving a brief reason, whether the series $\sum_{r=1}^{n} \dfrac{1}{\sqrt{r+2}+\sqrt{r}}$ converges.

© OCR AS Level Mathematics, Unit 4725 Further Pure Mathematics 1, June 2010

12 i Show that $\dfrac{1}{r}-\dfrac{1}{r+2} \equiv \dfrac{2}{r(r+2)}$.

ii Hence find an expression, in terms of n, for $\sum_{r=1}^{n} \dfrac{2}{r(r+2)}$.

iii Given that $\sum_{r=N+1}^{\infty} \dfrac{2}{r(r+2)} = \dfrac{11}{30}$, find the value of N.

© OCR AS Level Mathematics, Unit 4725 Further Pure Mathematics 1, June 2012

13 i Show that $\dfrac{1}{r^2}-\dfrac{1}{(r+2)^2} \equiv \dfrac{4(r+1)}{r^2(r+2)^2}$.

ii Hence find an expression, in terms of n, for $\sum_{r=1}^{n} \dfrac{4(r+1)}{r^2(r+2)^2}$.

iii Find $\sum_{r=5}^{\infty} \dfrac{4(r+1)}{r^2(r+2)^2}$, giving your answer in the form $\dfrac{p}{q}$ where p and q are integers.

© OCR AS Level Mathematics, Unit 4725/01 Further Pure Mathematics 1, June 2014

14 The sequence $u_1, u_2, u_3, \ldots$ is defined by $u_1 = 2$ and $u_{n+1} = \dfrac{u_n}{1+u_n}$ for $n \geqslant 1$.

i Find u_2 and u_3, and show that $u_4 = \dfrac{2}{7}$.

ii Hence suggest an expression for u_n.

iii Use induction to prove that your answer to part **ii** is correct.

© OCR AS Level Mathematics, Unit 4725/01 Further Pure Mathematics 1, January 2013

15 The roots of the equation $z^3 - 1 = 0$ are denoted by $1, \omega$ and ω^2.

i Sketch an Argand diagram to show these roots.

ii Show that $1 + \omega + \omega^2 = 0$.

iii Hence evaluate

   a $(2+\omega)(2+\omega^2)$,

   b $\dfrac{1}{2+\omega}+\dfrac{1}{2+\omega^2}$.

iv Hence find a cubic equation, with integer coefficients, which has roots $2, \dfrac{1}{2+\omega}$ and $\dfrac{1}{2+\omega^2}$.

© OCR A Level Mathematics, Unit 4727/01 Further Pure Mathematics 3, June 2008

16 i Write down, in Cartesian form, the roots of the equation $z^4 = 16$.

ii Hence solve the equation $w^4 = 16(1-w)^4$, giving your answers in Cartesian form.

© OCR A Level Mathematics, Unit 4727 Further Pure Mathematics 3, January 2010

17 The cube roots of 1 are denoted by 1, ω and ω^2, where the imaginary part of ω is positive.

i Show that $1+\omega+\omega^2=0$.

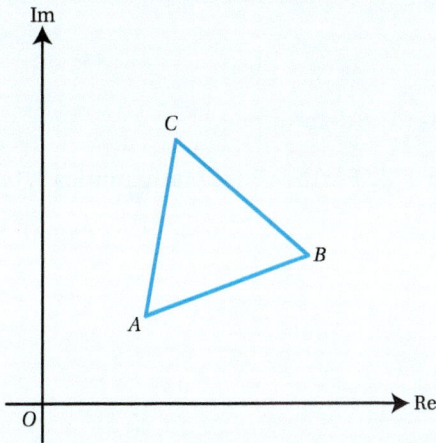

In the diagram, ABC is an equilateral triangle, labelled anticlockwise. The points A, B and C represent the complex numbers z_1, z_2 and z_3 respectively.

ii State the geometrical effect of multiplication by ω and hence explain why $z_1-z_3=\omega(z_3-z_2)$.

iii Hence show that $z_1+\omega z_2+\omega^2 z_3=0$.

© OCR A Level Mathematics, Unit 4727 Further Pure Mathematics 3, January 2011

18 **i** Solve the equation $z^5=1$, giving your answers in polar form.

ii Hence, by considering the equation $(z+1)^5=z^5$, show that the roots of

$$5z^4+10z^3+10z^2+5z+1=0$$

can be expressed in the form $\dfrac{1}{e^{i\theta}-1}$, stating the values of θ.

© OCR A Level Mathematics, Unit 4727/01 Further Pure Mathematics 3, January 2013

19 In an Argand diagram, the complex numbers 0, z and $ze^{\frac{1}{6}i\pi}$ are represented by the points O, A and B respectively.

i Sketch a possible Argand diagram showing the triangle OAB. Show that the triangle is isosceles and state the size of angle AOB.

The complex numbers $1 + i$ and $5 + 2i$ are represented by the points C and D respectively. The complex number w is represented by the point E, such that $CD = CE$ and angle $DCE = \dfrac{1}{6}\pi$.

ii Calculate the possible values of w, giving your answers exactly in the form $a + bi$.

© OCR A Level Mathematics, Unit 4727/01 Further Pure Mathematics 3, June 2015

20 i By expressing $\sin\theta$ and $\cos\theta$ in terms of $e^{i\theta}$ and $e^{-i\theta}$, prove that

$$\sin^3\theta\cos^2\theta \equiv -\frac{1}{16}(\sin 5\theta - \sin 3\theta - 2\sin\theta).$$

ii Hence show that all the roots of the equation

$$\sin 5\theta = \sin 3\theta + 2\sin\theta$$

are of the form $\theta = \dfrac{n\pi}{k}$, where n is any integer and k is to be determined.

© **OCR A Level Mathematics, Unit 4727 Further Pure Mathematics 3, June 2012**

21 i By expressing $\sin\theta$ in terms of $e^{i\theta}$ and $e^{-i\theta}$, show that

$$\sin^5\theta \equiv \frac{1}{16}(\sin 5\theta - 5\sin 3\theta + 10\sin\theta).$$

ii Hence solve the equation

$$\sin 5\theta + 4\sin\theta = 5\sin 3\theta$$

for $-\dfrac{1}{2}\pi \leqslant \theta \leqslant \dfrac{1}{2}\pi$.

© **OCR A Level Mathematics, Unit 4727/01 Further Pure Mathematics 3, June 2014**

22 The line l has equations $\dfrac{x-1}{2} = \dfrac{y-1}{3} = \dfrac{z+1}{2}$ and the point A is $(7, 3, 7)$. M is the point where the perpendicular from A meets l.

i Find, in either order, the coordinates of M and the perpendicular distance from A to l.

ii Find the coordinates of the point B on AM such that $\overrightarrow{AB} = 3\overrightarrow{BM}$.

© **OCR A Level Mathematics, Unit 4727 Further Pure Mathematics 3, January 2012**

23 The plane Π has equation $\mathbf{r} = \begin{pmatrix} 1 \\ 6 \\ 7 \end{pmatrix} + \lambda\begin{pmatrix} 2 \\ -1 \\ -1 \end{pmatrix} + \mu\begin{pmatrix} 2 \\ -3 \\ -5 \end{pmatrix}$ and the line l has equation $\mathbf{r} = \begin{pmatrix} 7 \\ 4 \\ 1 \end{pmatrix} + t\begin{pmatrix} 3 \\ 0 \\ -1 \end{pmatrix}$.

i Express the equation of Π in the form $\mathbf{r} \cdot \mathbf{n} = p$.

ii Find the point of intersection of l and Π.

iii The equation of Π may be expressed in the form $\mathbf{r} = \begin{pmatrix} 1 \\ 6 \\ 7 \end{pmatrix} + \lambda\begin{pmatrix} 2 \\ -1 \\ -1 \end{pmatrix} + \mu\mathbf{c}$, where $\mathbf{c}$ is perpendicular

to $\begin{pmatrix} 2 \\ -1 \\ -1 \end{pmatrix}$. Find $\mathbf{c}$.

© **OCR A Level Mathematics, Unit 4727 Further Pure Mathematics 3, January 2012**

24 The plane p has equation $\mathbf{r} \cdot (\mathbf{i} - 3\mathbf{j} + 4\mathbf{k}) = 4$ and the line l_1 has equation $\mathbf{r} = 2\mathbf{j} - \mathbf{k} + t(3\mathbf{i} + \mathbf{j} + 2\mathbf{k})$. The line l_2 is parallel to p and perpendicular to l_1, and passes through the point with position vector $\mathbf{i} + 4\mathbf{j} + 2\mathbf{k}$. Find the equation of l_2, giving your answer in the form $\mathbf{r} = \mathbf{a} + t\mathbf{b}$.

© **OCR A Level Mathematics, Unit 4727 Further Pure Mathematics 3, June 2012**

25 The lines l_1 and l_2 have equations

$$\mathbf{r} = \begin{pmatrix} 1 \\ 2 \\ 1 \end{pmatrix} + \lambda \begin{pmatrix} 2 \\ 3 \\ -1 \end{pmatrix} \text{ and } \mathbf{r} = \begin{pmatrix} 3 \\ 0 \\ 1 \end{pmatrix} + \mu \begin{pmatrix} 4 \\ -1 \\ -1 \end{pmatrix}$$

respectively.

i Find the shortest distance between the lines.

ii Find a Cartesian equation of the plane which contains l_1 and which is parallel to l_2.

© OCR A Level Mathematics, Unit 4727/01 Further Pure Mathematics 3, January 2013

26 **a** Integrate e^{kx} with respect to x.

b Show that, for $x \in \mathbb{R}$, the imaginary part of $e^{(1+3i)x}$ is $e^x \sin 3x$.

c Hence find the exact value of $\int_0^{\frac{\pi}{2}} e^x \sin 3x \, dx$.

© OCR A Level Mathematics, Unit 4727 Further Pure Mathematics 3, June 2012

27 **a** Show that $\cos z = \dfrac{e^{iz} + e^{-iz}}{2}$.

b Hence find possible complex numbers z for which $\cos z = 2$.

© OCR A Level Mathematics, Unit 4727 Further Pure Mathematics 3, June 2012

28 **a** If $z = \dfrac{1}{2}\cos\theta + \dfrac{i}{2}\sin\theta$, show that $|z| < 1$.

b Find an expression for $1 + \dfrac{1}{2}e^{i\theta} + \dfrac{1}{4}e^{2i\theta} + \dfrac{1}{8}e^{3i\theta} + \dots$.

c Hence show that $\dfrac{1}{2}\sin\theta + \dfrac{1}{4}\sin 2\theta + \dfrac{1}{8}\sin 3\theta + \dots = \dfrac{2\sin\theta}{5 - 4\cos\theta}$.

© OCR A Level Mathematics, Unit 4727 Further Pure Mathematics 3, June 2012

29 **a** Find the smallest positive integer N for which $N! > 2^N$.

b Prove that $n! > 2^n$ for all $n \geqslant N$.

30 **a** Prove using induction that

$$\sin\theta + \sin 3\theta + \dots + \sin(2n-1)\theta = \dfrac{\sin^2 n\theta}{\sin\theta} \text{ for integer } n \geqslant 1.$$

b Hence find the exact value of $\sin\dfrac{\pi}{7} + \sin\dfrac{3\pi}{7} + \dots + \sin\dfrac{13\pi}{7}$.

31 Prove, using induction, that for positive integer n,

$$\cos x \times \cos 2x \times \cos 4x \times \cos 8x \ \dots \ \times \cos(2^n x) = \dfrac{\sin(2^{n+1} x)}{2^{n+1} \sin x}$$

32 **i** Show that $\dfrac{1}{r}-\dfrac{2}{r+1}+\dfrac{1}{r+2}=\dfrac{2}{r(r+1)(r+2)}$.

ii Hence find an expression, in terms of n, for

$$\sum_{r=1}^{n}\frac{2}{r(r+1)(r+2)}.$$

iii Show that $\displaystyle\sum_{r=n+1}^{\infty}\frac{2}{r(r+1)(r+2)}=\frac{1}{(n+1)(n+2)}$.

© OCR AS Level Mathematics, Unit 4725 Further Pure Mathematics 1, January 2011

33 The integrals C and S are defined by

$$C=\int_0^{\frac{1}{2}\pi}e^{2x}\cos 3x\,\mathrm{d}x\quad\text{and}\quad S=\int_0^{\frac{1}{2}\pi}e^{2x}\sin 3x\,\mathrm{d}x.$$

By considering $C+\mathrm{i}S$ as a single integral, show that

$$C=-\frac{1}{13}\left(2+3e^{\pi}\right),$$

and obtain a similar expression for S.

(You may assume that the standard result for $\int e^{kx}\,\mathrm{d}x$ remains true when k is a complex constant,

so that $\int e^{(a+\mathrm{i}b)x}\mathrm{d}x=\dfrac{1}{a+\mathrm{i}b}e^{(a+\mathrm{i}b)x}$.)

© OCR A Level Mathematics, Unit 4727/01 Further Pure Mathematics 3, January 2008

34 **i** Use de Moivre's theorem to prove that

$$\tan 3\theta\equiv\frac{\tan\theta\left(3-\tan^2\theta\right)}{1-3\tan^2\theta}.$$

ii a By putting $\theta=\dfrac{1}{12}\pi$ in the identity in part **i**, show that $\tan\dfrac{1}{12}\pi$ is a solution of the equation

$t^3-3t^2-3t+1=0$.

b Hence show that $\tan\dfrac{1}{12}\pi=2-\sqrt{3}$.

iii Use the substitution $t=\tan\theta$ show that

$$\int_0^{2-\sqrt{3}}\frac{t(3-t^2)}{(1-3t^2)(1+t^2)}\,\mathrm{d}t=a\ln b,$$

where a and b are positive constants to be determined.

© OCR A Level Mathematics, Unit 4727 Further Pure Mathematics 3, June 2009

35 **i** Solve the equation $\cos 6\theta=0$, for $0<\theta<\pi$.

ii By using de Moivre's theorem, show that

$$\cos 6\theta\equiv\left(2\cos^2\theta-1\right)\left(16\cos^4\theta-16\cos^2\theta+1\right).$$

iii Hence find the exact value of

$$\cos\left(\frac{1}{12}\pi\right)\cos\left(\frac{5}{12}\pi\right)\cos\left(\frac{7}{12}\pi\right)\cos\left(\frac{11}{12}\pi\right),$$

justifying your answer.

© OCR A Level Mathematics, Unit 4727 Further Pure Mathematics 3, January 2010

36 **i** Use de Moivre's theorem to express $\cos 4\theta$ as a polynomial in $\cos\theta$.

    ii Hence prove that $\cos 4\theta \cos 2\theta \equiv 16\cos^6\theta - 24\cos^4\theta + 10\cos^2\theta - 1$.

    iii Use part **ii** to show that the only roots of the equation $\cos 4\theta \cos 2\theta = 1$ are $\theta = n\pi$, where n is an integer.

    iv Show that $\cos 4\theta \cos 2\theta = -1$ only when $\cos\theta = 0$.

© OCR A Level Mathematics, Unit 4727 Further Pure Mathematics 3, June 2011

37 **i** Use de Moivre's theorem to prove that
$$\tan 5\theta = \frac{5\tan\theta - 10\tan^3\theta + \tan^5\theta}{1 - 10\tan^2\theta + 5\tan^4\theta}.$$

    ii Solve the equation $\tan 5\theta = 1$, for $0 \leqslant \theta < \pi$.

    iii Show that the roots of the equation
$$t^4 - 4t^3 - 14t^2 - 4t + 1 = 0$$

    may be expressed in the form $\tan\alpha$, stating the exact values of α, where $0 \leqslant \alpha < \pi$.

© OCR A Level Mathematics, Unit 4727 Further Pure Mathematics 3, January 2012

38 Let $S = e^{i\theta} + e^{2i\theta} + e^{3i\theta} + \ldots + e^{10i\theta}$.

    i   **a**   Show that, for $\theta \neq 2n\pi$, where n is an integer,

$$S = \frac{e^{\frac{1}{2}i\theta}\left(e^{10i\theta} - 1\right)}{2i\sin\left(\frac{1}{2}\theta\right)}.$$

       b   State the value of S for $\theta = 2n\pi$, where n is an integer.

    ii Hence show that, for $\theta \neq 2n\pi$, where n is an integer,

$$\cos\theta + \cos 2\theta + \cos 3\theta + \ldots + \cos 10\theta = \frac{\sin\left(\frac{21}{2}\theta\right)}{2\sin\left(\frac{1}{2}\theta\right)} - \frac{1}{2}.$$

    iii Hence show that $\theta = \dfrac{1}{11}\pi$ is a root of $\cos\theta + \cos 2\theta + \cos 3\theta + \ldots + \cos 10\theta = 0$ and find

    another root in the interval $0 < \theta < \dfrac{1}{4}\pi$.

© OCR A Level Mathematics, Unit 4727/01 Further Pure Mathematics 3, January 2013

6 Hyperbolic functions

In this chapter you will learn how to:

- define the hyperbolic functions $\sinh x$, $\cosh x$ and $\tanh x$
- draw the graphs of hyperbolic functions, showing their domains and ranges
- write the inverse hyperbolic functions in terms of logarithms
- define the reciprocal hyperbolic functions $\operatorname{sech} x$, $\operatorname{cosec} x$ and $\coth x$
- solve equations and prove identities involving hyperbolic functions
- differentiate hyperbolic functions.

Before you start…

A Level Mathematics Student Book 2, Chapter 2	You should understand the terms domain and range of a function.	1	For the function $f(x) = \sqrt{x-3}$, state: a the largest possible domain b the corresponding range.
A Level Mathematics Student Book 2, Chapter 3	You should be able to draw a graph after two (or more) transformations.	2	The graph of $y = f(x)$ is shown. Sketch the graph of $y = 3 - f(2x)$, giving the new coordinates of the three points labelled on the original graph.
A Level Mathematics Student Book 2, Chapter 9	You should know how to differentiate and integrate the exponential function.	3	Find: a $\dfrac{d}{dx}\left(e^{2x}\right)$ b $\displaystyle\int e^{-x}\,dx$.
A Level Mathematics Student Book 2, Chapter 9	You should know how to differentiate and integrate trigonometric functions.	4	Find: a $\dfrac{d}{dx}(\tan x)$ b $\displaystyle\int \cos x\,dx$.
A Level Mathematics Student Book 2, Chapter 10	You should know how to use the chain rule, product rule and quotient rule for differentiation.	5	Find $f'(x)$ for these functions: a $f(x) = \sin^2 3x$ b $f(x) = x\cos x$ c $f(x) = \dfrac{\tan 2x}{x}$

What are hyperbolic functions?

Trigonometric functions are sometimes called **circular functions**. This is because of the definition that states: a point on the unit circle (with equation $x^2 + y^2 = 1$) defining a radius at an angle θ to the positive x-axis, has coordinates $(\cos\theta, \sin\theta)$.

Related to the circle is a curve with equation $x^2 - y^2 = 1$, called a **hyperbola**. Points on this hyperbola have coordinates $(\cosh\theta, \sinh\theta)$, although θ can no longer be interpreted as an angle.

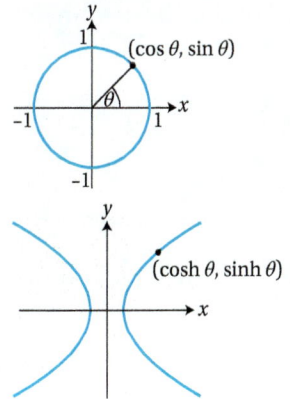

Section 1: Defining hyperbolic functions

Although the geometric definition of **hyperbolic functions** gives some helpful insight, a more useful definition is related to the number e.

> ### Key point 6.1
>
> - $\sinh x = \dfrac{e^x - e^{-x}}{2}$
>
> - $\cosh x = \dfrac{e^x + e^{-x}}{2}$

Tip

Cosh is pronounced as it reads, sinh is pronounced either 'sinch' or 'shine' and tanh is pronounced 'tanch' or 'than'.

You can define $\tanh x$ by analogy with the trigonometric definition of $\tan x$.

> ### Key point 6.2
>
> $$\tanh x \equiv \frac{\sinh x}{\cosh x} \equiv \frac{e^x - e^{-x}}{e^x + e^{-x}}$$

Did you know?

The tanh function is frequently used in physics, particularly in the context of special relativity and the study of entropy.

There are not many special values of these functions that you need to know, but, from Key points 6.1 and 6.2, you should be able to see that $\cosh 0 = 1$, $\sinh 0 = 0$ and $\tanh 0 = 0$.

As for trigonometric functions, you need to know the graphs of hyperbolic functions.

> ### Key point 6.3
>
> The graphs of $\sinh x$, $\cosh x$ and $\tanh x$ look like this.
>
> $y = \sinh x$ $y = \cosh x$ $y = \tanh x$
>
>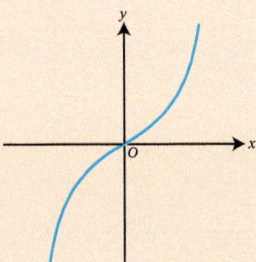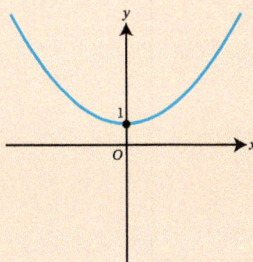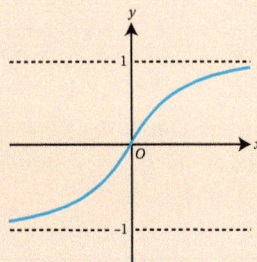

Did you know?

You may think that the graph of $\cosh x$ looks like a parabola, but it is slightly flatter. It is called a **catenary**, which is the shape formed by a hanging chain.

You can establish the domains and ranges of the hyperbolic functions from their graphs.

Key point 6.4

The domains and ranges of the hyperbolic functions $\sinh x$, $\cosh x$ and $\tanh x$ are:

Function	Domain	Range
$\sinh x$	$x \in \mathbb{R}$	$f(x) \in \mathbb{R}$
$\cosh x$	$x \in \mathbb{R}$	$f(x) \geqslant 1$
$\tanh x$	$x \in \mathbb{R}$	$-1 < f(x) < 1$

Rewind

You met the domain and range of a function in A Level Mathematics Student Book 2, Chapter 2.

WORKED EXAMPLE 6.1

Given that $f(x) = 3\tanh x + 2$ for $x \in \mathbb{R}$,

a sketch the graph of $y = f(x)$
b state the range of f.

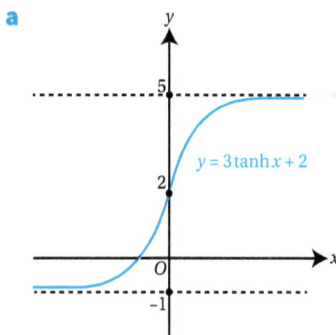

a

You need to apply two transformations to the graph of $y = \tanh x$:

Stretch by scale factor 3 parallel to the y-axis.

Translation by $\begin{pmatrix} 0 \\ 2 \end{pmatrix}$.

The horizontal asymptote at $y = 1$ moves to $y = 3(1) + 2 = 5$, and the one at $y = -1$ is still at -1 ($y = 3(-1) + 2 = -1$).

b $-1 < f(x) < 5$

From the graph, the function is bounded by the asymptotes at $y = -1$ and $y = 5$.

Rewind

You learnt about transformations of graphs in A Level Mathematics Student Book 1, Chapter 5.

EXERCISE 6A

1 For each hyperbolic function, sketch the graph of $y = f(x)$ and state the largest possible domain and the corresponding range.

a $f(x) = 3 - \sinh\left(\dfrac{x}{2}\right)$ **b** $f(x) = 2\sinh(x-1)$ **c** $f(x) = \cosh(2x+3)$

d $f(x) = 4 - \cosh x$ **e** $f(x) = 5 + 2\tanh x$ **f** $f(x) = 4\tanh(-x)$

2 The diagram shows the graph of $y = a\cosh(x+b) - 5$, where a and b are integers.

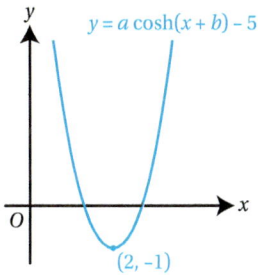

$y = a\cosh(x+b) - 5$

$(2, -1)$

Find the values of a and b.

3 The diagram shows the graph of $y = a\tanh(2x+b)$, where a and b are integers.

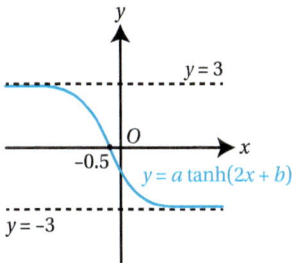

$y = 3$

-0.5

$y = a\tanh(2x+b)$

$y = -3$

Find the values of a and b.

> ⏮ **Rewind**
>
> You saw in A Level Mathematics Student Book 2, Chapter 2, how to form the graphs of inverse functions from their original function by reflection in the line $y = x$.

Section 2: Inverse hyperbolic functions

The **inverse functions** of the hyperbolic functions are called $\operatorname{arsinh} x$, $\operatorname{arcosh} x$ and $\operatorname{artanh} x$.

The graphs of these functions look like this.

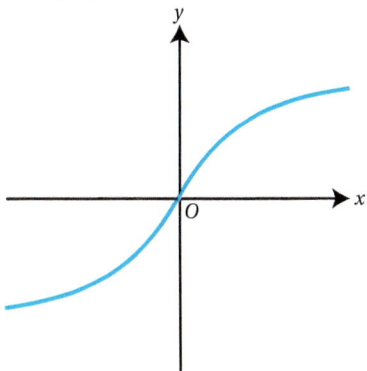

$y = \operatorname{arsinh} x$ $y = \operatorname{arcosh} x$ $y = \operatorname{artanh} x$

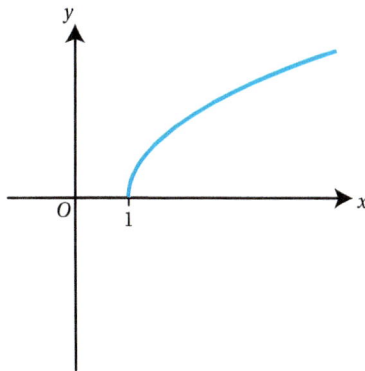

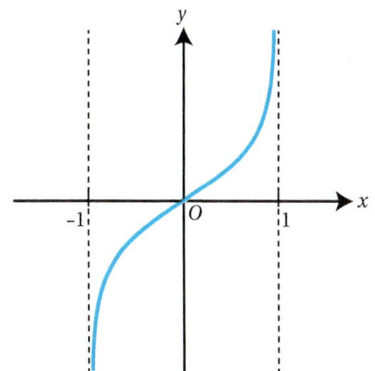

You need to know the domain and range of each function.

Key point 6.5

The domains and ranges of the inverse hyperbolic functions $\sinh^{-1}x$, $\cosh^{-1}x$ and $\tanh^{-1}x$ are:

Function	Domain	Range
$\sinh^{-1}x$	$x \in \mathbb{R}$	$f(x) \in \mathbb{R}$
$\cosh^{-1}x$	$x \geqslant 1$	$f(x) \geqslant 0$
$\tanh^{-1}x$	$-1 < x < 1$	$f(x) \in \mathbb{R}$

Tip

$\sinh^{-1}x$, $\cosh^{-1}x$ and $\tanh^{-1}x$ are alternative notations for $\operatorname{arsinh}x$, $\operatorname{arcosh}x$ and $\operatorname{artanh}x$.

WORKED EXAMPLE 6.2

Let $f(x) = 1 - \cosh^{-1}\left(\dfrac{x}{3}\right)$.

a State the largest possible domain of f.
b For the domain in part **a**, find the range of f.

a Domain: $x \geqslant 3$

The domain of $\cosh^{-1}x$ is $x \geqslant 1$, so $\dfrac{x}{3} \geqslant 1 \Rightarrow x \geqslant 3$.

b

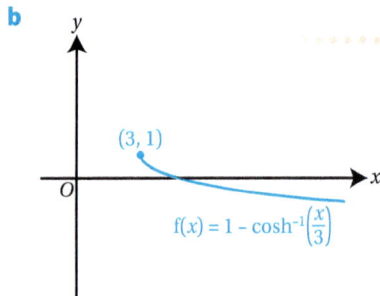

$f(x) = 1 - \cosh^{-1}\left(\frac{x}{3}\right)$

It is always a good idea to sketch the graph when finding the range.

You need to apply three transformations to the graph of $y = \cosh^{-1}x$:

Stretch by scale factor 3 parallel to the x-axis.

Reflection in the x-axis.

Translation by $\begin{pmatrix} 0 \\ 1 \end{pmatrix}$.

Range: $f(x) \leqslant 1$

You can use the inverse hyperbolic functions to solve simple equations involving hyperbolic functions. For example, if $\sinh x = 2$ then $x = \operatorname{arsinh} 2$, which you can evaluate, on a calculator, as $1.4436\ldots$.

However, you can use the definition of $\sinh x$ to derive a logarithmic form of this result. You can do this for all three inverse hyperbolic functions.

Key point 6.6

- $\operatorname{arsinh} x = \ln\left(x + \sqrt{x^2 + 1}\right)$
- $\operatorname{arcosh} x = \ln\left(x + \sqrt{x^2 - 1}\right)$
- $\operatorname{artanh} x = \frac{1}{2}\ln\left(\frac{1+x}{1-x}\right)$

These will be given in your formula book.

These results can all be proved in the same way. The proof for $\operatorname{arcosh} x$ is given here.

PROOF 3

Prove that $\operatorname{arcosh} x = \ln\left(x + \sqrt{x^2 - 1}\right)$.

Let $y = \operatorname{arcosh} x$

Let $y = \operatorname{arcosh} x$ and then look to find an expression for y.

Then $\cosh y = x$

Take cosh of both sides.

$$\frac{e^y + e^{-y}}{2} = x$$

Use the definition of $\cosh y$.

$$e^y + e^{-y} = 2x$$

Rearrange into a disguised quadratic in e^y.

$$e^y + \frac{1}{e^y} = 2x$$

$$(e^y)^2 + 1 = 2xe^y$$

$$(e^y)^2 - 2xe^y + 1 = 0$$

Use the quadratic formula.

So

$$e^y = \frac{2x \pm \sqrt{(2x)^2 - 4}}{2}$$

$$= \frac{2x \pm \sqrt{4x^2 - 4}}{2}$$

$$= \frac{2x \pm \sqrt{4(x^2 - 1)}}{2}$$

$$= \frac{2x \pm \sqrt{4}\sqrt{x^2 - 1}}{2}$$

Use the algebra of surds to simplify the expression.

$$= x \pm \sqrt{x^2 - 1}$$

But $\operatorname{arcosh} x$ is a function so it can only take one value.

Conventionally, you take the positive root, so this makes $e^y > 1$ and $y > 0$.

$$\therefore e^y = x + \sqrt{x^2 - 1}$$

$$y = \ln\left(x + \sqrt{x^2 - 1}\right)$$

But $y = \operatorname{arcosh} x$

So $\operatorname{arcosh} x = \ln\left(x + \sqrt{x^2 - 1}\right)$

WORKED EXAMPLE 6.3

Solve $\sinh x = 7$, giving your answer in the form $\ln\left(a+\sqrt{b}\right)$ for integers a and b.

$\sinh x = 7$

$\quad x = \operatorname{arsinh} 7$ .... Apply the inverse sinh function to find x.

$\quad\quad = \ln\left(7+\sqrt{7^2+1}\right)$ .... Use the logarithmic form $\operatorname{arsinh} x = \ln\left(x+\sqrt{x^2+1}\right)$.

$\quad\quad = \ln\left(7+5\sqrt{2}\right)$

When you are trying to solve equations involving hyperbolic cosines, using the inverse function does not give all the solutions: it just gives the positive one. This is because the cosh function is not one-to-one. As can be seen from the graph, there is a second, negative solution. (This is analogous to solving an equation like $x^2 = 6$, where taking the square root of both sides gives $x = \sqrt{6}$, but there are in fact two solutions, $x = \pm\sqrt{6}$.).

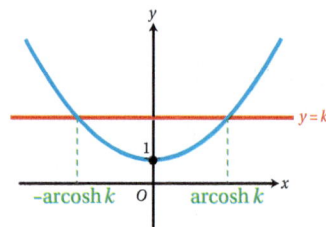

WORKED EXAMPLE 6.4

Given that $\cosh^2 x - \cosh x = 2$, express x in the form $\ln\left(a+\sqrt{b}\right)$ or $\ln\left(a-\sqrt{b}\right)$.

$\quad \cosh^2 x - \cosh x - 2 = 0$ .... The expression is a disguised quadratic, so rearrange it to make one side zero and then factorise it. You could also have used the quadratic formula.

$(\cosh x - 2)(\cosh x + 1) = 0$

$\quad \cosh x = 2$ or $\cosh x = -1$

But $\cosh x \geqslant 0$ so $\cosh x = -1$ has no solutions. ... Use the inverse cosh function to find x. Remember

$x = \pm \operatorname{arcosh} 2$ .... that you need a plus or minus ($\pm$).

$\quad = \pm \ln\left(2+\sqrt{2^2-1}\right)$ .... Use the logarithmic form $\operatorname{arcosh} x = \ln\left(x+\sqrt{x^2-1}\right)$.

$\quad = \pm \ln\left(2+\sqrt{3}\right)$

So

$x = \ln\left(2+\sqrt{3}\right)$ or $\ln\left(\dfrac{1}{2+\sqrt{3}}\right)$ .... Use the fact that $-\ln x \equiv \ln(x^{-1}) \equiv \ln\left(\dfrac{1}{x}\right)$.

$\ln\left(\dfrac{1}{2+\sqrt{3}}\right) = \ln\left(\dfrac{2-\sqrt{3}}{(2+\sqrt{3})(2-\sqrt{3})}\right)$ ...... Simplify the second solution by rationalising the denominator to produce the required form.

$\quad\quad = \ln\left(\dfrac{2-\sqrt{3}}{1}\right)$

So $x = \ln\left(2+\sqrt{3}\right)$ or $x = \ln\left(2-\sqrt{3}\right)$

You can generalise the method used at the end of Worked example 6.4 to write $-\ln\left(2+\sqrt{3}\right)$ as $\ln\left(2-\sqrt{3}\right)$, so the two solutions to $\cosh x = k$ for $k > 1$ can always be written as $x = \ln\left(k\pm\sqrt{k^2-1}\right)$.

EXERCISE 6B

1 For each inverse hyperbolic function, sketch the graph of $y = f(x)$ and state the largest possible domain and the corresponding range.

   a i $f(x) = 3\sinh^{-1} x + 2$ **ii** $f(x) = 2 + \sinh^{-1}(-x)$

   b i $f(x) = 1 + \cosh^{-1} 2x$ **ii** $f(x) = 2\cosh^{-1}\left(\dfrac{x}{3}\right)$

   c i $f(x) = 2\tanh^{-1}(x+1) + 1$ **ii** $f(x) = 3\tanh^{-1}(x-2) + 1$

2 Use your calculator to evaluate each expression where possible.

   a i $\cosh(-1)$ **ii** $\sinh 3$ **b i** $\tanh^2 1$ **ii** $\cosh^2 3$

   c i $3\sinh(0.2) + 1$ **ii** $5\tanh\left(\dfrac{1}{2}\right) + 8$ **d i** $\sinh^{-1} 0.5$ **ii** $\cosh^{-1} 2$

   e i $\tanh^{-1} 2$ **ii** $\cosh^{-1} 0$

3 Solve each equation, giving your answers to 3 significant figures.

   a i $\sinh x = -2$ **ii** $\sinh x = 0.1$ **b i** $2\cosh x = 5$ **ii** $\cosh x = 4$

   c i $4\tanh x = 3$ **ii** $\tanh x = 0.4$ **d i** $3\tanh x = 4$ **ii** $3\cosh x = 1$

   e i $\sinh^{-1} x = 5$ **ii** $\cosh^{-1} x = 4$

4 Without using your calculator, find the exact value of each expression.

   a i $\operatorname{arsinh} 1$ **ii** $\operatorname{arsinh} 2$ **b i** $\cosh^{-1} 2$ **ii** $\cosh^{-1} 3$

   c i $\operatorname{artanh} \dfrac{1}{3}$ **ii** $\operatorname{artanh} \dfrac{1}{4}$ **d i** $\operatorname{arcosh}(-3)$ **ii** $\operatorname{artanh}(-2)$

   e i $\sinh^{-1} \sqrt{2}$ **ii** $\tanh^{-1} \dfrac{1}{\sqrt{3}}$

5 Solve the equation $3\cosh(x-1) = 5$.

6 Find and simplify the exact value of $\cosh(\ln 2)$.

7 Find and simplify a rational expression for $\tanh(\ln 3)$.

8 The function f is given by $f(x) = \cosh^{-1}(x+a) + \tanh^{-1}(x+b)$, where a and b are constants.

   Find, in terms of a, the set of values of the constant b so that $f(x)$ has the largest domain possible.

9 Solve the equation $\sinh^2 3x = 5$.

10 Solve the equation $2\tanh^2 x + 2 = 5\tanh x$.

11 Find and simplify an expression for $\cosh(\sinh^{-1} x)$.

12 Prove that $\sinh^{-1} x = \ln(x + \sqrt{x^2 + 1})$.

13 Prove that $\tanh^{-1} x = \dfrac{1}{2}\ln\left(\dfrac{1+x}{1-x}\right)$.

14 In the derivation of $\cosh^{-1} x$ you found that two possible expressions were $\ln(x + \sqrt{x^2 - 1})$ and $\ln(x - \sqrt{x^2 - 1})$. Show that their sum is zero and hence explain why the expression chosen in Proof 3 is non-negative.

Section 3: Hyperbolic identities

Just as there is the identity $\sin^2 x + \cos^2 x \equiv 1$ linking the trigonometric functions $\sin^2 x$ and $\cos^2 x$, there is also an identity linking the hyperbolic functions $\sinh^2 x$ and $\cosh^2 x$.

▶◀ Rewind

The result in Key point 6.7 proves that $(\cosh x, \sinh x)$ lies on the hyperbola $x^2 - y^2 = 1$, as described in the introduction to this chapter.

🔑 Key point 6.7

$$\cosh^2 x - \sinh^2 x \equiv 1$$

This will be given in your formula book.

You can prove the identity in Key point 6.7 by using the definitions of sinh and cosh given in Key point 6.1.

WORKED EXAMPLE 6.5

Prove that $\cosh^2 x - \sinh^2 x \equiv 1$.

$\cosh^2 x = \left(\dfrac{e^x + e^{-x}}{2}\right)^2$

$\quad = \dfrac{e^{2x} + 2e^x e^{-x} + e^{-2x}}{4}$

$\quad = \dfrac{e^{2x} + 2 + e^{-2x}}{4}$

> Start from the definition of one of the hyperbolic functions. It doesn't matter which one. It is squared in the expression so square it and simplify.

> Since $e^x e^{-x} = 1$.

$\sinh^2 x = \left(\dfrac{e^x - e^{-x}}{2}\right)^2$

$\quad = \dfrac{e^{2x} - 2e^x e^{-x} + e^{-2x}}{4}$

$\quad = \dfrac{e^{2x} - 2 + e^{-2x}}{4}$

> Repeat with the $\sinh^2 x$ term.

> $e^x e^{-x} = 1$ again.

$\cosh^2 x - \sinh^2 x \equiv \dfrac{e^{2x} + 2 + e^{-2x}}{4} - \dfrac{e^{2x} - 2 + e^{-2x}}{4}$

$\quad \equiv \dfrac{e^{2x} + 2 + e^{-2x} - (e^{2x} - 2 + e^{-2x})}{4}$

$\quad \equiv \dfrac{e^{2x} + 2 + e^{-2x} - e^{2x} + 2 - e^{-2x}}{4}$

$\quad \equiv \dfrac{4}{4}$

$\quad \equiv 1$

> Combine the two terms and simplify.

You might be asked to prove other unfamiliar hyperbolic identities. To do this, always return to the definitions of the functions and follow a process similar to that in Worked examples 6.5 and 6.6.

WORKED EXAMPLE 6.6

Prove that $\sinh 2x = 2\sinh x \cosh x$.

$\text{LHS} \equiv \dfrac{e^{2x} - e^{-2x}}{2}$ ⋯⋯⋯⋯⋯ On the LHS use the definition of $\sinh x$ and replace each x with $2x$.

$\text{RHS} \equiv 2 \times \dfrac{e^{x} - e^{-x}}{2} \times \dfrac{e^{x} + e^{-x}}{2}$ ⋯⋯⋯⋯ Then work from the RHS. Substitute the definitions of $\sinh x$ and $\cosh x$.

$\equiv \dfrac{(e^{x} - e^{-x}) \times (e^{x} + e^{-x})}{2}$

$\equiv \dfrac{(e^{x})^{2} - (e^{-x})^{2}}{2}$ ⋯⋯⋯⋯⋯ Multiply out the brackets, using the difference of two squares.

$\equiv \dfrac{e^{2x} - e^{-2x}}{2}$ ⋯⋯⋯⋯⋯ Using the rules of indices.

$\equiv \text{LHS}$

EXERCISE 6C

1. Prove that $\cosh x - \sinh x \equiv e^{-x}$.

2. Simplify $\sqrt{1 + \sinh^{2} x}$.

3. Prove that $\cosh 2x \equiv \cosh^{2} x + \sinh^{2} x$.

4. Prove that $1 - \tanh^{2} x \equiv \dfrac{1}{\cosh^{2} x}$.

5. Prove that $\tanh 2x \equiv \dfrac{2\tanh x}{1 + \tanh^{2} x}$.

6. Prove that $\cosh x - 1 \equiv \dfrac{1}{2}\left(e^{0.5x} - e^{-0.5x}\right)^{2}$. Hence prove that $\cosh x \geqslant 1$.

7. Prove that $\cosh A + \cosh B \equiv 2\cosh\left(\dfrac{A+B}{2}\right)\cosh\left(\dfrac{A-B}{2}\right)$.

8. Use the binomial theorem to show that $\sinh^{3} x \equiv \dfrac{1}{4}\sinh 3x - \dfrac{3}{4}\sinh x$.

9. a Explain why $(\cosh x + \sinh x)^{3} \equiv \cosh 3x + \sinh 3x$ and $(\cosh x - \sinh x)^{3} \equiv \cosh 3x - \sinh 3x$.

   b Hence show that $\cosh 3x \equiv \cosh^{3} x + 3\cosh x \sinh^{2} x$.

   c Write $\cosh 3x$ in terms of $\cosh x$.

10. Given that $\tan y = \sinh x$ show that $\sin y = \pm\tanh x$.

Section 4: Solving harder hyperbolic equations

When you are solving equations involving hyperbolic functions you have several options:

- Rearrange to get a hyperbolic function that is equal to a constant and use inverse hyperbolic functions.
- Use the definition of hyperbolic functions to get an exponential function that is equal to a constant and use logarithms.
- Use an identity for hyperbolic functions to simplify the situation to one of the two preceding options.

It is only with experience that you will develop an instinct about which method will be most efficient.

WORKED EXAMPLE 6.7

Solve $\sinh x + \cosh x = 4$.

$$\sinh x + \cosh x = 4$$

$$\frac{e^x - e^{-x}}{2} + \frac{e^x + e^{-x}}{2} = 4$$

Use the definitions of $\sinh x$ and $\cosh x$.

$$\frac{2e^x}{2} = 4$$

$$e^x = 4$$

$$x = \ln 4$$

> **Tip**
>
> When you are dealing with the sum or difference of two hyperbolic functions, it is often useful to use the exponential form.

WORKED EXAMPLE 6.8

Solve $5\sinh x - 4\cosh x = 0$, giving your answer in the form $\ln a$.

$$5\sinh x = 4\cosh x$$

$$\frac{\sinh x}{\cosh x} = \frac{4}{5}$$

$$\tanh x = \frac{4}{5}$$

You could use the definitions of sinh and cosh, but it is easier to use the identity $\tanh x \equiv \dfrac{\sinh x}{\cosh x}$.

$$x = \operatorname{artanh} x$$

$$= \frac{1}{2}\ln\left(\frac{1 + \frac{4}{5}}{1 - \frac{4}{5}}\right)$$

Then use the logarithmic form of artanh.

$$= \frac{1}{2}\ln 9$$

$$= \ln 3$$

$\dfrac{1}{2}\ln 9 = \ln 9^{\frac{1}{2}} = \ln 3$

WORKED EXAMPLE 6.9

Solve $\cosh^2 x + 1 = 3\sinh x$, giving your answer in logarithmic form.

$\cosh^2 x + 1 = 3\sinh x$	The equation involves two types of function. You can use the identity $\cosh^2 x - \sinh^2 x \equiv 1$ to replace the $\cosh^2 x$ term.
$(1 + \sinh^2 x) + 1 = 3\sinh x$	
$\sinh^2 x - 3\sinh x + 2 = 0$	Solve the resulting quadratic.
$(\sinh x - 1)(\sinh x - 2) = 0$	
$\sinh x = 1$	
$\quad x = \text{arsinh } 1$	
$\quad x = \ln(1 + \sqrt{2})$	Use the logarithmic form of arsinh.
or	
$\sinh x = 2$	
$\quad x = \text{arsinh } 2$	
$\quad x = \ln(2 + \sqrt{5})$	

WORK IT OUT 6.1

Solve $\sinh 2x \cosh 2x = 6\sinh 2x$.

Which is the correct solution? Identify the errors made in the incorrect solutions.

Solution 1	Solution 2	Solution 3
Dividing by 2:	Dividing by $\sinh 2x$:	$\sinh 2x \cosh 2x = 6\sinh 2x$
$\sinh x \cosh x = 3\sinh x$	$\cosh 2x = 6$	$\sinh 2x \cosh 2x - 6\sinh 2x = 0$
$\sinh x \cosh x - 3\sinh x = 0$	$2x = \cosh^{-1} 6$	$\sinh 2x(\cosh 2x - 6) = 0$
$\sinh x(\cosh x - 3) = 0$	$\quad = \ln(6 + \sqrt{35})$	$\sinh 2x = 0$ or $\cosh 2x = 6$
$\sinh x = 0$ or $\cosh x = 3$	$x = \dfrac{1}{2}\ln(6 + \sqrt{35})$	$2x = \sinh^{-1} 0$ or $2x = \pm\cosh^{-1} 6$
$x = \sinh^{-1} 0$ or $x = \cosh^{-1} 3$		$x = 0$ or $x = \pm 1.24$
$x = 0$ or $x = 1.76$		

WORKED EXAMPLE 6.10

a Prove that $1-\tanh^2 x \equiv \dfrac{1}{\cosh^2 x}$.

b Solve the equation $\dfrac{3}{\cosh^2 x}+4\tanh x+1=0$.

a $1-\tanh^2 x \equiv 1-\dfrac{\sinh^2 x}{\cosh^2 x}$

Since the RHS of the required identity contains cosh, it is easiest to start from the definition of tanh in terms of sinh and cosh.

$\equiv \dfrac{\cosh^2 x-\sinh^2 x}{\cosh^2 x}$

Write as a single fraction.

$\equiv \dfrac{1}{\cosh^2 x}$

Use the identity from Key Point 6.7.

b $3-3\tanh^2 x+4\tanh x+1=0$

From **a**, $\dfrac{3}{\cosh^2 x}=3(1-\tanh^2 x)$.

$3\tanh^2 x-4\tanh x-4=0$

This is a quadratic equation in tanh x.

$(3\tanh x+2)(\tanh x-2)=0$

$\tanh x=-\dfrac{2}{3}\text{ or }2$

$\therefore x=\text{artanh}\left(-\dfrac{2}{3}\right)$

The range of tanh x is $(-1, 1)$, so $\tanh x=2$ is not possible.

$\therefore x=\dfrac{1}{2}\ln\left(\dfrac{1+\left(-\dfrac{2}{3}\right)}{1-\left(-\dfrac{2}{3}\right)}\right)$

Then use $\text{artanh}\,x=\dfrac{1}{2}\ln\left(\dfrac{1+x}{1-x}\right)$.

$=\dfrac{1}{2}\ln\left(\dfrac{\frac{1}{3}}{\frac{5}{3}}\right)$

$=\dfrac{1}{2}\ln\left(\dfrac{1}{5}\right)$

$=-\dfrac{1}{2}\ln 5$

EXERCISE 6D

1 Find the exact solution to $\cosh x = 5 - \sinh x$.

2 Solve $\cosh x - \sinh x = 2$, giving your answer in the form $\ln k$.

3 Solve $3(2\sinh x - 1)(\cosh x - 4) = 0$, giving your answers correct to 3 significant figures.

4 Solve $5\sinh x + 3\cosh x = 0$, giving your answer in the form $\ln k$, where k is a rational number.

5 Solve the equation $\dfrac{2}{\cosh x} = e^x$, giving your answer in the form $\ln k$.

6 Solve the equation $2\sinh 2x = 9\tanh x$, giving exact answers.

7 Find the exact solution to $2\sinh x = 1 + \cosh x$.

8 Solve $6\sinh x - 2\cosh x = 7$, giving your answers in logarithmic form.

9 Solve $2\cosh^2 x - 5\sinh x = 5$, giving your answers in exact form.

10 Solve $\sinh^2 x = \cosh x + 1$, giving your answer in logarithmic form.

11 Solve $\tanh x = \dfrac{1}{\cosh x}$, giving your answer in logarithmic form.

12 Solve the equation $6\tanh x - \dfrac{7}{\cosh x} = 2$, giving your answer in the form $\ln k$, where k is a rational number.

13 Using the identity $\dfrac{1}{\cosh^2 x} \equiv 1 - \tanh^2 x$, solve the equation $2\tanh^2 x = 4 - \dfrac{5}{\cosh x}$, giving your answers in the form $\ln k$.

14 $\sinh x + \sinh y = \dfrac{21}{8}$

$\cosh x + \cosh y = \dfrac{27}{8}$

   a Show that $e^x = 6 - e^y$ and $e^{-x} = 0.75 - e^{-y}$.

   b Hence find the exact solutions to the simultaneous equations.

15 Find a sufficient condition on p, q and r for $p^2\cosh x + q^2\sinh x = r^2$ to have at least one solution.

16 a Prove that $\sinh 3x \equiv 4\sinh^3 x + 3\sinh x$.

   b Hence solve the equation $\sinh 6x = 6 + \sinh 2x$, giving your answer in the form $a\ln b$.

17 a Show that, for any real number k,

   $(\cosh x + \sinh x)^k + (\cosh x - \sinh x)^k \equiv 2\cosh kx$.

   b Hence solve the equation

   $(\cosh x + \sinh x)^6 + (\cosh x - \sinh x)^6 = 6$

   giving your answers in the form $a\ln b$.

Section 5: Differentiation

> ### 🔑 Key point 6.8
>
> - $\dfrac{d}{dx}(\sinh x) = \cosh x$
>
> - $\dfrac{d}{dx}(\cosh x) = \sinh x$
>
> - $\dfrac{d}{dx}(\tanh x) = \dfrac{1}{\cosh^2 x}$
>
> **Only the final one of these will be given in your formula book.**

> ### 💡 Tip
>
> The first two of these formulae will *not* be given in your formula book.

You can derive the results for $\sinh x$ and $\cosh x$ by returning to the definitions of these functions.

WORKED EXAMPLE 6.11

Show that $\dfrac{d}{dx}(\sinh x) = \cosh x$.

$y = \sinh x$

$= \dfrac{e^x - e^{-x}}{2}$ .... Use the definition of $\sinh x$.

Differentiating:

$\dfrac{dy}{dx} = \dfrac{e^x - \left(-e^{-x}\right)}{2}$

$= \dfrac{e^x + e^{-x}}{2}$

$= \cosh x$

You can show the result for $\tanh x$ either from the definition again or by using $\tanh x \equiv \dfrac{\sinh x}{\cosh x}$ and the quotient rule.

WORKED EXAMPLE 6.12

Use the derivatives of $\sinh x$ and $\cosh x$ to show that $\dfrac{d}{dx}(\tanh x) = \dfrac{1}{\cosh^2 x}$.

$y = \tanh x$

$= \dfrac{\sinh x}{\cosh x}$ .... Use $\tanh x \equiv \dfrac{\sinh x}{\cosh x}$.

Differentiating using the quotient rule:

$\dfrac{dy}{dx} = \dfrac{\cosh x \cosh x - \sinh x \sinh x}{\cosh^2 x}$

$= \dfrac{\cosh^2 x - \sinh^2 x}{\cosh^2 x}$

$= \dfrac{1}{\cosh^2 x}$ .... Use $\cosh^2 x - \sinh^2 x \equiv 1$.

WORKED EXAMPLE 6.13

Given that $y = x\tanh(x^2)$, find $\dfrac{dy}{dx}$.

Let $u = x$ and $v = \tanh(x^2)$ .... Use the product rule.

Then $u' = 1$

and .... v is a composite function so use the chain rule to differentiate.

$v' = \dfrac{1}{\cosh^2 x} \times 2x$

$= \dfrac{2x}{\cosh^2 x}$

$\dfrac{dy}{dx} = 1 \times \tanh(x^2) + x \times \dfrac{2x}{\cosh^2 x}$ .... Now apply the product rule formula.

$= \tanh(x^2) + \dfrac{2x^2}{\cosh^2 x}$

EXERCISE 6E

1 Differentiate each function with respect to x.

a **i** $f(x) = \sinh 3x$ **ii** $f(x) = \sinh\dfrac{1}{2}x$

b **i** $f(x) = \cosh(4x+1)$ **ii** $f(x) = \cosh\dfrac{1}{3}x$

c **i** $f(x) = \tanh\dfrac{2}{3}x$ **ii** $f(x) = \tanh(1-2x)$

2 Differentiate each function with respect to x.

a $f(x) = x^2\tanh 3x$ **b** $f(x) = \dfrac{1}{\tanh^2 5x}$

3 Find the exact coordinates of the turning point on the curve $y = e^{\cosh x} - \cosh x$.

4 Find the exact coordinates of the minimum point on the curve $y = 3\sinh x + 5\cosh x$.

5 Show that the equation of the tangent to the curve $y = \tanh x$ at $x = \ln 2$ is $16x - 25y + 15 - 16\ln 2 = 0$.

6 Find the equation of the normal to the curve $y = 2\sinh x - \cosh x$ at $x = \ln 3$, giving your answer in the form $ax + by + c = 0$.

7 **a** Find the exact values of the x-coordinates of the turning points on the curve $y = \tanh 2x - x$.

b Show that the maximum point has y-coordinate $\dfrac{\sqrt{2} - \ln(\sqrt{2}+1)}{2}$.

8 Find the coordinates of the stationary point on the curve $y = e^{-x}\sinh\dfrac{1}{2}x$.

9 Show that the two points of inflection on the curve $y = \dfrac{1}{\cosh x}$ have x-coordinates $\pm\ln k$, stating the value of k.

10 **a** Find the exact value of the x-coordinates of the stationary points on the curve with equation $y = 8\sinh x - 27\tanh x$.

b Prove that one of the stationary points from part **a** is a local maximum and that one is a local minimum point.

Section 6: Integration

Key point 6.9

- $\int \sinh x\, dx = \cosh x + c$

- $\int \cosh x\, dx = \sinh x + c$

- $\int \tanh x\, dx = \ln\cosh x + c$

Tip

These results will *not* be given in your formula book.

As with differentiation, you can derive the results for the integrals of $\sinh x$ and $\cosh x$ by returning to the definitions of these functions.

WORKED EXAMPLE 6.14

a Show that $\int \sinh x\, dx = \cosh x + c$.

b Hence find the exact value of $\int_0^{\ln 5} \sinh 3x\, dx$.

a $\int \sinh x\, dx = \int \dfrac{e^x - e^{-x}}{2}\, dx$

$= \dfrac{e^x - (-e^{-x})}{2} + c$

$= \dfrac{e^x + e^{-x}}{2} + c$

$= \cosh x + c$

Use the definition of sinh x.

b $\int_0^{\ln 5} \sinh 3x\, dx = \left[\dfrac{1}{3}\cosh 3x \right]_0^{\ln 5}$

$= \dfrac{1}{3}\cosh(3\ln 5) - \dfrac{1}{3}\cosh(0)$

$= \dfrac{1}{3}\dfrac{\left(e^{3\ln 5} + e^{-3\ln 5}\right)}{2} - \dfrac{1}{3}$

$= \dfrac{125 + \frac{1}{125}}{6} - \dfrac{1}{3}$

$= \dfrac{7688}{375}$

Use the result from part **a** together with the reverse chain rule.

Use the definition of cosh in terms of e, and the fact that $\cosh(0) = 1$.

$e^{3\ln 5} = e^{\ln(5^3)} = 125$

You can now find the integral of $\tanh x$ either from the definition again or by using $\tanh x \equiv \dfrac{\sinh x}{\cosh x}$ and applying the reverse chain rule.

WORKED EXAMPLE 6.15

Show that $\int \tanh x \, dx = \ln \cosh x + c$.

$\int \tanh x \, dx = \int \dfrac{\sinh x}{\cosh x} \, dx$ .... Use $\tanh x \equiv \dfrac{\sinh x}{\cosh x}$.

$\qquad = \ln \cosh x + c$

This is of the form $\int \dfrac{f'(x)}{f(x)} \, dx$, so you can integrate it directly.

Tip

Look out for integrals of the form $\int f'(x)[f(x)]^n \, dx$ or $\int \dfrac{f'(x)}{f(x)} \, dx$ as you can integrate these without need for a substitution, by reversing the chain rule.

Rewind

See A Level Mathematics Student Book 2, Chapter 11, for a reminder of integrating trigonometric functions using the reverse chain rule, trigonometric identities and integration by parts.

Often you will need to use a hyperbolic identity before integrating.

Tip

When integrating hyperbolic functions, you can often use the same approach as with the corresponding trigonometric function.

WORKED EXAMPLE 6.16

Find $\int \cosh^2 x \, dx$.

$\cosh 2x \equiv 2 \cosh^2 x - 1$ .... Use the identity for $\cosh 2x$ in terms of $\cosh^2 x$.

$\Rightarrow \cosh^2 x \equiv \dfrac{\cosh 2x + 1}{2}$

$\therefore \int \cosh^2 x \, dx = \int \dfrac{\cosh 2x + 1}{2} \, dx$

$\qquad = \dfrac{1}{2}\left(\dfrac{1}{2}\sinh 2x + x\right) + c$ .... Remember to divide by the coefficient of x when integrating $\cosh 2x$.

Sometimes it's better to use the definition of the hyperbolic function, rather than the method you would have used with the corresponding trigonometric function.

WORKED EXAMPLE 6.17

Find $\int e^x \cosh x \, dx$.

$\int e^x \cosh x \, dx = \int e^x \left(\dfrac{e^x + e^{-x}}{2} \right) dx$ .... Use the definition of $\cosh x$.

$\qquad\qquad = \int \dfrac{e^{2x} + 1}{2} dx$

$\qquad\qquad = \dfrac{1}{2}\left(\dfrac{1}{2}e^{2x} + x \right) + c$

⏮ Rewind

If the integral in Worked example 6.17 had been $\int e^x \cos x \, dx$, you would have done integration by parts twice and rearranged.

EXERCISE 6F

1 Find:

a **i** $\int \sinh 3x \, dx$ **ii** $\int \sinh \dfrac{x}{2} \, dx$

b **i** $\int \cosh(2x+1) \, dx$ **ii** $\int \cosh 4x \, dx$

c **i** $\int \tanh(-2x) \, dx$ **ii** $\int \tanh(3x-2) \, dx$

2 Find the exact value of each integral.

a **i** $\int_0^2 \sinh x \, dx$ **ii** $\int_0^5 \cosh x \, dx$

b **i** $\int_0^1 \sinh 2x \, dx$ **ii** $\int_0^3 \cosh 2x \, dx$

c **i** $\int_{\ln 2}^{\ln 4} \cosh 3x \, dx$ **ii** $\int_{\ln 2}^{\ln 3} \sinh 2x \, dx$

3 Use an appropriate hyperbolic identity to find each integral.

a **i** $\int \sinh^2 2x \, dx$ **ii** $\int \cosh^2 3x \, dx$

b **i** $\int \tanh^2 \dfrac{x}{2} \, dx$ **ii** $\int \tanh^2 3x \, dx$

c **i** $\int \sinh x \cosh x \, dx$ **ii** $\int \sinh 3x \cosh 3x \, dx$

4 Use integration by parts to find each integral.

a **i** $\int x \sinh x \, dx$ **ii** $\int x \sinh 2x \, dx$

b **i** $\int 3x \cosh x \, dx$ **ii** $\int x \cosh \dfrac{x}{2} \, dx$

c **i** $\int x^2 \sinh x \, dx$ **ii** $\int x^2 \sinh 3x \, dx$

d **i** $\int x^2 \cosh 2x \, dx$ **ii** $\int 3x^2 \cosh x \, dx$

5 Use the definitions of $\sinh x$ and/or $\cosh x$ to find each integral.

a **i** $\int e^x \sinh 2x \, dx$ **ii** $\int e^{2x} \cosh x \, dx$

b **i** $\int \sinh x \sinh 4x \, dx$ **ii** $\int \cosh 2x \cosh 3x \, dx$

6 Find $\int \dfrac{\sinh x + \cosh x}{4 \cosh x} \, dx$.

7 By expressing $\sinh x$ and $\cosh x$ in terms of e^x, evaluate $\int_0^1 \dfrac{1}{\sinh x + \cosh x} \, dx$.

8 The diagram shows the region R, which is bounded by the curve
$y = \sinh x$, the x-axis and the line $x = \ln 3$.

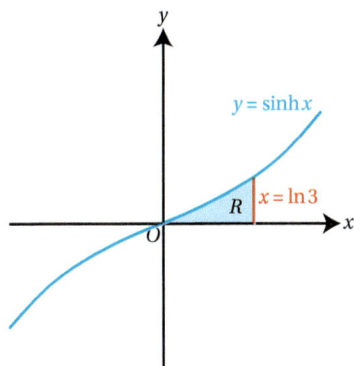

Show that the volume of the solid formed when the region R is
rotated through 2π radians about the x-axis is given by $\dfrac{\pi}{18}(20 - 9\ln 3)$.

9 Find:

a $\displaystyle\int \cosh^3 x \sinh^2 x \, dx$

b $\displaystyle\int \frac{\cosh^3 x}{\sinh^2 x} \, dx.$

10 Find $\displaystyle\int \frac{x}{\cosh^2 x} \, dx.$

11 The diagram shows the region R bounded by curve
$y = 4\cosh\dfrac{x}{2}$, for $x \geqslant 0$, the y-axis and the line $y = 3\sqrt{2}$.

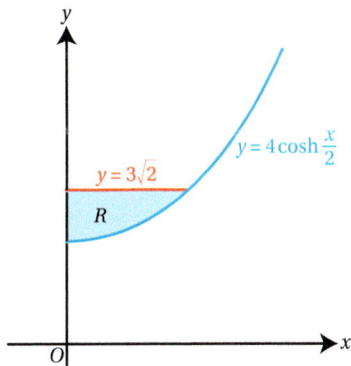

Find the exact volume of the solid formed when the region R is
rotated through 2 radians about the x-axis.

12 Using the substitution $u = \cosh x$, show that $\displaystyle\int_0^{\ln 2} \frac{\sinh^3 x}{\cosh x + 1} \, dx = \frac{1}{32}.$

13 Show that $\displaystyle\int_0^{\frac{\pi}{4}} \sin x \operatorname{artanh}(\sin x) \, dx = \frac{\pi - \sqrt{2}\ln(3 + 2\sqrt{2})}{4}$

14 **a** Given that $u = \operatorname{arsinh} x$, write down an expression for $\dfrac{du}{dx}$.

b Use integration by parts to show that
$\displaystyle\int \operatorname{arsinh} x \, dx = x\operatorname{arsinh} x - \sqrt{x^2 + 1} + c.$

📎 Checklist of learning and understanding

- Definitions of hyperbolic functions:

  - $\sinh x = \dfrac{e^x - e^{-x}}{2}$

  - $\cosh x = \dfrac{e^x + e^{-x}}{2}$

  - $\tanh x \equiv \dfrac{\sinh x}{\cosh x} \equiv \dfrac{e^x - e^{-x}}{e^x + e^{-x}}$

- Graphs of hyperbolic functions:

  - $y = \sinh x$

  - $y = \cosh x$

  - $y = \tanh x$

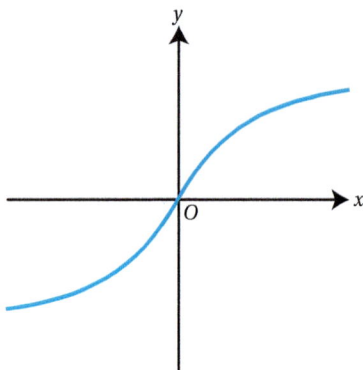 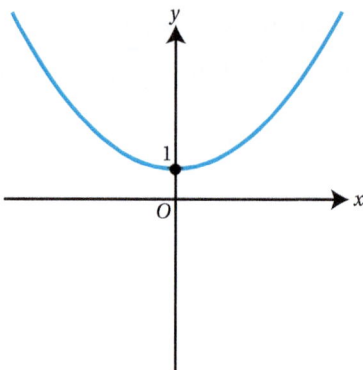 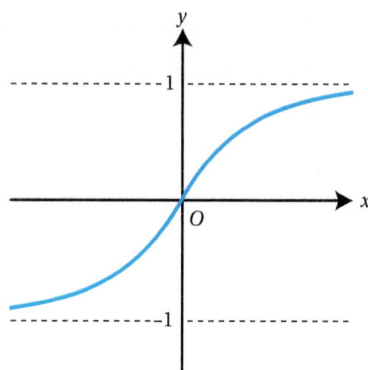

- Logarithmic form of inverse hyperbolic functions:

  - $\operatorname{arsinh} x = \ln\left(x + \sqrt{x^2 + 1}\right)$

  - $\operatorname{arcosh} x = \ln\left(x + \sqrt{x^2 - 1}\right)$

  - $\operatorname{artanh} x = \dfrac{1}{2}\ln\left(\dfrac{1+x}{1-x}\right)$

- Domain and range of hyperbolic, inverse hyperbolic, and reciprocal hyperbolic functions:

Function	Domain	Range
$\sinh x$	$x \in \mathbb{R}$	$f(x) \in \mathbb{R}$
$\cosh x$	$x \in \mathbb{R}$	$f(x) \geqslant 1$
$\tanh x$	$x \in \mathbb{R}$	$-1 < f(x) < 1$
$\sinh^{-1} x$	$x \in \mathbb{R}$	$f(x) \in \mathbb{R}$
$\cosh^{-1} x$	$x \geqslant 1$	$f(x) \geqslant 0$
$\tanh^{-1} x$	$-1 < x < 1$	$f(x) \in \mathbb{R}$

Continues on next page ...

- Identities:
  - $\cosh^2 x - \sinh^2 x \equiv 1$
  - To prove other identities, return to the definitions of the functions involved.
- Derivatives of hyperbolic and reciprocal hyperbolic functions:
  - $\dfrac{\mathrm{d}}{\mathrm{d}x}(\sinh x) = \cosh x$
  - $\dfrac{\mathrm{d}}{\mathrm{d}x}(\cosh x) = \sinh x$
  - $\dfrac{\mathrm{d}}{\mathrm{d}x}(\tanh x) = \dfrac{1}{\cosh^2 x}$
- Integrals of hyperbolic functions:
  - $\displaystyle\int \sinh x = \cosh x + c$
  - $\displaystyle\int \cosh x = \sinh x + c$
  - $\displaystyle\int \tanh x = \ln \cosh x + c$
- Many hyperbolic integrals can be done by using the same method that you would use for the corresponding trigonometric integral.

Mixed practice 6

1 Given that $f(x) = 3\tanh^2 x + 1$ for $x \in \mathbb{R}$,

    a sketch the graph of $y = f(x)$

    b state the range of f.

2 Solve $\sinh x + \cosh x = k$, giving your answer in terms of k.

3 Simplify $\tanh(1 + \ln p)$.

4 Solve $\cosh(x + 1) = 3$, giving your answer in terms of logarithms.

5 Solve $4\sinh 2x = \cosh 2x$, giving your answer correct to 3 significant figures.

6 Find the exact solutions to $16\cosh^2 x + 8\cosh x = 35$.

7 Solve the equation $\sinh x = \dfrac{1}{\cosh x}$, giving your answer in the form $a\ln b$.

8 Given $f(x) = \sinh^2 3x$, find $f''(x)$.

9 Show that the curve $y = e^x \cosh 2x$ has no points of inflection.

10 Given that $y = a\sinh nx + b\cosh nx$, show that $\dfrac{d^2 y}{dx^2} = n^2 y$.

11 By first expressing $\cosh x$ and $\sinh x$ in terms of exponentials, solve the equation $3\cosh x - 4\sinh x = 7$ giving your answer in an exact logarithmic form.

© OCR A Level Mathematics, Unit 4726/01 Further Pure Mathematics 2, January 2013

12 Find $\displaystyle\int \dfrac{\tanh 3x}{\cosh 3x}\,dx$.

13 Show that $\displaystyle\int_0^{\ln\sqrt{2}} e^{\cosh 4x}\sinh 4x\,dx = \dfrac{e}{4}\left(e^{\frac{9}{8}} - 1\right)$.

14 Solve the equation $\sinh 3x\cosh^2 3x = 5\sinh 3x$, giving your answers in exact form.

15 Solve $3\sinh^2 x - 13\cosh x + 7 = 0$, giving your answers in terms of natural logarithms.

16 Prove that $\cosh x > \sinh x$ for all x.

17 Find and simplify an expression for $\tanh(\operatorname{arsinh} x)$.

18 Use the binomial theorem to show that $\cosh^4 x \equiv \dfrac{1}{8}\cosh 4x + \dfrac{1}{2}\cosh 2x + \dfrac{3}{8}$.

19 **a** Sketch the graph of $y = \tanh x$.

    b Given that $u = \tanh x$, use the definitions of $\sinh x$ and $\cosh x$ in terms of e^x and e^{-x} to show that
$$x = \dfrac{1}{2}\ln\left(\dfrac{1+u}{1-u}\right).$$

    c **i** Show that the equation $\dfrac{3}{\cosh^2 x} + 7\tanh x = 5$ can be written as $3\tanh^2 x - 7\tanh x + 2 = 0$.

        ii Show that the equation $3\tanh^2 x - 7\tanh x + 2 = 0$ has only one solution for x.

        Find this solution in the form $\dfrac{1}{2}\ln a$ where a is an integer.

20 **a** Prove that $\dfrac{1}{\cosh^2 x} \equiv 1 - \tanh^2 x$.

    b Hence solve the equation $\dfrac{1}{\cosh^2 x} = 4 + \tanh x$, giving your answers in terms of natural logarithms.

21 i Given that $y = \sinh^{-1} x$, prove that $y = \ln\left(x + \sqrt{x^2 + 1}\right)$.

ii It is given that x satisfies the equation $\sinh^{-1} x - \cosh^{-1} x = \ln 2$. Use the logarithmic forms for $\sinh^{-1} x$ and $\cosh^{-1} x$ to show that $\sqrt{x^2 + 1} - 2\sqrt{x^2 - 1} = x$.

Hence, by squaring this equation, find the exact value of x.

© **OCR A Level Mathematics, Unit 4726 Further Pure Mathematics 2, January 2012**

22 Using the substitution $u = e^x$, find $\displaystyle\int \frac{1}{4\sinh x + 5\cosh x}\, dx$.

23 The diagram shows the graphs of $y = 5\cosh x$ and $y = \sinh x = 7$.

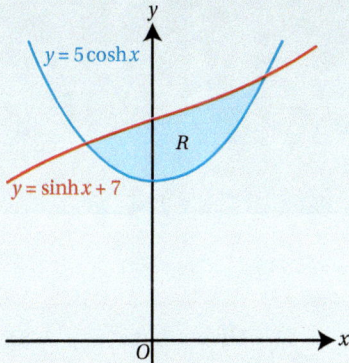

Find the exact value of the area of the shaded region.

24 a Prove the identity $\cosh 3x = 4\cosh^3 x - 3\cosh x$.

b If $48u^3 - 36u - 13 = 0$ and $u = \cosh x$ find the value of x.

c Hence find the exact real solution to $48u^3 - 36u - 13 = 0$, giving your answer in a form without logarithms.

25 Solve these simultaneous equations, giving your answers in exact logarithmic form.
$$\begin{cases} \sinh x + \sinh y = 3.15 \\ \cosh x + \cosh y = 3.85 \end{cases}$$

26 Using the logarithmic definition, prove that $\operatorname{arsinh}(-x) = -\operatorname{arsinh} x$.

27 Given that $\operatorname{artanh} x + \operatorname{artanh} y = \ln\sqrt{5}$, show that $y = \frac{3x-2}{2x-3}$.

28 Prove that $\frac{\sinh x + \cosh x - 1}{\sinh x + \cosh x + 1} = \tanh\left(\frac{1}{2}x\right)$.

7 Further calculus techniques

In this chapter you will learn how to:

- differentiate inverse trigonometric and inverse hyperbolic functions
- reverse those results to find integrals of the form $(a^2+x^2)^{-1}$, $(a^2-x^2)^{-\frac{1}{2}}$, $(a^2+x^2)^{-\frac{1}{2}}$ and $(x^2-a^2)^{-\frac{1}{2}}$
- use trigonometric and hyperbolic substitutions to find similar integrals
- integrate using partial fractions with a quadratic expression in the denominator.

Before you start…

A Level Mathematics Student Book 2, Chapter 10	You should know how to differentiate functions defined implicitly.	1 Given that $x^2 - y^3 = 5x$, find an expression for $\dfrac{dy}{dx}$.
A Level Mathematics Student Book 2, Chapter 11	You should be able to integrate using a substitution.	2 Use a suitable substitution to evaluate $\displaystyle\int_0^1 x^2\sqrt{1+2x^3}\,dx$.
A Level Mathematics Student Book 2, Chapter 11	You should know how to integrate rational functions by splitting them into partial fractions.	3 Find $\displaystyle\int \dfrac{2x^2 - 9x + 8}{(x-1)(x-2)^2}\,dx$.

Introduction

In this chapter you will extend your range of integration methods and the variety of functions you can integrate. Differentiation of inverse trigonometric functions leads to rules for integrating functions of the form $\dfrac{1}{a^2+x^2}$ and $\dfrac{1}{\sqrt{a^2-x^2}}$ and suggests that you can use trigonometric substitution to find other similar integrals. Likewise, differentiation of inverse hyperbolic functions leads to rules for integrating $\dfrac{1}{\sqrt{a^2+x^2}}$ and $\dfrac{1}{\sqrt{x^2-a^2}}$. You can use these results in combination with partial fractions to integrate many rational functions.

Section 1: Differentiation of inverse trigonometric functions

You already know how to differentiate $\sin x$, $\cos x$ and $\tan x$. To differentiate their inverse functions, you can use implicit differentiation.

⏮ **Rewind**

You met implicit differentiation in A Level Mathematics Student Book 2, Chapter 10.

WORKED EXAMPLE 7.1

Given that $y = \sin^{-1} x$, and that $|x| < 1$, find $\dfrac{dy}{dx}$ in terms of x.

$y = \sin^{-1} x$	You know how to differentiate sin, so express x in terms of y.
$\Rightarrow \sin y = x$	

Differentiating each term with respect to x:

$\cos y \dfrac{dy}{dx} = 1$	Remember the chain rule.
$\dfrac{dy}{dx} = \dfrac{1}{\cos y}$	
$\quad = \dfrac{1}{\sqrt{1 - \sin^2 y}}$	You want the answer in terms of x, so you need to change cos to sin.
$\quad = \dfrac{1}{\sqrt{1 - x^2}}$	

You should notice two important details in the derivation of the derivative of $\sin^{-1} x$ shown in Worked example 7.1. First, the $\sin^{-1} x$ function is defined for $|x| \leqslant 1$. However, you can see from the graph that the gradient at $x = \pm 1$ is not finite, so the condition $|x| < 1$ is required for the derivative to exist. (You can also see that the expression for $\dfrac{dy}{dx}$ is not defined when $x = \pm 1$.)

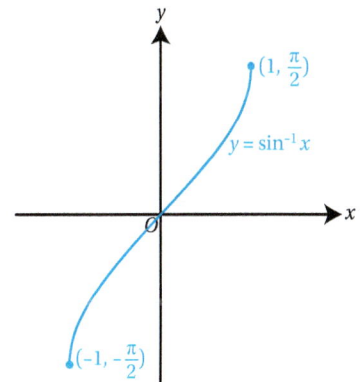

Second, you used $\cos^2 y = 1 - \sin^2 y$ to write $\cos y = \sqrt{1 - \sin^2 y}$. When taking a square root, you need to ask whether it should be positive or negative (or both). In this case, the range of the $\sin^{-1} x$ function is $-\dfrac{\pi}{2} \leqslant y \leqslant \dfrac{\pi}{2}$ and in this range, $\cos y > 0$; this justifies taking the positive square root.

You can establish the results for the inverse cos and tan functions similarly.

🔑 Key point 7.1

- $\dfrac{d}{dx}\left(\sin^{-1} x\right) = \dfrac{1}{\sqrt{1-x^2}}, |x| < 1$

- $\dfrac{d}{dx}\left(\cos^{-1} x\right) = \dfrac{-1}{\sqrt{1-x^2}}, |x| < 1$

- $\dfrac{d}{dx}\left(\tan^{-1} x\right) = \dfrac{1}{1+x^2}$

These will be given in your formula book.

Notice that the $\tan^{-1} x$ function is defined for all $x \in \mathbb{R}$, so there is no restriction on the domain of its derivative.

You can combine these results with other rules of differentiation.

WORKED EXAMPLE 7.2

Differentiate:

a $y = x^2 \tan^{-1} 4x$
b $y = \arccos\sqrt{x-3}$ and state the values for which the derivative is valid.

> **Tip**
> Remember that arccos x is alternative notation for $\cos^{-1} x$.

a Using the product rule:
$u = x^2, v = \tan^{-1} 4x$

$\dfrac{du}{dx} = 2x$

$\dfrac{dv}{dx} = \dfrac{1}{1+(4x)^2} \times 4$ .... Multiply by 4 (the derivative of $4x$) due to the chain rule.

$= \dfrac{4}{1+16x^2}$

$\therefore \dfrac{dy}{dx} = 2x\tan^{-1}4x + \dfrac{4x^2}{1+16x^2}$ ...... Use $\dfrac{dy}{dx} = \dfrac{du}{dx}v + u\dfrac{dv}{dx}$.

b Using the chain rule:
$\dfrac{dy}{dx} = \dfrac{-1}{\sqrt{1-(\sqrt{x-3})^2}} \times \dfrac{1}{2}(x-3)^{-\frac{1}{2}}$ ..... Multiply by $\dfrac{1}{2}(x-3)^{-\frac{1}{2}}$ (the derivative of $\sqrt{x-3}$).

$= \dfrac{-1}{\sqrt{1-(x-3)}} \times \dfrac{1}{2\sqrt{x-3}}$

$= \dfrac{-1}{2\sqrt{(4-x)(x-3)}}$

The derivative is valid when

$|x-3| < 1$ and $x \geqslant 3$ .... The derivative of arccos x is only defined for $|x| < 1$ and the square root is only defined when $x - 3 \geqslant 0$.

$3 \leqslant x < 4$

EXERCISE 7A

1 Find $\dfrac{dy}{dx}$ for each function.

a i $y = \cos^{-1} 3x$ **ii** $y = \cos^{-1} 2x$

b i $y = \tan^{-1}\left(\dfrac{x}{2}\right)$ **ii** $y = \tan^{-1}\left(\dfrac{2x}{5}\right)$

c i $y = x\arcsin x$ **ii** $y = x^2\arcsin x$

d i $y = \arctan(x^2+1)$ **ii** $y = \arcsin(1-x^2)$

2. Find the exact value of the gradient of the graph of $y = \cos^{-1}\left(\dfrac{x}{2}\right)$ at the point where $x = \dfrac{1}{3}$.

3. Find the exact value of the gradient of the graph of $y = x \arctan 3x$ at the point where $x = -\dfrac{1}{3}$.

4. Differentiate $\arctan(3x+2)$, simplifying your answer as far as possible.

5. Find the derivative of $\arcsin(x^2 - 3)$, stating the range of values of x for which your answer is valid.

6. **a** Given that $y = \tan^{-1} x$, show that $\dfrac{dy}{dx} = \dfrac{1}{1+x^2}$.

   b Hence differentiate $\tan^{-1}\dfrac{1}{x}$ with respect to x.

7. Given that $y = \sin^{-1}\left(\dfrac{3x}{2}\right)$, show that $\dfrac{dy}{dx} = \dfrac{3}{\sqrt{4-9x^2}}$ and state the values of x for which the derivative is valid.

8. Given that $x \arctan y = 1$, find an expression for $\dfrac{dy}{dx}$.

9. **a** Find $\dfrac{d}{dx}\left(x\sin^{-1} x\right)$.

   b Hence find $\int \sin^{-1} x\, dx$.

10. Show that the graph of $y = \arcsin(x^2)$ has no points of inflection.

Section 2: Differentiation of inverse hyperbolic functions

You know how to differentiate hyperbolic functions, and so again you can use implicit differentiation to differentiate their inverse functions.

> ⏪ **Rewind**
>
> See Chapter 6, Section 5, for differentiation of hyperbolic functions.

WORKED EXAMPLE 7.3

$y = \sinh^{-1} x$ Rewrite in terms of $\sinh y$.

$\Rightarrow \sinh y = x$

Differentiating with respect to y:

$\dfrac{dx}{dy} = \cosh y$

$\dfrac{dy}{dx} = \dfrac{1}{\cosh y}$ $\dfrac{dy}{dx} = \dfrac{1}{\left(\dfrac{dx}{dy}\right)}$, so take the reciprocal of both sides.

$= \dfrac{1}{\sqrt{\sinh^2 y + 1}}$ Use $\cosh^2 x - \sinh^2 x \equiv 1$.

$= \dfrac{1}{\sqrt{x^2 + 1}}$

You can establish the results for the inverse cosh and tanh functions similarly.

Key point 7.2

- $\dfrac{d}{dx}(\sinh^{-1} x) = \dfrac{1}{\sqrt{x^2 + 1}}$

- $\dfrac{d}{dx}(\cosh^{-1} x) = \dfrac{1}{\sqrt{x^2 - 1}}, \ x > 1$

- $\dfrac{d}{dx}(\tanh^{-1} x) = \dfrac{1}{1 - x^2}, \ |x| < 1$

These will be given in your formula book.

WORKED EXAMPLE 7.4

Find the value of the x-coordinate of the point on the curve $y = \operatorname{arcosh}\left(\dfrac{x}{2}\right)$ at which the tangent is parallel to the line $y = x$.

Differentiating using the chain rule:

$y' = \dfrac{1}{\sqrt{\left(\dfrac{x}{2}\right)^2 - 1}} \times \dfrac{1}{2}$

> Remember to multiply by the derivative of $\dfrac{x}{2}$.

$= \dfrac{1}{2\sqrt{\dfrac{x^2}{4} - 1}}$

> Simplify the denominator.

$= \dfrac{1}{2\sqrt{\dfrac{x^2 - 4}{4}}}$

$= \dfrac{1}{\sqrt{x^2 - 4}}$

So

$\dfrac{1}{\sqrt{x^2 - 4}} = 1$

> $y = x$ has gradient 1, so set $y' = 1$ and solve for x.

$\sqrt{x^2 - 4} = 1$

$x^2 - 4 = 1$

$x^2 = 5$

$\therefore x = \sqrt{5}$

> The domain of $y = \cosh^{-1}\left(\dfrac{x}{2}\right)$ is $x \geqslant 2$ so take the positive square root.

EXERCISE 7B

1 Differentiate each function with respect to x.

a i $f(x)=\operatorname{arsinh}2x$ **ii** $f(x)=\operatorname{arsinh}(x+2)$

b i $f(x)=\operatorname{arcosh}(-x)$ **ii** $f(x)=\operatorname{arcosh}3x$

c i $f(x)=\operatorname{artanh}4x$ **ii** $f(x)=\operatorname{artanh}(1-2x)$

2 Given that $f(x)=\operatorname{arsinh}(\cosh x)$, find $f'(x)$.

3 Given that $f(x)=x\operatorname{arcosh}(x^2)$, find $f'(x)$.

4 Given that $y=\operatorname{artanh}\left(\dfrac{2}{x}\right)$ for $x>2$, show that

$$\frac{dy}{dx}=\frac{a}{b-x^2}$$

where a and b are integers to be found.

5 Find the equation of the tangent to the curve $y=\operatorname{artanh}x$ at the point where $x=\dfrac{3}{5}$, giving your answer in the form $ax+by+c=0$.

6 The tangents at $x=0$ and $x=1$ to the curve $y=\operatorname{arsinh}x$ intersect at the point P. Show that the x-coordinate of P is $(2+\sqrt{2})\ln(1+\sqrt{2})-(1+\sqrt{2})$.

7 Find the coordinates of the point of inflection on the curve $y=\operatorname{arsinh}(x+1)$.

8 Show that $y=(\operatorname{arcosh}x)^2$ satisfies $(x^2-1)\dfrac{d^2y}{dx^2}+x\dfrac{dy}{dx}=2$.

9 Prove that the x-coordinate of the point on the curve $y=\operatorname{artanh}(e^x)$ at which the gradient is $\sqrt{2}$ is $x=a\ln2$, where a is a value to be found.

Section 3: Using inverse trigonometric and hyperbolic functions in integration

You can reverse the derivatives from Sections 1 and 2 to derive four more integration results:

$$\int\frac{1}{\sqrt{1-x^2}}\,dx=\sin^{-1}x+c$$

$$\int\frac{1}{1+x^2}\,dx=\tan^{-1}x+c$$

$$\int\frac{1}{\sqrt{1+x^2}}\,dx=\sinh^{-1}x+c$$

$$\int\frac{1}{\sqrt{x^2-1}}\,dx=\cosh^{-1}x+c$$

Tip

Notice that the results $\int\dfrac{-1}{\sqrt{1-x^2}}\,dx=\cos^{-1}x+c$ and $\int\dfrac{1}{1-x^2}\,dx=\tanh^{-1}x+c$ are not included in this list, as the first is just the negative of $\sin^{-1}x$ and the second can be done by partial fractions.

These results can be generalised slightly by making a linear substitution.

Key point 7.3

- $\int \frac{1}{\sqrt{a^2 - x^2}}\, dx = \sin^{-1}\left(\frac{x}{a}\right) + c, \quad |x| < a$

- $\int \frac{1}{a^2 + x^2}\, dx = \frac{1}{a}\tan^{-1}\left(\frac{x}{a}\right) + c$

- $\int \frac{1}{\sqrt{a^2 + x^2}}\, dx = \sinh^{-1}\left(\frac{x}{a}\right) + c$

- $\int \frac{1}{\sqrt{x^2 - a^2}}\, dx = \cosh^{-1}\left(\frac{x}{a}\right) + c, \quad x > a$

These will be given in your formula book.

Focus on ...

See Focus on ... Problem solving 2 for an example of using one of these integrals.

You also need to know how to derive these results using trigonometric or hyperbolic substitutions.

WORKED EXAMPLE 7.5

Use the substitution $x = a\sin u$ to prove the result $\int \frac{1}{\sqrt{a^2 - x^2}}\, dx = \sin^{-1}\left(\frac{x}{a}\right) + c$ when $|x| < a$.

$\frac{dx}{du} = a\cos u$

$\Rightarrow dx = a\cos u\, du$

Differentiate the substitution and express dx in terms of du.

$\frac{1}{\sqrt{a^2 - x^2}} = \frac{1}{\sqrt{a^2 - a^2\sin^2 u}}$

$= \frac{1}{\sqrt{a^2(1 - \sin^2 u)}}$

Express the integrand in terms of u.

$= \frac{1}{\sqrt{a^2\cos^2 u}}$

Use $\sin^2 u + \cos^2 u \equiv 1$.

$= \frac{1}{a\cos u}$

Since you are choosing the substitution, you can choose $a > 0$. For a given value of $\sin u$ there are two possible values of $\cos u$. You can choose the u that gives the positive value.

$\therefore \int \frac{1}{\sqrt{a^2 - x^2}}\, dx = \int \frac{1}{a\cos u} a\cos u\, du$

Make the substitution and integrate.

$= \int 1\, du = u + c$

$= \sin^{-1}\left(\frac{x}{a}\right) + c$

Write the answer in terms of x:

$x = a\sin u \Rightarrow u = \sin^{-1}\left(\frac{x}{a}\right)$.

You can derive the result $\int \frac{1}{a^2+x^2}\,dx = \frac{1}{a}\tan^{-1}\left(\frac{x}{a}\right)+c$ similarly, using the substitution $x = a\tan u$ and the identity $1+\tan^2 u \equiv \sec^2 u$.

Fast forward

You will be asked to derive this result in Question 8 in Exercise 7C.

WORKED EXAMPLE 7.6

Use the substitution $x = a\sinh u$ to prove the result $\int \frac{1}{\sqrt{a^2+x^2}}\,dx = \sinh^{-1}\left(\frac{x}{a}\right)+c$.

$\frac{dx}{du} = a\cosh u$

$\Rightarrow dx = a\cosh u\,du$

Differentiate the substitution and express dx in terms of du.

$\frac{1}{\sqrt{a^2+x^2}} = \frac{1}{\sqrt{a^2+a^2\sinh^2 u}}$

$= \frac{1}{\sqrt{a^2\left(1+\sinh^2 u\right)}}$

Express the integrand in terms of u.

$= \frac{1}{\sqrt{a^2\cosh^2 u}}$

Use $\cosh^2 u - \sinh^2 u \equiv 1$.

$= \frac{1}{a\cosh u}$

Since you are choosing the substitution, you can choose $a>0$.

$\therefore \int \frac{1}{\sqrt{a^2+x^2}}\,dx = \int \frac{1}{a\cosh u}a\cosh u\,du$

Make the substitution and integrate.

$= \int 1\,du = u+c$

$= \sinh^{-1}\left(\frac{x}{a}\right)+c$

Write the answer in terms of x:

$x = a\sinh u \Rightarrow u = \sinh^{-1}\left(\frac{x}{a}\right)$.

You can derive the result $\int \frac{1}{\sqrt{x^2-a^2}}\,dx = \cosh^{-1}\left(\frac{x}{a}\right)+c$ similarly, using the substitution $x = a\cosh u$ and the identity $\cosh^2 u - \sinh^2 u \equiv 1$.

You can combine these results with algebraic manipulation to integrate an even wider variety of functions.

Fast forward

You will be asked to derive this result in Question 9 in Exercise 7C.

WORKED EXAMPLE 7.7

Find $\int \dfrac{3}{9x^2+5}\,dx$.

$\int \dfrac{3}{9x^2+5}\,dx = \int \dfrac{3}{(3x)^2+(\sqrt{5})^2}\,dx$

$= \dfrac{1}{3}\dfrac{3}{\sqrt{5}}\tan^{-1}\left(\dfrac{x}{\sqrt{5}}\right)+c$

$= \dfrac{1}{\sqrt{5}}\tan^{-1}\left(\dfrac{x}{\sqrt{5}}\right)+c$

You can turn the integrand into the form $\dfrac{1}{u^2+a^2}$ (with $a=\sqrt{5}$) by making a substitution $u=3x$.

Since the substitution is linear, you can simply divide by the coefficient of x.

If the denominator is not in the form x^2+a^2 or $\sqrt{a^2-x^2}$, you might need to complete the square to write it in this form.

WORKED EXAMPLE 7.8

Find $\int \dfrac{1}{\sqrt{4x^2-12x-7}}\,dx$.

$4x^2-12x-7=(4x^2-12x)-7$

$=(2x-3)^2-9-7$

$=(2x-3)^2-16$

The expression in the denominator is quadratic, so you should check whether you can complete the square to write it in the form u^2-a^2.

Hence

$\int \dfrac{1}{\sqrt{4x^2-12x-7}}\,dx = \int \dfrac{1}{\sqrt{(2x-3)^2-16}}\,dx$

$= \dfrac{1}{2}\cosh^{-1}\left(\dfrac{2x-3}{4}\right)+c$

The integrand is of the form $\dfrac{1}{\sqrt{u^2-a^2}}$ with $u=2x-3$ and $a=4$. Remember to divide by the coefficient of x when integrating.

EXERCISE 7C

1 Find each indefinite integral.

a i $\int \dfrac{3}{x^2+4}\,dx$ ii $\int \dfrac{5}{x^2+36}\,dx$

b i $\int \dfrac{1}{9x^2+4}\,dx$ ii $\int \dfrac{4}{4x^2+25}\,dx$

c i $\int \dfrac{6}{2x^2+3}\,dx$ ii $\int \dfrac{10}{5x^2+2}\,dx$

d **i** $\int \dfrac{2}{\sqrt{9-x^2}}\,dx$ **ii** $\int \dfrac{5}{\sqrt{4-x^2}}\,dx$

e **i** $\int \dfrac{1}{\sqrt{9-4x^2}}\,dx$ **ii** $\int \dfrac{3}{\sqrt{25-9x^2}}\,dx$

f **i** $\int \dfrac{15}{\sqrt{5-3x^2}}\,dx$ **ii** $\int \dfrac{6}{\sqrt{7-12x^2}}\,dx$

2 Find each indefinite integral.

a **i** $\int \dfrac{3}{\sqrt{16+x^2}}\,dx$ **ii** $\int \dfrac{5}{\sqrt{25+x^2}}\,dx$

b **i** $\int \dfrac{10}{\sqrt{25+9x^2}}\,dx$ **ii** $\int \dfrac{3}{\sqrt{9+4x^2}}\,dx$

c **i** $\int \dfrac{4}{\sqrt{3+2x^2}}\,dx$ **ii** $\int \dfrac{6}{\sqrt{5+7x^2}}\,dx$

d **i** $\int \dfrac{2}{\sqrt{x^2-49}}\,dx$ **ii** $\int \dfrac{7}{\sqrt{x^2-36}}\,dx$

e **i** $\int \dfrac{1}{\sqrt{9x^2-16}}\,dx$ **ii** $\int \dfrac{15}{\sqrt{25x^2-36}}\,dx$

f **i** $\int \dfrac{5}{\sqrt{3x^2-7}}\,dx$ **ii** $\int \dfrac{2}{\sqrt{7x^2-11}}\,dx$

3 By first completing the square, find:

a **i** $\int \dfrac{1}{x^2+4x+5}\,dx$ **ii** $\int \dfrac{1}{x^2-6x+10}\,dx$

b **i** $\int \dfrac{1}{\sqrt{8x-x^2-15}}\,dx$ **ii** $\int \dfrac{1}{\sqrt{2x-x^2}}\,dx$

c **i** $\int \dfrac{6}{x^2+10x+27}\,dx$ **ii** $\int \dfrac{5}{\sqrt{-4x^2-12x}}\,dx$

d **i** $\int \dfrac{1}{\sqrt{x^2+6x+10}}\,dx$ **ii** $\int \dfrac{1}{\sqrt{x^2+4x+5}}\,dx$

e **i** $\int \dfrac{1}{\sqrt{x^2-4x-12}}\,dx$ **ii** $\int \dfrac{1}{\sqrt{x^2-2x}}\,dx$

f **i** $\int \dfrac{6}{\sqrt{4x^2-12x+4}}\,dx$ **ii** $\int \dfrac{3}{\sqrt{x^2+2x+5}}\,dx$

4 Find the exact value of $\displaystyle\int_0^4 \dfrac{1}{\sqrt{x^2+16}}\,dx$.

5 Find the exact value of $\displaystyle\int_0^{\frac{\sqrt{3}}{2}} \dfrac{3}{1+4x^2}\,dx$.

6 Find the exact value of $\displaystyle\int_{\frac{1}{\sqrt{3}}}^1 \dfrac{1}{\sqrt{4-3x^2}}\,dx$.

7 Find:

a $\int \dfrac{1}{1+9x^2}\,dx$ **b** $\int \dfrac{16}{16+x^2}\,dx$

8 a Use a trigonometric substitution to prove that $\int \dfrac{1}{a^2+x^2}\,dx = \dfrac{1}{a}\tan^{-1}\left(\dfrac{x}{a}\right)+c.$

   b Hence evaluate $\displaystyle\int_0^2 \dfrac{5}{4+x^2}\,dx.$

9 a Use a hyperbolic substitution to prove that $\int \dfrac{1}{\sqrt{x^2-a^2}}\,dx = \cosh^{-1}\left(\dfrac{x}{a}\right)+c, \ \ x>a.$

   b Hence evaluate $\displaystyle\int_3^6 \dfrac{2}{\sqrt{x^2-9}}\,dx$, giving your answer in terms of a natural logarithm.

10 a Write $2x^2+4x+11$ in the form $2(x+p)^2+q.$

    b Hence find $\int \dfrac{3}{2x^2+4x+11}\,dx.$

11 a Write $1+6x-3x^2$ in the form $a^2-3(x-b)^2.$

    b Hence find the exact value of $\displaystyle\int_1^2 \dfrac{1}{\sqrt{1+6x-3x^2}}\,dx.$

12 a Using a suitable substitution prove that, when $|x|<a,$

    $\int \dfrac{1}{\sqrt{a^2-x^2}}\,dx = \sin^{-1}\left(\dfrac{x}{a}\right)+c.$

    b Find $\int \dfrac{3}{\sqrt{-4x^2-4x+8}}\,dx.$

13 Use a suitable trigonometric substitution to show that

    $\displaystyle\int_{\frac{2}{5}}^{\frac{2\sqrt{3}}{5}} \dfrac{20}{25x^2+4}\,dx = \dfrac{\pi}{6}.$

14 Find $\int \dfrac{1}{x^2+2x+2}\,dx.$

15 Show that $\displaystyle\int_3^{5.5} \dfrac{10}{4x^2-24x+61}\,dx = \dfrac{\pi}{4}.$

16 Find $\int \dfrac{4x+5}{\sqrt{1-x^2}}\,dx.$

17 Find $\int \dfrac{x+1}{\sqrt{x^2-1}}\,dx.$

18 Find $\int \dfrac{6x-5}{x^2+9}\,dx.$

19 a Write $2x^2-8x+17$ in the form $a(x-p)^2+q.$

    b Hence find $\int \dfrac{2x+8}{2x^2-8x+17}\,dx.$

20 Use a suitable hyperbolic substitution to show that

    $\int \sqrt{x^2-9}\,dx = \dfrac{x}{2}\sqrt{x^2-9} - \dfrac{9}{2}\cosh^{-1}\left(\dfrac{x}{3}\right)+c.$

21 Use a suitable trigonometric substitution to show that

    $\int \sqrt{4-9x^2}\,dx = \dfrac{x}{2}\sqrt{4-9x^2} + \dfrac{2}{3}\sin^{-1}\left(\dfrac{3x}{2}\right)+c.$

22 a Given that $\tan u = x$ express $\cos u$ and $\sin u$ in terms of x.

b Use a suitable trigonometric substitution to show that
$$\int \frac{1}{1+2x^2+x^4}\,dx = \frac{x}{2(1+x^2)} + \frac{1}{2}\arctan x + c.$$

Section 4: Using partial fractions in integration

You have already used partial fractions to integrate rational expressions with linear and repeated linear factors in the denominator, such as $\int \frac{2x+1}{(x-1)(x+2)^2}\,dx$. You can now use the results from Section 3 to extend the range of rational functions you can integrate to include those with denominators of the form (x^2+q^2).

Rewind

You met partial fractions in A Level Mathematics Student Book 2, Chapter 5, and then used them in integration in Chapter 11.

In general, when there is a quadratic factor in the denominator, there are three possibilities:

- The quadratic factorises into two different linear factors, $(x-p)(x-q)$. The corresponding partial fractions are $\frac{A}{x-p} + \frac{B}{x-q}$.

- The quadratic is a perfect square, $(x-p)^2$. The corresponding partial fractions are $\frac{A}{x-p} + \frac{B}{(x-p)^2}$.

- The quadratic does not factorise (the quadratic factor is **irreducible**). For example, (x^2+1) or (x^2+2x+5). Then there is only one corresponding partial fraction, with a numerator of the form $Bx+C$.

Key point 7.4

If $f(x)$ is a polynomial of order less than or equal to 2, then
$$\frac{f(x)}{(x-p)(x^2+q^2)} = \frac{A}{x-p} + \frac{Bx+C}{x^2+q^2}$$

WORKED EXAMPLE 7.9

a Express $\frac{3x}{(x-1)(x^2+2)}$ in partial fractions.

b Hence find $\int \frac{3x}{(x-1)(x^2+2)}\,dx$.

a $\frac{3x}{(x-1)(x^2+2)} = \frac{A}{x-1} + \frac{Bx+C}{x^2+2}$ ⟶ Use the form from Key point 7.4.

$3x = A(x^2+2) + (Bx+C)(x-1)$ ⟶ Multiply through by the common denominator.

Continues on next page ...

$x = 1 : 3 = A(1+2) + 0 \Rightarrow A = 1$

$x = 0 : 0 = 1(0+2) + (C)(0-1) \Rightarrow C = 2$

Comparing coefficients of x^2:

$0 = 1 + B$

$B = -1$

Hence

$$\frac{3x}{(x-1)(x^2+2)} = \frac{1}{x-1} + \frac{-x+2}{x^2+2}$$

> Substitute in the values of x which make some of the terms zero.

> Look at the coefficient of x^2 to find B.

b $\displaystyle\int \frac{3x}{(x-1)(x^2+2)}\,dx = \int \left(\frac{1}{x-1} + \frac{-x}{x^2+2} + \frac{2}{x^2+2} \right) dx$

$$\int \frac{1}{x-1}\,dx = \ln|x-1|$$

$$\int -\frac{x}{x^2+2}\,dx = -\frac{1}{2}\ln|x^2+2|$$

$$\int \frac{2}{x^2+2}\,dx = \frac{2}{\sqrt{2}}\arctan\left(\frac{x}{\sqrt{2}} \right)$$

Hence

$$\int \frac{3x}{(x-1)(x^2+2)}\,dx = \ln|x-1| - \frac{1}{2}\ln|x^2+2| + \sqrt{2}\arctan\left(\frac{x}{\sqrt{2}} \right) + c$$

> Integrate each term separately before applying limits. You need to split the second integral into two in order to apply standard results.

> Here you can use a substitution $u = x^2 + 2$, or the reverse chain rule, as x is half the derivative of $x^2 + 2$.

> Use $\displaystyle\int \frac{1}{x^2+a^2}\,dx = \frac{1}{a}\arctan\left(\frac{x}{a} \right)$ with $a = \sqrt{2}$.

If there is a quadratic factor in the denominator, you first need to check whether it is irreducible or whether it can be factorised. If a quadratic factor is irreducible, you need to write it in completed square form before you can apply standard integration results.

WORKED EXAMPLE 7.10

Given that $\dfrac{dy}{dx} = \dfrac{16x+36}{(x^2-4)(x^2+4x+5)}$, find an expression for y in terms of x.

The denominator is

$(x-2)(x+2)(x^2+4x+5)$.

Hence

$$\frac{16x+36}{(x^2-4)(x^2+4x+5)} = \frac{A}{x-2} + \frac{B}{x+2} + \frac{(Cx+D)}{x^2+4x+5}$$

> You need to split the function into partial fractions before integrating.

> Check whether the quadratic factors factorise. The second one has the discriminant $4^2 - 20 < 0$ so it is irreducible.

Continues on next page ...

$$16x+36=A(x+2)(x^2+4x+5)+B(x-2)(x^2+4x+5)$$
$$+(Cx+D)(x-2)(x+2)$$

Multiply through by the denominator.

$$x=2:68=A(4)(17)\Rightarrow A=1$$

Substitute in suitable values of x.

$$x=-2:4=B(-4)(1)\Rightarrow B=-1$$

$$x=0:36=1(2)(5)-1(-2)(5)+D(-2)(2)\Rightarrow D=-4$$

$$x=1:52=1(3)(10)-1(-1)(10)+(C-4)(-1)(3)\Rightarrow C=0$$

$$y=\int\left(\frac{1}{x-2}-\frac{1}{x+2}-\frac{4}{x^2+4x+5}\right)dx$$

$$\int\frac{4}{(x+2)^2+1}\,dx=4\arctan(x+2)$$

For the third integral, you need to complete the square and then use $\int\frac{1}{x^2+a^2}\,dx=\arctan\left(\frac{x}{a}\right).$

$$\therefore y=\ln\left|\frac{x-2}{x+2}\right|-4\arctan(x+2)+c$$

EXERCISE 7D

1 Use partial fractions to find each integral.

a i $\displaystyle\int\frac{2x^2+x+7}{(x^2+2)(x+3)}\,dx$ **ii** $\displaystyle\int\frac{-x^2+2x-5}{(x^2+1)(x-2)}\,dx$

b i $\displaystyle\int\frac{2x^2+13x+21}{(x+1)(x^2+6x+10)}\,dx$ **ii** $\displaystyle\int\frac{x^2-2x+13}{(x-2)(x^2+2x+5)}\,dx$

c i $\displaystyle\int\frac{-x^2+3x-2}{(x^2+1)(x+1)}\,dx$ **ii** $\displaystyle\int\frac{-3x-2}{(x^2+4)(x-1)}\,dx$

d i $\displaystyle\int\frac{x^3+2x^2+x+8}{(x+1)^2(x^2+3)}\,dx$ **ii** $\displaystyle\int\frac{x^3+x^2-7x+7}{(x-2)^2(x^2+1)}\,dx$

2 Use partial fractions to find the exact value of $\displaystyle\int_0^1\frac{-2x^2+x-1}{(x+1)(x^2+1)}\,dx$

3 a Write $\dfrac{x^2-x+11}{(x-2)(x^2+9)}$ in partial fractions.

b Given that $\dfrac{dy}{dx}=\dfrac{x^2-x+11}{(x-2)(x^2+9)}$, and that $y=0$ when $x=0$, find y in terms of x.

4 Use partial fractions to integrate:

a $\dfrac{3x^2+x-5}{(x-2)(x^2+2x+1)}$ **b** $\dfrac{x^2+4x-2}{(x-2)(x^2+2x+2)}.$

5 Let $f(x) = \dfrac{2x^3 + x^2 + 8x - 4}{(x^2 - 4)(x^2 + 4)}$.

     a Write $f(x)$ in partial fractions.   **b** Hence find the exact value of $\displaystyle\int_0^{2\sqrt{3}} f(x)\,dx$.

6 Let $f(x) = \dfrac{x^3 + 4x^2 + 3x + 4}{(x+1)^2(x^2+1)}$. Use partial fractions to evaluate $\displaystyle\int_0^1 f(x)\,dx$.

7 Use partial fractions to find $\displaystyle\int \dfrac{2x^2 + 4x + 18}{(x^2 + 2x - 3)(x^2 + 2x + 3)}\,dx$.

8 Show that

$$\int \frac{2x^2 + 3x - 3}{(x^2 + 2x + 5)(x+1)}\,dx = P\ln(x^2 + 2x + 5) + Q\arctan\left(\frac{x+1}{2}\right) + R\ln|x+1| + c$$

where P, Q and R are constants to be found.

✎ Checklist of learning and understanding

- Derivatives of inverse trigonometric functions:

  - $\dfrac{d}{dx}(\sin^{-1} x) = \dfrac{1}{\sqrt{1-x^2}}$, $|x| < 1$

  - $\dfrac{d}{dx}(\cos^{-1} x) = \dfrac{-1}{\sqrt{1-x^2}}$, $|x| < 1$

  - $\dfrac{d}{dx}(\tan^{-1} x) = \dfrac{1}{1+x^2}$

- Derivatives of inverse hyperbolic functions:

  - $\dfrac{d}{dx}(\sinh^{-1} x) = \dfrac{1}{\sqrt{x^2+1}}$

  - $\dfrac{d}{dx}(\cosh^{-1} x) = \dfrac{1}{\sqrt{x^2-1}}$, $x > 1$

  - $\dfrac{d}{dx}(\tanh^{-1} x) = \dfrac{1}{1-x^2}$, $|x| < 1$

- You can derive the corresponding integrals using a trigonometric substitution ($x = a\sin u$ or $x = a\tan u$) or a hyperbolic substitution ($x = a\sinh u$ or $x = a\cosh u$):

  - $\displaystyle\int \dfrac{1}{\sqrt{a^2 - x^2}}\,dx = \sin^{-1}\left(\dfrac{x}{a}\right) + c$, $|x| < a$

  - $\displaystyle\int \dfrac{1}{a^2 + x^2}\,dx = \dfrac{1}{a}\tan^{-1}\left(\dfrac{x}{a}\right) + c$

  - $\displaystyle\int \dfrac{1}{\sqrt{a^2 + x^2}}\,dx = \sinh^{-1}\left(\dfrac{x}{a}\right) + c$

  - $\displaystyle\int \dfrac{1}{\sqrt{x^2 - a^2}}\,dx = \cosh^{-1}\left(\dfrac{x}{a}\right) + c$, $x > a$

- You might need to write a quadratic expression in completed square form in order to apply one of the results shown.

- When splitting an expression into partial fractions, if the denominator has an irreducible quadratic factor $x^2 + px + q$, then the corresponding partial fraction is $\dfrac{Ax + B}{x^2 + px + q}$.

Mixed practice 7

1. Differentiate $f(x) = \arctan(e^x)$.

2. Find $\dfrac{dy}{dx}$ when $y = x^2 \sin^{-1} x$.

3. Differentiate $f(x) = \cos^{-1}(1 - x^2)$.

4. Find the x-coordinates of the points on the curve $y = \tanh^{-1}\left(\dfrac{x}{2}\right)$ where the gradient is 2.

5. By first completing the square, find the exact value of $\displaystyle\int_{\frac{1}{2}}^{1} \dfrac{1}{\sqrt{2x - x^2}}\, dx$.

© OCR A Level Mathematics, Unit 4726/01 Further Pure Mathematics 2, June 2015

6. It is given that $f(x) = \dfrac{x^2 + 9x}{(x - 1)(x^2 + 9)}$.

   i Express $f(x)$ in partial fractions.

   ii Hence find $\int f(x)\, dx$.

© OCR A Level Mathematics, Unit 4726/01 Further Pure Mathematics 2, June 2007

7. Given that $y = \tan^{-1}(x^2)$, find $\dfrac{d^2 y}{dx^2}$.

8. Show that $y = (\operatorname{arsinh} x)^2$ satisfies $(1 + x^2)\dfrac{d^2 y}{dx^2} + x\dfrac{dy}{dx} - 2 = 0$.

9. Find $\displaystyle\int \dfrac{6x + 4}{x^2 + 4}\, dx$.

10. Find $\displaystyle\int \dfrac{1}{2 - 2x + x^2}\, dx$.

In questions 11 and 12 you must show detailed reasoning.

11. Given that
$$\int_{0}^{1} \dfrac{1}{\sqrt{16 + 9x^2}}\, dx + \int_{0}^{2} \dfrac{1}{\sqrt{9 + 4x^2}}\, dx = \ln a,$$
find the exact value of a.

© OCR A Level Mathematics, Unit 4726 Further Pure Mathematics 2, June 2009

> **Tip**
>
> Remember that 'show detailed reasoning' means that you need to show full algebraic working, rather than using your calculator to evaluate indefinite integrals.

12. i Given that
$$y = x\sqrt{1 - x^2} - \cos^{-1} x,$$
find $\dfrac{dy}{dx}$ in a simplified form.

   ii Hence, or otherwise, find the exact value of $\displaystyle\int_{0}^{1} 2\sqrt{1 - x^2}\, dx$.

© OCR A Level Mathematics, Unit 4726/01 Further Pure Mathematics 2, June 2007

13. i Express $\dfrac{4}{(1 - x)(1 + x)(1 + x^2)}$ in partial fractions.

   ii Show that $\displaystyle\int_{0}^{\frac{1}{\sqrt{3}}} \dfrac{4}{1 - x^4}\, dx = \ln\left(\dfrac{\sqrt{3} + 1}{\sqrt{3} - 1}\right) + \dfrac{1}{3}\pi$.

© OCR A Level Mathematics, Unit 4726/01 Further Pure Mathematics 2, January 2010

14 a Show that $\sqrt{\dfrac{1-3x}{1+3x}} = \dfrac{1-3x}{\sqrt{1-9x^2}}$.

b Hence find $\displaystyle\int \sqrt{\dfrac{1-3x}{1+3x}}\,\mathrm{d}x$.

15 Use the substitution $x = 9\sinh^2\theta$ to show that

$$\int_0^1 \sqrt{\dfrac{x+9}{x}}\,\mathrm{d}x = 9\sinh^{-1}\left(\dfrac{1}{3}\right) + \sqrt{A}$$

and state the value of the integer A.

16 a Split $\dfrac{4-3x}{(x+2)(x^2+1)}$ into partial fractions.

b Hence find $\displaystyle\int \dfrac{4-3x}{(x+2)(x^2+1)}\,\mathrm{d}x$.

c Find the exact value of $\displaystyle\int_0^{\frac{\sqrt{3}}{2}} \dfrac{4-3x}{\sqrt{1-x^2}}\,\mathrm{d}x$.

17 a Show that $\displaystyle\int \sqrt{k^2-x^2}\,\mathrm{d}x = \dfrac{k^2}{2}\arcsin\left(\dfrac{x}{k}\right) + \dfrac{x}{2}\sqrt{k^2-x^2} + c$.

b Hence show that the area enclosed by the ellipse with equation $\dfrac{x^2}{a^2} + \dfrac{y^2}{b^2} = 1$ is πab.

18 i Prove that $\dfrac{\mathrm{d}}{\mathrm{d}x}\left(\cosh^{-1}x\right) = \dfrac{1}{\sqrt{x^2-1}}$.

ii Hence, or otherwise, find $\displaystyle\int \dfrac{1}{\sqrt{4x^2-1}}\,\mathrm{d}x$.

iii By means of a suitable substitution, find $\displaystyle\int \sqrt{4x^2-1}\,\mathrm{d}x$.

© OCR A Level Mathematics, Unit 4726/01 Further Pure Mathematics 2, January 2008

8 Applications of calculus

In this chapter you will learn how to:

- find infinite series expansions (called Maclaurin series) of functions
- use given results to find the Maclaurin series of more complicated functions
- understand for which values of x these series are valid
- find the value of definite integrals in certain cases where a limiting process is required (improper integrals)
- find the volume of a shape formed by rotating a curve around the x-axis or the y-axis
- find the mean value of a function.

Before you start…

A Level Mathematics Student Book 2, Chapter 10	You should know how to differentiate functions using the chain rule.	1 For $f(x) = (x^2 + 3)^5$, find $f'(x)$.

A Level Mathematics Student Book 2, Chapter 9	You should know how to differentiate exponential, logarithmic and trigonometric functions.	2 For each function, find:
		   i $f'(x)$ **ii** $f''(x)$.
		   a $f(x) = e^{-2x}$
		   b $f(x) = \ln x$
		   c $f(x) = \cos 3x$

Chapter 6	You should know how to differentiate hyperbolic functions.	3 For $f(x) = \tanh x$, find: **a** $f'(x)$ **b** $f''(x)$.
Chapter 6	You should know how to differentiate inverse hyperbolic functions.	4 For $f(x) = \cosh^{-1} x$, find: **a** $f'(x)$ **b** $f''(x)$.
Chapter 7	You should know how to differentiate inverse trigonometric functions.	5 For $f(x) = \sin^{-1} x$, find: **a** $f'(x)$ **b** $f''(x)$.

Introduction

In the first part of this chapter you will see that, with certain restrictions, you can write many functions as infinite series in ascending positive integer powers of x. Being able to take a number of terms of the infinite series as an approximation to a given function has many uses; for example, calculators and computers use these series to evaluate a function at a particular value or to produce an approximation for definite integrals of functions that can't be integrated using standard functions.

You will also look at definite integrals where either the integrand is not defined at one or more points in the range of integration or where the range of integration extends to infinity. These are known as improper integrals.

Finally, you will see two further applications of calculus: finding volumes and finding the mean value of a function.

Section 1: Maclaurin series

You know from A Level Mathematics Student Book 2, Chapter 6, that functions such as $(1-3x)^{-2}$ and $\sqrt{1+x}$ can be written as infinite series using the binomial expansion. In general, for some function f(x) such a series will be of the form

$f(x) = a_0 + a_1x + a_2x^2 + a_3x^3 + \ldots$ where a_0, a_1, a_2, etc. are real constants.

Differentiating this series several times:

$$f'(x) = a_1 + 2a_2x + 3a_3x^2 + 4a_4x^3 + \ldots$$
$$f''(x) = 2a_2 + (3 \times 2)a_3x + (4 \times 3)a_4x^2 + \ldots$$
$$f'''(x) = (3 \times 2)a_3 + (4 \times 3 \times 2)a_4x + \ldots$$
$$\vdots$$

Substituting $x = 0$ into f(x) and each of its derivatives to find expressions for a_0, a_1, a_2, etc:

$$f(0) = a_0$$
$$f'(0) = a_1$$
$$f''(0) = 2!a_2$$
$$f'''(0) = 3!a_3$$
$$\vdots$$

Substituting these expressions for a_0, a_1, a_2, etc. back into the expression for f(x) gives the **Maclaurin series** formula for any function.

ⓘ Did you know?

Maclaurin series are named after the 18th-century mathematician Colin Maclaurin, who also developed some of Newton's work on calculus, algebra and gravitation theory.

⏭ Fast forward

You will meet the interval of convergence for some standard functions in Section 2.

You are assuming here that you can differentiate an infinite series term by term in the same way as a finite series. In fact, this is only possible for values of x within the interval of convergence of the series.

You can use Key point 8.1 to find the first few terms of a Maclaurin series. You can sometimes spot a pattern in the derivatives, which enables you to also write down the **general term**, $\dfrac{f^{(r)}(0)}{r!}x^r$.

> ### 🔑 Key point 8.1
>
> The Maclaurin series of a function $f(x)$ is given by
>
> $$f(x) = f(0) + f'(0)x + \frac{f''(0)}{2!}x^2 + \ldots + \frac{f^{(r)}(0)}{r!}x^r + \ldots$$

WORKED EXAMPLE 8.1

a Find the first three non-zero terms in the Maclaurin series of $\sin x$.

b Write down a conjecture for the general term (you need not prove your conjecture).

a $f(x) = \sin x \Rightarrow f(0) = 0$ | Find $f(0)$.

$f'(x) = \cos x \Rightarrow f'(0) = 1$ | Then differentiate and evaluate each derivative at $x = 0$.

$f''(x) = -\sin x \Rightarrow f''(0) = 0$ | Notice that you need to go as far as the fifth derivative to get three non-zero terms.

$f'''(x) = -\cos x \Rightarrow f'''(0) = -1$

$f^{(4)}(x) = \sin x \Rightarrow f^{(4)}(0) = 0$

$f^{(5)}(x) = \cos x \Rightarrow f^{(5)}(0) = 1$

So

$f(x) = 0 + 1x + \dfrac{0}{2!}x^2 + \dfrac{-1}{3!}x^3 + \dfrac{0}{4!}x^4 + \dfrac{1}{5!}x^5 \ldots$ | Substitute these values into the Maclaurin series formula:

$= x - \dfrac{x^3}{3!} + \dfrac{x^5}{5!} + \ldots$ | $f(x) = f(0) + f'(0)x + \dfrac{f''(0)}{2!}x^2 + \ldots$

b The general term is

$\dfrac{(-1)^n \cos(0)}{(2n+1)!}x^{2n+1} \ldots$ | The series will only contain terms with odd powers because even derivatives are $\pm \sin x$, and $\sin(0) = 0$.

$= \dfrac{(-1)^n}{(2n+1)!}x^{2n+1} \ldots$ | Because of the pattern of differentiating $\sin x$, the signs will alternate between $+$ and $-$.

Not every function has a Maclaurin series. For example, for $f(x) = \ln x$, $f(0)$ doesn't exist (nor do any of the derivatives of $\ln x$ at $x = 0$). However, $f(x) = \ln(1 + x)$ does have a Maclaurin series as now $f(0)$, and all the derivatives at $x = 0$, do exist.

WORKED EXAMPLE 8.2

a Find the Maclaurin series of $\ln(1+x)$ up to and including the term in x^4.

b Prove by induction that the general term is $\dfrac{(-1)^{n-1}}{n}x^n$.

a $f(x)=\ln(1+x)\Rightarrow f(0)=0$ | Find f(0).

$f'(x)=(1+x)^{-1}\Rightarrow f'(0)=1$

$f''(x)=-(1+x)^{-2}\Rightarrow f''(0)=-1$ | Then differentiate and evaluate each derivative at $x=0$.

$f'''(x)=2(1+x)^{-3}\Rightarrow f'''(0)=2$

$f^{(4)}(x)=-6(1+x)^{-4}\Rightarrow f^{(4)}(0)=-6$

So

$f(x)=0+1x-\dfrac{1}{2!}x^2+\dfrac{2}{3!}x^3-\dfrac{6}{4!}x^4+...$ | Substitute these values into the Maclaurin series formula:

$=x-\dfrac{x^2}{2}+\dfrac{x^3}{3}-\dfrac{x^4}{4}+...$

$$f(x)=f(0)+f'(0)x+\frac{f''(0)}{2!}x^2+...$$

b To prove:

$f^{(n)}(x)=(-1)^{n-1}(n-1)!(1+x)^{-n}$ | To find the general term you need an expression for the nth derivative, which you then divide by $n!$ to get the coefficient of x_n. You can conjecture the expression by looking at the first four derivatives you found in part **a**.

When $n=1$:

$f'(x)=(1+x)^{-1}$

$=(-1)^0(0!)(1+x)^{-1}$ | Show that the expression is correct when $n=1$.

Hence the expression is correct for $n=1$.

Suppose that the expression is correct

for the kth derivative: | Now suppose that the expression is correct for some k and show that it is still correct for $k+1$.

$f^{(k)}(x)=(-1)^{k-1}(k-1)!(1+x)^{-k}$

Then the $(k+1)$th derivative is: | Differentiate $f^{(k)}$ to get $f^{(k+1)}$.

$f^{(k+1)}(x)=(-1)^{k-1}(k-1)!(-k)(1+x)^{-(k+1)}$

$=(-1)^k k!(1+x)^{-(k+1)}$ | The negative sign in $-k$ changes the sign of the whole expression.

Hence the expression is correct for the

$(k+1)$th derivative. | $(k-1)!\,k=k!$

The expression is correct for $n=1$ and, if it

is correct for $n=k$, then it is also correct for | Write a conclusion.

$n=k+1$. It is therefore true for all $n\geqslant 1$,

by the principle of mathematical induction.

One use of Maclaurin series is to approximate definite integrals of functions that can't be integrated by standard methods.

Fast forward

In Section 2 you will see how to use the Maclaurin series of certain standard functions to find the series of more complicated functions.

Tip

The question will make it clear whether you are required to find the Maclaurin series of a function from first principles or whether you can use one of the standard results in the formula book to find the series you need.

WORKED EXAMPLE 8.3

a Use the Maclaurin series of $\sin x$ to find the first three non-zero terms in the Maclaurin series of $\sin(x^2)$.

b Hence find an approximate value for $\int_0^{\frac{\pi}{3}} \sin(x^2)\,dx$, giving your answer to three decimal places.

a $\sin(x^2) = (x^2) - \dfrac{(x^2)^3}{3!} + \dfrac{(x^2)^5}{5!} - \cdots$

Substitute x^2 into the series for $\sin x$.

$= x^2 - \dfrac{x^6}{3!} + \dfrac{x^{10}}{5!} - \cdots$

b $\int_0^{\frac{\pi}{3}} \sin(x^2)\,dx \approx \int_0^{\frac{\pi}{3}} \left(x^2 - \dfrac{x^6}{3!} + \dfrac{x^{10}}{5!}\right) dx$

$= \left[\dfrac{x^3}{3} - \dfrac{x^7}{7\times3!} + \dfrac{x^{11}}{11\times5!}\right]_0^{\frac{\pi}{3}}$

Integrate the polynomial as usual.

≈ 0.351

EXERCISE 8A

1 Using Key point 8.1, find the first four non-zero terms of the Maclaurin series of these functions. Also conjecture the general term for each series (you need not prove your conjecture).

a i e^x **ii** e^{-3x}

b i $\sin(-x)$ **ii** $\sin 2x$

c i $\cos x$ **ii** $\cos 3x$

d i $\ln(1-x)$ **ii** $\ln(1+2x)$

e i $\sinh x$ **ii** $\sinh 2x$

f i $\cosh x$ **ii** $\cosh(-x)$

2 Find the Maclaurin series of $f(x) = \sqrt{3+e^x}$ up to and including the term in x^2.

3 a Show that the first two non-zero terms in the Maclaurin series of $\tan x$ are $x + \dfrac{1}{3}x^3$.

 b Hence find, to three decimal places, an approximation to $\tan\dfrac{\pi}{10}$.

4 It is given that $f(x) = e^{-x^2}$.

 a i Find the first four derivatives of $f(x)$.

  ii Hence find the Maclaurin series of $f(x)$, up to and including the term in x^4.

b Use your result from part **a ii** to find an approximation to $\int_0^1 e^{-x^2}\,dx$.

Give your answer in the form $\dfrac{a}{b}$ where a and b are integers.

5 **a** Find the Maclaurin series of $\cos^2 x$ up to and including the term in x^6.

b Hence state the Maclaurin series of $\sin^2 x$ up to and including the term in x^6.

6 It is given that $f(x) = \ln(1 + \sin x)$, $-\dfrac{\pi}{2} < x < \dfrac{3\pi}{2}$.

a **i** Show that $f'(x) = \dfrac{\cos x}{1 + \sin x}$ and $f''(x) = -\dfrac{1}{1 + \sin x}$

ii Find the third and fourth derivatives of f.

b Hence find the Maclaurin series of $\ln(1 + \sin x)$ up to and including the term in x^4.

c Use your series from part **b** to find an approximation to $\int_0^{\frac{\pi}{6}} \ln(1 + \sin x)\,dx$, giving your answer to three decimal places.

7 The function f is defined by $f(x) = \ln\left(\dfrac{1}{1-x}\right)$, $-1 < x < 1$.

a **i** Find the first three derivatives of $f(x)$.

ii Hence show that the Maclaurin series for $f(x)$ up to and including the x^3 term is $x + \dfrac{x^2}{2} + \dfrac{x^3}{3}$.

b Use the series from part **a ii** to find an approximate value for $\ln 3$. Give your answer in the form $\dfrac{a}{b}$, where a and b are integers.

8 **a** Given that $f(x) = x\,e^x$, use induction to show that $f^{(n)}(x) = (n + x)e^x$.

b Hence find the general term in the Maclaurin series for $f(x)$.

9 **a** Given that $f(x) = \dfrac{1}{1 - 5x}$, prove by induction that $f^{(n)}(x) = \dfrac{5^n n!}{(1 - 5x)^{n+1}}$.

b Hence find the general term in the Maclaurin series for $\dfrac{1}{1 - 5x}$.

10 **a** Find the first three non-zero terms in the Maclaurin series for $\arcsin x$.

b **i** Let $f(x) = \arcsin x$ and $g(x) = \arccos x$. State the relationship between $f^{(n)}(x)$, the nth derivative of $f(x)$, and $g^{(n)}(x)$, the nth derivative of $g(x)$, for any integer $n > 0$.

ii Hence show that $\arcsin x + \arccos x = k$, where k is a constant to be determined.

11 The diagram shows part of the graph of $y = f(x)$.

Explain why neither of these can be Maclaurin series of the function $f(x)$:

a $\dfrac{1}{2} + \dfrac{x}{2} + \dfrac{x^2}{8} + \dots$

b $1 - 3x - \dfrac{x^2}{4} + \dots$

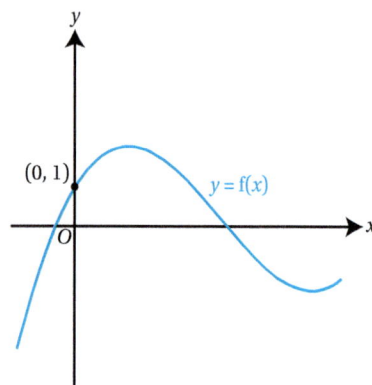

Section 2: Using standard Maclaurin series

The Maclaurin series of a few standard functions are given in your formula book. These can often be used, without needing to be derived, to find the series for more complicated functions.

> ### Key point 8.2
>
> Maclaurin series for some common functions and the values of x for which they are valid:
>
> - $e^x = \exp(x) = 1 + x + \dfrac{x^2}{2!} + \ldots + \dfrac{x^r}{r!} + \ldots$ for all x
>
> - $\ln(1+x) = x - \dfrac{x^2}{2} + \dfrac{x^3}{3} - \ldots + (-1)^{r+1}\dfrac{x^r}{r} + \ldots \; -1 < x \leqslant 1$
>
> - $\sin x = x - \dfrac{x^3}{3!} + \dfrac{x^5}{5!} - \ldots + (-1)^r \dfrac{x^{2r+1}}{(2r+1)!} + \ldots$ for all x
>
> - $\cos x = 1 - \dfrac{x^2}{2!} + \dfrac{x^4}{4!} - \ldots + (-1)^r \dfrac{x^{2r}}{(2r)!} + \ldots$ for all x
>
> - $(1+x)^n = 1 + nx + \dfrac{n(n-1)}{2!}x^2 + \ldots + \dfrac{n(n-1)\ldots(n-r+1)}{r!}x^r + \ldots \; |x|<1, n \in \mathbb{R}$
>
> **These will be given in your formula book.**

> **Tip**
>
> Don't overlook the information on the values of x for which these series are valid; this is a very important part of each result.

> **Rewind**
>
> Note that the last result in Key point 8.2 is the binomial expansion, which is covered in A Level Mathematics Student Book 2, Chapter 6.
>
> Note also that replacing x by ix in the series for e^x gives the same series as adding $\cos x + i \sin x$, which agrees with Euler's formula from Chapter 2, Section 2.

WORKED EXAMPLE 8.4

a Use the Maclaurin series for $\cos x$ to find the first four terms in the series for $\cos(2x^3)$.
b State the values of x for which the series is valid.

a $\cos(2x^3) = 1 - \dfrac{(2x^3)^2}{2!} + \dfrac{(2x^3)^4}{4!} - \dfrac{(2x^3)^6}{6!} + \ldots$

Substitute $2x^3$ into the series for $\cos x$:
$\cos x = 1 - \dfrac{x^2}{2!} + \dfrac{x^4}{4!} - \dfrac{x^6}{6!} + \cdots$

$= 1 - \dfrac{4x^6}{2!} + \dfrac{16x^{12}}{4!} - \dfrac{64x^{18}}{6!} + \ldots$

Expand and simplify.

$= 1 - 2x^6 + \dfrac{2}{3}x^{12} - \dfrac{4}{45}x^{18} + \ldots$

b Valid for all $x \in \mathbb{R}$.

Both $2x^3$ and $\cos x$ are valid for all $x \in \mathbb{R}$.

This process can be more complicated if it involves finding two separate Maclaurin series and then combining them.

WORKED EXAMPLE 8.5

a Use the Maclaurin series for $\sin x$ and that for e^x to find the series for $e^{\sin x}$ as far as the term in x^4.

b State the values of x for which your series is valid.

a $e^{\sin x} = e^{x - \frac{x^3}{3!} + \cdots}$

Start by replacing $\sin x$ with its series.

$\approx e^x \times e^{-\frac{x^3}{3!}}$

Now split this into a product of two terms.

$= \left(1 + x + \frac{x^2}{2!} + \frac{x^3}{3!} + \frac{x^4}{4!} + \cdots\right)\left(1 + \left(-\frac{x^3}{3!}\right) + \cdots\right)$

Then form the series for each of them. For $e^{-\frac{x^3}{3!}}$ substitute $-\frac{x^3}{3!}$ into the series for e^x.

$= 1 + x + \frac{x^2}{2} + \frac{x^3}{6} + \frac{x^4}{24} - \frac{x^3}{6} - \frac{x^4}{6} + \cdots$

$= 1 + x + \frac{x^2}{2} - \frac{x^4}{8} + \cdots$

Expand term by term and simplify.

b Valid for all $x \in \mathbb{R}$.

The series for both e^x and $\sin x$ are valid for all $x \in \mathbb{R}$.

WORKED EXAMPLE 8.6

a Find the first three terms in the Maclaurin series for $\ln(2 - 3x)$.

b Hence find the Maclaurin series up to the term in x^3 for $\ln\left(\frac{\sqrt{1+2x}}{2-3x}\right)$.

c State the interval in which the expansion is valid.

a $\ln(2 - 3x) = \ln\left[2\left(1 - \frac{3x}{2}\right)\right]$

You know the series expansion for $\ln(1 + x)$ so you need to write $\ln(2 - 3x)$ in this form. Start by factorising 2.

$= \ln\left[2\left(1 + \frac{-3x}{2}\right)\right]$

$= \ln 2 + \ln\left(1 + \frac{-3x}{2}\right)$

Separate the 2, using $\ln(ab) = \ln a + \ln b$

$= \ln 2 + \left(\frac{-3x}{2}\right) - \frac{\left(\frac{-3x}{2}\right)^2}{2} + \frac{\left(\frac{-3x}{2}\right)^3}{3} + \cdots$

Then substitute $\frac{-3x}{2}$ into the series of $\ln(1 + x)$:

$\ln(1 + x) = x - \frac{x^2}{2} + \frac{x^3}{3} - \frac{x^4}{4} + \cdots$

$= \ln 2 + \frac{-3x}{2} - \frac{\frac{9x^2}{4}}{2} + \frac{\frac{-27x^3}{8}}{3} + \cdots$

Expand and simplify.

$= \ln 2 - \frac{3x}{2} - \frac{9x^2}{8} - \frac{9x^3}{8} + \cdots$

Continues on next page

b $\ln\left(\dfrac{\sqrt{1+2x}}{2-3x}\right)=\ln(\sqrt{1+2x})-\ln(2-3x)$

$\qquad\qquad\qquad$ Again, you need everything in the form of $\ln(1+x)$. First, use the laws of logs.

$\qquad\qquad=\dfrac{1}{2}\ln(1+2x)-\ln(2-3x)$

$\dfrac{1}{2}\ln(1+2x)=\dfrac{1}{2}\left((2x)-\dfrac{(2x)^2}{2}+\dfrac{(2x)^3}{3}+\cdots\right)$

$\qquad\qquad\qquad$ For the first term, substitute $2x$ into the series for $\ln(1+x)$.

$\qquad\qquad=\dfrac{1}{2}\left(2x-2x^2+\dfrac{8x^3}{3}+\cdots\right)$

$\qquad\qquad=x-x^2+\dfrac{4x^3}{3}+\cdots$

So:

$\qquad\qquad\qquad$ You know the series expansion for the second term from part **a**.

$\ln\left(\dfrac{\sqrt{1+2x}}{2-3x}\right)=\left(x-x^2+\dfrac{4x^3}{3}+\cdots\right)$

$\qquad\qquad\qquad$ Now put both series together.

$\qquad\qquad-\left(\ln 2-\dfrac{3x}{2}-\dfrac{9x^2}{8}-\dfrac{9x^3}{8}+\cdots\right)$

$\qquad=x-x^2+\dfrac{4x^3}{3}+\cdots-\ln 2+\dfrac{3x}{2}+\dfrac{9x^2}{8}+\dfrac{9x^3}{8}+\cdots$

$\qquad=-\ln 2+\dfrac{5x}{2}+\dfrac{x^2}{8}+\dfrac{59x^3}{24}+\cdots$

c Since $\ln(1+x)$ is valid when $-1<x\leqslant 1$:

$\qquad\qquad\qquad$ Find the interval of validity separately for each function.

- $\ln\left(1+\dfrac{-3x}{2}\right)$ is valid when $-1<\dfrac{-3x}{2}\leqslant 1$

  This is when $-\dfrac{2}{3}\leqslant x<\dfrac{2}{3}$.

- $\ln(1+2x)$ is valid when $-1<2x\leqslant 1$

  This is when $-\dfrac{1}{2}<x\leqslant\dfrac{1}{2}$.

Therefore, $\ln\left(\dfrac{\sqrt{1+2x}}{2-3x}\right)$ is valid

when $-\dfrac{1}{2}<x\leqslant\dfrac{1}{2}$.

$\qquad\qquad\qquad$ For both to be valid, you need the smaller interval.

1 Find the first three non-zero terms and the general term of the Maclaurin series for each expression.

   a i e^{-3x} **ii** e^{x^3} **b i** $\ln(1+3x)$ **ii** $\ln(1-2x)$

   c i $\sin\left(-\dfrac{x}{2}\right)$ **ii** $\sin(3x^2)$ **d i** $\cos\left(\dfrac{x^2}{3}\right)$ **ii** $\cos(-2x)$

   e i $(1-4x)^{\frac{1}{2}}$ **ii** $\left(1+\dfrac{x}{3}\right)^{-4}$

2 By first manipulating it into an appropriate form, find the first three non-zero terms of the Maclaurin series for each expression.

   a i $\ln(3+x)$ **ii** $\ln\left(\dfrac{1}{2}-x\right)$

   b i $(2-3x)^{-3}$ **ii** $\left(\dfrac{1}{4}+2x\right)^{-\frac{1}{2}}$

   c i $(8x-27)^{\frac{1}{3}}$ **ii** $(3x-4)^{-2}$

3 By combining Maclaurin series of different functions, find the series expansion as far as the term in x^4 for each expression.

   a i $\ln(1+x)\sin 2x$ **ii** $\ln(1-x)\cos 3x$

   b i $\dfrac{e^x}{1+x}$ **ii** $\dfrac{\sin x}{1-2x}$

   c i $\ln(1+\sin x)$ **ii** $\ln(\cos x)$

4 Find the Maclaurin series for $\ln(1+4x^2+4x)$ and state the interval in which the series is valid.

5 Find the Maclaurin series as far as the term in x^4 for $e^{3x}\sin 2x$.

6 Show that $\sqrt{1+x^2}\,e^{-x}=1-x+x^2-\dfrac{2}{3}x^3+\dfrac{1}{6}x^4+\dots$.

7 **a** Find the first two non-zero terms of the Maclaurin series for $\tan x$.

   b Hence find the Maclaurin series of $e^{\tan x}$ up to and including the term in x^4.

8 **a** By using the Maclaurin series for $\cos x$, find the series expansion for $\ln(\cos x)$ up to the term in x^4.

   b Hence find the first two non-zero terms of the expansion of $\ln(\sec x)$.

   c Use your result from **b** to find the first two non-zero terms of the series for $\tan x$.

9 **a** Find the first four terms of the Maclaurin series for $f(x)=\ln[(2+x)^3(1-3x)]$.

   b Find the equation of the tangent to $f(x)$ at $x=0$.

10 **a** Find the Maclaurin series for $\ln\sqrt{\dfrac{1+x}{1-x}}$, stating the interval in which the series is valid.

   b Use the first three terms of this series to estimate the value of $\ln 2$, stating the value of x used.

Section 3: Improper integrals

Integrals where the range of integration extends to infinity

You are by now very familiar with evaluating definite integrals

$$\int_a^b f(x)dx = g(b) - g(a) \text{ where } g'(x) = f(x).$$

In the examples you have encountered so far, the limits were often convenient, relatively small numbers such as $0, 1, \pi$. However, there is nothing to stop them from being very large numbers; this would make no difference to the method for evaluating the integral.

If you continue along this line and let $b \to \infty$, you can still find a finite value for the integral in certain cases. In much the same way that you have seen that a sequence can either converge to a finite limit or diverge to infinity, so can an integral. Integrals of the form $\int_a^\infty f(x)dx$ are known as **improper integrals**.

To evaluate an improper integral, you need to replace the infinite limit by b, find the value of the integral in terms of b and then consider what happens when $b \to \infty$.

> ### 🔑 Key point 8.3
>
> The value of the improper integral $\int_a^\infty f(x)dx$ is
>
> $$\lim_{b \to \infty} \int_a^b f(x)dx = \lim_{b \to \infty}\{I(b)\} - I(a),$$
>
> if this limit exists and is finite.
>
> If this limit is infinite you say that the improper integral diverges (does not have a value).

WORKED EXAMPLE 8.7

a Explain why $\int_0^\infty e^{-3x}\,dx$ is an improper integral.

b Evaluate $\int_0^\infty e^{-3x}\,dx$.

a The integral is improper because the range of integration extends to infinity.

b $\int_0^\infty e^{-3x}\,dx = \lim_{b \to \infty} \int_0^b e^{-3x}\,dx$

  Integrate as normal, but replace the upper limit with b and take the limit as $b \to \infty$ after you have completed the integration.

$$= \lim_{b \to \infty}\left[-\frac{1}{3}e^{-3x}\right]_0^b$$

$$= \lim_{b \to \infty}\left(-\frac{1}{3}e^{-3b} + \frac{1}{3}\right)$$

$$= \lim_{b \to \infty}\left(-\frac{1}{3}e^{-3b}\right) + \frac{1}{3}$$

$$= \frac{1}{3}$$

  As $b \to \infty$, $e^{-3b} \to 0$. Therefore the integral converges.

WORKED EXAMPLE 8.8

Explain why the improper integral $\int_2^\infty \frac{1}{x}\,dx$ diverges.

$\int_2^\infty \frac{1}{x}dx = \lim_{b\to\infty}\int_2^b \frac{1}{x}dx$ .... Integrate as normal, but replace the upper limit with b and consider the limit as $b\to\infty$ after you have completed the integration.

$\qquad = \lim_{b\to\infty}\left[\ln x\right]_2^b$

When $b\to\infty$, $\ln x$ tends to infinity.

Therefore the integral diverges.

When evaluating improper integrals, you might need to use some more complicated limits. These will be given in each question.

WORKED EXAMPLE 8.9

Evaluate $\int_0^\infty xe^{-x}\,dx$, showing clearly the limiting process used.

[You can use without proof the result $\lim_{x\to\infty} x^k e^{-x} = 0$, where $k > 0$.]

$\int_0^\infty xe^{-x}\,dx = \lim_{b\to\infty}\int_0^b xe^{-x}\,dx$ .... Integrate as normal, but replace the upper limit with b and consider the limit as $b\to\infty$ after you have completed the integration.

$u = x, \dfrac{dv}{dx} = e^{-x}$ .... Use integration by parts.

$\Rightarrow \dfrac{du}{dx} = 1, v = -e^{-x}$

$\int_0^b xe^{-x}\,dx = \left[-xe^{-x}\right]_0^b - \int_0^b -e^{-x}\,dx$

$\qquad = \left[-xe^{-x} - e^{-x}\right]_0^b$

$\qquad = \left(-be^{-b} - e^{-b}\right) - (0 - 1)$

$\qquad = -be^{-b} - e^{-b} + 1$

$\lim_{b\to\infty}\int_0^b xe^{-x}\,dx = \lim_{b\to\infty}\left(-be^{-b} - e^{-b} + 1\right)$ .... Now take the limit as $b\to\infty$.

$\qquad = 0 + 0 + 1$ .... Use the given result: $\lim_{x\to\infty} x^k e^{-x} = 0$, with $k = 1$.

$\qquad = 1$

$\therefore \int_0^\infty xe^{-x}\,dx = 1$

Integrals where the integrand is undefined at a point within the range of integration

There is another type of improper integral, where the range of integration is finite but the integrand is not defined at a point within the range of integration (which could be at an end point or inside the range).

Examples of such integrals are $\int_0^2 x^3 \ln x \, dx$, which isn't defined at $x = 0$,

and $\int_0^5 \dfrac{1}{\sqrt{x-3}} \, dx$, which isn't defined at $x = 3$.

To evaluate the first of these integrals, you need to replace 0 by b as the lower limit, find the value of the integral in terms of b and then consider the limit $b \to 0$.

🔑 Key point 8.4

If $f(x)$ is not defined at $x = k$, then

$$\int_a^k f(x)\,dx = \lim_{b \to k} \int_a^b f(x)\,dx$$

and

$$\int_k^c f(x)\,dx = \lim_{b \to k} \int_b^c f(x)\,dx.$$

If the limit is not finite, then the improper integral diverges (does not have a value).

WORKED EXAMPLE 8.10

Evaluate $\int_0^2 x^3 \ln x \, dx$, showing clearly the limiting process used.

[You may use without proof the result $\lim\limits_{x \to 0} x^k \ln x = 0$, where $k > 0$.]

$\int_0^2 x^3 \ln x \, dx = \lim\limits_{b \to 0} \int_b^2 x^3 \ln x \, dx$

$\ln x$ is not defined at $x = 0$. Integrate as normal, but replace the lower limit with b and consider the limit as $b \to 0$ after you have completed the integration.

$u = \ln x, \dfrac{dv}{dx} = x^3$

$\Rightarrow \dfrac{du}{dx} = \dfrac{1}{x}, v = \dfrac{1}{4}x^4$

Use integration by parts.

$\int_b^2 x^3 \ln x \, dx = \left[\dfrac{1}{4}x^4 \ln x\right]_b^2 - \int_b^2 \dfrac{1}{4}x^3 \, dx$

Remember that integrals with ln are an exception where you take $\ln x = u$.

$= \left[\dfrac{1}{4}x^4 \ln x - \dfrac{1}{16}x^4\right]_b^2$

$= (4\ln 2 - 1) - \left(\dfrac{1}{4}b^4 \ln b - \dfrac{1}{16}b^4\right)$

$\lim\limits_{b \to 0} \int_b^2 x^3 \ln x \, dx = \lim\limits_{b \to 0}\left(4\ln 2 - 1 - \dfrac{1}{4}b^4 \ln b + \dfrac{1}{16}b^4\right)$

Now take the limit as $b \to 0$.

Use the given result: $\lim\limits_{x \to 0} x^k \ln x = 0$, with $k = 4$.

$= 4\ln 2 - 1 - 0 + 0$

$= 4\ln 2 - 1$

$\therefore \int_0^2 x^3 \ln x \, dx = 4\ln 2 - 1$

If the point where the integrand is not defined is not an end point, you need to split the integral into two.

> ### 🔑 Key point 8.5
>
> If $f(x)$ is undefined at $x = k \in (a, c)$, then
>
> $$\int_a^c f(x)\,dx = \lim_{b \to k} \int_a^b f(x)\,dx + \lim_{b \to k} \int_b^c f(x)\,dx.$$
>
> If either limit is not finite, then the improper integral diverges (does not have a value).

WORKED EXAMPLE 8.11

Find the exact value of $\displaystyle\int_0^5 \frac{1}{\sqrt[3]{x-3}}\,dx$.

$$\int_0^5 \frac{1}{\sqrt[3]{x-3}}\,dx = \int_0^3 \frac{1}{\sqrt[3]{x-3}}\,dx + \int_3^5 \frac{1}{\sqrt[3]{x-3}}\,dx$$

The integrand is not defined at $x = 3$, so you need to split the integral in two.

$$\int_0^3 \frac{1}{\sqrt[3]{x-3}}\,dx = \lim_{b \to 3} \int_0^b \frac{1}{\sqrt[3]{x-3}}\,dx$$

For each integral replace 3 by b, evaluate the integral and then find the limit when $b \to 3$.

$$= \lim_{b \to 3} \int_0^b (x-3)^{-\frac{1}{3}}\,dx$$

$$= \lim_{b \to 3} \left[\frac{3}{2}(x-3)^{\frac{2}{3}} \right]_0^b$$

Integrate as usual.

$$= \lim_{b \to 3} \left[\frac{3}{2}(b-3)^{\frac{2}{3}} - \frac{3}{2}(0-3)^{\frac{2}{3}} \right]$$

$$= -\frac{3}{2}\sqrt[3]{9}$$

As $b \to 3$, $(b-3)^{\frac{2}{3}} \to 0$.

$$\int_3^5 \frac{1}{\sqrt[3]{x-3}}\,dx = \lim_{b \to 3} \int_b^5 \frac{1}{\sqrt[3]{x-3}}\,dx$$

$$= \lim_{b \to 3} \left[\frac{3}{2}(5-3)^{\frac{2}{3}} - \frac{3}{2}(b-3)^{\frac{2}{3}} \right]$$

Repeat the same process for the second integral.

$$= \frac{3}{2}\sqrt[3]{4}$$

$$\therefore \int_0^5 \frac{1}{\sqrt[3]{x-3}}\,dx = \frac{3}{2}\left(\sqrt[3]{4} - \sqrt[3]{9} \right)$$

EXERCISE 8C

1 Determine which of these improper integrals converge. Evaluate the ones that do converge.

a $\displaystyle\int_0^\infty \frac{1}{(1+x)^2}\,dx$

b $\displaystyle\int_0^\infty e^{-\frac{x}{4}}\,dx$

c $\displaystyle\int_0^\infty \frac{1}{\sqrt{1+x}}\,dx$

d $\displaystyle\int_0^\infty x e^{-x^2}\,dx$

2 For what values of p do each of these improper integrals converge?

a $\displaystyle\int_0^\infty e^{px}\,dx$

b $\displaystyle\int_1^\infty \frac{\ln x}{x^p}\,dx$

3 Explain why $\displaystyle\int_0^4 \frac{1}{\sqrt{x}}\,dx$ is an improper integral and find its value.

4 Evaluate $\displaystyle\int_0^\infty e^{-x}\,dx$.

5 Evaluate $\displaystyle\int_0^\infty \frac{1}{x^2+1}\,dx$.

6 Evaluate the improper integral $\displaystyle\int_0^\infty \left(\frac{2x}{x^2+5} - \frac{6}{3x+2}\right)dx$, giving your answer in the form $\ln k$, where k is a constant.

Section 4: Volumes of revolution

In A Level Mathematics Student Book 1, Chapter 15, you saw that the area between a curve and the x-axis from $x = a$ to $x = b$ is given by $\int_a^b y\,dx$, as long as $y > 0$. In this section, you will use a similar formula to find the volume of a shape formed by rotating the curve about either the x-axis or the y-axis.

If a curve is rotated about the x-axis or the y-axis, the resulting shape is called a **solid of revolution** and the volume of that shape is referred to as the **volume of revolution**.

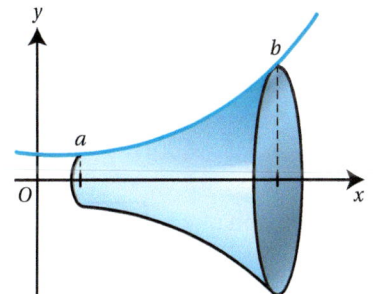

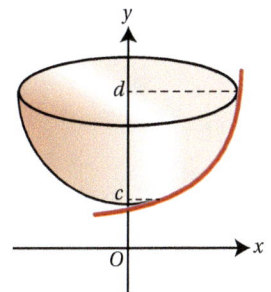

🔑 Key point 8.6

- When the curve $y = f(x)$ between $x = a$ and $x = b$ is rotated 360° about the x-axis, the volume of revolution is given by $V = \pi\displaystyle\int_a^b y^2\,dx$.

- When the curve $y = f(x)$ between $y = c$ and $y = d$ is rotated 360° about the y-axis, the volume of revolution is given by $V = \pi\displaystyle\int_c^d x^2\,dy$.

The proof of these results is very similar. The proof for rotation about the x-axis is given in Proof 4.

PROOF 4

The solid can be split into small cylinders.

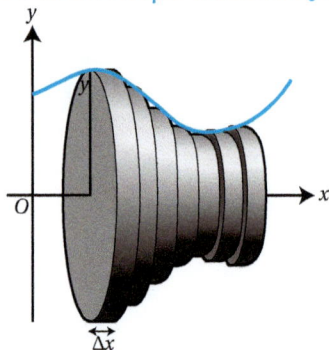

Draw an outline of a representative function to illustrate the argument.

The volume of each cylinder is $\pi y^2 \Delta x$.

The radius of each cylinder is the y-coordinate and the height is Δx.

The total volume is approximately:

$$V \approx \sum_a^b \pi y^2 \Delta x$$

You are starting at $x = a$ and stopping at $x = b$.
It is only approximate because the volume of revolution is not exactly the same as the total volume of the cylinders.

$$V = \lim_{\Delta x \to 0} \sum_a^b \pi y^2 \Delta x$$

$$= \int_a^b \pi y^2 \, dx$$

$$= \pi \int_a^b y^2 \, dx$$

However, as you make the cylinders smaller the volume gets more and more accurate. The sum then becomes an integral. You can leave π out of the integration and multiply by it at the end.

WORKED EXAMPLE 8.12

The graph of $y = \sqrt{x^2 + 1}$, $0 \leqslant x \leqslant 3$, is rotated $360°$ about the x-axis.

Find, in terms of π, the volume of the solid generated.

$$V = \pi \int_0^3 \left(x^2 + 1\right) dx$$

Use the formula: $V = \pi \int_a^b y^2 \, dx$.

$$= \pi \left[\frac{x^3}{3} + x \right]_0^3$$

Evaluate the definite integral.

$$= \pi \left[\left(\frac{3^3}{3} + 3 \right) - 0 \right]$$

$$= 12\pi$$

To find the volume of revolution about the y-axis you will often have to rearrange the equation of the curve to find x in terms of y.

Tip

Remember that the limits of the integration need to be in terms of y and not x.

WORKED EXAMPLE 8.13

The part of the curve $y = \dfrac{1}{x}$ between $x = 1$ and $x = 4$ is rotated $360°$ about the y-axis. Find the exact value of the volume of the solid generated.

When $x = 1$, $y = \dfrac{1}{1} = 1$ .... Find the limits in terms of y.

When $x = 4$, $y = \dfrac{1}{4}$

$y = \dfrac{1}{x} \Rightarrow x = \dfrac{1}{y}$ .... Express x in terms of y.

$V = \pi \displaystyle\int_a^b x^2 \, dy$ .... Use the formula $V = \pi \displaystyle\int_a^b x^2 \, dy$, substituting in $x = \dfrac{1}{y}$.

$\quad = \pi \displaystyle\int_{\frac{1}{4}}^{1} \left(\dfrac{1}{y}\right)^2 dy$

$\quad = \pi \displaystyle\int_{\frac{1}{4}}^{1} y^{-2} \, dy$

$\quad = \pi \left[-y^{-1}\right]_{\frac{1}{4}}^{1}$

$\quad = \pi \left[(-1) - (-4)\right]$

$\quad = 3\pi$

You might also be asked to find a volume of revolution of an area between two curves.

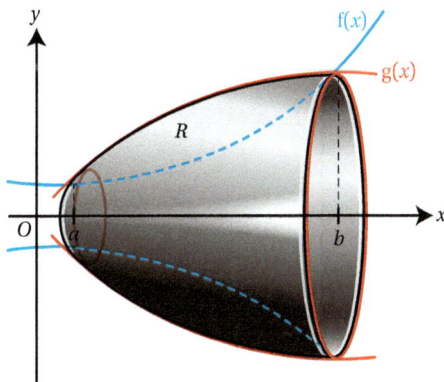

From the diagram you can see that the volume formed when the region R is rotated around the x-axis is given by the volume of revolution of $g(x)$ minus the volume of revolution of $f(x)$.

Key point 8.7

The volume of revolution of the region between curves $g(x)$ and $f(x)$ is:

$$V = \pi \int_a^b \left(g(x)^2 - f(x)^2 \right) dx$$

where the curve of $g(x)$ is above $f(x)$ and the curves intersect at $x = a$ and $x = b$.

Tip

Make sure that you square each term within the brackets and do not make the mistake of squaring the whole expression inside the brackets: the formula is **not** $\pi \int_a^b \left(g(x) - f(x) \right)^2 dx$.

WORKED EXAMPLE 8.14

Find the volume formed when the region enclosed by $y = x^2 + 6$ and $y = 8x - x^2$ is rotated through $360°$ about the x-axis.

For points of intersection:

$$x^2 + 6 = 8x - x^2$$
$$2x^2 - 8x + 6 = 0$$
$$x^2 - 4x + 3 = 0$$
$$(x - 1)(x - 3) = 0$$
$$x = 1 \text{ or } x = 3$$

First find the x-coordinates of the points where the curves meet, by equating the RHS of both equations and solving. This will give you the limits of integration.

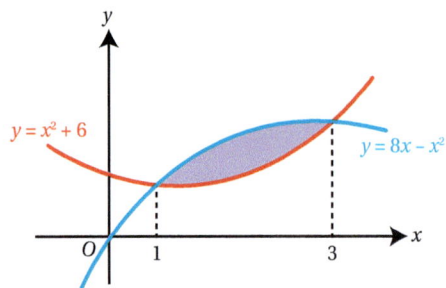

Sketch the graphs in the region concerned.

$y = 8x - x^2$ is above $y = x^2 + 6$.

$$V = \pi \int_1^3 \left((8x - x^2)^2 - (x^2 + 6)^2 \right) dx$$

$$= \frac{176}{3} \pi$$

Apply the formula

$$V = \pi \int_a^b \left(g(x)^2 - f(x)^2 \right) dx.$$

Use a calculator to evaluate the integral.

Tip

Don't forget that you can use your calculator to evaluate definite integrals. However, look out for the instruction to 'show detailed reasoning' which means that you must show full integration and evaluation.

(i) Did you know?

There are also formulae to find the surface area of a solid formed by rotating a region around an axis. Some particularly interesting examples arise if you allow one end of the region to tend to infinity; for example, rotating the region formed by the lines $y = \frac{1}{x}$, $x = 1$ and the x-axis results in a solid called Gabriel's horn or Torricelli's trumpet.

Areas and volumes can also be calculated using improper integrals, and it turns out that it is possible to have a solid of finite volume but infinite surface area!

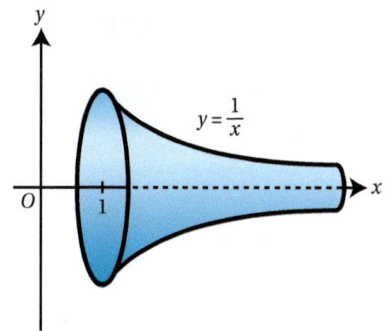

You can also use the formulae for the volume of revolution about the coordinate axes in the case where the curve is defined parametrically.

(⏪) Rewind

You learnt how to find the area defined by a parametric curve in A Level Mathematics Student Book 2, Chapter 12.

(🔑) Key point 8.8

When the part of a curve with parametric equations $x = f(t)$, $y = g(t)$, between points with parameter values t_1 and t_2, is rotated about one of the coordinate axes, the resulting volume of revolution is

$$\pi \int_{t_1}^{t_2} y^2 \frac{\mathrm{d}x}{\mathrm{d}t} \, \mathrm{d}t \text{ for rotation about the } x\text{-axis}$$

$$\pi \int_{t_1}^{t_2} x^2 \frac{\mathrm{d}y}{\mathrm{d}t} \, \mathrm{d}t \text{ for rotation about the } y\text{-axis.}$$

WORKED EXAMPLE 8.15

The curve shown in the diagram has parametric equations $x = 5t^2$, $y = 2t^3$ for $t \in \mathbb{R}$.

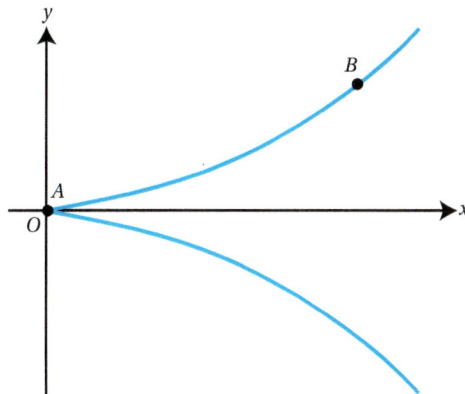

The part of the curve between the points $A(0, 0)$ and $B(20, 16)$ is rotated about the x-axis. Find the exact value of the resulting volume of revolution.

Continues on next page ...

When $x = 0$, $t = 0$.

When $x = 20$, $t = \pm 2$; but $y > 0$ so $t = 2$.

$$V = \pi \int_0^2 (2t^3)^2 (10t)\, dt$$

$$= \pi \int_0^2 40t^7\, dt$$

$$= 1280\pi$$

Find the values of t corresponding to the end points.

Use the formula $\pi \int_{t_1}^{t_2} y^2 \dfrac{dx}{dt}\, dt$ for the volume of revolution about the x-axis.

You can use your calculator to evaluate the integral (unless the question asks you to 'show detailed reasoning').

EXERCISE 8D

In this exercise, whenever a question asks for an exact volume, you must show detailed reasoning.

1 The part of the curve $y = f(x)$ for $a \leqslant y \leqslant b$ is rotated 360° about the x-axis. Find the exact volume of revolution formed in each case.

    a   **i** $f(x) = x^2$; $a = -1$, $b = 1$ **ii** $f(x) = x^3$; $a = 0$, $b = 2$

    b   **i** $f(x) = x^2 + 6$; $a = -1$, $b = 3$ **ii** $f(x) = 2x^3 + 1$; $a = 0$, $b = 1$

    c   **i** $f(x) = \dfrac{1}{x}$; $a = 1$, $b = 2$ **ii** $f(x) = \dfrac{1}{x^2}$; $a = 1$, $b = 4$

2 Find the exact volume of revolution formed when each curve, for $a \leqslant x \leqslant b$, is rotated through 2π radians about the x-axis.

    a   **i** $y = e^x$; $a = 0$, $b = 1$ **ii** $y = e^{-x}$; $a = 0$, $b = 3$

    b   **i** $y = e^{2x} + 1$; $a = 0$, $b = 1$ **ii** $y = e^{-x} + 2$; $a = 0$, $b = 2$

    c   **i** $y = \sqrt{\sin x}$; $a = 0$, $b = \pi$ **ii** $y = \sqrt{\cos x}$; $a = 0$, $b = \dfrac{\pi}{2}$

3 The part of the curve for $a \leqslant y \leqslant b$ is rotated 360° about the y-axis.

Find the exact volume of revolution formed in each case.

    a   **i** $y = 4x^2 + 1$; $a = 1$, $b = 17$ **ii** $y = \dfrac{x^2 - 1}{3}$; $a = 0$, $b = 5$

    b   **i** $y = x^3$; $a = 0$, $b = 8$ **ii** $y = x^4$; $a = 2$, $b = 8$

    c   **i** $y = \dfrac{1}{x^3}$; $a = 8$, $b = 27$ **ii** $y = \dfrac{1}{x^5}$; $a = 1$, $b = 32$

4 The part of the curve $y = f(x)$ for $a \leqslant y \leqslant b$ is rotated 360° about the y-axis.

Find the exact volume of revolution formed in each case.

    a   **i** $f(x) = \ln x + 1$; $a = 1$, $b = 3$ **ii** $f(x) = \ln(2x - 1)$; $a = 0$, $b = 4$

    b   **i** $f(x) = \dfrac{1}{x^2}$; $a = 1$, $b = 2$ **ii** $f(x) = \dfrac{1}{x^2} + 2$; $a = 3$, $b = 5$

    c   **i** $f(x) = \arcsin x$; $a = -\dfrac{\pi}{2}$, $b = \dfrac{\pi}{2}$ **ii** $f(x) = \arcsin x$; $a = -\dfrac{\pi}{4}$, $b = \dfrac{\pi}{4}$

5 The part of the curve $x = f(t), y = g(t)$ for $a \leqslant t \leqslant b$ is rotated through $360°$ about the x-axis. Find the exact volume of revolution formed in each case.

a i $x = 4t^2, y = 3t^3; a = 0, b = 2$

ii $x = \dfrac{1}{2}t^2, y = 2t^3; a = 0, b = 1$

b i $x = t^2 + 1, y = t + \dfrac{1}{t}; a = 1, b = 2$

ii $x = t^3 + 2, y = 2t + \dfrac{1}{t}; a = 1, b = 2$

c i $x = \cos t, y = \sqrt{\sin t}; a = 0, b = \dfrac{\pi}{2}$

ii $x = \cos t, y = \dfrac{1}{\sqrt{\sin t}}; a = \dfrac{\pi}{6}, b = \dfrac{\pi}{2}$

6 The part of the curve $x = f(t), y = g(t)$ for $a \leqslant t \leqslant b$ is rotated through $360°$ about the y-axis. Find the exact volume of revolution formed in each case.

a i $x = 4t^2, y = 3t^3; a = 0, b = 2$

ii $x = \dfrac{1}{2}t^2, y = 2t^3; a = 0, b = 1$

b i $x = t^2 + 1, y = t + \dfrac{1}{t}; a = \dfrac{1}{2}, b = 1$

ii $x = t^3 + 2, y = 2t + \dfrac{1}{t}; a = 1, b = 2$

c i $x = \sqrt{\cos t}, y = \sin t; a = 0, b = \dfrac{\pi}{2}$

ii $x = \cos t, y = \sin^2 t; a = 0, b = \dfrac{\pi}{2}$

7 The diagram shows the region, R, bounded by the curve $y = \sqrt{x} - 2$, the x-axis and the line $x = 9$.

a Find the coordinates of the point A where the curve crosses the x-axis.

This region is rotated $360°$ about the x-axis.

b Find the exact volume of the solid generated.

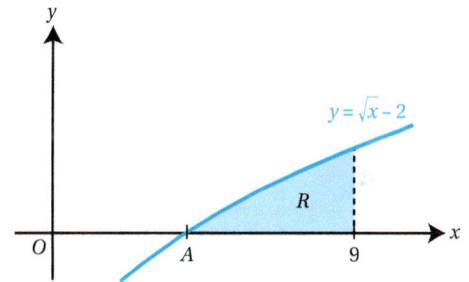

8 The curve $y = 3x^2 + 1$, for $0 \leqslant x \leqslant 2$, is rotated through $360°$ about the y-axis.

Find the volume of revolution generated, correct to 3 significant figures.

9 The part of the curve $y^2 = \sin x$ between $x = 0$ and $x = \dfrac{\pi}{2}$ is rotated through 2π radians about the x-axis.

Find the exact volume of the solid generated.

10 The curve $y = x^2$, for $0 < x < a$, is rotated through $180°$ about the x-axis.

The resulting volume is $\dfrac{16\pi}{5}$.

Find the value of a.

11 The region enclosed by the curve $y = x^2 - a^2$ and the x-axis is rotated $90°$ about the x-axis.

Find an expression, in terms of a, for the volume of revolution formed.

12 The part of the curve $y = \sqrt{\dfrac{3}{x}}$ between $x = 1$ and $x = a$ is rotated through 2π radians about the x-axis. The volume of the resulting solid is $\pi \ln \dfrac{64}{27}$.

Find the exact value of a.

13 a Find the coordinates of the points of intersection of curves $y = x^2 + 3$ and $y = 4x + 3$.

b Find the volume of revolution generated when the region between the curves $y = x^2 + 3$ and $y = 4x + 3$ is rotated through $360°$ about the x-axis.

14 The region bounded by the curves $y = x^2 + 6$ and $y = 8x - x^2$ is rotated through $360°$ about the x-axis. Find the volume of the resulting solid.

15 a Find the coordinates of the points of intersection of the curves $y = 4\sqrt{x}$ and $y = x + 3$.

b The region between the curves $y = 4\sqrt{x}$ and $y = x + 3$ is rotated through $360°$ about the y-axis. Find the volume of the solid generated.

16 By rotating the circle $x^2 + y^2 = r^2$ around the x-axis, prove that the volume of a sphere of radius r is given by $\frac{4}{3}\pi r^3$.

17 The part of the curve with parametric equations $x = t^2 + 3$, $y = 3t + 1$, between the points $(4, 4)$ and $(7, 7)$, is rotated through $360°$ about the x-axis. Find the exact volume generated.

18 The diagram shows a part of the curve with parametric equations

$x = \dfrac{1}{1+t}, y = t^2$.

The section of the curve between points $P\left(\dfrac{1}{2}, 1\right)$ and $Q\left(\dfrac{1}{3}, 4\right)$ is

rotated a full turn about the y-axis. Find the exact volume of the resulting solid.

19 The diagram shows the curve with parametric equations
$x = \sin\theta, y = \sin 2\theta$.

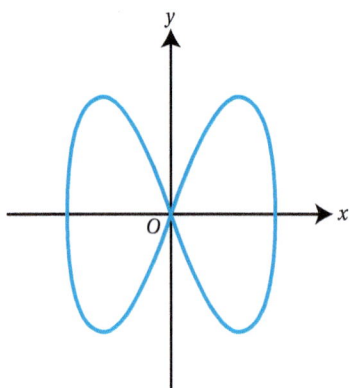

One of the loops of the curve is rotated 2π radians about the x-axis. Find the exact value of the volume of revolution.

20 By choosing a suitable function to rotate around the x-axis, prove that the volume of a circular cone with base radius r and height h is $\dfrac{\pi r^2 h}{3}$.

21 Find the volume of revolution when the region enclosed by the graphs of $y = e^x$, $y = 1$ and $x = 1$ is rotated through $360°$ about the line $y = 1$.

Section 5: Mean value of a function

Suppose an object travels between time $t = 0$ and $t = 3$ with a velocity given by $v = t$. Its velocity–time graph looks like this.

Its average velocity can be found from:

$$\frac{\text{initial velocity} + \text{final velocity}}{2} = \frac{0+3}{2}$$
$$= 1.5$$

Suppose, instead, the object has velocity given by $v = \dfrac{t^2}{3}$. Then you can compare the two velocity–time graphs.

The formula $\dfrac{\text{initial velocity} + \text{final velocity}}{2}$ would give the same mean velocity for the two graphs, which can't be correct because the red curve is underneath the blue line everywhere other than at the end points.

You need a method of calculating the mean that takes into account the value of the function everywhere.

One possibility is to use $\dfrac{\text{total distance}}{\text{time taken}}$.

You can then use the fact that total distance is the integral of velocity with respect to time.

For the blue line this gives:

$$\text{average velocity} = \frac{\int_0^3 t\,dt}{3}$$
$$= \frac{1}{3}\left[\frac{t^2}{2}\right]_0^3$$
$$= 1.5$$

For the red curve this gives:

$$\text{average velocity} = \frac{\int_0^3 \dfrac{t^2}{3}\,dt}{3}$$
$$= \frac{1}{3}\left[\frac{t^3}{9}\right]_0^3$$
$$= 1$$

This process can be generalised for any function.

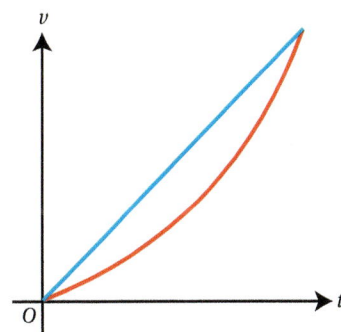

Key point 8.9

The mean value of a function $f(x)$ between a and b is:

$$\frac{1}{b-a}\int_a^b f(x)\,dx$$

WORKED EXAMPLE 8.16

You are given that $f(x) = x^2 - x$. Find the mean value of $f(x)$ between 3 and 4.

Mean value $= \dfrac{1}{4-3} \displaystyle\int_3^4 \left(x^2 - x\right) dx$ · · · · · · · · Use the formula for the mean value of a function: $\dfrac{1}{b-a} \displaystyle\int_a^b f(x) dx$.

$= \dfrac{1}{4-3} \left[\dfrac{x^3}{3} - \dfrac{x^2}{2} \right]_3^4$

$= \dfrac{37}{3} - \dfrac{7}{2}$

$= \dfrac{53}{6}$ · · · · · · · · · · · · Notice that $x^2 - x$ varies between 6 and 12, so a mean of around 9 seems reasonable.

EXERCISE 8E

1 Find the mean value of each function between the given values of x.

   a **i** x^2 for $0 < x < 1$ **ii** x^2 for $1 < x < 3$

   b **i** $\sqrt{x}$ for $0 < x < 4$ **ii** $\dfrac{1}{x^2}$ for $1 < x < 5$

   c **i** $x^3 + 1$ for $0 < x < 4$ **ii** $x^4 - x$ for $0 < x < 10$

2 Find the mean value of each function over the domain given.

   a **i** $\sin x$ for $0 < x < \pi$ **ii** $\cos x$ for $0 < x < \pi$

   b **i** e^x for $0 < x < 1$ **ii** $\dfrac{1}{x}$ for $1 < x < e$

   c **i** $\sqrt{x+1}$ for $3 < x < 8$ **ii** $x \sin(x^2)$ for $0 < x < \sqrt{\pi}$

3 The velocity of a rocket is given by $v = 30\sqrt{t}$ where t is time, in seconds, and v is velocity, in metres per second.

   Find the mean velocity in the first T seconds.

4 The mean value of the function $f(x) = x^2 - x$ for $0 < x < a$ is zero.

   Find the value of a.

5 $f(x) = x^2$ for $x \geq 0$.

   a f_{mean} is the mean value of $f(x)$ between 0 and a. Find an expression for f_{mean} in terms of a.

   b Given that $f(c) = f_{mean}$ find an expression for c in terms of a.

6 Show that the mean value of $\dfrac{1}{x^2}$ between 1 and a is inversely proportional to a.

7 An alternating current has time period 2. The power dissipated by the current through a resistor is given by $P = P_0 \sin^2(\pi t)$.

Find the ratio of the mean power of one complete period to the maximum power.

8 The mean value of f(x) between a and b is F.

Prove that the mean value of f$(x) + 1$ between a and b is $F + 1$.

9 **a** Sketch the graph of f$(x) = \dfrac{1}{2\sqrt{x}}$.

b Use the graph to explain why the mean value of f(x) between a and b is less than the mean of f(a) and f(b).

c Hence prove that, if $0 < a < b$, $\sqrt{b} - \sqrt{a} < \dfrac{1}{3}\left(\dfrac{b}{\sqrt{a}} - \dfrac{a}{\sqrt{b}}\right)$.

10 If f_{mean} is the mean value of f(x) for $a < x < b$ and f$(a) <$ f(b), then f$(a) < f_{mean} <$ f(b).

Either prove this statement or disprove it using a counterexample.

Checklist of learning and understanding

- The Maclaurin series for a function f(x) is given by

$$f(x) = f(0) + f'(0)x + \frac{f''(0)}{2!}x^2 + \dots + \frac{f^{(r)}(0)}{r!}x^r + \dots$$

- Maclaurin series for some common functions and the values of x for which they are valid:

  - $e^x = \exp(x) = 1 + x + \dfrac{x^2}{2!} + \dots + \dfrac{x^r}{r!} + \dots$ for all x

  - $\ln(1+x) = x - \dfrac{x^2}{2} + \dfrac{x^3}{3} - \dots + (-1)^{r+1}\dfrac{x^r}{r} + \dots \quad -1 < x \leqslant 1$

  - $\sin x = x - \dfrac{x^3}{3!} + \dfrac{x^5}{5!} - \dots + (-1)^r \dfrac{x^{2r+1}}{(2r+1)!} + \dots$ for all x

  - $\cos x = 1 - \dfrac{x^2}{2!} + \dfrac{x^4}{4!} - \dots + (-1)^r \dfrac{x^{2r}}{(2r)!} + \dots$ for all x

  - $(1+x)^n = 1 + nx + \dfrac{n(n-1)}{2!}x^2 + \dots + \dfrac{n(n-1)\dots(n-r+1)}{r!}x^r + \dots \quad |x| < 1, n \in \mathbb{R}$

- Improper integrals are definite integrals where either:
  - the range of integration is infinite or
  - the integrand isn't defined at every point in the range of integration.
- The value of the improper integral $\int_a^\infty f(x)\,dx$ is

$$\lim_{b\to\infty}\int_a^b f(x)\,dx = \lim_{b\to\infty}\{I(b)\} - I(a)$$

if this limit exists and is finite.
- If f(x) is not defined at $x = k$, then

$$\int_a^k f(x)\,dx = \lim_{b\to k}\int_a^b f(x)\,dx \quad \text{and} \quad \int_k^c f(x)\,dx = \lim_{b\to k}\int_b^c f(x)\,dx$$

if the limits exist and are finite.

Continues on next page ...

- If $f(x)$ is undefined at $x = k \in (a,c)$, then

$$\int_a^c f(x)\,dx = \lim_{b \to k} \int_a^b f(x)\,dx + \lim_{b \to k} \int_b^c f(x)\,dx$$

  if the limits exist and are finite.
- The volume of a shape formed by rotating a curve about the x-axis or the y-axis is known as the volume of revolution.
  - When the curve $y = f(x)$ between $x = a$ and $x = b$ is rotated $360°$ about the x-axis, the volume of revolution is given by

$$V = \pi \int_a^b y^2\,dx$$

  - When the curve $y = f(x)$ between $y = c$ and $y = d$ is rotated $360°$ about the y-axis, the volume of revolution is given by

$$V = \pi \int_c^d x^2\,dy$$

  - The volume of revolution of the region between curves $g(x)$ and $f(x)$ is:

$$V = \pi \int_a^b \left(g(x)^2 - f(x)^2 \right) dx$$

    where the curve of $g(x)$ is above $f(x)$ and the curves intersect at $x = a$ and $x = b$.

  - When the part of a curve with parametric equations $x = f(t)$, $y = g(t)$, between points with parameter values t_1 and t_2, is rotated about one of the coordinate axes, the resulting volume of revolution is

$$\pi \int_{t_1}^{t_2} y^2 \frac{dx}{dt}\,dt \text{ for rotation about the } x\text{-axis}$$

$$\pi \int_{t_1}^{t_2} x^2 \frac{dy}{dt}\,dt \text{ for rotation about the } y\text{-axis}$$

- The mean value of a function $f(x)$ between a and b is:

$$\frac{1}{b-a} \int_a^b f(x)\,dx$$

Mixed practice 8

1. **a** Show that the first four terms in the Maclaurin series of e^x are $1 + x + \dfrac{x^2}{2} + \dfrac{x^3}{6}$.

    b Use the series in part **a** to show that $\sqrt{e} \approx \dfrac{79}{48}$.

2. **a** Find the first four derivatives of $f(x) = \ln(1+x)$.

    b Hence find the first four non-zero terms of the Maclaurin series of $f(x)$.

    c Using this expansion, find the exact value of the infinite series $1 - \dfrac{1}{2} + \dfrac{1}{3} - \dfrac{1}{4} + \dots$.

3. **a** Explain why $\displaystyle\int_0^4 \dfrac{5-x}{\sqrt{x^3}}\,dt$ is an improper integral.

    b Either find the value of the integral $\displaystyle\int_0^4 \dfrac{5-x}{\sqrt{x^3}}\,dx$ or explain why it does not have a finite value.

4. Given that $f(x) = \ln(\cos 3x)$, find $f'(0)$ and $f''(0)$. Hence show that the first term in the Maclaurin series for $f(x)$ is ax^2, where the value of a is to be found.

    © OCR A Level Mathematics, Unit 4726 Further Pure Mathematics 2, January 2012

5. The curve $y = \sqrt{x}$ between 0 and a is rotated through $360°$ about the x-axis. The resulting solid has a volume of 18π.

    Find the value of a.

6. The curve $x = \dfrac{y^2 - 1}{3}$, with $1 \leqslant y \leqslant 4$, is rotated through $360°$ about the y-axis.

    Find the volume of revolution generated, correct to 3 significant figures.

7. For $0 < x < a$, the mean value of x is equal to the mean value of x^2.

    Find the value of a.

8. The mean value of $\dfrac{1}{\sqrt{x}}$ from 0 to b is 1.

    Find the value of b.

9. Find the set of values of x for which the Maclaurin series of the function $f(x) = \sqrt{1 + e^x}$ is valid.

10. **a** Find the Maclaurin series of $\ln(\cos x)$ up to and including the term in x^4.

    b Hence show that $\ln 2 \approx \dfrac{\pi^2}{16}\left(1 + \dfrac{\pi^2}{96}\right)$.

11. It is given that $f(x) = \ln(1 + \sin x)$, $-\dfrac{\pi}{2} < x < \dfrac{3\pi}{2}$.

    a Find the first three derivatives of $f(x)$.

    b Hence find the Maclaurin series of $f(x)$ up to and including the term in x^3.

    c Hence find $\displaystyle\lim_{x \to 0} \dfrac{\ln(1 + \sin x) - x}{x^2}$.

12 You are given that $f(x) = e^{-x} \sin x$.

   i Find $f(0)$ and $f'(0)$.

   ii Show that $f''(x) = -2f'(x) - 2f(x)$ and hence, or otherwise, find $f''(0)$.

   iii Find a similar expression for $f'''(x)$ and hence, or otherwise, find $f'''(0)$.

   iv Find the Maclaurin series for $f(x)$ up to and including the term in x^3.

© OCR A Level Mathematics, Unit 4726/01 Further Pure Mathematics 2, January 2013

13 The region bounded by the curve $y = ax - x^2$ and the x-axis is rotated one full turn about the x-axis. Find, in terms of a, the resulting volume of revolution.

In questions 14 to 20 you must show detailed reasoning.

14 The diagram shows the curve $y = \ln x$ and the line $y = -\frac{1}{e}x + 2$.

   a Show that the two graphs intersect at $(e, 1)$.

   The shaded region is rotated through 360° about the y-axis.

   b Find the exact value of the volume of revolution.

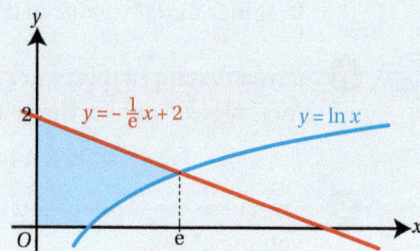

15 The region enclosed by $y = (x-1)(x-2) + 1$ and the line $y = 1$ is rotated through $180°$ about the line $y = 1$. Find the exact value of the resulting volume.

16 The diagram shows the curve $y = \sqrt{\dfrac{3}{4x+1}}$ for $0 \leqslant x \leqslant 20$. The point

P on the curve has coordinates $\left(20, \frac{1}{9}\sqrt{3}\right)$. The shaded region R is

enclosed by the curve and the lines $x = 0$ and $y = \frac{1}{9}\sqrt{3}$.

   i Find the exact area of R.

   ii Find the exact volume of the solid obtained when R is rotated completely about the x-axis.

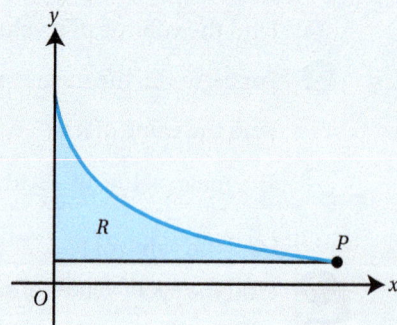

© OCR A Level Mathematics, Unit 4723/01 Core Mathematics 3, June 2014

17 **i** Show that the derivative with respect to y of

$$y\ln(2y) - y$$

   is $\ln(2y)$.

   ii The diagram shows the curve with equation $y = \frac{1}{2}e^{x^2}$. The point

$P\left(2, \frac{1}{2}e^4\right)$ lies on the curve. The shaded region is bounded

by the curve and the lines $x = 0$ and $y = \frac{1}{2}e^4$. Find the exact volume of the solid produced when the shaded region is rotated completely about the y-axis.

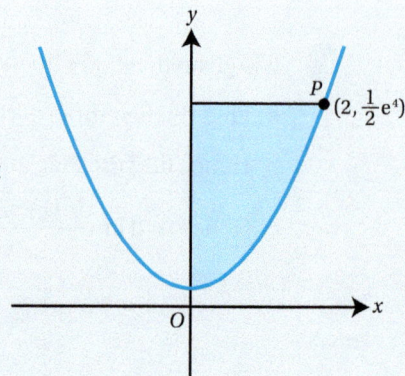

   iii Hence find the volume of the solid produced when the region bounded by the curve and the lines $x = 0$, $x = 2$ and $y = 0$ is rotated completely about the y-axis.

© OCR A Level Mathematics, Unit 4723 Core Mathematics 3, June 2012

18 The region bounded by the curves $y = \sqrt{x}$ and $y = 2\sqrt{x-3}$ is rotated one full turn, about the x-axis. Find the resulting volume of revolution.

19 The shape of a rugby ball can be modelled as a solid obtained by rotating an ellipse about one of its axes of symmetry.

An ellipse has parametric equations $x = a\cos\theta$, $y = b\sin\theta$. Find the volume of the solid generated when the ellipse is rotated through $360°$ about the x-axis.

20 The diagram shows a part of the curve with parametric equations $x = e^t$, $y = t^2 - 1$.

The part of the curve between points $A(1, -1)$ and $B\left(\dfrac{1}{e}, 0\right)$ is

rotated a full turn about the y-axis. Find the exact volume of the resulting solid.

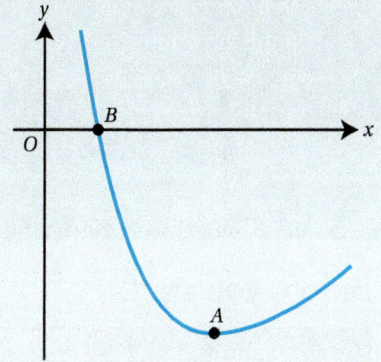

21 **a** Using the series for e^x, or otherwise, find the Maclaurin series of $x e^x$, stating the first four non-zero terms and the general term.

  b Hence find a series expansion of $\displaystyle\int_0^x t e^t \,dt$.

  c Hence show that $\dfrac{1}{2} + \dfrac{1}{3} + \dfrac{1}{4(2!)} + \dfrac{1}{5(3!)} + \ldots = 1$.

22 The part of the curve $y = x^2 + 3$ between $y = 3$ and $y = k$ $(k > 0)$ is rotated $360°$ about the y-axis. The volume of revolution formed is 25π.

Find the value of k.

23 Consider two curves with equations $y = x^2 - 8x + 12$ and $y = 12 + x - x^2$.

  a Find the coordinates of the points of intersection of the two curves.

The region enclosed by the curves is rotated through $360°$ about the x-axis.

  b Write down an integral expression for the volume of the solid generated.

  c Evaluate the volume, giving your answer to the nearest integer.

24 The region enclosed by $y = x^2$ and $y = \sqrt{x}$ is to be labelled R.

  a Draw a sketch showing R.

  b Find the volume when R is rotated through $360°$ about the x-axis.

  c Hence find the volume when R is rotated through $360°$ about the y-axis.

9 Polar coordinates

In this chapter you will learn how to:

- use polar coordinates to represent curves
- establish various properties of those curves
- convert between polar and Cartesian equations of a curve
- find the area enclosed by a polar curve, or between two curves.

Before you start…

A Level Mathematics Student Book 2, Chapter 7	You should be able to use radians.	1 a	Express $\dfrac{7\pi}{6}$ radians in degrees.
		b	State the exact value of $\sin \dfrac{4\pi}{3}$.
A Level Mathematics Student Book 1, Chapter 10	You should be familiar with graphs of trigonometric functions.	2 a	Find the set of values of θ, between 0 and 2π, for which $\cos\theta < 0$.
		b	State the greatest possible value of $5 - 2\cos\theta$.
A Level Mathematics Student Book 2, Chapter 11	You should be able to integrate trigonometric functions.	3	Evaluate:
		a	$\displaystyle\int_0^{\frac{\pi}{6}} \cos 3\theta \, d\theta$ $\qquad$ b $\displaystyle\int_0^{\frac{\pi}{2}} \sin^2\theta \, d\theta$

What are polar coordinates?

You are familiar with describing positions of points in the plane by using Cartesian coordinates, which represent the distance of a point from the x- and y-axes. But you are also familiar with bearings, which determine a direction in terms of an angle from a fixed line. If you know that a point lies on a certain bearing, you can describe its exact position by also specifying the distance from the origin. For example, this diagram shows that P is 4 cm from O on a bearing of 250°.

Polar coordinates use a similar idea: positions of points are described in terms of a direction and the distance from the origin. They can be used to describe curves that cannot easily be represented in Cartesian coordinates. In this chapter, you will learn about equations of curves such as these.

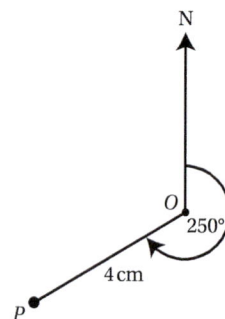

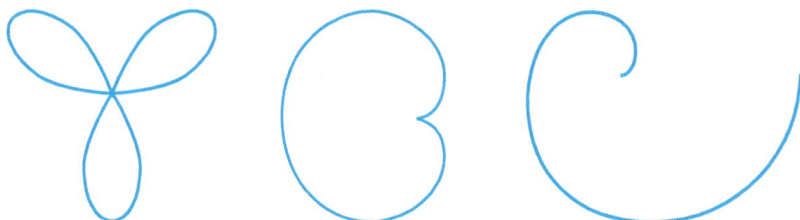

Because the distance from the origin explicitly features as a variable, polar coordinates are often used to describe quantities that vary with distance from a point, such as the strength of a gravitational field.

Section 1: Curves in polar coordinates

Polar coordinates describe the position of a point by specifying its distance from the origin (also called the **pole**) and the angle relative to a fixed line (called the **initial line**). By convention the initial line is drawn horizontally, in the direction of the positive x-axis, and the angle is measured anticlockwise.

You write polar coordinates as (r, θ), where r is the distance from the pole and θ is the angle. For example, point A in the diagram has polar coordinates $\left(3, \dfrac{\pi}{6}\right)$, point B $(5, \pi)$ and point $C\left(3, \dfrac{7\pi}{4}\right)$.

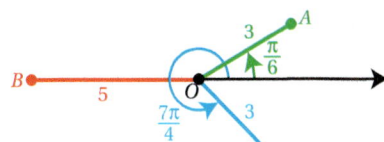

Notice that polar coordinates of a point are not uniquely defined. For example, you could also say that the polar coordinates of B are $(5, -\pi)$ or $(5, 3\pi)$. Conventionally, θ is taken to be between 0 and 2π (i.e. $0 \leqslant \theta < 2\pi$).

WORKED EXAMPLE 9.1

Points M and N have polar coordinates $\left(12, \dfrac{\pi}{4}\right)$ and $\left(9, \dfrac{5\pi}{6}\right)$. Find:

a the length MN
b the area of the triangle MON, where O is the pole.

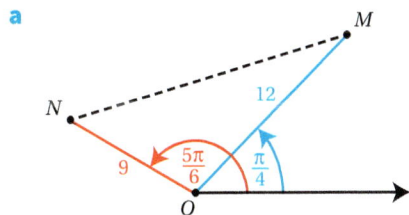

a

Polar coordinates give information about lengths and angles.

In triangle MON, $OM = 12$, $ON = 9$.

$\angle MON = \dfrac{5\pi}{6} - \dfrac{\pi}{4} = \dfrac{7\pi}{12}$

So:

You can use the cosine rule in triangle MON.

$MN^2 = 12^2 + 9^2 - 2 \times 12 \times 9 \cos\dfrac{7\pi}{12}$

$MN = \sqrt{280.9} = 16.8$

b Area $= \dfrac{1}{2}(12)(9)\sin\dfrac{7\pi}{12} = 52.2$

Use area of triangle $= \dfrac{1}{2}ab\sin C$.

In Cartesian coordinates, an equation of a curve gives a relationship between the x- and y-coordinate of any point on the curve. Similarly, a polar equation of a curve is a relationship between r and θ that holds for any point on the curve.

WORKED EXAMPLE 9.2

a Make a table of values for the curve with polar equation $r = \sqrt{\theta}$ for $0 \leqslant \theta < 2\pi$.
b Hence sketch the curve.

a

θ	O	$\frac{\pi}{4}$	$\frac{\pi}{2}$	$\frac{3\pi}{4}$	π	$\frac{5\pi}{4}$	$\frac{3\pi}{2}$	$\frac{7\pi}{4}$	2π
r	O	0.89	1.25	1.53	1.77	1.98	2.17	2.34	2.51

Use $r = \sqrt{\theta}$ to calculate r for various values of θ.

b

Plot the points and join them up.

There might be values of θ for which r is not defined. This happens when the expression for r has a negative value; r is a distance, so it must be

Tip

If you use a graphical calculator or graphing software to plot polar curves, you will find that some of them allow negative values of r, showing parts of a curve where we claim the curve is not defined. In this course, we require that r takes non-negative values.

WORKED EXAMPLE 9.3

Sketch the curve with equation $r = \sin 2\theta$ for $0 \leqslant \theta < 2\pi$.

Finding disallowed values of θ:
$\sin 2\theta \geqslant 0$

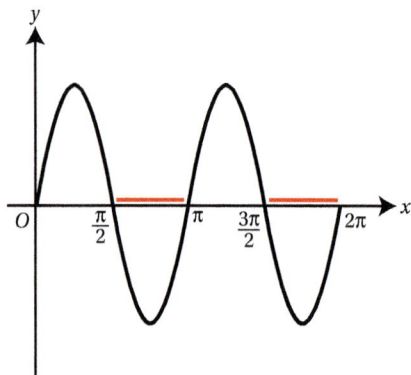

The curve is not defined when $\sin 2\theta < 0$. Sketch the graph of $y = \sin 2x$ to see where it is negative.

Continues on next page ...

r is not defined for:

$$\frac{\pi}{2}<\theta<\pi \text{ or } \frac{3\pi}{2}<\theta<2\pi$$

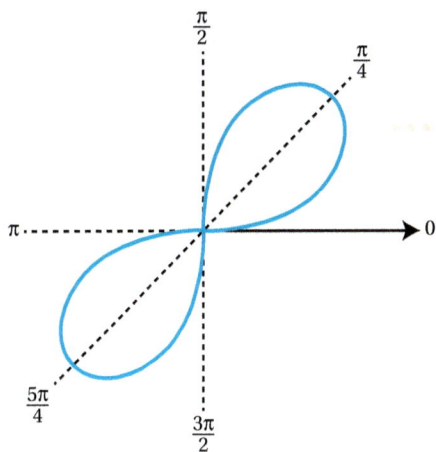

There should be no points in the sections $\frac{\pi}{2}<\theta<\pi$

and $\frac{3\pi}{2}<\theta<2\pi$.

You can also see from the graph that r increases for

$0<\theta<\frac{\pi}{4}$, then decreases for $\frac{\pi}{4}<\theta<\frac{\pi}{2}$, then repeats

the same values between π and $\frac{3\pi}{2}$.

non-negative. An effective way to identify such values is to sketch the graph of r against θ.

EXERCISE 9A

1 Plot the points with the given polar coordinates.

a i $\left(3,\frac{\pi}{4}\right)$ **ii** $\left(4,\frac{2\pi}{3}\right)$ **b i** $\left(5,\frac{7\pi}{6}\right)$ **ii** $\left(3,\frac{5\pi}{4}\right)$

c i $(2,\pi)$ **ii** $\left(1,\frac{3\pi}{2}\right)$

2 For points A and B with given polar coordinates, find the distance AB and the area of the triangle AOB.

a i $A\left(5,\frac{\pi}{6}\right),B\left(7,\frac{\pi}{4}\right)$ **ii** $A\left(2,\frac{\pi}{2}\right),B\left(5,\frac{2\pi}{3}\right)$

b i $A\left(10,\frac{\pi}{4}\right),B\left(8,\frac{7\pi}{6}\right)$ **ii** $A\left(4,\frac{3\pi}{4}\right),B\left(5,\frac{5\pi}{3}\right)$

c i $A\left(1,\frac{2\pi}{3}\right),B\left(2,\frac{11\pi}{6}\right)$ **ii** $A\left(6,\frac{7\pi}{4}\right),B\left(4,\frac{\pi}{4}\right)$

3 For each equation, make a table of values (for $0\leqslant\theta<2\pi$) and sketch the curve.

a i $r=2\theta$ **ii** $r=\theta^2$

b i $r=\theta^2-5\theta+6$ **ii** $r=\theta^2-2\theta$

c i $r=2\cos2\theta$ **ii** $r=\sin3\theta$

4 Shade the region described by each inequality. They are given in polar coordinates.

     a $r \leqslant 2$ **b** $1 \leqslant r \leqslant 3$ **c** $\dfrac{\pi}{4} < \theta \leqslant \pi$

5 A curve has polar equation $r = \cos 2\theta, 0 \leqslant \theta < 2\pi$.

     a State the values of θ for which the curve is not defined.

     b Hence sketch the curve.

6 A curve has polar equation $r = 4 \sin \theta$.

     a Find the set of values of θ for which r is not defined.

     b Show that the points A and B, with polar coordinates $\left(2, \dfrac{\pi}{6}\right)$ and $\left(4, \dfrac{\pi}{2}\right)$ lie on the curve.

     c Sketch the curve.

     d Find the exact length of AB.

Section 2: Some features of polar curves

When sketching curves in Cartesian coordinates you normally mark the axis intercepts, maximum and minimum points. For polar curves, there are similar features that you can deduce from the equation.

Minimum and maximum values of r

Since r is a function of θ, you can use differentiation to find its minimum and maximum values.

> ### 🔑 Key point 9.1
>
> The minimum and maximum values of r occur where $\dfrac{\mathrm{d}r}{\mathrm{d}\theta} = 0$.

WORKED EXAMPLE 9.4

A curve has polar equation $r = 150 + 9\theta^2 - 2\theta^3$ for $0 \leqslant \theta < 2\pi$.

a Find the minimum and maximum values of r, and the values of θ for which they occur.
b Explain why there is a point on the curve corresponding to every value of θ.
c Sketch the curve.

a $\dfrac{\mathrm{d}r}{\mathrm{d}\theta} = 18\theta - 6\theta^2$ The maximum and minimum values of r occur where $\dfrac{\mathrm{d}r}{\mathrm{d}\theta} = 0$.

     When $\dfrac{\mathrm{d}r}{\mathrm{d}\theta} = 0$:

     $18\theta - 6\theta^2 = 0$

     $6\theta(3 - \theta) = 0$

     $\theta = 0$ or $\theta = 3$

Continues on next page …

$$\frac{d^2r}{d\theta^2} = 18 - 12\theta$$

When $\theta = 0$:

$18 - 12(0) = 18 \, (> 0)$ so there is a minimum at $(150, 0)$.

When $\theta = 3$:

$18 - 12(3) = -18 \, (< 0)$ so there is maximum at $(177, 3)$.

When $\theta = 2\pi$, $r = 9.21$.

Hence the minimum value of $r = 9.21$ occurs when $\theta = 2\pi$.

The maximum value of $r = 177$ occurs when $\theta = 3$.

b r is always positive for $0 \leqslant \theta \leqslant 2\pi$, so the curve exists for all θ.

c

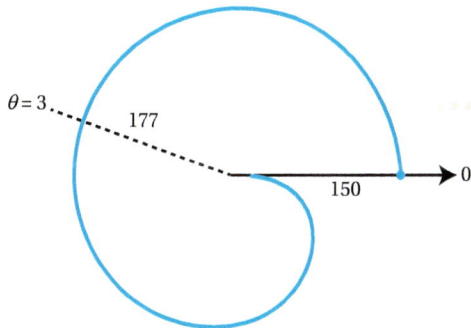

> When $\theta = 0$, $r = 150$.
>
> When $\theta = 3$, $r = 177$.
>
> The minimum value could occur at the end of the domain. You have already checked $\theta = 0$, so you just need to check $\theta = 2\pi$.
>
> For the curve to be defined, r needs to be positive.
>
> As θ increases from 0 to 2π, r increases from 150 to 177 (when $\theta = 3$, which is just below π) and then decreases to 9.21.

Polar equations often involve trigonometric functions. You might be able to use trigonometric graphs, rather than differentiation, to find the maximum and minimum values of r.

WORKED EXAMPLE 9.5

A curve has polar equation $r = 5 - 2\cos\theta$ for $0 \leqslant \theta < 2\pi$.

a   Find the largest and smallest values of r.
b   Hence sketch the curve.

a $-1 \leqslant \cos\theta \leqslant 1$

so $3 \leqslant 5 - 2\cos\theta \leqslant 7$

The largest value is $r = 7$ when $\theta = \pi$.
The smallest value is $r = 3$ when $\theta = 0, 2\pi$.

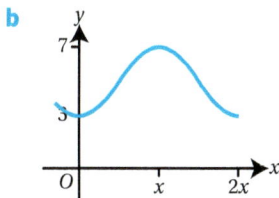

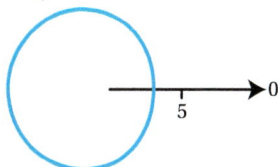

b

> Start by considering the minimum and maximum values of $\cos\theta$: -1, when $\theta = \pi$, and 1, when $\theta = 0$ and 2π.
>
> Sketch the graph of $y = 5 - 2\cos x$:
>
> You can see that r increases from 3 to 7 and decreases to 3 again.

Tangents at the pole

In Worked example 9.3 you saw a curve ($r = \sin 2\theta$) that is only defined for certain values of θ. This happens because we do not allow negative values of r.

The value of $r = \sin 2\theta$ changes from positive to negative, or vice versa, when $\theta = 0, \frac{\pi}{2}, \pi, \frac{3\pi}{2}, 2\pi$. Each of those θ values corresponds to a half-line, shown in red in the diagram. As the curve approaches each of the lines, r gets closer to zero (so points on the curve get closer and closer to the pole). This means that each of the lines $\theta = 0, \frac{\pi}{2}, \pi, \frac{3\pi}{2}$ is a tangent to the curve at the pole.

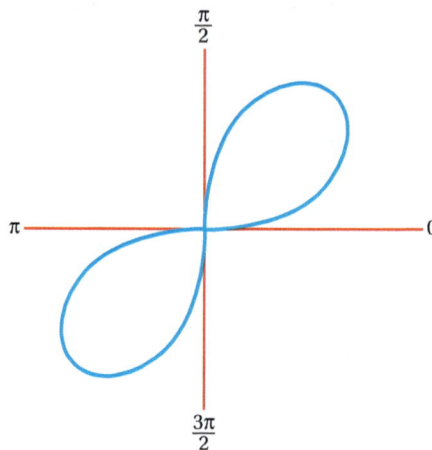

Key point 9.2

For a curve with polar equation $r = f(\theta)$, the line $\theta = \alpha$ is a **tangent at the pole** if $f(\alpha) = 0$ but $f(\alpha) > 0$ on one side of the line.

WORKED EXAMPLE 9.6

For the curve with polar equation $r = 2\cos 3\theta$, find the tangents at the pole and hence sketch the curve.

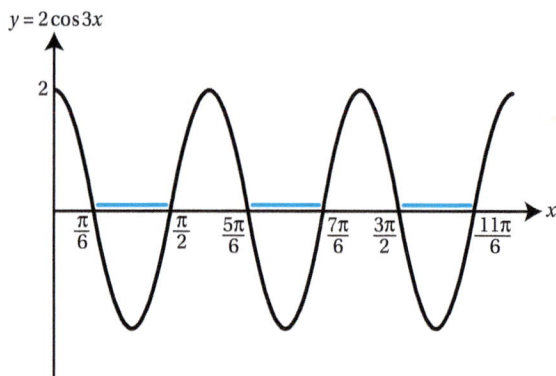

Sketch the graph of $y = 2\cos 3x$ to see which values of θ would produce negative values of r.

The tangents at the pole are:
$\theta = \frac{\pi}{6}, \frac{\pi}{2}, \frac{5\pi}{6}, \frac{7\pi}{6}, \frac{3\pi}{2}$ and $\frac{11\pi}{6}$.

The value of $2\cos 3\theta$ passes through zero when $\theta = \frac{\pi}{6}, \frac{\pi}{2}, \frac{5\pi}{6}, \frac{7\pi}{6}, \frac{3\pi}{2}$ and $\frac{11\pi}{6}$.

Continues on next page ...

Between the tangents where the curve is defined, the value of r increases from 0 to the maximum value of 2 and then decreases back to 0.

EXERCISE 9B

1 For each curve, find the maximum and minimum possible value of r, and the corresponding values of θ. Hence sketch the curve. (In all cases, $0 \leqslant \theta < 2\pi$.)

  a   **i** $r = 3 + 2\sin\theta$ **ii** $r = 5 + \cos\theta$

  b   **i** $r = 7 - 3\cos 2\theta$ **ii** $r = 5 - 2\sin 2\theta$

2 Find the equations of the tangents at the pole for each curve. Hence sketch the curve. (In all cases, $0 \leqslant \theta < 2\pi$.)

  a   **i** $r = 2\sin 3\theta$ **ii** $r = 3\cos 2\theta$

  b   **i** $r = 2 + 4\cos 3\theta$ **ii** $r = \sqrt{2} - 2\sin 4\theta$

3 Consider the curve with polar equation $r = 3\cos 4\theta$, $r \geqslant 0$ for $0 \leqslant \theta < 2\pi$.

  a   Find the equations of the tangents at the pole.

  b   State the set of values of θ for which the curve is not defined.

  c   Hence sketch the curve.

4   **a**   Sketch the curve with polar equation $r = 3 - 2\cos^2\theta$, $0 \leqslant \theta < 2\pi$.

  b   State the largest and smallest values of r.

5 Consider the curve with polar equation $r = 1 + 2\sin 2\theta$, $0 \leqslant \theta < 2\pi$.

  a   Find the range of values of θ for which the curve exists.

  b   Sketch the curve, labelling the tangents at the pole and indicating the points where r has maximum value.

6 A curve has polar equation $r = \dfrac{3}{2 - \sin\theta}$, $0 \leqslant \theta < 2\pi$.

    a Find the largest and smallest value of r and the values of θ at which they occur.

    b Hence sketch the graph.

7 **a** Sketch the graph of $y = x\left(x - \dfrac{2\pi}{3}\right)\left(x - \dfrac{3\pi}{2}\right)$ for $0 \leqslant x < 2\pi$.

    b Sketch the curve with polar equation $r = \theta\left(\theta - \dfrac{2\pi}{3}\right)\left(\theta - \dfrac{3\pi}{2}\right)$.

8 **a** Find the smallest and largest values of $y = 3x^2 - 6\pi x + 4\pi^2$ for $x \in [0, 2\pi]$.

    b Sketch the curve with the polar equation $r = 3\theta^2 - 6\pi\theta + 4\pi^2$, for $0 \leqslant x < 2\pi$.

9 Sketch the curve with equation $r = -\theta(\theta - \pi)^2(\theta - 2\pi)$ for $0 \leqslant \theta < 2\pi$.

10 Sketch the curve with equation $r = \tan\theta$ for $0 \leqslant \theta < 2\pi$.

Section 3: Changing between polar and Cartesian coordinates

You can use trigonometry to find the Cartesian coordinates of a point with given polar coordinates. Usually, the origin of the Cartesian coordinates is taken to be the pole, and the x-axis to be the initial line.

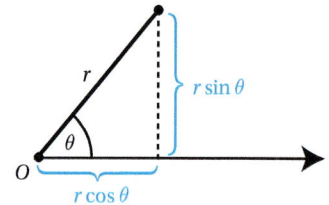

🔑 Key point 9.3

A point with polar coordinates (r, θ) has Cartesian coordinates $(r\cos\theta, r\sin\theta)$.

WORKED EXAMPLE 9.7

Points P and Q have polar coordinates $\left(4, \dfrac{\pi}{3}\right)$ and $\left(2, \dfrac{7\pi}{6}\right)$.

a Show points P and Q on the same diagram.

b Find the Cartesian coordinates of P and Q.

a

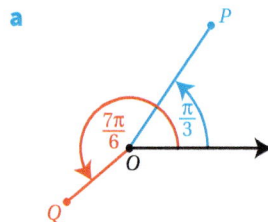

The first coordinate is the distance from the origin and the second coordinate is the angle.

$\dfrac{\pi}{3}$ is $60°$ and $\dfrac{7\pi}{6}$ is $210°$.

b For P:

Use $x = r\cos\theta$ and $y = r\sin\theta$.

$x = 4\cos\dfrac{\pi}{3} = 2$

$y = 4\sin\dfrac{\pi}{3} = 2\sqrt{3}$

So the Cartesian coordinates of P are $(2, 2\sqrt{3})$.

Continues on next page ...

For Q:

$x = 2\cos\dfrac{7\pi}{6} = -\sqrt{3}$

$y = 2\sin\dfrac{7\pi}{6} = -1$

So the Cartesian coordinates of Q are $(-\sqrt{3}, -1)$.

To change from Cartesian to polar coordinates, consider the same diagram again.

The value of r is the distance from the origin, so $r^2 = x^2 + y^2$. Remember that we require r to be positive. You need to be a little careful when finding the angle. Since $x = r\cos\theta$ and $y = r\sin\theta$, you can divide the two equations to get $\tan\theta = \dfrac{y}{x}$. However, there are two values $\theta \in [0, 2\pi)$ with the same value of $\tan\theta$; you need to consider the position of the point to decide which one is correct.

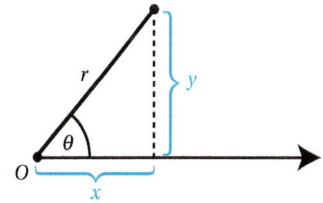

🔑 Key point 9.4

For a point with Cartesian coordinates (x, y) the polar coordinates satisfy:

- $r = \sqrt{x^2 + y^2}$

- $\tan\theta = \dfrac{y}{x}$

⏪ Rewind

This should remind you of the modulus and argument of a complex number – see Pure Core Student Book 1, Chapter 4, and Chapter 2 of this book.

WORKED EXAMPLE 9.8

Find the polar coordinates of the points $A(-3, 5)$ and $B(4, -1)$.

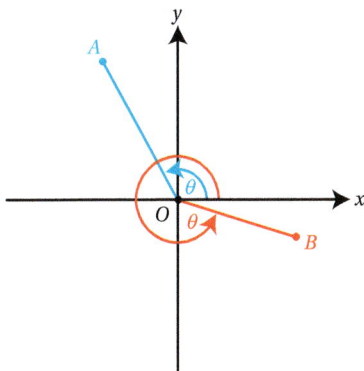

Start by plotting the points to see which angle to use.

For A:

$r = \sqrt{(-3)^2 + 5^2}$

$\quad = \sqrt{34}$

Use $r = \sqrt{x^2 + y^2}$.

Continues on next page ...

$\tan\theta = \dfrac{5}{-3}$

$\Rightarrow \theta = 2.11$ or $\theta = 5.25$

Find $\tan^{-1}\left(-\dfrac{5}{3}\right)$, then add π to get the two possible values between 0 and 2π.

Hence the polar coordinates of A are $(\sqrt{34}, 2.11)$.

The angle for A is smaller than π.

For B:

$r = \sqrt{4^2 + (-1)^2}$

$ = \sqrt{17}$

$\tan\theta = -\dfrac{1}{4}$

$\Rightarrow \theta = 2.90$ or $\theta = 6.04$

Hence the polar coordinates of B are $(\sqrt{17}, 6.04)$.

The angle for B is greater than π.

You can now convert equations of curves between polar and Cartesian forms.

WORKED EXAMPLE 9.9

Find the Cartesian equation of the curve with polar equation $r = 2\sin\theta$.

$r = \sqrt{x^2 + y^2}$

$\sin\theta = \dfrac{y}{r}$

$\therefore \sin\theta = \dfrac{y}{\sqrt{x^2 + y^2}}$

Use $r^2 = x^2 + y^2$ and $y = r\sin\theta$.

Hence:

$\sqrt{x^2 + y^2} = 2\left(\dfrac{y}{\sqrt{x^2 + y^2}}\right)$

Substitute for $\sin\theta$ in the equation of the curve.

$x^2 + y^2 = 2y$

Simplify if possible. In this case, multiply both sides by the 'square root' term.

WORKED EXAMPLE 9.10

Find the polar equation of the curve $(x^2 + y^2)^3 = 3xy$.

$(r^2)^3 = 3(r\cos\theta)(r\sin\theta)$

$r^6 = 3r^2\sin\theta\cos\theta$

$r^4 = 3\sin\theta\cos\theta$

Use $x = r\cos\theta$ and $y = r\sin\theta$.

You also know that $x^2 + y^2 = r^2$.

EXERCISE 9C

1 Each point is given in polar coordinates. Find the Cartesian coordinates.

a i $\left(5,\dfrac{\pi}{4}\right)$ **ii** $\left(3,\dfrac{\pi}{3}\right)$

b i $\left(\sqrt{2},\dfrac{3\pi}{4}\right)$ **ii** $\left(\sqrt{3},\dfrac{5\pi}{6}\right)$

c i $(6,4.1)$ **ii** $(3,5.7)$

2 Each point is given in Cartesian coordinates. Find the polar coordinates. Take $\theta \in [0,2\pi)$.

a i $(5,2)$ **ii** $(3,4)$

b i $(0,2)$ **ii** $(0,-3)$

c i $(-1,-5)$ **ii** $(4,-1)$

3 Find the polar equation of each curve.

a i $x^2+y^2=3xy$ **ii** $(x^2+y^2)^2=2x^2y$

b i $\dfrac{1}{x}+\dfrac{1}{y}=\dfrac{1}{5}$ **ii** $x^3+y^3=3$

c i $y=3x+1$ **ii** $x^2+y^2=6$

4 Find the Cartesian equation of each curve.

a i $r=2\theta$ **ii** $r=3\theta^2$

b i $r=4\sin\theta$ **ii** $r=2\cos\theta$

c i $r=2\tan\theta$ **ii** $r^2=\tan\theta$

5 A curve has polar equation $r=3\cos\theta$. Point Q, with polar coordinates $(2,\alpha)$ where $\pi<\alpha<2\pi$, lies on the curve.

a Find the value of α.

b Find Cartesian coordinates of Q.

c Find the Cartesian equation of the curve.

6 Find the polar equation of the circle $(x-1)^2+(y-1)^2=2$, giving your answer in the form $r=f(\theta)$.

7 Find the Cartesian equation of the curve with polar equation $r=3\tan\theta$.

8 A curve has polar equation $r=\dfrac{3}{2+\cos\theta}$.

Show that the Cartesian equation of the curve can be written as $3x^2+4y^2+6x=9$.

9 Find the Cartesian equation of the curve with polar equation:

a $r=\cos^3\theta$ **b** $r=\dfrac{1}{\cos^3\theta}$.

Section 4: Area enclosed by a polar curve

Finding the area bounded by a polar curve is similar to finding the area bounded by a Cartesian curve, except that rather than being the area between the curve, the x-axis and vertical lines $x = a$ and $x = b$, now it is the area between the curve, the pole and lines from the pole $\theta = \alpha$ and $\theta = \beta$.

> ### 🔑 Key point 9.5
>
> The area enclosed between a polar curve and the half-lines $\theta = \alpha$ and $\theta = \beta$ is
>
> $$A = \int_{\alpha}^{\beta} \frac{1}{2} r^2 \, d\theta.$$

PROOF 5

Consider a curve $r = f(\theta)$, where $r \geqslant 0$ and $\alpha < \beta$.

You can split the region into small sectors of angle $\Delta\theta$ and area ΔA.

The polar coordinates of the point P are (r, θ) and the polar coordinates of the nearby point Q are $(r + \Delta r, \theta + \Delta\theta)$.

The area of each sector is approximately the same as the area of a sector of a circle with angle ΔA and radius r:

$$\Delta A = \frac{1}{2} r^2 \Delta\theta$$

The area of a sector of a circle is $A = \frac{1}{2} r^2 \theta$.

The total area is approximately:

$$A \approx \sum_{\theta = \alpha}^{\theta = \beta} \frac{1}{2} r^2 \Delta\theta$$

Summing all these sectors between $\theta = \alpha$ and $\theta = \beta$ gives the approximate total area.

$$A = \lim_{\Delta\theta \to 0} \sum_{\theta = \alpha}^{\theta = \beta} \frac{1}{2} r^2 \Delta\theta$$

The approximation becomes more and more accurate as the angle gets smaller.

$$= \int_{\alpha}^{\beta} \frac{1}{2} r^2 \, d\theta$$

In the limit as $\Delta\theta \to 0$ the sum becomes an integral.

WORKED EXAMPLE 9.11

In this question you must show detailed reasoning.

The diagram shows the curve with polar equation $r = 3 - 2\sin\theta$.

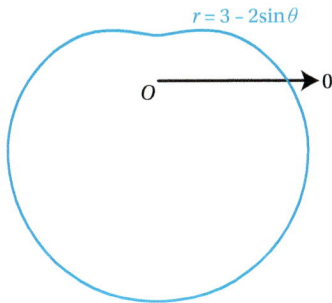

Find the area enclosed by the curve.

$A = \displaystyle\int_0^{2\pi} \frac{1}{2}(3 - 2\sin\theta)^2 \, d\theta$ .... Use the formula $A = \displaystyle\int_\alpha^\beta \frac{1}{2}r^2 \, d\theta$.

$= \dfrac{1}{2}\displaystyle\int_0^{2\pi} (9 - 12\sin\theta + 4\sin^2\theta) \, d\theta$ .... Expand the brackets.

$= \dfrac{1}{2}\displaystyle\int_0^{2\pi} (9 - 12\sin\theta + 2 - 2\cos2\theta) \, d\theta$ .... To integrate $\sin^2\theta$, use the $\cos2\theta$ identity: $4\sin^2\theta = 2 - 2\cos2\theta$.

$= \dfrac{1}{2}\displaystyle\int_0^{2\pi} (11 - 12\sin\theta - 2\cos2\theta) \, d\theta$

$= \dfrac{1}{2}\Big[11\theta + 12\cos\theta - \sin2\theta\Big]_0^{2\pi}$

$= \dfrac{1}{2}\Big[(22\pi + 12 - 0) - (0 + 12 - 0)\Big]$

$= 11\pi$

EXERCISE 9D

1 Find the area enclosed between these polar curves and half-lines.

a i $r^2 = \cos2\theta,\, a = -\dfrac{\pi}{4},\, b = \dfrac{\pi}{4}$ **ii** $r^2 = \sin3\theta,\, a = 0,\, b = \dfrac{\pi}{3}$

b i $r = 2\theta,\, a = 0,\, b = \pi$ **ii** $r = \theta^2,\, a = 0,\, b = \pi$

c i $r = 2e^\theta,\, a = 0,\, b = 2\pi$ **ii** $r = e^{\frac{\theta}{2}},\, a = 0,\, b = 2\pi$

d i $r = \cos\theta,\, a = 0,\, b = \dfrac{\pi}{2}$ **ii** $r = \sin\theta,\, a = \dfrac{\pi}{4},\, b = \dfrac{\pi}{2}$

e i $r = 1 + \sin\theta,\, a = -\dfrac{\pi}{2},\, b = \dfrac{\pi}{2}$ **ii** $r = 1 - \cos\theta,\, a = 0,\, b = 2\pi$

2 a Write down the polar equation of a circle of radius a with centre at the pole.

b Using your answer to part **a**, show that the area of the circle is πa^2.

> **Focus on …**
>
> Some areas can only be calculated exactly using polar coordinates. Focus on … Proof 2 looks at one important example.

3 Find the exact value of the area enclosed between the curve $r = \tan\theta$, the initial line and the half-line $\theta = \dfrac{\pi}{4}$, clearly showing all your working.

4 The diagram shows the curve with polar equation $r = 2 + \cos\theta$, $0 \leqslant \theta < 2\pi$.

$r = 2 + \cos\theta$

Show that the exact area enclosed by the curve is $\dfrac{9\pi}{2}$.

5 The diagram shows the curve with polar equation $r = 5 + 2\sin\theta$, $0 \leqslant \theta < 2\pi$.

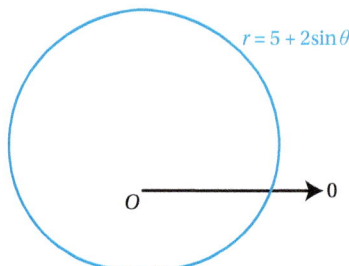

$r = 5 + 2\sin\theta$

Find the exact area enclosed by the curve, clearly showing all your working.

6 **a** Sketch the curve with polar equation $r = \theta^{-\frac{1}{2}}$, $0 \leqslant \theta < 2\pi$.

   b Show that the area enclosed between the lines $\theta = a$ and $\theta = 2a$, where $0 < a \leqslant \pi$, is independent of a.

7 **a** Sketch the curve C with polar equation $r = 3\sin 2\theta$, $0 \leqslant \theta < 2\pi$.

   b Find the total area of the region enclosed by C, clearly showing all your working.

8 The curve C has polar equation $r = a\cos 2\theta$, for $r \geqslant 0$ and $0 \leqslant \theta < 2\pi$.

   a Sketch C, giving the equations of any tangents at the pole.

   b Find, in terms of a, the total area of the region enclosed by C, clearly showing all your working.

9 The area of the region enclosed between the curve with polar equation $r = a(1 + \tan\theta)$, the initial line and the half-line $\theta = k$ is $a^2\left(\ln 2 + \dfrac{\sqrt{3}}{2}\right)$.

Find the value of the positive constant k.

10 The diagram shows the curve with polar equation $r = a\sin\theta \sin 2\theta$, $0 \leqslant \theta < \dfrac{\pi}{2}$.

$r = a\sin\theta \sin 2\theta$

Show that the area of the region enclosed by the curve is $\dfrac{\pi a^2}{16}$.

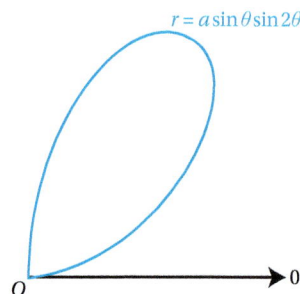

Section 5: Area between two curves

To find the area enclosed between two polar curves, find the intersection points of the curves and calculate the part of the area bounded by each curve separately.

WORKED EXAMPLE 9.12

Two curves have polar equations:

$$C_1 : r = 3 + \cos\theta$$

$$C_2 : r = 7\cos\theta$$

for $0 \leqslant \theta < 2\pi$.

a Find the polar coordinates of the points of intersection of C_1 and C_2.
b Find the exact value of the area of the finite region enclosed between C_1 and C_2.

a The curves intersect where

$$3 + \cos\theta = 7\cos\theta$$

$$\cos\theta = \frac{1}{2}$$

> Equate the equations of the two curves.

$$\theta = \frac{\pi}{3}, \frac{5\pi}{3}$$

> Solve for θ. There are two values between 0 and 2π.

From C_2: when $\cos\theta = \frac{1}{2}$, $r = \frac{7}{2}$.

> Find the corresponding values of r.

The points of intersection are $\left(\frac{7}{2}, \frac{\pi}{3}\right)$ and $\left(\frac{7}{2}, \frac{5\pi}{3}\right)$.

b Sketching the curves:

> It is a good idea to sketch the curves to see where the required region is.

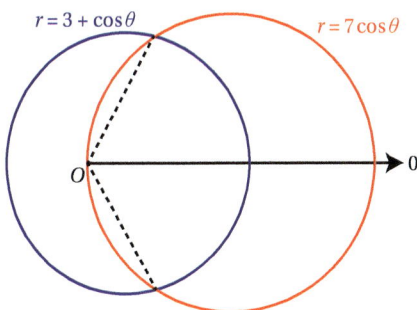

$$\text{Area} = 2\left[\int_0^{\frac{\pi}{3}} \frac{1}{2}(3 + \cos\theta)^2 \, d\theta + \int_{\frac{\pi}{3}}^{\frac{\pi}{2}} \frac{1}{2}(7\cos\theta)^2 \, d\theta\right]$$

> The required region above the initial line is made up of two parts: one bounded by C_1 between $\theta = 0$ and $\theta = \frac{\pi}{3}$ and one bounded by C_2 between $\theta = \frac{\pi}{3}$ and $\theta = \frac{\pi}{2}$.
>
> By symmetry about the initial line, the full area is double this.

Continues on next page ...

$$= \int_0^{\frac{\pi}{3}} \left(9 + 6\cos\theta + \cos^2\theta \right) d\theta + \int_{\frac{\pi}{3}}^{\frac{\pi}{2}} 49\cos^2\theta \, d\theta \quad \cdots\cdots\cdots \quad \boxed{\text{Expand and use } \cos^2\theta = \frac{\cos 2\theta + 1}{2}.}$$

$$= \int_0^{\frac{\pi}{3}} \left(9 + 6\cos\theta + \frac{\cos 2\theta + 1}{2} \right) d\theta + \int_{\frac{\pi}{3}}^{\frac{\pi}{2}} \frac{49}{2} \left(\cos 2\theta + 1 \right) d\theta$$

$$= \left[9\theta + 6\sin\theta + \frac{1}{4}\sin 2\theta + \frac{1}{2}\theta \right]_0^{\frac{\pi}{3}} + \frac{49}{2}\left[\frac{1}{2}\sin 2\theta + \theta \right]_{\frac{\pi}{3}}^{\frac{\pi}{2}}$$

$$= \left[\left(\frac{19\pi}{6} + 3\sqrt{3} + \frac{\sqrt{3}}{8} \right) - (0) \right] + \frac{49}{2}\left[\left(\frac{\pi}{2} \right) - \left(\frac{\sqrt{3}}{4} + \frac{\pi}{3} \right) \right]$$

$$= \frac{29\pi}{4} - 3\sqrt{3}$$

💡 **Tip**

In more complicated questions where the region is between two curves, remember that in polar coordinates you are finding the area of a sector bounded by the curve and two half-lines from the pole; not a region bounded by two vertical lines as in Cartesian coordinates.

EXERCISE 9E

1 The diagram shows the curve with polar equation $r = 2$ and the line with polar equation $r = \sqrt{3}\sec\theta$, both defined for $0 \leqslant \theta < \frac{\pi}{2}$.

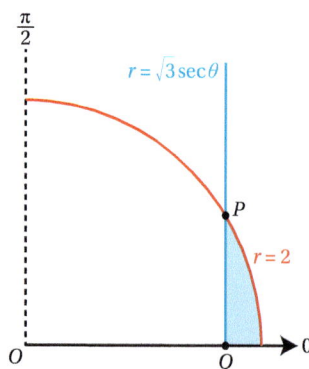

The line intersects the curve at the point P and the initial line at the point Q.

a Find the polar coordinates of P.

b **i** Find the exact area of the triangle OPQ.
　　ii Hence show that the area of the shaded region is $\dfrac{2\pi - 3\sqrt{3}}{6}$.

2 The diagram shows the curve with polar equation $r = a(1 + \cos\theta)$, $0 \leqslant \theta < \pi$.

Find, in terms of a, the exact area of the shaded region R.

3 The diagram shows the curves with polar equations $r = a$ and $r = 2a\sin 2\theta$ for $0 \leqslant \theta < \dfrac{\pi}{2}$.

a Find the polar coordinates of the points of intersection of the two curves.

b Find the exact area of the shaded region enclosed within both curves.

4 The diagram shows the curves with polar equations $r = \dfrac{1}{2}$ and $r = 1 - \sin\theta$, $0 \leqslant \theta < 2\pi$.

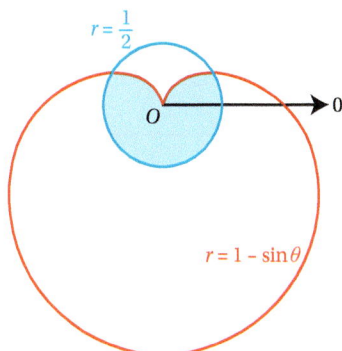

a Find the polar coordinates of the points of intersection of the two curves.

b Find the exact area enclosed inside both curves which is shaded on the diagram.

5 The diagram shows the curves C_1 and C_2 with polar equations $r = 6 - 6\cos\theta$ and $r = 2 + 2\cos\theta$, $0 \leqslant \theta < 2\pi$.

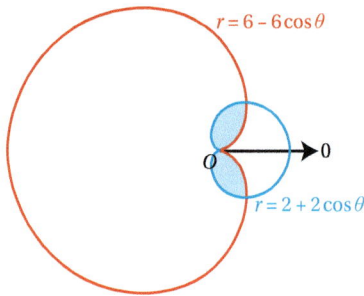

$r = 6 - 6\cos\theta$

O

0

$r = 2 + 2\cos\theta$

   a Find the polar coordinates of the points of intersection of the two curves.

   b Show that the area of the shaded region is $2\left(11\pi - 18\sqrt{3}\right)$.

6 **a** On the same axes, sketch the curves with polar equations $r = 2$ and $r = 3 - 2\cos\theta$ for $0 \leqslant \theta < 2\pi$.

   b Show that the exact value of the area inside $r = 3 - 2\cos\theta$ but outside $r = 2$ is $\dfrac{33\sqrt{3} + 28\pi}{6}$.

Checklist of learning and understanding

- Polar coordinates (r, θ) describe the position of a point in terms of its distance from the pole and the angle measured anticlockwise from the initial line.
- The connection between polar and Cartesian coordinates is:
  - $x = r\cos\theta$, $y = r\sin\theta$
  - $r = \sqrt{x^2 + y^2}$, $\tan\theta = \dfrac{y}{x}$
- For a curve with equation given in polar coordinates:
  - r cannot be negative, so not all values of θ are possible
  - there might be one or more tangents at the pole, given by the values of θ for which $r = 0$.
- The area enclosed between a polar curve and the half-lines $\theta = \alpha$ and $\theta = \beta$ is $A = \displaystyle\int_\alpha^\beta \frac{1}{2} r^2 \, d\theta$.
- To find the area enclosed between two polar curves, find the intersection points of the curves and calculate the part of the area bounded by each curve separately.

Mixed practice 9

1. Find the greatest distance from the pole of any point on the curve $r = 2(5 - 3\sin\theta)$, $0 \leqslant \theta < 2\pi$.

2. Find the polar equation of the curve $x^2 + y^2 = a(x - y)$.

3. Points A and B have polar coordinates $\left(7, \dfrac{\pi}{4}\right)$ and $\left(4, \dfrac{5\pi}{6}\right)$. Find:
   a the distance AB
   b the area of the triangle AOB.

4. Sketch the curve with polar equation $r = 1 - \cos 2\theta$, $0 \leqslant \theta < 2\pi$.

5. Find the area of the region enclosed between the initial line and the curve with polar equation
   $r = \sqrt{2a}\, e^{\frac{\theta}{2}}$, $0 \leqslant \theta < 2\pi$.

6. a Sketch the curve with polar equation $r = \theta$, $0 \leqslant \theta < 2\pi$.
   b Show that the area bounded by the curve and the initial line is $\dfrac{4\pi^3}{3}$.

7. a Sketch the curve with polar equation $r = 5 - 4\cos\theta$, $0 \leqslant \theta < 2\pi$.
   b Find the exact area enclosed by the curve, clearly showing your working.

8. The Cartesian equation of a curve is $x^2 + y^2 - 2x = \sqrt{x^2 + y^2}$, $r \geqslant 0$.
   a Show that the polar equation of the curve is $r = 1 + 2\cos\theta$.
   b Find the equation of any tangents at the pole.
   c Hence sketch the curve for $0 \leqslant \theta < 2\pi$.
   d Show that the area enclosed by the curve is $2\pi + \dfrac{3\sqrt{3}}{2}$.

9. Find a Cartesian equation for the curve $r = 3\cos^2\theta$, $0 \leqslant \theta < 2\pi$.

10. A curve has polar equation $r = 2 + 4\sin\theta$, $0 \leqslant \theta < 2\pi$.
    a Find the equations of the tangents at the pole.
    b Sketch the curve.
    c Find the Cartesian equation of the curve.

11. a Sketch the curve with polar equation $r = 5 - 3\cos\theta$, $0 \leqslant \theta < 2\pi$.
    b Find the Cartesian coordinates of the point that is furthest away from the pole.

12. A curve has polar equation $r = 2 - 4\sin 2\theta$, $0 \leqslant \theta < 2\pi$.
    a Find the equations of the tangents at the pole.
    b State the polar coordinates of the points at greatest distance from the pole.
    c Hence sketch the graph.

13. A curve is defined by the polar equation $r = 3\theta$ for $0 \leqslant \theta < 2\pi$.
    a Sketch the curve.
    b Find the Cartesian coordinates of the point where the curve intersects the line $\theta = \dfrac{2\pi}{3}$.

14 Sketch the curve with polar equation $r = 5\cos 3\theta, 0 \leqslant \theta < 2\pi$. Indicate the equations of the tangents at the pole, and give the polar coordinates of the point where the curve crosses the initial line.

15 The diagram shows the curve with polar equation $r = 2 + 4\cos\theta, 0 \leqslant \theta < 2\pi$.

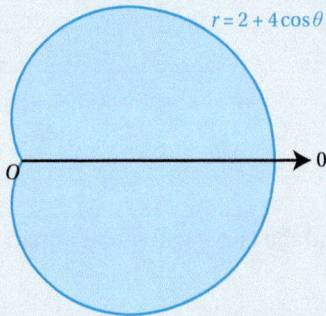

Find the exact value of the shaded area, clearly showing your working.

16 The diagram shows the curve C with polar equation $r = 10 - 10\cos\theta, 0 \leqslant \theta < 2\pi$.

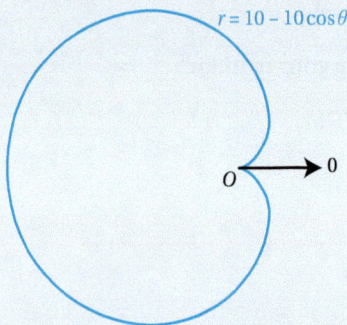

a Show that the area of the region bounded by C is 150π.

The circle $x^2 + y^2 = 25$ intersects C at the points A and B.

b Find the polar coordinates of A and B.

c Find the area enclosed between the circle and C.

17 The diagram shows the curve with polar equation $r = \cos\theta + \cos 3\theta, 0 \leqslant \theta < \pi$.

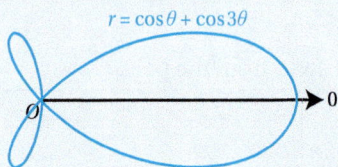

a i Show that $\cos 3\theta \equiv 4\cos^3\theta - 3\cos\theta$.

  ii Hence find the equations of the tangents at the pole.

b Show that the area enclosed in the large loop is $\dfrac{3\pi + 8}{12}$.

18 The equation of a curve is $x^2 + y^2 - x = \sqrt{x^2 + y^2}$.

  i Find the polar equation of this curve in the form $r = f(\theta)$.

  ii Sketch the curve.

  iii The line $x + 2y = 2$ divides the region enclosed by the curve into two parts. Find the ratio of the two areas.

© OCR A Level Mathematics, Unit 4726/01 Further Pure Mathematics 2, June 2013

19

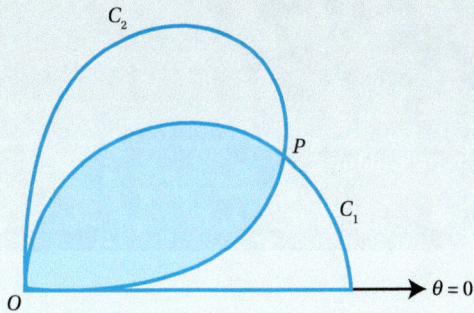

The diagram shows two curves, C_1 and C_2, which intersect at the pole O and at the point P. The polar equation of C_1 is $r = \sqrt{2}\cos\theta$ and the polar equation of C_2 is $r = \sqrt{2\sin 2\theta}$. For both curves, $0 \leqslant \theta \leqslant \frac{1}{2}\pi$. The value of θ at P is α.

  i Show that $\tan\alpha = \frac{1}{2}$.

  ii Show that the area of the region common to C_1 and C_2, shaded in the diagram, is $\frac{1}{4}\pi - \frac{1}{2}\alpha$.

© OCR A Level Mathematics, Unit 4726 Further Pure Mathematics 2, January 2012

10 Differential equations

In this chapter you will learn how to:

- understand and use the language associated with differential equations
- solve differential equations of the form $\dfrac{dy}{dx} + P(x)y = Q(x)$
- solve differential equations of the form $a\dfrac{d^2 y}{dx^2} + b\dfrac{dy}{dx} + cy = f(x)$
- use substitutions to turn differential equations into the required form.

Before you start…

A Level Mathematics Student Book 2, Chapter 13	You should know how to solve separable differential equations.	1	Solve $\dfrac{dy}{dx} = xy$.
A Level Mathematics Student Book 2, Chapter 10	You should know how to differentiate expressions, including using the product and chain rules.	2	Differentiate $y = xe^{3x}$ with respect to x.
A Level Mathematics Student Book 2, Chapter 11	You should be able to use various methods to integrate expressions.	3	Integrate xe^x with respect to x.

Introduction

In many academic areas such as Physics and Economics it is important to describe situations in terms of rates of change. This produces differential equations. In this chapter you will extend the types of differential equations you can solve. You will then see in Chapter 11 how these methods can be applied in many real-life situations.

⏮ **Rewind**

You met differential equations in A Level Mathematics Student Book 2, Chapter 13.

ⓘ **Did you know?**

If you read other books about differential equations you will see that the equations covered in this book are referred to as ordinary differential equations (ODEs). This is in contrast to another type of differential equation called partial differential equations (PDEs) which use a different type of differentiation.

Section 1: Terminology of differential equations

Differential equations have an **independent variable** (which is the variable on the bottom of the derivatives) and at least one **dependent variable** (which is the variable on the top of the derivatives). For example, in the equation $\dfrac{dy}{dx} = x^2 + y$ the independent variable is x and the dependent variable is y.

There are many different types of differential equation. To decide which technique to use when solving differential equations you need to be able to categorise them.

- The **order of a differential equation** is the largest number of times the dependent variable is differentiated. For example: $\dfrac{d^3y}{dx^3} + 3\dfrac{dy}{dx} + y = x^2$ is a third-order differential equation.

- A **linear differential equation** is one in which the dependent variable (y in these examples) only appears to the power of 1 (or not at all) in any expression. For example: $x^2\dfrac{dy}{dx} + 3\dfrac{d^2y}{dx^2} + \sin x = 0$ is a linear differential equation, but any differential equation involving y^2, $\sin y$ or even $y\dfrac{dy}{dx}$ is non-linear.

- A **homogeneous differential equation** is one where every term involves the dependent variable. For example: $\dfrac{d^2y}{dx^2} + x\dfrac{dy}{dx} + y^2 = 0$ is homogeneous, but $\dfrac{dy}{dx} = y + 2$ is a **non-homogeneous differential equation**. Every non-homogeneous differential equation has a homogeneous differential equation associated with it, formed by removing all the terms not involving the dependent variable.

> **Tip**
>
> In some books homogeneous is used to refer to a different property of differential equations. The definition given here is the only one relevant to this course.

WORKED EXAMPLE 10.1

Consider the differential equation $x\dfrac{dy}{dx} + 3y = \ln x$.

a State the order of the differential equation.
b Is the equation linear? Explain your answer.
c Explain why the equation is non-homogeneous, and write down the associated homogeneous equation.

a The equation is first order.

> The highest derivative of y in the equation is $\dfrac{dy}{dx}$, which is a first derivative.

b The equation is linear because $3y$ and $\dfrac{dy}{dx}$ are both linear terms.

> The only occurrences of the dependent variable are y and $\dfrac{dy}{dx}$, and they are both linear.
>
> (Note that it doesn't matter that $\dfrac{dy}{dx}$ is multiplied by x.)

Continues on next page ...

c The equation is non-homogeneous because the term $\ln x$ does not contain y.

> Non-homogeneous equations include terms which do not contain the dependent variable.

The associated homogeneous equation is

> Remove all terms without y to get a homogeneous equation.

$$x\frac{dy}{dx} + 3y = 0$$

To solve a differential equation, you need to find y as a function of x (if y is the dependent variable and x is the independent variable). When solving a differential equation, because the process is effectively integration, there will be arbitrary constants involved. The solution containing all the arbitrary constants is called the **general solution**.

🔑 Key point 10.1

The general solution to an nth-order differential equation has n arbitrary constants.

To fix the arbitrary constants you either use initial conditions (values of y, $\frac{dy}{dx}$, etc. at one value of x) or boundary conditions (values of y, $\frac{dy}{dx}$, etc. at several values of x). You need one piece of information for each arbitrary constant, and then you normally use simultaneous equation techniques to find the values of each constant. When the constants in the general solution have the values required to fit the conditions, the result is called the **particular solution**.

WORKED EXAMPLE 10.2

a Show that $y = Ax^3 + \dfrac{B}{x^2} + 3x^2$ is the general solution of the differential equation

$$x^2\frac{d^2y}{dx^2} - 6y + 12x^2 = 0.$$

b Find the particular solution that satisfies $y = 7$ and $\dfrac{dy}{dx} = 3$ when $x = 1$.

a $y = Ax^3 + \dfrac{B}{x^2} + 3x^2$

$\dfrac{dy}{dx} = 3Ax^2 - \dfrac{2B}{x^3} + 6x$

$\dfrac{d^2y}{dx^2} = 6Ax + \dfrac{6B}{x^4} + 6$

Continues on next page ...

Then:

$$x^2 \frac{d^2 y}{dx^2} - 6y + 12x^2$$

Substitute y and $\frac{d^2 y}{dx^2}$ into the given differential equation.

$$= \left(6Ax^3 + \frac{6B}{x^2} + 6x^2 \right) - \left(6Ax^3 + \frac{6B}{x^2} + 18x^2 \right) + 12x^2$$

$$= 0$$

Hence the given function is the general solution.

The given function satisfies the differential equation for all values of A and B. Notice that it also contains two arbitrary constants, as stated in Key point 10.1.

b When $x = 1$:

$$y = A + B + 3 = 7$$

$$\frac{dy}{dx} = 3A - 2B + 6 = 3$$

Substitute the given values of x, y and $\frac{dy}{dx}$ into the general solution.

$$\Leftrightarrow \begin{cases} A + B = 4 \\ 3A - 2B = -3 \end{cases}$$

Solve the simultaneous equations (you can use a calculator).

$$\therefore A = 1, B = 3$$

The particular solution is

$$y = x^3 + \frac{3}{x^2} + 3x^2$$

Non-homogeneous linear equations

In general, homogeneous equations are easier to solve than non-homogeneous ones. Fortunately, it turns out that, for linear equations, a general solution of a non-homogeneous equation can be found by using the general solution of the associated homogeneous equation, called the **complementary function**.

Once you have the complementary function, all you need is one solution of the full non-homogeneous equation. This is called a **particular integral**. Combining this with the complementary function gives the general solution of the full non-homogeneous equation.

For example, the differential equation $x^2 \frac{d^2 y}{dx^2} - 6y + 12x^2 = 0$ from Worked example 10.2 is non-homogeneous. You can check that $y_P = 3x^2$ satisfies this equation, so it can be used as a particular integral.

The associated homogeneous equation is $x^2 \frac{d^2 y}{dx^2} - 6y = 0$ (obtained by removing the term without y). You can check that $y_c = Ax^3 + \frac{B}{x^2}$

satisfies this homogeneous equation for all values of A and B; hence this is the complementary function. The general solution of the non-homogeneous equation is then $y = Ax^3 + \frac{B}{x^2} + 3x^2$, which is the sum of the particular integral and the complementary function.

> **Tip**
>
> The complementary function will contain arbitrary constants.

> **Fast forward**
>
> In Section 4 you will learn how to 'guess' the form of a particular integral in some specific cases.

> **Tip**
>
> Particular integrals are not unique. For example, $y = 3x^2 + 5x^3$, $y = 3x^2 + \frac{2}{x^2}$ and $y = 3x^2 + 5x^3 + \frac{2}{x^2}$ can all be taken as particular integrals for our equation. You may wish to investigate how using different particular integrals affects the values of the constants A and B for the particular solution in part **b** of Example 10.2.

For a linear differential equation, the general solution is given by

$$y = y_c + y_p$$

where y_c is the complementary function and y_p is a particular integral.

> 💡 **Tip**
>
> Don't confuse particular integrals with the particular solution. A particular integral is *any* solution of a differential equation. The particular solution needs to satisfy given initial conditions.

The proof of the result in Key point 10.2 is shown in Proof 6. It generalises the observations made in the discussion above. You do not need to learn (or even understand) this proof, but it might help to explain the result. It also shows you an example of a type of proof often seen in advanced mathematics.

Any linear differential equation can be written as $L[y] = f(x)$ where L is called a linear differential operator. For example: if $\dfrac{d^2 y}{dx} + x\dfrac{dy}{dx} + 3y = x^3$ then $L[y]$ is $\dfrac{d^2 y}{dx^2} + x\dfrac{dy}{dx} + 3y$ and $f(x) = x^3$.

Proof 6 uses the fact that $L[y_1 + y_2] = L[y_1] + L[y_2]$. This is because the derivative of a sum equals the sum of the derivatives.

PROOF 6

If y_c is the complementary function and y_p is the particular integral of a linear differential equation $L[y] = f(x)$, then $y = y_c + y_p$ will be a solution to the differential equation.

The complementary function, y_c, is defined by:

$L[y_c] = 0$ (1) y_c is the solution of the associated homogeneous equation.

The particular integral is any function, y_p, satisfying

$L[y_p] = f(x)$ (2) y_p is a solution of the full equation.

Then if $y = y_c + y_p$,

$L[y] = L[y_c + y_p]$ Use the given fact about linear differential operators.

$\quad = L[y_c] + L[y_p]$

$\quad\quad = 0 + f(x)$ Use properties (1) and (2).

So y also satisfies $L[y] = f(x)$.

WORKED EXAMPLE 10.3

The differential equation $x\dfrac{dy}{dx} + y = \ln x$ is satisfied for $x > 0$.

a Find the complementary function.
b A particular integral has the form $y = a\ln x + b$. Find the values of a and b.
c Hence find the particular solution with initial condition $y = 3$ when $x = 1$.

a The associated homogenous differential equation is

$$x\dfrac{dy}{dx} + y = 0$$

$$\dfrac{dy}{dx} = -\dfrac{y}{x}$$

$$\dfrac{1}{y}\dfrac{dy}{dx} = -\dfrac{1}{x}$$

$$\int \dfrac{1}{y}\,dy = \int -\dfrac{1}{x}\,dx$$

$$\ln y = -\ln x + c$$

$$= \ln\left(\dfrac{1}{x}\right) + \ln A$$

$$= \ln\left(\dfrac{A}{x}\right)$$

$$\therefore y_c = \dfrac{A}{x}$$

> The given equation is non-homogeneous. The complementary function is the solution of the associated homogeneous equation, obtained by removing any terms which do not contain y.

> You need to be able to recognise that this differential equation is separable.

> Write c as $\ln A$ in order to use rules of logarithms.

> Remember that the complementary function should contain one arbitrary constant (since the equation is first order).

b If $y = a\ln x + b$, then $\dfrac{dy}{dx} = \dfrac{a}{x}$.

Substituting: $x \times \dfrac{a}{x} + a\ln x + b = \ln x$

So, $a + b + a\ln x = \ln x$

Comparing coefficients of $\ln x$: $a = 1$

> Frequently when finding the particular integral you look at the coefficients on both sides of the equation.

Comparing coefficients of the constant term and using the fact that $a = 1: a + b = 0$

$$b = -1$$

So the particular integral is $\ln x - 1$.

c The general solution is

$$y = \dfrac{A}{x} + \ln x - 1.$$

Using the initial condition, substituting in $x = 1$ when $y = 3$:

$$3 = A + \ln 1 - 1$$

$$A = 4$$

> The general solution is the sum of the complementary function and the particular integral.

> To find the particular solution, use the initial condition to find the value of the constant A.

Therefore the particular solution is $y = \dfrac{4}{x} + \ln x - 1$.

EXERCISE 10A

1 Write an example of each type of differential equation.

a i Linear   **ii** Non-linear

b i Second order   **ii** Third order

c i Homogeneous   **ii** Non-homogeneous

2 Classify each differential equation, stating its order, whether or not it is linear, and whether it is homogeneous or non-homogeneous.

a i $\dfrac{d^2y}{dx^2}+3\dfrac{dy}{dx}+4y=\sin x$ **ii** $\dfrac{d^2y}{dx^2}+3\dfrac{dy}{dx}+4x=0$

b i $\dfrac{d^2y}{dx^2}+3y\dfrac{dy}{dx}+4y=0$ **ii** $\left(\dfrac{dy}{dx}\right)^2+4y=x^2$

c i $5\dfrac{d^2z}{dt^2}+\dfrac{dz}{dt}+\sin z+e^t=0$ **ii** $t^3\dfrac{d^3z}{dt^3}+t^2\dfrac{d^2z}{dt^2}+t\dfrac{dz}{dt}+z=0$

d i $\dfrac{d}{dx}\left(x\dfrac{dy}{dx}\right)=y$ **ii** $\dfrac{d}{dx}(x+y)=y^2$

3 Given these solutions to differential equations and the initial or boundary conditions, find the particular solution to each differential equation.

a i $y=Ax^2+4$; $y=12$ when $x=2$ **ii** $y=A\cos x-2$; $y=6$ when $x=0$

b i $y=Ae^{-x}+3x$; $\dfrac{dy}{dx}=2$ when $x=0$ **ii** $y=A\ln x-2x^2$; $\dfrac{dy}{dx}=2$ when $x=1$

c i $y=Ax^2+Bx$; $y=1$ when $x=1$; $y=8$ when $x=2$

ii $y=A\sin x+B\cos x$; $y=5$ when $x=0$; $y=10$ when $x=\dfrac{\pi}{2}$

d i $y=Ae^{2x}+Be^x$; $y=2$ and $\dfrac{dy}{dx}=5$ when $x=0$

ii $y=A\sin x+B\cos 2x$; $y=-2$ and $\dfrac{dy}{dx}=10$ when $x=0$

4 The differential equation $2x\dfrac{dy}{dx}+y=\ln x$ is defined for $x>0$.

a Find the complementary function.

b A particular integral has the form $y=a\ln x+b$. Find the values of a and b.

c Hence find the particular solution with initial condition $y=3$ when $x=1$.

5 The differential equation $\cos x\dfrac{dy}{dx}+2\sin 2x=y\sin x$ is defined for $-\dfrac{\pi}{2}<x<\dfrac{\pi}{2}$.

a Find the complementary function.

b A particular integral exists of the form $y=a\sin x+b\cos x$. Find the values of a and b.

c Hence find the general solution to the differential equation $\cos x\dfrac{dy}{dx}+2\sin 2x=y\sin x$ defined for $-\dfrac{\pi}{2}<x<\dfrac{\pi}{2}$.

6 The differential equation $y\dfrac{dy}{dx}+y^2=e^x$ is defined for $y>0$.

a Show that this differential equation has a particular integral of the form $y=Ae^{Bx}$, stating the values of A and B.

b Solve the associated homogenous differential equation.

c Explain why the general solution cannot be written as the sum of the complementary function and the particular integral.

d By considering the expression $\dfrac{d}{dx}\left(e^{2x}y^2\right)$, or otherwise, find the general solution of the differential equation.

7 The differential equation $e^x\dfrac{dy}{dx}+y=2$ is defined for all x.

    a Find the complementary function.

    b Find a particular integral.

    c Hence find the general solution.

Section 2: The integrating factor method for first order equations

You already have the necessary tools to solve a differential equation such as:

$$x^2\frac{dy}{dx}+2xy=e^x$$

because the left-hand side is of a convenient form. Notice that $2x$ is the derivative of x^2 and that $\dfrac{dy}{dx}$ is the derivative of y, which means you have an expression that has resulted from the differentiation of a product (x^2y) using the product rule. Therefore you can write the equation equivalently as

$$\frac{d}{dx}\left(x^2y\right)=e^x$$

Now, integrating both sides and rearranging:

$$x^2y=\int e^x\,dx$$

$$y=\frac{e^x+c}{x^2}$$

When faced with a differential equation like this where you cannot separate the variables, it will not often be the case that the left-hand side is quite so convenient. However, this method does suggest a way forward in such cases.

Consider, for example, the equation $\dfrac{dy}{dx}-\dfrac{y}{x}=x$. This can be rewritten as $\dfrac{1}{x}\dfrac{dy}{dx}-\dfrac{y}{x^2}=1$ by dividing throughout by x. This is $\dfrac{d}{dx}\left(\dfrac{y}{x}\right)=1$.

Consider, in general, a similar linear first order differential equation:

$$\frac{dy}{dx}+P(x)y=Q(x)$$

where $P(x)$ and $Q(x)$ are just functions of x. Note that if there is a function in front of $\dfrac{dy}{dx}$, you can divide through the equation by that function to get it in this form.

To make the left-hand side the derivative of a product as before, you can multiply through the equation by a function $I(x)$:

$$I(x)\frac{dy}{dx} + I(x)P(x)y = I(x)Q(x)$$

and then notice that if $I(x)$ is chosen such that $I'(x) = I(x)P(x)$ you have the left-hand side in the required form. From here you can proceed exactly as before:

$$\frac{d}{dx}(I(x)y) = I(x)Q(x)$$

$$y = \frac{1}{I(x)}\int I(x)Q(x)dx$$

The only remaining question is to decide on the function $I(x)$ to make this work.

You need

$$I'(x) = I(x)P(x)$$

$$\frac{I'(x)}{I(x)} = P(x)$$

$$\int\frac{I'(x)}{I(x)}dx = \int P(x)dx$$

$$\ln|I(x)| = \int P(x)dx$$

$$I(x) = e^{\int P(x)dx}$$

This function $I(x)$ is known as the **integrating factor**.

🔑 Key point 10.3

Given a first order linear differential equation

$$\frac{dy}{dx} + P(x)y = Q(x)$$

multiply through by the integrating factor:

$$I(x) = e^{\int P(x)dx}$$

then the general solution will be:

$$y = \frac{1}{I(x)}\int I(x)Q(x)dx.$$

💡 Tip

When calculating $\int P(x)dx$ you do not need to include the '$+c$'. It turns out that it does not matter whether there is a constant, as it would cancel later in the process.

WORKED EXAMPLE 10.4

Solve the differential equation

$$\cos x\frac{dy}{dx} - 2y\sin x = 3 \text{ for } -\frac{\pi}{2} < x < \frac{\pi}{2}$$

where $y = 1$ when $x = 0$.

Continues on next page ...

$$\cos x \frac{dy}{dx} - 2y \sin x = 3$$

You can't write the LHS as the derivative of a product: always check for this first. (Note that if the LHS had been $\cos x \frac{dy}{dx} - y \sin x$, i.e. without the 2, you could have written it as $\frac{d}{dx}(y \cos x)$.)

$$\frac{dy}{dx} - 2y \frac{\sin x}{\cos x} = \frac{3}{\cos x}$$

$$\frac{dy}{dx} - (2 \tan x)y = 3 \sec x$$

Therefore, start by dividing through by $\cos x$ to get the equation in the correct form for applying the integrating factor.

$$I(x) = e^{\int -2 \tan x \, dx}$$

$$= e^{-2 \ln |\sec x|}$$

$$= e^{\ln(\sec x)^{-2}}$$

$$= e^{\ln \cos^2 x}$$

$$= \cos^2 x$$

Find the integrating factor $I(x) = e^{\int P(x) \, dx}$, making sure not to miss the $-$ sign on $P(x)$.

So

$$\cos^2 x \frac{dy}{dx} - (2 \cos^2 x \tan x)y = 3 \cos^2 x \sec x$$

$$\cos^2 x \frac{dy}{dx} - (2 \cos x \sin x)y = 3 \cos x$$

$$\frac{d}{dx}(y \cos^2 x) = 3 \cos x$$

Now multiply through by $\cos^2 x$ and check that the LHS is of the form $\frac{d}{dx}(y \cos^2 x) = \cos^2 x \frac{dy}{dx} - 2 \cos x \sin x.$

$$y \cos^2 x = \int 3 \cos x \, dx$$

$$= 3 \sin x + c$$

You can now integrate both sides.

Since $x = 0$, $y = 1$:

$$1 \cos^2 0 = 3 \sin 0 + c$$

$$c = 1$$

Finally, you need to find the constant c and rearrange into the form $y = f(x)$.

$$\therefore y \cos^2 x = 3 \sin x + 1$$

$$y = \sec^2 x (3 \sin x + 1)$$

You can transform some differential equations into the required form by using a substitution. In an examination, you would generally be given the required substitution.

WORKED EXAMPLE 10.5

Show that the substitution $z = y^3$ transforms the differential equation

$$3y^2 \frac{dy}{dx} + \frac{y^3}{x} = \frac{e^x}{x}$$

into a linear differential equation. Hence find the general solution of the given differential equation.

If $z = y^3$ then $\frac{dz}{dx} = 3y^2 \frac{dy}{dx}$.

Continues on next page ...

Substituting this turns the given differential equation into:

$$\frac{dz}{dx} + \frac{z}{x} = \frac{e^x}{x}$$

which is a linear differential equation.

The integrating factor is $e^{\int\left(\frac{1}{x}\right)dx} = e^{\ln x} = x.$

> Use Key point 10.3.

Therefore $z = \frac{1}{x}\int x \times \frac{e^x}{x}dx$

> Use Key point 10.3 again.

$$= \frac{1}{x}\left(e^x + c\right)$$

> Notice that the $+ c$ is in the brackets.

Therefore $y^3 = \frac{1}{x}\left(e^x + c\right)$ or $y = \sqrt[3]{\frac{1}{x}\left(e^x + c\right)}$

> You need to give the solution in terms of y and x. If you can easily rewrite it to make y the subject, this is the conventional thing to do.

EXERCISE 10B

1 Use an integrating factor to find the general solution to each linear differential equation.

a i $\dfrac{dy}{dx} + 2y = e^x$ **ii** $\dfrac{dy}{dx} - 4y = e^x$

b i $\dfrac{dy}{dx} + y\cot x = 1$ **ii** $\dfrac{dy}{dx} - (\tan x)y = \sec x$

c i $\dfrac{dy}{dx} + \dfrac{y}{x} = \dfrac{1}{x^2}$ **ii** $\dfrac{dy}{dx} + \dfrac{y}{x} = \dfrac{1}{x^3}$

2 Find the particular solution of the linear differential equation $\dfrac{dy}{dx} + y = e^x$ which has $y = e$ when $x = 1$.

3 Find the general solution to the differential equation $x^2\dfrac{dy}{dx} - 2xy = \dfrac{x^4}{x-3}$.

4 Find the general solution of the differential equation $\dfrac{dy}{dx} + y\sin x = e^{\cos x}$.

5 Find the particular solution of the linear differential equation $x^2\dfrac{dy}{dx} + xy = \dfrac{2}{x}$ that passes through the point $(1,1)$.

6 Given that $\cos x\dfrac{dy}{dx} + y\sin x = \cos^2 x$ and that $y = 2$ when $x = 0$, find y in terms of x.

7 Prove that when finding $\int P(x)dx$ in Key point 10.3, it does not matter whether or not the constant of integration is included.

8 Find the general solution to the differential equation $x\dfrac{dy}{dx} + 2y = 1 + \dfrac{1}{x}\dfrac{dy}{dx}$.

9 **a** Use the substitution $z = \dfrac{1}{y}$ to transform the equation

$$\frac{dy}{dx} + xy = xy^2$$

into a linear differential equation in x and z.

     b Solve the resulting equation, writing z in terms of x.

     c Find the particular solution to the original equation that has $y = 1$ when $x = 1$.

10 **a** Using the substitution $z = y^2$, or otherwise, solve the equation:

$$2y\frac{dy}{dx} + \frac{y^2}{x} = x^2$$

given that when $x = 4$, $y = -5$. Give your answer in the form $y = f(x)$.

b Use another substitution to find the general solution to the equation $\cos y\frac{dy}{dx} + \tan x \sin y = \sin x$.

Section 3: Homogeneous second order linear differential equations with constant coefficients

A differential equation of the form

$$a\frac{d^2 y}{dx^2} + b\frac{dy}{dx} + cy = 0$$

is called a homogeneous second order linear differential equation with constant coefficients.

To find the solution to this type of differential equation, you need to create an **auxiliary equation**.

> ### 🔑 Key point 10.4
>
> The auxiliary equation of the differential equation
>
> $$a\frac{d^2 y}{dx^2} + b\frac{dy}{dx} + cy = 0$$
>
> is
>
> $$a\lambda^2 + b\lambda + c = 0.$$

> ### 💡 Tip
>
> How the auxiliary equation arises is shown in Proof 7.

Solving the auxiliary equation gives you important information about the solution of the differential equation, as set out in Proofs 7, 8 and 9. However, in most instances you will be able to just quote the results, which will be summarised in Key point 10.5.

If the auxiliary equation has real, distinct roots, λ_1 and λ_2, then the solution to the differential equation is $y = Ae^{\lambda_1 x} + Be^{\lambda_2 x}$.

PROOF 7

If $b^2 - 4ac > 0$ and $a\frac{d^2 y}{dx^2} + b\frac{dy}{dx} + cy = 0$

then $y = Ae^{\lambda_1 x} + Be^{\lambda_2 x}$, where λ_1 and λ_2 are the roots of the equation.

Continues on next page ...

If $y = e^{\lambda x}$

Then

$\dfrac{dy}{dx} = \lambda e^{\lambda x}$

and

$\dfrac{d^2 y}{dx^2} = \lambda^2 e^{\lambda x}.$

One possible solution to the differential equation could occur if you had a function whose derivative and second derivative are proportional to the original function, allowing everything to 'cancel' and result in zero. $e^{\lambda x}$ is one example of a function which has this property (although you will see later that there are others).

Substituting these into the differential equation:

$a\lambda^2 e^{\lambda x} + b\lambda e^{\lambda x} + c e^{\lambda x} = 0$

$e^{\lambda x}(a\lambda^2 + b\lambda + c) = 0$ .... Taking out a factor of $e^{\lambda x}$.

$a\lambda^2 + b\lambda + c = 0$ .... Since $e^{\lambda x}$ can never be zero. This is the auxiliary equation.

If $b^2 - 4ac > 0$ then the auxiliary equation will have two real solutions. Call these λ_1 and λ_2.

Therefore two possible solutions to the differential equation are $y = e^{\lambda_1 x}$ and $y = e^{\lambda_2 x}$.

$y = Ae^{\lambda_1 x} + Be^{\lambda_2 x}$ will also be a solution, since any linear combination of solutions is also a solution. .... This can be proved in a similar way to Proof 6.

This is a solution with two arbitrary constants, therefore it is the general solution. .... Use Key point 10.1 to justify that your 'guess' gives the complete solution.

If the roots of the auxiliary equation are complex, you could still write the solution as $y = Ae^{\lambda_1 x} + Be^{\lambda_2 x}$. However, you can then rewrite this in terms of trigonometric functions.

⏮ **Rewind**

You learnt in Chapter 2, Section 2, that $e^{p+q} = e^p(\cos q + i \sin q)$.

PROOF 8

If $y = Ce^{\lambda_1 x} + De^{\lambda_2 x}$ with $\lambda_1 = \alpha + \beta i$ and $\lambda_2 = \alpha - \beta i$, then y can be written in the form $y = e^{\alpha x}(A\cos(\beta x) + B\sin(\beta x))$.

Substituting in the given information:

$y = Ce^{(\alpha + \beta i)x} + De^{(\alpha - \beta i)x}$ .... Take out a factor of $e^{\alpha x}$.

$= e^{\alpha x}(Ce^{\beta x i} + De^{-\beta x i})$ .... Rewrite the complex exponential into polar form.

$= e^{\alpha x}(C(\cos(\beta x) + i\sin(\beta x)) + D(\cos(\beta x) - i\sin(\beta x)))$

$= e^{\alpha x}(A\cos(\beta x) + B\sin(\beta x))$ .... Separate out the cosine and sine terms.

with $A = C + D$ and $B = i(C - D)$

If the auxiliary equation has equal roots, $\lambda_1 = \lambda_2 = \lambda$, the general solution becomes $y = A\,e^{\lambda x} + B\,e^{\lambda x} = C\,e^{\lambda x}$, which contains only one arbitrary constant. According to Key point 10.1, you need to find another complementary function. Proof 9 shows you how to do this.

PROOF 9

If $b^2 - 4ac = 0$ and $a\dfrac{d^2 y}{dx^2} + b\dfrac{dy}{dx} + cy = 0$, then $y = x e^{\lambda x}$ is a possible solution for a suitably chosen λ.

If $y = x e^{\lambda x}$

then

$\dfrac{dy}{dx} = (1 + \lambda x)e^{\lambda x}$

and

$\dfrac{d^2 y}{dx^2} = \lambda(2 + \lambda x)e^{\lambda x}$.

Substituting into the differential equation:

$a\lambda(2 + \lambda x)e^{\lambda x} + b(1 + \lambda x)e^{\lambda x} + cx e^{\lambda x} = 0$

$2a\lambda + a\lambda^2 x + b + b\lambda x + cx = 0$ Divide through by $e^{\lambda x}$ (which can never be 0) and tidy up.

$2a\lambda + b + x(a\lambda^2 + b\lambda + c) = 0$

Comparing coefficients:

$x^0\colon 2a\lambda + b = 0$

$\qquad \lambda = -\dfrac{b}{2a}$

$x^1\colon a\lambda^2 + b\lambda + c = 0$

Checking $\lambda = -\dfrac{b}{2a}$ in the second equation:

$a\left(-\dfrac{b}{2a}\right)^2 + b\left(-\dfrac{b}{2a}\right) + c = \dfrac{b^2}{4a} - \dfrac{b^2}{2a} + c$

$\qquad\qquad = \dfrac{-b^2 + 4ac}{4a} = 0$ You are given that $b^2 - 4ac = 0$.

So $y = x e^{\left(-\frac{b}{2a}x\right)}$ is a solution to the differential equation in this case.

To get to the general solution a linear combination of the two possible values is required, leading to $y = A e^{\lambda x} + Bx e^{\lambda x}$.

When solving differential equations normally you will just be able to write down the solution without going through the Proofs 7, 8 and 9.

Key point 10.5

Solution to auxiliary equation	General solution to differential equation
Two distinct roots, λ_1 and λ_2	$y = Ae^{\lambda_1 x} + Be^{\lambda_2 x}$
Repeated root, λ	$y = (A + Bx)e^{\lambda x}$
Complex roots, $\alpha + i\beta$	$y = e^{\alpha x}(A\sin\beta x + B\cos\beta x)$

WORKED EXAMPLE 10.6

Solve the differential equation $\dfrac{d^2 y}{dx^2} - 3\dfrac{dy}{dx} + 2y = 0$

given that $y = 1$ and $\dfrac{dy}{dx} = 0$ when $x = 0$.

The auxiliary equation is
$\lambda^2 - 3\lambda + 2 = 0$.

This has roots $\lambda = 1$ and $\lambda = 2$, so the general solution to the differential equation is
$y = Ae^x + Be^{2x}$.

> Since the roots are real and distinct you can write the solution to the differential equation in exponential form using Key point 10.5.

$\dfrac{dy}{dx} = Ae^x + 2Be^{2x}$

> You need to differentiate the expression to make use of the initial conditions.

Using the initial conditions, when $x = 0$:
$y = 1 = A + B$
$\dfrac{dy}{dx} = 0 = A + 2B$

Solving gives $A = 2$ and $B = -1$.

> You can use technology to solve these types of simultaneous equations.

So the particular solution is $y = 2e^x - e^{2x}$.

EXERCISE 10C

1   a   Write the auxiliary equation associated with the differential equation $\dfrac{d^2 y}{dx^2} + 5\dfrac{dy}{dx} + 6y = 0$.

    b   Hence find the general solution of the differential equation.

2   a   Write the auxiliary equation associated with the differential equation $\dfrac{d^2 y}{dx^2} + 4y = 0$.

    b   Hence find the general solution of the differential equation.

3   a   Write the auxiliary equation associated with the differential equation $y'' + 2y' + y = 0$.

    b   Hence find the general solution of the differential equation.

Tip

Remember that y' is another notation for $\dfrac{dy}{dx}$.

4 **a** Find the general solution to the differential equation $\dfrac{d^2y}{dx^2} + 8y = 6\dfrac{dy}{dx}$.

   b Find the particular solution that satisfies $y = 5$ and $\dfrac{dy}{dx} = 12$ when $x = 0$.

5 **a** Find the general solution to the differential equation $y'' + 4y' + 4y = 0$.

   b Find the particular solution that satisfies $y(0) = 1$ and $y'(0) = 0$.

6 **a** Find the general solution to the differential equation $\dfrac{d^2x}{dt^2} - 2\dfrac{dx}{dt} + 2x = 0$.

   b Find the particular solution that satisfies $x(0) = 1$ and $\dot{x}(0) = 0$.

> **Tip**
>
> Remember $\ddot{x}$ that means $\dfrac{dx}{dt}$.

7 **a** Find the general solution to the differential equation $\ddot{x} + 3x = 4\dot{x}$.

   b Given that $x(0) = 1$ and $x(1) = e$, find $x(2)$.

8 **a** Find the general solution to the differential equation $\dfrac{d^2y}{dt^2} - 6\dfrac{dy}{dt} + 9y = 0$.

   b Find the particular solution that satisfies $y(0) = 0$ and $y(1) = p$, writing your answer in terms of p.

9 Find the general solution to the differential equation $y''' - 5y'' + 9y' = 5y$.

10 Find the general solution to the differential equation $y''' + 3y'' + 3y' + y = 0$.

11 By using the trial function $y = x^n$, find the general solution to the differential equation $x^2\dfrac{d^2y}{dx^2} + x\dfrac{dy}{dx} - 9y = 0$.

12 **a** Use the substitution $y = x^2$ to turn the differential equation $x\dfrac{d^2x}{dt^2} + \left(\dfrac{dx}{dt}\right)^2 + x\dfrac{dx}{dt} = 0$

   into a second order differential equation with constant coefficients involving y and t.

   b Solve the differential equation to find y as a function of t.

   c Hence solve the original differential equation given that when $t = 0$, $x = 2$ and $\dfrac{dx}{dt} = \dfrac{1}{4}$.

Section 4: Non-homogeneous second order linear differential equations with constant coefficients

The second order differential equations you need to solve can all be written in the form

$$a\dfrac{d^2y}{dx^2} + b\dfrac{dy}{dx} + cy = f(x).$$

To solve these differential equations you use the method in Key point 10.2. You first of all solve the associated homogeneous equation to find a complementary function (using Key point 10.5) and then find a particular integral.

The form of the particular integral will depend on $f(x)$. Key point 10.6 gives trial functions for some common situations. This trial function needs to be substituted into the differential equation to find the unknown constants.

Key point 10.6

f(x)	Trial function
$ax+b$	$px+q$
Polynomial	General polynomial of the same order
$a\mathrm{e}^{bx}$	$c\mathrm{e}^{bx}$
$a\cos bx$ $a\sin bx$	$p\sin bx + q\sin bx$

Tip

You need to learn these forms of trial functions. If a different trial function is needed, it will be given in the question.

WORKED EXAMPLE 10.7

Find the general solution to the differential equation

$$\frac{\mathrm{d}^2 y}{\mathrm{d}x^2}+7\frac{\mathrm{d}y}{\mathrm{d}x}+12y=24x+60\mathrm{e}^{2x}.$$

The associated homogenous equation is

$$\frac{\mathrm{d}^2 y}{\mathrm{d}x^2}+7\frac{\mathrm{d}y}{\mathrm{d}x}+12y=0.$$

This has auxiliary equation

$$\lambda^2+7\lambda+12=0$$

which has roots -3 and -4.

Therefore, the complementary function is
$y=A\mathrm{e}^{-3x}+B\mathrm{e}^{-4x}$.

The trial function associated with $60\mathrm{e}^{2x}$ is $y=c\mathrm{e}^{2x}$.

Differentiating twice:

$$\frac{\mathrm{d}y}{\mathrm{d}x}=2c\mathrm{e}^{2x}$$

$$\frac{\mathrm{d}^2 y}{\mathrm{d}x^2}=4c\mathrm{e}^{2x}.$$

Substituting into the left-hand side of the differential equation and comparing to the exponential part of the right-hand side:

$$4c\mathrm{e}^{2x}+7\times 2c\mathrm{e}^{2x}+12\times c\mathrm{e}^{2x}=60e^{2x}$$
$$30c\mathrm{e}^{2x}=60e^{2x}$$
$$c=2$$

The trial function associated with $24x$ is $y=px+q$.

Differentiating twice:

$$\frac{\mathrm{d}y}{\mathrm{d}x}=p$$

$$\frac{\mathrm{d}^2 y}{\mathrm{d}x^2}=0.$$

First solve the associated homogenous equation to find the complementary function.

You can find the particular integral in two stages. Notice that the coefficient of x in the power of the trial function mirrors the original function.

Notice that although the expression on the right only involves a term in x, you need the general linear expression in the trial function.

Continues on next page ...

Substituting these into the left-hand side of the differential equation and comparing to the linear part of the right-hand side:

$$0 + 7 \times p + 12(px + q) = 24x$$

$$12px + 7p + 12q = 24x$$

Comparing the coefficient of x:

$$12p = 24$$

$$p = 2$$

Comparing the constant term:

$$7p + 12q = 0$$

$$14 + 12q = 0$$

$$q = -\frac{7}{6}$$

The particular integral is $2e^{2x} + 2x - \dfrac{7}{6}$.

The general solution is

$$y = Ae^{-3x} + Be^{-4x} + 2e^{2x} + 2x - \frac{7}{6}.$$

> Compare coefficients to find p and q.

> You can use the fact that $p = 2$ to solve for q.

> The general solution is the sum of the complementary function and the particular integral, from Key point 10.2.

Sometimes the trial function given in Key point 10.6 already appears as a part of the complementary function. In that case, you need to modify the particular integral.

Key point 10.7

If your trial function is already part of the complementary function, try multiplying the trial function by x.

WORKED EXAMPLE 10.8

Find the general solution to the differential equation

$$\frac{\mathrm{d}^2 y}{\mathrm{d}x^2} + y = 16\sin x.$$

Continues on next page ...

The associated homogenous equation is

$\dfrac{d^2y}{dx^2} + y = 0.$

This has auxiliary equation

$\lambda^2 + 1 = 0$

which has roots i and −i.

Therefore, the complementary function is

$y = A\sin x + B\cos x.$

The trial function associated with
$16\sin x$ is $y = p\sin x + q\cos x.$

This already appears in the complementary function,
so try $y = x(p\sin x + q\cos x).$

Differentiating twice:

$\dfrac{dy}{dx} = p\sin x + q\cos x + x(p\cos x - q\sin x)$

$\dfrac{d^2y}{dx^2} = p\cos x - q\sin x + p\cos x - q\sin x + x(-p\sin x - q\cos x).$

Substituting into the left-hand side of the
differential equation:

$2p\cos x - 2q\sin x = 16\sin x$

$p = 0, q = -8$

The particular integral is $-8x\cos x.$

The general solution is

$y = A\sin x + B\cos x - 8x\cos x.$

> First solve the associated homogenous equation to find the complementary function.

> From Key point 10.5.

> Although the right-hand side of the equation involves only sin x, you need to include both sin and cos in the trial function.

> You need to adjust the trial function according to Key point 10.7.

> You need to use the product rule to differentiate.

> Notice that the terms containing x sin x and x cos x all cancel. This will always happen in the situation described in Key point 10.7.

> The general solution is the sum of the complementary function and the particular integral, from Key point 10.2.

EXERCISE 10D

1 For the differential equation $\dfrac{d^2y}{dx^2} - 4\dfrac{dy}{dx} - 5y = e^{2x}$:

   a find the complementary function

   b find the particular integral

   c hence write down the general solution.

2 For the differential equation $y'' + 9y' + 20y = 60x$:

   a find the complementary function

   b find the particular integral

   c hence write down the general solution.

3 For the differential equation $\dfrac{d^2y}{dx^2} + \dfrac{dy}{dx} + \sin x = 0$:

    a find the complementary function

    b find the particular integral

    c hence find the general solution.

4 For the differential equation $\dfrac{d^2y}{dx^2} + 9y = 20\,e^{-x}$:

    a find the complementary function

    b hence find the general solution

    c find the particular solution given that $y = 7$ and $y' = 10$ when $x = 0$.

5 For the differential equation $\dfrac{d^2y}{dx^2} + 4\dfrac{dy}{dx} + 4y = 12x + 25\sin x$, find:

    a the general solution

    b the particular solution given that $y = 0$ and $y' = 10$ when $x = 0$.

6 For the differential equation $y'' - 4y' + 8y = 32t^2$, find:

    a the general solution

    b the particular solution given that $y = 0$ and $y' = 0$ when $t = 0$.

7 For the differential equation $\dfrac{d^2y}{dx^2} - 10\dfrac{dy}{dx} + 25y = e^{5x}$:

    a find the complementary function

    b show that there is a particular integral of the form qx^2e^{5x}

    c hence find the general solution

    d find the particular solution given that $y = 4$ and $y' = 2$ when $x = 0$.

8 The function $f(x)$ satisfies the differential equation $f''(x) - 2f'(x) = 4e^x \sin x$ with boundary conditions $f(0) = f(\pi) = 3$. Given that there is a particular integral of the form $pe^x \sin x$, find the particular solution of the differential equation.

9 Find the general solution of the differential equation $\dfrac{d^2y}{dx^2} + 3\dfrac{dy}{dx} - 4y = e^x$.

10 For the differential equation $\dfrac{d^2y}{dx^2} + 4y = 12\cos 2x$, find

    a the general solution

    b the particular solution which satisfies $y = 5$ and $\dfrac{dy}{dx} = 6$ when $x = 0$.

Checklist of learning and understanding

- The general solution to an nth-order differential equation has n arbitrary constants.
- For a linear differential equation, the general solution is given by $y = y_c + y_p$, where y_c is the complementary function and y_p is a particular integral.
    - The complementary function is the solution of the associated homogeneous equation, obtained from the original equation by removing the terms that do not contain y.
- Given a first order linear differential equation $\dfrac{dy}{dx} + P(x)y = Q(x)$, multiply through by the integrating factor $I(x) = e^{\int P(x)dx}$. The solution will be $y = \dfrac{1}{I(x)} \int I(x)Q(x)dx$.
- The auxiliary equation to the homogeneous differential equation $a\dfrac{d^2y}{dx^2} + b\dfrac{dy}{dx} + cy = 0$ is $a\lambda^2 + b\lambda + c = 0$.
    - The solution to the auxiliary equation gives the form of the general solution of the homogeneous equation:

Solution to auxiliary equation	General solution to differential equation
Two distinct roots, λ_1 and λ_2	$y = Ae^{\lambda_1 x} + Be^{\lambda_2 x}$
Repeated root, λ	$y = (A + Bx)e^{\lambda x}$
Complex roots, $\alpha + i\beta$	$y = e^{\alpha x}(A\sin\beta x + B\cos\beta x)$

- The form of the particular integral for the homogeneous differential equation $a\dfrac{d^2y}{dx^2} + b\dfrac{dy}{dx} + cy = f(x)$ depends on $f(x)$:

$f(x)$	Trial function
$ax + b$	$px + q$
Polynomial	General polynomial of the same order
ae^{bx}	pe^{bx}
$a\cos bx$ \ $a\sin bx$	$p\sin bx + q\sin bx$

- If the trial function is already part of the complementary function, multiply the trial function by x.

Mixed practice 10

1 Find the complementary function of the differential equation $4\dfrac{d^2y}{dx^2}+9y=\sin 3x$.

2 What is the integrating factor for the differential equation $\dfrac{dy}{dx}-\dfrac{3}{x}y=x^2$?

3 **a** Solve the quadratic equation $\lambda^2+6\lambda+5=0$.

   b Hence write down the general solution of the differential equation $\dfrac{d^2y}{dx^2}+6\dfrac{dy}{dx}+5y=0$.

4 **a** Show that the integrating factor for the differential equation $\dfrac{dy}{dx}+2xy=x\,e^{-x^2}$ is e^{x^2}.

   b Hence find the general solution of the differential equation.

5 The variables x and y satisfy the differential equation

$$\dfrac{dy}{dx}+4y=5\cos 3x.$$

   i Find the complementary function.

   ii Hence, or otherwise, find the general solution.

   iii Find the approximate range of values of y when x is large and positive.

© OCR A Level Mathematics, Unit 4727 Further Pure Mathematics 3, June 2011

6 The differential equation $y''+7y'+10y=e^x$ is defined for all x.

   a By considering the associated homogeneous differential equation, find the complementary function.

   b Show that a function of the form $q\,e^x$ forms a particular integral, and find the value of q.

   c Hence write down the general solution of the differential equation.

   d Find the particular solution with initial conditions $y(0)=0$ and $y'(0)=6$.

7 **a** Find the value of the constant q for which $q\cos x$ is a particular integral of the differential equation
   $$\dfrac{d^2y}{dx^2}+4y=\cos x.$$

   b Hence find the general solution of the differential equation.

8 The differential equation $\dfrac{d^2y}{dx^2}+6\dfrac{dy}{dx}+25y=50x$ is defined for all x.

   a By considering the associated homogeneous differential equation, find the complementary function.

   b Show that a function of the form $px+q$ forms a particular integral, and find the values of p and q.

   c Hence write down the general solution of the differential equation.

   d Find the particular solution with initial conditions $y=-\dfrac{12}{25}$ and $y'=6$ when $x=0$.

9 Find the particular solution of the differential equation

$$x\dfrac{dy}{dx}+3y=x^2+x$$

   for which $y=1$ when $x=1$, giving y in terms of x.

© OCR A Level Mathematics, Unit 4727/01 Further Pure Mathematics 3, June 2015

10 Solve the differential equation

$$\frac{d^2y}{dx^2} + 5\frac{dy}{dx} + 6y = e^{-x}$$

subject to the conditions $y = \frac{dy}{dx} = 0$ when $x = 0$.

© OCR A Level Mathematics, Unit 4727/01 Further Pure Mathematics 3, June 2014

11 Solve the differential equation $x\frac{dy}{dx} - 3y = x^4e^{2x}$ for y in terms of x, given that $y = 0$ when $x = 1$.

© OCR A Level Mathematics, Unit 4727/01 Further Pure Mathematics 3, January 2013

12 The variables x and y satisfy the differential equation

$$2\frac{d^2y}{dx^2} + 3\frac{dy}{dx} - 2y = 5e^{-2x}.$$

i Find the complementary function of the differential equation.

ii Given that there is a particular integral of the form $y = px\,e^{-2x}$, find the constant p.

iii Find the solution of the equation for which $y = 0$ and $\frac{dy}{dx} = 4$ when $x = 0$.

© OCR A Level Mathematics, Unit 4727 Further Pure Mathematics 3, January 2012

13 **i** Find the general solution of the differential equation

$$3\frac{d^2y}{dx^2} + 5\frac{dy}{dx} - 2y = -2x + 13$$

ii Find the particular solution for which $y = -\frac{7}{2}$ and $\frac{dy}{dx} = 0$ when $x = 0$.

iii Write down the function to which y approximates when x is large and positive.

© OCR A Level Mathematics, Unit 4727 Further Pure Mathematics 3, January 2011

14 **a** Find the general solution of $x\frac{du}{dx} + 3u = x$ for $x > 0$.

b Show that the substitution $u = \frac{dy}{dx}$ transforms the differential equation $x\frac{d^2y}{dx^2} + 3\frac{dy}{dx} = x$ into

$$x\frac{du}{dx} + 3u = x.$$

c Hence find the general solution of the differential equation $x\frac{d^2y}{dx^2} + 3\frac{dy}{dx} = x$ for $x > 0$.

15 **a** Find the general solution to the differential equation $\cos x\frac{dy}{dx} + y\sin x = \sin 2x$.

b Find the particular solution with $y(0) = 5$.

16 Find the general solution of the differential equation $y'' - 2y' = 5$.

11 Applications of differential equations

In this chapter you will learn how to:

- use differential equations in modelling, in kinematics and in other contexts
- solve the equation for simple harmonic motion and relate the solution to the motion
- model damped oscillations using second order differential equations and interpret their solution
- solve coupled first order differential equations and use them to model situations with two dependent variables.

Before you start…

Chapter 10	You should know how to solve second order differential equations.	1	Find the general solution to $y'' + 5y' + 4y = x$.
A Level Mathematics Student Book 1, Chapter 21	You should be able to use Newton's second law.	2	A falling object of mass 10 kg is subjected to a constant air resistance of 50 N. Find the acceleration of the object if $g = 9.8$ m s^{-2}.

Real world modelling

In reality, nearly everything of interest – be it the effect of a medicine or the price of a share – changes over time. The tool that mathematicians use to model these situations is a differential equation. In this chapter you will look at some common situations modelled by differential equations and how you can use the methods from Chapter 10 to solve them and interpret their solutions in context.

Section 1: Forming differential equations

When modelling real-life situations, it is often the case that the description can be interpreted in terms of differential equations.

You have already met many examples of setting up differential equations in A Level Mathematics Student Book 2, Chapter 13. In this section you will see further types of situations where differential equations arise, but this time you will often need to use methods from Chapter 10 to solve them. You will also look at the type of assumptions which are made when writing these differential equations.

⏮ **Rewind**

See A Level Mathematics Student Book 1, Chapter 21, for a reminder of using $F = ma$, and A Level Mathematics Student Book 2, Chapter 13, for its use in setting up differential equations.

WORKED EXAMPLE 11.1

A car, of mass m kg, is moving along a straight horizontal road. At time t seconds, the car has speed v m s^{-1}. The only force acting is a resistance, which is modelled as being proportional to $m(144+v^2)$ newtons.

Initially the car is moving with speed 12 m s^{-1} and deceleration 2 m s^{-2}.

Find the time taken for the car to come to rest.

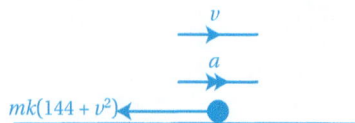

The resistance force is $mk(144+v^2)$ for some constant k.

Using $F = ma$ and $a = \dfrac{dv}{dt}$:

$$-mk(144+v^2) = m\frac{dv}{dt}$$

The resistance force is negative as the car is moving in the opposite direction to which this force acts.

$$-k(144+v^2) = \frac{dv}{dt}$$

Initially:

$$-k(144+12^2) = -2$$

$$k = \frac{1}{144}$$

Use the initial conditions $\left(v = 12, \dfrac{dv}{dt} = -2\right)$ to find the value of k.

Separating the variables and integrating:

$$\int \frac{1}{144+v^2}\,dv = -\int \frac{1}{144}\,dt$$

$$\frac{1}{12}\arctan\left(\frac{v}{12}\right) = -\frac{1}{144}t + c$$

This is a standard arctan integral.

When $t = 0$, $v = 12$:

$$\frac{1}{12}\arctan\left(\frac{12}{12}\right) = 0 + c$$

Use the initial condition again to find c.

$$c = \frac{\pi}{48}$$

$$\therefore \frac{1}{12}\arctan\left(\frac{v}{12}\right) = -\frac{1}{144}t + \frac{\pi}{48}$$

$$\arctan\left(\frac{v}{12}\right) = -\frac{1}{12}t + \frac{\pi}{4}$$

When $v = 0$:

$$\arctan 0 = -\frac{1}{12}t + \frac{\pi}{4}$$

The car will come to rest when $v = 0$.

$$t = 3\pi \text{ seconds}$$

Tip

You can shortcut having to find c and then setting $v = 0$ by using definite instead of indefinite integration. After separating variables, integrate with limits 12 and 0 for v, and 0 and T for t:

$$\int_{12}^{0} \frac{1}{144+v^2}\,dv = -\int_{0}^{T} \frac{1}{144}\,dt$$

Then T will be the time taken for the car to stop.

EXERCISE 11A

1 A stone of mass m falls vertically downwards under gravity. At time t, the stone has speed v, and it experiences air resistance of magnitude kmv, where k is a constant.

  a Find an expression for $\dfrac{\mathrm{d}v}{\mathrm{d}t}$ in terms of v, g and k.

  b The initial speed of the stone is u. Find an expression for v at time t.

2 A car of mass m kg is moving along a straight horizontal road. At time t seconds, the car has speed v m s^{-1}. The magnitude of the resistance force, in newtons, is modelled by $3mv^{\frac{3}{2}}$. No other horizontal force acts on the car. The initial speed of the car is 9 m s^{-1}.

  a Show that, according to this model, $v = \left(\dfrac{6}{9t+2}\right)^2$.

  b A student performs an experiment to measure the speed of the car. She finds that the speed of the car after half a second is 0.8 m s^{-1} and the speed after two seconds is 0.3 m s^{-1}. Comment on the suitability of the model.

3 The current (I) at time t in a circuit with resistance ($2R$), capacitance (C) and inductance (L) is modelled by the differential equation:
$$L\frac{\mathrm{d}^2 I}{\mathrm{d}t^2} + 2R\frac{\mathrm{d}I}{\mathrm{d}t} + \frac{1}{C}I = 0$$
Solve to find I as a function of t and sketch the solution in each situation.

  a $R=0$ **b** $R^2 < \dfrac{L}{C}$ **c** $R^2 > \dfrac{L}{C}$

4 The rate of immigration into a country is modelled as exponentially decreasing. The initial rate is 200 000 per year. One year later the rate is 50 000 per year.

  a Write a differential equation for the population (Y), assuming that changes in the population are due only to immigration.

  b Given that the initial population is 12 million, find the long-term population predicted by the model.

  c The model is refined by adding the term $0.02Y$ to the rate of change. Suggest what this term represents.

5 A chicken is to be cooked and is placed into an oven. The temperature of the oven, T_{oven} °C, follows the rule $T_{\text{oven}} = 25 + 20t$, where t is the time in minutes after the chicken is put into the oven. The rate of increase of the temperature of the chicken (T °C) is modelled as proportional to the difference between the chicken's temperature and the oven's temperature.

  a Write a differential equation for the temperature of the chicken.

  b If the temperature of the chicken is originally 5 °C and increasing at a rate of 10 °Cs^{-1}, find the particular solution of the differential equation.

  c Find an estimate of the chicken's temperature after 10 minutes, giving your answer to an appropriate degree of accuracy.

  d Describe one way in which the model is a simplification of the chicken's temperature.

6 A school has N students. The rate of spread of a rumour in a school is thought to be proportional to both the number of students who know the rumour (R) and the number who do not know the rumour.

  a Write this information as a differential equation.

  b Find the number of students who know the rumour when the rumour is spreading fastest.

  c Write down two assumptions that are being made in this situation.

7   A bacterium is modelled as a sphere. According to one biological model the volume of the bacterium (V) follows this differential equation:

$$\frac{dV}{dt} = 2V^{\frac{2}{3}} - V$$

a   Explain the biological significance of the $V^{\frac{2}{3}}$ term.

b   By using the substitution $u = V^{\frac{1}{3}}$, solve the differential equation given that initially $V = 1$.

c   Sketch the solution and hence find the long-term volume of the bacterium.

Did you know?

The model in question 7 is called Von Bertalanffy growth. It is very important in mathematical biology.

Section 2: Simple harmonic motion

In Chapter 10, you saw that some differential equations have solutions involving sines and cosines. These describe **oscillating behaviour**.

The differential equation which has pure sinusoidal behaviour is called **simple harmonic motion**. It occurs in a surprisingly wide range of physical situations.

Tip

Simple harmonic motion is often abbreviated to SHM.

Key point 11.1

The differential equation for simple harmonic motion is

$$\frac{d^2x}{dt^2} = -\omega^2 x$$

Tip

Using the dot notation to represent differentiation with respect to time, you can also write this equation as $\ddot{x} = -\omega^2 x$.

In the equation in Key point 11.1, x represents the displacement of an object from its **equilibrium position** (the position where the acceleration of the object is zero).

To solve this differential equation you find the auxiliary equation:

$$\lambda^2 = -\omega^2$$

This has solutions $\lambda = \pm i\omega$, which lead to the general solution to the differential equation.

Key point 11.2

The general solution to the simple harmonic motion differential equation is
$$x = A\sin\omega t + B\cos\omega t$$

Rewind

The general form of solutions for second order differential equations was given in Key point 10.5.

There is some terminology that is useful in describing these solutions:

- The average position around which the object oscillates (corresponding to $x = 0$) is the **equilibrium position**.
- The maximum distance from the equilibrium position is called the **amplitude**.
- The motion repeats itself after time, T, which is called the **period**.
- The value ω is called the **angular frequency**.

If initially the object is:

- at the equilibrium position, then the solution will be $x = a\sin\omega t$
- at the maximum displacement from the equilibrium position, then the solution will be $x = a\cos\omega t$.

In both of these cases the amplitude is given by a.

You know from A Level Mathematics Student Book 2, Chapter 8 that $A\sin\omega t + B\cos\omega t$ can also be written in the form $R\sin(\omega t + \varphi)$.

Key point 11.3

The general solution to the simple harmonic motion can also be written as $R\sin(\omega t + \varphi)$.

In the equation in Key point 11.3, the amplitude is R and φ is called the phase shift. $R\sin\varphi$ gives the initial displacement from the equilibrium position.

Since $\sin\theta$ and $\cos\theta$ repeat when θ gets to 2π, one full period, T, occurs when $\omega T = 2\pi$.

Key point 11.4

The period, T, of a particle moving with simple harmonic motion is
$$T = \frac{2\pi}{\omega}.$$

The object has its maximum speed as it is going through the equilibrium position, and it is instantaneously at rest when it reaches the maximum displacement, a.

Key point 11.5

The relationship between velocity and displacement for a particle moving with simple harmonic motion is
$$v^2 = \omega^2\left(a^2 - x^2\right).$$

PROOF 10

Prove that $v^2 = \omega^2(a^2 - x^2)$.

If $x = 0$ when $t = 0$, then	Since you are only looking for a relationship between v and x, choose to start the time when the object moves through the equilibrium position.
$x = a\sin\omega t$.	
$v = \dfrac{dx}{dt} = a\omega\cos\omega t$	Use the fact that $v = \dfrac{dx}{dt}$ from kinematics.
$v^2 = \omega^2 a^2 \cos^2\omega t$	Use $\cos^2\theta \equiv 1 - \sin^2\theta$.
$\quad = \omega^2 a^2\left(1 - \sin^2\omega t\right)$	
$\quad = \omega^2\left(a^2 - \left(a\sin\omega t\right)^2\right)$	Group the terms together to make a link with the expression for displacement.
$\quad = \omega^2\left(a^2 - x^2\right)$	

A common context for simple harmonic motion is the situation with springs. One of the forces acting on the object is the tension, T, in the spring. The tension is always directed back towards the equilibrium position. There is a standard model in physics for the magnitude of the tension, but in this course you will always be told the form to use.

If the only force acting on the object is the tension from the spring (for example, when the object is moving on a smooth horizontal table) then in the equilibrium position the spring is neither extended nor compressed (it is at its natural length).

However, if there are additional forces, this spring might be extended or compressed at the equilibrium position.

Explore

The tension in an elastic spring can be modelled using Hooke's Law. You will learn about it if you study the Further Mechanics option.

WORKED EXAMPLE 11.2

A spring of natural length 10 cm is attached to a hook in the ceiling. A particle of mass 0.5 kg is attached to the other end of the spring. When the extension of the spring from its natural length is x m, the tension in the spring has magnitude $100x$ N.

Use $g = 10$ m s^{-2}, giving your final answers to an appropriate degree of accuracy.

a Show that, in the equilibrium position, the length of the spring is 15 cm.
b Show that, if the spring is displaced from the equilibrium, the particle will perform simple harmonic motion and find the time period of oscillations about this equilibrium.
c The spring is stretched 2 cm from the equilibrium position and then released. Find the maximum speed of the particle.

a

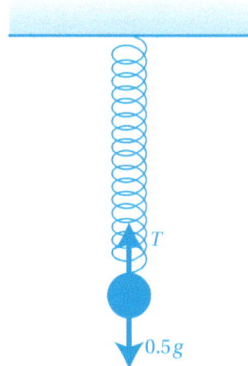

Draw a diagram to help visualise the situation.

Only tension and weight are acting on the mass. These must balance in equilibrium.

When the spring is in equilibrium:

$100x = 0.5g$

$x = \dfrac{5}{100}$

To keep consistent units you need to use metres for x.

So the extension is 5 cm and the equilibrium length is 15 cm.

Continues on next page ...

b

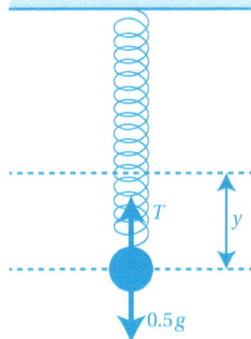

If y is the extension below the equilibrium position then:

$$m\frac{d^2y}{dt^2}=mg-100(0.05+y)$$

$$=0.5\times10-100\times0.05-100y$$

$$=-100y$$

$$\therefore \frac{d^2y}{dt^2}=-200y$$

So, $\frac{d^2y}{dt^2}=-\omega^2 y$, where $\omega^2=200$.

This is the equation for simple harmonic motion.

Then $T=\frac{2\pi}{\omega}$

$T=\frac{2\pi}{\omega}=0.44\text{ s}$ (2 d.p.)

c If the spring is stretched an additional 2 cm, then the amplitude is 0.02 m.

$$v^2=\omega^2(a^2-x^2)=200(0.02^2-x^2).$$

This is maximised when $x=0$, so the maximum speed is

$$\sqrt{200\times0.02^2}=0.28\text{ms}^{-1}\text{ (2 d.p.)}$$

This is Newton's second law vertically. The acceleration is $\frac{d^2y}{dt^2}$.

The total extension of the spring is $0.05+y$ so the magnitude of the tension is $100(0.05+y)$.

If y is positive when below the equilibrium position, then you need to use down as positive, so the resultant force is weight − tension.

Rearrange into the standard form for simple harmonic motion.

Use $T=\frac{2\pi}{\omega}$ from Key point 11.4.

It is important to be consistent with units, in this case working only in metres.

Use $v^2=\omega(a^2-x^2)$ from Key point 11.5.

EXERCISE 11B

1 State the amplitude and the period of the simple harmonic motion described by each equation. Also state whether the particle is at rest, passing through the equilibrium position, or neither, when $t=0$.

a   i $x=4.5\cos3t$ ii $x=3\sin4t$

b   i $x=2.6\sin\dfrac{t}{3}$ ii $x=5\cos\dfrac{t}{4}$

c **i** $x = 3.2 \sin \dfrac{2\pi t}{3}$ **ii** $x = 10.4 \cos \dfrac{3\pi t}{5}$

d **i** $x = 2.6 \left(\sin \dfrac{t}{3} + 0.2\right)$ **ii** $x = 5 \sin\left(\dfrac{t}{4} + 1.4\right)$

e **i** $x = 3.2 \sin \dfrac{5\pi t}{4} + 1.4 \cos \dfrac{5\pi t}{4}$ **ii** $x = 3 \sin \dfrac{2\pi t}{7} + 8 \cos \dfrac{2\pi t}{7}$

2 For each description of simple harmonic motion write an equation for x in terms of t (where x is in metres and t is in seconds).

  a **i** Amplitude 0.6 m, period $\dfrac{\pi}{5}$ seconds; at rest when $t = 0$.

     ii Amplitude 3.4 m, period $\dfrac{\pi}{7}$ seconds; at rest when $t = 0$.

  b **i** Amplitude 0.7 m, period 6π seconds; in equilibrium when $t = 0$.

     ii Amplitude 1.3 m, period 10π seconds; at in equilibrium when $t = 0$.

  c **i** Amplitude 12.1 m, period 2.5 seconds; in equilibrium when $t = 0$.

     ii Amplitude 0.3 m, period 0.6 seconds; at rest when $t = 0$.

3 Each differential equation models a particle performing simple harmonic motion. Find the period of the motion.

  a **i** $\dfrac{d^2 x}{dt^2} + 25x = 0$ **ii** $\dfrac{d^2 x}{dt^2} + 9x = 0$

  b **i** $\ddot{x} + 2x = 0$ **ii** $\ddot{x} + 8x = 0$

  c **i** $3\dfrac{d^2 x}{dt^2} + 9x = 0$ **ii** $5\dfrac{d^2 x}{dt^2} + 45x = 0$

  d **i** $4\ddot{x} = -4x$ **ii** $3\ddot{x} = -15x$

4 A particle performs simple harmonic motion with amplitude 0.2 m and angular frequency 5 s⁻¹. The particle passes through the equilibrium position when $t = 0$.

  a Find the distance of the particle from the equilibrium position when $t = 4$ s.

  b Find the maximum speed of the particle.

5 A small ball is attached to one end of an elastic spring. When $t = 0$ the ball is released from rest 0.6 m from the equilibrium position and performs simple harmonic motion with angular frequency 12 s⁻¹.

  a Find the displacement of the ball from the equilibrium position after 3 seconds.

  b Find the time when the ball first passes through the equilibrium position, and the speed of the ball at this time.

6 A small ball attached to the end of a spring performs simple harmonic motion with amplitude 8 cm and angular frequency 15 s⁻¹.

  a Find the maximum speed of the ball.

  b The ball is at rest when $t = 0$. Find the speed of the ball 5 seconds later. Find also the magnitude of acceleration of the ball at this time.

7 A particle performs simple harmonic motion with amplitude 12 cm. Its speed as it passes through the equilibrium position is 0.08 m s⁻¹.

  a Find the angular frequency of the simple harmonic motion.

  b Find the speed of the particle when its displacement from the equilibrium position is 8 cm.

8 A particle performs simple harmonic motion with amplitude 0.6 m and angular frequency 10 s^{-1}. The particle passes through the equilibrium position when $t = 0$ with positive displacement immediately after $t = 0$.

   a Find the displacement of the particle when $t = 3.6$ s.

   b Find the time when the particle is first 0.3 m from the equilibrium position.

   c Find the speed of the particle at that point. Is it moving towards or away from the equilibrium position?

9 A particle is attached to one end of an elastic spring. It is displaced from its equilibrium position and performs simple harmonic motion with amplitude 0.5 m. When its displacement from the equilibrium position is 0.2 m the speed of the particle is 0.4 m s^{-1}.

   a Find the angular frequency of the simple harmonic motion.

   b Hence find the distance from the equilibrium position when the speed of the particle is 0.05 m s^{-1}.

10 A particle moves in a straight line between points A and B which are 0.6 m apart. The midpoint of AB is O and the displacement of the particle from O at time t seconds is x metres.

   The motion of the particle is described by the equation $\dfrac{d^2 x}{dt^2} + 0.16x = 0$. When $t = 0$ the particle is at A.

   a Write down the amplitude and the period of the simple harmonic motion.

   b Write down an expression for x in terms of t.

   c Point C is between A and B, and $AC = 0.4$ m. Find the time when the particle first passes through C.

   d The mass of the particle is 0.2 kg. Find the magnitude of the force acting on the particle when it passes through C.

11 A cart of mass 300 kg is moving in a straight line with a speed of 12 m s^{-1} when it hits a buffer which is attached to a fixed wall by a light spring.

   At time t seconds after the impact the compression of the spring is x metres and the force in the spring is given by $T = 192x$ newtons. Any other forces acting on the cart can be ignored.

   a Show that the cart performs simple harmonic motion as long as it remains in contact with the buffer.

   b Find the maximum compression of the spring and the magnitude of the force acting on the cart at that point.

   c Find the time taken to reach the point of maximum compression.

12 A particle of mass m kg is attached to one end of a light spring and rests on a smooth horizontal table. The string is horizontal and its other end is attached to a fixed wall.

   The particle is displaced away from the wall so that the extension of the spring is 0.6 m and then released. When the extension of the spring is e the elastic force in the spring is $T = mq^2 e$, where q is a constant. All other forces on the particle can be ignored.

   a Show that the particle performs simple harmonic motion and find, in terms of q, the period of the motion.

   b Find the maximum speed and the maximum acceleration of the particle.

   c Find the extension of the spring at the moment when the speed of the particle equals half of its maximum speed.

13 A particle of mass 0.5 kg rests on a smooth horizontal table. The particle is attached to two light springs and the other ends of the springs are attached to fixed points A and B, which are 0.8 m apart. The natural length of each spring is 0.3 m and the magnitude of the tension in each spring is given by $1.2e$, where e is the extension of the spring.

The particle is released from rest 0.04 m from O, the midpoint of AB.

At time t the displacement of the particle from O is x.

a Find the magnitude of the resultant force on the particle at time t.

b Hence show that the particle performs simple harmonic motion.

c Find an expression for x in terms of t.

14 A light spring is attached to a fixed point A. A particle of mass 0.2 kg is attached to the other end of the spring and hangs vertically below A.

When the extension of the spring is e metres, the magnitude of the tension in the spring is $T = 2ge$ N.

a The particle hangs in equilibrium at point B. Find the extension of the spring.

The particle is displaced x m downward from the equilibrium position.

b Write down the extension of the spring. Hence show that, as long as $|x| < 0.1$, the magnitude of the resultant force on the particle is $2gx$.

c Hence show that the particle performs simple harmonic motion and find the period of the motion.

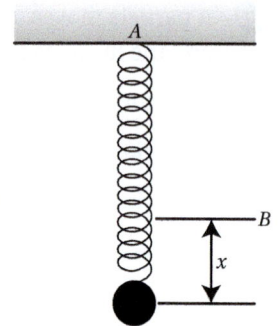

Section 3: Damping and damped oscillations

When objects are moving they are usually subjected to resistive forces such as air resistance or drag in water. There are several ways in which you can model this situation. One common model is to say that the drag force, D, is proportional to the speed, acting in the opposite direction.

Key point 11.6

The drag force on an object moving with speed v is given by
$$D = -Kv$$
where K is a constant.

If you add this to the standard equation for simple harmonic motion, the differential equation becomes:

$$\frac{d^2x}{dt^2} + k\frac{dx}{dt} + \omega^2 x = 0$$

where $k = \dfrac{K}{m}$ is a positive constant. **Damped** oscillations result, or **damped simple harmonic motion**.

As you saw in Chapter 10, the solutions to this differential equation depend upon how many solutions there are to the auxiliary equation, $\lambda^2 + k\lambda + \omega^2 = 0$, and each case is given a different name.

🔑 Key point 11.7

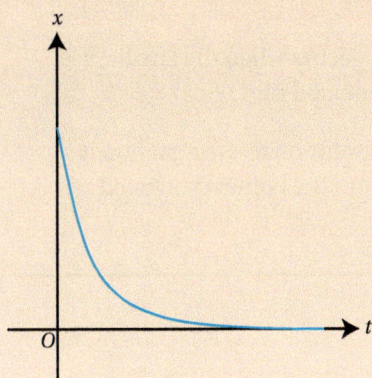

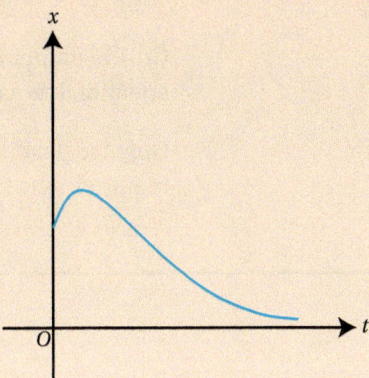

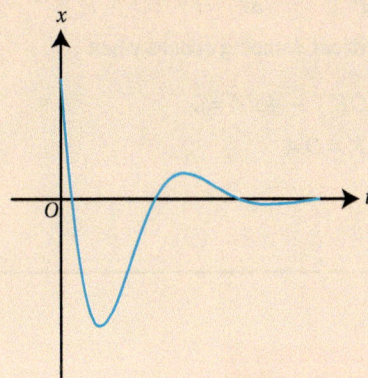

$k^2 - 4\omega^2 > 0$

Overdamping

$x = A\mathrm{e}^{\alpha t} + B\mathrm{e}^{\beta t}$

$k^2 - 4\omega^2 = 0$

Critical damping

$x = (A + Bt)\mathrm{e}^{-\frac{k}{2}t}$

$k^2 - 4\omega^2 < 0$

Underdamping

$x = \mathrm{e}^{-\frac{k}{2}t}(A\sin qt + B\cos qt)$

ℹ️ Did you know?

In physical situations, such as the suspension of a car, critical damping is often desirable as it minimises vibrations without too much jerkiness.

WORKED EXAMPLE 11.3

A bob of mass 0.1 kg is connected to a spring. In air the bob is found to follow simple harmonic motion with period π seconds. The bob is then placed into oil where there is a drag force of magnitude Kv. Find the value of K which produces critical damping.

If the period is T, then the value of ω is given by $= \dfrac{2\pi}{T} = 2$.

Rearrange the formula in Key point 11.4 to express ω in terms of T.

Therefore, in air the differential equation is:
$$\frac{d^2 x}{dt^2} + 4x = 0$$

The force is given by:
$$F = ma = m\frac{d^2 x}{dt^2} = -4mx$$

In oil, there must be the additional drag force:
$$F = m\frac{d^2 x}{dt^2} = -4mx - K\frac{dx}{dt}$$

$$\therefore \frac{d^2 x}{dt^2} + \frac{K}{m}\frac{dx}{dt} + 4x = 0$$

Divide through by m.

Continues on next page …

$$\frac{d^2x}{dt^2} + 10K\frac{dx}{dt} + 4x = 0$$

Since $m = 0.1$.

Critical damping occurs when

Critical damping occurs when the auxiliary equation has a repeated root.

$$(10K)^2 - 4 \times 4 = 0$$

$$\therefore K = 0.4$$

Only the positive solution to the equation is required, since from the context you need $K > 0$.

EXERCISE 11C

1 Each differential equation describes damped harmonic motion. In each case determine whether the damping is critical, underdamping or overdamping.

Tip

Remember that $\ddot{x}$ is an alternative notation for $\frac{d^2x}{dt^2}$, and $\dot{x}$ is an alternative notation for $\frac{dx}{dt}$.

a i $\dfrac{d^2x}{dt^2} + 2\dfrac{dx}{dt} + 5x = 0$

ii $\dfrac{d^2x}{dt^2} + 8\dfrac{dx}{dt} + 3x = 0$

b i $\ddot{x} + 4\dot{x} + 4x = 0$

ii $\ddot{x} + 3\dot{x} + 4x = 0$

c i $3\dfrac{d^2x}{dt^2} + 3\dfrac{dx}{dt} + 9x = 0$

ii $5\dfrac{d^2x}{dt^2} + 30\dfrac{dx}{dt} + 45x = 0$

d i $4\ddot{x} = -8\dot{x} - 4x$

ii $3\ddot{x} = -6\dot{x} - 15x$

2 A particle performs damped oscillations described by the differential equation $\dfrac{d^2x}{dt^2} + 3n\dfrac{dx}{dt} + 6nx = 0$. Given that the damping is critical, find the value of n.

3 A particle of mass 0.2 kg is attached to one end of a light spring and rests on a horizontal table, with the spring horizontal. When the extension of the spring is x metres the tension in the spring has magnitude $3.6x$ N. The resistance force acting on the particle has magnitude kv N, where v m s^{-1} is the speed of the particle.

a Show that the equation of motion of the particle is $\ddot{x} + 5k\dot{x} + 18x = 0$.

b Given that the motion of the particle is critically damped, find the exact value of k.

c Name the type of damping that occurs when $k = 1.8$.

4 A particle P of mass m kg is attached to one end of a spring. When the displacement of P from its equilibrium position is x metres, the magnitude of the tension in the spring is $4nx$ N and the resistance force on P has magnitude $5c\dot{x}$ N.

a Write down a differential equation that models the motion of the particle.

b Given that the motion of the particle is critically damped, express n in terms of m and c.

5 A particle is attached to one end of a spring and moves under the action of a tension and a resistance force. The motion of the particle is described by the differential equation $\dfrac{d^2x}{dt^2} + 4\dfrac{dx}{dt} + 13x = 0$, when $t = 0$, $x = 0$ and $\dfrac{dx}{dt} = 2.7$.

a Find an expression for x in terms of t.

b Name the type of damping that occurs and sketch the graph of x as a function of t.

6 A particle of mass 0.2 kg is attached to one end of a spring and moves in a straight line on a horizontal table. When the displacement of the particle from a fixed point O is x m the tension in the spring has magnitude $1.2x$ N. The resistance force acting on the particle has magnitude $1.4v$ N, where v m s^{-1} is the speed of the particle.

   a Show that the equation of motion for the particle is $\dfrac{d^2x}{dt^2} + 7\dfrac{dx}{dt} + 6x = 0$.

   When $t = 0$ the particle is at rest, 1 m from O.

   b Find an equation for x in terms of t.

   c Show that the particle never reaches O.

   d Name the type of damping that occurs in this case.

7 A particle of mass 3 kg moves in a straight line under the action of two forces. When the particle's displacement from a fixed point O is x m there is a force towards O of magnitude $2.43x$ N as well as a resistance force of magnitude $12kv$ m s^{-1} (where v is the speed of the particle). When $t = 0$ the particle is at rest 0.8 m from O.

   a Show that the motion of the particle is described by the equation $\ddot{x} + 4k\dot{x} + 0.81x = 0$.

   b Given that the motion of the particle is critically damped, find the value of k.

   c In this case, find an expression for x in terms of t.

8 A particle P of mass 0.16 kg is suspended by a light elastic string, and the other end of the string is attached to a fixed point A vertically above P. The natural length of the string is 1.2 m. When the extension of the string is d m, the magnitude of the tension in the string is $T = 4d$ N.

   a P hangs in equilibrium at a point B. Taking $g = 10$ m s^{-2}, find the extension of the string at this point.

   P is held at rest with the string at its natural length, and then released. When the speed of P is v m s^{-1} the resistance force acting on P has magnitude $1.28v$ N.

   b Show that the subsequent motion of P can be modelled by the differential equation $\ddot{x} + 8\dot{x} + 25x = 0$.

   c Name the type of damping that occurs in this case and find an expression for x in terms of t.

   d According to this model, what will the length of the string be in the long term?

   e Find the speed of P when it passes through B for the first time.

9 A particle is attached to one end of a light spring and performs damped oscillations described by the differential equation $\ddot{x} + k\dot{x} + c^2x = 0$, where x is the extension of the spring beyond the equilibrium position. It is given that $k = \dfrac{8c}{5}$.

   a Determine the type of damping that occurs.

   At $t = 0$, $x = 0$ and $\dot{x} = u$.

   b Find an expression for x in terms of t.

   c Show that the maximum extension of the spring is approximately $\dfrac{0.424u}{c}$.

Section 4: Linear systems

There are many situations where two variables are linked by coupled differential equations; for example, each variable might change with time, but the rate of change might depend of the value of the other variable. If both of these differential equations are linear and first order, then you can eliminate one of the variables to form a second order linear differential equation.

Rewind

You learnt how to solve linear second order differential equations in Chapter 10, Sections 3 and 4.

WORKED EXAMPLE 11.4

In a population of foxes (f thousands) and rabbits (r thousands), the foxes have a birth rate $3r$ and a death rate $6f$. The rabbits have a birth rate of $4r$ and a death rate of $8f$.

a Write this information in the form of a pair of differential equations.
b Rewrite these differential equations as a second order differential equation for f.
c Solve this second order differential equation given that initially $f = 2$ and $\dfrac{df}{dt} = 2$.
d Hence find the solution for r, given that the initial population of rabbits is five thousand.
e What is the long-term population of foxes and rabbits?

a $\dfrac{df}{dt} = 3r - 6f$ $\qquad$ (1)

$\dfrac{dr}{dt} = 4r - 8f$ $\qquad$ (2)

The rate of change of the fox population will be (birth rate – death rate), and likewise for the rabbit population.

b Differentiating (1) with respect to t:

$\dfrac{d^2 f}{dt^2} = 3\dfrac{dr}{dt} - 6\dfrac{df}{dt}$

Substituting in $\dfrac{dr}{dt}$ from (2):

$\dfrac{d^2 f}{dt^2} = 12r - 24f - 6\dfrac{df}{dt}$

Rearranging (1):

$3r = \dfrac{df}{dt} + 6f$

Substituting this in:

$\dfrac{d^2 f}{dt^2} = 4 \times \left(\dfrac{df}{dt} + 6f\right) - 24f - 6\dfrac{df}{dt}$

$\qquad = -2\dfrac{df}{dt}$

So $\dfrac{d^2 f}{dt^2} + 2\dfrac{df}{dt} = 0$

c Auxiliary equation is:

$\lambda^2 + 2\lambda = 0$

$\lambda(\lambda + 2) = 0$

$\lambda = 0 \text{ or } -2$

$\therefore f = A + Be^{-2t}$

Continues on next page ...

$$\frac{df}{dt} = -2Be^{-2t}$$

When $t=0$, $\frac{df}{dt}=2$ so $B=-1$.

When $t=0$, $f=2$ so $A=3$.

So the solution is $f=3-e^{-2t}$.

d Substituting the solution from part **c** into (2):

$$\frac{dr}{dt} = 4r - 24 + 8e^{-2t}$$

$$\frac{dr}{dt} - 4r = 8e^{-2t} - 24$$

> This is a first order linear differential equation, so you can write it in an appropriate form to use integrating factors.

The integrating factor is e^{-4t}, so:

$$re^{-4t} = \int e^{-4t}\left(8e^{-2t} - 24\right)dt$$

$$= \int 8e^{-6t} - 24e^{-4t}\,dt$$

$$= -\frac{4}{3}e^{-6t} + 6e^{-4t} + c$$

$$\therefore r = -\frac{4}{3}e^{-2t} + 6 + ce^{4t}$$

When $t=0$, $r=5$ so $c = \frac{1}{3}$.

$$\therefore r = -\frac{4}{3}e^{-2t} + 6 + \frac{1}{3}e^{4t}$$

e As t gets very large e^{-2t} gets very small, but e^{4t} gets very large. The population of foxes tends towards 3000, but the population of rabbits grows without limit.

EXERCISE 11D

1 Write each pair of differential equations as a single second order equation for x. Hence find the general solution for x and y in terms of t.

a i $\dfrac{dx}{dt} = 5x - 2y, \dfrac{dy}{dt} = x + 2y$ **ii** $\dfrac{dx}{dt} = x + 2y, \dfrac{dy}{dt} = 2x + y$

b i $\dot{x} = 4y - 3x, \dot{y} = y - 2x$ **ii** $\dot{x} = 3x - y, \dot{y} = 8x - y$

c i $\dot{x} = 4y - 5x + e^{-3t}, \dot{y} = 2y - 3x + 2e^{-3t}$ **ii** $\dot{x} = 2y - 3x + 5, \dot{y} = 2y - 2x - 8$

2 Find the general solution of $\dfrac{dy}{dt} = x + \cos t, \dfrac{dx}{dt} = y + \sin t$.

3 Find the general solution for x and y in terms of t for this system of differential equations:

$$\frac{dx}{dt} = -2y, \quad \frac{dy}{dt} = -8x.$$

4　The variables x and y satisfy the differential equations

$$\frac{\mathrm{d}x}{\mathrm{d}t}=35-5y,\ 5\frac{\mathrm{d}y}{\mathrm{d}t}=16x-192.$$

When $t=0$, $x=17$ and $y=7$.

Find expressions for x and y in terms of t.

5　Consider the system of differential equations

$$\frac{\mathrm{d}x}{\mathrm{d}t}=3x+y,\ \frac{\mathrm{d}y}{\mathrm{d}t}=6x-2y.$$

a　Find a second order differential equation for x.

When $t=0$, $x=1$ and $y=15$.

b　Find expressions for x and y in terms of t.

6　Three identical cylindrical buckets, each with cross-sectional area 0.25 m², are placed vertically above each other. A hole is drilled in the base of each of the top two buckets so that water can flow from the top bucket to the middle one and from the middle one to the bottom one.

For each of the top two buckets, when the height of water in the bucket is h m, the rate of flow of water out of the bucket is $0.5h$ m³ s⁻¹. Initially, the height of water in the top bucket is 30 cm and the middle bucket is empty.

Let x be the height of water in the top bucket and y be the height of water in the middle bucket at time t.

The time taken for water to fall between buckets can be ignored.

a　Show that $x=0.3\mathrm{e}^{-2t}$ and write a differential equation for y in terms of t.

b　Find an expression for y in terms of t and show that this model predicts that the second bucket never empties.

c　Find the maximum height of water in the second bucket.

7　A system contains sharks (S thousand) and fish (F million). The sharks have a birth rate given by $0.1F+1$ and a death rate given by $0.2S$. The fish have a birth rate given by $0.2F+4$ and a death rate given by $0.5S$.

> **Focus on …**
>
> You can find out how this model can be improved in Focus on … Modelling 2.

a　Write this information in the form of a pair of differential equations.

b　Rewrite these differential equations as a second order differential equation for S.

c　Solve this second order differential equation given that initially $S=17$ and $F=28$.

d　By writing the solution for S in the form $S=A+B\cos(kt-\alpha)$, where $\alpha\in[0,\pi)$, find the time of the first peak in the shark population. Find the equivalent time at which the fish population first peaks.

e　Describe the long-term behaviour of the two populations.

8　A predator-prey system is of the form $\dfrac{\mathrm{d}x}{\mathrm{d}t}=ax+by,\ \dfrac{\mathrm{d}y}{\mathrm{d}t}=cx+dy.$

Prove that the system will only oscillate if $(a+d)^2<4(ad-bc)$.

Checklist of learning and understanding

- The differential equation for simple harmonic motion is $\dfrac{d^2x}{dt^2} = -\omega^2 x$.
- The general solution to the simple harmonic motion differential equation is $x = A\sin\omega t + B\cos\omega t$, which can also be written as $x = R\sin(\omega t + \varphi)$.
- Period of the solution: $T = \dfrac{2\pi}{\omega}$
- Speed, v, is given by: $v^2 = \omega^2(a^2 - x^2)$.
- Drag force is given by $D = -Kv$.
- If there is a drag force there can be underdamping, overdamping or critical damping depending on the number of solutions to the auxiliary equation.
- You can rewrite coupled pairs of linear first order differential equations as a second order differential equation in one variable.

Mixed practice 11

1. Find the period of the oscillations of a particle whose motion is modelled by the differential equation
$$\frac{d^2 y}{dt^2} + 4y = 0.$$

2. Find the value of q which would result in critical damping in the system modelled by
$$\frac{d^2 x}{dt^2} + 16\frac{dx}{dt} + qx = 0.$$

3. A particle of mass 5 kg is acted on by a force $F = 10\sin t$ newtons, where t is the time measured in seconds.

   a Write down a differential equation satisfied by the displacement, x metres, of the particle from its initial position.

   b Given that the particle is initially at rest, find its displacement after 3 seconds.

4. A ball is attached to one end of an elastic string and performs simple harmonic motion with amplitude 0.3 m and angular frequency 6 s^{-1}.

   a Find the maximum speed of the ball.

   b Find the speed of the ball when its displacement from the equilibrium position is 0.2 m.

   c The mass of the ball is 0.3 kg. Find the magnitude of the maximum force acting on the ball during the motion.

5. A particle of mass 2 kg moves in a straight line so that, when the displacement of the particle from the origin is x m, the force acting on the particle is directed towards the origin and has magnitude $18x$ N.

   a Show that the displacement of the particle satisfies the differential equation $\frac{d^2 x}{dt^2} = -9x$.

   b Verify that, for some value of ω which you should state, $x = A\cos\omega t + B\sin\omega t$, where A and B are constants, satisfies this differential equation.

   c The particle is initially at rest 0.3 m from the origin. Find the value of the constants A and B.

   d Hence find the maximum speed of the particle.

6. A particle P of mass 0.2 kg moves on a smooth horizontal plane. Initially it is projected with velocity 0.8 m s^{-1} from a fixed point O towards another fixed point A. At time t s after projection, P is x m from O and is moving with velocity v m s^{-1}, with the direction OA being positive. A force of $(1.5t - 1)$ N acts on P in the direction parallel to OA.

   i Find an expression for v in terms of t.

   ii Find the time when the velocity of P is next 0.8 m s^{-1}.

   iii Find the times when P subsequently passes through O.

   iv Find the distance P travels in the third second of its motion.

   © OCR A Level Mathematics, Unit 4730/01 Mechanics 3, June 2013

7. One end of a light elastic spring is attached to a fixed wall and a small ball is attached to the other end. The ball rests on a smooth horizontal table.

   At $t = 0$ the ball is given the velocity of 15 m s^{-1} away from the equilibrium position. When the displacement of the ball from the equilibrium position is x m the force acting on the particle is $5x$ N.

a Given that the mass of the ball is 0.2 kg, show that the equation of motion of the ball is $\dfrac{d^2x}{dt^2}=-25x$.

b Show that $x=R\sin(5t+\varphi)$ satisfies the equation and find the value of the constants R and φ.

c Find the time when the particle first returns to the equilibrium position.

8 The spread of a disease through a population is modelled using the following differential equations:

$$\begin{cases}\dfrac{dS}{dt}=-2I\\[2mm]\dfrac{dI}{dt}=2I\end{cases}$$

where S is the number of uninfected individuals and I is the number of infected individuals in the population at time t months.

Initially there are 199 uninfected individuals and 1 infected individual. According to this model, how long will it take for half the population to become infected?

9 Solve this system of differential equations:

$$\begin{cases}\dot{x}=3x-5y\\ \dot{y}=5x-3y\end{cases}$$

given that $x(0)=1$ and $y(0)=1$.

10 Two particles, P and Q, each have a mass of 1 kg and are initially at rest. P moves under the action of a force F_p newtons, modelled by $F_p(t)=t+1$, where t is the time measured in seconds. Q moves under the action of a force F_Q newtons, modelled by $F_Q(x)=x+1$, where x metres is the displacement from the initial position. Which particle travels further in the first five seconds?

11 A particle moves with simple harmonic motion in a straight line between points A and B, which are 1.2 m apart. The midpoint of AB is O.

The motion of the particle satisfies the differential equation $\dfrac{d^2x}{dt^2}=-\omega^2x$, where x is the displacement of the particle from O.

a Show that $x=0.6\cos\omega t$ satisfies the differential equation. Hence show that $v^2=\omega^2(0.36-x^2)$.

b Given that the particle takes four seconds to travel from A to B, find the value of ω.

c Given that the mass of the particle is 400 grams, find the maximum force acting on the particle.

12 One end of a light spring is attached to a fixed wall. A ball of mass 0.25 kg is attached to the other end of the spring and rests on a smooth horizontal table. The ball is displaced 0.2 m from the equilibrium position and then released. When the extension of the spring is x m, the magnitude of the tension in the spring is given by $T=64x$ N.

a Show that the equation of motion of the ball can be written as $\dfrac{d^2x}{dt^2}=-256x$.

b Find the maximum speed of the ball.

13 One end of a light spring is attached to a fixed wall. A particle P of mass m kg is attached to the other end of the spring and rests on a smooth horizontal table with the spring horizontal. P is displaced 30 cm from its equilibrium position and released from rest. When the displacement of P from equilibrium is x m the tension in the spring has magnitude $6.25mx$ N.

a Show that $\dfrac{d^2x}{dt^2}-6.25x=0$.

b Show that $x=A\sin2.5t+B\cos2.5t$ satisfies the differential equation and find the values of A and B. Hence show that $v^2=6.25(0.09-x^2)$.

c Find the maximum speed of the particle.

14 A cart of mass 13 kg is attached to one end of a horizontal spring. When the extension of the spring is x m the tension in the spring is $13x$ N. Initially the cart is displaced 5 m from its equilibrium position along the axis of the spring. It is held at rest and then released.

In a simple model the only force acting on the cart is the tension in the spring.

a Find an expression for x in terms of time.

b How long does it take for the cart to reach the equilibrium position for the first time?

In an improved model there is also a resistance force on the cart of magnitude $10v$ N, where $v\,\text{ms}^{-1}$ is the speed of the cart.

c Find an expression for x in terms of t for the second model.

d Which model predicts the cart reaching the equilibrium position later?

15 Find the general solution of this system of differential equations:
$$\frac{dx}{dt} = x - y, \frac{dy}{dt} = 2x + y$$

16 Snakes and badgers are in competition for resources on a plain. There are no other types of animals on this plain. The populations snakes (S) and badgers (B) at time t months are modelled by these differential equations:
$$\frac{dB}{dt} = B - S, \frac{dS}{dt} = S - B$$

Initially there are 1000 badgers and 3000 snakes on the plain. Find the total number of animals on the plain after three months.

17 A particle P starts from rest from a point A and moves in a straight line with simple harmonic motion about a point O. At time t seconds after the motion starts the displacement of P from O is x m towards A. The particle P is next at rest when $t = 0.25\pi$ having travelled a distance of 1.2 m.

i Find the maximum velocity of P.

ii Find the value of x and the velocity of P when $t = 0.7$.

iii Find the other values of t, for $0 < t < 1$, at which P's speed is the same as when $t = 0.7$. Find also the corresponding values of x.

© OCR A Level Mathematics, Unit 4730/01 Mechanics 3, June 2015

18 O is a fixed point on a horizontal plane. A particle P of mass 0.25 kg is released from rest at O and moves in a straight line on the plane. At time t s after release the only horizontal force acting on P has magnitude
$$\frac{1}{2400}\left(144 - t^2\right) \text{ N for } 0 \leqslant t \leqslant 12$$
and
$$\frac{1}{2400}\left(t^2 - 144\right) \text{ N for } t \geqslant 12.$$

The force acts in the direction of P's motion. P's velocity at time t s is $v\,\text{ms}^{-1}$.

i Find an expression for v in terms of t, valid for $t \geqslant 12$, and hence show that v is three times greater when $t = 24$ than it is when $t = 12$.

ii Sketch the (t, v) graph for $0 \leqslant t \leqslant 24$.

© OCR A Level Mathematics, Unit 4730 Mechanics 3, June 2010

19 A particle of mass 3 kg is attached to two identical springs, each of natural length 1 m. The magnitude of the tension in each spring is $24e$, where e is the extension of the spring.

The other ends of the springs are attached to points A and B, which are 2.6 m apart on a smooth horizontal surface. The midpoint of AB is C.

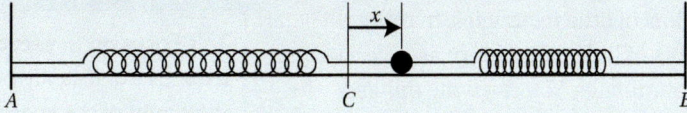

The particle is released from rest 0.15 m from C.

a Show that, when the displacement of the particle from C is x, the magnitude of the force acting on the particle is $48x$.

b Hence show that the particle performs simple harmonic motion, and find the period of the motion.

c Find the speed of the particle when it is 0.05 m from C.

20 In a strongman competition the competitors pull a truck (initially at rest) for 20 seconds. The winner is the person who pulls the truck furthest.

The truck has a mass of 2000 kg and is subject to a constant resistance force of 2000 N. Brawny Bill initially pulls the truck with a force of 3000 N, but by the end of the 20 seconds he is pulling it with a force of 1000 N.

a State one assumption needed to model this force as a linear function of time. Comment on the appropriateness of this assumption.

b Given the assumption from part **a**, write down a differential equation satisfied by the displacement, x, of the truck from its initial position.

c Solve your differential equation and hence find the displacement of the truck at the end of the 20 seconds.

d Muscly Mike's force, F newtons, is modified as $F = 3000 \times \left(\dfrac{1}{3}\right)^{\frac{t}{20}}$ at time t seconds. Determine who wins the competition.

Elements of area and Gaussian integrals

You have seen that you can find areas under a curve using an integral which you thought of as summing up lots of little rectangles. In more advanced work it is useful to sum up lots of little elements of area instead and do a double sum over all coordinates. You can write this as:

$$A = \iint dA$$

where, in Cartesian coordinates, $dA = dy\,dx$.

> **Tip**
>
> This Focus on ... section extends significantly beyond the scope of the specification, but it will be of interest to anyone wanting to go on to study Mathematics, Physics, Chemistry, Engineering or Theoretical Economics.

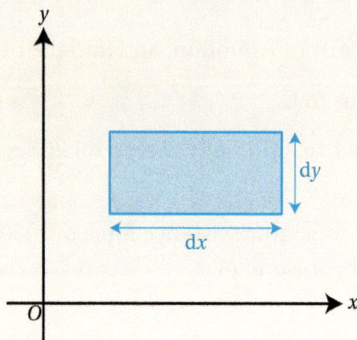

The double integrals become

$$A = \int_{x=a}^{x=b} \left(\int_{y=c}^{y=d} dy \right) dx$$

If you are looking for the area between a curve and the x-axis then the limits on y are from 0 to $f(x)$ so the area is:

$$A = \int_{x=a}^{x=b} \left(\int_{y=0}^{y=f(x)} dy \right) dx$$
$$= \int_{x=a}^{x=b} \left[y \right]_0^{f(x)} dx$$
$$= \int_{x=a}^{x=b} f(x) dx$$

which is the formula you are used to using.

In Chapter 9 you found that the area between two half-lines in polar coordinates is given by $A = \int_\alpha^\beta \frac{1}{2} r^2 \, d\theta$.

You can derive this using a similar method to the one shown for Cartesian coordinates. In the diagram the shaded area is approximately a rectangle with one dimension $r\,d\theta$ (using the formula for arc length in radians) and the other dr.

The area element in polar coordinates is therefore:

$$dA = r \, dr \, d\theta$$

Question

1 Prove that the area bounded by the lines $\theta = \alpha$, $\theta = \beta$ and $r = r(\theta)$ is given by the formula $A = \int_{\alpha}^{\beta} \frac{1}{2} r(\theta)^2 \, d\theta$.

These area elements have some lovely consequences, including allowing you to evaluate otherwise impossible integrals.

Consider the integral $I = \int_{x=-\infty}^{x=\infty} e^{-x^2} \, dx$.

You cannot find the indefinite integral of e^{-x^2} using standard functions, however over this range you can evaluate the integral exactly. The x in the integral is just a dummy variable. You could also write:

$$I = \int_{y=-\infty}^{y=\infty} e^{-y^2} \, dy$$

Multiplying the two expressions together:

$$I^2 = \int_{x=-\infty}^{x=\infty} e^{-x^2} \, dx \int_{y=-\infty}^{y=\infty} e^{-y^2} \, dy$$

It turns out that you can combine these two integrals into one double integral:

$$I^2 = \int_{x=-\infty}^{x=\infty} \int_{y=-\infty}^{y=\infty} e^{-y^2} e^{-x^2} \, dy \, dx$$

$$= \int_{x=-\infty}^{x=\infty} \int_{y=-\infty}^{y=\infty} e^{-y^2 - x^2} \, dy \, dx$$

But $dx \, dy = dA$ is just an element of area, so you could rewrite it as $r \, dr \, d\theta$. You can recast the whole expression in terms of polar coordinates, noting that $x^2 + y^2 = r^2$ and that the limits represent the whole plane:

$$I^2 = \int_{\theta=0}^{\theta=2\pi} \left(\int_{r=0}^{r=\infty} e^{-r^2} r \, dr \right) d\theta$$

Questions

2 Complete the proof to evaluate I.

3 Hence evaluate $\int_{-\infty}^{\infty} e^{\frac{x^2}{2\sigma^2}} \, dx$ where σ is a constant.

⏪ Rewind

This is an example of an improper integral which you met in Chapter 8.

⏪ Rewind

The integral in question 3 is of vital importance in working with the normal distribution, which you met in A Level Mathematics Student Book 2, Chapter 17.

Finding the shape of a hanging chain

Consider this problem:

> A uniform chain is suspended from two fixed points at the
> same height and hangs under its own weight. Find the shape
> of the chain.

The first step is to express the question in a mathematical form.
If you set up the coordinate axes so that the two end points have the
same y-coordinate, then you can describe the shape of the chain by
a function $y = f(x)$. The task is thens to find an expression for $f(x)$.

It is clear that the shape of the chain will be symmetrical, with the
lowest point halfway between the end points. Note that the position
of the x-axis is irrelevant, since the shape of the chain does not
change if the end points are moved vertically.

Next you need to introduce some parameters: what could the exact
shape of the chain depend on? It seems reasonable to consider
these factors:

- thc mass of the chain (M)
- the length of the chain (L)
- the distance between the end points ($2D$).

As already noted, the height of the end points does not affect the
shape of the chain.

The shape of the chain is determined by the forces acing on it. As
well as the mass of the chain, there is a tension force acting along
the chain. At each point the tension acts along the tangent to the
chain. So, if you can determine the direction of the tension at each
point, you will know the gradient of the tangent, which is $\dfrac{dy}{dx}$.
Knowing the gradient will enable you to find the equation for y
in terms of x.

Consider the part of the chain between the lowest point and
another point with a variable x-coordinate. The forces acting on this
part of the chain are shown in the diagram (m is the mass of this
part of the chain). The force T_0 is fixed (it is the force from the left
half of the chain on the right half of the chain), but T changes with x.

Resolving forces horizontally and vertically gives:

$$T\cos\theta = T_0, \quad T\sin\theta = mg$$

and therefore $\tan\theta = \dfrac{mg}{T_0}$. But, since the force T is directed along the
tangent to the curve, $\tan\theta$ equals the gradient of the curve at that
point. Hence $\dfrac{dy}{dx} = \dfrac{mg}{T_0}$.

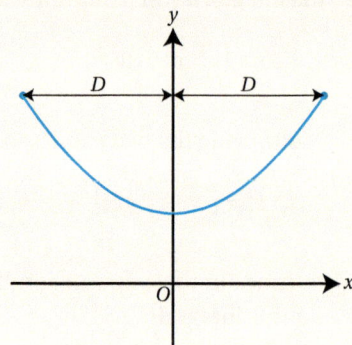

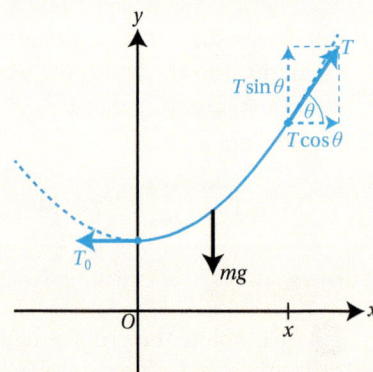

⏮ Rewind

For a reminder of resolving
forces, see A Level
Mathematics Student Book 2,
Chapter 21.

In the expression for the gradient, g and T_0 are constants, but m (the mass of this part of the chain) depends on the x-coordinate. If you can express m in terms of x, you can then integrate $\dfrac{dy}{dx}$ to obtain your required equation for the shape of the chain.

Since the chain is uniform, the mass of a part of the chain is proportional to the length of that part. The whole chain has length L and mass M, so $m = \dfrac{Ms}{L}$ where s is the length of the section of the chain between x-coordinates 0 and x.

To find an expression for the length of the curve in terms of x, consider a small section of the curve between points with coordinates (x, y) and $(x + \Delta x, y + \Delta y)$, and denote the length of this small section Δs. The small section of the curve is close to a straight line, so

$$\Delta s \approx \sqrt{(\Delta x)^2 + (\Delta y)^2} = (\Delta x)\sqrt{1 + \left(\frac{\Delta y}{\Delta x}\right)^2}$$

$$\Rightarrow \frac{\Delta s}{\Delta x} \approx \sqrt{1 + \left(\frac{\Delta y}{\Delta x}\right)^2}$$

As $\Delta x \to 0$, $\dfrac{\Delta s}{\Delta x} \to \dfrac{ds}{dx}$ and $\dfrac{\Delta y}{\Delta x} \to \dfrac{dy}{dx}$, and so

$$\frac{ds}{dx} = \sqrt{1 + \left(\frac{dy}{dx}\right)^2}$$

You now have the equation $\dfrac{dy}{dx} = \dfrac{mg}{T_0} = \dfrac{Mg}{T_0 L} s$, since $m = \dfrac{Ms}{L}$.

Differentiating this gives $\dfrac{d^2y}{dx^2} = \dfrac{gM}{T_0 L} \dfrac{ds}{dx}$, and so

$$\frac{d^2y}{dx^2} = \frac{gM}{T_0 L}\sqrt{1 + \left(\frac{dy}{dx}\right)^2}$$

You can now proceed to solve this differential equation.

Questions

1 Make a substitution $u = \dfrac{dy}{dx}$ and show that
$$\int \frac{1}{\sqrt{1+u^2}}\,du = \frac{gM}{T_0 L}\int 1\,dx.$$

2 Explain why the constant of integration is zero. Hence show that $u = \sinh\left(\dfrac{gM}{T_0 L}x\right)$.

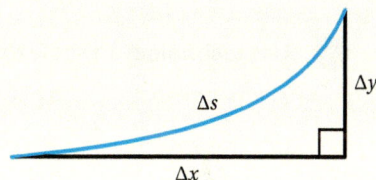

⏮ **Rewind**

You met integrals of this type in Chapter 7, Section 3.

3 Hence find an expression for y in terms of x. Explain why the constant of integration can be taken to be zero.

In the expression you found in question 3, g is a constant and M and L are fixed properties of the chain. However, you don't yet know what T_0 is; you defined it as the magnitude of the tension acting at the lowest point of the chain. You should also notice that you have not yet used the condition that the end points of the chain are a distance $2D$ apart. It seems reasonable that the tension in the chain will depend on how far apart the end points are.

Questions

4 Show that the length of the curve $y = k\cosh\left(\dfrac{x}{k}\right)$ between points with coordinates $x = a$ and $x = b$ is

$$k\left(\sinh\left(\frac{b}{k}\right) - \sinh\left(\frac{a}{k}\right)\right).$$

5 Use the fact that the total length of the chain is L, and that the end points are at $x = -D$ and $x = D$, to show that

$$\frac{2T_0}{gM}\sinh\left(\frac{gMD}{LT_0}\right) = 1.$$

6 Use technology to show that this equation has a solution for T_0 whenever $\dfrac{D}{L} < \dfrac{1}{2}$. Explain why this condition always holds in this problem.

In summary, you have found that a chain suspended freely from two fixed points hangs in the shape of a cosh curve, $y = k\cosh\left(\dfrac{x}{k}\right)$, where k is a constant depending on the mass and the length of the chain and the distance between the end points.

> **Tip**
>
> You have seen that the length of the curve satisfies
> $$\frac{ds}{dx} = \sqrt{1 + \left(\frac{dy}{dx}\right)^2}.$$ Integrating this expression, you find that
> $$s = \int_a^x \sqrt{1 + \left(\frac{dy}{dx}\right)^2}\, dx.$$

> **Did you know?**
>
> The cosh curve is called a **catenary**, meaning 'relating to a chain'.

The Lotka–Volterra model and phase planes

During World War I the marine biologist Umberto D'Ancona noticed something puzzling about fish in the Adriatic Sea. Although they were being fished less (and so their natural death rate decreased), the numbers of small fish were actually decreasing while the numbers of predator fish were increasing. His father-in-law, Vito Volterra, applied the work of Alfred Lotka to try to explain this observation.

Consider a population of a species of fish (F million) and sharks (S thousand).

The natural net birth rate of the fish (i.e. the birth rate minus the death rate) is proportional to the number of fish, with constant of proportionality a. There is also a death rate due to predation which is proportional to both the number of fish and the number of sharks, with constant of proportionality b. This means that:

$$\frac{dF}{dt} = aF - bFS$$

A similar differential equation governs the population of sharks:

$$\frac{dS}{dt} = cFS - kS$$

where the cFS term represents the growth in the shark population due to their predation on the fish and the $-kS$ term is the natural net death rate of the sharks.

Question

1 Describe some modelling assumptions that have been made in creating this model.

When analysing systems like this, it is often the case that solving the differential equation is less important than finding fixed points of the system (values of the population where there is no change in the population i.e. places where $\frac{dF}{dt} = 0$ and $\frac{dS}{dt} = 0$.)

Questions

2 Find all fixed points of the differential equations in the Lotka-Volterra model. Which correspond to the biological equilibrium values if the populations do not go extinct?

3 When trawler fishing is reduced, the net birth rate of the fish will increase and the net death rate of the sharks will decrease. Use the Lokta–Volterra model to explain this.

A common way to visualise these systems of equations is to use a phase plane. These plot the 'flow' of the system at each value of F and S. You can find phase plane plotters online. For $a = b = c = k = 1$, the phase plane for Lotka–Volterra is shown in the diagram.

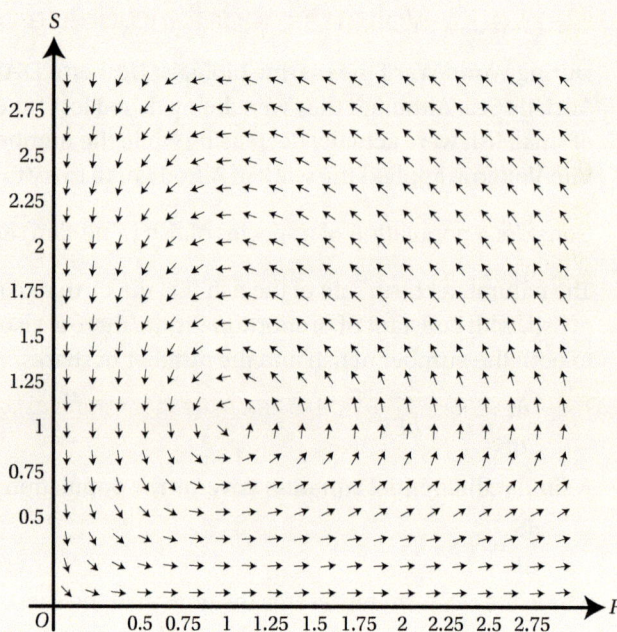

Questions

4 Sketch the curve in the phase plane above, corresponding to the initial values $F = 0.5$ and $S = 1$. Use this curve to estimate the maximum fish population.

5 Hence sketch the behaviour of F against t and of S against t for these initial conditions.

6 The effect of competition amongst the fish can be included in the model by adding another term in F^2 to the original differential equation:

$$\frac{dF}{dt} = aF - eF^2 - bFS$$

$$\frac{dS}{dt} = cFS - kS$$

Assuming that all parameters are positive, explain why this adaptation introduces a competition effect into the differential equations.

7 By using online technology, investigate this system with $a = b = c = k = 1$ and $e = 0.5$. How has the introduction of competition changed the behaviour of the system?

CROSS-TOPIC REVIEW EXERCISE 2

1. The curve C has polar equation $r^2 = a \sin 4\theta$, where $0 \leqslant \theta < 2\pi$ and $a > 0$.

   a Sketch C, clearly stating the range of values of θ for which it is defined.

   b Find the total area enclosed by C.

2. Solve the equation $3 \sinh x = 2 \cosh x$, giving your answer in the form $x = \ln \sqrt{a}$.

3. Solve the equation $3 \sinh^2 x + 2 \sinh x - 8 = 0$, giving your answer in terms of natural logarithms.

4. The Cartesian equation of a circle is $(x+4)^2 + (y-7)^2 = 65$.

   Using the origin O as the pole and the positive x-axis as the initial line, find the polar equation of this circle, giving your answer in the form $r = p \sin \theta + q \cos \theta$.

5. The shaded region in the diagram is bounded by the curve with equation $y = 4x^{-\frac{3}{2}} + 1$, the x-axis and lines $x = 1$ and $x = 4$.

   Calculate the volume of revolution when the shaded region is rotated $360°$ about the x-axis.

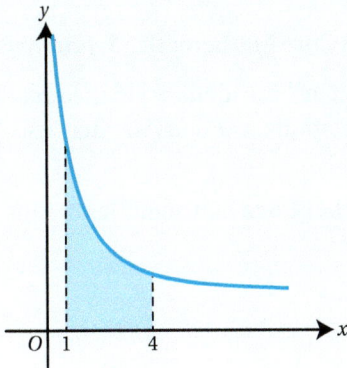

6. Express $\sinh 2x$ in terms of exponentials and hence, by using the substitution $u = e^{2x}$, find $\displaystyle\int \frac{1}{\sinh 2x}\, dx$.

7. By first completing the square in the denominator, find the exact value of

   $$\int_{\frac{1}{2}}^{\frac{3}{2}} \frac{1}{4x^2 - 4x + 5}\, dx.$$

   © OCR A Level Mathematics, Unit 4726 Further Pure Mathematics 2, January 2012

8

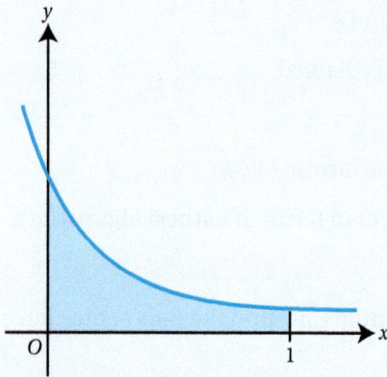

The diagram shows part of the curve $y = \dfrac{6}{(2x+1)^2}$. The shaded region is bounded by the curve and the lines $x=0$, $x=1$ and $y=0$. Find the exact volume of the solid produced when this shaded region is rotated completely about the x-axis.

© OCR A Level Mathematics, Unit 4723 Core Mathematics 3, January 2012

9 Given that the first three terms of the Maclaurin series for $(1+\sin x)e^{2x}$ are identical to the first three terms of the binomial series for $(1+ax)^n$, find the values of the constants a and n. (You may use appropriate results given in the List of Formulae (MFl).)

© OCR A Level Mathematics, Unit 4726 Further Pure Mathematics 2, June 2010

10 **a** By using the substitution $u = \sinh x$, show that

$$\int \frac{1}{\cosh x}\,dx = \arctan(\sinh x) + c$$

The part of the curve $y = \sqrt{\dfrac{1}{\cosh x}}$ between $x = 0$ and $x = k$, where k is a positive constant, is rotated through $360°$ about the x-axis. The volume of the solid formed is $\dfrac{\pi^2}{6}$.

b Find the exact value of k.

11 Show that $\sinh(\ln \sin x) = -\dfrac{\cos x}{2\tan x}$.

12 **a** Show that $9\sinh x - \cosh x = 4e^x - 5e^{-x}$.

b Given that $9\sinh x - \cosh x = 8$, find the exact value of $\tanh x$.

13 **i** Use the definitions of hyperbolic function in terms of exponentials to prove that

$$8\sinh^4 x \equiv \cosh 4x - 4\cosh 2x + 3.$$

ii Solve the equation

$$\cosh 4x - 3\cosh 2x + 1 = 0,$$

giving your answer in logarithmic form.

© OCR A Level Mathematics, Unit 4726 Further Pure Mathematics 2, January 2011

14 **i** Show that $\dfrac{\mathrm{d}}{\mathrm{d}x}(\sinh^{-1}x)=\dfrac{1}{\sqrt{x^2+1}}$.

ii Given that $y=\cosh(a\sinh^{-1}x)$, where a is a constant, show that

$$(x^2+1)\frac{\mathrm{d}^2y}{\mathrm{d}x^2}+x\frac{\mathrm{d}y}{\mathrm{d}x}-a^2y=0.$$

© OCR A Level Mathematics, Unit 4726 Further Pure Mathematics 2, June 2010

15 **i** Prove that, if $y=\sin^{-1}x$, then $\dfrac{\mathrm{d}y}{\mathrm{d}x}=\dfrac{1}{\sqrt{1-x^2}}$.

ii Find the Maclaurin series for $\sin^{-1}x$, up to and including the term in x^3.

iii Use the result of part **ii** and the Maclaurin series for $\ln(1+x)$ to find the Maclaurin series for $(\sin^{-1}x)\ln(1+x)$ up to and including the term in x^4.

© OCR A Level Mathematics, Unit 4726 Further Pure Mathematics 2, June 2011

16 It is given that $\mathrm{f}(x)=\tanh^{-1}\left(\dfrac{1-x}{3+x}\right)$ for $x>-1$.

i Show that $\mathrm{f}''(x)=\dfrac{1}{2(x+1)^2}$.

ii Hence find the Maclaurin series for $\mathrm{f}(x)$ up to and including the term in x^2.

© OCR A Level Mathematics, Unit 4726/01 Further Pure Mathematics 2, June 2013

17 The equation of a curve in polar coordinates is $r=2\sin3\theta$ for $0\leqslant\theta\leqslant\dfrac{1}{3}\pi$.

i Sketch the curve.

ii Find the area of the region enclosed by this curve.

iii By expressing $\sin3\theta$ in terms of $\sin\theta$, show that a Cartesian equation for the curve is

$$\left(x^2+y^2\right)^2=6x^2y-2y^3.$$

© OCR A Level Mathematics, Unit 4726/01 Further Pure Mathematics 2, June 2015

18 **i** Find the general solution of the differential equation
$$\frac{\mathrm{d}^2y}{\mathrm{d}x^2}+2\frac{\mathrm{d}y}{\mathrm{d}x}+17y=17x+36.$$

ii Show that, when x is large and positive, the solution approximates to a linear function, and state its equation.

© OCR A Level Mathematics, Unit 4727 Further Pure Mathematics 3, June 2010

19 Find the solution of the differential equation $\dfrac{\mathrm{d}^2y}{\mathrm{d}x^2}+2\dfrac{\mathrm{d}y}{\mathrm{d}x}+5y=\mathrm{e}^{-x}$ for which $y=\dfrac{\mathrm{d}y}{\mathrm{d}x}=0$ when $x=0$.

© OCR A Level Mathematics, Unit 4727/01 Further Pure Mathematics 3, June 2013

20 Find the solution of the differential equation
$$\frac{\mathrm{d}y}{\mathrm{d}x}+y\cot x=2x$$
for which $y=2$ when $x=\dfrac{1}{6}\pi$. Give your answer in the form $y=\mathrm{f}(x)$.

© OCR A Level Mathematics, Unit 4727 Further Pure Mathematics 3, June 2012

21 At time $t=0$ s a particle P, of mass 0.3 kg, is 1 m away from a point O on a smooth horizontal plane and is moving away from O with speed $\sqrt{5}$ m s^{-1}. The only horizontal force acting on P has magnitude $1.5x$ N, where x is the distance OP, and acts away from O.

   i Show that the speed of P, v m s^{-1}, is given by $v=\sqrt{5}x$.

   ii Find an expression for v in terms of t.

© OCR A Level Mathematics, Unit 4730/01 Mechanics 3, January 2013

22 A particle P starts from rest at a point A and moves in a straight line with simple harmonic motion. At time t s after the motion starts, P's displacement from a point O on the line is x m towards A. The particle P returns to A for the first time when $t=0.4\pi$. The maximum speed of P is 4 m s^{-1} and occurs when P passes through O.

   i Find the distance OA.

   ii Find the values of x and the velocity of P when $t=1$.

   iii Find the number of occasions in the interval $0<t<1$ at which P's speed is the same as that when $t=1$, and find the corresponding values of x and t.

© OCR A Level Mathematics, Unit 4730 Mechanics 3, January 2012

23 A particle P of mass 0.25 kg is projected horizontally with speed 5 m s^{-1} from a fixed point O on a smooth horizontal surface and moves in a straight line on the surface. The only horizontal force acting on P has magnitude $0.2v^2$ N, where v m s^{-1} is the velocity of P at time t s after it is projected from O. This force is directed towards O.

   i Find an expression for v in terms of t.

The particle P passes through a point X with speed 0.2 m s^{-1}.

   ii Find the average speed of P for its motion between O and X.

© OCR A Level Mathematics, Unit 4730 Mechanics 3, June 2011

24 Find the set of values of k for which

$$2\sinh x+3\cosh x=k$$

has at least one solution.

25 The mean value of the function $f(x)=2-\dfrac{1}{2\sqrt{x}}$ between 1 and k is $\dfrac{8}{5}$.

Find the value of k.

26 **a** Find, up to the term in x^3, the Maclaurin series for $\ln\left(\dfrac{2+x}{2-x}\right)$.

   b Find the set of x values for which the expansion is valid.

   c By evaluating the series in part **a** at an appropriate value of x, find a rational approximation to $\ln 3$.

27 Prove that if $y=\ln(\tan x)$, then $\tanh y=-\cos 2x$.

28 **a** Use the substitution $x=\sinh\theta$ to show that

$$\int\frac{1}{x^2\sqrt{1+x^2}}\,dx=-\frac{\sqrt{1+x^2}}{x}+c$$

   b Hence find $\displaystyle\int\frac{\sqrt{1+x^2}}{x^2}\,dx$.

29 Evaluate the improper integral $\displaystyle\int_0^\infty \frac{6x-4}{(3x^2+2)(x+1)}\,dx$, clearly showing the limiting process used.

30 i Using the definition of $\cosh x$ in terms of e^x and e^{-x}, show that

$4\cosh^3 x - 3\cosh x \equiv \cosh 3x.$

ii Use the substitution $u = \cosh x$ to find, in terms of $5^{\frac{1}{3}}$, the real root of the equation

$20u^3 - 15u - 13 = 0.$

© OCR A Level Mathematics, Unit 4726 Further Pure Mathematics 2, June 2010

31 i Use the substitution $x = \cosh^2 u$ to find $\displaystyle\int \sqrt{\frac{x}{x-1}}\,dx$, giving your answer in the form $f(x) + \ln\big(g(x)\big)$.

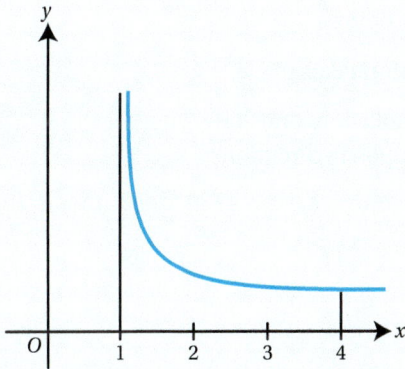

ii Hence calculate the exact area of the region between the curve $y = \sqrt{\dfrac{x}{x-1}}$, the x-axis and the lines $x=1$ and $x=4$ (see diagram).

iii What can you say about the solid of revolution obtained when the region defined in part **ii** is rotated completely about the x-axis? Justify your answer.

© OCR A Level Mathematics, Unit 4726 Further Pure Mathematics 2, June 2011

32 A curve has polar equation $r = 5\sin 2\theta$ for $0 \leqslant \theta \leqslant \frac{1}{2}\pi$.

i Sketch the curve, indicating the line of symmetry and stating the polar coordinates of the point P on the curve which is furthest away from the pole.

ii Calculate the area enclosed by the curve.

iii Find the Cartesian equation of the tangent to the curve at P.

iv Show that a Cartesian equation of the curve is $\left(x^2 + y^2\right)^3 = (10xy)^2$.

© OCR A Level Mathematics, Unit 4726/01 Further Pure Mathematics 2, January 2013

33 The differential equation $\dfrac{d^2 y}{dx^2} + 4y = \sin kx$ is to be solved, where k is a constant.

i In the case $k=2$, by using a particular integral of the form $ax\cos 2x + bx\sin 2x$, find the general solution.

ii Describe briefly the behaviour of y when $x \to \infty$.

iii In the case $k \neq 2$, explain whether y would exhibit the same behaviour as in part **ii** when $x \to \infty$.

© OCR A Level Mathematics, Unit 4727/01 Further Pure Mathematics 3, January 2013

34

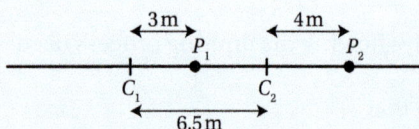

Particles P_1 and P_2 are each moving with simple harmonic motion along the same straight line. P_1's motion has centre C_1, period 2π s and amplitude 3 m; P_2's motion has centre C_2, period $\frac{4}{3}\pi$ s and amplitude 4 m. The points C_1 and C_2 are 6.5 m apart. The displacements of P_1 and P_2 from their centres of oscillation at time t s are denoted by x_1 m and x_2 m respectively. The diagram shows the positions of the particles at time $t = 0$, when $x_1 = 3$ and $x_2 = 4$.

i State expressions for x_1 and x_2 in terms of t, which are valid until the particles collide.

The particles collide when $t = 5.99$, correct to 3 significant figures.

ii Find the distance travelled by P_1 and P_2 before the collision takes place.

iii Find the velocities of P_1 and P_2 immediately before the collision, and state whether the particles are travelling in the same direction or in opposite directions.

© OCR A Level Mathematics, Unit 4730 Mechanics 3, June 2010

35 **i** Given that $y = \cosh^{-1} x$, show that $y = \ln\left(x + \sqrt{x^2 - 1}\right)$.

ii Show that $\dfrac{\mathrm{d}}{\mathrm{d}x}\left(\cosh^{-1} x\right) = \dfrac{1}{\sqrt{x^2 - 1}}$

iii Solve the equation $\cosh x = 3$, giving your answer in logarithmic form.

© OCR A Level Mathematics, Unit 4726/01 Further Pure Mathematics 2, June 2014

36 A function is defined by $f(x) = \sinh^{-1} x + \sinh^{-1}\left(\dfrac{1}{x}\right)$, for $x = 0$.

i When $x > 0$, show that the value of $f(x)$ for which $f'(x) = 0$ is $2\ln\left(1 + \sqrt{2}\right)$.

ii

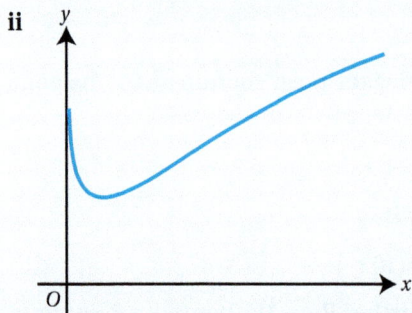

The diagram shows the graph of $y = f(x)$ for $x > 0$. Sketch the graph of $y = f(x)$ for $x < 0$ and state the range of values that $f(x)$ can take for $x \neq 0$.

© OCR A Level Mathematics, Unit 4726 Further Pure Mathematics 2, June 2012

37 **a** Find $\int \arcsin x \, \mathrm{d}x$.

   b Show that $\int \sin^2 x \, \mathrm{d}x = \dfrac{1}{2}(x - \sin x \cos x) + c$.

The area A is bounded by the curve with equation $y = \sin^2 x$, the y-axis and the line $y = p$ as shown.

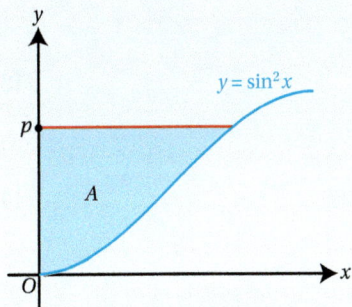

   c **i** Find the area A in terms of p.

   ii Hence state $\int \arcsin \sqrt{x} \, \mathrm{d}x$, $0 < x < 1$.

38

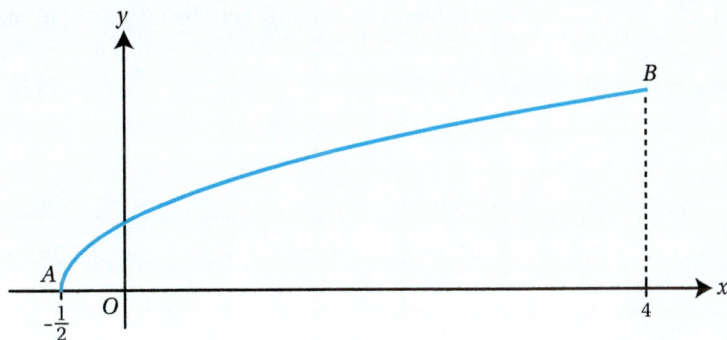

The diagram shows the curve with equation $y = \sqrt{2x+1}$ between the points $A\left(-\frac{1}{2}, 0\right)$ and $B(4, 3)$.

i Find the area of the region bounded by the curve, the x-axis and the line $x = 4$. Hence find the area of the region bounded by the curve and the lines OA and OB, where O is the origin.

ii Show that the curve between B and A can be expressed in polar coordinates as

$$r = \frac{1}{1 - \cos\theta}, \text{ where } \tan^{-1}\left(\tfrac{3}{4}\right) \leqslant \theta \leqslant \pi.$$

iii Deduce from parts **i** and **ii** that $\displaystyle\int_{\tan^{-1}\left(\frac{3}{4}\right)}^{\pi} \operatorname{cosec}^4\left(\tfrac{1}{2}\theta\right)\mathrm{d}\theta = 24$.

© OCR A Level Mathematics, Unit 4726 Further Pure Mathematics 2, June 2010

1 hour 30 minutes, 75 marks

1 Let $\mathbf{a} = \begin{pmatrix} 3 \\ -1 \\ 2 \end{pmatrix}$, $\mathbf{b} = \begin{pmatrix} 1 \\ 1 \\ k \end{pmatrix}$ and $\mathbf{c} = \begin{pmatrix} 2 \\ 22 \\ 8 \end{pmatrix}$.

Find the value of k such that $\mathbf{c}$ is perpendicular to $\mathbf{a}$ and $\mathbf{b}$. **[3 marks]**

2 Use the definitions of $\sinh x$ and $\cosh x$ to prove that

$\cosh(x+y) \equiv \cosh x \cosh y + \sinh x \sinh y$. **[3 marks]**

3 **In this question you must show detailed reasoning.**

Evaluate $\int_0^9 \dfrac{1}{\sqrt{x}}\,dx$, explaining clearly why the integral converges. **[3 marks]**

4 **In this question you must show detailed reasoning.**

Given that α, β and γ are roots of the equation $2x^3 - x + 5 = 0$, find the value of $\alpha^2 + \beta^2 + \gamma^2$. **[4 marks]**

5 The region bounded by the curve $y = \dfrac{1}{\sqrt[4]{x}} + 1$, the x-axis and the lines $x = 1$ and $x = 4$ is

rotated through $360°$ about the x-axis.

Show that the volume generated is $\dfrac{(a + b\sqrt{2})\pi}{c}$, where a, b and c are integers to be found. **[5 marks]**

6 **a** Write down the 7th roots of unity in the form $\operatorname{cis}\theta$. **[3 marks]**

b

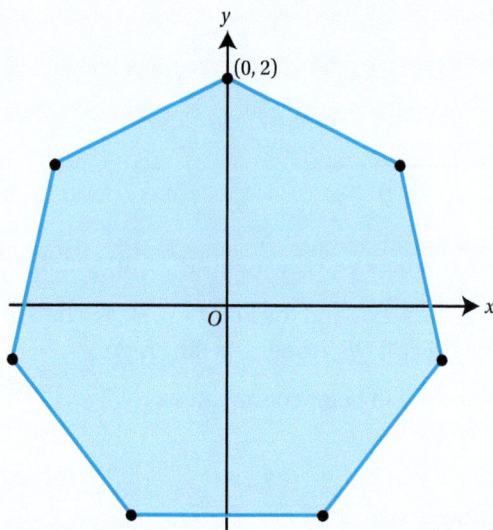

The diagram shows a regular seven-sided polygon with the centre at the origin and one vertex at $(0, 2)$, whose vertices represent the roots of the equation $z^n = w$, with $n \in \mathbb{N}$ and $w \in \mathbb{C}$. Find the values of n and w. **[2 marks]**

7 Find the general solution of the system of differential equations:

$$\begin{cases} \dfrac{dx}{dt} = 2y \\ \dfrac{dy}{dt} = 2x \end{cases}$$

[5 marks]

8 **a** Write $\dfrac{2x^2-x+3}{(x-1)(x^2+1)}$ in partial fractions. **[4 marks]**

    b Hence find $\displaystyle\int \dfrac{2x^2-x+3}{(x-1)(x^2+1)}\,\mathrm{d}x.$ **[3 marks]**

9 Plane Π contains the points $A(-1, 1, 2)$, $B(3, 5, -1)$ and $C(2, 2, 1)$.

    a Find the Cartesian equation of Π. **[5 marks]**

    b Point D has coordinates $(3, -1, 1)$. Find the angle between Π and the line AD, giving your answer to the nearest $0.1°$. **[4 marks]**

10 A curve C has polar equation $r(1-\cos\theta)=3$.

    a Find its Cartesian equation in the form $y^2 = \mathrm{f}(x)$.

    The curve C intersects the line $3r = \dfrac{4}{\cos\theta}$. **[4 marks]**

    b Find the value of r at the points of intersection. **[3 marks]**

11 Consider the matrix $\mathbf{A}=\begin{pmatrix} a & 2 & 1 \\ 1 & a & 3 \\ 1 & 1 & 1 \end{pmatrix}$.

    a Show that $\mathbf{A}$ is non-singular for all values of a. **[2 marks]**

    b Find $\mathbf{A}^{-1}$ in terms of a. **[5 marks]**

    c Hence solve the simultaneous equations **[3 marks]**
$$\begin{cases} ax+2y+z=1 \\ x+ay+3z=0 \\ x+y+z=0 \end{cases}$$

12 Given that $y=1$ when $x=0$, solve the differential equation
$$\left(1-x^2\right)\frac{\mathrm{d}y}{\mathrm{d}x}-xy=2, |x|<1.$$
Give your answer in the form $y=\mathrm{f}(x)$. **[6 marks]**

13 **a** Prove by induction that, for $n\in\mathbb{Z}^+$,
$$\cos\theta+\cos(3\theta)+\cos(5\theta)+\ldots+\cos\big((2n-1)\theta\big)=\frac{\sin(2n\theta)}{2\sin\theta}.$$ **[6 marks]**

    b Hence find the exact value of $\cos\dfrac{\pi}{7}+\cos\dfrac{3\pi}{7}+\ldots+\cos\dfrac{13\pi}{7}$. **[2 marks]**

1 hour 30 minutes, 75 marks

1 In this question you must show detailed reasoning.

Let $z = \dfrac{1 - 2i}{1 + 3i}$.

  a Find z in the form $a + bi$.   **[2 marks]**

  b Find the modulus and argument of z.   **[3 marks]**

2 A transformation is described by the matrix $\mathbf{M} = \begin{pmatrix} \dfrac{1}{2} & -\dfrac{\sqrt{3}}{2} \\ \dfrac{\sqrt{3}}{2} & \dfrac{1}{2} \end{pmatrix}$.

  a The image of point A is the point $(1, -1)$. Find the coordinates of A.   **[2 marks]**

  b Describe the transformation fully.   **[2 marks]**

3 Find the Cartesian equation of the curve with polar equation $r^3 = 8\cos\theta$.   **[2 marks]**

4 Use the formulae for $\displaystyle\sum_{r=1}^{n} r$ and $\displaystyle\sum_{r=1}^{n} r^2$ to show that $\displaystyle\sum_{r=1}^{n}(2r-1)^2 = \dfrac{n}{3}(2n-1)(2n+1)$.   **[5 marks]**

5 In this question you must show detailed reasoning.

By first completing the square, find the exact value of $\displaystyle\int_{0}^{8} \dfrac{1}{\sqrt{x^2 - 8x + 25}}\, dx$.   **[6 marks]**

6 **a** Show that the Maclaurin series of the function $\ln(1 + \sin x)$ up to the

    term in x^4 is $x - \dfrac{x^2}{2} + \dfrac{x^3}{6} - \dfrac{x^4}{12} + \ldots$.   **[6 marks]**

  b A student claims that this series is valid for all $x \in \mathbb{R}$. Show by means of a counterexample that he is wrong.   **[2 marks]**

7 Matrix $\mathbf{A}$ is given by $\mathbf{A} = \begin{pmatrix} 1 & 5 \\ 0 & 1 \end{pmatrix}$.

  a Find $\mathbf{A}^2$, $\mathbf{A}^3$ and $\mathbf{A}^4$.   **[2 marks]**

  b Conjecture an expression for $\mathbf{A}^n$ for $n \in \mathbb{N}$.   **[1 mark]**

  c Use mathematical induction to prove your conjecture.   **[5 marks]**

8 Show that the lines with Cartesian equations $\dfrac{x-2}{5} = \dfrac{y+1}{-1} = \dfrac{z-1}{1}$ and

$\dfrac{x+1}{2} = \dfrac{y-1}{2} = \dfrac{z-2}{7}$ are skew and find the shortest distance between them.   **[7 marks]**

9 Show that the mean value of the function $f(x) = \dfrac{3x^2 + 2x - 5}{(2x+1)(x^2+5)}$ between 0 and 4 is $a\ln b$,

where a and b are constants to be found.   **[7 marks]**

10 **a** If $z = \cos\theta + i\sin\theta$, show that $\dfrac{1}{z} = \cos\theta - i\sin\theta$.   **[1 mark]**

  b Show that $\cos(n\theta) = \dfrac{1}{2}\left(z^n + \dfrac{1}{z^n}\right)$ for $n \in \mathbb{N}$.   **[4 marks]**

  c Hence solve $z^4 - 3z^3 + 4z^2 - 3z + 1 = 0$.   **[3 marks]**

11 Three planes have equations $\Pi_1 : x - 2y + z = 0$, $\Pi_2 : 3x - z = 4$, $\Pi_3 : x + y - z = k$.

   a Show that, for all values of k, the planes do not intersect at a unique point. **[2 marks]**

   b Find the value of k for which the intersection of the three planes is a line. **[4 marks]**

12 A particle P of mass m is attached to one end of a light horizontal spring. The other end of the spring is attached to a fixed point.

The magnitude of the tension in the spring is given by $2mk^2x$, where x is the extension in the spring at time t seconds and $k > 0$ is a constant.

The particle experiences a resistance to motion of magnitude $3kmv$, where v is the speed of the particle at time t seconds.

   a Show that $\dfrac{d^2x}{dt^2} + 3k\dfrac{dx}{dt} + 2k^2x = 0$. **[3 marks]**

   b Given that when $t = 0$, $x = 4$ and $v = -3k$:

      i find x in terms of t and k **[5 marks]**

      ii state whether the damping is underdamping, overdamping or critical. **[1 mark]**

FORMULAE

Learners will be given the following formulae in the Formulae Booklet in each assessment.

Pure Mathematics

Arithmetic series

$$S_n = \frac{1}{2}n(a+l) = \frac{1}{2}n\{2a+(n-1)d\}$$

Geometric series

$$S_n = \frac{a(1-r^n)}{1-r}$$

$$S_\infty = \frac{a}{1-r} \text{ for } |r|<1$$

Binomial series

$$(a+b)^n = a^n + {}^nC_1 a^{n-1}b + {}^nC_2 a^{n-2}b^2 + \ldots + {}^nC_r a^{n-r}b^r + \ldots + b^n \ (n\in\mathbb{N}),$$

where $ {}^nC_r = \binom{n}{r} = \frac{n!}{r!(n-r)!}$

$$(1+x)^n = 1+nx+\frac{n(n-1)}{2!}x^2+\ldots+\frac{n(n-1)\ldots(n-r+1)}{r!}x^r+\ldots \ (|x|<1,\ n\in\mathbb{R})$$

Series

$$\sum_{r=1}^{n} r^2 = \frac{1}{6}n(n+1)(2n+1), \quad \sum_{r=1}^{n} r^3 = \frac{1}{4}n^2(n+1)^2$$

Maclaurin series

$$\mathrm{f}(x) = \mathrm{f}(0) + \mathrm{f}'(0)x + \frac{\mathrm{f}''(0)}{2!}x^2 + \ldots + \frac{\mathrm{f}^{(r)}(0)}{r!}x^r + \ldots$$

$$\mathrm{e}^x = \exp(x) = 1 + x + \frac{x^2}{2!} + \ldots + \frac{x^r}{r!} + \ldots \text{ for all } x$$

$$\ln(1+x) = x - \frac{x^2}{2} + \frac{x^3}{3} - \ldots + (-1)^{r+1}\frac{x^r}{r} + \ldots \ (-1<x\leqslant 1)$$

$$\sin x = x - \frac{x^3}{3!} + \frac{x^5}{5!} - \ldots + (-1)^r \frac{x^{2r+1}}{(2r+1)!} + \ldots \text{ for all } x$$

$$\cos x = 1 - \frac{x^2}{2!} + \frac{x^4}{4!} - \ldots + (-1)^r \frac{x^{2r}}{(2r)!} + \ldots \text{ for all } x$$

$$(1+x)^n = 1+nx+\frac{n(n-1)}{2!}x^2+\ldots+\frac{n(n-1)\ldots(n-r+1)}{r!}x^r+\ldots \ (|x|<1,\ n\in\mathbb{R})$$

Matrix transformations

Reflection in the line $y = \pm x$: $\begin{pmatrix} 0 & \pm 1 \\ \pm 1 & 0 \end{pmatrix}$

Rotations through θ about the coordinate axes. The direction of positive rotation is taken to be anticlockwise when looking towards the origin from the positive side of the axis of rotation.

$$\mathbf{R}_x = \begin{pmatrix} 1 & 0 & 0 \\ 0 & \cos\theta & \sin\theta \\ 0 & -\sin\theta & \cos\theta \end{pmatrix}$$

$$\mathbf{R}_y = \begin{pmatrix} \cos\theta & 0 & -\sin\theta \\ 0 & 1 & 0 \\ \sin\theta & 0 & \cos\theta \end{pmatrix}$$

$$\mathbf{R}_z = \begin{pmatrix} \cos\theta & \sin\theta & 0 \\ -\sin\theta & \cos\theta & 0 \\ 0 & 0 & 1 \end{pmatrix}$$

Differentiation

$f(x)$	$f'(x)$
$\tan kx$	$k \sec^2 kx$
$\sec x$	$\sec x \tan x$
$\cot x$	$-\text{cosec}^2 x$
$\text{cosec}\, x$	$-\text{cosec}\, x \cot x$
$\arcsin x$ or $\sin^{-1} x$	$\dfrac{1}{\sqrt{1-x^2}}$
$\arccos x$ or $\cos^{-1} x$	$-\dfrac{1}{\sqrt{1-x^2}}$
$\arctan x$ or $\tan^{-1} x$	$\dfrac{1}{1+x^2}$

Quotient rule $y = \dfrac{u}{v}$, $\dfrac{dy}{dx} = \dfrac{v\dfrac{du}{dx} - u\dfrac{dv}{dx}}{v^2}$

Differentiation from first principles

$$f'(x) = \lim_{h \to 0} \frac{f(x+h) - f(x)}{h}$$

Integration

$$\int \frac{f'(x)}{f(x)}\, dx = \ln|f(x)| + c$$

$$\int f'(x)(f(x))^n \, dx = \frac{1}{n+1}(f(x))^{n+1} + c$$

Integration by parts $\int u \dfrac{dv}{dx} dx = uv - \int v \dfrac{du}{dx} dx$

The mean value of f(x) on the interval [a, b] is $\dfrac{1}{b-a}\int_a^b f(x)dx$

Area of sector enclosed by polar curve is $\dfrac{1}{2}\int r^2 d\theta$

f(x)	$\int f(x)dx$
$\dfrac{1}{\sqrt{a^2-x^2}}$	$\sin^{-1}\left(\dfrac{x}{a}\right)(|x|<a)$
$\dfrac{1}{a^2+x^2}$	$\dfrac{1}{a}\tan^{-1}\left(\dfrac{x}{a}\right)$
$\dfrac{1}{\sqrt{a^2+x^2}}$	$\sinh^{-1}\left(\dfrac{x}{y}\right)$ or $\ln\left(x+\sqrt{x^2+a^2}\right)$
$\dfrac{1}{\sqrt{x^2-a^2}}$	$\cosh^{-1}\left(\dfrac{x}{a}\right)$ or $\ln\left(x+\sqrt{x^2-a^2}\right)(x>a)$

Numerical methods

Trapezium rule: $\int_a^b y \, dx \approx \dfrac{1}{2}h\{(y_0+y_n)+2(y_1+y_2+...+y_{n-1})\}$, where $h = \dfrac{b-a}{n}$

The Newton-Raphson iteration for solving f$(x)=0$: $x_{n+1} = x_n - \dfrac{f(x_n)}{f'(x_n)}$

Complex numbers

Circles: $|z-a|=k$

Half-lines: $\arg(z-a)=\alpha$

Lines: $|z-a|=|z-b|$

De Moivre's theorem: $\{r(\cos\theta+i\sin\theta)\}^n = r^n(\cos n\theta+i\sin n\theta)$

Roots of unity: The roots of $z^n=1$ are given by $z=\exp\left(\dfrac{2\pi k}{n}i\right)$ for $k=0,1,2,...,n-1$

Vectors and 3D coordinate geometry

Cartesian equation of the line through the point A with position vector $\mathbf{a}=a_1\mathbf{i}+a_2\mathbf{j}+a_3\mathbf{k}$ in direction $\mathbf{u}=u_1\mathbf{i}+u_2\mathbf{j}+u_3\mathbf{k}$ is $\dfrac{x-a_1}{u_1}=\dfrac{y-a_2}{u_2}=\dfrac{z-a_3}{u_3}(=\lambda)$

Cartesian equation of a plane is $n_1x+n_2y+n_3z+d=0$

Vector product: $\mathbf{a}\times\mathbf{b}=\begin{pmatrix}a_1\\a_2\\a_3\end{pmatrix}\times\begin{pmatrix}b_1\\b_2\\b_3\end{pmatrix}=\begin{vmatrix}\mathbf{i}&a_1&b_1\\\mathbf{j}&a_2&b_2\\\mathbf{k}&a_3&b_3\end{vmatrix}=\begin{pmatrix}a_2b_3-a_3b_2\\a_3b_1-a_1b_3\\a_1b_2-a_2b_1\end{pmatrix}$

The distance between skew lines is $D=\dfrac{|(\mathbf{b}-\mathbf{a})\cdot\mathbf{n}|}{|\mathbf{n}|}$, where $\mathbf{a}$ and $\mathbf{b}$ are position vectors of points on each line and $\mathbf{n}$ is a mutual perpendicular to both lines

The distance between a point and a line is $D = \dfrac{|ax_1 + by_1 - c|}{\sqrt{a^2 + b^2}}$, where the coordinates of the point are (x_1, y_1) and the equation of the line is given by $ax + by = c$

The distance between a point and a plane is $D = \dfrac{|\mathbf{b}\cdot\mathbf{n} - p|}{|\mathbf{n}|}$, where $\mathbf{b}$ is the position vector of the point and the equation of the plane is given by $\mathbf{r}\cdot\mathbf{n} = p$

Small angle approximations

$\sin\theta \approx \theta$, $\cos\theta \approx 1 - \dfrac{1}{2}\theta^2$, $\tan\theta \approx \theta$ where θ is small and measured in radians

Trigonometric identities

$\sin(A \pm B) = \sin A \cos B \pm \cos A \sin B$

$\cos(A \pm B) = \cos A \cos B \mp \sin A \sin B$

$\tan(A \pm B) = \dfrac{\tan A \pm \tan B}{1 \mp \tan A \tan B}\left(A \pm B \neq \left(k + \dfrac{1}{2}\right)\pi\right)$

Hyperbolic functions

$\cosh^2 x - \sinh^2 x = 1$

$\sinh^{-1} x = \ln\left[x + \sqrt{(x^2 + 1)}\right]$

$\cosh^{-1} x = \ln\left[x + \sqrt{(x^2 - 1)}\right]$, $x \geq 1$

$\tanh^{-1} x = \dfrac{1}{2}\ln\left(\dfrac{1+x}{1-x}\right)$, $-1 < x < 1$

Simple harmonic motion

$x = A\cos(\omega t) + B\sin(\omega t)$

$x = R\sin(\omega t + \varphi)$

Answers

Chapter 1

Before you start...

1 Proof

2 $u_1 = 3$, $u_2 = 9$, $u_3 = 17$

3 $3n^2(n+1)$

4 62

5 $(5+6x)e^{3x}$

6 $\dfrac{1}{r} - \dfrac{1}{r+1}$

Exercise 1A

1, 2 Proof

3 a $4, 24, 124, 624$ **b** $u_n = 5^n - 1$; Proof

4 a $\begin{pmatrix} 1 & 0 \\ 2 & 1 \end{pmatrix}, \begin{pmatrix} 1 & 0 \\ 3 & 1 \end{pmatrix}, \begin{pmatrix} 1 & 0 \\ 4 & 1 \end{pmatrix}$

  b $\begin{pmatrix} 1 & 0 \\ n & 1 \end{pmatrix}$; Proof

5 a $4, 24, 124, 624$

  b 4   **c** Proof

6, 7 Proof

8 a $\begin{pmatrix} 2 & 2 \\ 2 & 2 \end{pmatrix}, \begin{pmatrix} 4 & 4 \\ 4 & 4 \end{pmatrix}, \begin{pmatrix} 8 & 8 \\ 8 & 8 \end{pmatrix}$

  b $\begin{pmatrix} 2^{n-1} & 2^{n-1} \\ 2^{n-1} & 2^{n-1} \end{pmatrix}$; Proof

9 a 5   **b** Proof

10–15 Proof

Exercise 1B

1–10 Proof

Work it out 1.1

The correct answer is Solution 2.

Exercise 1C

1 a i 9455   **ii** 44 100

  b i 1 379 609   **ii** 4750

2 a i $2n(4n+1)$ **ii** $\dfrac{n}{2}(3n+1)(6n+1)$

  b i $\dfrac{n}{6}(n-1)(2n-1)$ **ii** $\dfrac{(n+1)^2(n+2)^2}{4}$

3, 4 Proof

5 a $3n^2 + 10n$ **b** 27

6 a Proof   **b** 25 225

7 Proof; $k = 3$

8 a Proof   **b** $n(2n+1)(n-1)(2n+3)$

9 a Proof   **b** $\ln 3^{2660}$

10 Proof; $a = 4$

Exercise 1D

1, 2 Proof

3 a Proof   **b i** Proof   **ii** $\dfrac{7}{78}$

4 a $\dfrac{1}{2r-1} - \dfrac{1}{2r+1}$ **b** Proof

  c 1

5 Proof

6 a Proof   **b** $\dfrac{3}{2}$

7 a Proof   **b** $\dfrac{n}{3(2n+3)}$

  c Proof

8 $\dfrac{n^2 + 5n}{12(n+2)(n+3)}$

9 a Proof

  b Proof; $a = 2$, $b = 3$, $c = 4$

  c $\dfrac{5}{1848}$

10 a $\ln(n+1)$ **b** Proof

Mixed practice 1

1 a $16, 88, 736, 6568$; 8

  b Proof

2 Proof

3 a $3e^{3x}, 9e^{3x}, 27e^{3x}$; $3^n e^{3x}$

  b Proof

4 $3n^2 + 2n$

5 Proof

6 1 278 270

7 a Proof   **b** Proof; $a = 1$, $b = 5$

8 $2n^2(4n+3)$

9–11 Proof

12 a Proof   **b** 603 330

13 a Proof   **b** Proof

14 a Proof   **b** $1 - \dfrac{1}{(n+1)!}$

  c 1

15 i 6, 27, 129   **ii** 3   **iii** Proof

16–18 Proof

19 7; Proof

20, 21 Proof

22 i Proof   **ii** $\dfrac{n}{n+1}$ **iii** $\dfrac{1}{n+1}$

23 a $\dfrac{1}{r-1}-\dfrac{1}{r+2}$ **b** Proof; $a=99,\,b=144,\,c=49$

24 Proof; $a=5,\,b=9,\,c=2$

Chapter 2

Before you start...

1 5; 2.21 radians

2 $-3+2i$; $-3i$

3 a $1-3i$ **b** $\dfrac{4}{5}-\dfrac{7}{5}i$

4 a $20\left(\cos\left(-\dfrac{7\pi}{12}\right)+i\sin\left(-\dfrac{7\pi}{12}\right)\right)$

  b $5\left(\cos\dfrac{\pi}{12}+i\sin\dfrac{\pi}{12}\right)$

5 a $-3-5i$ **b** $3\left(\cos\left(-\dfrac{\pi}{4}\right)+i\sin\left(-\dfrac{\pi}{4}\right)\right)$

6 a Reflection in the real axis.

  b Translation by $\begin{pmatrix}2\\1\end{pmatrix}$.

Exercise 2A

1 a i $64\left(\cos\left(\dfrac{-4\pi}{5}\right)+i\sin\left(\dfrac{-4\pi}{5}\right)\right)$

    ii $81\left(\cos\dfrac{2\pi}{3}+i\sin\dfrac{2\pi}{3}\right)$

  b i $\cos\left(\dfrac{-11\pi}{12}\right)+i\sin\left(\dfrac{-11\pi}{12}\right)$

    ii $\cos\left(\dfrac{-5\pi}{6}\right)+i\sin\left(\dfrac{-5\pi}{6}\right)$

  c i $\cos\left(\dfrac{-\pi}{2}\right)+i\sin\left(\dfrac{-\pi}{2}\right)$

    ii $\cos\dfrac{\pi}{2}+i\sin\dfrac{\pi}{2}$

2 a $z^2=\cos\dfrac{\pi}{3}+i\sin\dfrac{\pi}{3}$; $z^3=\cos\dfrac{\pi}{2}+i\sin\dfrac{\pi}{2}$;

    $z^4=\cos\dfrac{2\pi}{3}+i\sin\dfrac{2\pi}{3}$

b

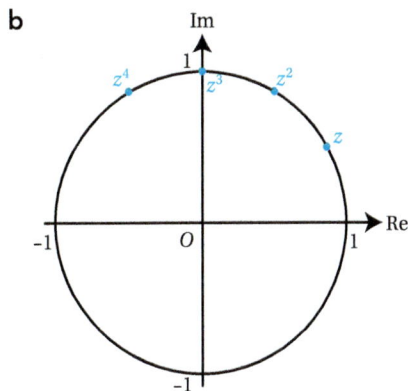

3 a i $z^2=\cos\left(-\dfrac{2\pi}{3}\right)+i\sin\left(-\dfrac{2\pi}{3}\right)$; $z^3=1$,

    $z^4=\cos\dfrac{2\pi}{3}+i\sin\dfrac{2\pi}{3}$

  ii

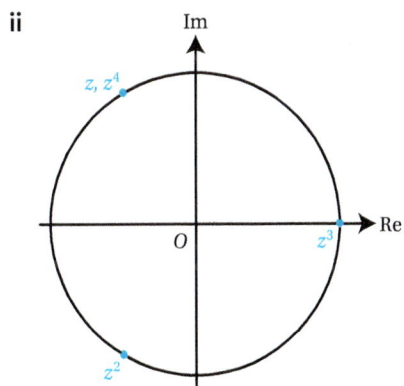

  b $n=1+3k,\ k\in\mathbb{Z}^+$

4 a $\text{mod}=2,\ \arg=\dfrac{\pi}{3}$

  b $32\left(\cos\dfrac{5\pi}{3}+i\sin\dfrac{5\pi}{3}\right)$ **c** $16-16\sqrt{3}i$

5 a $2\text{cis}\left(\dfrac{3\pi}{4}\right)$ **b** $64i$

6 12

7 14

Exercise 2B

1 a i $\dfrac{3\sqrt{3}}{2}+\dfrac{3}{2}i$ **ii** $2\sqrt{2}+2\sqrt{2}i$

  b i -4 **ii** 5

  c i $-\dfrac{1}{2}+\dfrac{\sqrt{3}}{2}i$ **ii** $-2i$

2 a i $5\sqrt{2}\,e^{i\frac{\pi}{4}}$ **ii** $4e^{-i\frac{\pi}{6}}$

  b i $\dfrac{1}{\sqrt{2}}e^{i\frac{3\pi}{4}}$ **ii** $\sqrt{13}\,e^{0.983i}$

  c i $4e^{-i\frac{\pi}{2}}$ **ii** $5e^{i\pi}$

3 a i $20e^{i\frac{5\pi}{12}}$　　**ii** $\frac{1}{2}e^{i\frac{\pi}{2}}$

b i $\frac{8}{25}e^{i\frac{\pi}{12}}$　　**ii** $2e^{-i\frac{\pi}{2}}$

4

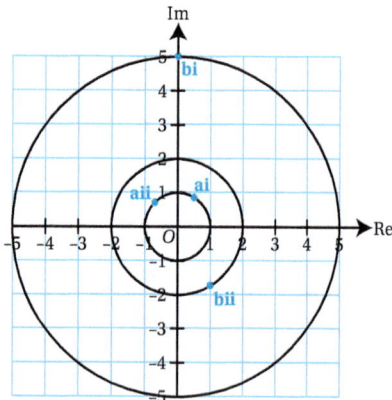

5 Proof

6 a $\frac{2}{3}i$　　**b** $-432\sqrt{3}-432i$

7 $-54.1+7.70i$

8 $\frac{e^2}{2}-\frac{e^2\sqrt{3}}{2}i$

9 $a=-3,\ b=3,\ c=-2$

10 $x^4-\left(2+\sqrt{3}\right)x^3+\left(5+2\sqrt{3}\right)x^2-\left(2+4\sqrt{3}\right)x+4=0$

11 $\cos(\ln 5)+i\sin(\ln 5)$

12 $9\cos(\ln 3)-9i\sin(\ln 3)$

Exercise 2C

1 a i $3,\ 3e^{\frac{2\pi i}{3}},\ 3e^{-\frac{2\pi i}{3}}$

　　ii $\sqrt[3]{100},\ \sqrt[3]{100}\,e^{\frac{2\pi i}{3}},\ \sqrt[3]{100}\,e^{-\frac{2\pi i}{3}}$

b i $2e^{\frac{\pi i}{6}},\ 2e^{\frac{5\pi i}{6}},\ 2e^{\frac{3\pi i}{2}}$　　**ii** $e^{\frac{\pi i}{6}},\ e^{\frac{5\pi i}{6}},\ e^{\frac{3\pi i}{2}}$

c i $2^{\frac{1}{6}}e^{\frac{\pi i}{12}},\ 2^{\frac{1}{6}}e^{\frac{3\pi i}{4}},\ 2^{\frac{1}{6}}e^{\frac{17\pi i}{12}}$

　　ii $7^{\frac{1}{6}}e^{-0.238i},\ 7^{\frac{1}{6}}e^{1.86i},\ 7^{\frac{1}{6}}e^{-2.33i}$

2 a i $2\operatorname{cis}\frac{\pi}{4},\ 2\operatorname{cis}\frac{3\pi}{4},\ 2\operatorname{cis}\frac{5\pi}{4},\ 2\operatorname{cis}\frac{7\pi}{4}$;

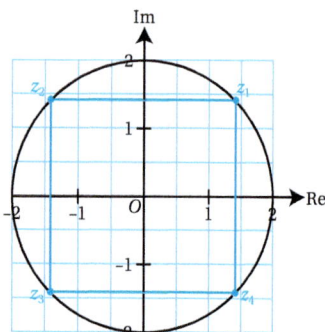

ii $3\operatorname{cis}\frac{\pi}{8},\ 3\operatorname{cis}\frac{5\pi}{8},\ 3\operatorname{cis}\frac{9\pi}{8},\ 3\operatorname{cis}\frac{13\pi}{8}$;

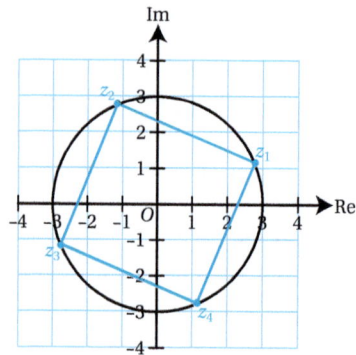

b i $2\operatorname{cis}\frac{\pi}{16},\ 2\operatorname{cis}\frac{9\pi}{16},\ 2\operatorname{cis}\frac{17\pi}{16},\ 2\operatorname{cis}\frac{25\pi}{16}$;

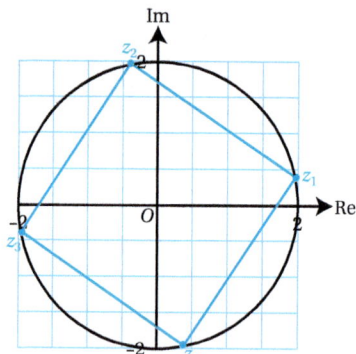

ii $\operatorname{cis}\frac{\pi}{12},\ \operatorname{cis}\frac{7\pi}{12},\ \operatorname{cis}\frac{13\pi}{12},\ \operatorname{cis}\frac{19\pi}{12}$;

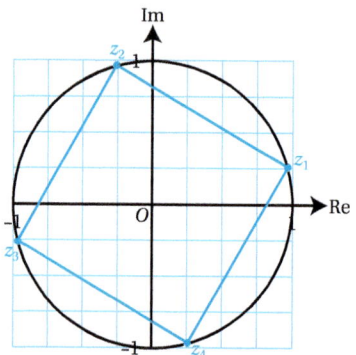

3 $-2,\ 1+\sqrt{3}i,\ 1-\sqrt{3}i$

4 a $16;\ -\frac{\pi}{6}$

b $2\operatorname{cis}\left(-\frac{\pi}{24}\right),\ 2\operatorname{cis}\left(\frac{11\pi}{24}\right),\ 2\operatorname{cis}\left(\frac{23\pi}{24}\right),\ 2\operatorname{cis}\left(-\frac{13\pi}{24}\right)$

5 $\frac{\sqrt{2}+\sqrt{6}}{2}+\left(\frac{\sqrt{2}-\sqrt{6}}{2}\right)i,\ \frac{\sqrt{2}-\sqrt{6}}{2}+\left(\frac{\sqrt{2}+\sqrt{6}}{2}\right)i,$
$-\sqrt{2}-\sqrt{2}i$

6 $3\left(\cos\left(\dfrac{3\pi}{8}\right)+i\sin\left(\dfrac{3\pi}{8}\right)\right), 3\left(\cos\left(\dfrac{7\pi}{8}\right)+i\sin\left(\dfrac{7\pi}{8}\right)\right),$

$3\left(\cos\left(\dfrac{11\pi}{8}\right)+i\sin\left(\dfrac{11\pi}{8}\right)\right), 3\left(\cos\left(\dfrac{15\pi}{8}\right)+i\sin\left(\dfrac{15\pi}{8}\right)\right)$

7 a $8e^{i\frac{\pi}{3}}$

b $8^{\frac{1}{4}}e^{i\frac{\pi}{12}}, 8^{\frac{1}{4}}e^{i\frac{7\pi}{12}}, 8^{\frac{1}{4}}e^{i\frac{13\pi}{12}}, 8^{\frac{1}{4}}e^{i\frac{19\pi}{12}}$

c

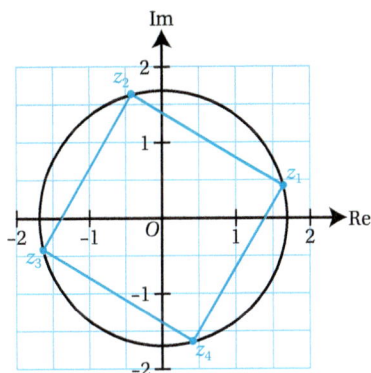

8 $n=4, w=-64$

9 a $\sqrt{2}\pm\sqrt{2}i, -\sqrt{2}\pm\sqrt{2}i$

b $\left(z^2+2\sqrt{2}z+4\right)\left(z^2-2\sqrt{2}z+4\right)$

10 a $2i, -\sqrt{3}-i, \sqrt{3}-i$

b $\dfrac{4}{5}-\dfrac{2}{5}i, \dfrac{14-3\sqrt{3}}{13}+\dfrac{\left(5-2\sqrt{3}\right)}{13}i, \dfrac{14+3\sqrt{3}}{13}+\dfrac{\left(5+2\sqrt{3}\right)}{13}i$

11 a $2\left(\cos\left(\dfrac{\pi}{4}\right)+i\sin\left(\dfrac{\pi}{4}\right)\right),$

$2\left(\cos\left(\dfrac{11\pi}{12}\right)+i\sin\left(\dfrac{11\pi}{12}\right)\right),$

$2\left(\cos\left(\dfrac{19\pi}{12}\right)+i\sin\left(\dfrac{19\pi}{12}\right)\right)$

b $1; \dfrac{7\pi}{12}$ **c** $\dfrac{\sqrt{2}}{2}-\dfrac{\sqrt{2}}{2}i$

12 a $-1, e^{\frac{i\pi}{3}}, e^{-\frac{i\pi}{3}}$ **b** $x^3+6x^2+12x+8$

c $-3, -\dfrac{3}{2}+\dfrac{\sqrt{3}}{2}i, -\dfrac{3}{2}-\dfrac{\sqrt{3}}{2}i$

Exercise 2D

1 a i $\cos 0+i\sin 0, \cos\left(\dfrac{2\pi}{3}\right)+i\sin\left(\dfrac{2\pi}{3}\right),$

$\cos\left(\dfrac{4\pi}{3}\right)+i\sin\left(\dfrac{4\pi}{3}\right)$

ii $\cos 0+i\sin 0, \cos\pi+i\sin\pi$

b i $\cos 0+i\sin 0, \cos\left(\dfrac{\pi}{3}\right)+i\sin\left(\dfrac{\pi}{3}\right),$

$\cos\left(\dfrac{2\pi}{3}\right)+i\sin\left(\dfrac{2\pi}{3}\right), \cos\pi+i\sin\pi,$

$\cos\left(\dfrac{4\pi}{3}\right)+i\sin\left(\dfrac{4\pi}{3}\right),$

$\cos\left(\dfrac{5\pi}{3}\right)+i\sin\left(\dfrac{5\pi}{3}\right)$

ii $\cos 0+i\sin 0, \cos\left(\dfrac{\pi}{2}\right)+i\sin\left(\dfrac{\pi}{2}\right),$

$\cos\pi+i\sin\pi, \cos\left(\dfrac{3\pi}{2}\right)+i\sin\left(\dfrac{3\pi}{2}\right)$

2 a i $1, -\dfrac{1}{2}\pm i\dfrac{\sqrt{3}}{2}$ **ii** ± 1

b i $\pm 1, \dfrac{1}{2}\pm i\dfrac{\sqrt{3}}{2}, -\dfrac{1}{2}\pm i\dfrac{\sqrt{3}}{2}$ **ii** $\pm 1, \pm i$

3 a $1, e^{\frac{2\pi i}{5}}, e^{\frac{4\pi i}{5}}, e^{\frac{6\pi i}{5}}, e^{\frac{8\pi i}{5}}$

b

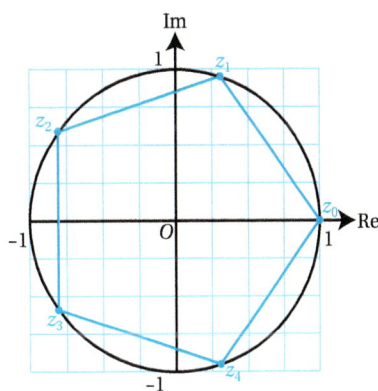

4 a 5   **b** -1

c A, B, C

5 a $1 \text{ (or } \omega^0), \omega, \omega^2, \omega^3, \omega^4, \omega^5, \omega^6$

b No. Consider $\omega^7=1$, or an Argand diagram.

c 3   **d** 5

6 a, b Proof

7 a $1, -\dfrac{1}{2}\pm\dfrac{\sqrt{3}}{2}i$ **b** $\dfrac{-1\pm\sqrt{3}i}{2}$

8 a^2+b^2+ab

9 a $(1), \omega, \omega^2, \omega^3, \omega^4$

b $\dfrac{1+\omega^k}{1-\omega^k}$ for $k=1, 2, 3, 4$

c–e Proof

10 a

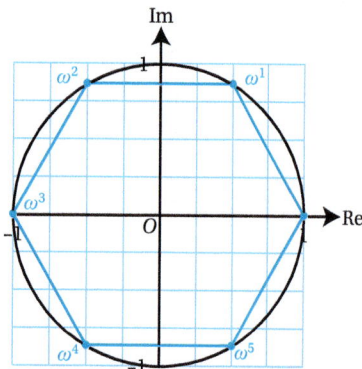

b i $\dfrac{\sqrt{3}}{2}e^{i\frac{\pi}{6}}$ **ii** $\dfrac{\sqrt{3}}{2}e^{i\frac{7\pi}{6}}$

11 a Proof

  b i Proof   **ii** $-\dfrac{1}{2}$

  c Proof

Exercise 2E

1 a $2e^{i\frac{\pi}{4}}, 2e^{i\frac{3\pi}{4}}, 2e^{i\frac{5\pi}{4}}, 2e^{i\frac{7\pi}{4}}$

  b $\sqrt{2}+i\sqrt{2}, \sqrt{2}-i\sqrt{2}, -\sqrt{2}+i\sqrt{2}, -\sqrt{2}-i\sqrt{2}$

  c $\left(z^2+2\sqrt{2}z+4\right)\left(z^2-2\sqrt{2}z+4\right)$

2 $\left(z^2-2z+2\right)\left(z^2+2z+2\right)\left(z^2-2\right)\left(z^2+2\right)=0$

3 Proof; $\theta=\dfrac{2\pi}{5}$, $\phi=\dfrac{4\pi}{5}$

4 a $1, \omega, \omega^2, \omega^3, \omega^4$ **b** 31

5 a Proof   **b** $e^{\frac{i\pi}{6}}, e^{\frac{2i\pi}{3}}, e^{\frac{7i\pi}{6}}, e^{\frac{5i\pi}{3}}$

  c $\left(z^2+z+1\right)\left(z^2-z+1\right)\left(z^2+\sqrt{3}z+1\right)\left(z^2-\sqrt{3}z+1\right)$

6 a $\omega, \omega^2, \omega^3, \omega^4, \omega^5, \omega^6$

  (or $\omega, \omega^2, \omega^3, \omega^*, \omega^{*2}, \omega^{*3}$)

  b, c Proof

Exercise 2F

1 a $\sqrt{17}$, 0.245; $\sqrt{34}$, 0.540

  b $\sqrt{2}$, 0.295 $(16.9°)$

2 a Proof

  b Rotation 1.76 radians $(101°)$ about the origin.

3 $\dfrac{3\sqrt{3}-2}{2}+i\dfrac{3+2\sqrt{3}}{2}$

4 $(7,-3), (2,-5)$

5 $b=az, c=az^3, d=\dfrac{a}{z^2}$

6 a $\sqrt{60}$

  b $\left(6-\sqrt{3}, 3+2\sqrt{3}\right)$

7 a

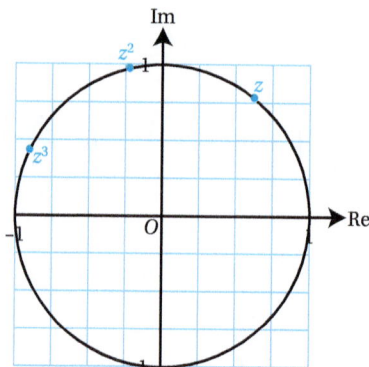

b Rotation through 1.85 radians about the origin.

8 a $\dfrac{\pi}{3}$ **b** $\dfrac{\pi}{6}$

  c $\left(\dfrac{\sqrt{6}-\sqrt{2}}{2}, \dfrac{\sqrt{6}+\sqrt{2}}{2}\right)$

9 a $4-i$

  b $B\left(\dfrac{\sqrt{3}-4}{2}, \dfrac{4\sqrt{3}+1}{2}\right), C\left(-\dfrac{4+\sqrt{3}}{2}, -\dfrac{4\sqrt{3}-1}{2}\right)$

10 $2\operatorname{cis}\left(\dfrac{2\pi}{5}\right), 2\operatorname{cis}\left(\dfrac{4\pi}{5}\right), 2\operatorname{cis}\left(\dfrac{6\pi}{5}\right), 2\operatorname{cis}\left(\dfrac{8\pi}{5}\right)$

11 a iz^* **b** Proof

12 a Proof   **b** $\left(\dfrac{3-4\sqrt{3}}{2}, \dfrac{2-\sqrt{3}}{2}\right)$

13 a $c-a=(b-a)e^{i\theta}$ **b** $0.935+4.89i$

14 $\dfrac{1-\sqrt{3}}{4}+\dfrac{1+\sqrt{3}}{4}i$

Mixed practice 2

1 $\dfrac{1}{r}e^{i\theta}$

2 a $2; \dfrac{2\pi}{3}$ **b** $-16-16\sqrt{3}i$

3 a $e^{\frac{2k\pi i}{5}}$ for $k=0,1,2,3,4$

  b

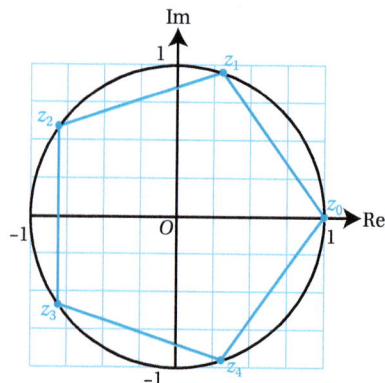

4 $e^{\frac{7\pi i}{12}}$

5 a $8\sqrt{2}; \dfrac{7\pi}{4}$

   b $2^{\frac{7}{8}}e^{i\left(\frac{(8k-1)\pi}{16}\right)}$ for $k=0,1,2,3$

6 a $\sqrt{2}, \dfrac{\pi}{4}$ **b** $z^6=-8i$

7 i $e^{\frac{i\pi}{3}}$ **ii** 6

8 $\dfrac{\sqrt{3}}{3}$

9 $-\dfrac{1}{64}$

10 a $\cos\left(-\dfrac{\pi}{6}\right)+i\sin\left(-\dfrac{\pi}{6}\right)$

   b Proof; $c=1$
   c $m=6, n=4$

11, 12 Proof

13 θ

14 Proof

15 a $e^{i\frac{\pi}{2}}$ **b** $e^{-\frac{\pi}{2}}$

16 i

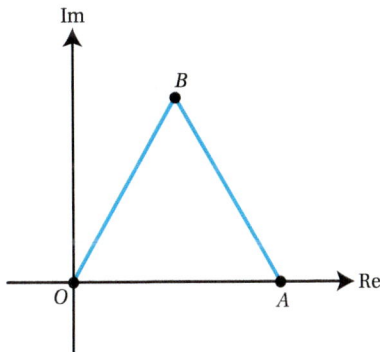

Proof

  ii $3e^{-i\frac{\pi}{3}}$ **iii** $\dfrac{243}{2}+\dfrac{243}{2}i\sqrt{3}$

17 a $\omega^2=e^{\frac{4\pi i}{5}}, \omega^3=e^{\frac{6\pi i}{5}}, \omega^4=e^{\frac{8\pi i}{5}}$

   b, c Proof

   d $4\cos^2\alpha+2\cos\alpha-1=0\left(\text{where }\alpha=\dfrac{2\pi}{5}\right)$

18 a Proof

   b i 1   **ii** 1

   c $z^3-4z^2+4z-3=0$

19 a $1+2i$

   b $\left(\sqrt{3}-2, 2\sqrt{3}+1\right), \left(\sqrt{3}+2, 2\sqrt{3}-1\right)$

20 Proof; it equals $2|a|\cos\dfrac{\theta}{2}$.

21 i Proof

  ii $1, e^{i\frac{2\pi}{7}}, e^{i\frac{4\pi}{7}}, e^{i\frac{6\pi}{7}}, e^{i\frac{8\pi}{7}}, e^{i\frac{10\pi}{7}}, e^{i\frac{12\pi}{7}};$

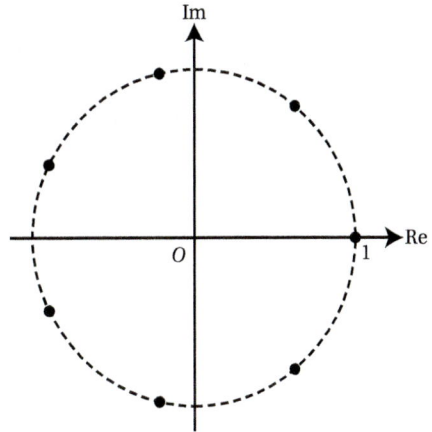

  iii $(z-1)\left(z^2-\left(2\cos\dfrac{2\pi}{7}\right)z+1\right)$

      $\times\left(z^2-\left(2\cos\dfrac{4\pi}{7}\right)z+1\right)\times\left(z^2-\left(2\cos\dfrac{6\pi}{7}\right)z+1\right)$

22 a $\dfrac{3-\sqrt{3}}{2}+\dfrac{1+3\sqrt{3}}{2}i$ **b** 5.35

23 a Proof

   b i a **ii** $b(\cos\theta+i\sin\theta)$

    iii $AB=\sqrt{b^2\sin^2\theta+(a-b\cos\theta)^2}$

    iv Proof

Chapter 3

Before you start...

1 $32\left(\cos\dfrac{5\pi}{7}+i\sin\dfrac{5\pi}{7}\right)$

2 a $2-i2\sqrt{3}$ **b** $2+e^{-3i}$

3 $x^5-10x^4y+40x^3y^2-80x^2y^3+80xy^4-32y^5$

4 $\text{Re}=0; \text{Im}=-\dfrac{\sin x}{1+\cos x}$

5 a $\dfrac{e^x\left(1-e^{nx}\right)}{1-e^x}$ **b** $x<0$

Exercise 3A

1 a $4\cos^3\theta\sin\theta-4\cos\theta\sin^3\theta$
   b Proof

2 $\sin3\theta=3\sin\theta-4\sin^3\theta$

3 a $\cos^5\theta+5i\cos^4\theta\sin\theta-10\cos^3\theta\sin^2\theta$
      $-10i\cos^2\theta\sin^3\theta+5\cos\theta\sin^4\theta+i\sin^5\theta$
   b $16\sin^5\theta-20\sin^3\theta+5\sin\theta$

4 a Proof   **b** $\dfrac{\pi}{2}, \dfrac{3\pi}{2}$

5 a $A=16$, $B=20$, $C=5$ **b** $0,\pm\dfrac{\sqrt{2}}{2}$

6 a Real: $\cos^4\theta-6\sin^2\theta\cos^2\theta+\sin^4\theta$;
Imaginary: $4\cos^3\theta\sin\theta-4\cos\theta\sin^3\theta$

b $\dfrac{4\tan\theta-4\tan^3\theta}{1-6\tan^2\theta+\tan^4\theta}$

7 a Proof   **b** $\dfrac{\pi}{24},\dfrac{5\pi}{24},\dfrac{3\pi}{8}$

8 a Real: $\cos^5\theta-10\cos^3\theta\sin^2\theta+5\cos\theta\sin^4\theta$;
Imaginary: $5\cos^4\theta\sin\theta-10\cos^2\theta\sin^3\theta+\sin^5\theta$

b Proof   **c** $5-20\theta^2$

Exercise 3B

1 a $2\cos^2\theta-1$ **b** $2c^2-\left(1+\dfrac{\sqrt{3}}{2}\right)=0$

c $\dfrac{\sqrt{2+\sqrt{3}}}{2}$

2 a Proof   **b** $\dfrac{\pi}{8},\dfrac{5\pi}{8}$

c $-1-\sqrt{2}$

3 a $\dfrac{\pi}{10},\dfrac{3\pi}{10},\dfrac{7\pi}{10},\dfrac{9\pi}{10}$ **b** Proof

4 a Proof

b $\dfrac{\pi}{18},\dfrac{13\pi}{18},\dfrac{25\pi}{18}$ or $\dfrac{5\pi}{18},\dfrac{17\pi}{18},\dfrac{29\pi}{18}$

c Proof; $\sin\dfrac{5\pi}{18},\sin\dfrac{25\pi}{18}$

5 a Proof

b $\tan3\theta=\dfrac{3t-t^3}{1-3t^2}$; $\tan4\theta=\dfrac{4t-4t^3}{1-6t^2+t^4}$; proof

6 a Proof   **b** $\dfrac{\sqrt{7}}{8}$

Exercise 3C

1 a i $2\cos3\theta+6\cos\theta$
  ii $2\cos4\theta+8\cos2\theta+6$
b i $2\cos4\theta-8\cos2\theta+6$
  ii $2i\sin5\theta-10i\sin3\theta+20i\sin\theta$
2 a Proof
  b Proof; $A=2$, $B=10$, $C=20$

3 a Proof   **b** $\dfrac{5\pi}{48}-\dfrac{9\sqrt{3}}{64}$

4 a i, ii Proof
b i $z^5-5z^3+10z-\dfrac{10}{z}+\dfrac{5}{z^3}-\dfrac{1}{z^5}$
  ii $a=1$, $b=-5$, $c=10$

c $\dfrac{1}{16}\left(-\dfrac{1}{10}\cos10x+\dfrac{5}{6}\cos6x-5\cos2x\right)+c$

5 a Proof
b $\left(z+z^{-1}\right)^6=z^6+6z^4+15z^2+20+15z^{-2}+6z^{-4}+z^{-6}$
$\left(z-z^{-1}\right)^6=z^6-6z^4+15z^2-20+15z^{-2}-6z^{-4}+z^{-6}$
c Proof

6 a $\sin x=\dfrac{e^{ix}-e^{-ix}}{2i}$, $\cos x=\dfrac{e^{ix}+e^{-ix}}{2}$

b $\dfrac{4}{35}$

Exercise 3D

1 a $\dfrac{3}{3-e^{i\theta}}$ or $\dfrac{3\left(3-e^{-i\theta}\right)}{10-6\cos\theta}$
b $\dfrac{9-3\cos1}{10-6\cos1}(\approx1.09)$

2 a Proof; $\dfrac{2}{2-e^{i\theta}}$ or $\dfrac{2\left(2-e^{-i\theta}\right)}{5-4\cos\theta}$
b Proof

3 $\dfrac{2\sin1}{5+4\cos2}(\approx0.505)$

4, 5 Proof

6 a $\dfrac{e^{i\theta}\left(1-e^{2ni\theta}\right)}{1-e^{2i\theta}}$ **b** Proof

c $\theta=\dfrac{\pi}{6},\dfrac{2\pi}{6},\dfrac{3\pi}{6},\dfrac{4\pi}{6},\dfrac{5\pi}{6}$

Mixed practice 3

1 a $\cos^4\theta+4i\cos^3\theta\sin\theta-6\cos^2\theta\sin^2\theta$
$-4i\cos\theta\sin^3\theta+\sin^4\theta$
b $A=4$, $B=8$

2 Proof; $\dfrac{1}{5},-\dfrac{1}{5}$

3 a $A=\dfrac{1}{16}$, $B=\dfrac{5}{16}$, $C=\dfrac{5}{8}$ **b** $\dfrac{8}{15}$

4–6 Proof
7 i Proof   **ii** $\dfrac{\pi}{3},\dfrac{\pi}{2},\dfrac{2\pi}{3}$
8 a, b Proof
  c i Proof   **ii** $\dfrac{2}{3}\pm\dfrac{\sqrt{5}}{3}i,\dfrac{-1}{2}\pm\dfrac{\sqrt{3}}{2}i$

9 a $\cos3\theta=4\cos^3\theta-3\cos\theta$
$\sin3\theta=3\sin\theta-4\sin^3\theta$
b, c Proof
d Proof; $-1,2\pm\sqrt{3}$
e Proof
f $2-\sqrt{3}$

10 i Proof   **ii** $\cos\left(\dfrac{\pi}{18}\right)$
11 i Proof
  ii Proof; $S=\dfrac{2\sin\theta}{5-4\cos\theta}$

Chapter 4

Before you start...

1 a $r = \begin{pmatrix} 3 \\ -1 \\ 2 \end{pmatrix} + \lambda \begin{pmatrix} 1 \\ 1 \\ 3 \end{pmatrix}$ **b** $\dfrac{x-3}{1} = \dfrac{y+1}{1} = \dfrac{z-2}{3}$

2 $(4, 0, 5)$

3 $42.4°$

4 $-2i - 10j + 4k$ (or $i + 5j - 2k$)

Exercise 4A

1 a i $r = \begin{pmatrix} 1 \\ 0 \\ 2 \end{pmatrix} + \lambda \begin{pmatrix} -1 \\ 5 \\ 2 \end{pmatrix} + \mu \begin{pmatrix} 1 \\ -2 \\ 3 \end{pmatrix}$

　　ii $r = \begin{pmatrix} 0 \\ 2 \\ 0 \end{pmatrix} + \lambda \begin{pmatrix} 0 \\ 4 \\ -1 \end{pmatrix} + \mu \begin{pmatrix} 5 \\ 3 \\ 0 \end{pmatrix}$

b i $r = (j + k) + \lambda(3i + j - 3k) + \mu(i - 3j)$

　ii $r = (i - 6j + 2k) + \lambda(5i - 6j) + \mu(-i + 3j - k)$

2 a i $r = \begin{pmatrix} 3 \\ -1 \\ 3 \end{pmatrix} + \lambda \begin{pmatrix} -2 \\ 2 \\ -1 \end{pmatrix} + \mu \begin{pmatrix} 1 \\ 0 \\ -1 \end{pmatrix}$

　　ii $r = \begin{pmatrix} -1 \\ -1 \\ 5 \end{pmatrix} + \lambda \begin{pmatrix} 5 \\ 2 \\ -3 \end{pmatrix} + \mu \begin{pmatrix} -6 \\ 2 \\ -4 \end{pmatrix}$

b i $r = \begin{pmatrix} 9 \\ 0 \\ 0 \end{pmatrix} + \lambda \begin{pmatrix} -11 \\ 1 \\ 0 \end{pmatrix} + \mu \begin{pmatrix} -8 \\ -1 \\ 2 \end{pmatrix}$

　　ii $r = \begin{pmatrix} 11 \\ -7 \\ 3 \end{pmatrix} + \lambda \begin{pmatrix} -10 \\ 21 \\ -1 \end{pmatrix} + \mu \begin{pmatrix} -16 \\ 17 \\ -3 \end{pmatrix}$

3 a i $r = \begin{pmatrix} -1 \\ 4 \\ 3 \end{pmatrix} + \lambda \begin{pmatrix} 4 \\ 1 \\ 2 \end{pmatrix} + \mu \begin{pmatrix} 2 \\ -1 \\ 2 \end{pmatrix}$

　　ii $r = \begin{pmatrix} 11 \\ 12 \\ 13 \end{pmatrix} + \lambda \begin{pmatrix} 6 \\ -3 \\ 1 \end{pmatrix} + \mu \begin{pmatrix} 2 \\ 15 \\ 6 \end{pmatrix}$

b i $r = \begin{pmatrix} -3 \\ 1 \\ 0 \end{pmatrix} + \lambda \begin{pmatrix} 0 \\ 0 \\ 1 \end{pmatrix} + \mu \begin{pmatrix} -7 \\ -3 \\ -1 \end{pmatrix}$

　　ii $r = \begin{pmatrix} 4 \\ 0 \\ 2 \end{pmatrix} + \lambda \begin{pmatrix} 2 \\ 1 \\ 1 \end{pmatrix} + \mu \begin{pmatrix} 4 \\ 0 \\ 2 \end{pmatrix}$

4 a i $r \cdot \begin{pmatrix} 3 \\ -5 \\ 2 \end{pmatrix} = -4$　**ii** $r \cdot \begin{pmatrix} 6 \\ -1 \\ 2 \end{pmatrix} = 19$

b i $r \cdot \begin{pmatrix} 3 \\ -1 \\ 0 \end{pmatrix} = -9$　**ii** $r \cdot \begin{pmatrix} 4 \\ 0 \\ -5 \end{pmatrix} = -10$

5 a i $\begin{pmatrix} 10 \\ 13 \\ -12 \end{pmatrix}$　**ii** $\begin{pmatrix} 10 \\ 4 \\ 3 \end{pmatrix}$

b i $\begin{pmatrix} 1 \\ 5 \\ 0 \end{pmatrix}$　**ii** $\begin{pmatrix} 1 \\ 20 \\ 7 \end{pmatrix}$

6 a i $r \cdot \begin{pmatrix} 10 \\ 13 \\ -12 \end{pmatrix} = 38$　**ii** $r \cdot \begin{pmatrix} 10 \\ 4 \\ 3 \end{pmatrix} = 3$

b i $r \cdot \begin{pmatrix} 1 \\ 5 \\ 0 \end{pmatrix} = 22$　**ii** $r \cdot \begin{pmatrix} 1 \\ 20 \\ 7 \end{pmatrix} = 152$

7 a i $10x + 13y - 12z = 38$

　　ii $10x + 4y + 3z = 3$

　b i $x + 5y = 22$　**ii** $x + 20y + 7z = 152$

8 a i $x + y + z = 10$　**ii** $z = 2$

　b i $40x + 5y + 8z = 580$

　　ii $x + y + z = 1$

9 a–c Proof

10 $6x + y - 3z = 1$

11 $r = \begin{pmatrix} 5 \\ 1 \\ 5 \end{pmatrix} + \lambda \begin{pmatrix} 8 \\ 0 \\ 3 \end{pmatrix} + \mu \begin{pmatrix} 5 \\ 0 \\ 0 \end{pmatrix}$

12 a $\begin{pmatrix} -1 \\ -4 \\ -17 \end{pmatrix}$　**b** $x + 4y + 17z = 37$

13 a $r \cdot \begin{pmatrix} -11 \\ 7 \\ 13 \end{pmatrix} = 39$　**b** No

14 a $\begin{pmatrix} 13 \\ -11 \\ -8 \end{pmatrix}$　**b** $-13x + 11y + 8z = 38$

15 a $\begin{pmatrix} -2 \\ 3 \\ -1 \end{pmatrix}$

　b i Proof　**ii** $(1, -3, 14)$

　c $2x - 3y + z = 25$

16 No

17 a $\begin{pmatrix} 1 \\ -3 \\ 4 \end{pmatrix}$ **b** $p = -\dfrac{1}{4}, q = \dfrac{3}{4}$

c $\mathbf{r} = \begin{pmatrix} 0 \\ 0 \\ 4 \end{pmatrix} + \lambda \begin{pmatrix} 4 \\ 0 \\ -1 \end{pmatrix} + \mu \begin{pmatrix} 0 \\ 4 \\ 3 \end{pmatrix}$

18 a $8x - 10y + 17z = 75$

b For example: $\mathbf{r} = \begin{pmatrix} 0 \\ -1 \\ 4 \end{pmatrix} + \lambda \begin{pmatrix} 5 \\ 2 \\ 1 \end{pmatrix} + \mu \begin{pmatrix} -1 \\ 1 \\ 4 \end{pmatrix}$

19 For example: $\mathbf{r} = \begin{pmatrix} 0 \\ 0 \\ -5 \end{pmatrix} + \lambda \begin{pmatrix} 1 \\ 0 \\ 0 \end{pmatrix} + \mu \begin{pmatrix} 0 \\ 1 \\ 3 \end{pmatrix}$

Exercise 4B

1 a i $(7, 1, 1)$ **ii** $(-19, -5, 7)$

b i $\left(-\dfrac{4}{3}, -\dfrac{7}{3}, -4\right)$ **ii** $(8, -3, 2)$

2 a–d Proof

3 $(15, 5, 8)$

4 a $\begin{pmatrix} -1 \\ 2 \\ 1 \end{pmatrix}$ **b** $(0.2, 2.6, 3.8)$

5 $(5, 3, 0)$

6 a $(5, 0, 0), (0, -20, 0), (0, 0, 12)$

b 133 (3 s.f.)

Exercise 4C

1 a i $46.4°$ **ii** $17.5°$

b i $47.6°$ **ii** $10.8°$

2 a $75.8°$ **b** $60°$

3 a $\begin{pmatrix} 2 \\ 3 \\ -1 \end{pmatrix}$ **b** $32.8°$

4 a $\begin{pmatrix} 3 \\ -1 \\ 1 \end{pmatrix}$ **b** $57°$

5 a $\begin{pmatrix} 4 \\ 2 \\ 3 \end{pmatrix}$ **b** $28.3°$

6 a $\begin{pmatrix} 13 \\ 7 \\ -8 \end{pmatrix}$ **b** $3.46°$

7 Proof

8 a $\begin{pmatrix} -5 \\ 0 \\ 1 \end{pmatrix}$ **b** $64.4°$

9 a Proof **b** $\dfrac{\sqrt{7}}{3}$

c $3\sqrt{5}$ **d** $\sqrt{35}$

Exercise 4D

1 a i $\dfrac{19}{\sqrt{6}}$ **ii** $\dfrac{12}{\sqrt{37}}$

b i $\dfrac{\sqrt{14}}{7}$ **ii** $\dfrac{12}{\sqrt{17}}$

2 a i $\dfrac{27}{\sqrt{73}}$ **ii** $\dfrac{24}{\sqrt{10}}$

b i $\dfrac{23}{\sqrt{17}}$ **ii** $\dfrac{19}{\sqrt{10}}$

3 a i 0.894 **ii** 0.596

b i 1.23 **ii** 3.77

4 a i Proof; 1.94 **ii** Proof; 3

b i Proof; 4.60 **ii** Proof; 3.11

5 1.52

6 $\dfrac{8}{\sqrt{11}}$

7 $\sqrt{2}$

8 $\dfrac{16}{\sqrt{41}}$

9 Proof; 5.30

10 $\dfrac{47}{\sqrt{122}}; \dfrac{47}{2}$

11 Proof; 2.57

12 a, b Proof

c $\dfrac{6\sqrt{11}}{11}$

13 a Proof **b** 0

c $6\sqrt{2}$

14 a $\mathbf{r} = \begin{pmatrix} -3 \\ -3 \\ 4 \end{pmatrix} + \lambda \begin{pmatrix} 2 \\ 2 \\ -1 \end{pmatrix}$ **b** $(3, 3, 1)$

c 9

15 a $\left(\dfrac{96}{41}, -\dfrac{32}{41}, \dfrac{16}{41}\right)$ **b** $\dfrac{16\sqrt{41}}{41}$

16 $\left(\dfrac{64}{9}, \dfrac{4}{9}, \dfrac{19}{9}\right); \dfrac{\sqrt{29}}{3}$

17 a, b Proof

c $\left(\dfrac{184}{11}, \dfrac{-32}{11}, \dfrac{-1}{11}\right)$ **d** 6.99

18 a Proof   **b** $11\mu - 6\lambda = 21$

   c $\sqrt{30}$

Mixed practice 4

1 $\dfrac{6\sqrt{10}}{5}$

2 $\dfrac{2}{\sqrt{19}}$

3 a $3x + y - z = 6$ **b** $\dfrac{5}{4}$

4 a $\begin{pmatrix} 0 \\ 5 \\ 5 \end{pmatrix}$ **b** $2x - y + z = 0$

  c Proof   **d** $47.1°\,(3\text{ s.f.})$

5 i $\mathbf{r} = \begin{pmatrix} 1 \\ 6 \\ 2 \end{pmatrix} + \lambda\begin{pmatrix} 4 \\ -4 \\ -1 \end{pmatrix} + \mu\begin{pmatrix} 0 \\ 3 \\ 2 \end{pmatrix}$

  ii $5x + 8y - 12z = 29$

6 i $2\mathbf{i} - \mathbf{j} - \mathbf{k}$ **ii** $\dfrac{7\sqrt{6}}{6}$

7 a $\begin{pmatrix} -2 \\ 7 \\ -3 \end{pmatrix}$ **b** $(3,3,8)$

  c $\begin{pmatrix} -2 \\ 7 \\ -3 \end{pmatrix}$ **d** $2x - 7y + 3z = 9$

8 a $\mathbf{r} = \begin{pmatrix} 4 \\ -1 \\ 2 \end{pmatrix} + \lambda\begin{pmatrix} 1 \\ -2 \\ 1 \end{pmatrix}$ **b** $(6, -5, 4)$

  c $2\sqrt{6}$

9 a Proof
  b $\mathbf{r} = (5\mathbf{i} + \mathbf{j} + 3\mathbf{k}) + t(4\mathbf{i} - 8\mathbf{j} + 3\mathbf{k})$
  c $\mathbf{r} = (9\mathbf{i} - 7\mathbf{j} + 6\mathbf{k}) + t(\mathbf{i} - 4\mathbf{j} + 2\mathbf{k})$
  d $(7, 1, 2)$
  e $2\sqrt{21}$

10 $\sqrt{14}$

11 a $-2, 2$ **b** 1.58
  c 5.91

12 a $(10, 11, -6)$ **b** $\begin{pmatrix} 7 \\ -9 \\ -5 \end{pmatrix}$

  c $7x - 9y - 5z = 1$

13 a $\dfrac{x-3}{3} = \dfrac{y-1}{-1} = \dfrac{z+4}{-1}$

  b $(0, 2, -3)$ **c** $(-3, 3, -2)$

  d Proof   **e** $3\sqrt{2}$

14 i $(2, 1, -3)$ **ii** $12x + 13y - 8z = 61$

15 i $(1, -3, -2);\ 6$ **ii** $41.8°$

16 a $\mathbf{r} = \begin{pmatrix} -1 \\ 1 \\ 4 \end{pmatrix} + \lambda\begin{pmatrix} 6 \\ 1 \\ 5 \end{pmatrix}$ **b** Proof

  c $\begin{pmatrix} 13 \\ -38 \\ -8 \end{pmatrix}$ **d** $13x - 38y - 8z = -83$

17 $k = 8$

18 $7x + 2y - 3z = 3$

19 a $\begin{pmatrix} -1 \\ 4 \\ 7 \end{pmatrix}$ **b, c** Proof

  d $\mathbf{r} = \begin{pmatrix} 1 \\ 1 \\ 2 \end{pmatrix} + \lambda\begin{pmatrix} -1 \\ 4 \\ 7 \end{pmatrix}$

20 a Proof   **b** $\mathbf{r}\cdot\begin{pmatrix} -1 \\ 5 \\ 2 \end{pmatrix} = -5$

  c $\sqrt{30}$ **d** $(8, -5, -1)$

21 a $\begin{pmatrix} -3 \\ -10 \\ 2 \end{pmatrix}$ **b** 5.32

  c $3x + 10y - 2z = 16$ **d** $\mathbf{r} = \begin{pmatrix} -7 \\ -28 \\ 11 \end{pmatrix} + \lambda\begin{pmatrix} -3 \\ -10 \\ 2 \end{pmatrix}$

  e $(2, 2, 5);\ 31.9\,(3\text{ s.f.})$ **f** 56.5

22 i Proof
  ii Symmetry in the plane $y = 0$.

  iii $\dfrac{1}{3}$

23 i, ii Proof

  iii $\mathbf{r} = \dfrac{1}{3}(\mathbf{u} + \mathbf{v} + \mathbf{w}) + t(\mathbf{u} - \mathbf{v})\times(\mathbf{u} - \mathbf{w})$

  iv $\dfrac{\sqrt{3}}{3}$

Chapter 5

Before you start...

1 -5

2 $\begin{pmatrix} 2 & -3 \\ 5 & -8 \end{pmatrix}\begin{pmatrix} x \\ y \end{pmatrix} = \begin{pmatrix} 2 \\ 3 \end{pmatrix};\ \mathbf{M}^{-1} = \begin{pmatrix} 8 & -3 \\ 5 & -2 \end{pmatrix};\ x = 7, y = 4$

3 -5

4 $x = 7, y = 4$

5 a $4x - y + 2z = 3$ **b** $3\mathbf{i} - 4\mathbf{k}$

Exercise 5A

1 a i Unique   **ii** Unique
  b i No solutions   **ii** Infinitely many
  c i Infinitely many   **ii** No solutions

2 a i Unique; $x = \dfrac{36}{5}, y = 7, z = \dfrac{122}{5}$

    ii Unique; $x = 3, y = -1, z = -2$

  b i Unique; $x = \dfrac{5}{3}, y = \dfrac{16}{3}, z = -\dfrac{7}{3}$

    ii Not unique; inconsistent

  c i Not unique; consistent

    ii Not unique; consistent

3 Proof

4 a $2, -3$

  b No solutions for both

5 a Proof   **b** $\pm 1, -2$

  c $x = 2, y = -1, z = 5$ **d** Inconsistent

6 a, b Proof

7 $k = 2, c = 1$

Exercise 5B

1 a i Consistent; unique solution:
     $x = 3, y = 2, z = 2$

    ii Consistent; unique solution:
     $x = 1, y = 1, z = 4$

  b i Inconsistent; two parallel planes with a
     single intersecting plane.

    ii Consistent; line intersection of three
     distinct planes (sheaf)

  c i Consistent; unique solution;
     $x = -1, y = -\dfrac{1}{3}, z = \dfrac{13}{3}$

    ii Inconsistent; two parallel planes with a
     single intersecting plane

  d i Consistent; unique solution;
     $x = \dfrac{1}{6}, y = \dfrac{1}{6}, z = \dfrac{7}{6}$

    ii Consistent; unique solution:
     $x = 0.5, y = 0.5, z = 4$

2 a Proof $(\det \mathbf{M} \neq 0)$; $x = \dfrac{5}{3}, y = \dfrac{16}{3}, z = -\dfrac{7}{3}$

  b The planes intersect at a single point.

3 a Proof

  b The planes intersect at a single point.

4 $\left(\dfrac{3}{2}, -\dfrac{11}{6}, -\dfrac{1}{6} \right)$

5 a, b Proof

  c Triangular prism

6 Π_1 and Π_3 are parallel, Π_2 intersects them.

7 a $k = 25$ **b** Triangular prism

8 a Proof   **b** -2

  c Intersect along a line (sheaf)

9 a Proof

  b Intersect along a line (sheaf)

10 a $\begin{pmatrix} 0.5 & 0.5 & 0.5 \\ -0.5 & 0.5 & 0.5 \\ 0.5 & 0.5 & -0.5 \end{pmatrix}$ **b** $\left(\dfrac{5+d}{2}, \dfrac{d-3}{2}, \dfrac{5-d}{2} \right)$

11 a 2

  b Inconsistent; triangular prism

12 a -4 **b** 2

  c Intersect in a line (form a sheaf)

13 $a = -3, b = 15$

Mixed practice 5

1 $(6, 1, 0)$

2 $k = 4, -4$

3 a Proof   **b** $k = 9$

4 a $a = 3, -3$

  b $\left(\dfrac{-9(2a+3)}{a^2 - 9}, \dfrac{3a}{a^2 - 9}, \dfrac{4a^2 - 9}{a^2 - 9} \right)$

  c $k = 1.5$

5 i $a^2 - 7a + 6$

  ii $1, 6$

  iii Infinitely many

6 i $a^3 - 4a$

  ii a Unique

     b Not unique, inconsistent

     c Not unique, consistent

Focus on ... 1

Focus on ... Proof 1

1, 2 Proof

Focus on ... Problem solving 1

1 $\begin{pmatrix} -\dfrac{1}{2}x - \dfrac{\sqrt{3}}{2}y \\ \dfrac{\sqrt{3}}{2}x - \dfrac{1}{2}y \end{pmatrix}$

2 The velocity is parallel to $S_2 - S_1$; proof

3–7 Proof

8 $t = \dfrac{2}{\sqrt{3}v}$; $\theta \to \infty$

9 Proof; $\dfrac{dy}{dt} = \dfrac{dr}{dt}\sin\theta + r\dfrac{d\theta}{dt}\cos\theta$

10 $\begin{cases} \dfrac{dr}{dt}\cos\theta - r\dfrac{d\theta}{dt}\sin\theta = -\dfrac{v\sqrt{3}}{2}\cos\theta - \dfrac{v}{2}\sin\theta \\ \dfrac{dr}{dt}\sin\theta + r\dfrac{d\theta}{dt}\cos\theta = \dfrac{v}{2}\cos\theta - \dfrac{v\sqrt{3}}{2}\sin\theta \end{cases}$; proof

11 $\dfrac{dr}{d\theta} = -r\sqrt{3}$; proof

Focus on ... Modelling 1

1 Proof

2 $j_{n+1} = 3a_n$

3 $\begin{pmatrix} 0.9 & 0.8 \\ 3 & 0 \end{pmatrix}$

4 a 180   **b** 179 108

5 2.06

6 1.45

7 0.125

8 For example: all adults are the same; there is no reference to gender; the average of 3 might not give a good prediction with small numbers; no randomness; there are no limiting factors such as the size of the island; there are no direct effects of predators.

9 Investigation

Cross-topic review exercise 1

1 a Proof   **b** 3.76

2 a $8e^{-i\frac{\pi}{2}}$

  b $\sqrt{2}e^{-i\frac{\pi}{12}}, \sqrt{2}e^{i\frac{\pi}{4}}, \sqrt{2}e^{i\frac{7\pi}{12}}, \sqrt{2}e^{i\frac{11\pi}{12}}, \sqrt{2}e^{-i\frac{5\pi}{12}}, \sqrt{2}e^{-i\frac{3\pi}{4}}$

3 a $6x - 2y + 3z = 4$ **b** $180° - 99.2° = 80.8°$

4 $a = 4, b = -4$

5 $n^3(n+1)$

6 i Proof

  ii $\left(z + e^{\frac{i\pi}{6}}\right)\left(z - e^{\frac{i\pi}{6}}\right)\left(z + e^{\frac{5i\pi}{6}}\right)\left(z - e^{\frac{5i\pi}{6}}\right)$

7 i $z_1 = \sqrt{2}\operatorname{cis}\left(\dfrac{\pi}{12}\right)$ $z_2 = \sqrt{2}\operatorname{cis}\left(\dfrac{7\pi}{12}\right)$

    $z_3 = \sqrt{2}\operatorname{cis}\left(\dfrac{13\pi}{12}\right)$ and $z_4 = \sqrt{2}\operatorname{cis}\left(\dfrac{19\pi}{12}\right)$

  ii

8 $\dfrac{6\sqrt{6}}{5}$

9 $\lambda = \dfrac{1}{3}$ or 2

10 a Proof   **b** $5\arctan\left(\dfrac{1}{a}\right)$

  c Proof

11 i Proof

  ii $\dfrac{1}{2}\left(\sqrt{n+2} + \sqrt{n+1} - 1 - \sqrt{2}\right)$

  iii Does not converge;

    as $n \to \infty, \displaystyle\sum_{r=1}^{n} \dfrac{1}{\sqrt{r+2} + \sqrt{r}} \to \infty$

12 i Proof

  ii $\dfrac{3}{2} - \dfrac{2n+3}{(n+1)(n+2)}$

  iii $N = 4$

13 i Proof

  ii $\dfrac{5}{4} - \dfrac{1}{(n+1)^2} - \dfrac{1}{(n+2)^2}$

  iii $\dfrac{61}{900}$

14 i $u_2 = \dfrac{2}{3}, u_3 = \dfrac{2}{5}$; proof

  ii $u_n = \dfrac{2}{2n-1}$

  iii Proof

15 i

  ii Proof
  iii a 3   **b** 1
  iv $3z^3 - 9z^2 + 7z - 2 = 0$

16 i $z = \pm 2$ or $\pm 2i$

  ii $\dfrac{2}{3}, 2, \dfrac{4+2i}{5}, \dfrac{4-2i}{5}$

17 i Proof

  ii Rotation of a point in the Argand plane by $\dfrac{2\pi}{3}$ about the origin; proof

  iii Proof

18 i $z = 1, e^{\frac{2\pi i}{5}}, e^{\frac{4\pi i}{5}}, e^{\frac{6\pi i}{5}}$ or $e^{\frac{8\pi i}{5}}$

  ii Proof; $\theta = \dfrac{2\pi}{5}, \dfrac{4\pi}{5}, \dfrac{6\pi}{5}$ or $\dfrac{8\pi}{5}$

19 i

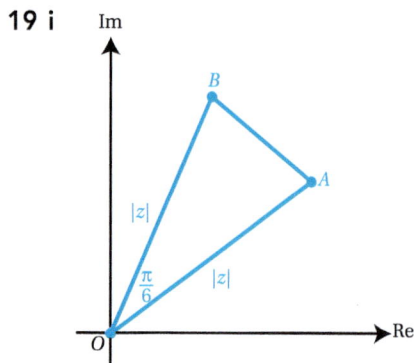

ii $\omega = \dfrac{1}{2} + 2\sqrt{3} + i\left(3 + \dfrac{\sqrt{3}}{2}\right)$ or

$\omega = \dfrac{3}{2} + 2\sqrt{3} + i\left(-1 + \dfrac{\sqrt{3}}{2}\right)$

20 i Proof
   ii Proof; $k = 2$
21 i Proof
   ii $\theta = 0$ or ± 0.899
22 i $M(5, 7, 3)$; 6
   ii $(5.5, 6, 4)$

23 i $\mathbf{r} \cdot \begin{pmatrix} 1 \\ 4 \\ -2 \end{pmatrix} = 11$

   ii $(1, 4, 3)$

   iii $\mathbf{c} = k\begin{pmatrix} 2 \\ 1 \\ 3 \end{pmatrix}$

24 $\mathbf{r} = \begin{pmatrix} 1 \\ 4 \\ 2 \end{pmatrix} + t\begin{pmatrix} -1 \\ 1 \\ 1 \end{pmatrix}$

25 i $\dfrac{\sqrt{6}}{9}$

   ii $2x + y + 7z = 11$

26 a $\dfrac{1}{k}e^{kx} + c$ **b** Proof

   c $\dfrac{3 - e^{\frac{\pi}{2}}}{10}$

27 a Proof   **b** $-i\ln\left(2 \pm \sqrt{3}\right)$

28 a Proof   **b** $\dfrac{1}{1 - \frac{1}{2}e^{i\theta}}$

   c Proof
29 a 4 **b** Proof
30 a Proof   **b** 0

31 Proof
32 a Proof   **b** $\dfrac{1}{2} - \dfrac{1}{n+1} + \dfrac{1}{n+2}$
   c Proof
33 Proof; $S = \dfrac{3 - 2e^{\pi}}{13}$
34 i, ii Proof
   iii Proof; $a = \dfrac{1}{6}$, $b = 2$. So $a = \dfrac{1}{6}$, $b = 2$
35 i $\theta = \dfrac{\pi}{12}, \dfrac{\pi}{4}, \dfrac{5\pi}{12}, \dfrac{7\pi}{12}, \dfrac{3\pi}{4}, \dfrac{11\pi}{12}$
   ii Proof
   iii $\dfrac{1}{16}$
36 i $8\cos^4\theta - 8\cos^2\theta + 1$
   ii–iv Proof
37 i Proof
   ii $\theta = \dfrac{\pi}{20}, \dfrac{5\pi}{20} = \dfrac{\pi}{4}, \dfrac{9\pi}{20}, \dfrac{13\pi}{20}, \dfrac{17\pi}{20}$
   iii $t = \tan\left(\dfrac{\pi}{20}\right), \tan\left(\dfrac{9\pi}{20}\right), \tan\left(\dfrac{13\pi}{20}\right), \tan\left(\dfrac{17\pi}{20}\right)$
38 i a Proof   **b** $S = 10$
   ii Proof
   iii Proof; $\theta = \dfrac{\pi}{5}$

Chapter 6

Before you start...

1 a $x \geqslant 3$
   b $f(x) \geqslant 0$

2

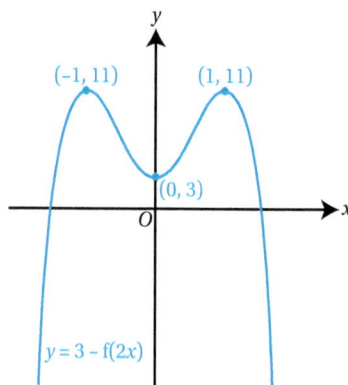

3 a $2e^{2x}$ **b** $-e^{-x} + c$
4 a $\sec^2 x$ **b** $\sin x + c$
5 a $6\sin 3x \cos 3x$ **b** $\cos x - x\sin x$
   c $\dfrac{2x\sec^2 2x - \tan 2x}{x^2}$

Exercise 6A

1 a

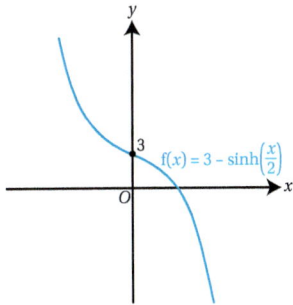

Domain: $x \in \mathbb{R}$; range: $f(x) \in \mathbb{R}$

b

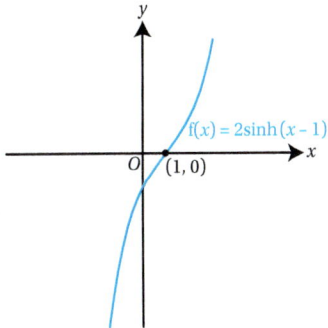

Domain: $x \in \mathbb{R}$; range: $f(x) \in \mathbb{R}$

c

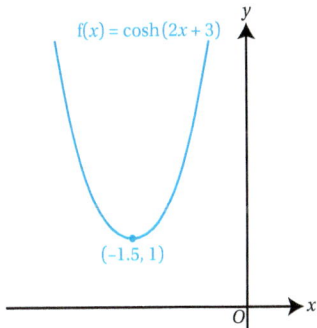

Domain: $x \in \mathbb{R}$; range: $f(x) \geqslant 1$

d

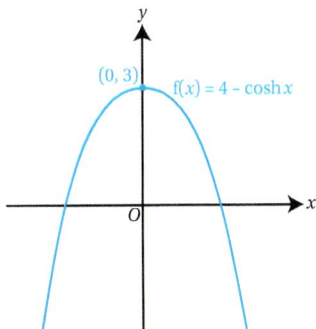

Domain: $x \in \mathbb{R}$; range: $f(x) \leqslant 3$

e

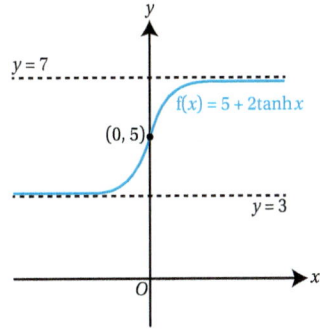

Domain: $x \in \mathbb{R}$; range: $3 < f(x) < 7$

f

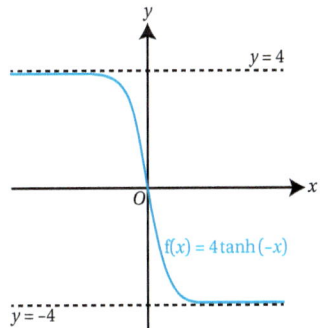

Domain: $x \in \mathbb{R}$; range: $-4 < f(x) < 4$

2 $a = 4$, $b = -2$

3 $a = -3$, $b = 1$

Exercise 6B

1 a i

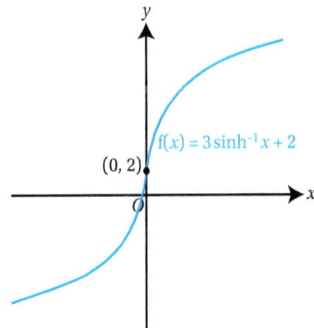

Domain: $x \in \mathbb{R}$; range: $f(x) \in \mathbb{R}$

ii

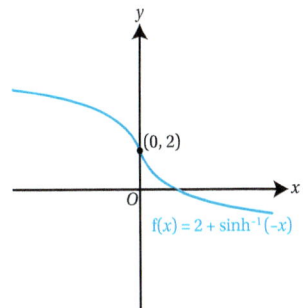

Domain: $x \in \mathbb{R}$; range: $f(x) \in \mathbb{R}$

b i

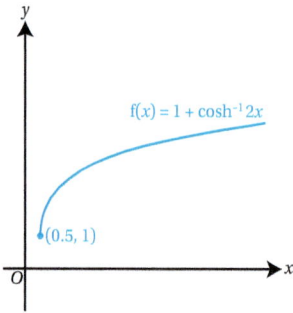

Domain: $x \geqslant 0.5$; range: $f(x) \geqslant 1$

ii

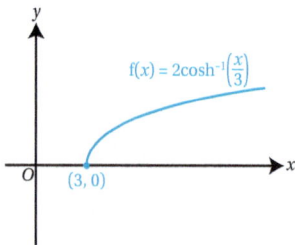

Domain: $x \geqslant 3$; range: $f(x) \geqslant 0$

c i

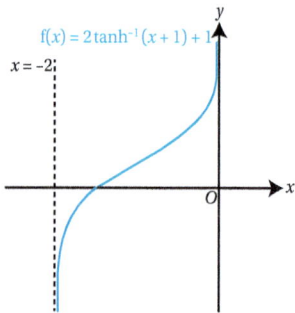

Domain: $-2 < x < 0$; range: $f(x) \in \mathbb{R}$

ii

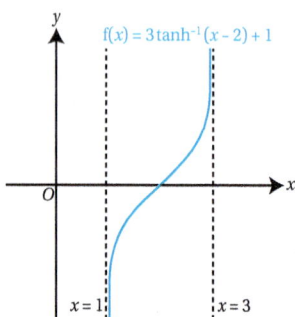

Domain: $1 < x < 3$; range: $f(x) \in \mathbb{R}$

2 a i 1.54   **ii** 10.0
  b i 0.580   **ii** 101
  c i 1.60   **ii** 10.3
  d i 0.481   **ii** 1.32
  e i Not possible   **ii** Not possible
3 a i -1.44 **ii** 0.0998
  b i ± 1.57 **ii** ± 2.06
  c i 0.973   **ii** 0.424
  d i No solution   **ii** No solution
  e i 74.2   **ii** 27.3

4 a i $\ln\left(1+\sqrt{2}\right)$ **ii** $\ln\left(2+\sqrt{5}\right)$
  b i $\ln\left(2+\sqrt{3}\right)$ **ii** $\ln\left(3+\sqrt{8}\right)$
  c i $\frac{1}{2}\ln 2$ **ii** $\frac{1}{2}\ln\left(\frac{5}{3}\right)$
  d i Doesn't exist   **ii** Doesn't exist
  e i $\ln\left(\sqrt{2}+\sqrt{3}\right)$ **ii** $\frac{1}{2}\ln\left(2+\sqrt{3}\right)$

5 2.10 or -0.0986

6 $\frac{5}{4}$

7 $\frac{4}{5}$

8 $b \leqslant a-2$

9 ± 0.515

10 0.549

11 $\sqrt{1+x^2}$

12–14 Proof

Exercise 6C

1 Proof

2 $\cosh x$

3–8 Proof

9 a, b Proof   **c** $4\cosh^3 x - 3\cosh x$

10 Proof

Work it out 6.1

Solution 3 is correct.

Exercise 6D

1 $\ln 5$

2 $\ln\dfrac{1}{2}$

3 $0.481, 2.06, -2.06$

4 $\ln\dfrac{1}{2}$

5 $\ln\sqrt{3}$

6 $0, \ln\left(\dfrac{3\pm\sqrt{5}}{2}\right)$

7 $\ln 3$

8 $\ln 4$

9 $\ln\left(\dfrac{\sqrt{5}-1}{2}\right)$ or $\ln\left(3+\sqrt{10}\right)$

10 $\ln(2\pm\sqrt{3})$

11 $\ln(1+\sqrt{2})$

12 $\ln 4$

13 $\ln\left(2\pm\sqrt{3}\right)$

14 a Proof

   b $x = \ln 2,\ y = \ln 4$ or $x = \ln 4,\ y = \ln 2$

15 $r^4 \geqslant p^4 - q^4$

16 a Proof   **b** $\dfrac{1}{2}\ln\left(1+\sqrt{2}\right)$

17 a Proof   **b** $\dfrac{1}{6}\ln\left(3\pm2\sqrt{2}\right)$

Exercise 6E

1 a i $3\cosh 3x$ **ii** $\dfrac{1}{2}\cosh\dfrac{1}{2}x$

  b i $4\sinh(4x+1)$ **ii** $\dfrac{1}{3}\sinh\dfrac{1}{3}x$

  c i $\dfrac{2}{3}\operatorname{sech}^2\dfrac{2}{3}x$ **ii** $-2\operatorname{sech}^2(1-2x)$

2 a $2x\tanh 3x + \dfrac{3x^2}{\cosh^2 3x}$ **b** $-\dfrac{10}{\tanh 5x\sinh^2 5x}$

3 $(0,\ e-1)$

4 $(-\ln 2,\ 4)$

5 Proof

6 $x+2y-\ln 3 - 2 = 0$

7 a $\pm\dfrac{1}{2}\ln\left(\sqrt{2}+1\right)$ **b** Proof

8 $\left(\ln 3,\ \dfrac{\sqrt{3}}{9}\right)$

9 Proof; $k = 1+\sqrt{2}$

10 a $\pm\ln\left(\dfrac{3+\sqrt{5}}{2}\right)$ **b** Proof

Exercise 6F

1 a i $\dfrac{1}{3}\cosh 3x + c$ **ii** $2\cosh\dfrac{x}{2} + c$

  b i $\dfrac{1}{2}\sinh(2x+1)+c$ **ii** $\dfrac{1}{4}\sinh 4x + c$

  c i $-\dfrac{1}{2}\ln\cosh(2x)+c$ **ii** $\dfrac{1}{3}\ln\cosh(3x-2)+c$

2 a i $\dfrac{e^4-2e^2+1}{2e^2}$ **ii** $\dfrac{e^{10}-1}{2e^5}$

  b i $\dfrac{e^4-2e^2+1}{4e^2}$ **ii** $\dfrac{e^{12}-1}{4e^6}$

  c i $\dfrac{1197}{128}$ **ii** $\dfrac{175}{144}$

3 a i $\dfrac{1}{8}\sinh 4x - \dfrac{1}{2}x + c$ **ii** $\dfrac{1}{12}\sinh 6x + \dfrac{1}{2}x + c$

  b i $x - 2\tanh\dfrac{x}{2} + c$ **ii** $x - \dfrac{1}{3}\tanh 3x + c$

c i $\dfrac{1}{4}\cosh 2x + c$ **ii** $\dfrac{1}{12}\cosh 6x + c$

4 a i $x\cosh x - \sinh x + c$

  ii $\dfrac{1}{2}x\cosh 2x - \dfrac{1}{4}\sinh 2x + c$

  b i $3x\sinh x - 3\cosh x + c$

  ii $2x\sinh\dfrac{x}{2} - 4\cosh\dfrac{x}{2} + c$

  c i $x^2\cosh x - 2x\sinh x + 2\cosh x + c$

  ii $\dfrac{1}{3}x^2\cosh 3x - \dfrac{2}{9}x\sinh 3x + \dfrac{2}{27}\cosh 3x + c$

  d i $\dfrac{1}{2}x^2\sinh 2x - \dfrac{1}{2}x\cosh 2x + \dfrac{1}{4}\sinh 2x + c$

  ii $3x^2\sinh x - 6x\cosh x + 6\sinh x + c$

5 a i $\dfrac{1}{6}e^{3x} + \dfrac{1}{2}e^{-x} + c$ **ii** $\dfrac{1}{6}e^{3x} + \dfrac{1}{2}e^{x} + c$

  b i $\dfrac{1}{20}e^{5x} - \dfrac{1}{20}e^{-5x} - \dfrac{1}{12}e^{3x} + \dfrac{1}{12}e^{-3x} + c$

  ii $\dfrac{1}{20}e^{5x} - \dfrac{1}{20}e^{-5x} + \dfrac{1}{4}e^{x} - \dfrac{1}{4}e^{-x} + c$

6 $\dfrac{1}{4}(\ln(\cosh x) + x) + c$

7 $1 - e^{-1}$

8 Proof

9 a $\dfrac{1}{5}\sinh^5 x + \dfrac{1}{3}\sinh^3 x + c$

  b $\sinh x - \dfrac{1}{\sinh x} + c$

10 $x\tanh x - \ln(\cosh x) + c$

11 $2\pi(5\ln 2 - 3)$

12, 13 Proof

14 a $\dfrac{1}{\sqrt{x^2+1}}$ **b** Proof

Mixed practice 6

1 a

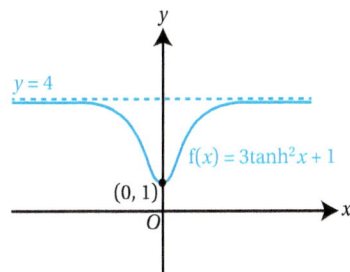

b $1 \leqslant f(x) < 4$

2 $\ln k$

3 $\dfrac{e^2 p^2 - 1}{e^2 p^2 + 1}$

4 $-1\pm\ln\left(3+\sqrt{8}\right)$

5 0.128

6 $\pm\ln 2$

7 $x=\dfrac{1}{2}\ln\left(2+\sqrt{5}\right)$

8 $18\cosh 6x$

9, 10 Proof

11 $\ln\left(2\sqrt{14}-7\right)$

12 $-\dfrac{1}{3\cosh 3x}+c$

13 Proof

14 0 or $\dfrac{1}{3}\ln\left(\sqrt{5}\pm 2\right)$

15 $\ln\left(4\pm\sqrt{15}\right)$

16 Proof

17 $\dfrac{x}{\sqrt{1+x^2}}$

18 Proof

19 a

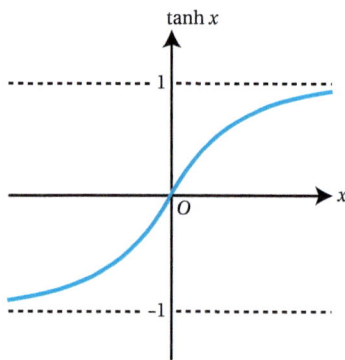

b Proof

c i Proof   **ii** Proof; $\dfrac{1}{2}\ln 2$

20 a Proof   **b** $\ln\sqrt{3},\ -\ln\sqrt{5}$

21 i Proof   **ii** Proof; $\dfrac{5\sqrt{6}}{12}$

22 $\dfrac{2}{3}\tan^{-1}(3e^x)+c$

23 $7\ln 6-10$

24 a Proof   **b** $\pm\dfrac{1}{3}\ln\dfrac{3}{2}$ **c** $\dfrac{1}{2}\left(\sqrt[3]{\dfrac{3}{2}}+\sqrt[3]{\dfrac{2}{3}}\right)$

25 $x=\ln 2,\ y=\ln 5$ or $x=\ln 5,\ y=\ln 2$

26–28 Proof

Chapter 7

Before you start...

1 $\dfrac{2x-5}{3y^2}$

2 $\dfrac{1}{9}\left(3\sqrt{3}-1\right)$

3 $\ln\left|(x-1)(x-2)\right|+\dfrac{2}{x-2}+c$

Exercise 7A

1 a i $\dfrac{-3}{\sqrt{1-9x^2}}$ **ii** $\dfrac{-2}{\sqrt{1-4x^2}}$

  b i $\dfrac{2}{4+x^2}$ **ii** $\dfrac{10}{25+4x^2}$

  c i $\arcsin x+\dfrac{x}{\sqrt{1-x^2}}$ **ii** $2x\arcsin x+\dfrac{x^2}{\sqrt{1-x^2}}$

  d i $\dfrac{2x}{\left(2+2x^2+x^4\right)}$ **ii** $-\dfrac{2}{\sqrt{2-x^2}}$

2 $-\dfrac{3}{\sqrt{35}}$

3 $\dfrac{3\pi-2}{4}$

4 $\dfrac{3}{9x^2+12x+5}$

5 $\dfrac{2x}{\sqrt{-x^4+6x^2-8}};\ x\in\left(-2,-\sqrt{2}\right)\cup\left(\sqrt{2},2\right)$

6 a Proof   **b** $-\dfrac{1}{x^2+1}$

7 Proof; $\left|x\right|<\dfrac{2}{3}$

8 $-\dfrac{1+\tan^2\left(\dfrac{1}{x}\right)}{x^2}$

9 a $\sin^{-1}x+\dfrac{x}{\sqrt{1-x^2}}$ **b** $x\sin^{-1}x+\sqrt{1-x^2}+c$

10 Proof

Exercise 7B

1 a i $\dfrac{2}{\sqrt{4x^2+1}}$ **ii** $\dfrac{1}{\sqrt{(x+2)^2+1}}$

  b i $\dfrac{-1}{\sqrt{x^2-1}}$ **ii** $\dfrac{3}{\sqrt{9x^2-1}}$

  c i $\dfrac{4}{1-16x^2}$ **ii** $\dfrac{1}{2x(x-1)}$

2 $\dfrac{\sinh x}{\sqrt{\cosh^2 x+1}}$

3 $\operatorname{arcosh}\left(x^2\right)+\dfrac{2x^2}{\sqrt{x^4-1}}$

4 Proof; $a = 2$, $b = 4$

5 $25x - 16y + 16\ln 2 - 15 = 0$

6 Proof

7 $(-1, 0)$

8 Proof

9 Proof; $a = -\dfrac{1}{2}$

Exercise 7C

1 a i $\dfrac{3}{2}\arctan\left(\dfrac{x}{2}\right) + c$ **ii** $\dfrac{5}{6}\arctan\left(\dfrac{x}{6}\right) + c$

   b i $\dfrac{1}{6}\arctan\left(\dfrac{3x}{2}\right) + c$ **ii** $\dfrac{2}{5}\arctan\left(\dfrac{2x}{5}\right) + c$

   c i $\sqrt{6}\arctan\left(\dfrac{\sqrt{6}x}{3}\right) + c$

     ii $\sqrt{10}\arctan\left(\dfrac{\sqrt{10}x}{2}\right) + c$

   d i $2\arcsin\left(\dfrac{x}{3}\right) + c$ **ii** $5\arcsin\left(\dfrac{x}{2}\right) + c$

   e i $\dfrac{1}{2}\arcsin\left(\dfrac{2x}{3}\right) + c$ **ii** $\arcsin\left(\dfrac{3x}{5}\right) + c$

   f i $5\sqrt{3}\arcsin\left(\dfrac{\sqrt{15}x}{5}\right) + c$

     ii $\sqrt{3}\arcsin\left(\dfrac{\sqrt{84}x}{7}\right) + c$

2 a i $3\sinh^{-1}\left(\dfrac{x}{4}\right) + c$ **ii** $5\sinh^{-1}\left(\dfrac{x}{5}\right) + c$

   b i $\dfrac{10}{3}\sinh^{-1}\left(\dfrac{3x}{5}\right) + c$ **ii** $\dfrac{3}{2}\sinh^{-1}\left(\dfrac{2x}{3}\right) + c$

   c i $2\sqrt{2}\sinh^{-1}\left(\sqrt{\dfrac{2}{3}}x\right) + c$

     ii $\dfrac{6\sqrt{7}}{7}\sinh^{-1}\left(\sqrt{\dfrac{7}{5}}x\right) + c$

   d i $2\cosh^{-1}\left(\dfrac{x}{7}\right) + c$ **ii** $7\cosh^{-1}\left(\dfrac{x}{6}\right) + c$

   e i $\dfrac{1}{3}\cosh^{-1}\left(\dfrac{3x}{4}\right) + c$ **ii** $3\cosh^{-1}\left(\dfrac{5x}{6}\right) + c$

   f i $\dfrac{5\sqrt{3}}{3}\cosh^{-1}\left(\sqrt{\dfrac{3}{7}}x\right) + c$

     ii $\dfrac{2\sqrt{7}}{7}\cosh^{-1}\left(\sqrt{\dfrac{7}{11}}x\right) + c$

3 a i $\arctan(x+2) + c$ **ii** $\arctan(x-3) + c$

   b i $\arcsin(x-4) + c$ **ii** $\arcsin(x-1) + c$

   c i $3\sqrt{2}\arctan\left(\dfrac{x+5}{\sqrt{2}}\right) + c$

     ii $\dfrac{5}{2}\arcsin\left(\dfrac{2x+3}{3}\right) + c$

   d i $\sinh^{-1}(x+3) + c$ **ii** $\sinh^{-1}(x+2) + c$

   e i $\cosh^{-1}\left(\dfrac{x-2}{4}\right) + c$ **ii** $\cosh^{-1}(x-1) + c$

   f i $3\cosh^{-1}\left(\dfrac{2x-3}{\sqrt{5}}\right) + c$

     ii $3\sinh^{-1}\left(\dfrac{x+1}{2}\right) + c$

4 $\ln\left(1+\sqrt{2}\right)$

5 $\dfrac{\pi}{2}$

6 $\dfrac{\pi\sqrt{3}}{18}$

7 a $\dfrac{1}{3}\arctan(3x) + c$ **b** $4\arctan\left(\dfrac{x}{4}\right) + c$

8 a Proof   **b** $\dfrac{5\pi}{8}$

9 a Proof   **b** $2\ln(2+\text{sqrt}(3))$

10 a $2(x+1)^2 + 9$

   b $\dfrac{1}{\sqrt{2}}\arctan\left(\dfrac{\sqrt{2}(x+1)}{3}\right) + c$

11 a $2^2 - 3(x-1)^2$ **b** $\dfrac{\sqrt{3}\pi}{9}$

12 a Proof   **b** $\dfrac{3}{2}\sin^{-1}\left(\dfrac{2x+1}{3}\right) + c$

13 Proof

14 $\arctan(x+1) + c$

15 Proof

16 $-4\sqrt{1-x^2} + 5\arcsin x + c$

17 $\sqrt{x^2-1} + \cosh^{-1}x + c$

18 $3\ln(x^2+9) - \dfrac{5}{3}\arctan\left(\dfrac{x}{3}\right) + c$

19 a $2(x-2)^2 + 9$

   b $\dfrac{1}{2}\ln\left|2x^2 - 8x + 17\right| + 2\sqrt{2}\arctan\left(\dfrac{\sqrt{2}(x-2)}{3}\right) + c$

20, 21 Proof

22 a $\cos u = \sqrt{\dfrac{1}{1+x^2}}$, $\sin u = \dfrac{x}{\sqrt{1+x^2}}$

   b Proof

Exercise 7D

1 a i $\dfrac{1}{\sqrt{2}}\arctan\left(\dfrac{x}{\sqrt{2}}\right)+2\ln|x+3|+c$

ii $2\arctan x-\ln|x-2|+c$

b i $2\ln|x+1|+\arctan(x+3)+c$

ii $\ln|x-2|+2\arctan\left(\dfrac{2}{x+1}\right)+c$

c i $\ln|x^2+1|+\arctan x-3\ln|x+1|+c$

ii $\dfrac{1}{2}\ln|x^2+4|-\arctan\left(\dfrac{x}{2}\right)-\ln|x-1|+c$

d i $\ln|x+1|-\dfrac{2}{x+1}-\dfrac{1}{\sqrt{3}}\arctan\left(\dfrac{x}{\sqrt{3}}\right)+c$

ii $\ln|x-2|-\dfrac{1}{x-2}+2\arctan x+c$

2 $\dfrac{\pi}{4}-\ln 4$

3 a $\dfrac{1}{x-2}-\dfrac{1}{x^2+9}$

b $y=\ln\left|\dfrac{x-2}{2}\right|-\dfrac{1}{3}\arctan\left(\dfrac{x}{3}\right)$

4 a $\ln|x-2|+2\ln|x+1|-\dfrac{1}{x+1}+c$

b $\ln|x-2|+2\arctan(x+1)+c$

5 a $\dfrac{1}{x-2}+\dfrac{1}{x+2}+\dfrac{1}{x^2+4}$

b $\ln 2+\dfrac{\pi}{6}$

6 $\ln 2+1+\dfrac{\pi}{4}$

7 $\ln\left|\dfrac{x-1}{x+3}\right|-\sqrt{2}\arctan\left(\dfrac{x+1}{\sqrt{2}}\right)+c$

8 Proof; $P=\dfrac{3}{2}, Q=-\dfrac{1}{2}, R=-1$

Mixed practice 7

1 $\dfrac{e^x}{1+e^{2x}}$

2 $2x\sin^{-1}x+\dfrac{x^2}{\sqrt{1-x^2}}$

3 $\dfrac{2x}{\sqrt{1-\left(1-x^2\right)^2}}$

4 $\pm\sqrt{3}$

5 $\dfrac{\pi}{6}$

6 i $\dfrac{1}{x-1}+\dfrac{9}{x^2+9}$ **ii** $\ln|x-1|+3\arctan\left(\dfrac{x}{3}\right)+c$

7 $\dfrac{2\left(1-3x^4\right)}{\left(1+x^4\right)^2}$

8 Proof

9 $3\ln\left(x^2+4\right)+2\arctan\left(\dfrac{x}{2}\right)+c$

10 $\arctan(x-1)+c$

11 $108^{\frac{1}{6}}$

12 i $2\sqrt{1-x^2}$ **ii** $\dfrac{\pi}{2}$

13 i $\dfrac{1}{1-x}+\dfrac{1}{1+x}+\dfrac{2}{1+x^2}$ **ii** Proof

14 a Proof

b $\dfrac{1}{3}\left(\arcsin 3x+\sqrt{1-9x^2}\right)+c$

15 Proof; $A=10$

16 a $\dfrac{2}{x+2}-\dfrac{2x-1}{x^2+1}$

b $2\ln|x+2|-\ln|x^2+1|+\arctan x+c$

c $\dfrac{4\pi}{3}-\dfrac{3}{2}$

17 a, b Proof

18 i Proof

ii $\dfrac{1}{2}\cosh^{-1}2x+c$

iii $\dfrac{x}{2}\sqrt{4x^2-1}-\dfrac{1}{4}\cosh^{-1}2x+c$

Chapter 8

Before you start...

1 $10x\left(x^2+3\right)^4$

2 a i $-2e^{-2x}$ **ii** $4e^{-2x}$

b i $\dfrac{1}{x}$ **ii** $-\dfrac{1}{x^2}$

c i $-3\sin 3x$ **ii** $-9\cos 3x$

3 a $\operatorname{sech}^2 x$ **b** $-2\operatorname{sech}^2 x\tanh x$

4 a $\dfrac{1}{\sqrt{x^2-1}}$ **b** $-x\left(x^2-1\right)^{-\frac{3}{2}}$

5 a $\dfrac{1}{\sqrt{1-x^2}}$ **b** $x\left(1-x^2\right)^{-\frac{3}{2}}$

Exercise 8A

1 a i $1+x+\dfrac{x^2}{2!}+\dfrac{x^3}{3!}+\ldots+\dfrac{x^n}{n!}$

ii $1-3x+\dfrac{9}{2}x^2-\dfrac{9}{2}x^3+\ldots+\dfrac{(-3)^n}{n!}x^n$

b i $-x+\dfrac{x^3}{3!}-\dfrac{x^5}{5!}+\dfrac{x^7}{7!}+\ldots+\dfrac{(-1)^{n+1}x^{2n+1}}{(2n+1)!}$

ii $2x-\dfrac{4}{3}x^3+\dfrac{4}{15}x^5-\dfrac{8}{315}x^7+\ldots$
$+\dfrac{(-1)^n\,2^{2n+1}}{(2n+1)!}x^{2n+1}$

c i $1-\dfrac{x^2}{2!}+\dfrac{x^4}{4!}-\dfrac{x^6}{6!}+\ldots+\dfrac{(-1)^n x^{2n}}{(2n)!}$

ii $1-\dfrac{9}{2}x^2+\dfrac{27}{8}x^4-\dfrac{81}{80}x^6+\ldots+\dfrac{(-1)^n 3^{2n}}{(2n)!}x^{2n}$

d i $-x-\dfrac{x^2}{2}-\dfrac{x^3}{3}-\dfrac{x^4}{4}+\ldots+-\dfrac{x^n}{n}$

ii $2x-2x^2+\dfrac{8}{3}x^3-4x^4+\ldots+\dfrac{-(-2)^n x^n}{n}$

e i $x+\dfrac{x^3}{3!}+\dfrac{x^5}{5!}+\dfrac{x^7}{7!}+\ldots+\dfrac{x^{2n+1}}{(2n+1)!}$

ii $2x+\dfrac{8x^3}{3!}+\dfrac{32x^5}{5!}+\dfrac{128x^7}{7!}+\ldots+\dfrac{2^{2n+1}x^{2n+1}}{(2n+1)!}$

f i $1+\dfrac{x^2}{2!}+\dfrac{x^4}{4!}+\dfrac{x^6}{6!}+\ldots+\dfrac{x^{2n}}{(2n)!}$

ii $1+\dfrac{x^2}{2!}+\dfrac{x^4}{4!}+\dfrac{x^6}{6!}+\ldots+\dfrac{x^{2n}}{(2n)!}$

2 $2+\dfrac{1}{4}x+\dfrac{7}{64}x^2$

3 a Proof   **b** 0.324

4 a i $f'(x)=-2xe^{-x^2}; f''(x)=-2e^{-x^2}+4x^2e^{-x^2};$
$f'''(x)=12xe^{-x^2}-8x^3e^{-x^2};$
$f^{(4)}(x)=\left(16x^4-48x^2+12\right)e^{-x^2}$

ii $1-x^2+\dfrac{1}{2}x^4$

b $\dfrac{23}{30}$

5 a $1-x^2+\dfrac{1}{3}x^4-\dfrac{2}{45}x^6$

b $x^2-\dfrac{1}{3}x^4+\dfrac{2}{45}x^6$

6 a i Proof

ii $f'''(x)=\dfrac{\cos x}{(1+\sin x)^2};$
$f^{(4)}(x)=-\dfrac{1+\sin x+\cos^2 x}{(1+\sin x)^3}$

b $x-\dfrac{1}{2}x^2+\dfrac{1}{6}x^3-\dfrac{1}{12}x^4$

c 0.116

7 a i $f'(x)=\dfrac{1}{1-x}; f''(x)=\dfrac{1}{(1-x)^2};$
$f'''(x)=\dfrac{2}{(1-x)^3}$

ii Proof

b $\dfrac{80}{81}$

8 a Proof   **b** $\dfrac{x^n}{(n-1)!}$

9 a Proof   **b** $5^n x^n$

10 a $x+\dfrac{1}{6}x^3+\dfrac{3}{40}x^5$

b i $f^{(n)}(x)=-g^{(n)}(x)$ **ii** Proof; $k=\dfrac{\pi}{2}$

11 a Not equal to f(0).

b f(x) is an increasing function at $x=0$, but first derivative of series is negative at $x=0$.

Exercise 8B

1 a i $1-3x+\dfrac{9x^2}{2}+\ldots+\dfrac{(-3)^n}{n!}x^n$

ii $1+x^3+\dfrac{x^6}{2}+\ldots+\dfrac{x^{3n}}{n!}$

b i $3x-\dfrac{9x^2}{2}+9x^3+\ldots+\dfrac{(-1)^{(n+1)}3^n}{n}x^n$

ii $-2x-2x^2-\dfrac{8}{3}x^3+\ldots+\dfrac{-2^n}{n}x^n$

c i $-\dfrac{x}{2}+\dfrac{x^3}{48}-\dfrac{x^5}{3840}+\ldots+\dfrac{(-1)^n}{2^{2n+1}(2n+1)!}x^{2n+1}$

ii $3x^2-\dfrac{9x^6}{2}+\dfrac{81x^{10}}{40}+\ldots+\dfrac{(-1)^n 3^{2n+1}}{(2n+1)!}x^{4n+2}$

d i $1-\dfrac{x^4}{18}+\dfrac{x^8}{1944}+\ldots+\dfrac{(-1)^n}{3^{2n}(2n)!}x^{4n}$

ii $1-2x^2+\dfrac{2x^4}{3}+\ldots+\dfrac{(-1)^n 2^{2n}}{(2n)!}x^{2n}$

e i $1-2x-2x^2+\ldots+\dfrac{\left(\dfrac{1}{2}\right)\left(-\dfrac{1}{2}\right)\ldots\left(\dfrac{3}{2}-n\right)}{n!}(-4)^n x^n$

ii $1-\dfrac{4x}{3}+\dfrac{10x^2}{9}+\ldots+\dfrac{(-4)(-5)\ldots(-3+n)}{3^n n!}x^n$

2 a i $\ln 3+\dfrac{x}{3}-\dfrac{x^2}{18}$ **ii** $-\ln 2-2x-2x^2$

b i $\dfrac{1}{8}+\dfrac{9x}{16}+\dfrac{27x^2}{16}$ **ii** $2-8x+48x^2$

c i $-3+\dfrac{8x}{27}+\dfrac{64x^2}{2187}$ **ii** $\dfrac{1}{16}+\dfrac{3x}{32}+\dfrac{27x^2}{256}$

3 a i $2x^2 - x^3 - \dfrac{2}{3}x^4$

ii $-x - \dfrac{1}{2}x^2 + \dfrac{25}{6}x^3 + 2x^4$

b i $1 + \dfrac{1}{2}x^2 - \dfrac{1}{3}x^3 + \dfrac{3}{8}x^4$

ii $x + 2x^2 + \dfrac{23}{6}x^3 + \dfrac{23}{3}x^4$

c i $x - \dfrac{x^2}{2} + \dfrac{x^3}{6} - \dfrac{x^4}{12}$

ii $-\dfrac{x^2}{2} - \dfrac{x^4}{12}$

4 $4x - 4x^2 + \dfrac{16}{3}x^3 - 8x^4 + \ldots; \ -\dfrac{1}{2} < x \leqslant \dfrac{1}{2}$

5 $2x + 6x^2 + \dfrac{23}{3}x^3 + 5x^4$

6 Proof

7 a $x + \dfrac{x^3}{3}$

b $1 + x + \dfrac{x^2}{2} + \dfrac{x^3}{2} + \dfrac{3}{8}x^4$

8 a $-\dfrac{x^2}{2} - \dfrac{x^4}{12}$

b $\dfrac{x^2}{2} + \dfrac{x^4}{12}$

c $x + \dfrac{x^3}{3}$

9 a $\ln 8 - \dfrac{3}{2}x - \dfrac{39}{8}x^2 - \dfrac{71}{8}x^3$

b $y = \ln 8 - 1.5x$

10 a $\displaystyle\sum_{k=1}^{\infty} \dfrac{x^{2k-1}}{2k-1}$; convergence for $|x| < 1$

b $x = \dfrac{3}{5}; \ \ln 2 \approx 0.688$

Exercise 8C

1 a 1

b 4

c Integral diverges.

d $\dfrac{1}{2}$

2 a $p < 0$

b $p > 1$

3 Proof; 4

4 1

5 $\dfrac{\pi}{2}$

6 $\ln\dfrac{4}{45}$

Exercise 8D

1 a i 0.4π

ii $\dfrac{128\pi}{7}$

b i 304.8π

ii $\dfrac{18\pi}{7}$

c i $\dfrac{\pi}{2}$

ii $\dfrac{21\pi}{64}$

2 a i $\dfrac{\pi}{2}(e^2 - 1)$

ii $\dfrac{\pi}{2}(1 - e^{-6})$

b i $\pi\left(\dfrac{e^4}{4} + e^2 - \dfrac{1}{4}\right)$

ii $\pi\left(\dfrac{25}{2} - 4e^{-2} - \dfrac{e^{-4}}{2}\right)$

c i 2π

ii π

3 a i 32π

ii $\dfrac{85}{2}\pi$

b i $\dfrac{96}{5}\pi$

ii $\dfrac{28\sqrt{2}}{3}\pi$

c i 3π

ii $\dfrac{35}{3}\pi$

4 a i $\dfrac{\pi}{2}(e^4 - 1)$

ii $\dfrac{\pi}{8}(e^8 + 4e^4 + 3)$

b i $\pi \ln 2$

ii $\pi \ln 3$

c i $\dfrac{\pi^2}{2}$

ii $(\pi - 2)\dfrac{\pi}{4}$

5 a i 2304π

ii 0.5π

b i $\left(\dfrac{27}{2} + 2\ln 2\right)\pi$

ii $\dfrac{527}{5}\pi$

c i $\dfrac{\pi^2}{4}$

ii $\dfrac{\pi^2}{3}$

6 a i $\dfrac{18432\pi}{7}$

ii $\dfrac{3\pi}{14}$

b i $\dfrac{487\pi}{480}$

ii $\dfrac{2103\pi}{35}$

c i $\dfrac{\pi^2}{4}$

ii $\dfrac{\pi}{2}$

7 a $(4,0)$

b $\dfrac{11\pi}{6}$

8 75.4

9 π

10 2

11 $\dfrac{4\pi a^5}{15}$

12 $\dfrac{4}{3}$

13 a $(0,3),(4,19)$

b 630 (3 s.f.)

14 184 (3 s.f.)

15 a $(1,4),(9,12)$

b $\dfrac{736\pi}{15}$

16 Proof

17 $\dfrac{197}{2}\pi$

18 $2\pi\left(\ln\dfrac{3}{2} - \dfrac{1}{6}\right)$

19 $\dfrac{8}{15}\pi$

20 Proof; use $y = \dfrac{rx}{h}$.

21 $\pi\left(\dfrac{1}{2}e^2 - 2e + \dfrac{5}{2}\right)$

Exercise 8E

1 a i $\dfrac{1}{3}$ **ii** $\dfrac{13}{3}$

  b i $\dfrac{4}{3}$ **ii** $\dfrac{1}{5}$

  c i 17   **ii** 1995

2 a i $\dfrac{2}{\pi}$ **ii** 0

  b i $e-1$ **ii** $\dfrac{1}{e-1}$

  c i $\dfrac{38}{15}$ **ii** $\dfrac{1}{\sqrt{\pi}}$

3 $20\sqrt{T}$

4 1.5

5 a $\dfrac{a^2}{3}$ **b** $\dfrac{a}{\sqrt{3}}$

6 Proof

7 $1:2$

8 Proof

9 a

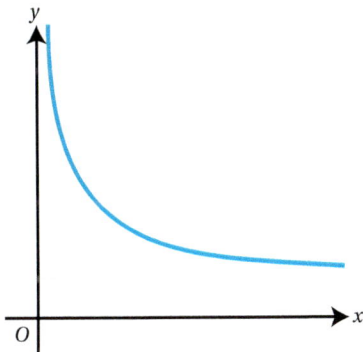

  b Curve is concave up.

  c Proof

10 Not true. For example: $f(x)=x^2-1$ between -1 and 1.

Mixed practice 8

1 a, b Proof

2 a $f'(x)=(1+x)^{-1}, f''(x)=-(1+x)^{-2},$ $f'''(x)=2(1+x)^{-3}, f^{(4)}(x)=-6(1+x)^{-4}$

  b $f(x)\approx x-\dfrac{1}{2}x^2+\dfrac{1}{3}x^3-\dfrac{1}{4}x^4$

  c $\ln 2$

3 a The integrand is not defined at $x=0$.

  b $\displaystyle\lim_{a\to 0}\left\{\int_a^4 \dfrac{5-x}{\sqrt{x^3}}dx\right\}$ does not converge to a finite value.

4 $-\dfrac{9}{2}x^2$

5 6

6 57.8

7 $\dfrac{3}{2}$

8 4

9 $x<0$

10 a $-\dfrac{1}{2}x^2-\dfrac{1}{12}x^4$ **b** Proof

11 a $f'(x)=\dfrac{\cos x}{1+\sin x}, f''(x)=-\dfrac{1}{1+\sin x},$

     $f'''(x)=\dfrac{\cos x}{(1+\sin x)^2}$

  b $x-\dfrac{1}{2}x^2+\dfrac{1}{6}x^3$ **c** $-\dfrac{1}{2}$

12 i $f(0)=0, f'(0)=1$

  ii Proof; $f''(0)=-2$

  iii $f'''(x)=-2f''(x)-2f'(x); f'''(0)=2$

  iv $f(x)=x-x^2+\dfrac{1}{3}x^3$

13 $\dfrac{|\pi a^5|}{30}$

14 a Proof   **b** $\pi\left(\dfrac{5e^2}{6}-\dfrac{1}{2}\right)$

15 $\dfrac{\pi}{60}$

16 i $\dfrac{16\sqrt{3}}{9}$

  ii $\left(3\ln 3-\dfrac{20}{27}\right)\pi$

17 i Proof

  ii $\dfrac{3e^4+1}{2}\pi$ **iii** $\dfrac{\pi}{2}(e^4-1)$

18 6π

19 $\dfrac{4}{3}\pi ab^2$

20 $\dfrac{\pi}{2}\left(1-\dfrac{3}{e^2}\right)$

21 a $x+\dfrac{x^2}{1!}+\dfrac{x^3}{2!}+\dfrac{x^4}{3!}+\ldots+\dfrac{x^n}{(n-1)!}+\ldots$

  b $\dfrac{x^2}{2}+\dfrac{x^3}{3(1!)}+\dfrac{x^4}{4(2!)}+\ldots+\dfrac{x^n}{n(n-2)!}+\ldots$

  c Proof

22 $3+5\sqrt{2}$

23 a $(0,12)$ and $(4.5,-3.75)$

b $\pi\int_0^{4.5}(14x^3-111x^2+216x)\,dx$

c 787

24 a

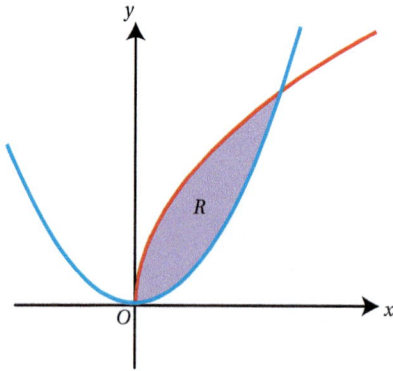

b $\dfrac{3\pi}{10}$ **c** $\dfrac{3\pi}{10}$

Chapter 9

Before you start…

1 a 210° **b** $-\dfrac{\sqrt{3}}{2}$

2 a $\theta\in\left(\dfrac{\pi}{2},\dfrac{3\pi}{2}\right)$ **b** 7

3 a $\dfrac{1}{3}$ **b** $\dfrac{\pi}{4}$

Exercise 9A

1 a–c

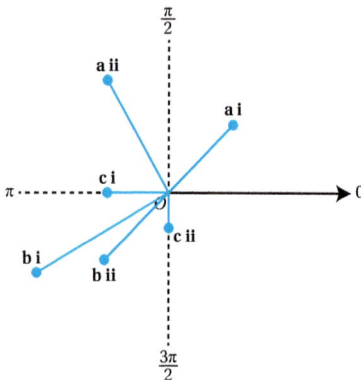

2 a i Distance $AB=2.53$; area $AOB=4.53$
  ii Distance $AB=3.42$; area $AOB=2.5$
 b i Distance $AB=17.8$; area $AOB=10.4$
  ii Distance $AB=8.92$; area $AOB=2.59$

c i Distance $AB=2.91$; area $AOB=0.5$
 ii Distance $AB=7.21$; area $AOB=12$

3 a i

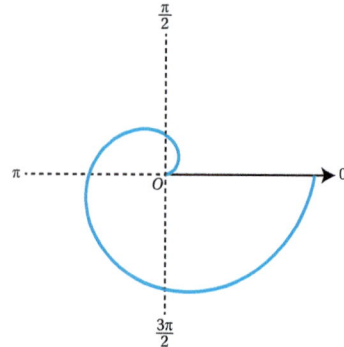

ii

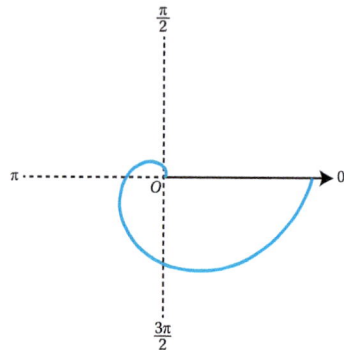

b i

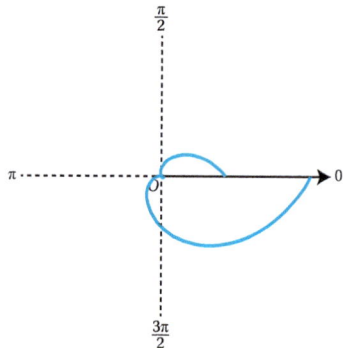

ii

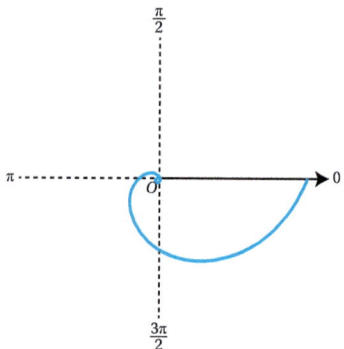

c i

ii

4 a

b

c

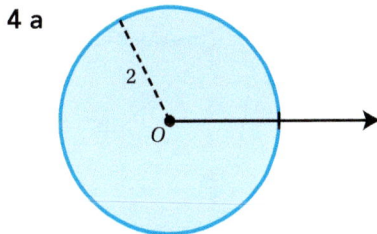

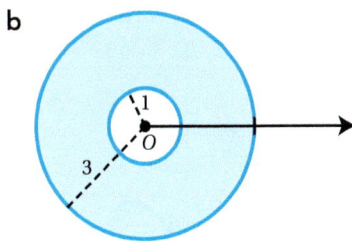

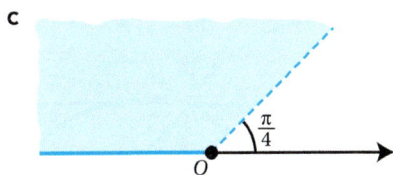

5 a $\dfrac{\pi}{4} < \theta < \dfrac{3\pi}{4}$ and $\dfrac{5\pi}{4} < \theta < \dfrac{7\pi}{4}$

b

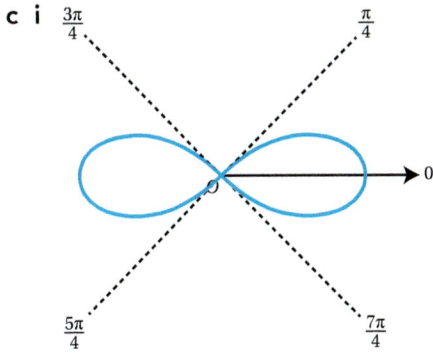

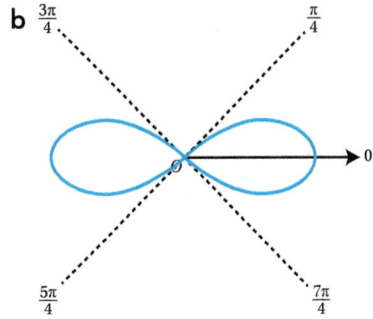

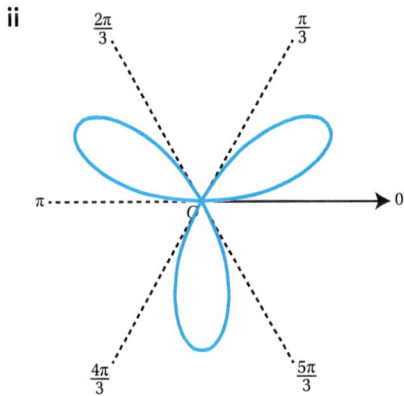

6 a $\pi < \theta < 2\pi$ **b** Proof

c

d $2\sqrt{3}$

Exercise 9B

1 a i Maximum $\left(5, \dfrac{\pi}{2}\right)$; minimum $\left(1, \dfrac{3\pi}{2}\right)$

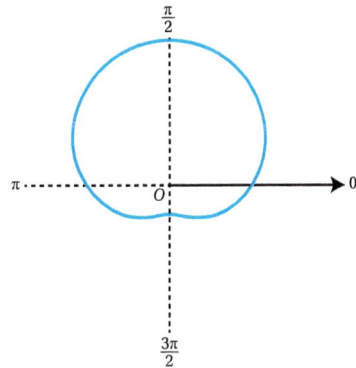

ii Maximum $(6,0)$; minimum $(4,\pi)$

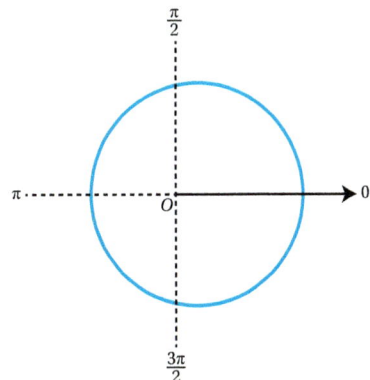

b i Maxima $\left(10, \dfrac{\pi}{2}\right)$ and $\left(10, \dfrac{3\pi}{2}\right)$;

minima $(4, 0)$ and $(4, \pi)$

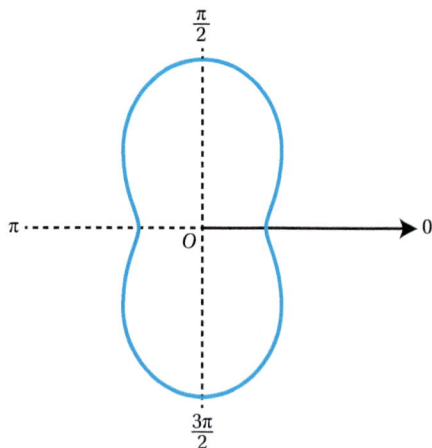

ii Maxima $\left(7, \dfrac{3\pi}{4}\right)$ and $\left(7, \dfrac{7\pi}{4}\right)$;

minima $\left(3, \dfrac{\pi}{4}\right)$ and $\left(3, \dfrac{5\pi}{4}\right)$

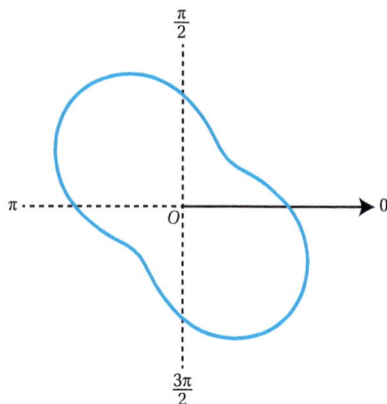

2 a i $\theta = 0, \dfrac{\pi}{3}, \dfrac{2\pi}{3}, \pi, \dfrac{4\pi}{3}, \dfrac{5\pi}{3}$

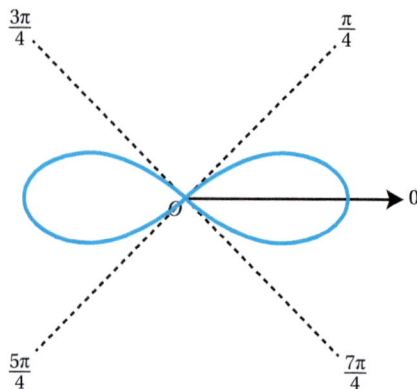

ii $\theta = \dfrac{\pi}{4}, \dfrac{3\pi}{4}, \dfrac{5\pi}{4}, \dfrac{7\pi}{4}$

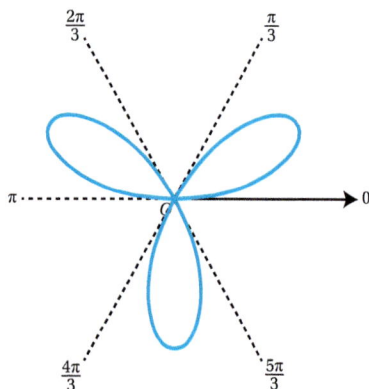

b i $\theta = \dfrac{2\pi}{9}, \dfrac{4\pi}{9}, \dfrac{8\pi}{9}, \dfrac{10\pi}{9}, \dfrac{14\pi}{9}, \dfrac{16\pi}{9}$

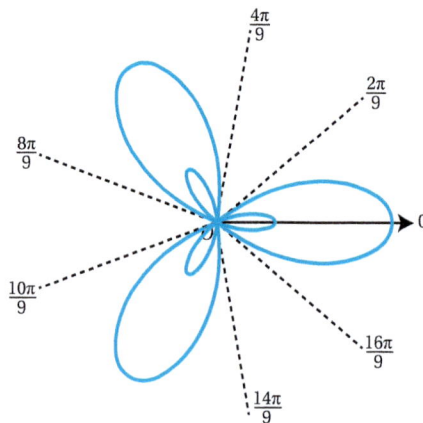

ii $\theta = \dfrac{\pi}{16}, \dfrac{3\pi}{16}, \dfrac{9\pi}{16}, \dfrac{11\pi}{16}, \dfrac{17\pi}{16}, \dfrac{19\pi}{16},$

$\dfrac{25\pi}{16}, \dfrac{27\pi}{16}$

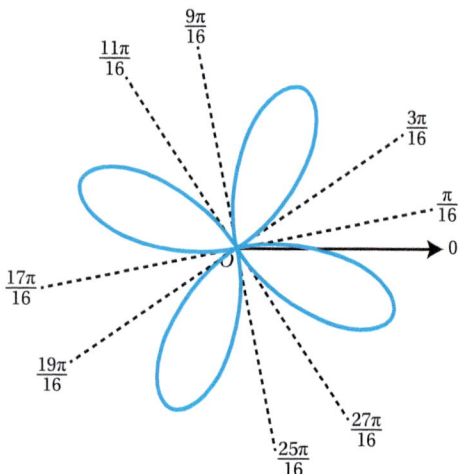

3 a $\theta = \dfrac{\pi}{8}, \dfrac{3\pi}{8}, \dfrac{5\pi}{8}, \dfrac{7\pi}{8}, \dfrac{9\pi}{8}, \dfrac{11\pi}{8}, \dfrac{13\pi}{8}, \dfrac{15\pi}{8}$

b $\dfrac{\pi}{8} < \theta < \dfrac{3\pi}{8}, \dfrac{5\pi}{8} < \theta < \dfrac{7\pi}{8}, \dfrac{9\pi}{8} < \theta < \dfrac{11\pi}{8}$ and

$\dfrac{13\pi}{8} < \theta < \dfrac{15\pi}{8}$

c

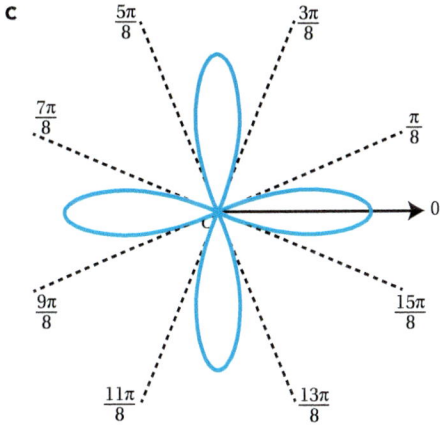

4 a

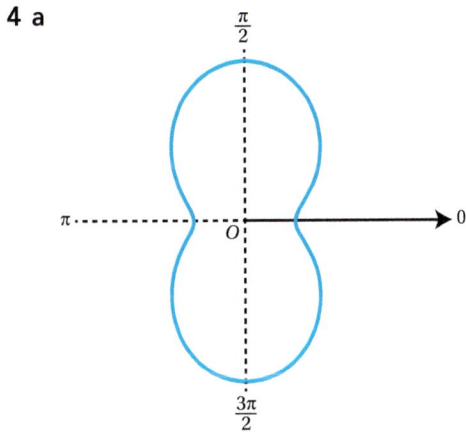

b $1 \leqslant r \leqslant 3$

5 a $\theta \in \left[0, \dfrac{7\pi}{12}\right] \cup \left[\dfrac{11\pi}{12}, \dfrac{19\pi}{12}\right] \cup \left[\dfrac{23\pi}{12}, 2\pi\right]$

b $\theta = \dfrac{7\pi}{12}, \dfrac{11\pi}{12}, \dfrac{19\pi}{12}, \dfrac{23\pi}{12}$; maximum r at $\left(3, \dfrac{\pi}{4}\right)$

and $\left(3, \dfrac{5\pi}{4}\right)$

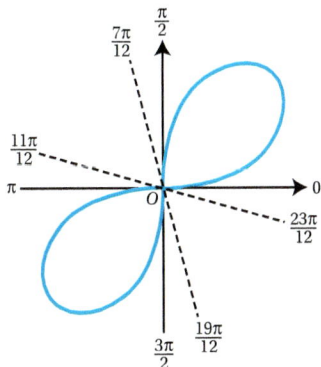

6 a Largest $r = 3$ when $\theta = \dfrac{\pi}{2}$; smallest $r = 1$

when $\theta = \dfrac{3\pi}{2}$.

b

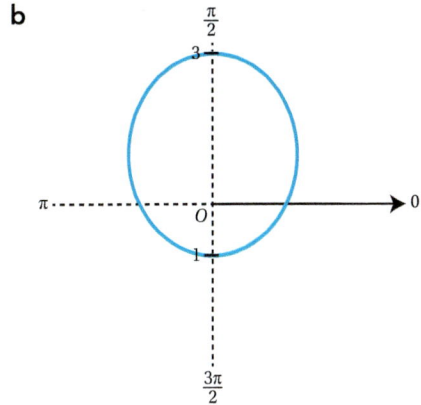

7 a

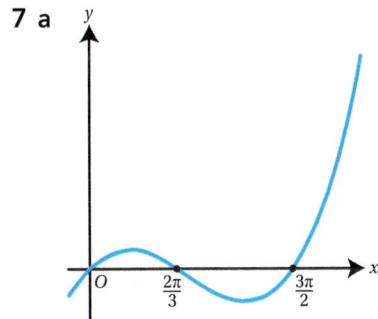

b

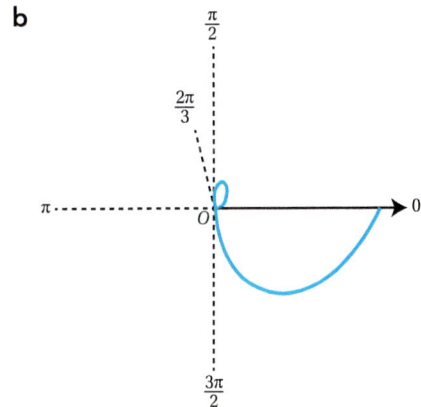

8 a Maximum y is $4\pi^2$; minimum y is π^2

b

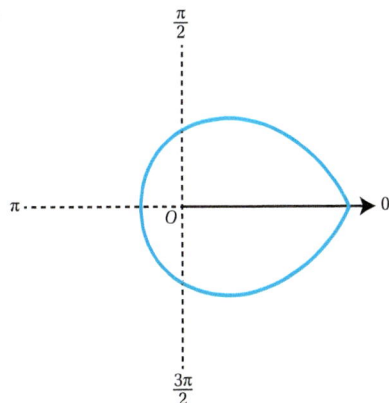

9

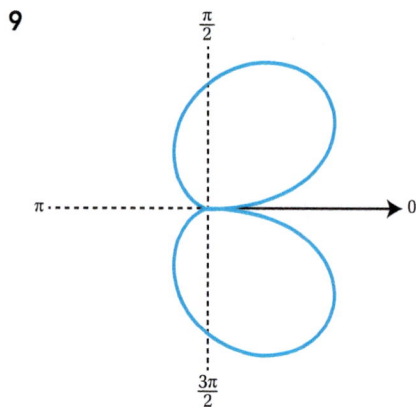

10

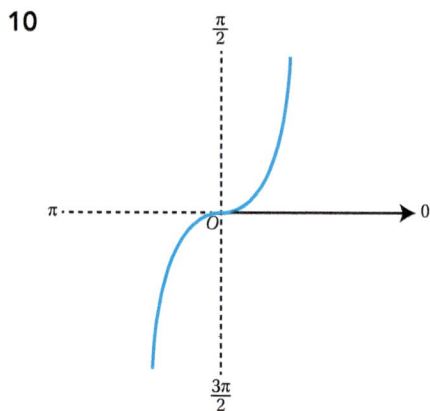

Exercise 9C

1 a i $\left(\dfrac{5}{\sqrt{2}},\dfrac{5}{\sqrt{2}}\right)$ **ii** $\left(\dfrac{3}{2},\dfrac{3\sqrt{3}}{2}\right)$

b i $(-1,1)$ **ii** $\left(-\dfrac{3}{2},\dfrac{\sqrt{3}}{2}\right)$

c i $(-3.45,-4.91)$ **ii** $(2.50,-1.65)$

2 a i $(\sqrt{29},0.381)$ **ii** $(5,0.927)$

b i $\left(2,\dfrac{\pi}{2}\right)$ **ii** $\left(3,\dfrac{3\pi}{2}\right)$

c i $(\sqrt{26},4.51)$ **ii** $(\sqrt{17},6.04)$

3 a i $\sin 2\theta=\dfrac{2}{3}$ **ii** $r=2\cos^2\theta\sin\theta$

b i $r=5(\sec\theta+\csc\theta)$

    ii $r=\left(\dfrac{3}{\sin^3\theta+\cos^3\theta}\right)^{\frac{1}{3}}$

c i $r=\dfrac{1}{\sin\theta-3\cos\theta}$ **ii** $r=\sqrt{6}$

---

4 a i $x^2+y^2=4\left(\tan^{-1}\dfrac{y}{x}\right)^2$

    ii $x^2+y^2=9\left(\tan^{-1}\dfrac{y}{x}\right)^4$

b i $x^2+y^2=4y$ **ii** $y^2=2x-x^2$

c i $x^4+x^2y^2-4y^2=0$ **ii** $x^2+y^2=\dfrac{y}{x}$

5 a 0.841 **b** $\left(\dfrac{4}{3},\dfrac{2\sqrt{5}}{3}\right)$

    c $x^2+y^2=3x$

6 $r=2(\cos\theta+\sin\theta)$

7 $x^4+x^2y^2-9y^2=0$

8 Proof

9 a $y=\sqrt{x\sqrt{x}-x^2}$ **b** $y^2=x^3-x^2$

Exercise 9D

1 a i $\dfrac{1}{2}$ **ii** $\dfrac{1}{3}$

b i $\dfrac{2\pi^3}{3}$ **ii** $\dfrac{\pi^5}{10}$

c i $e^{4\pi}-1$ **ii** $\dfrac{e^{2\pi}-1}{2}$

d i $\dfrac{\pi}{8}$ **ii** $\dfrac{2+\pi}{16}$

e i $\dfrac{3\pi}{4}$ **ii** $\dfrac{3\pi}{2}$

2 a $r=a$ **b** Proof

3 $\dfrac{4-\pi}{8}$

4 Proof

5 27π

6 a

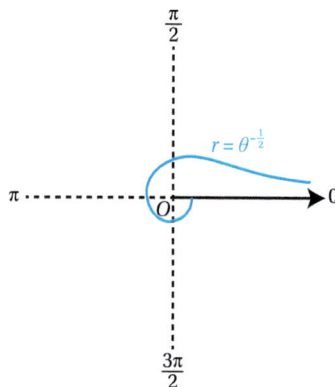

b Proof

7 a

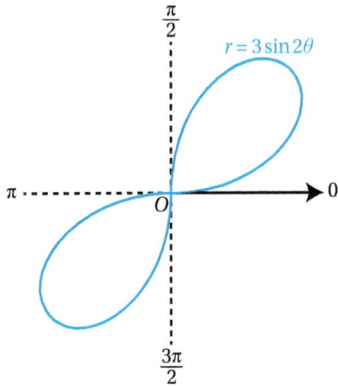

$r = 3\sin 2\theta$

b $\dfrac{9\pi}{4}$

8 a

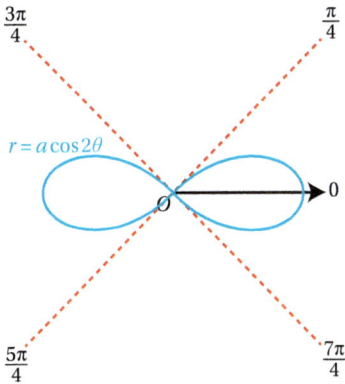

$r = a\cos 2\theta$

Tangents: $\theta = \dfrac{\pi}{4}, \dfrac{3\pi}{4}, \dfrac{5\pi}{4}, \dfrac{7\pi}{4}$

b $\dfrac{a^2\pi}{4}$

9 $\dfrac{\pi}{3}$

10 Proof

Exercise 9E

1 a $\left(2, \dfrac{\pi}{6}\right)$

b i $\dfrac{\sqrt{3}}{2}$ **ii** Proof

2 $\dfrac{3\pi a^2}{8}$

3 a $\left(a, \dfrac{\pi}{12}\right)$ and $\left(a, \dfrac{5\pi}{12}\right)$

b $\dfrac{a^2}{12}\left(4\pi - 3\sqrt{3}\right)$

4 a $\left(\dfrac{1}{2}, \dfrac{\pi}{6}\right), \left(\dfrac{1}{2}, \dfrac{5\pi}{6}\right)$ **b** $\dfrac{16\pi - 21\sqrt{3}}{24}$

5 a $\left(3, \dfrac{\pi}{3}\right), \left(3, \dfrac{5\pi}{3}\right)$ **b** Proof

6 a

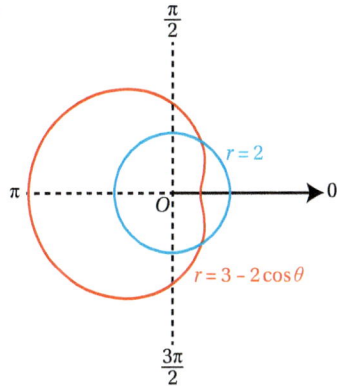

$r = 2$

$r = 3 - 2\cos\theta$

b Proof

Mixed practice 9

1 16

2 $r = a(\cos\theta - \sin\theta)$

3 a 8.92   **b** 13.5

4

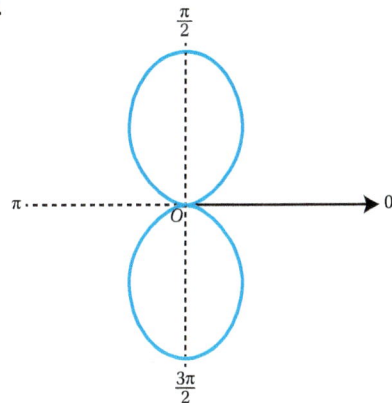

5 $a\left(e^{2\pi} - 1\right)$

6 a

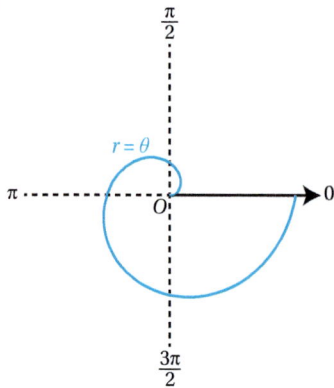

$r = \theta$

b Proof

7 a

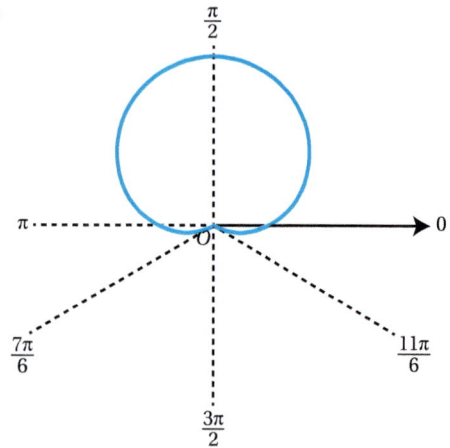

$r = 5 - 4\cos\theta$

b 33π

8 a Proof   **b** $\theta = \dfrac{2\pi}{3}$ or $\dfrac{4\pi}{3}$

c

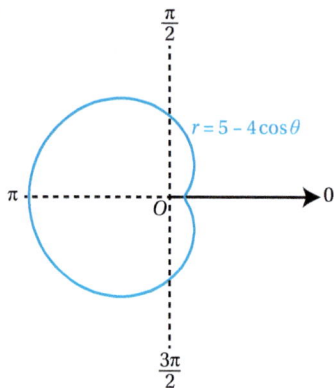

$r = 1 + 2\cos\theta$

d Proof

9 $\left(x^2 + y^2\right)^{\frac{3}{2}} = 3x^2$

10 a $\theta = \dfrac{7\pi}{6}$ and $\theta = \dfrac{11\pi}{6}$

b

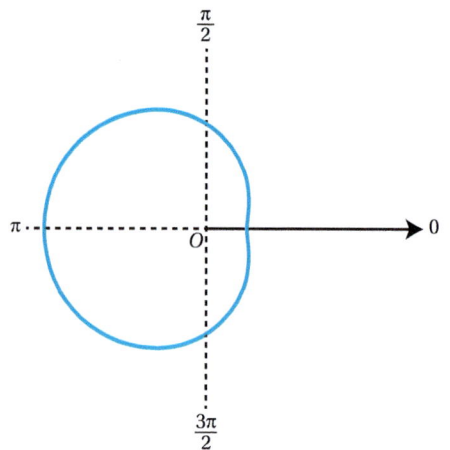

c $(x^2 + y^2 - 4y)^2 = 4(x^2 + y^2)$

11 a

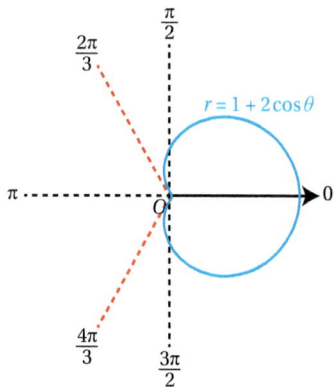

b $(-8, 0)$

12 a $\theta = \dfrac{\pi}{12}, \dfrac{5\pi}{12}, \dfrac{13\pi}{12}, \dfrac{17\pi}{12}$

b $\left(6, \dfrac{3\pi}{4}\right)$ and $\left(6, \dfrac{7\pi}{4}\right)$

c

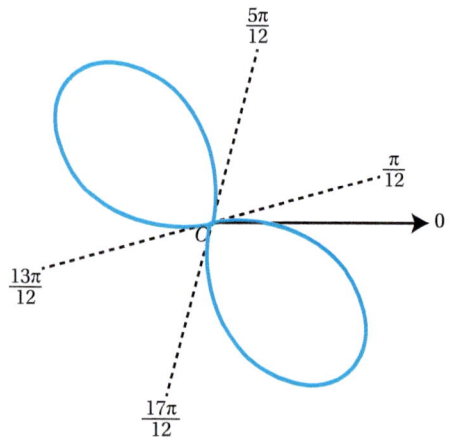

13 a

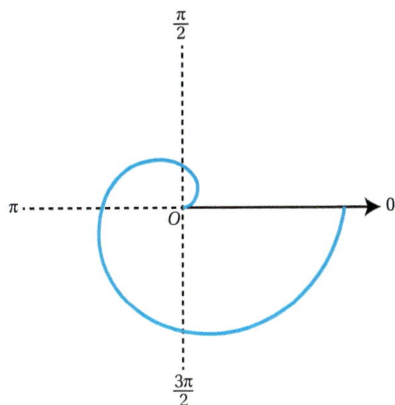

b $(-\pi, \pi\sqrt{3})$

14

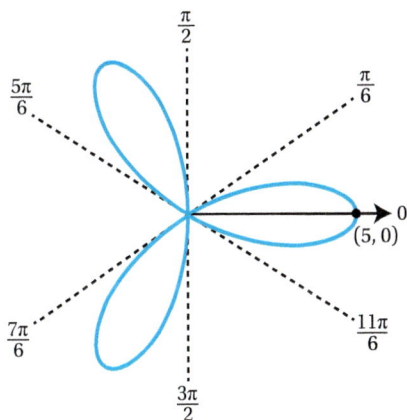

15 $8\pi + 6\sqrt{3}$

16 a Proof

b $\left(5, \dfrac{\pi}{3}\right)$ and $\left(5, \dfrac{5\pi}{3}\right)$

c $\dfrac{200\pi}{3} - \dfrac{175\sqrt{3}}{2}$

17 a i Proof

ii $\theta = \dfrac{\pi}{4}, \dfrac{\pi}{2}, \dfrac{3\pi}{4}, \dfrac{5\pi}{4}, \dfrac{3\pi}{2}, \dfrac{7\pi}{4}$

b Proof

18 i $r = 1 + \cos\theta$

ii

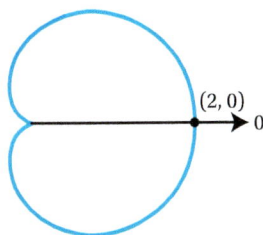

iii $1:3$

19 i, ii Proof

Chapter 10

Before you start...

1 $y = A e^{\frac{1}{2}x^2}$

2 $\dfrac{\mathrm{d}y}{\mathrm{d}x} = \mathrm{e}^{3x} + 3x\,\mathrm{e}^{3x}$

3 $\mathrm{e}^x(x-1) + c$

Exercise 10A

1 a i For example: $\dfrac{\mathrm{d}^2 y}{\mathrm{d}x^2} + 2\dfrac{\mathrm{d}y}{\mathrm{d}x} + y = 0$

   ii For example: $\dfrac{\mathrm{d}^2 y}{\mathrm{d}x^2} + 2\dfrac{\mathrm{d}y}{\mathrm{d}x} + \cos y = 0$

b i For example: $\dfrac{\mathrm{d}^2 y}{\mathrm{d}x^2} + 2\dfrac{\mathrm{d}y}{\mathrm{d}x} + y = 0$

   ii For example: $\dfrac{\mathrm{d}^3 y}{\mathrm{d}x^3} + 2\dfrac{\mathrm{d}y}{\mathrm{d}x} + y = 0$

c i For example: $\dfrac{\mathrm{d}^2 y}{\mathrm{d}x^2} + 2\dfrac{\mathrm{d}y}{\mathrm{d}x} + y = 0$

   ii For example: $\dfrac{\mathrm{d}^2 y}{\mathrm{d}x^2} + 2\dfrac{\mathrm{d}y}{\mathrm{d}x} + y = x$

2 a i Second order linear non-homogeneous

   ii Second order linear non-homogeneous

b i Second order non-linear homogeneous

   ii First order non-linear non-homogeneous

c i Second order non-linear non-homogeneous

   ii Third order linear homogeneous

d i Second order linear homogeneous

   ii First order non-linear non-homogeneous

3 a i $y = 2x^2 + 4$

   ii $y = 8\cos x - 2$

b i $y = \mathrm{e}^{-x} + 3x$

   ii $y = 6\ln x - 2x^2$

c i $y = 3x^2 - 2x$

   ii $y = 10\sin x + 5\cos x$

d i $y = 3\mathrm{e}^{2x} - \mathrm{e}^x$

   ii $y = 10\sin x - 2\cos x$

4 a $y = \dfrac{A}{\sqrt{x}}$ **b** $a = 1, b = -2$

c $y = \dfrac{5}{\sqrt{x}} + \ln x - 2$

5 a $y = \dfrac{A}{\cos x}$ **b** $a = 0, b = 2$

c $y = \dfrac{A}{\cos x} + 2\cos x$

6 a Proof; $A = \dfrac{1}{\sqrt{2}}, B = 1$ **b** $y = C\mathrm{e}^{-x}$

c The equation is not linear.

d $y = \sqrt{\frac{2}{3}e^x + ce^{-2x}}$

7 a $y = Ae^{e^{-x}}$ **b** 2

  c $y = Ae^{e^{-x}} + 2$

Exercise 10B

1 a i $y = \frac{1}{3}e^x + ce^{-2x}$ **ii** $y = -\frac{1}{3}e^x + ce^{4x}$

  b i $y = -\cot x + c\,\mathrm{cosec}\,x$

    ii $y = \frac{x+c}{\cos x}$

  c i $y = \frac{\ln|x|}{x} + \frac{c}{x}$ **ii** $-\frac{1}{x^2} + \frac{c}{x}$

2 $y = \frac{1}{2}e^x + \frac{1}{2}e^{2-x}$

3 $y = x^2 \ln|x - 3| + cx^2$

4 $y = e^{\cos x}(x + c)$

5 $y = -\frac{2}{x^2} + \frac{3}{x}$

6 $y = (x + 2)\cos x$

7 Proof

8 $y = \frac{x^2 + c}{2(x^2 - 1)}$

9 a $\frac{dz}{dx} - xz = -x$

  b $z = 1 + ce^{\frac{x^2}{2}}$ **c** 1

10 a $y = -\sqrt{\frac{x^3}{4} + \frac{36}{x}}$

  b $\sin y = \cos x(\ln \sec x + c)$

Exercise 10C

1 a $\lambda^2 + 5\lambda + 6 = 0$ **b** $y = Ae^{-3x} + Be^{-2x}$

2 a $\lambda^2 + 4 = 0$

  b $y = A\cos(2x) + B\sin(2x)$

3 a $\lambda^2 + 2\lambda + 1 = 0$ **b** $y = (Ax + B)e^{-x}$

4 a $y = Ae^{4x} + Be^{2x}$ **b** $y = e^{4x} + 4e^{2x}$

5 a $y = (A + Bx)e^{-2x}$ **b** $y = (1 + 2x)e^{-2x}$

6 a $x = e^t(A\cos t + B\sin t)$

  b $x = e^t(\cos t - \sin t)$

7 a $x = Ae^t + Be^{3t}$ **b** e^2

8 a $y = (A + Bt)e^{3t}$ **b** $y = pt\,e^{3(t-1)}$

9 $y = Ae^x + e^{2x}(B\sin x + C\cos x)$

10 $y = (A + Bx + Cx^2)e^{-x}$

11 $y = Ax^3 + \frac{B}{x^3}$

12 a $\frac{d^2 y}{dt^2} + \frac{dy}{dt} = 0$ **b** $y = A + Be^{-t}$

  c $x = \sqrt{5 - e^{-t}}$

Exercise 10D

1 a $Ae^{5x} + Be^{-x}$ **b** $-\frac{1}{9}e^{2x}$

  c $y = Ae^{5x} + Be^{-x} - \frac{1}{9}e^{2x}$

2 a $Ae^{-5x} + Be^{-4x}$ **b** $3x - \frac{27}{20}$

  c $y = Ae^{-5x} + Be^{-4x} + 3x - \frac{27}{20}$

3 a $Ae^{-x} + B$ **b** $\frac{1}{2}\sin x + \frac{1}{2}\cos x$

  c $y = Ae^{-x} + B + \frac{1}{2}\sin x + \frac{1}{2}\cos x$

4 a $A\sin 3x + B\cos 3x$

  b $y = A\sin 3x + B\cos 3x + 2e^{-x}$

  c $y = 4\sin 3x + 5\cos 3x + 2e^{-x}$

5 a $y = (A + Bx)e^{-2x} + 3x - 3 + 3\sin x - 4\cos x$

  b $y = (7 + 18x)e^{-2x} + 3x - 3 + 3\sin x - 4\cos x$

6 a $y = e^{2t}(A\sin 2t + B\cos 2t) + 4t^2 + 4t + 1$

  b $y = e^{2t}(-\sin 2t - \cos 2t) + 4t^2 + 4t + 1$

7 a $(A + Bx)e^{5x}$

  b Proof

  c $y = (A + Bx)e^{5x} + \frac{1}{2}x^2 e^{5x}$

  d $y = (4 - 18x)e^{5x} + \frac{1}{2}x^2 e^{5x}$

8 $f(x) = 3 - 2e^x \sin x$

9 $Ae^{-4x} + Be^x + 0.2xe^x$

10 a $A\sin 2x + B\cos 2x + 3x\sin 2x$

  b $3\sin 2x + 5\cos 2x + 3x\sin 2x$

Mixed practice 10

1 $A\cos\left(\frac{3}{2}x\right) + B\sin\left(\frac{3}{2}x\right)$

2 $\frac{1}{x^3}$

3 a $\lambda = -5, -1$ **b** $y = Ae^{-5x} + Be^{-x}$

4 a Proof **b** $y = e^{-x^2}\left(\dfrac{x^2}{2} + c\right)$

5 i $y = Ae^{-4x}$

ii $y = Ae^{-4x} + \dfrac{1}{5}(4\cos 3x + 3\sin 3x)$

iii For large x, the function oscillates approximately between -1 and $+1$.

6 a $Ae^{-2x} + Be^{-5x}$

b Proof; $\dfrac{1}{18}$

c $y = Ae^{-2x} + Be^{-5x} + \dfrac{e^x}{18}$

d $y = \dfrac{34}{18}e^{-2x} - \dfrac{35}{18}e^{-5x} + \dfrac{e^x}{18}$

7 a $\dfrac{1}{3}$

b $y = A\cos 2x + B\sin 2x + \dfrac{\cos x}{3}$

8 a $e^{-3x}(A\cos 4x + B\sin 4x)$

b Proof; $p = 2, q = -\dfrac{12}{25}$

c $y = e^{-3x}(A\cos 4x + B\sin 4x) + 2x - \dfrac{12}{25}$

d $y = e^{-3x}\sin 4x + 2x - \dfrac{12}{25}$

9 $y = \dfrac{1}{20}(4x^2 + 5x + 11x^{-3})$

10 $y = \dfrac{1}{2}e^{-x} - e^{-2x} + \dfrac{1}{2}e^{-3x}$

11 $y = \dfrac{1}{2}x^3(e^{2x} - e^2)$

12 i $y = Ae^{\frac{x}{2}} + Be^{-2x}$

ii $p = -1$

iii $y = 2e^{\frac{x}{2}} - (2 + x)e^{-2x}$

13 i $y = Ae^{\frac{x}{3}} + Be^{-2x} + x - 4$

ii $y = \dfrac{1}{2}e^{-2x} + x - 4$

iii $y = x - 4$

14 a $\dfrac{A}{x^3} + \dfrac{x}{4}$

b Proof

c $y = \dfrac{B}{x^2} + C + \dfrac{x^2}{8}$

15 a $y = A\cos x - 2\cos x \ln|\cos x|$

b $y = 5\cos x - 2\cos x \ln|\cos x|$

16 $A + Be^{2x} - 2.5x$

Chapter 11

Before you start...

1 $y = Ae^{-4x} + Be^{-x} + \dfrac{x}{4} - \dfrac{5}{16}$

2 $4.8\ \text{m s}^{-2}$

Exercise 11A

1 a $\dfrac{dv}{dt} = g - kv$

b $v = \dfrac{1}{k}\left(g - (g - ku)e^{-kt}\right)$

2 a Proof

b The model seems suitable initially, but is not accurate for later times.

3 a

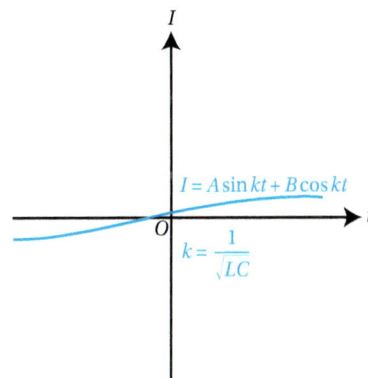

b

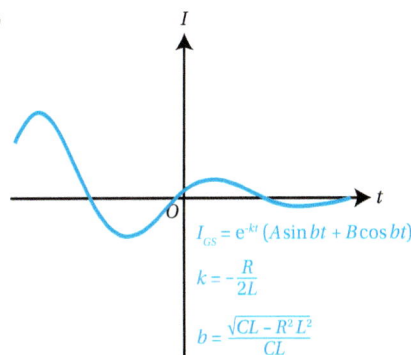

c

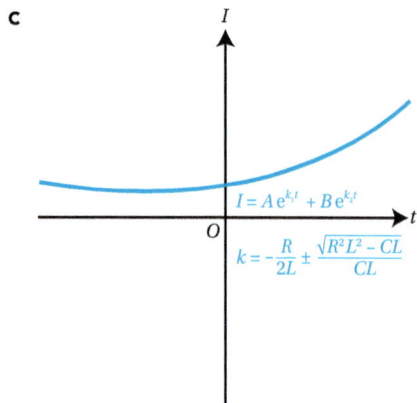

The graph shows $I = A e^{k_1 t} + B e^{k_2 t}$ with $k = -\dfrac{R}{2L} \pm \dfrac{\sqrt{R^2 L^2 - CL}}{CL}$

4 a $\dfrac{dY}{dt} = 200000 \times e^{-t \ln 4}$

b 12 144 270

c A natural net birth rate of the population

5 a $\dfrac{dT}{dt} = k(25 + 20t - T)$

b $T = 20t - 15 + 20 e^{-\frac{t}{2}}$

c 185 °C

d Different parts of the chicken are likely to have different temperatures.

6 a $\dfrac{dR}{dt} = kR(N - R)$ **b** $\dfrac{N}{2}$

c For example: interest in the rumour remains constant; the number of students who know the rumour is modelled as a continuous variable; there are no people outside the school spreading the rumour.

7 a This is proportional to the surface area of the bacterium. Larger surface areas make the bacterium more efficient in taking up nutrients so it will grow faster.

b $V = \left(2 - e^{-\frac{t}{3}}\right)^3$

c

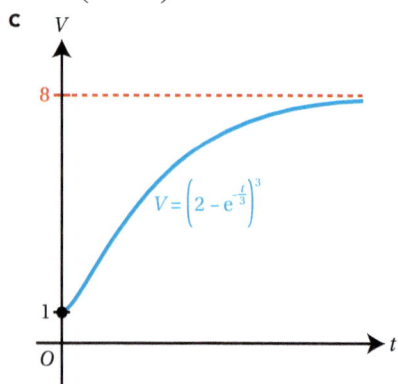

The graph shows $V = \left(2 - e^{-\frac{t}{3}}\right)^3$

Exercise 11B

1 a i $A = 4.5,\ T = \dfrac{2\pi}{3}$; at rest

    ii $A = 3,\ T = \dfrac{\pi}{2}$; equilibrium

b i $A = 2.6,\ T = 6\pi$; equilibrium

    ii $A = 5,\ T = 8\pi$; at rest

c i $A = 3.2,\ T = 3$; equilibrium

    ii $A = 10.4,\ T = \dfrac{10}{3}$; at rest

d i $A = 2.6,\ T = 6\pi$; neither

    ii $A = 5,\ T = 8\pi$; neither

e i $A = 3.49,\ T = 1.6$; neither

    ii $A = 8.54,\ T = 7$; neither

2 a i $x = 0.6 \cos 10t$

    ii $x = 3.4 \cos 14t$

b i $x = 0.7 \sin \dfrac{t}{3}$

    ii $x = 1.3 \sin \dfrac{t}{5}$

c i $x = 12.1 \sin \dfrac{4\pi t}{5}$

    ii $x = 0.3 \cos \dfrac{10\pi t}{3}$

3 a i $\dfrac{2\pi}{5}$ **ii** $\dfrac{2\pi}{3}$

b i $\pi\sqrt{2}$ **ii** $\dfrac{\pi}{\sqrt{2}}$

c i $\dfrac{2\pi}{\sqrt{3}}$ **ii** $\dfrac{2\pi}{3}$

d i 2π **ii** $\dfrac{2\pi}{\sqrt{5}}$

4 a 0.183 m   **b** $1\ \text{m s}^{-1}$

5 a -0.0768 m   **b** 0.131 s; $7.2\ \text{m s}^{-1}$

6 a $1.2\ \text{m s}^{-1}$ **b** $0.465\ \text{m s}^{-1}$; $16.6\ \text{m s}^{-2}$

7 a $0.667\ \text{s}^{-1}$ **b** $0.0596\ \text{m s}^{-1}$

8 a -0.595 m   **b** 0.0524 s

    c $5.20\ \text{m s}^{-1}$; away

9 a $0.873\ \text{s}^{-1}$ **b** 0.497 m

10 a 0.3 m; 15.7 s   **b** $x = 0.3 \cos 0.4t$

    c 4.78 s   **d** 0.0032 N

11 a Proof   **b** 15 m; 2880 N

    c 1.96 s

12 a Proof; $\dfrac{2\pi}{q}$ **b** $0.6q\ \text{m s}^{-1}$; $0.6q^2\ \text{m s}^{-2}$

    c 0.520 m

13 a $2.4x$ **b** Proof

    c $x = 0.04 \cos 2.19t$

14 a 0.1 m　　　　**b** $0.1 + x$; proof

　c Proof; $T = \dfrac{2\pi}{\sqrt{10g}}$ s

Exercise 11C

1 a i Underdamping　**ii** Overdamping
　b i Critical　　　　**ii** Underdamping
　c i Underdamping　**ii** Critical
　d i Critical　　　　**ii** Underdamping

2 $\dfrac{8}{3}$

3 a Proof　　　　**b** $\dfrac{6\sqrt{2}}{5}$

　c Overdamping

4 a $m\dfrac{d^2x}{dt^2} + 5c\dfrac{dx}{dt} + 4nx = 0$

　b $n = \dfrac{25c^2}{16m}$

5 a $x = 0.9e^{-2t}\sin 3t$　　**b** Underdamping

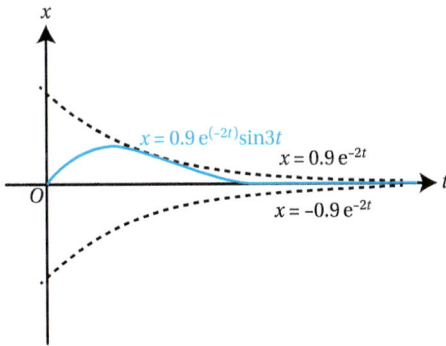

6 a Proof　　　**b** $x = 1.2e^{-t} - 0.2e^{-6t}$

　c Proof　　　**d** Overdamping

7 a Proof　　　**b** 0.45

　c $x = (0.8 + 0.72t)e^{-0.9t}$

8 a 0.4 m　　　**b** Proof

　c Underdamping;

　　$x = -0.4e^{-4t}\left(\dfrac{4}{3}\sin 3t + \cos 3t\right)$

　d 0.4 m　　　**e** 0.0714 m s^{-1}

9 a Underdamping　**b** $x = \dfrac{5u}{3c}e^{-\frac{4ct}{5}}\sin\dfrac{3ct}{5}$

　c Proof

Exercise 11D

1 a i $\dfrac{d^2x}{dt^2} - 7\dfrac{dx}{dt} + 12x = 0$;

　　$x = Ae^{3t} + Be^{4t}, y = Ae^{3t} + \dfrac{B}{2}e^{4t}$

ii $\dfrac{d^2x}{dt^2} - 2\dfrac{dx}{dt} - 3x = 0$;

　$x = Ae^{3t} + Be^{-t}, y = Ae^{3t} - Be^{-t}$

b i $\ddot{x} + 2\dot{x} + 5x = 0$;

　$x = e^{-t}(A\cos 2t + B\sin 2t)$,

　$y = 2e^{-t}((A+B)\cos 2t - (A-B)\sin 2t)$

ii $\ddot{x} - 2\dot{x} + 5x = 0$; $x = e^{t}(A\cos 2t + B\sin 2t)$,

　$y = 2e^{t}((A+B)\cos 2t + (B-A)\sin 2t)$

c i $\ddot{x} + 3\dot{x} + 2x = 3e^{-3t}$; $x = Ae^{-t} + Be^{-2t} + \dfrac{3}{2}e^{-3t}$,

　$y = Ae^{-t} + \dfrac{3B}{4}e^{-2t} + \dfrac{1}{2}e^{-3t}$

ii $\ddot{x} + \dot{x} - 2x = -26$;

　$x = Ae^{t} + Be^{-2t} + 13, y = 2Ae^{t} + \dfrac{B}{2}e^{-2t} + 17$

2 $x = Ae^{t} + Be^{-t} - \cos t, y = Ae^{t} - Be^{-t}$

3 $x = Ae^{4t} + Be^{-4t}, y = -2Ae^{4t} + 2Be^{-4t}$

4 $x = 12 + 5\cos 4t, y = 7 + 4\sin 4t$

5 a $\dfrac{d^2x}{dt^2} - \dfrac{dx}{dt} - 12x = 0$

　b $x = 3e^{4t} - 2e^{-3t}, y = 3e^{4t} + 12e^{-3t}$

6 a Proof; $\dfrac{dy}{dt} + 2y = 0.6e^{-2t}$

　b $y = 0.6te^{-2t}$; proof

　c 11.0 cm

7 a $\dfrac{dS}{dt} = 0.1F - 0.2S + 1$

　$\dfrac{dF}{dt} = 0.2F - 0.5S + 4$

　b $\dfrac{d^2S}{dt^2} + 0.01S = 0.2$

　c $S = 20 - 3\cos 0.1t + 4\sin 0.1t$

　　$F = 30 - 2\cos 0.1t + 11\sin 0.1t$

　d $S = 20 + 5\cos(0.1t - 2.21)$; shark peak at
　　$t = 22.1$, fish peak at $t = 17.5$

　e The populations oscillate with the same
　　period (62.8 time units), and with a
　　phase delay of 4.6 time units between fish
　　population peaking and shark population
　　peaking.

8 Proof

Mixed practice 11

1 π

2 64

3 a $\ddot{x} = 2\sin t$ **b** 5.72 m

4 a 1.8 m s^{-1} **b** 1.34 m s^{-1}

  c 3.24 N

5 a Proof   **b** Proof; $\omega = 3$

  c $A = 0.3, B = 0$ **d** 0.9 m s^{-1}

6 i $v = \dfrac{15}{4}t^2 - 5t + 0.8$ **ii** $\dfrac{4}{3}$ s

  iii 0.4 s, 1.6 s   **iv** 12.1 m

7 a Proof   **b** Proof; $R = 3, \varphi = 0$

  c $\dfrac{\pi}{5}$ seconds

8 $2.30 \approx 2$ months

9 $x = \cos 4t - \dfrac{1}{2}\sin 4t, y = \cos 4t + \dfrac{1}{2}\sin 4t$

10 $Q\,(x_P = 33.3 \text{ m}, x_Q = 74.2 \text{ m})$

11 a Proof   **b** 0.785 s^{-1}

  c 0.148 N

12 a Proof   **b** 3.2 m s^{-1}

13 a Proof

  b $A = 0, B = 0.3$; proof   **c** 0.75 m s^{-1}

14 a $x = 5\cos t$ **b** 1.57 s

  c $x = e^{-\frac{5}{13}t}\left(5\cos\left(\dfrac{12}{13}t\right) + \dfrac{25}{12}\sin\left(\dfrac{12}{13}t\right)\right)$

  d The second model (after 2.13 s)

15 $x = e^t\left(A\cos\left(t\sqrt{2}\right) + B\sin\left(t\sqrt{2}\right)\right),$

  $y = e^t\sqrt{2}\left(A\sin\left(t\sqrt{2}\right) - B\cos\left(t\sqrt{2}\right)\right)$

16 4000

17 i 2.4 m s^{-1}

  ii $x = -0.565$ m, $v = -0.804$ m s^{-1}

  iii $t = 0.0854$ s ($x = 0.565$ m, $v = -0.804$ m s^{-1}),
    $t = 0.871$ s ($x = -0.565$ m, $v = 0.804$ m s^{-1})

18 i $v = \dfrac{t^3 - 432t + 6912}{1800}$; proof

  ii v

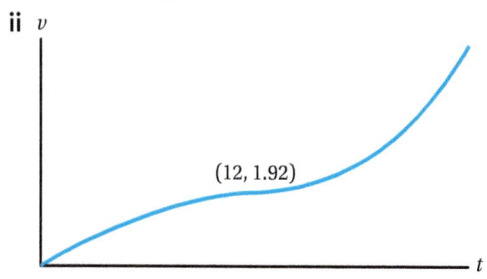

19 a Proof   **b** Proof; $\dfrac{\pi}{2}$

  c 0.566 m s^{-1}

20 a The force decreases at a constant rate. In reality the force will vary over each stride; he might be more motivated towards the end.

  b $2000\ddot{x} = 1000 - 100t$

  c $x = \dfrac{1}{4}t^2 - \dfrac{1}{120}t^3$; 33.3 m

  d Bill (Mike's distance is 14.7 m but you don't need to find this. A sketch of both forces makes it clear that Mike is never pulling harder.)

Focus on ... 2

Focus on ... Proof 2

1 Proof

2 $\sqrt{\pi}$

3 $\sqrt{2\pi\sigma}$

Focus on ... Problem solving 2

1 Proof

2 The tangent to the chain at $x = 0$ is horizontal; proof

3 $y = \dfrac{T_0 L}{gM}\cosh\left(\dfrac{gM}{T_0 L}x\right)$; moving the chain vertically does not change its shape.

4 Proof

5 Proof

6 Investigation (using graphing software); the distance between the end-points ($2D$) has to be smaller than the length of the chain (L).

Focus on ... Modelling 2

1 For example: all fish and sharks are treated as equivalent so that the effects of age or disease average out over the population; there is no randomness, which might be acceptable if the populations are large enough; there are no external populations (i.e. no other predators or sources of food); there is no seasonality so the birth rate stays constant over time.

2 $F = 0, S = 0$ or $F = \dfrac{k}{c}, S = \dfrac{a}{b}$; the second solution is the biologically relevant one.

3 The equilibrium value of the fish goes down when k decreases. The equilibrium value of the sharks goes up when a increases.

4

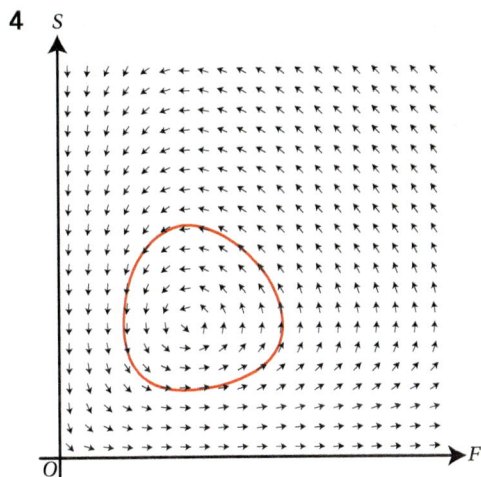

S vs F direction field with closed orbit curve.

Maximum fish population is about 1.75 million.

5

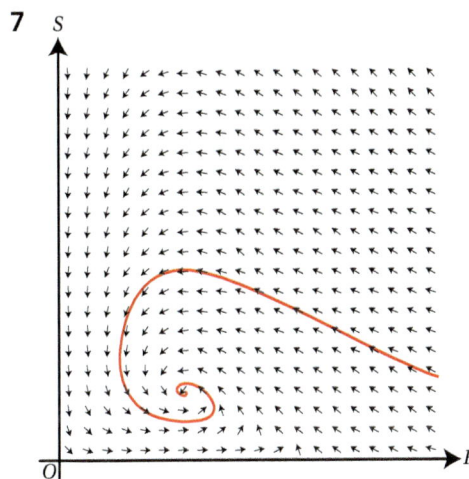

F vs t oscillating curve.

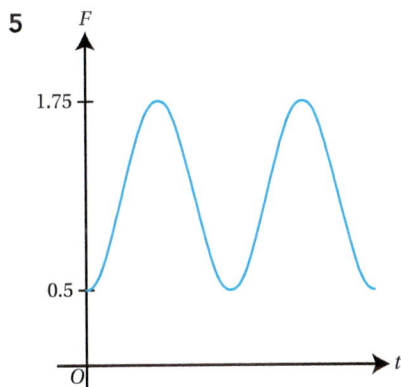

S vs t oscillating curve.

6 When F is small the aF term will be relatively more important. When F is large the eF^2 term will dominate, meaning that when the population is too large there is a net death, as would be expected with internal competition.

7

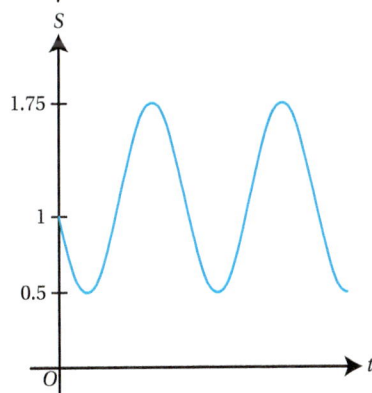

S vs F direction field with spiral curve.

The system tends to an equilibrium at $F = 1$, $S = 0.5$.

Cross-topic review exercise 2

1 a $0 \leqslant \theta \leqslant \dfrac{\pi}{4}, \dfrac{\pi}{2} \leqslant \theta \leqslant \dfrac{3\pi}{4}, \pi \leqslant \theta \leqslant \dfrac{5\pi}{4}, \dfrac{3\pi}{2} \leqslant \theta \leqslant \dfrac{7\pi}{4}$

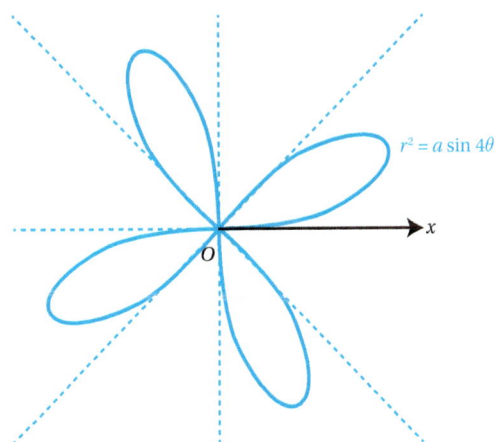

$r^2 = a\sin 4\theta$

b a

2 $\ln \sqrt{5}$

3 $\ln 3$ or $\ln(\sqrt{5} - 2)$

4 $r = 14\sin\theta - 8\cos\theta$

5 $\dfrac{37\pi}{2}$

6 $\dfrac{1}{2}\ln\dfrac{e^x - e^{-x}}{e^x + e^{-x}} + c = \dfrac{1}{2}\ln(\tanh x) + c$

7 $(2x-1)^2 + 4;\ \dfrac{\pi}{16}$

8 $\dfrac{52\pi}{9}$

9 $a=\dfrac{1}{3}, n=9$

10 a Proof   **b** $\ln\sqrt{3}$

11 Proof

12 a Proof   **b** $\dfrac{21}{29}$

13 i Proof   **ii** $\ln\left(\dfrac{\sqrt{5}\pm1}{2}\right)$

14 i, ii Proof

15 i Proof   **ii** $x+\dfrac{1}{6}x^3$

     iii $x^2-\dfrac{1}{2}x^3+\dfrac{1}{2}x^4$

16 i Proof   **ii** $\ln\sqrt{2}-\dfrac{1}{2}x+\dfrac{1}{4}x^2$

17 i

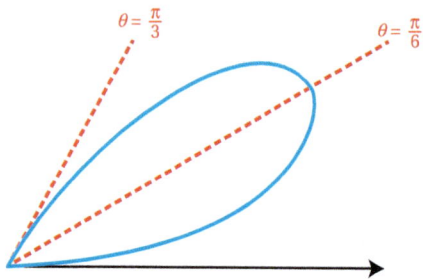

     ii $\dfrac{\pi}{3}$

     iii Proof

18 i $y=e^{-x}(A\cos4x+B\sin4x)+x+2$

     ii Proof; $y=x+2$

19 $y=\dfrac{e^{-x}}{4}(1-\cos2x)$

20 $y=2-2x\cot x+\dfrac{\pi\sqrt{3}}{6}\operatorname{cosec}x$

21 i Proof

     ii $v=\sqrt{5}\,e^{\sqrt{5}t}$

22 i $0.8\,\text{m}$

     ii $x=0.227$, velocity $3.84\,\text{ms}^{-1}$

     iii 3; $t=0.257$ and $x=0.227$, $t=0.372$ and $x=-0.227$, $t=0.885$ and $x=-0.227$

23 i $v=\dfrac{1}{0.8t+0.2}=\dfrac{5}{4t+1}$

     ii $0.671\,\text{ms}^{-1}$ (3 s.f.)

24 $k\geqslant\sqrt{5}$

25 $\dfrac{9}{4}$

26 a $x+\dfrac{x^3}{12}$ **b** $-2<x<2$

     c $\dfrac{13}{12}$

27 Proof

28 a Proof

     b $\operatorname{arsinh}x-\dfrac{\sqrt{1+x^2}}{x}+c$

29 $\ln\dfrac{3}{2}$

30 i Proof

     ii $u=\dfrac{1}{2}\left(5^{\frac{1}{3}}+\dfrac{1}{5^{\frac{1}{3}}}\right)$

31 i $\sqrt{x(x-1)}+\ln\left(\sqrt{x}+\sqrt{x^2-1}\right)+c$

     ii $2\sqrt{3}+\ln\left(2+\sqrt{3}\right)$

     iii The volume of revolution is infinite.

32 i

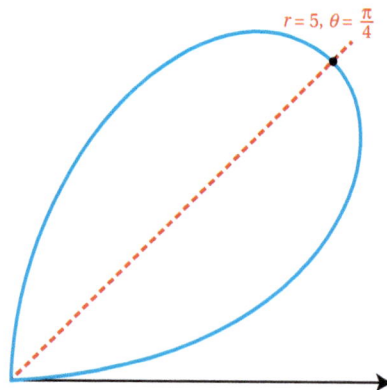

     ii $\dfrac{25\pi}{8}$

     iii $y=5\sqrt{2}-x$

     iv Proof

33 i $y=\left(A-\dfrac{1}{4}x\right)\cos2x+B\sin2x$

     ii y oscillates increasingly widely, with amplitude proportional to x.

     iii y oscillates with a stable amplitude throughout all values of x, and does not grow without limit.

34 i $x_1=3\cos t$, $x_2=4\cos1.5t$

     ii The period of P_1 is $2\pi>5.99$ so the particle has not completed a full cycle. $23.6\,\text{m}$

     iii $v_{P_1}=-0.867\,\text{ms}^{-1}$, $v_{P_2}=2.55\,\text{ms}^{-1}$. The particles are travelling in opposite directions.

35 i, ii Proof

   iii $\pm\ln\left(3+2\sqrt{2}\right)$

36 i Proof

   ii

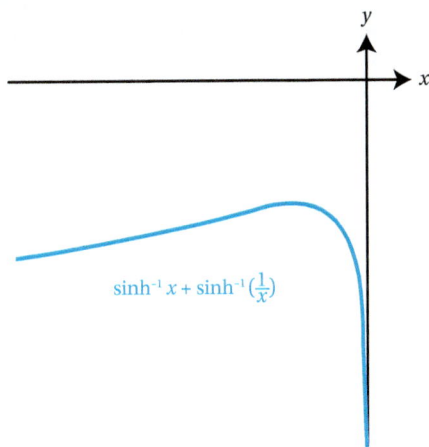

$$\left(-\infty,-2\ln\left(1+\sqrt{2}\right)\right]\cup\left[2\ln\left(1+\sqrt{2}\right),\infty\right)$$

37 a $x\arcsin x+\sqrt{1-x^2}+c$

   b Proof

   c i $p\arcsin\sqrt{p}-\dfrac{1}{2}\arcsin\sqrt{p}+\dfrac{1}{2}\sqrt{p-p^2}$

   ii $x\arcsin\sqrt{x}-\dfrac{1}{2}\arcsin\sqrt{x}+\dfrac{1}{2}\sqrt{x-x^2}+c$

38 i $9;3$

   ii, iii Proof

Practice paper 1

1 $k=-3$

2 Proof

3 6; proof

4 1

5 Proof; $a=7, b=16, c=3$

6 a $1,\operatorname{cis}\left(\dfrac{2\pi}{7}\right),\operatorname{cis}\left(\dfrac{4\pi}{7}\right),\operatorname{cis}\left(\dfrac{6\pi}{7}\right),$

   $\operatorname{cis}\left(\dfrac{8\pi}{7}\right),\operatorname{cis}\left(\dfrac{10\pi}{7}\right),\operatorname{cis}\left(\dfrac{12\pi}{7}\right)$

   b $n=7, w=-128i$

7 $x=Ae^{2t}+Be^{-2t}, y=Ae^{2t}-Be^{-2t}$

8 a $\dfrac{2}{x-1}-\dfrac{1}{x^2+1}$

   b $2\ln|x-1|-\arctan x+c$

9 a $x+5y+8z=20$ **b** $18.8°$

10 a $y^2=6x+9$ **b** $r=\dfrac{13}{3}$

11 a Proof

   b $\dfrac{1}{a^2-4a+5}\begin{pmatrix} a-3 & -1 & 6-a \\ 2 & a-1 & 1-3a \\ 1-a & 2-a & a^2-2 \end{pmatrix}$

   c $x=\dfrac{a-3}{a^2-4a+5}, y=\dfrac{2}{a^2-4a+5}, z=\dfrac{1-a}{a^2-4a+5}$

12 $y=\dfrac{2\arcsin x+1}{\sqrt{1-x^2}}$

13 a Proof   **b** 0

Practice paper 2

1 a $-\dfrac{1}{2}-\dfrac{1}{2}i$ **b** $\dfrac{\sqrt{2}}{2},\dfrac{5\pi}{4}$

2 a $\left(\dfrac{1-\sqrt{3}}{2},\dfrac{-\sqrt{3}-1}{2}\right)$

   b Rotation through $60°$ anticlockwise about the origin

3 $\left(x^2+y^2\right)^2=8x$

4 Proof

5 $\ln 9$

6 a Proof

   b For example, $x=\dfrac{3\pi}{2}$

7 a $\begin{pmatrix} 1 & 10 \\ 0 & 1 \end{pmatrix},\begin{pmatrix} 1 & 15 \\ 0 & 1 \end{pmatrix},\begin{pmatrix} 1 & 20 \\ 0 & 1 \end{pmatrix}.$

   b $\begin{pmatrix} 1 & 5n \\ 0 & 1 \end{pmatrix}$ **c** Proof

8 Proof; 0.745

9 Proof; $a=\dfrac{1}{4}, b=\dfrac{7}{5}$

10 a, b Proof

   c $1,\dfrac{1}{2}\pm i\dfrac{\sqrt{3}}{2}$

11 a Proof   **b** $k=2$

12 a Proof

   b i $x=5e^{-kt}-e^{-2kt}$ **ii** Overdamping

Glossary

amplitude (of an object moving with SHM): The maximum distance from the equilibrium position.

angular frequency: The constant ω in the simple harmonic motion equation $\frac{d^2x}{dt^2} = -\omega^2 x$.

auxiliary equation: The quadratic equation $a\lambda^2 + b\lambda + c = 0$ associated with the second order differential equation $a\frac{d^2y}{dx^2} + b\frac{dy}{dx} + c = 0$.

circular functions: A name sometimes given to trigonometric functions, related to points on a unit circle (with equation $x^2 + y^2 = 1$).

complementary function: The solution to the homogeneous differential equation associated with a non-homogenous differential equation.

consistent (system of equations): A set of simultaneous equations that have (a) solution(s).

converges: When a series approaches a finite value as more terms are added.

damped harmonic motion: Simple harmonic motion in which a resistive force proportional to the object's velocity is added, causing the amplitude to decay with time.

De Moivre's theorem: A theorem for finding powers of complex numbers: $z^n = r^n(\cos n\theta + i\sin n\theta)$ for any integer n.

dependent variable (in a differential equation): The variable on the top of the derivatives.

equilibrium position: The position where the acceleration of an object is zero.

exponential form (of a complex number): A way of expressing a complex number, z, in terms of its modulus, r, and argument, θ: $z = re^{i\theta}$

finite series: The sum of a finite number of terms in a sequence.

general solution (of a differential equation): The solution involving all the necessary arbitrary constants.

general term (in a Maclaurin series): The expression for the general term in a Maclaurin series is $\frac{f^{(r)}(0)}{r!}x^r$.

homogeneous differential equation: A differential equation in which every term involves the dependent variable.

hyperbola: A curve consisting of two branches. Coordinates of points on the curve are of the form $(\cosh\theta, \sinh\theta)$.

hyperbolic functions: The functions $\sinh x$, $\cosh x$ and $\tanh x$.

improper integral: An integral where either the range of integration is infinite or the integrand is undefined at a point within the range of integration.

inconsistent (system of equations): A set of simultaneous equations that have no solution.

independent variable (in a differential equation): The variable on the bottom of the derivatives.

infinite series: The sum of the terms of an infinite sequence (this may have a finite sum).

initial line: In polar coordinates, a fixed line from the pole from which the angle to a point is measured.

integrating factor: A function that is multiplied through a first order linear differential equation so that the side containing the dependent variable can be expressed as the derivative of a product.

inverse functions: A function denoted by f^{-1} that reverses the effect of function f. The inverse hyperbolic functions are $\operatorname{arshin} x$, $\operatorname{arcosh} x$ and $\operatorname{artanh} x$.

irreducible (quadratic): A quadratic that does not factorise.

linear differential equation: A differential equation in which the dependent variable only appears to the power 1 in any expression.

Maclaurin series: An expansion of a function as a series in powers of x.

non-homogeneous differential equation: A differential equation in which at least one term doesn't involve the dependent variable.

normal vector (of a plane): A vector that is perpendicular to the plane.

order (of a differential equation): The largest number of times the dependent variable is differentiated.

oscillating behaviour: To vary in magnitude or position in a regular manner about a central (equilibrium) point.

particular integral: Any solution of a differential equation.

particular solution: The general solution with the values of all constants found to fit a set of specific conditions.

period (of an object moving with SHM): The time after which the motion repeats itself.

polar coordinates: A way of describing the position of a point by means of its distance from the origin and the angle relative to a fixed line.

pole: In polar coordinates, an origin from which the distance of a point is measured.

roots of unity: The n solutions of the equation $z^n = 1$, where z is a complex number.

scalar product equation: A form of writing an equation of a plane, using the scalar (dot) product, denoted $\mathbf{r} \cdot \mathbf{n} = \mathbf{a} \cdot \mathbf{n}$, where $\mathbf{n}$ is the normal to the plane and $\mathbf{a}$ is the position vector of a point in the plane.

series: The sum of the terms of a sequence.

sigma notation: A shorter way of writing a series, using a large Greek sigma with limits.

simple harmonic motion: Oscillating motion where the acceleration of the object is proportional to the displacement from a central (equilibrium) position and is in the direction opposite to that of the displacement.

solid of revolution: A 3D shape formed by rotating a curve, usually around the x- or y-axis.

vector product: A way of multiplying two vectors $\mathbf{a}$ and $\mathbf{b}$, denoted as $\mathbf{a} \times \mathbf{b}$ where $\times$ is the cross product. The direction of the product is perpendicular to both $\mathbf{a}$ and $\mathbf{b}$.

volume of revolution: The volume of a solid of revolution.

Index

Acknowledgements

The authors and publishers acknowledge the following sources of copyright material and are grateful for the permissions granted. While every effort has been made, it has not always been possible to identify the sources of all the material used, or to trace all copyright holders. If any omissions are brought to our notice, we will be happy to include the appropriate acknowledgements on reprinting.

Thanks to the following for permission to reproduce images:

Cover image: huskyomega/Getty Images

Back cover: Fabian Oefner www.fabianoefner.com

Mondadori Portfolio/Contributor; Sakkmesterke/Getty Images; Laguna Design/Getty Images; Chris Clor/Getty Images; Phillippe Bourseiller/Getty Images; Universal Images Group/Contributor/Getty Images; Hani Alahmadi/EyeEm/Getty Images; Wangwukong/Getty Images; Alter Your Reality/Getty Images; Anna Bliokh/Getty Images; Nobi Prizue/Getty Images; Cuppuppycake/Getty Images

TGCh Hanfodol

CBAC
WJEC

ar gyfer CBAC

Safon U2 | Stephen Doyle

DREF WEN

Mae'r deunydd hwn wedi cael ei gymeradwyo gan CBAC ac mae'n cynnig cymorth o safon uchel ar gyfer cymwysterau CBAC. Er bod y deunydd hwn wedi mynd drwy broses sicrhau ansawdd CBAC, y cyhoeddwr sy'n parhau'n llwyr gyfrifol am y cynnwys.

Golygydd:	Geoff Tuttle
Datblygiad y project:	Rick Jackman (Jackman Publishing Solutions Ltd) a Claire Hart
Dylunio cysyniadol:	Patricia Briggs
Gosodiad tudalen:	GreenGate Publishing Services
Darluniau:	GreenGate Publishing Services
Dyluniad y clawr:	Jump To! www.jumpto.co.uk
Delwedd y clawr:	Drwy garedigrwydd Chris Harvey/Fotolia.com

Cyhoeddwyd gyntaf yn y Saesneg yn 2009 gan Folens Limited.

Cyhoeddwyd gyntaf yn y Gymraeg yn 2011 gan Wasg y Dref Wen.

Cyfieithwyd gan Cymen

Noddwyd gan Lywodraeth Cynulliad Cymru.

Cyhoeddwyd dan nawdd Cynllun Adnoddau Addysgu a Dysgu CBAC.

Data Catalogio mewn Cyhoeddiad y Llyfrgell Brydeinig.

Mae cofnod catalogio'r cyhoeddiad hwn ar gael gan y Llyfrgell Brydeinig.

ISBN 978-1-85596-907-0

Cynnwys

Uned IT3 Y Defnydd o TGCh a'i Heffaith

Cyflwyniad i'r Unedau U2

Mae Technoleg Gwybodaeth a Chyfathrebu U2 yn cynnwys dwy uned:

- Uned IT3 Y Defnydd o TGCh a'i Heffaith
- Uned IT4 Cronfeydd Data Perthynol

Ar gyfer safon U2, mae cydbwysedd y marciau ar gyfer y ddwy uned fel a ganlyn:

Uned IT3 Y Defnydd o TGCh a'i Heffaith   60%
Uned IT4 Cronfeydd Data Perthynol 40%

Uned IT3
Y Defnydd o TGCh a'i Heffaith

Asesiad ar gyfer Uned IT3 Y Defnydd o TGCh a'i Heffaith

Mae'r asesiad yn cynnwys papur cwestiynau ac ysgrifennir yr atebion mewn llyfr ateb sy'n cael ei farcio'n allanol gan CBAC. Mae'r papur ysgrifenedig yn 2 awr 30 munud o hyd.
Mae dwy adran yn y papur:

Adran A

- Dylech ateb yr holl gwestiynau yn yr adran hon.
- Mae'n cynnwys cwestiynau strwythuredig sy'n asesu hyd a lled eich gwybodaeth o'r fanyleb IT3.
- Gall ansawdd eich cyfathrebu ysgrifenedig (eglurder yr ateb, gramadeg, sillafu, ac atalnodi) gael ei asesu mewn rhai cwestiynau.

Adran B

- Mae gofyn i chi ateb un o ddau gwestiwn.
- Mae'n cynnwys cwestiynau strwythuredig sy'n asesu hyd a lled eich gwybodaeth o'r fanyleb IT3.
- Gall ansawdd eich cyfathrebu ysgrifenedig gael ei asesu mewn rhai cwestiynau.

Trefniadaeth Uned IT3 Y Defnydd o TGCh a'i Heffaith

Mae Uned IT3 wedi'i rhannu'n dopigau fel a ganlyn:

1. Rhwydweithiau
2. Y Rhyngrwyd
3. Rhyngwyneb cyfrifiadur-dyn
4. Gweithio gyda TGCh
5. Polisïau diogelwch
6. Systemau cronfa ddata
7. Rheoli newid
8. Systemau gwybodaeth rheoli
9. Cylchred oes datblygu system

1 Rhwydweithiau

Buoch yn ymdrin â hanfodion rhwydweithiau yn Nhopig 8 o'r Uned UG IT1 ac yn yr uned hon byddwch yn edrych yn fanylach ar sut i ddewis rhwydwaith a'r gwahanol fathau o rwydwaith sydd ar gael. Yn y topig hwn byddwch yn dysgu am y gwahaniaethau rhwng rhwydweithiau cymar wrth gymar a rhwydweithiau cleient/gweinydd a'u manteision ac anfanteision cymharol. Rhoddwn sylw hefyd i dopolegau rhwydwaith, hynny yw, y gwahanol ffyrdd y gall rhwydwaith gael ei gysylltu gan ddefnyddio gwifrau neu'n ddiwifr.

2 Y Rhyngrwyd

Y Rhyngrwyd yw'r rhwydwaith mwyaf. Rhwydwaith o rwydweithiau ydyw. Yn y topig hwn byddwch yn dysgu am agweddau ar y Rhyngrwyd megis protocol trosi ffeiliau, e-fasnach, cronfeydd data ar-lein a chyfrifiadura gwasgaredig. Byddwch yn dod i ddeall y gwahaniaeth rhwng deialu a band llydan, a rhwng cysylltiad cebl â'r Rhyngrwyd a mynediad symudol i'r Rhyngrwyd.
Byddwch hefyd yn dysgu am y materion moesol, cymdeithasol a moesegol sy'n gysylltiedig â'r Rhyngrwyd.

3 Rhyngwyneb cyfrifiadur-dyn

Buoch yn ystyried hanfodion rhyngwynebau cyfrifiadur-dyn yn Nhopig 9 o'r Uned UG IT1. Yn yr uned hon fe roddwn sylw manylach i'r ffactorau y mae angen eu cymryd i ystyriaeth wrth ddylunio rhyngwyneb defnyddiwr da.

4 Gweithio gyda TGCh

Mae'r topig hwn yn ymdrin ag agweddau ar weithio gyda TGCh megis telegymudo (teleweithio a fideo-gynadledda), codau ymddygiad, cynnwys cod ymddygiad a'r gwahaniaeth rhwng materion cyfreithiol a moesol mewn perthynas â chodau ymddygiad.

5 Polisïau diogelwch

Yn y topig hwn byddwch yn dysgu am y bygythiadau posibl i ddata, canlyniadau camddefnyddio data, a'r angen i wneud copïau wrth gefn.
Byddwch yn dysgu sut y gellir mabwysiadu gweithdrefnau a fydd yn helpu i osgoi camddefnydd bwriadol a damweiniol. Byddwch yn dysgu hefyd am ddadansoddi risg a sut y gall corff leihau risgiau i'r eithaf.

6 Systemau cronfa ddata

Mae Topig 11 yn Uned UG IT1 yn rhoi cyflwyniad byr i gronfeydd data, ac yn enwedig i'r gwahaniaethau rhwng cronfeydd data perthynol a ffeiliau fflat.
Yn y topig hwn byddwch yn edrych yn fanwl ar sut mae cronfeydd data perthynol yn cael eu creu a sut mae strwythur ar gyfer y data'n cael ei baratoi. Rhoddir sylw hefyd i'r defnydd o gronfeydd data gwasgaredig a'u manteision ac anfanteision cymharol.

7 Rheoli newid

Yn y topig hwn byddwch yn ystyried sut y gall newid gael ei reoli pan gaiff system TGCh newydd ei chyflwyno i gorff. Byddwch yn edrych ar yr hyfforddiant sydd ei angen i ddysgu sgiliau newydd a'r newidiadau posibl y bydd angen eu gwneud i'r strwythur trefniadaethol, patrymau gwaith, gweithdrefnau mewnol a'r gweithlu.

8 Systemau gwybodaeth rheoli

Yn y topig hwn byddwch yn dysgu bod systemau gwybodaeth rheoli yn gasgliadau o bobl, gweithdrefnau ac adnoddau sy'n helpu rheolwyr i wneud penderfyniadau.

Byddwch yn dysgu am nodweddion systemau gwybodaeth rheoli ac am y defnydd sy'n cael ei wneud ohonynt. Byddwch yn dysgu hefyd am y ffactorau a all arwain at system gwybodaeth rheoli wael.

9 Cylchred oes datblygu system

Mae'r topig hwn yn ymdrin â phrif gydrannau'r gylchred oes datblygu system a sut y gellir eu cymhwyso wrth ddatblygu system TGCh.

Byddwch yn dysgu am yr holl gamau yn y gylchred oes datblygu system, megis ymchwilio i'r system, dadansoddi'r system, dylunio'r system, gweithredu'r system, cynnal y system a gwerthuso'r system.

Uned IT4 Cronfeydd Data Perthynol

Asesiad ar gyfer Uned IT4 Cronfeydd Data Perthynol

Mae'r uned hon yn gofyn i chi gynhyrchu project wedi'i seilio ar gronfa ddata berthynol. Er nad oes rhaid i'r project a ddewiswch fod â chyd-destun masnachol gwirioneddol, mae angen iddo fod yn realistig.

Rhaid i chi ystyried yn ofalus beth sydd ei angen ar gyfer y project cyn dechrau, er mwyn i chi allu cynhyrchu project digon heriol a fydd yn ymdrin â'r holl feini prawf asesu.

Cydran 1: Gofynion y defnyddiwr

Ar gyfer y gydran hon rhaid i chi roi'r broblem y byddwch chi'n ei datrys yn ei chyd-destun drwy roi disgrifiad cyffredinol o'r corff yr ydych wedi'i ddewis.

Cydran 2: Manyleb ddylunio

Cydran hanfodol yw hon sy'n cyfrif am gyfran fawr o'r marciau ar gyfer y project cyfan. Heb ddylunio gofalus ac ystyriol, ni fydd y datrysiad yn un effeithlon ac mae'n bosibl na fydd yn gweithio'n iawn. Mae angen i chi weithio drwy'r holl gamau dylunio'n ofalus a sicrhau eich bod chi wedi cynhyrchu tystiolaeth ar gyfer pob cam.

Cydran 3: Gweithredu

Ar gyfer y gydran hon rydych chi'n cymryd eich dyluniad ac yn defnyddio'r offer meddalwedd i gynhyrchu datrysiad sy'n gweithio. Os oes angen i chi newid y dyluniad mewn rhyw ffordd er mwyn ei weithredu, dylid cofnodi hyn.

Cydran 4: Profi

Ar gyfer y gydran hon mae angen i chi ddatblygu cynllun profi sy'n cymharu'r canlyniadau gwirioneddol â'r canlyniadau disgwyliedig, ac mae unrhyw waith addasu y mae angen ei gwblhau yn cael ei gofnodi. Rhaid profi pob agwedd ar y system a rhaid i chi ddarparu tystiolaeth bod hyn wedi'i wneud.

Cydran 5: Dogfennaeth defnyddiwr

Mae dogfennaeth defnyddiwr yn cael ei chynhyrchu i helpu'r defnyddiwr i ddefnyddio'r system yn gywir. Bydd dogfennaeth dda yn galluogi'r defnyddiwr i ddefnyddio'r datrysiad gyda chyn lleied o hyfforddiant â phosibl. Dylai'r ddogfennaeth defnyddiwr gwmpasu pob agwedd ar y system, a dylech ddarparu sgrinluniau i'ch helpu i esbonio sut mae'r gronfa ddata'n gweithio.

Cydran 6: Gwerthuso

Ar gyfer y gydran hon mae angen i chi werthuso'ch datrysiad yn erbyn gofynion y defnyddiwr i weld pa mor dda mae'n cwrdd â nhw. Hefyd mae angen i chi roi sylwadau ar y problemau y daethoch ar eu traws wrth weithredu'ch datrysiad a disgrifio'r strategaethau a ddefnyddiwyd gennych i'w goresgyn.

Cydran 7: Cynllunio'r project

Mae'r gydran hon yn berthnasol drwy gydol y project ac yma mae angen i chi ddangos eich bod chi wedi rheoli'ch gwaith yn effeithiol ac wedi defnyddio cynllun amser yn llwyddiannus. Rhaid i chi ddarparu tystiolaeth eich bod chi wedi cynllunio a rheoli'ch project yn llwyddiannus.

Cyflwyniad i nodweddion llyfr y myfyriwr

Yr athroniaeth sy'n sail i lyfr y myfyriwr

Mae'r llyfr hwn wedi'i seilio ar ymchwil helaeth o ysgolion a cholegau i'r gwahanol ffyrdd y caiff TGCh ei ddysgu, a manteisiwyd ar yr holl wybodaeth hon wrth ddatblygu'r gyfrol. Gan mai manyleb newydd yw hon, bydd llawer o athrawon/darlithwyr wrthi'n ymgyfarwyddo â hi, a nod y llyfr yw ymdrin yn fanwl â'r holl ddeunydd ar gyfer Unedau IT3 ac IT4. Mae'r llyfr hwn yn cwmpasu'r holl ddeunydd ar gyfer safon U2 TGCh.

Dylai athrawon/darlithwyr ddefnyddio'r llyfr hwn ar y cyd â'r deunyddiau atodol i athrawon. Wrth gwrs, gellir defnyddio'r llyfr hwn ar ei ben ei hun, ond os ydych yn athro mae llawer o adnoddau ar gael yn y deunyddiau atodol i athrawon i helpu'ch myfyrwyr i lwyddo ac ennill y marciau mwyaf posibl. Mae'r *CD-ROM* i athrawon yn cynnwys nifer o adnoddau nad ydynt yn rhai digidol: Atebion i'r Cwestiynau, Gweithgareddau ac Astudiaethau Achos, ac mae hefyd yn darparu Cwestiynau ac Astudiaethau Achos ychwanegol.

Mae'r *CD-ROM* hefyd yn cynnwys cyfoeth o ddeunyddiau digidol megis cyflwyniadau PowerPoint, cwestiynau dewis lluosog, tasgau gair coll a thasgau testun rhydd. Bydd pob un o'r rhain yn helpu'ch myfyrwyr i gyfnerthu eu dealltwriaeth o'r topigau.

Strwythur llyfr y myfyriwr

Mae U2 CBAC yn cynnwys dwy uned ac mae pob uned wedi'i rhannu'n dopigau. Yn y llyfr hwn mae pob topig wedi'i rannu ymhellach yn daeniadau (*spreads*). Mae hyn yn ei gwneud hi'n bosibl i rannu pob topig yn ddarnau o ddeunydd sy'n hawdd eu cymryd i mewn. Er mwyn cysondeb, ac i wneud llyfr y myfyriwr yn hawdd ei ddefnyddio, mae'r holl dopigau wedi'u strwythuro yn yr un ffordd.

Taeniadau topig

Tudalennau cyflwyno topig

Mae tudalen gyntaf pob topig yn cyflwyno'r deunydd yn y topig ac yn cynnwys y nodweddion canlynol:

Cyflwyniad i'r topig: ychydig o baragraffau sy'n cyflwyno'r myfyrwyr i gynnwys y topig.

Cysyniadau allweddol: mae hyn yn rhestru'r cysyniadau allweddol sy'n cael sylw yn y topig. Mae'r cysyniadau allweddol yr un fath â'r rheiny ym manyleb U2 CBAC.

Cynnwys: mae'r cynnwys yn rhestru'r taeniadau a ddefnyddir i ymdrin â'r topig ac mae pob taeniad yn ymdrin â chysyniadau allweddol.

Cartwnau: mae cartwnau perthnasol gan y cartwnydd Randy Glasbergen yn ychwanegu tipyn o hiwmor a hwyl at y topigau.

Cyflwyniad: mae'n cyflwyno'r cynnwys ar y taeniadau.

Byddwch yn dysgu: mae hyn yn dweud wrthych beth y byddwch chi'n ei ddysgu o gynnwys y taeniadau.

Y cynnwys: mae'r cynnwys yn nodi'r hyn y bydd angen i chi ei ddysgu, ac mae'r deunydd hwn wedi cael ei ysgrifennu i roi'r wybodaeth hanfodol sydd ei hangen i ateb cwestiynau arholiad.

Geiriau allweddol: termau arbenigol yw'r rhain a ddefnyddir yn y cynnwys ac mae'n bwysig i chi gofio'r geiriau hyn a gallu eu defnyddio'n hyderus wrth ddisgrifio agweddau ar systemau TGCh. Hefyd mae geirfa yng nghefn y llyfr sy'n rhoi ystyr llawer o'r termau.

Diagramau a ffotograffau: mae'r topig yn cael ei fywiocáu gan ddelweddau perthnasol sy'n gynnyrch ymchwil gofalus.

Yn yr arholiad: cynghorion defnyddiol wedi'u seilio ar y problemau sy'n wynebu myfyrwyr wrth ateb cwestiynau ar y topigau.

Taeniadau Cwestiynau, Gweithgareddau ac Astudiaethau achos

Cynhwysir y rhain fel rheol ar ddiwedd y taeniadau cynnwys a chânt eu defnyddio i gyfnerthu'r dysgu. Weithiau caiff Gweithgareddau neu Gwestiynau eu cynnwys o fewn y taeniadau cynnwys. Mae pob bloc o gwestiynau'n ymdrin â nifer penodol o dudalennau a byddant ar ffurf taeniad dwbl fel rheol. Mae hyn yn caniatáu i chi edrych ar y taeniadau ac yna ymarfer y cwestiynau. Mae'r atebion i'r holl gwestiynau ar gael yn y deunyddiau atodol i athrawon. Mae'r rhain i'w cael ar wahân ar *CD-ROM* ac ategant destun y myfyriwr.

Cwestiynau 1   tt. 2–3

1 Mae sefydliad yn ystyried rhwydweithio ei gyfrifiaduron arunig i ffurfio rhwydwaith.
  (a) Eglurwch **dri** ffactor a fydd yn dylanwadu ar ei ddewis o rwydwaith. (6 marc)
  (b) Ysgrifennwch restr o **bum** cwestiwn y byddai angen i chi eu gofyn cyn gallu penderfynu ar rwydwaith addas i'r sefydliad. Ar gyfer pob cwestiwn, dylech ddisgrifio'n fyr pam y mae angen yr atebion. (5 marc)
2 Wrth ddewis rhwydwaith i gwmni, rhaid cymryd i ystyriaeth y perfformiad sydd ei angen o'r rhwydwaith. Nodwch ac eglurwch **dri** mater perfformiad hollol

wahanol y bydd angen rhoi ystyriaeth iddynt. (6 marc)
3 Mae ysgol yn uwchraddio ei rhwydwaith. Mae'r prifathro'n poeni y gallai'r problemau diogelwch a gododd gyda'i hen rwydwaith godi gyda'r un newydd hefyd.
  Amlinellwch **dri** mater diogelwch y dylai'r rhwydwaith newydd roi sylw iddynt. (3 marc)
4 Mae sefydliad yn prynu rhwydwaith newydd. Disgrifiwch **dair** eitem wahanol o galedwedd y bydd angen eu prynu ac, ar gyfer pob un, eglurwch pam y mae eu hangen. (3 marc)

Gweithgaredd: Topolegau rhwydwaith

Mae ymgynghorydd rhwydwaith yn ymweld â busnes bach ac yn darganfod bod ganddo ddeg cyfrifiadur arunig. Ar ôl mynd drwy broses darganfod ffeithiau ac ymchwilio i'r llifoedd gwybodaeth yn y busnes, mae'r ymgynghorydd wedi dweud wrth y busnes bod angen rhwydwaith ardal leol arno. Mae wedi dweud hefyd fod yna nifer o dopolegau rhwydwaith i ddewis o'u plith.
1 Eglurwch beth yw topoleg rhwydwaith.
2 Mae'r ymgynghorydd wedi dweud bod y topolegau rhwydwaith canlynol ar gael:
  Seren
  Bws
  Cylch
  Gan ddefnyddio'r 10 cyfrifiadur yn y swyddfa, lluniwch ddiagram ar gyfer pob topoleg sydd yn y rhestr uchod i ddangos sut y dylid cysylltu'r cyfrifiaduron.
3 Mae'r ymgynghorydd yn dweud, 'Mae'n bwysig cael y lled band yn gywir'. Beth yw ystyr y gosodiad hwn?

Astudiaeth achos 4   tt. 76–77

Ymosodiad gan derfysgwyr

Ar ddydd Sadwrn yn ystod mis Mehefin 1996 derbyniodd yr heddlu rybudd wedi'i godio fod bom wedi cael ei blannu yng Nghanolfan Arndale, prif ganolfan siopa Manceinion. Ychydig dros awr ar ôl i'r ganolfan a'r ardal o'i chwmpas gael ei chlirio fe ffrwydrodd y bom, gan anafu dros ddau gant o bobl a dinistrio rhan helaeth o'r ganolfan. Achosodd y bom, y mwyaf o'i fath ym Mhrydain ar adeg o heddwch, ddifrod helaeth i swyddfeydd cwmni yswiriant Royal and Sun Alliance lle y cafodd rhai o'r staff eu niweidio.

Gan fod prif gyfrifiadur y cwmni yn yr adeilad hwn, yr oedd ofn y byddai gweithgareddau pob dydd y cwmni yn dioddef yn enbyd. Roedd gan y cwmni swyddfa yn Lerpwl, ac roedd terfynellau yma a oedd wedi'u cysylltu â'r prif gyfrifiadur. Darganfu'r staff fod peth bywyd o hyd yn y system er bod difrod helaeth i'r adeilad lle'r oedd y prif gyfrifiadur. Roedd peth gobaith y gellid

adfer y system, ond oherwydd bod perygl o ffrwydradau gan fod y bibell nwy wedi'i thorri, fe dorrodd y dynion tân y cyflenwad trydan i ffwrdd. Roedd y rhan fwyaf o'r caledwedd wedi cael ei ddinistrio i bob pwrpas yn y ffrwydrad, ond, fel pob cwmni doeth, roedd gan hwn gynllun wrth gefn. Roedd ganddo gytundeb gyda chwmni adfer data arbenigol a oedd yn berchen ar galedwedd cyffelyb a chopïau o'r un math o feddalwedd ag a oedd yn cael ei ddefnyddio gan Royal and Sun Alliance. Oherwydd bod angen staff arnynt a oedd yn deall y busnes yswiriant i weithredu'r cyfrifiadur, cafodd staff y Royal eu cludo i swyddfeydd y cwmni adfer ac aethant ati i adfer y data o'r cyfryngau storio wrth gefn a oedd yn cael eu cadw oddi ar y safle. Erbyn bore Llun roedd yr holl ddata wedi cael eu hadfer ac roedd switsfwrdd dros dro wedi'i sefydlu, ac nid oedd yr un diwrnod o fasnachu wedi cael ei golli.

1 Mae gan gwmni yswiriant Royal and Sun Alliance bolisi diogelwch TGCh. Fel rhan o'r polisi byddent wedi gwneud dadansoddiad risg ac wedi sefydlu rhaglen adfer o drychineb.
  (a) Eglurwch beth sy'n cael ei wneud mewn dadansoddiad risg. (2 farc)
  (b) Disgrifiwch y technegau a ddefnyddir mewn rhaglen adfer o drychineb i adfer y data a ddefnyddir gan y systemau TGCh. (6 marc)
2 Mae gweithdrefnau gwneud copïau wrth gefn a gweithdrefnau adfer yn hanfodol i sicrhau diogelwch systemau TGCh.
  (a) Eglurwch y gwahaniaeth rhwng gweithdrefnau gwneud copïau wrth gefn a gweithdrefnau adfer. (2 farc)
  (b) Rhowch **dri** pheth y mae'n rhaid eu hystyried wrth ddewis gweithdrefnau gwneud copïau wrth gefn. (3 marc)
  (c) Eglurwch pam mae gweithdrefnau adfer yn hanfodol os yw corff fel y cwmni yswiriant hwn yn dymuno sicrhau diogelwch ei ddata. (2 farc)

Cwestiynau: cynhwysir y rhain ar ddiwedd pob topig a chyfeiriant at y cynnwys yn y taeniadau. Maen nhw wedi'u labelu'n glir fel y gallwch eu gwneud ar ôl pob taeniad dwbl neu i gyd ar unwaith ar ddiwedd y topig. Mae'r cwestiynau'n debyg i gwestiynau arholiad U2 ac mae ganddynt farciau i ddangos i'r myfyrwyr sut y caiff yr atebion eu marcio.

Cynhwysir yr atebion i'r cwestiynau yn y *CD-ROM* i athrawon.

Gweithgareddau: mae'r rhain yn cynnig pethau diddorol i chi eu gwneud a fydd yn ategu ac yn atgyfnerthu cynnwys y taeniadau.

Astudiaethau achos: cynhwysir astudiaethau achos o fywyd go iawn sy'n ymwneud yn uniongyrchol â'r deunydd yn y topig. Mae astudiaethau achos yn rhoi cyd-destun ar gyfer ateb y cwestiynau arholiad. Mae llawer o gwestiynau arholiad ar TGCh yn gofyn nid yn unig am ddiffiniad neu esboniad ond hefyd am enghraifft. Mae astudiaethau achos yn cynyddu'ch gwybodaeth o sut mae'r theori a ddysgwch yn cael ei defnyddio'n ymarferol.

Cwestiynau ar yr astudiaethau achos: bydd y rhain yn rhoi cyfle i chi ateb cwestiynau'n ymwneud â sefyllfaoedd go iawn. Mae'r cwestiynau wedi'u llunio'n ofalus i fod yn debyg i'r cwestiynau arholiad a gewch, ac maen nhw'n ymwneud yn uniongyrchol â'r astudiaeth achos a'r deunydd arall yn y taeniadau cynnwys.

Mae'r atebion i'r cwestiynau hyn yn y *CD-ROM* i athrawon.

Cymorth gyda'r arholiad

Enghreifftiau: mae'r rhain yn nodwedd bwysig gan eu bod yn rhoi amcan i chi o sut y caiff cwestiynau arholiad eu marcio. Ar safon U2, mae'n bosibl cael marciau gwael hyd yn oed os yw'r wybodaeth gennych gan nad ydych chi'n cyfleu'r hyn a wyddoch yn effeithiol. Mae'n hollbwysig i chi ddeall yr hyn a ddisgwylir gennych wrth ateb cwestiynau ar y lefel hon.

Atebion myfyrwyr: gwelwch gwestiwn arholiad sydd wedi cael ei ateb gan ddau fyfyriwr gwahanol. Rhoddir sylwadau Arholwr ar bob ateb.

Sylwadau'r arholwr: mae'r rhain yn dangos i chi sut mae arholwyr yn marcio atebion myfyrwyr. Y prif bwrpas yw dangos y camgymeriadau y mae rhai myfyrwyr yn eu gwneud i sicrhau na fyddwch chi'n eu gwneud hefyd. Drwy ddadansoddi'r ffordd y caiff atebion eu hateb, byddwch chi'n gallu ennill mwy o farciau am y cwestiynau a atebwch drwy beidio â gwneud camgymeriadau cyffredin.

Atebion yr arholwr: cynigiant rai o'r llu o atebion posibl a rhoddant amcan o sut y caiff y marciau eu dosbarthu. Dylid cofio bod llawer o atebion cywir i rai cwestiynau a bod unrhyw gynllun marcio'n dibynnu ar allu'r marcwyr i'w ddehongli a rhoi marciau am atebion nad ydynt i'w cael ynddo.

Mapiau meddwl cryno

Mae mapiau meddwl yn hwyl i'w cynhyrchu ac yn ffordd wych o adolygu. Yn y gyfrol hon maen nhw'n crynhoi'r deunydd yn y topig. Weithiau bydd un map meddwl yn unig, ac ar adegau eraill bydd sawl un – gan ddibynnu ar sut mae'r deunydd yn y topig yn cael ei rannu.

Yn ogystal â defnyddio'r mapiau meddwl hyn i'ch helpu i adolygu, dylech lunio eich rhai eich hun.

Beth am eu llunio ar y cyfrifiadur? Mae'n ddigon hawdd cael gafael ar feddalwedd sy'n creu mapiau meddwl.

Enghraifft 1

1 Wrth ddylunio unrhyw system TGCh, mae rhyngwyneb cyfrifiadur-dyn yn rhan bwysig ohoni. Enwch **bedwar ffactor** y dylid eu cymryd i ystyriaeth wrth ddylunio rhyngwyneb cyfrifiadur-dyn ac, ar gyfer pob ffactor, disgrifiwch pam mae'n bwysig. **(8 marc)**

Ateb myfyriwr 1

1 Y gallu i newid maint y testun. Er enghraifft, mewn ffeil pdf gallwch addasu maint y ddogfen drwy ei chwyddo fel y gall pobl â golwg gwael ei gweld.

Dylai'r botwm Nesaf neu'r botymau eraill a ddefnyddir i lywio fod yn yr un lle ar y sgrin fel bod defnyddwyr yn gwybod ble maen nhw ar unwaith ar bob sgrin.

Dylid darparu cymorth sgrin fel bod defnyddwyr yn gallu teipio ymadrodd neu frawddeg a gadael i'r meddalwedd wneud y dasg iddynt.

Dylai'r defnyddiwr allu addasu'r cyfuniadau lliw o destun a chefndir ar y sgrin.

Sylwadau'r arholwr

1 Mae'r ffactor a'r rheswm ar gyfer y pwynt cyntaf yn gywir. Mae'r ail ffactor yn cael un marc yn unig gan nad yw'r myfyriwr wedi dweud pam mae'n bwysig. Dylai fod wedi nodi pam mae'n bwysig i Nesaf fod yn yr un lle ar bob sgrin.

Mae'r trydydd ffactor yn cael un marc am grybwyll cymorth sgrin ond nid yw'r rheswm yn gwneud synnwyr. Mae'r ffactor olaf yn gywir ond ni roddir rheswm dros ei phwysigrwydd. Efallai bod y myfyriwr yn meddwl bod y rheswm yn amlwg, ond er hynny rhaid ei gynnwys i ennill y marc. **(5 marc allan o 8)**

Ateb myfyriwr 2

1 Mae maint y ffont yn bwysig oherwydd bod angen ffont mawr ar bobl oedrannus a phlant ifanc iawn. Ond mae ffont llai yn addas ar gyfer defnyddwyr eraill er mwyn iddynt allu gweld mwy o wybodaeth ar y sgrin.

Dylai fod strwythur llywio clir fel y gall y defnyddiwr symud yn ôl ac ymlaen o sgrin i sgrin heb wastraffu amser.

Dylai'r rhyngwyneb fod yn reddfol fel ei fod yn hawdd ei ddefnyddio. Bydd dewisiadau lliw da ar gyfer y sgrin a'r cefndir yn gwneud y sgrin yn hawdd ei gweld, sy'n bwysig i ddefnyddwyr â golwg gwael neu sy'n ddall i liwiau, ac mewn rhai pecynnau neu wefannau gallwch newid y cyfuniadau lliw yn hawdd.

Sylwadau'r arholwr

1 Mae angen esboniad llawn o bob ffactor a pham mae'n bwysig i ennill y ddau farc ar gyfer pob ffactor. Mae'r ffactor gyntaf, maint y ffont, yn cael ei hesbonio'n ddigon da i ennill dau farc.

Rhoddir dau esboniad ar gyfer yr ail ffactor, felly dau farc am hyn.

Er bod y trydydd ffactor yn gywir, nid oes esboniad digon manwl, felly ni roddir unrhyw farciau.

Mae'r ffactor olaf am liw wedi cael ei hesbonio'n dda ac mae'n werth dau farc. **(6 marc allan o 8)**

Ateb yr arholwr

1 Nid yw nodi'r ffactor yn unig yn ennill marciau, felly rhaid i'r myfyrwyr roi disgrifiad pellach o'r ffactor a/neu fanylu ar beth sy'n gwneud y ffactor yn berthnasol i ryngwyneb defnyddiwr da i ennill y ddau farc.

Dylai unrhyw bedwar o'r ffactorau canlynol gael eu disgrifio'n fanwl:

Maint y ffont – bydd rhai defnyddwyr am weld mwy ar y sgrin, felly dylai maint y ffont fod yn fach (1) neu bydd angen ffont mwy o faint ar ddefnyddwyr ifanc a defnyddwyr hŷn sydd â golwg gwael (1).

Cymorth ar y sgrin – mae'n bwysig os nad yw cymorth ar gael o ffynonellau eraill (1) gan ei fod bob amser ar gael o fewn y rhaglen (1).

Dyluniad sy'n briodol i'r dasg. Gall fod yn well gan arbenigwyr deipio gorchmynion i mewn yn hytrach na defnyddio llygoden (1). Bydd yn well gan ddefnyddwyr llai profiadol ddefnyddio'r llygoden a rhyngwyneb defnyddiwr graffigol (1) oherwydd nad oes angen iddynt ddysgu gorchmynion.

Strwythur llywio clir. Dylai fod yn amlwg i ddefnyddwyr sut i fynd i'r cam neu sgrin nesaf (1) a dylai nodweddion llywio (e.e. Nesaf, saethau Ymlaen ac Yn ôl) gael eu gosod yn yr un lle (1).

Dewis o liw. Dylid defnyddio cyfuniadau sy'n darparu cyferbyniad rhwng y testun a'r cefndir (1) a dylai defnyddwyr allu newid y lliwiau (1) fel y gallant osgoi'r cyfuniad coch/gwyrdd os ydynt yn ddall i liwiau (1).

Cysondeb arwyddbostio a gwybodaeth naidlen. Fel bod popeth ar y rhyngwyneb i'w weld yn syth (1), sy'n gwneud y meddalwedd yn haws ei ddysgu (1).

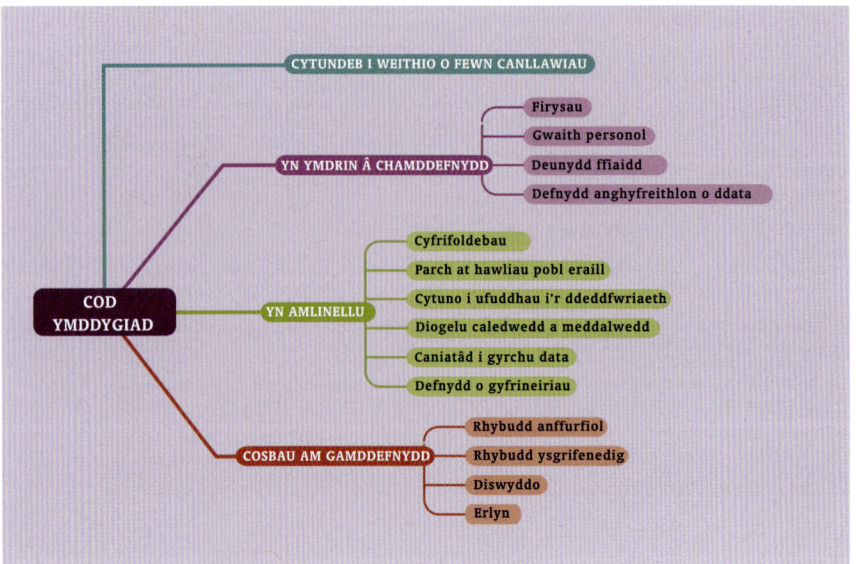

TOPIG 1: Rhwydweithiau

Cafodd hanfodion rhwydweithiau sylw yn Nhopig 8 o fewn Uned UG IT1. Mae'r topig hwn yn ymhelaethu ar y wybodaeth honno drwy edrych ar y ffactorau y mae angen eu hystyried wrth ddewis rhwydwaith ar gyfer cwmni neu sefydliad. Mae'r topig hwn yn ymdrin â'r ddau brif fath o rwydwaith – rhwydweithiau cymar wrth gymar a rhwydweithiau cleient/gweinydd – a hefyd yn edrych ar y gwahanol ffyrdd y gall cyfrifiaduron gael eu trefnu, a manteision ac anfanteision cymharol pob math. Yn ogystal, mae'r cydrannau meddalwedd sy'n sicrhau diogelwch y system ac yn ei gwneud hi'n bosibl i weinyddu'r rhwydwaith yn cael eu trafod.

▼ Y cysyniadau allweddol sy'n cael sylw yn y topig hwn yw:

▶ Dewis rhwydwaith ar gyfer cwmni

▶ Y mathau o rwydwaith sydd ar gael a'r defnydd o galedwedd cysylltiedig

▶ Cydrannau meddalwedd

CYNNWYS

Dewis rhwydwaith ar gyfer cwmni

▼ **Byddwch yn dysgu**

► Am y ffactorau sy'n dylanwadu ar y dewis o rwydwaith

Cyflwyniad

Mae sefydliadau a chwmnïau yn amrywio o ran maint, ac felly mae ganddynt ofynion rhwydweithio gwahanol. Er enghraifft, mae cwmnïau mawr fel cwmnïau hedfan yn gweithredu mewn llawer o wahanol wledydd ac mae eu gofynion rhwydweithio yn hollol wahanol i ofynion cwmni bach dylunio ceginau sy'n cyflogi ychydig o staff yn unig yn yr un swyddfa.

Ffactorau sy'n dylanwadu ar y dewis o rwydwaith

Mae'r ffactorau sy'n dylanwadu ar y dewis o rwydwaith yn cynnwys:
- cost y rhwydwaith
- maint y sefydliad
- sut y caiff y system ei defnyddio
- y systemau presennol
- y perfformiad sydd ei angen
- materion diogelwch.

Cost rhwydwaith

Gall hyn gynnwys costau'r gweinydd (os yw'n berthnasol), costau ceblau, costau meddalwedd a chostau gwasanaethau cyfathrebu trydydd parti.

Mae cost yn ffactor pwysig wrth ystyried unrhyw rwydwaith. Nid yw cost yn broblem fel rheol i gwmnïau mawr, ond i gwmnïau llai mae cyflogi rhywun i reoli'r rhwydwaith, ymdrin ag unrhyw broblemau a chadw'r rhwydwaith i fynd yn gost sylweddol. Mae costau cyflog yn uchel o'u cymharu â chostau caledwedd a meddalwedd, a gall hyn fod yn broblem i gwmni bach. Hefyd gall fod yn anodd dod o hyd i rywun â chymwysterau addas am bris y gallwch ei fforddio.

Mae'r diagram canlynol yn dangos costau eraill:

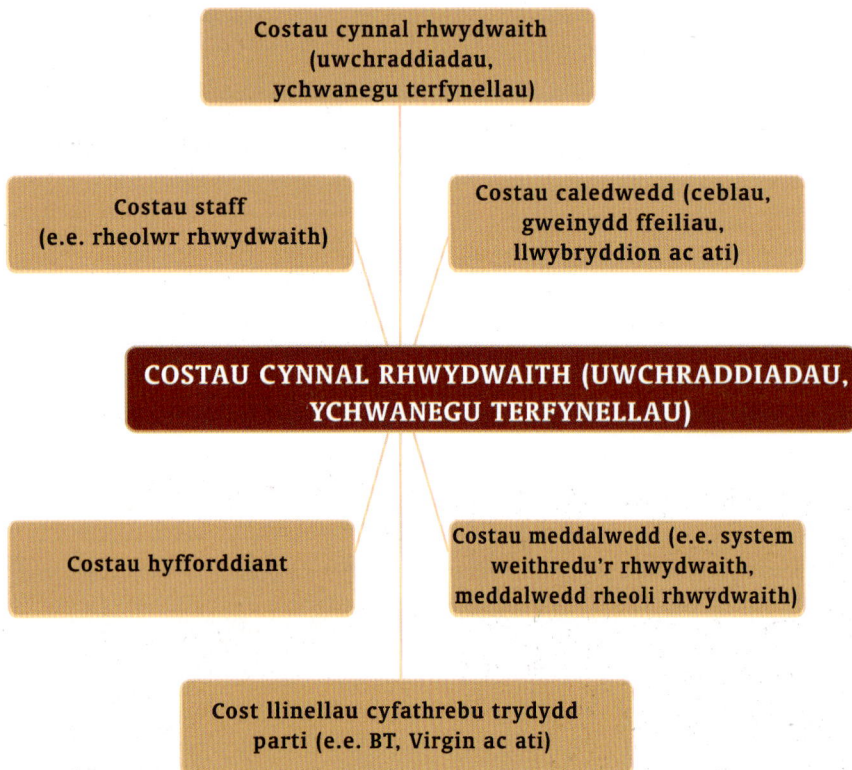

Maint y sefydliad

Hyn sy'n penderfynu cymhlethdod y rhwydwaith. Mae gan lawer o sefydliadau mawr ddefnyddwyr ar lawer safle ym mhedwar ban byd, felly gall costau cyfathrebu fod yn uchel oherwydd bod angen defnyddio cysylltau lloeren. Bydd maint y corff hefyd yn penderfynu a oes angen Rhwydwaith Ardal Leol (*RhAL/LAN: Local Area Network*) neu Rhwydwaith Ardal Eang (RhAE/*WAN: Wide Area Network*). Mae angen mesurau diogelwch cymhleth ar gyfer rhwydweithiau mawr ac amrywiaeth o staff arbenigol i gadw'r rhwydwaith i fynd.

Sut y caiff y system ei defnyddio

Bydd hyn yn effeithio ar faint y rhwydwaith. Bydd adwerthwyr mawr yn gorfod sefydlu rhwydweithiau ardal leol ym mhob siop, a chaiff pob un ei gysylltu i ffurfio rhwydwaith ardal eang. Bydd hyn yn galluogi pob siop i wneud peth prosesu lleol ond hefyd yn caniatáu cyfathrebu data â'r brif swyddfa. Mae rhai rhwydweithiau'n caniatáu i staff weithio gartref gan ddefnyddio cronfeydd data ar-lein, fideo-gynadledda, ac ati.

Bydd rhai tasgau, e.e. prosesu archebion neu archebu tocynnau, yn cael eu gwneud yn rhyngweithiol trwy gynnal deialog gyda'r defnyddiwr. Mae angen i rwydweithiau a ddefnyddir yn y ffordd hon fod yn gyflym er mwyn lleihau'r amser ymateb.

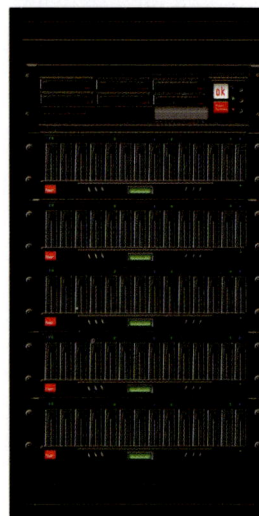

Costau cynnal rhwydwaith (uwchraddiadau, ychwanegu terfynellau)

Costau staff (e.e. rheolwr rhwydwaith)

Costau caledwedd (ceblau, gweinydd ffeiliau, llwybryddion ac ati)

COSTAU CYNNAL RHWYDWAITH (UWCHRADDIADAU, YCHWANEGU TERFYNELLAU)

Costau hyfforddiant

Costau meddalwedd (e.e. system weithredu'r rhwydwaith, meddalwedd rheoli rhwydwaith)

Cost llinellau cyfathrebu trydydd parti (e.e. BT, Virgin ac ati)

Costau'n gysylltiedig â rhwydwaith

Mae gweinydd ffeiliau yn rhan hanfodol o'r mwyafrif o rwydweithiau

Mae rhai tasgau – e.e. cynhyrchu rhestr gyflogau, biliau trydan neu filiau cerdyn credyd – yn defnyddio swp-brosesu, sy'n golygu bod yr holl fewnbynnau'n cael eu casglu ynghyd dros gyfnod o amser ac yna eu prosesu i gyd gyda'i gilydd. Mae swp-brosesu'n gofyn am lawer o rym prosesu, a bydd hyn yn arafu'r rhwydwaith i ddefnyddwyr eraill. Felly mae'n well i'r swp-brosesu ddigwydd pan nad yw defnyddwyr eraill yn defnyddio'r rhwydwaith yn rhyngweithiol. O ganlyniad, caiff swp-brosesu ei wneud ar adegau llai prysur, yn y nos fel arfer.

Y systemau presennol

Pan gaiff rhwydweithiau eu cyflwyno neu eu hehangu, yn aml mae'n rhaid iddynt weithio gyda'r systemau presennol. Yn aml bydd angen i'r rhwydwaith ar ei newydd wedd weithio gyda'r caledwedd a'r meddalwedd presennol. Gan fod y staff yn deall y systemau'n barod, ni fydd hyn yn cyflwyno gormod o bethau newydd iddynt eu dysgu. Mae ailddefnyddio rhai o'r systemau presennol yn lleihau costau hefyd.

Y perfformiad sydd ei angen

Bydd hyn yn penderfynu pa ffordd y dylai'r cyfrifiaduron a chaledwedd arall gael eu cysylltu â'i gilydd (h.y. y dopoleg i'w defnyddio). Bydd hefyd yn penderfynu pa fath o weinydd sydd ei angen. Un mesur o berfformiad yw pa mor dda mae'r rhwydwaith yn cefnogi defnyddwyr, er enghraifft, y cymorth ar-lein sydd ar gael a pha mor hawdd ydyw i ddefnyddwyr wneud eu gwaith gan ddefnyddio'r rhyngwyneb defnyddiwr. Bydd angen i systemau e-fasnach fod yn gyflym neu bydd y defnyddwyr yn diflasu ac yn symud

Mae'r rhwydwaith bach hwn wedi'i gysylltu â'r Rhyngrwyd drwy gyfrwng cyswllt cyflym iawn o'r enw asgwrn cefn

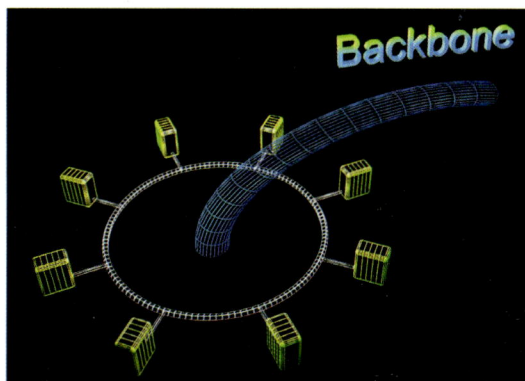

Cyflymder

Dibynadwyedd

MESURAU O BERFFORMIAD RHWYDWAITH

Cost

Defnyddioldeb

Mesurau o berfformiad rhwydwaith

at gwmni arall. Mae'n hollbwysig i'r rhwydwaith fod yn ddibynadwy ac ar gael i ddefnyddwyr ar bob adeg, felly mae dibynadwyedd yn fesur o berfformiad hefyd.

I grynhoi, bydd rhwydwaith perfformiad uchel yn:

- gyflym
- dibynadwy
- mor rhad â phosibl
- hawdd ei ddefnyddio.

Materion diogelwch

Mae'r mwyafrif o rwydweithiau'n cynnig ffordd o ganiatáu i ddefnyddwyr gyrchu'r Rhyngrwyd. Mewn llawer o gyrff, mae'r Rhyngrwyd yn rhan hanfodol o'r rhwydwaith, felly mae angen cymryd camau i gadw hacwyr allan a sicrhau nad yw manylion ariannol cwsmeriaid yn cael eu rhyng-gipio neu eu datgelu. Os yw data personol yn cael eu cadw, yna mae angen i fesurau diogelwch priodol fod ar waith i rwystro mynediad heb ei awdurdodi. Mae problemau eraill yn cynnwys cyflwyno firysau, defnyddio cyfrifiaduron i gopïo deunydd sydd wedi'i warchod gan hawlfraint, anfon negeseuon e-bost cas, ac ati.

Bydd gofynion diogelwch un sefydliad yn wahanol i ofynion sefydliad arall. Er enghraifft, bydd ysgol yn poeni am ddisgyblion yn cyrchu rhai gwefannau neu'n defnyddio ystafelloedd sgwrsio, tra bydd busnesau'n poeni fwy am wneud taliadau diogel.

Dyma rai materion diogelwch y bydd angen eu hystyried wrth ddewis rhwydwaith:

- mynediad at ddeunydd anghyfreithlon (e.e. pornograffi, gwefannau sy'n hyrwyddo casineb hiliol, ac ati)
- a yw defnyddwyr yn rhydd i ddefnyddio unrhyw eiriau mewn peiriannau chwilio neu a ddylid cyfyngu ar rai geiriau
- sut i gadw hacwyr allan (e.e. defnyddio muriau gwarchod, ac ati)
- sut y gall manylion taliadau gael eu diogelu wrth eu hanfon ar hyd llinellau cyfathrebu (e.e. amgryptio)
- problemau pan fydd defnyddwyr yn cysylltu â rhwydweithiau'n ddiwifr
- yr angen i osgoi firysau
- yr angen i gyfyngu ar y ffeiliau y gall staff eu cyrchu
- atal deunydd sydd wedi'i warchod gan hawlfraint rhag cael ei lwytho i lawr (e.e. ffilmiau, cerddoriaeth, ac ati).

Backbone

Y mathau o rwydwaith sydd ar gael a'r defnydd o galedwedd cysylltiedig

Cyflwyniad

Yn yr adran hon byddwn yn ystyried sut mae'r terfynellau mewn rhwydwaith wedi'u cysylltu. Er ein bod ni'n defnyddio'r gair 'cysyllt', rhaid cofio bod llawer o rwydweithiau'n ddiwifr.

Bydd yr adran hon hefyd yn ymdrin â'r ddau brif fath o rwydwaith: cymar wrth gymar a chleient/gweinydd.

Cymar wrth gymar neu gleient/gweinydd

Gallwn rannu rhwydweithiau yn ddau brif fath:

- cymar wrth gymar
- cleient/gweinydd.

Rhwydweithiau cymar wrth gymar

Mewn rhwydweithiau cymar wrth gymar mae gan bob cyfrifiadur yr un statws, a gallant gyfathrebu â'i gilydd ar lefel gyfartal. Mae hyn yn golygu y gall pob cyfrifiadur ar y rhwydwaith gyrchu holl adnoddau unrhyw gyfrifiadur arall ar y rhwydwaith.

Defnyddir rhwydweithiau cymar wrth gymar ar gyfer rhwydweithiau cartref. Os ydych am rannu ffeiliau ac argraffyddion rhwng sawl cyfrifiadur yn eich cartref fe fydd rhwydwaith cymar wrth gymar yn iawn.

Mae'r rhwydweithiau hyn bellach yn boblogaidd iawn ymhlith defnyddwyr Rhyngrwyd ar gyfer rhannu ffeiliau. Mae pob defnyddiwr yn gallu cysylltu â chyfrifiadur defnyddiwr arall dros y rhwydwaith, felly nid oes unrhyw reolaeth ganolog. Mae hyn wedi achosi llawer o broblemau, yn enwedig wrth i bobl rannu cerddoriaeth, fideos, delweddau, ac ati sydd wedi'u gwarchod gan hawlfraint.

Mae systemau meddalwedd cymar wrth gymar, megis Kazaa a Napster, yn rhaglenni hynod o boblogaidd gyda defnyddwyr.

Mewn rhwydweithiau cymar wrth gymar, mae gan bob cyfrifiadur yr un statws, a gallant gyrchu adnoddau ei gilydd.

Manteision rhwydweithio cymar wrth gymar:

- arbed costau – nid oes angen gweinydd, felly gall yr holl gyfrifiaduron fod yr un fath
- nid oes angen rheolwr rhwydwaith – mae'r holl ddefnyddwyr yn gyfrifol am y rhwydwaith
- hawdd eu gosod – y rhain yw'r rhwydweithiau cyfrifiadurol symlaf, felly gall unrhyw un eu gosod
- nid yw'r rhwydwaith yn dibynnu ar weinydd – felly nid oes angen poeni y bydd y gweinydd yn torri
- costau gweithredu is – costau gosod a chynnal is
- pawb yn gyfrifol – y defnyddwyr sy'n penderfynu pa adnoddau y gall defnyddwyr eraill eu defnyddio ar eu cyfrifiaduron.

Anfanteision rhwydweithio cymar wrth gymar:

- ni ellir gwneud ffeiliau wrth gefn yn ganolog – mae hyn yn rhoi cyfrifoldeb ar bob defnyddiwr i wneud copïau o'u data eu hunain;

Mae cymar wrth gymar yn caniatáu i gyfrifiaduron rannu ffeiliau

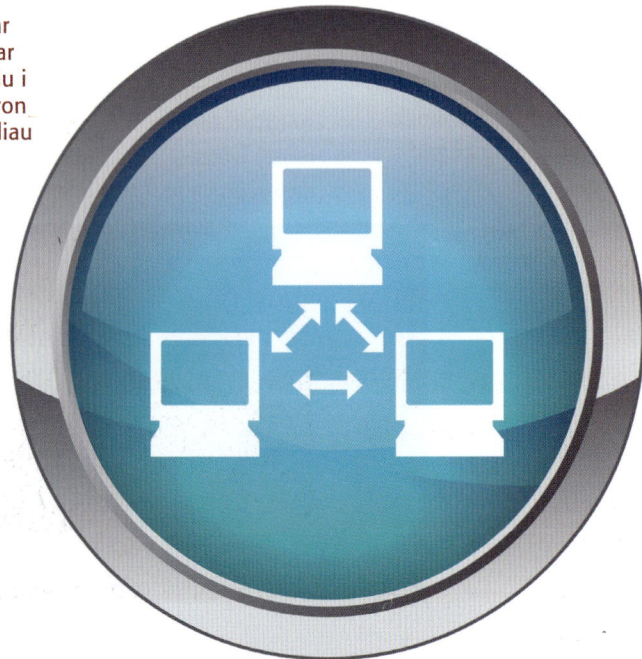

ni allwch fod yn siŵr y bydd pob defnyddiwr yn gwneud hyn
- mae angen mwy o wybodaeth TG ar y defnyddwyr – gan mai nhw fydd yn gyfrifol am y ffeiliau ar eu cyfrifiaduron eu hunain
- diogelwch llai effeithiol – caiff adnoddau eu rhannu, felly rhaid i ddefnyddwyr benderfynu pa rai o'u hadnoddau y gall defnyddwyr eraill y rhwydwaith eu defnyddio
- mae'n bosibl y bydd rhai cyfrifiaduron yn rhedeg yn araf – os oes adnoddau ar gyfrifiadur a ddefnyddir gan bob defnyddiwr, bydd yn rhedeg yn araf iawn
- efallai y bydd defnyddwyr yn cael trafferth dod o hyd i ffeiliau gan na chânt eu trefnu a'u storio'n ganolog
- yn addas ar gyfer rhwydweithiau bach iawn yn unig, gyda llai na 15 o gyfrifiaduron wedi'u rhwydweithio fel rheol.

Rhwydweithiau cleient/gweinydd

Rhwydweithiau cleient/gweinydd sydd orau ar gyfer rhwydweithiau mawr. Mewn rhwydwaith o'r fath, nid oes gan bob cyfrifiadur yr un statws. Mae cyfrifiadur canolog mwy pwerus yn cael ei ddefnyddio i storio ffeiliau a rhaglenni, a'r enw a roddir ar hwn yw'r gweinydd. Mae'r cyfrifiaduron eraill ar y rhwydwaith yn cael eu galw'n gleientiaid.

Gan fod y gweinydd yn rhan mor bwysig o rwydwaith cleient/gweinydd, bydd rheolwr rhwydwaith yn cael ei benodi fel rheol i ofalu amdano.

Manteision rhwydweithiau cleient/gweinydd:

- diogelwch gwell – gan ei fod wedi'i ganoli a bod rhywun yn gyfrifol amdano
- data wedi'u canoli – caiff yr holl ddata eu cadw ar y gweinydd ffeiliau sy'n golygu bod pob defnyddiwr yn gallu cyrchu'r un set o ddata
- caiff ffeiliau wrth gefn eu cadw'n ganolog – bydd rheolwr y rhwydwaith yn gwneud copïau wrth gefn yn rheolaidd, felly mae'n llai tebygol y caiff data a rhaglenni eu colli
- gellir cyrchu rhaglenni a ffeiliau'n gyflymach – defnyddir gweinyddion, sef cyfrifiaduron nerthol, felly mae'r rhwydwaith cyfan yn gweithio'n gyflymach
- gweinyddiaeth ganolog – mae holl waith gweinyddol y rhwydwaith (e.e. rhoi enwau defnyddwyr a chyfrineiriau, cymorth, ac ati) yn cael ei wneud yn ganolog, felly nid oes angen i ddefnyddwyr boeni am hyn.

Anfanteision rhwydweithiau cleient/gweinydd:

- drutach – mae gweinyddion yn ddrud
- mae angen gwybodaeth arbenigol – rhaid cael rhywun sy'n deall manylion technegol y rhwydwaith i ofalu amdano
- mae'r meddalwedd yn soffistigedig a drud – mae systemau gweithredu rhwydwaith yn gostus. Mae'n bosibl y bydd angen prynu meddalwedd rheoli rhwydwaith ar gyfer rhwydweithiau cleient/gweinydd mawr
- os bydd y gweinydd yn torri, fydd neb yn gallu defnyddio'r rhwydwaith nes iddo gael ei drwsio.

Mewn rhwydwaith cleient/gweinydd, mae cyfrifiadur mwy pwerus yn gofalu am y cyfleusterau rhwydweithio

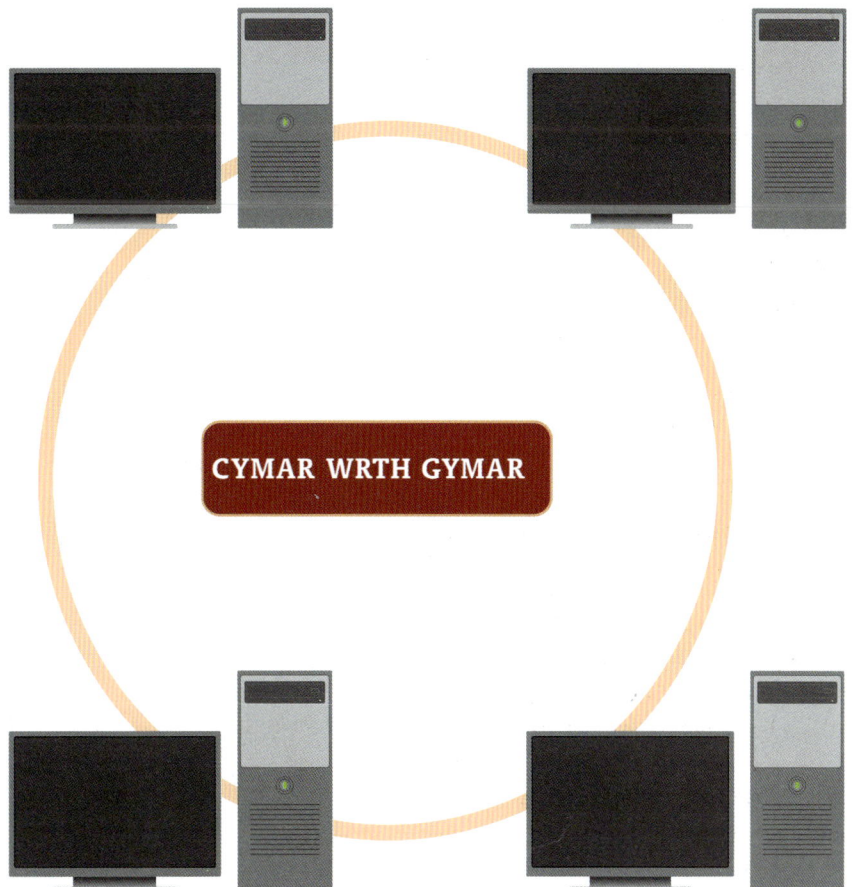

CYMAR WRTH GYMAR

Mewn rhwydwaith cymar wrth gymar, mae gan yr holl gyfrifiaduron yr un statws

5

Y mathau o rwydwaith sydd ar gael a'r defnydd o galedwedd cysylltiedig (parhad)

Topolegau rhwydwaith

Gall y dyfeisiau mewn rhwydwaith gael eu gosod mewn llawer gwahanol ffordd. Topoleg yw'r enw a roddwn ar y ffordd mae hyn yn cael ei wneud.

Mae topoleg rhwydwaith yn dangos sut mae'r cyfrifiaduron wedi'u cysylltu pan ddefnyddir gwifrau, ac os defnyddir diwifr (radio, isgoch a microdonnau) mae'n dangos sut mae'r dyfeisiau ar y rhwydwaith yn cyfathrebu â'i gilydd.

Topoleg cylch

Rhwydwaith cylch yw hwn. Mae hefyd yn rhwydwaith cymar wrth gymar gan nad oes gweinydd.

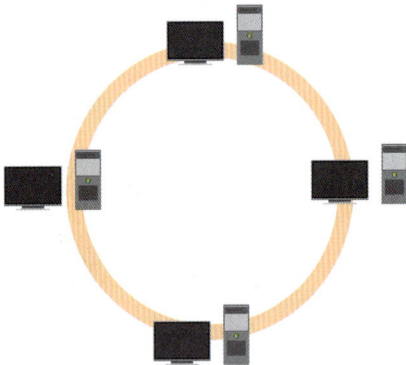

Mewn topoleg cylch:

- mae'r holl gyfrifiaduron wedi'u trefnu mewn cylch
- mae data sy'n cael eu hanfon gan un cyfrifiadur yn teithio o amgylch y cylch nes cyrraedd y cyfrifiadur cywir.

Manteision rhwydweithiau cylch:

- nid yw'r rhwydwaith yn dibynnu ar gyfrifiadur canolog
- mae gan bob cyfrifiadur yr un lefel o fynediad â'r lleill, felly ni all un cyfrifiadur gadw'r rhwydwaith iddo'i hun.

Anfanteision rhwydweithiau cylch:

- os bydd toriad yn y cysylltiad (gwifr neu ddiwifr), bydd y rhwydwaith cyfan yn methu
- mae'n anodd dod o hyd i namau
- mae'n amhosibl cadw'r rhwydwaith i fynd tra mae cyfarpar yn cael ei ychwanegu neu ei adnewyddu gan nad oes ond un llwybr i'r data ei ddilyn.

Topoleg bws

Mewn topoleg bws, mae'r holl ddyfeisiau ar y rhwydwaith wedi'u cysylltu â chebl cyffredin sy'n cael ei alw'n asgwrn cefn. Caiff signalau eu trosglwyddo i'r naill gyfeiriad neu'r llall ar hyd yr asgwrn cefn.

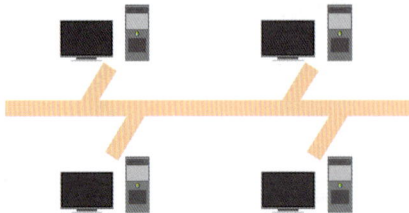

Manteision rhwydweithiau topoleg bws:

- cost-effeithiol gan nad oes angen llawer o gebl
- mae rhediadau cebl syml yn ei gwneud yn hawdd ei osod
- mae'n hawdd ychwanegu dyfeisiau ychwanegol at y rhwydwaith.

Anfanteision rhwydweithiau topoleg bws:

- os caiff mwy na 12 dyfais eu cysylltu â'r rhwydwaith, bydd ei berfformiad yn dioddef
- os bydd toriad yn y cebl asgwrn cefn, ni fydd yn bosibl defnyddio'r rhwydwaith.

Topoleg seren

Mae topoleg seren yn defnyddio man cysylltu canolog i gysylltu'r holl ddyfeisiau mewn rhwydwaith â'i gilydd. Gall y man cysylltu canolog fod yn foth, yn switsh neu'n llwybrydd.

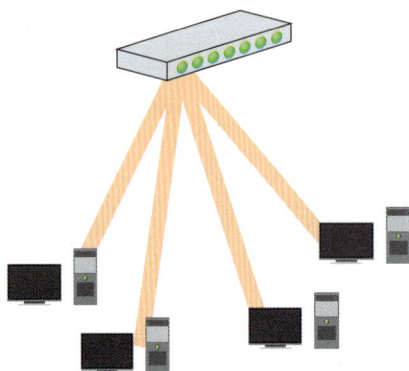

Manteision rhwydweithiau seren:

- gallant oddef namau – os bydd un o'r ceblau'n methu, bydd y cyfrifiaduron eraill yn parhau i weithio
- gallant oddef llwyth – gall rhagor o gyfrifiaduron gael eu hychwanegu heb effeithio rhyw lawer ar berfformiad oherwydd bod gan bob cyfrifiadur ei lwybr ei hun i'r both
- mae'n hawdd ychwanegu cyfrifiaduron ychwanegol – heb darfu ar y rhwydwaith.

Anfanteision rhwydweithiau seren:

- cost uwch – mae'n dopoleg ddrud oherwydd yr holl gebl sydd ei angen
- dibyniaeth ar y both, switsh neu lwybrydd canolog – os bydd y ddyfais yng nghanol y rhwydwaith yn methu, bydd y rhwydwaith cyfan yn methu.

Rhwydweithio diwifr

Erbyn hyn mae cyfrifiaduron yn gallu cysylltu â'r Rhyngrwyd neu gyfathrebu â chyfrifiaduron eraill mewn rhwydwaith ardal leol yn ddiwifr. Gyda chyfathrebu diwifr, y cyfrwng trosglwyddo data yw'r aer y mae'r tonnau radio yn teithio drwyddo.

Mae rhwydweithiau diwifr yn galluogi pobl i gysylltu â'r Rhyngrwyd neu rwydwaith ardal leol yn ddiwifr. Mae hyn yn golygu y gallant weithio lle bynnag y mae signal radio ar gael ar gyfer eu rhwydwaith.

Mae llawer o bobl, yn enwedig pobl sy'n teithio cryn dipyn, yn gorfod cyrchu'r Rhyngrwyd yn rheolaidd. Mae llawer o leoedd cyhoeddus lle gall pobl ddefnyddio eu gliniadur neu ddyfais gludadwy arall fel ffôn symudol neu Gynorthwyydd Digidol Cludadwy (*PDA: Portable Digital Assistant*) i gyrchu'r Rhyngrwyd.

Mannau poeth yw'r enw a roddir ar y lleoedd hynny lle gallwch ddefnyddio Wi-Fi i gyrchu'r Rhyngrwyd.

I osod rhwydwaith Wi-Fi bach, byddai angen:

- cysylltiad band llydan â'r Rhyngrwyd
- llwybrydd
- cyfrifiaduron â gallu Wi-Fi (mae addasydd diwifr wedi'i osod yn y mwyafrif o gyfrifiaduron). Gallwch brynu addasyddion diwifr ar gyfer cyfrifiaduron hŷn.

Sut mae Wi-Fi yn gweithio

1. Mae'r llwybrydd yn cael ei gysylltu â'r Rhyngrwyd drwy gysylltiad band llydan cyflym.
2. Mae'r llwybrydd yn derbyn data o'r Rhyngrwyd.
3. Mae'n trosglwyddo'r data fel signal radio, gan ddefnyddio antena.
4. Mae addasydd diwifr y cyfrifiadur yn codi'r signal radio ac yn ei droi'n ddata y gall y cyfrifiadur eu deall.

Wrth anfon data, mae'r prosesau uchod yn gweithio'r ffordd groes.

Manteision ac anfanteision rhwydweithiau diwifr

Manteision Wi-Fi:

- gall rhwydweithiau ardal leol rhad gael eu gosod heb geblau
- mae'n rhoi rhyddid i bobl weithio lle bynnag y gellir derbyn signal
- mae'n ddelfrydol ar gyfer rhwydweithiau mewn hen adeiladau rhestredig lle na fyddai'n bosibl cael caniatâd i osod ceblau
- set fyd-eang o safonau – gallwch ddefnyddio Wi-Fi ar hyd a lled y byd.

Anfanteision Wi-Fi:

- mae'n defnyddio llawer o drydan – sy'n golygu bod batrïau gliniaduron yn dadwefru'n gyflym
- mae'n bosibl bod problemau iechyd yn gysylltiedig â defnyddio Wi-Fi
- gall fod problemau diogelwch, hyd yn oed pan ddefnyddir amgryptio
- mae gan rwydweithiau cartref gyrhaeddiad cyfyngedig iawn (e.e. 150 troedfedd)
- gall ymyriant fod yn broblem os bydd signalau rhwydweithiau diwifr yn gorgyffwrdd.

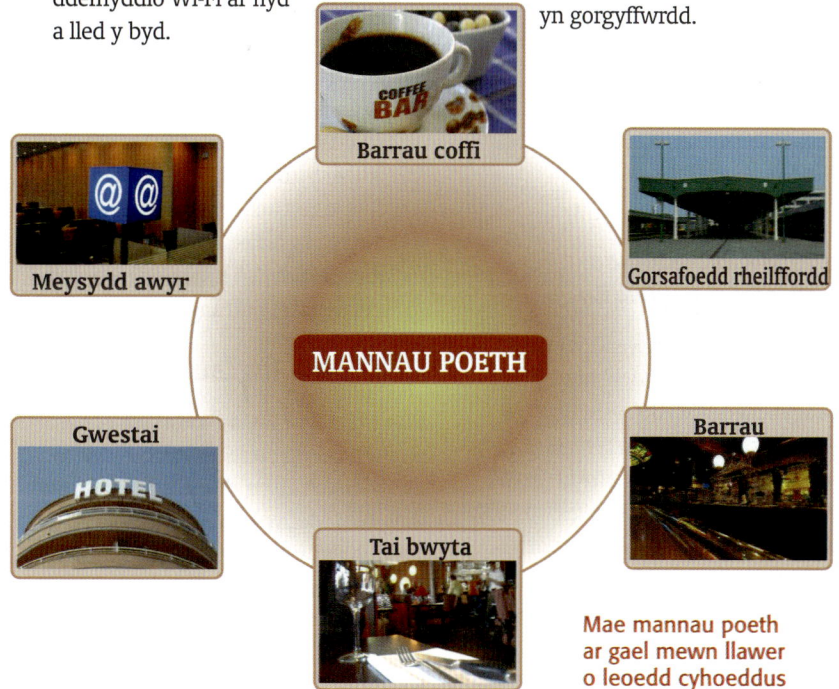

Barrau coffi

Gorsafoedd rheilffordd

Meysydd awyr

MANNAU POETH

Barrau

Gwestai

Tai bwyta

Mae mannau poeth ar gael mewn llawer o leoedd cyhoeddus

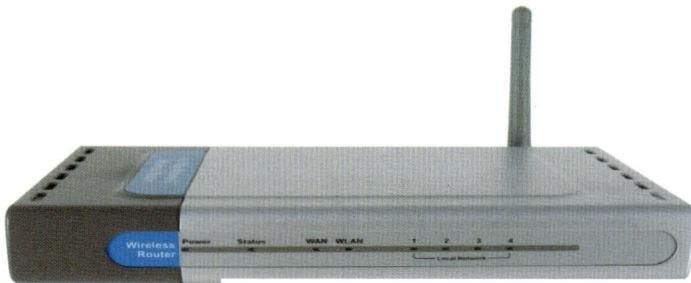

Llwybrydd diwifr a ddefnyddir i osod rhwydwaith diwifr bach yn y cartref neu'r swyddfa

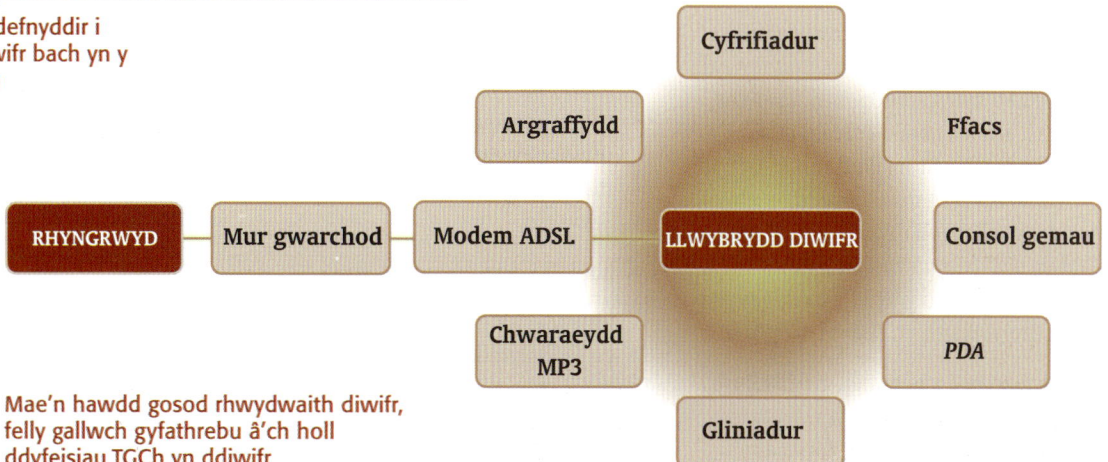

RHYNGRWYD — Mur gwarchod — Modem ADSL — LLWYBRYDD DIWIFR

Cyfrifiadur

Argraffydd

Ffacs

Consol gemau

Chwaraeydd MP3

PDA

Gliniadur

Mae'n hawdd gosod rhwydwaith diwifr, felly gallwch gyfathrebu â'ch holl ddyfeisiau TGCh yn ddiwifr

7

Cydrannau meddalwedd

Cyflwyniad

Buom yn edrych ar galedwedd y rhwydwaith yn yr adrannau blaenorol. Yn yr adran hon byddwn yn rhoi sylw i'r meddalwedd sy'n caniatáu i'r holl gydrannau caledwedd weithio'n gywir ac i bob cyfrifiadur gyfathrebu â'i gilydd.

Meddalwedd rhwydweithio

Mae angen meddalwedd ar rwydweithiau i ddweud wrth y dyfeisiau cysylltiedig sut i gyfathrebu â'i gilydd.

Meddalwedd systemau gweithredu rhwydwaith

Gellir defnyddio meddalwedd Windows i redeg rhwydweithiau bach, ond mae angen meddalwedd systemau gweithredu rhwydwaith arbenigol ar gyfer rhwydweithiau cleient/gweinydd mawr.

Mae systemau gweithredu rhwydwaith yn fwy cymhleth oherwydd bod angen iddynt gydgysylltu gweithgareddau'r holl gyfrifiaduron a dyfeisiau eraill sydd wedi'u cysylltu â'r rhwydwaith.

Enghreifftiau o feddalwedd systemau gweithredu rhwydwaith yw:

- UNIX
- Linux
- Novell Netware – mae hon yn system weithredu boblogaidd iawn ar gyfer rhwydweithiau cleient/gweinydd.

Meddalwedd rheoli rhwydwaith

Os byddech chi'n rheolwr rhwydwaith ac yn gyfrifol am rwydwaith yn cynnwys cannoedd o gyfrifiaduron, byddai angen cymorth arnoch i ofalu amdanynt a chadw'r rhwydwaith i fynd.

Yn ffodus mae meddalwedd o'r enw meddalwedd rheoli rhwydwaith ar gael a fydd yn eich helpu i wneud hyn.

Rhai o'r tasgau y bydd y meddalwedd rheoli rhwydwaith yn helpu i'w cyflawni:

- Sicrhau bod gan yr holl gyfrifiaduron feddalwedd cyfoes gyda'r clytiau diogelwch diweddaraf, fel na all hacwyr fynd i mewn i'r rhwydwaith.
- Cadw trefn ar y meddalwedd sy'n cael ei ddefnyddio ar bob cyfrifiadur a sicrhau bod trwyddedau ar gyfer yr holl feddalwedd.
- Diweddaru'r holl feddalwedd rhaglenni.
- Darparu cyfleusterau rheoli o bell fel y gall staff desg gymorth ddatrys problemau defnyddwyr drwy weld beth yn union sydd ar eu sgrin.
- Gwirio bod y lled band yn cael ei ddefnyddio'n gywir.
- Darganfod a yw defnyddiwr wedi gosod meddalwedd didrwydded ar gyfrifiadur ar y rhwydwaith heb gael caniatâd.
- Gwirio'r cof a chyflymder y prosesydd sydd gan gyfrifiadur penodol ar y rhwydwaith. Gall hyn helpu i adnabod cyfrifiaduron sydd angen uwchraddiadau.

Cyfrifon a logiau defnyddwyr

Mae cyfrifon a logiau defnyddwyr yn rhan bwysig o feddalwedd y rhwydwaith ac yn rhwystro defnyddwyr rhag camddefnyddio'r rhwydwaith. Os bydd rhywun yn ei gamddefnyddio, mae'n hawdd iawn darganfod pwy ydyw drwy ddefnyddio'r meddalwedd hwn.

Cyfrifon defnyddwyr

Bydd pawb sy'n defnyddio rhwydwaith yn cael cyfrif defnyddiwr, sy'n cael ei sefydlu gan weinyddwr neu reolwr y rhwydwaith.

Yr un adeg ag y caiff y cyfrif ei greu, bydd y gweinyddwr neu reolwr rhwydwaith yn rhoi hawliau penodol i'r defnyddiwr, sef y pethau y gall y defnyddiwr eu gwneud wrth ddefnyddio'r rhwydwaith. Gall holl ddefnyddwyr rhwydwaith:

- newid eu cyfrinair
- newid gosodiadau eu cyfrifiadur (h.y. personoli'r rhyngwyneb defnyddiwr)
- rheoli eu ffeiliau eu hunain yn eu lle storio.

Cyfrifoldeb y person sy'n rheoli'r rhwydwaith yw penderfynu:

- pa feddalwedd y gall y defnyddiwr ei gyrchu
- pa ffeiliau sy'n cael eu rhannu y gall y defnyddiwr eu cyrchu
- a yw'r defnyddiwr yn cael copïo ffeiliau

▼ Byddwch yn dysgu

- ▶ Am feddalwedd rhwydweithio
- ▶ Am feddalwedd systemau gweithredu rhwydwaith
- ▶ Am feddalwedd rheoli rhwydwaith
- ▶ Am gyfrifon a logiau defnyddwyr
- ▶ Am strategaethau diogelwch
- ▶ Am reoli ffurfweddiad
- ▶ Am reoli o bell
- ▶ Am gynllunio ar gyfer trychineb (h.y. copïau wrth gefn ac adfer)
- ▶ Am archwilio (cadw logiau)

GEIRIAU ALLWEDDOL

Mewngofnodi – dweud wrth y rhwydwaith pwy ydych er mwyn cael mynediad

Allgofnodi – hysbysu'r rhwydwaith eich bod am gau mynediad i'r cyfleusterau rhwydwaith tan y mewngofnodi nesaf

Cyfrinair – dilyniant o nodau (sy'n cael ei gelu rhag pobl eraill) y mae'r defnyddiwr yn ei deipio i gael mynediad i'r rhwydwaith

Enw defnyddiwr – cyfres unigryw o nodau y mae'r defnyddiwr neu reolwr y rhwydwaith yn ei dewis. Enw'r defnyddiwr ei hun yw hwn yn aml, neu lysenw efallai.

- a yw'r defnyddiwr yn cael gosod meddalwedd.

Logiau defnyddwyr

Pan fydd defnyddiwr yn mewngofnodi bydd yn rhoi ei enw defnyddiwr a'i gyfrinair. Yna bydd y system yn neilltuo adnoddau rhwydwaith i'r defnyddiwr. Ar ôl gorffen gweithio ar y rhwydwaith, gall y defnyddiwr allgofnodi. Yn ogystal â sicrhau mai defnyddwyr awdurdodedig yn unig sy'n cael mynediad i'r rhwydwaith, gall y drefn fewngofnodi roi trywydd archwilio sy'n dangos pwy sydd wedi defnyddio pa adnoddau ar y system. Mae hyn yn cael ei wneud drwy ddarparu log defnyddiwr, sy'n gofnod o ymdrechion llwyddiannus ac aflwyddiannus i fewngofnodi a hefyd o'r adnoddau sydd wedi cael eu defnyddio gan y rheiny sydd â'r hawl i gyrchu adnoddau'r rhwydwaith.

Sgrin fewngofnodi ar gyfer defnyddiwr

Strategaethau diogelwch

Bydd sefydliadau yn wynebu nifer o fygythiadau diogelwch wrth ddefnyddio rhwydweithiau, felly mae angen datblygu strategaethau i leihau'r bygythiadau hyn.

Gall meddalwedd helpu gyda strategaeth diogelwch mewn sawl ffordd:

- Defnyddio cyfrineiriau a rhifau adnabod – i ddilysu defnyddwyr. Mae mynediad yn cael ei gyfyngu i ddefnyddwyr awdurdodedig yn unig.
- Gwirwyr firysau – mae gan bob cyfrifiadur y meddalwedd gwirio firysau diweddaraf sy'n sganio'r cyfrifiadur a'r holl ffeiliau, e-byst a negeseuon sydyn am firysau. Os byddant yn canfod firysau, gallant gael eu gwaredu'n awtomatig.
- Muriau gwarchod – gallant fod yn feddalwedd, yn galedwedd neu'r ddau ac maent yn amddiffyn y rhwydwaith rhag hacwyr.
- Amgryptio – defnyddir hyn i gadw data'n gyfrinachol wrth eu hanfon dros rwydweithiau, er enghraifft, wrth anfon manylion ariannol (manylion banc, manylion cerdyn credyd, ac ati) dros y Rhyngrwyd.

Rheoli ffurfweddiad

Ar ôl creu rhwydwaith, mae'n rhaid ei ffurfweddu i gael y perfformiad gorau posibl. Rheoli ffurfweddiad rhwydwaith yw'r broses o drefnu a rheoli'r holl wybodaeth am rwydwaith.

Pan fydd angen addasu, trwsio, ehangu neu uwchraddio rhwydwaith, bydd y rheolwr neu weinyddwr rhwydwaith yn cyfeirio at y gronfa ddata ffurfweddiad rhwydwaith cyn penderfynu beth i'w wneud. Mae'r gronfa ddata ffurfweddiad rhwydwaith yn cynnwys lleoliadau a chyfeiriadau rhwydwaith yr holl ddyfeisiau caledwedd a ddefnyddir ar y rhwydwaith a gwybodaeth am y rhaglenni, a fersiynau a diweddariadau'r meddalwedd ar y cyfrifiaduron ar y rhwydwaith.

Manteision defnyddio meddalwedd rheoli ffurfweddiad yw:

- Mae'n llawer haws trwsio, ehangu neu uwchraddio'r rhwydwaith.
- Bydd y rhwydwaith yn cael ei optimeiddio ac felly bydd yn rhedeg yn gyflymach.
- Bydd y rhwydwaith yn methu'n llai aml gan ei fod yn cael ei reoli'n well.
- Caiff diogelwch y rhwydwaith ei optimeiddio.
- Os bydd y newidiadau sy'n cael eu gwneud yn cael effaith andwyol ar berfformiad y rhwydwaith, bydd yn bosibl mynd yn ôl i ffurfweddiad blaenorol.
- Mae'n cadw cofnod o'r holl newidiadau sy'n cael eu gwneud, felly nid oes angen i chi gofnodi'r gosodiadau newydd.

Rheoli o bell

Mae llawer o dasgau y gall rheolwr neu weinyddwr rhwydwaith eu gwneud ar ei derfynell. Gall ddefnyddio rheoli o bell i:

- weld pa ddefnyddwyr sy'n defnyddio'r rhwydwaith
- cadw golwg ar e-byst sy'n cael eu hanfon yn amser y cwmni
- cadw golwg ar y gwefannau mae staff yn eu cyrchu
- cadw golwg ar y caledwedd (disgyrwyr, prosesyddion, ac ati) i weld a oes angen uwchraddiadau
- sicrhau nad oes gormod o ddefnyddwyr yn defnyddio meddalwedd ac yn torri telerau'r drwydded
- tywys defnyddwyr drwy'r problemau maen nhw'n eu cael
- sicrhau nad oes gan ddefnyddwyr feddalwedd heb ei awdurdodi ar eu cyfrifiaduron
- allgofnodi defnyddiwr o'i gyfrifiadur os bydd yn gadael ei gyfrifiadur ac yntau'n dal wedi'i fewngofnodi
- gweld a yw cydrannau ar y rhwydweithiau'n methu
- cau gweithfannau nad ydynt yn gweithio'n iawn
- ailadeiladu gweithfannau drwy ychwanegu meddalwedd.

Cynllunio ar gyfer trychineb (copïau wrth gefn ac adfer)

Yn hwyr neu'n hwyrach bydd y mwyafrif o gwmnïau'n gorfod ymdopi â sefyllfa a fydd yn achosi colli caledwedd, meddalwedd, data neu wasanaethau cyfathrebu neu gyfuniad o'r rhain. Rhai o achosion posibl y colledion hyn yw:

- caledwedd yn methu
- namau yn y meddalwedd
- trychinebau naturiol (e.e. llifogydd, tân, daeargrynfeydd, corwyntoedd, ac ati)
- difrod bwriadol (difrod maleisus gan staff, firysau, hacio, fandaliaeth, bomiau terfysgwyr, ac ati)
- difrod damweiniol (e.e. dileu ffeiliau'n ddamweiniol, difrodi cyfarpar yn ddamweiniol, ac ati).

Mae angen cynllunio ar gyfer trychinebau er mwyn:

- lleihau tarfu ar waith y cwmni
- cael y systemau ar eu traed eto mor fuan â phosibl
- sicrhau bod yr holl staff yn gwybod beth i'w wneud i adfer data, rhaglenni ac ati.

Cydrannau meddalwedd (parhad)

Copïau wrth gefn

Nid yw colli rhaglenni a chaledwedd yn ormod o broblem gan ei bod hi'n bosibl cael rhai newydd. Ond os caiff data eu colli, ac os nad oes copi wrth gefn, gall fod yn anodd iawn cael y data'n ôl. Gallai colled o'r fath achosi cymaint o drafferth ariannol fel bod y cwmni'n mynd i'r wal.

Mae gwneud copi o'r data ar rwydwaith yn bwysig iawn. Dylai copïau wrth gefn gael eu:

- gwneud yn rheolaidd neu hyd yn oed yn awtomatig
- cadw dan glo i ffwrdd o'r cyfrifiadur mewn diogell wrthdan neu, yn well byth, oddi ar y safle
- gwneud gan ddefnyddio'r egwyddor taid-tad-mab yn achos systemau swp-brosesu
- defnyddio gyda system arae ddiangen o ddisgiau rhad (*RAID: random array of inexpensive disks*).

Adfer

Does dim pwynt gwneud copïau wrth gefn o ffeiliau heb sicrhau y gellir defnyddio'r copïau i adfer y ffeiliau gwreiddiol. Mae'n bwysig i'r staff ddeall yr hyn y mae angen iddynt ei wneud i adfer data sydd wedi'u colli neu eu difrodi. Dylai sefydliad wneud adferiad ffug o bryd i'w gilydd i sicrhau bod y system yn gweithio.

Gall adferiad gael ei wneud ar raddfa fach. Er enghraifft, gall chwalfa gyfrifiadurol olygu bod y copi presennol o ffeil wedi'i lygru. Mae gan y rhan fwyaf o feddalwedd ffordd o adfer ffeiliau o'r fath. Hefyd mae'n bosibl ail-greu ffeiliau sydd wedi cael eu dileu drwy ddamwain.

Archwilio (cadw logiau)

Pwrpas archwilio yw cadw cofnod o bwy sydd wedi gwneud beth ar rwydwaith. Mae archwilio'n cadw cofnod o:

- enwau defnyddwyr
- amserau mewngofnodi ac allgofnodi
- manylion y rhaglenni a ddefnyddiwyd
- manylion y ffeiliau sydd wedi cael eu cyrchu
- manylion y newidiadau sydd wedi cael eu gwneud.

Mae archwilio yn darganfod achosion o gamddefnyddio'r system gan ddefnyddwyr awdurdodedig ac achosion o fynediad heb awdurdod gan hacwyr.

Hawlfraint 2005 gan Randy Glasbergen
www.glasbergen.com

"Rydyn ni'n gwneud copïau wrth gefn o'n data ar nodion glynu gan nad ydyn nhw byth yn chwalu."

▶ Cwestiynau

1 Mae busnes lleol yn defnyddio rhwydwaith ardal leol ac mae'n rhan annatod o'u busnes.
   (a) Mae staff sy'n defnyddio'r rhwydwaith yn cael cyfrifon defnyddwyr. Disgrifiwch swyddogaeth cyfrifon defnyddwyr. (2 farc)
   (b) Mae rheolwr rhwydwaith yn gofalu am y rhwydwaith. Mae'n gyfrifol am reoli defnyddwyr y rhwydwaith. Mae'n defnyddio rheoli ffurfweddiad a rheoli o bell i helpu i reoli'r rhwydwaith. Disgrifiwch **ddwy** o swyddogaethau rheoli ffurfweddiad a **dwy** o swyddogaethau rheoli o bell. (4 marc)
   (c) Mae'n hanfodol i'r rhwydwaith gael ei ddiogelu. Eglurwch y rhan sy'n cael ei chwarae gan archwilio wrth sicrhau diogelwch y data sy'n cael eu storio ar rwydwaith. (3 marc)

2 Mae deintyddfa leol yn defnyddio rhwydwaith i reoli cofnodion y cleifion, apwyntiadau, cyflog y staff a'r holl swyddogaethau ariannol. Mae'r rheolwr gweinyddol yn poeni am gyfrinachedd cofnodion y cleifion.
   (a) Eglurwch pam y dylai fod gan y ddeintyddfa strategaeth ddiogelwch a rhowch **ddwy** enghraifft o'r hyn y dylai'r strategaeth ddiogelwch hon ei gynnwys. (4 marc)
   (b) Trafodwch y defnydd o gyfrifon a logiau defnyddwyr fel ffordd o sicrhau cyfrinachedd cofnodion cleifion. (3 marc)

3 Mae cwmni'n cyflwyno rhwydwaith newydd ac yn ystyried y materion diogelwch a fydd yn codi o hyn. Trafodwch y materion sy'n gysylltiedig â sicrhau diogelwch y rhwydwaith. (10 marc)

Cwestiynau

▶ Cwestiynau 1 — tt. 2–3

1 Mae sefydliad yn ystyried rhwydweithio ei gyfrifiaduron arunig i ffurfio rhwydwaith.

  (a) Eglurwch **dri** ffactor a fydd yn dylanwadu ar ei ddewis o rwydwaith. (6 marc)

  (b) Ysgrifennwch restr o **bum** cwestiwn y byddai angen i chi eu gofyn cyn gallu penderfynu ar rwydwaith addas i'r sefydliad. Ar gyfer pob cwestiwn, dylech ddisgrifio'n fyr pam y mae angen yr atebion. (5 marc)

2 Wrth ddewis rhwydwaith i gwmni, rhaid cymryd i ystyriaeth y perfformiad sydd ei angen o'r rhwydwaith. Nodwch ac eglurwch **dri** mater perfformiad hollol wahanol y bydd angen rhoi ystyriaeth iddynt. (6 marc)

3 Mae ysgol yn uwchraddio ei rhwydwaith. Mae'r prifathro'n poeni y gallai'r problemau diogelwch a gododd gyda'r hen rwydwaith godi gyda'r un newydd hefyd.

  Amlinellwch **dri** mater diogelwch y dylai'r rhwydwaith newydd roi sylw iddynt. (3 marc)

4 Mae sefydliad yn prynu rhwydwaith newydd. Disgrifiwch **dair** eitem wahanol o galedwedd y bydd angen eu prynu ac, ar gyfer pob un, eglurwch pam y mae eu hangen. (3 marc)

▶ Cwestiynau 2 — tt. 4–5

1 (a) Eglurwch **ddau** brif wahaniaeth rhwng rhwydwaith cleient/gweinydd a rhwydwaith cymar wrth gymar. (2 farc)

  (b) Mae rhwydwaith mawr yn cael ei adeiladu ar gyfer coleg. Nodwch pa un o'r **ddau** fath o rwydwaith sy'n cael eu crybwyll yn rhan (a) fyddai orau a rhowch ddau reswm dros eich dewis. (3 marc)

2 Mae cwmni'n ystyried rhwydweithio ei gyfrifiaduron arunig.

  (a) Rhowch **ddwy** o fanteision rhwydweithio'r cyfrifiaduron. (2 farc)

  (b) Trafodwch rinweddau cymharol rhwydweithiau'n seiliedig ar weinydd a rhwydweithiau cymar wrth gymar. (6 marc)

3 Mae chwech o staff yn cael eu cyflogi mewn swyddfa cwmni dylunio ceginau. Mae gan bob aelod o staff gyfrifiadur arunig ac argraffydd. Hoffai'r cwmni rwydweithio'r cyfrifiaduron hyn. Mae arbenigwr rhwydwaith wedi nodi y byddai'n fwy effeithlon pe byddai'r chwe chyfrifiadur yn cael eu cynnwys mewn rhwydwaith cymar wrth gymar.

  (a) Nodwch **dair** mantais i'r cwmni o rwydweithio eu systemau cyfrifiadurol fel system cymar wrth gymar yn hytrach na system wedi'i seilio ar weinydd. (3 marc)

  (b) Pa galedwedd ychwanegol fyddai ei angen i gysylltu'r chwe chyfrifiadur arunig fel rhwydwaith cymar wrth gymar? Nodwch pam y mae angen pob eitem o galedwedd. (4 marc)

▶ Cwestiynau 3 — tt. 6–7

1 Mae cwmni'n gosod rhwydwaith newydd mewn hen adeilad ac mae wedi penderfynu defnyddio rhwydwaith diwifr.

  (a) Disgrifiwch **ddwy** fantais a **dwy** anfantais defnyddio rhwydwaith diwifr. (4 marc)

  (b) Bydd y rhwydwaith diwifr yn rhwydwaith ardal leol (RhAL).

    (i) Nodwch beth yw ystyr rhwydwaith ardal leol a rhowch **ddau** reswm pam y mae'r rhwydwaith sydd wedi cael ei ddewis yn rhwydwaith ardal leol. (2 farc)

    (ii) Enwch **un** dopoleg sy'n addas ar gyfer rhwydwaith ardal leol a rhowch **un** fantais sydd i'r dopoleg yr ydych wedi'i dewis. (2 farc)

2 Dau fath o dopoleg rhwydwaith yw cylch a seren.

  (a) Enwch **un** math arall o dopoleg rhwydwaith, a lluniwch ddiagram neu eglurwch mewn geiriau sut y mae'r cyfrifiaduron wedi'u trefnu yn y dopoleg. (2 farc)

  (b) Cymharwch a chyferbynnwch fanteision ac anfanteision cymharol y topolegau cylch a seren. (6 marc)

Astudiaeth achos a Gweithgaredd

▶ Astudiaeth achos 1 — tt. 8–10

Gweithdrefnau ar gyfer rhwydwaith prifysgol

Mae pob prifysgol, coleg ac ysgol yn defnyddio rhwydweithiau ac mae cael set o weithdrefnau i'r staff a'r myfyrwyr yn helpu i osgoi problemau.

Mae'r ddogfen hon yn nodi'r gweithdrefnau ar gyfer rhwydwaith. Darllenwch nhw'n ofalus ac yna atebwch y cwestiynau sy'n dilyn.

1 **Enwau defnyddwyr a chyfrineiriau**
 (a) Mae angen enw defnyddiwr a chyfrinair ar yr holl staff a myfyrwyr i fewngofnodi i'r rhwydwaith.
 (b) Enw defnyddiwr yw'r rhif gweithiwr yn achos aelodau staff a'r rhif myfyriwr yn achos myfyrwyr.
 (c) Dylai fod gan gyfrineiriau o leiaf 7 nod a dylent gynnwys o leiaf 3 rhif.
 (ch) Dylai cyfrineiriau gael eu newid bob tri mis.

2 **Defnydd heb ei awdurdodi**
 (a) Peidiwch â rhannu eich cyfrinair â neb.
 (b) Os bydd unrhyw un heb awdurdod yn defnyddio'r rhwydwaith hwn, rhaid hysbysu rheolwr y rhwydwaith fel y gall chwilio am unrhyw ddifrod i'r rhwydwaith.

3 **Cyfrinachedd**
 (a) Ni ddylai staff ganiatáu i fyfyrwyr fewnbynnu neu weld graddau neu gofnodion electronig myfyrwyr eraill.
 (b) Caiff cyfeiriaduron eu cynnal o bryd i'w gilydd gan staff gweinyddol y rhwydwaith i helpu i sicrhau perfformiad a chyfanrwydd y rhwydwaith. Byddant yn edrych ar wybodaeth yn ymwneud â maint ffeiliau, nifer y ffeiliau ac ati yn unig.
 (c) Bydd staff gweinyddol y rhwydwaith yn cael caniatâd y defnyddiwr yn gyntaf pan fydd angen iddynt ddatrys problem sydd gan ddefnyddiwr neu pan fydd angen edrych ar ffeiliau defnyddiwr.

4 **Defnydd gan fyfyrwyr**
 (a) Rhaid monitro/goruchwylio myfyrwyr pan fyddant yn defnyddio'r rhwydwaith.
 (b) Dylai myfyrwyr ddefnyddio eu henw defnyddiwr a'u cyfrinair eu hunain yn unig i fewngofnodi i'r rhwydwaith.
 (c) Rhaid i'r holl ddata gael eu cadw yn adran y myfyriwr ar y rhwydwaith. Rhaid peidio â storio data ar y gyriant disg caled lleol.
 (ch) Ni chaiff myfyrwyr osod meddalwedd ar y gyriannau disg caled na'r rhwydwaith.

5 **Allgofnodi**
 (a) Pan fydd staff yn gadael eu gweithfan, dylent allgofnodi o'r rhwydwaith.
 (b) Dylai pob cyfrifiadur gael ei ddiffodd ar derfyn dydd.
 (c) Allgofnodwch bob amser cyn diffodd eich cyfrifiadur.

6 **Firysau**
 (a) Os dewch â chyfryngau o'ch cartref neu o rywle arall, sganiwch nhw am firysau.
 (b) Bydd pob gyriant disg caled yn cael ei sganio am firysau gan staff technegol cyn cael eu cysylltu â'r rhwydwaith.
 (c) Wrth ddefnyddio gliniaduron ar y rhwydwaith, sicrhewch fod copi cyfoes o sganiwr firysau wedi'i osod arnynt a bod hwn yn cael ei ddiweddaru'n rheolaidd.
 (ch) Os caiff ffeiliau eu llwytho i lawr o'r Rhyngrwyd, sicrhewch eu bod yn cael eu llwytho i lawr fel ffeiliau dros dro ac yna sganiwch nhw cyn eu hagor.

7 **Copïau wrth gefn**
 (a) Mae copi wrth gefn yn cael ei wneud o'r rhwydwaith bob dydd. Hefyd dylech wneud eich copïau eich hun o'ch ffeiliau personol ar gyfryngau symudadwy.

8 **Meddalwedd**
 (a) Mae'r lle storio sydd gennych yn gyfyngedig. Felly dylech gynnal eich ffeiliau drwy eu copïo i gyfryngau symudadwy a'u dileu o'r rhwydwaith.
 (b) Gall staff, ond nid myfyrwyr, osod meddalwedd cyfreithiol a gwreiddiol ar yriant caled eu cyfrifiaduron.
 (c) Dim ond rheolwr y rhwydwaith all ychwanegu meddalwedd at y rhwydwaith.

9 **Hawlfraint**
 (a) Rhaid i'r holl staff a myfyrwyr gydymffurfio â'r deddfau hawlfraint presennol.
 (b) Ni ddylai meddalwedd ar y rhwydwaith gael ei gopïo.
 (c) Ni ddylai ffeiliau awdio (e.e. MP3) a fideo sydd heb eu gwarchod gan hawlfraint gael eu storio ar y rhwydwaith.

10 **Caledwedd**
 (a) Rhaid i'r holl galedwedd ar y rhwydwaith aros ar y safle a rhaid sicrhau ei fod wedi'i gysylltu â'r rhwydwaith bob amser.
 (b) Rhaid peidio â benthyca cyfrifiaduron a chyfarpar arall a mynd â nhw adref.
 (c) Rhaid i galedwedd gael ei addasu neu ei atgyweirio gan dechnegwyr y rhwydwaith yn unig.

11 **Cyfryngau symudadwy**
 (a) Rhaid bod yn ofalus wrth roi cof pin i mewn i gyfrifiadur a'i dynnu allan.
 (b) Cadwch gyfryngau magnetig i ffwrdd o feysydd magnetig (top y sgrin, cypyrddau metel, ac ati).

12 **Datrys problemau**
 (a) Dylech wneud pob cais am gymorth i'r ddesg gymorth.
 (b) Mae cymorth ar gael 7 diwrnod yr wythnos o 8 a.m. hyd 6 p.m.

Gweler y dudalen nesaf am gwestiynau.

Cwestiynau: Gweithdrefnau ar gyfer rhwydwaith prifysgol

Mae'r holl gwestiynau sy'n dilyn yn cyfeirio at yr astudiaeth achos.

1 Mae'r cwestiwn hwn yn ymwneud â'r enwau defnyddwyr a chyfrineiriau sydd eu hangen i fewngofnodi i'r rhwydwaith hwn.

   (a) Eglurwch pam y caiff enwau defnyddwyr eu harddangos ond pam na chaiff cyfrineiriau byth eu harddangos ar y sgrin. (1 marc)

   (b) Eglurwch bwrpas:
       (i) enw defnyddiwr (1 marc)
       (ii) cyfrinair. (1 marc)

   (c) Mae rheolwr y rhwydwaith wedi dweud bod angen i ddefnyddwyr y rhwydwaith gael mwy o arweiniad ar ddewis cyfrineiriau. Disgrifiwch **bedwar** peth gwahanol y dylai defnyddwyr eu cadw mewn cof wrth ddewis cyfrineiriau. (4 marc)

   (ch) I fod yn effeithiol, mae angen newid cyfrineiriau'n rheolaidd. Ar gyfer y rhwydwaith hwn, dylai'r cyfrineiriau gael eu newid bob tri mis. Rhowch **un** rheswm pam y gallai rheolwr y rhwydwaith fod yn erbyn cynnig i wneud i ddefnyddwyr newid eu cyfrinair bob wythnos. (1 marc)

2 (a) Rhowch **un** rheswm pam y mae rheolwr y rhwydwaith yn caniatáu i fyfyrwyr storio data ar y rhwydwaith yn unig ac nid ar yriant disg caled lleol y cyfrifiadur y maen nhw'n gweithio arno. (2 farc)

   (b) Eglurwch pam y mae'r gweithdrefnau'n mynnu bod myfyrwyr yn cynnal eu lle storio eu hunain ar y rhwydwaith. (2 farc)

3 Nid yw rhai rhwydweithiau mewn busnesau yn caniatáu i gyfryngau symudadwy gael eu defnyddio yng nghyfrifiaduron y rhwydwaith.

   (a) Rhowch **un** rheswm dros hyn. (1 marc)

   (b) Eglurwch pam nad yw hyn yn bosibl mewn gwirionedd mewn rhwydwaith coleg neu brifysgol. (1 marc)

4 Eglurwch pam y dylai meddalwedd gael ei ychwanegu at y rhwydwaith gan reolwr y rhwydwaith yn unig. (2 farc)

5 Mae myfyrwyr yn cael defnyddio cyfryngau symudadwy i storio eu gwaith at ddibenion gwneud copïau wrth gefn.

   (a) Rhowch enwau **tair** enghraifft o gyfrwng symudadwy y gallai'r myfyrwyr eu defnyddio. (3 marc)

   (b) Gall myfyrwyr ddod â chofion pin a chyfryngau eraill gyda nhw i'r brifysgol. Rhowch **un** perygl i'r rhwydwaith sy'n deillio o hyn a nodwch beth y gellir ei wneud i leihau'r risg. (2 farc)

▶ **Gweithgaredd: Topolegau rhwydwaith**

Mae ymgynghorydd rhwydwaith yn ymweld â busnes bach ac yn darganfod bod ganddo ddeg cyfrifiadur arunig. Ar ôl mynd drwy broses darganfod ffeithiau ac ymchwilio i'r llifoedd gwybodaeth yn y busnes, mae'r ymgynghorydd wedi dweud wrth y busnes bod angen rhwydwaith ardal leol arno. Mae wedi dweud hefyd fod yna nifer o dopolegau rhwydwaith i ddewis o'u plith.

1   Eglurwch beth yw topoleg rhwydwaith.

2   Mae'r ymgynghorydd wedi dweud bod y topolegau rhwydwaith canlynol ar gael:
     Seren
     Bws
     Cylch
     Gan ddefnyddio'r 10 cyfrifiadur yn y swyddfa, lluniwch ddiagram ar gyfer pob topoleg sydd yn y rhestr uchod i ddangos sut y dylid cysylltu'r cyfrifiaduron.

3   Mae'r ymgynghorydd yn dweud, 'Mae'n bwysig cael y lled band yn gywir'. Beth yw ystyr y gosodiad hwn?

Cymorth gyda'r arholiad

Enghraifft 1

1 (a) Heblaw am gost y rhwydwaith a maint y corff, eglurwch **bedwar ffactor sy'n dylanwadu ar gwmni wrth ddewis rhwydwaith. (8 marc)**

   (b) Mae dau brif fath o rwydwaith y gallai cwmni eu dewis: cymar wrth gymar a chleient/gweinydd. Cymharwch a chyferbynnwch y **ddau** fath hyn o rwydwaith. (4 marc)

Ateb myfyriwr 1

1 (a) Faint o weithwyr sydd gan gwmni, gan y bydd hyn yn penderfynu faint o gyfrifiaduron sydd eu hangen.

      Faint o arian sydd i'w wario ar rwydwaith. Os nad oes llawer o arian, bydd angen rhwydwaith syml.

      Faint o bobl sydd wedi defnyddio rhwydwaith o'r blaen, gan y bydd hyn yn penderfynu pa mor gymhleth y gall y rhwydwaith fod.

      Cynllun y safle neu safleoedd. Os yw'r cwmni wedi'i leoli ar fwy nag un safle byddai angen RhAE, ond os yw ar un safle gallent ddefnyddio RhAL.

   (b) Mae rhwydwaith cymar wrth gymar yn rhatach gan nad oes angen gweinydd. Mae rhwydweithiau cymar wrth gymar yn symlach i'w gweithredu, felly nid oes angen un person arbennig i reoli'r rhwydwaith. Maen nhw'n llawer symlach i'w sefydlu na rhwydweithiau cleient/gweinydd.

      Mae rhwydweithiau cleient/gweinydd yn defnyddio data wedi'u canoli, sy'n golygu y gall yr holl ddefnyddwyr gyrchu'r un data. Mae hyn yn gwella cysondeb data. Gyda rhwydweithiau cleient/gweinydd mae copïau wrth gefn yn cael eu gwneud yn ganolog, felly mae'n fwy tebygol y cânt eu gwneud gan fod rhywun yn gyfrifol am hyn. Un o anfanteision rhwydwaith cleient/gweinydd yw y bydd y rhwydwaith yn methu os bydd y gweinydd yn torri i lawr.

Sylwadau'r arholwr

1 (a) Ni roddir unrhyw farciau am yr ateb cyntaf gan ei fod yn ymwneud â maint y sefydliad.

      Mae'r ail ateb yn sôn am gost. Ni roddir unrhyw farciau am hyn chwaith. Nid yw'r myfyriwr wedi darllen a deall y cwestiwn.

      Mae'r trydydd ateb yn ddilys ac yn ennill dau farc, gan ei fod yn llawer haws gweithredu rhwydwaith os yw'r defnyddwyr wedi defnyddio un yn barod.

      Mae'r pedwerydd ateb yn dda ac yn cael dau farc.

   (b) Mae'r holl bwyntiau hyn yn gywir ac yn cyfeirio at nodweddion y ddwy system, felly rhoddir marciau llawn. **(8 marc allan o 12)**

Ateb myfyriwr 2

1 (a) Bydd y perfformiad sydd ei angen o'r rhwydwaith yn penderfynu beth fydd maint y gweinydd sydd ei angen i redeg y rhwydwaith.

      Sut y caiff y system ei defnyddio. Os defnyddir y rhwydwaith yn rhyngweithiol gan yr holl ddefnyddwyr, bydd yn rhaid i'r rhwydwaith gael topoleg a chydrannau caledwedd sy'n gallu darparu cyflymder.

      A oes angen i'r rhwydwaith weithio gyda'r systemau presennol? Mae'n bosibl bod rhwydwaith yn ei le yn barod neu efallai bod yna raglenni y mae angen iddynt weithio gyda'r rhwydwaith newydd neu rwydwaith ychwanegol.

      Materion diogelwch. Mae angen rhwydwaith diogel iawn ar gyfer trafodion ariannol.

   (b) Mae rhwydweithiau cymar wrth gymar yn llawer rhatach i'w sefydlu gan nad oes angen gweinydd canolog drud na meddalwedd rhwydwaith arbenigol. Gyda rhwydwaith cymar wrth gymar nid yw'r data'n cael eu storio'n ganolog, ond gyda rhwydwaith cleient/gweinydd caiff yr holl raglenni a data eu storio'n ganolog.

      Mae rhwydweithiau cleient/gweinydd yn defnyddio gweinydd nerthol yng nghanol y system sy'n gwneud i'r rhwydwaith weithio'n llawer cyflymach na'r mwyafrif o rwydweithiau cymar wrth gymar.

      Mewn rhwydwaith cleient/gweinydd caiff yr holl weithgareddau diogelwch a gwneud copïau eu cynnal yn ganolog gan reolwr y rhwydwaith, sy'n golygu bod y broses yn fwy sicr na phe bai'n cael ei gadael i unigolion.

Sylwadau'r arholwr

1 (a) Ni roddir unrhyw farciau am grybwyll y ffactor ond hyd at ddau am esboniad mwy manwl. Felly, er enghraifft, pe bai'r myfyriwr wedi sôn am 'berfformiad' yn unig ni châi farc. Ond mae esboniad fel 'Bydd y perfformiad sydd ei angen o'r rhwydwaith...' yn cael y marc. Rhoddir dau farc yr un am y tri phwynt cyntaf ond mae'r pwynt olaf yn cael un marc yn unig gan fod angen rhagor o wybodaeth.

   (b) Dyma restr dda o gymariaethau. Sylwch ar y ffordd y mae'r myfyriwr wedi cymharu'r nodweddion. Mae'n llawer gwell gwneud hyn na dim ond cynhyrchu rhestr o'r nodweddion sydd gan bob math o rwydwaith, gan fod perygl o ailadrodd wedyn. Rhoddir mwy na phedwar pwynt perthnasol yma ac mae'r myfyriwr wedi cyfeirio at y ddwy system, felly caiff farciau llawn am yr ateb hwn. **(11 marc allan o 12)**

Ateb yr arholwr

1 (a) Dau farc am esboniad llawn o bob ffactor. Dim marc am ffactor ar ei phen ei hun a rhaid i'r ffactorau beidio â chynnwys cost na maint y sefydliad. Ffactorau megis:

Sut y defnyddir y system – bydd angen Rhwydwaith Ardal Leol neu Rwydwaith Ardal Eang (1) gan ddibynnu ar ba mor gymhleth y mae angen i'r rhwydwaith fod (1).

Sut y bydd angen i'r rhwydwaith gyd-fynd â'r systemau presennol – er enghraifft, gall rhwydwaith gael ei ehangu neu gall fod angen iddo gyfathrebu â rhwydwaith arall (1). Efallai y bydd y sefydliad am i'r rhwydwaith weithio gyda'r caledwedd neu feddalwedd presennol (1).

Pa berfformiad sydd ei angen o'r rhwydwaith – bydd hyn yn penderfynu pa galedwedd sydd ei angen a sut mae'n cael ei gysylltu (1).

Bydd angen gweinydd grymus iawn os yw traffig rhwydwaith trwm yn cael ei ragweld (1).

Materion diogelwch – pa mor ddiogel y mae angen i'r rhwydwaith fod (1). Mae hyn yn bwysig os bydd data ariannol neu bersonol yn cael eu trosglwyddo (1). Bydd angen dulliau fel amgryptio i gadw'r data'n ddiogel rhag hacwyr (1).

(b) Un marc yr un am y pwyntiau canlynol hyd at uchafswm o bedwar:

Mae gan rwydwaith cleient/gweinydd weinydd sy'n gyfrifiadur mwy nerthol (1) ond mae gan yr holl gyfrifiaduron mewn rhwydwaith cymar wrth gymar yr un statws (1).

Mae rhwydweithiau cymar wrth gymar yn symlach na rhwydweithiau cleient/gweinydd (1), felly fel rheol nid oes angen cael rheolwr rhwydwaith i oruchwylio'r rhwydwaith (1).

Mae rhwydweithiau cleient/gweinydd yn defnyddio storfeydd canolog o ddata a rhaglenni (1), ond caiff y data mewn rhwydwaith cymar wrth gymar eu storio yn yr holl gyfrifiaduron (1).

Mae rhwydweithiau cleient/gweinydd yn gwneud copïau wrth gefn yn ganolog, gan wella diogelwch ffeiliau (1), ond mewn rhwydweithiau cymar wrth gymar mae pob defnyddiwr yn gyfrifol am wneud copïau wrth gefn (1).

Gyda rhwydweithiau cleient/gweinydd mae'r cyfan yn dibynnu ar y gweinydd (1), ond os bydd un cyfrifiadur yn torri i lawr mewn rhwydwaith cymar wrth gymar, mae'n bosibl na fydd yn effeithio ar y rhwydwaith (1).

Enghraifft 2

2 Mae cwmni'n gosod rhwydwaith newydd. Rhaid i'r cwmni benderfynu rhwng topoleg cylch a thopoleg seren.

(a) Eglurwch ystyr y term topoleg. (2 farc)

(b) Bydd wyth cyfrifiadur yn y rhwydwaith. Gwnewch frasluniau i ddangos sut y caiff y cyfrifiaduron eu trefnu mewn rhwydwaith cylch a rhwydwaith seren. (2 farc)

(c) Eglurwch fanteision ac anfanteision cymharol rhwydwaith seren a rhwydwaith cylch. (6 marc)

Ateb myfyriwr 1

2 (a) Diagram sy'n dangos y rhwydwaith yw topoleg.

(b) Rhwydwaith cylch   Rhwydwaith seren

(c) Un o fanteision y dopoleg seren yw os bydd un o'r cyfrifiaduron yn methu yna gall y lleill barhau i gyrchu data.

Gyda thopoleg seren, mae'n hawdd iawn ychwanegu rhagor o gyfrifiaduron, ond mae hyn yn anoddach gyda thopoleg cylch.

Gyda thopoleg seren gallwch newid y llinellau cyfathrebu rhwng pob cyfrifiadur a'r gweinydd canolog. Felly, os defnyddir un cyfrifiadur ar y rhwydwaith ar gyfer ymholiadau rhyngweithiol, gallai cyswllt cyflym iawn i'r gweinydd gael ei ddefnyddio.

Un o anfanteision rhwydwaith seren yw bod angen rhagor o gebl ac mae cebl yn ddrud.

Sylwadau'r arholwr

2 (a) Nid yw'r ateb hwn yn ddigon manwl. Nid yw'n dweud bod topoleg yn dangos sut mae'r cyfrifiaduron/terfynellau neu ddyfeisiau eraill ar y rhwydwaith yn cyfathrebu â'i gilydd. Ni roddir unrhyw farciau.

(b) Mae'r cwestiwn yn nodi'n glir bod wyth cyfrifiadur yn cael eu rhwydweithio, felly dylai'r diagramau ddangos wyth. Ond mae'r topolegau'n gywir, felly rhoddir un marc.

(c) Rhoddir chwe marc yma, felly mae angen gwneud chwe phwynt hollol wahanol. Rhoddir un marc am y frawddeg gyntaf. Mae cymhariaeth rhwng y ddau rwydwaith yn yr ail frawddeg ac mae'n werth dau farc. Mae'r ddwy frawddeg nesaf yn werth dau farc gan fod yr ail frawddeg yn ychwanegu manylion pellach at y frawddeg gyntaf. Mae'r frawddeg olaf yn werth un marc. **(7 marc allan o 10)**

Cymorth gyda'r arholiad (parhad)

Ateb myfyriwr 2

2 (a) Mae'r term topoleg yn cyfeirio at gynllun y dyfeisiau cysylltiedig ar rwydwaith. Mae'n dangos sut mae'r dyfeisiau ar y rhwydwaith yn cyfathrebu â'i gilydd.

(b) Rhwydwaith cylch

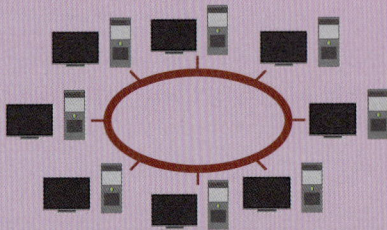

Rhwydwaith seren

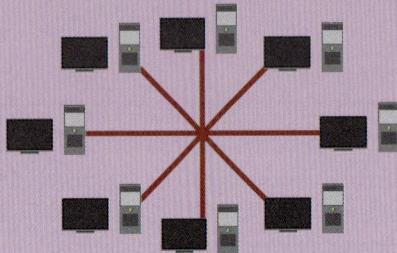

(c) Gyda thopoleg seren mae yna ddyfais ganolog sy'n rheoli'r data sy'n mynd heibio, ac os bydd hon yn methu bydd y rhwydwaith cyfan yn methu.

Mae rhwydwaith cylch yn symlach ond bydd hwn hefyd yn methu os bydd cebl neu un o'r cyfrifiaduron yn torri.

Bydd topoleg seren yn parhau i weithio hyd yn oed os bydd un o'r ceblau'n torri.

Mae'n hawdd ehangu rhwydwaith seren, drwy ychwanegu cyfrifiadur at y ddyfais ganolog. Mae'n anoddach ehangu rhwydweithiau cylch gan na ellir defnyddio'r rhwydwaith wrth i'r dyfeisiau ychwanegol gael eu cysylltu.

Sylwadau'r arholwr

2 (a) Mae'r ateb yn rhoi dau bwynt cywir, felly mae'n werth dau farc.

(b) Mae'r ddau ddiagram sy'n dangos y topolegau rhwydwaith yn gywir ac nid yw'r myfyriwr wedi cael ei gosbi am beidio â dangos cyfrifiadur, llwybrydd neu switsh yng nghanol y rhwydwaith seren, felly rhoddir dau farc.

(c) Dylai'r frawddeg gyntaf fod wedi rhoi enw'r ddyfais ganolog (cyfrifiadur, llwybrydd neu switsh) ond nid yw hyn yn cael ei gosbi yma. Un marc am y frawddeg hon.

Rhoddir un marc yr un am y pedair brawddeg arall.

(9 marc allan o 10)

Ateb yr arholwr

2 (a) Dau farc am esboniad clir, e.e.:

Mae topoleg yn fap o rwydwaith sy'n dangos sut mae'r gwahanol ddyfeisiau wedi'u cysylltu â'i gilydd. Dangosant y ffordd mae'r dyfeisiau'n cyfathrebu â'i gilydd.

(b) I ennill dau farc rhaid i bob topoleg fod yn gywir, gan ddangos y nifer cywir o gyfrifiaduron.

Braslun cywir o rwydwaith cylch yn dangos cyfrifiaduron wedi'u cysylltu mewn cylch.

Braslun cywir o rwydwaith seren yn dangos y cysylltiadau'n ymestyn allan o bwynt canolog.

(c) Un marc am bob pwynt ar gyfer y naill dopoleg neu'r llall.

Rhwydweithiau seren:

Mae angen dyfais galedwedd ychwanegol yn y canol ar rwydweithiau seren (cyfrifiadur, llwybrydd neu switsh).

Os bydd y ddyfais ganolog mewn rhwydwaith seren yn methu, bydd y rhwydwaith cyfan yn methu.

Os bydd cebl yn torri mewn rhwydwaith seren, bydd y terfynellau eraill yn parhau i weithio.

Gyda rhwydwaith seren, gallwch ddefnyddio llinellau cyflym rhwng terfynellau sydd angen mwy o gyflymder.

Rhwydweithiau cylch:

Nid yw rhwydweithiau cylch yn gallu ymdopi â namau cystal, oherwydd y bydd toriad mewn un cebl yn peri i'r rhwydwaith cyfan fethu.

Mae'n rhatach i'w osod oherwydd bod angen llai o gebl.

Rhaid cau'r rhwydwaith pan fydd rhagor o derfynellau'n cael eu hychwanegu.

Mae gan bob cyfrifiadur yr un mynediad i'r rhwydwaith, felly ni all un cyfrifiadur gadw'r rhwydwaith iddo'i hun.

Mapiau meddwl cryno

Ffactorau sy'n effeithio ar ddewis rhwydwaith

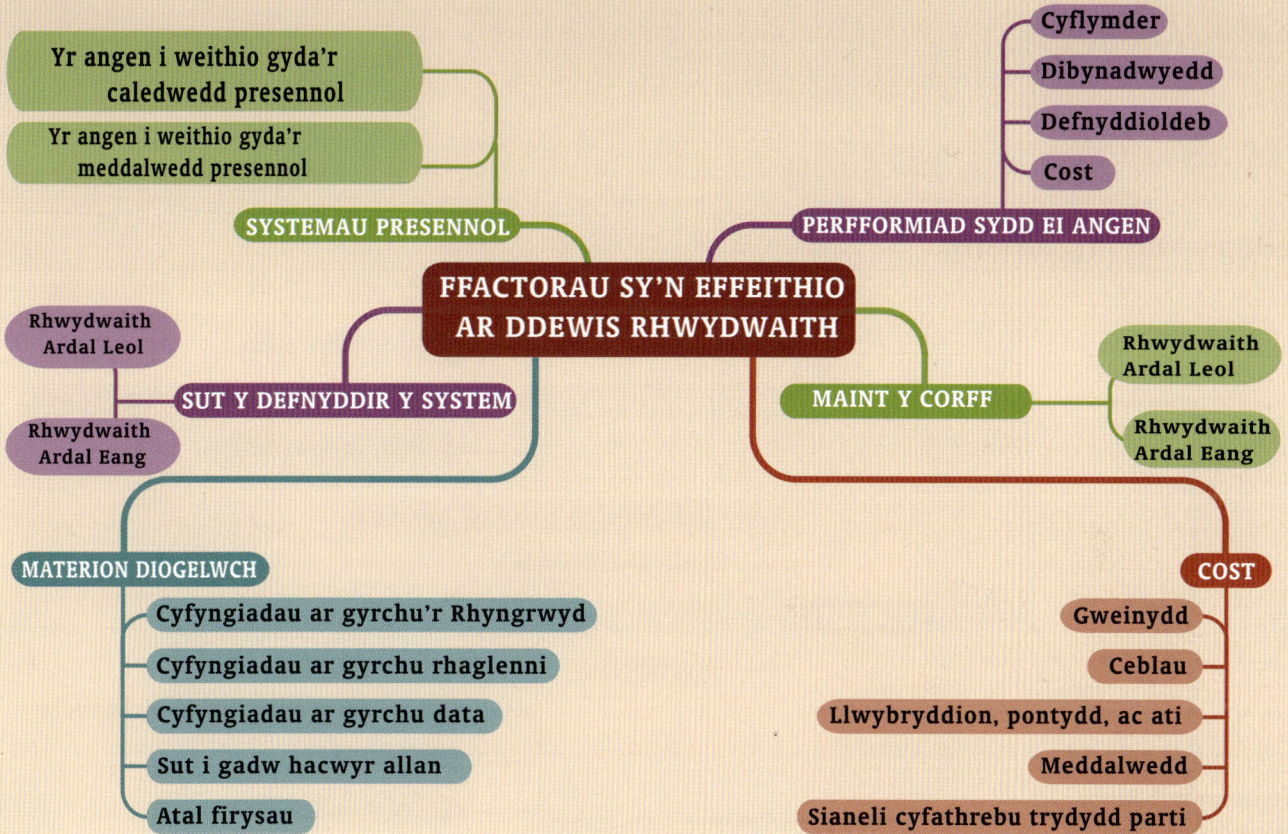

Yr angen i weithio gyda'r caledwedd presennol

Yr angen i weithio gyda'r meddalwedd presennol

SYSTEMAU PRESENNOL

Cyflymder

Dibynadwyedd

Defnyddioldeb

Cost

PERFFORMIAD SYDD EI ANGEN

FFACTORAU SY'N EFFEITHIO AR DDEWIS RHWYDWAITH

Rhwydwaith Ardal Leol

Rhwydwaith Ardal Eang

SUT Y DEFNYDDIR Y SYSTEM

MAINT Y CORFF

Rhwydwaith Ardal Leol

Rhwydwaith Ardal Eang

MATERION DIOGELWCH

Cyfyngiadau ar gyrchu'r Rhyngrwyd

Cyfyngiadau ar gyrchu rhaglenni

Cyfyngiadau ar gyrchu data

Sut i gadw hacwyr allan

Atal firysau

COST

Gweinydd

Ceblau

Llwybryddion, pontydd, ac ati

Meddalwedd

Sianeli cyfathrebu trydydd parti

Y ddau fath o rwydwaith

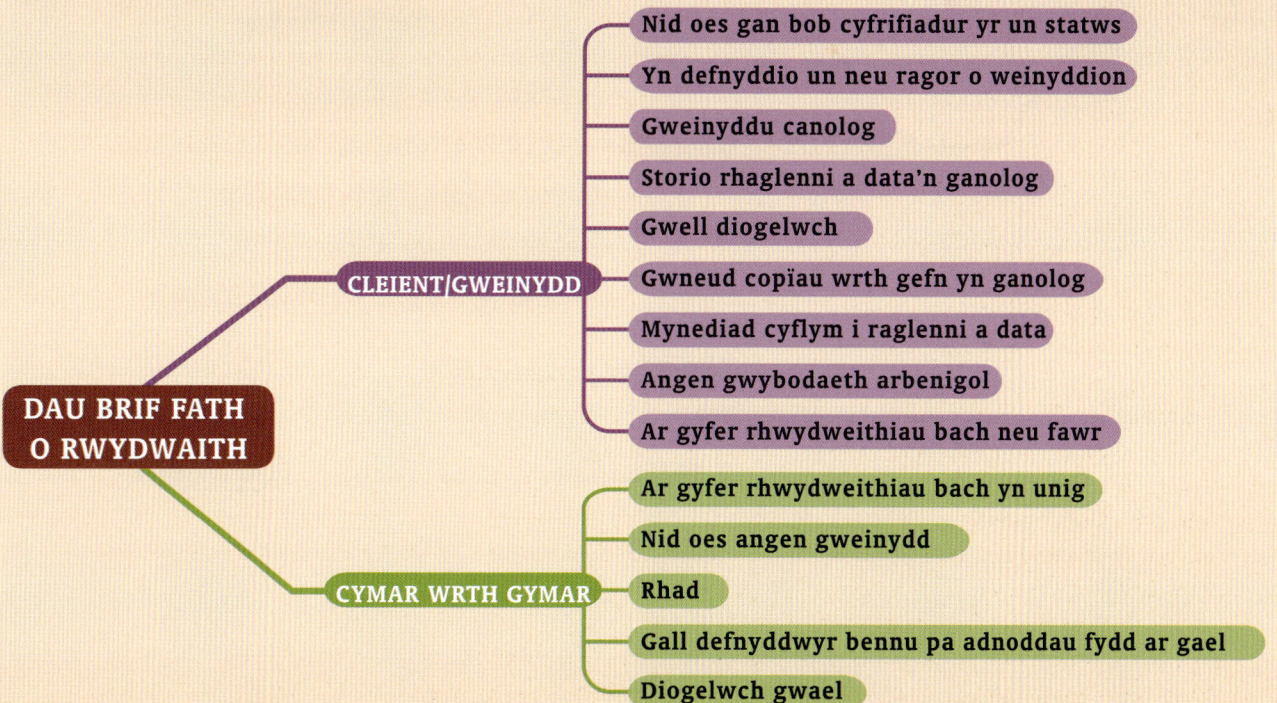

Nid oes gan bob cyfrifiadur yr un statws

Yn defnyddio un neu ragor o weinyddion

Gweinyddu canolog

Storio rhaglenni a data'n ganolog

Gwell diogelwch

Gwneud copïau wrth gefn yn ganolog

Mynediad cyflym i raglenni a data

Angen gwybodaeth arbenigol

Ar gyfer rhwydweithiau bach neu fawr

CLEIENT/GWEINYDD

DAU BRIF FATH O RWYDWAITH

Ar gyfer rhwydweithiau bach yn unig

Nid oes angen gweinydd

Rhad

Gall defnyddwyr bennu pa adnoddau fydd ar gael

Diogelwch gwael

CYMAR WRTH GYMAR

Meddalwedd rhwydwaith

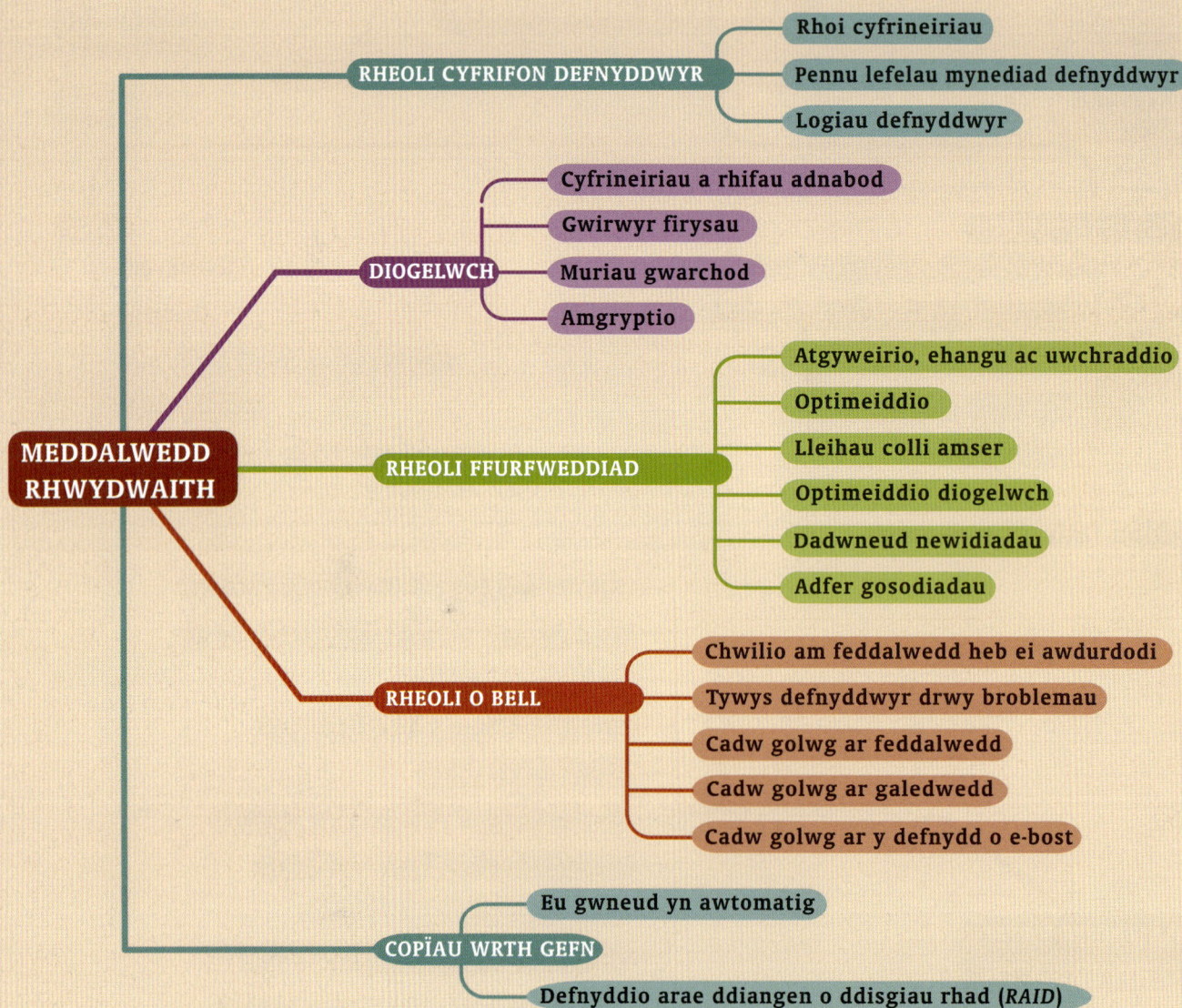

MEDDALWEDD RHWYDWAITH

RHEOLI CYFRIFON DEFNYDDWYR
- Rhoi cyfrineiriau
- Pennu lefelau mynediad defnyddwyr
- Logiau defnyddwyr

DIOGELWCH
- Cyfrineiriau a rhifau adnabod
- Gwirwyr firysau
- Muriau gwarchod
- Amgryptio

RHEOLI FFURFWEDDIAD
- Atgyweirio, ehangu ac uwchraddio
- Optimeiddio
- Lleihau colli amser
- Optimeiddio diogelwch
- Dadwneud newidiadau
- Adfer gosodiadau

RHEOLI O BELL
- Chwilio am feddalwedd heb ei awdurdodi
- Tywys defnyddwyr drwy broblemau
- Cadw golwg ar feddalwedd
- Cadw golwg ar galedwedd
- Cadw golwg ar y defnydd o e-bost

COPÏAU WRTH GEFN
- Eu gwneud yn awtomatig
- Defnyddio arae ddiangen o ddisgiau rhad (*RAID*)

TOPIG 2: Y Rhyngrwyd

Dros y blynyddoedd mae'r Rhyngrwyd wedi chwyldroi'r ffordd yr ydym yn gweithio a chwarae. Mae llawer yn ystyried mai mynediad i'r Rhyngrwyd yw'r allwedd i ddatblygu economaidd ac mae'n ffynhonnell incwm bwysig iawn i gwmnïau.

Yn y topig hwn byddwch chi'n edrych ar agweddau ar y Rhyngrwyd, y dulliau o gysylltu â'r Rhyngrwyd a'r materion moesol, cymdeithasol a moesegol sy'n gysylltiedig â defnyddio'r Rhyngrwyd.

▼ Y cysyniadau allweddol sy'n cael sylw yn y topig hwn yw:

▶ Effaith y Rhyngrwyd ar fusnes

▶ Cysylltu â'r Rhyngrwyd

▶ Materion moesol, cymdeithasol a moesegol sy'n gysylltiedig â'r Rhyngrwyd

CYNNWYS

Effaith y Rhyngrwyd ar fusnes

▼ **Byddwch yn dysgu**

▶ Beth yw'r Rhyngrwyd

▶ Am brotocol trosglwyddo ffeiliau

▶ Am e-fasnach

▶ Am gronfeydd data ar-lein

Cyflwyniad

Rydym wedi gweld newidiadau enfawr yn y ffordd mae busnesau'n cael eu rhedeg yn sgil dyfodiad y Rhyngrwyd. Mae'r Rhyngrwyd wedi rhoi cychwyn i lawer o fusnesau newydd sbon ac mae llawer o fusnesau traddodiadol wedi gorfod ymateb yn gyflym i'r her o weithio dros y Rhyngrwyd.

Yn yr adran hon byddwch yn dysgu am nodweddion y Rhyngrwyd sy'n fanteisiol i fusnes.

Beth yw'r Rhyngrwyd?

Dylech wybod am hyn o'ch gwaith UG, ond dyma ddiffiniad y gallwch ei ddefnyddio.

Grŵp enfawr o rwydweithiau wedi'u cysylltu â'i gilydd yw'r Rhyngrwyd. Mae pob un o'r rhwydweithiau hyn yn cynnwys llawer o rwydweithiau llai. Mae hyn yn golygu bod y Rhyngrwyd yn cynnwys caledwedd.

Protocol trosglwyddo ffeiliau (*FTP*)

Os ydych am drosglwyddo ffeil o un cyfrifiadur i gyfrifiadur arall gallwch, wrth gwrs, atodi'r ffeil wrth e-bost a'i hanfon fel atodiad ffeil. Mae'r dull hwn yn araf, a chan fod y lle storio ar weinydd eich darparwr gwasanaeth Rhyngrwyd (*ISP: Internet Service Provider*) neu weinydd Rhyngrwyd arall yn gyfyngedig, gallech fynd dros yr uchafswm a gallai'r trosglwyddiad gael ei wrthod.

Defnyddir *FTP* i drawsyrru unrhyw fath o ffeil: rhaglenni cyfrifiadurol, ffeiliau testun, graffigau, ac ati, drwy broses sy'n casglu'r data yn becynnau. Caiff pecyn o ddata ei anfon, ac mae'r system sy'n ei dderbyn yn gwirio'r pecyn i sicrhau nad oes unrhyw wallau wedi cael eu cyflwyno yn ystod y trawsyriad. Yna caiff neges ei hanfon i'r system anfon i roi gwybod iddi fod y pecyn yn iawn a'i bod yn barod i dderbyn y pecyn nesaf o ddata.

Pan drosglwyddwch ffeil gerddoriaeth drwy ei llwytho i lawr o'r Rhyngrwyd i'ch cyfrifiadur, rydych chi'n defnyddio *FTP*. Gallwch ddefnyddio *FTP* i lwytho ffeil i lawr o weinydd drwy'r Rhyngrwyd yn ogystal ag ar gyfer llwytho ffeil i fyny i weinydd. Un enghraifft o lwytho ffeil i weinydd yw rhoi tudalen we neu wefan ar weinydd. Defnyddir *FTP* gan gwmnïau i ddosbarthu gwybodaeth rhwng busnesau a phobl eraill megis cwsmeriaid a chyflenwyr.

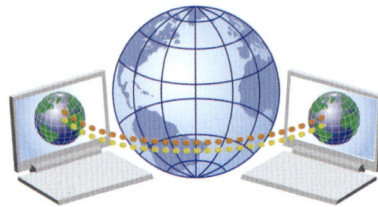

Mae *FTP* yn caniatáu i ffeiliau gael eu trosglwyddo drwy'r Rhyngrwyd

Prif fantais defnyddio *FTP* yn hytrach nag atodi ffeiliau wrth e-byst i'w hanfon yw eich bod chi'n gallu trosglwyddo ffeiliau eithriadol o fawr gyda phrotocol trosglwyddo ffeiliau, ond mae maint y ffeiliau y gallwch eu hanfon fel atodiadau e-bost wedi'i gyfyngu.

E-fasnach

Bydd llawer o bobl yn defnyddio e-fasnach i brynu nwyddau megis bwyd, llyfrau, cryno ddisgiau, nwyddau trydanol, ac ati. Er mwyn i gwmni weithredu gwasanaeth siopa ar-lein rhyngweithiol, rhaid cyflawni nifer o ofynion.

Y gofynion ar gyfer gwasanaeth siopa ar-lein rhyngweithiol

Prif ofynion gwasanaeth siopa ar-lein yw:

- Staff hyfforddedig i greu a chynnal y wefan – mae angen arbenigedd technegol a medrau dylunio i greu gwefan e-fasnach dda a diweddaru'r safle gyda chynhyrchion newydd, prisiau newydd, nodweddion newydd, ac ati.
- Catalog/cronfa ddata electronig o stoc – cronfa ddata o'r stoc gyda rhyngwyneb defnyddiwr sy'n rhoi gwybodaeth i'r defnyddiwr am gynhyrchion, prisiau, ac ati.
- Gall cwsmeriaid ddefnyddio'r cyfleuster chwilio i fynd yn syth at y cynnyrch sydd o ddiddordeb iddynt neu gallant bori.
- Dulliau talu diogel/troli siopa – rhaid i gwsmeriaid allu ychwanegu nwyddau at eu troli ac yna mynd at y ddesg dalu ar-lein i dalu amdanynt. Mae angen cadw manylion cardiau credyd a debyd yn ddiogel wrth eu trawsyrru dros y Rhyngrwyd, a defnyddir amgryptio i sicrhau na all hacwyr gyrchu'r wybodaeth hon.

Gofynion ar gyfer gwasanaeth siopa ar-lein

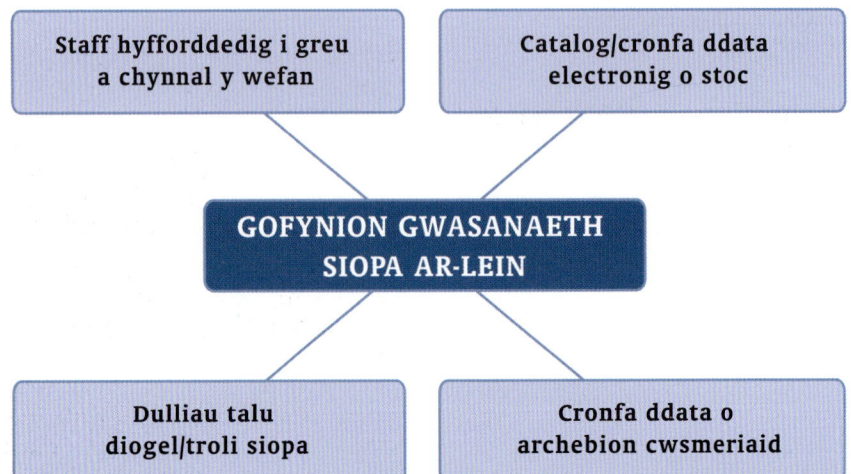

Staff hyfforddedig i greu a chynnal y wefan	Catalog/cronfa ddata electronig o stoc

GOFYNION GWASANAETH SIOPA AR-LEIN

Dulliau talu diogel/troli siopa	Cronfa ddata o archebion cwsmeriaid

- Cronfa ddata o archebion cwsmeriaid
  – mae'r gronfa ddata hon yn cadw
  manylion y nwyddau mae cwsmeriaid
  wedi'u harchebu a defnyddir hi os caiff
  nwyddau eu dychwelyd. Gall gael ei
  defnyddio hefyd ar gyfer archebion
  pellach gan yr un cwsmer; nid oes
  angen iddo roi'i holl fanylion i mewn
  eto ac mae hyn yn arbed amser. Gall y
  gronfa ddata helpu gyda strategaeth
  farchnata'r siop hefyd gan ei bod yn
  dangos beth mae cwsmeriaid ei eisiau.

Rhai o'r botymau a ddefnyddir mewn
siop ar-lein

GEIRIAU ALLWEDDOL

Protocol trosglwyddo ffeiliau *(FTP: file
transfer protocol)* – protocol (ffordd
o wneud pethau) safonol sy'n
cynnig ffordd syml o drosglwyddo
ffeiliau rhwng cyfrifiaduron gan
ddefnyddio'r Rhyngrwyd

Manteision ac anfanteision e-fasnach

Mae gan e-fasnach fanteision ac
anfanteision i gwsmeriaid a busnesau
ac mae'r rhain wedi'u crynhoi yn y
diagramau sy'n dilyn.

Manteision i gwsmeriaid

Y gallu i archebu nwyddau 24/7

Dewis ehangach o nwyddau o farchnad fyd-eang

Nwyddau rhatach yn golygu arbedion i'r cwsmer

MANTEISION E-FASNACH I GWSMERIAID

Dim angen teithio – nwyddau'n cael eu harchebu o'r cartref

Gall pobl anabl wneud eu siopa eu hunain

Anfanteision i gwsmeriaid

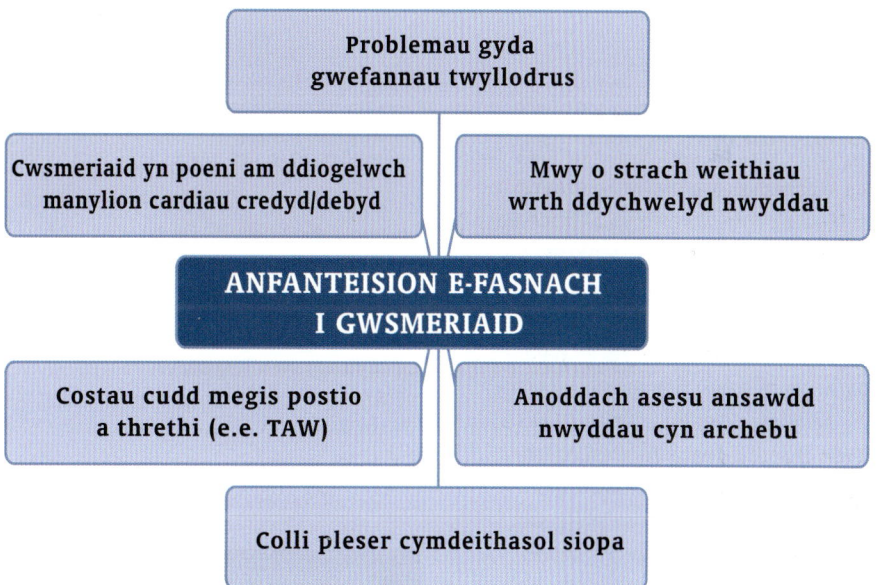

Problemau gyda gwefannau twyllodrus

Cwsmeriaid yn poeni am ddiogelwch manylion cardiau credyd/debyd

Mwy o strach weithiau wrth ddychwelyd nwyddau

ANFANTEISION E-FASNACH I GWSMERIAID

Costau cudd megis postio a threthi (e.e. TAW)

Anoddach asesu ansawdd nwyddau cyn archebu

Colli pleser cymdeithasol siopa

Manteision i fusnesau

Haws diweddaru catalogau ar-lein

Costau cychwyn a rhedeg is o'u cymharu â siopau traddodiadol

Angen llai o staff

MANTEISION E-FASNACH I FUSNESAU

Prisio hyblyg – haws newid prisiau o ddydd i ddydd

Y gallu i gyrraedd cwsmeriaid ar unrhyw adeg o'r dydd (24/7)

Marchnad fyd-eang – cwsmeriaid o bedwar ban byd

Effaith y Rhyngrwyd ar fusnes *(parhad)*

Anfanteision i fusnesau

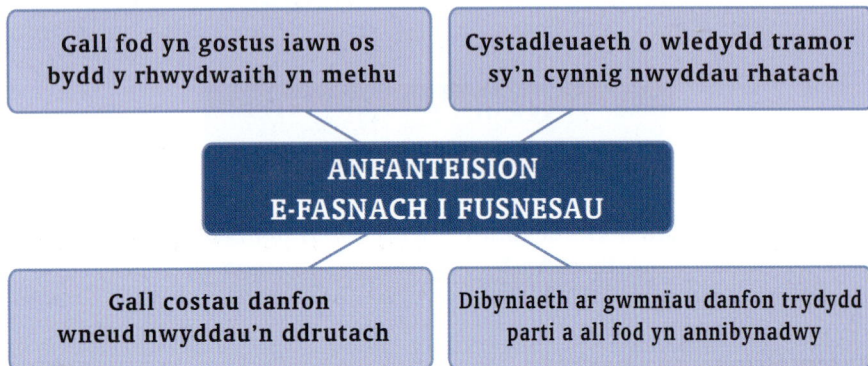

Gall fod yn gostus iawn os bydd y rhwydwaith yn methu	Cystadleuaeth o wledydd tramor sy'n cynnig nwyddau rhatach

ANFANTEISION E-FASNACH I FUSNESAU

Gall costau danfon wneud nwyddau'n ddrutach	Dibyniaeth ar gwmnïau danfon trydydd parti a all fod yn annibynadwy

Cronfeydd data ar-lein

Un o'r manteision mawr i gwmni sy'n defnyddio rhwydwaith yn hytrach na chyfres o gyfrifiaduron arunig yw bod yr holl ddefnyddwyr yn gallu cyrchu'r un gronfa ddata, gan osgoi gorfodi dyblygu data. Gan fod y data'n cael eu haddasu'n barhaus, byddant bob amser yn gyfoes.

Felly mae'r gallu i ryngweithio gyda chronfa ddata yn wasanaeth pwysig sy'n cael ei ddarparu gan unrhyw rwydwaith. Mae systemau e-fasnach yn defnyddio cronfeydd data ar-lein sy'n cynnwys manylion cynhyrchion, cwsmeriaid ac archebion. Mae ymgyfnewid data electronig (*EDI: electronic data interchange*), lle mae cwmnïau'n cyfnewid data'n awtomatig, yn cael ei ddefnyddio gan adwerthwyr mawr i archebu nwyddau'n awtomatig gan eu cyflenwyr. Mae'r system hon yn archebu, yn gwirio ac yn gwneud taliadau yn awtomatig heb fod angen unrhyw waith papur.

Sut i gyrchu gwybodaeth ar-lein

Mae tair prif ffordd o ddod o hyd i wybodaeth ar y Rhyngrwyd:

1 Defnyddio'r Lleolydd Adnoddau

Unffurf (*URL: Uniform Resource Locator*): os gwyddoch beth yw cyfeiriad gwe (*URL*) gwefan gallwch ei deipio i mewn. Os na wyddoch beth yw cyfeiriadau gwefannau sydd o ddiddordeb i chi, gallwch brynu llyfrau (cyfeiriaduron) neu gylchgrawn Rhyngrwyd sy'n eu cynnwys. Bellach mae cyfeiriadau Rhyngrwyd yn gymaint rhan o hunaniaeth cwmnïau â'u rhifau ffôn neu ffacs, ac maen nhw i'w gweld bob amser bron yn eu hysbysebion.

2 Pori neu syrffio'r Rhyngrwyd drwy ddilyn hypergysylltiadau: gallwch ddefnyddio hypergysylltiadau ar y Rhyngrwyd i symud o un maes sydd o ddiddordeb i faes arall. Mae'r cysylltau hyperdestun hyn naill ai ar ffurf testun wedi'i danlinellu neu mewn lliw gwahanol y cliciwch arno i symud i'r safle hwnnw. Yn y modd hwn gallwch symud o safle i safle. Os ydych yn ddylunydd gwefannau mae'n rhaid i chi sicrhau bod digon o gysylltau â'ch gwefan gan y bydd hyn yn denu mwy o 'borwyr' neu 'syrffwyr'.

3 Defnyddio peiriant chwilio: gallwch ddefnyddio rhaglen arbennig o'r enw peiriant chwilio. Rydych chi'n teipio geiriau allweddol neu enwau pynciau i mewn a bydd y rhaglen yn chwilio am wefannau sy'n cynnwys y geiriau hyn. Oherwydd y swm enfawr o ddeunydd sydd ar y Rhyngrwyd, gall fod yn anodd gwybod sut i lunio chwiliadau i'w gwneud yn fwy penodol.

Mae nifer o beiriannau chwilio ar gael, gan gynnwys:

Google: http://www.google.com
Alta Vista: http://www.altavista.com
Yahoo: http://www.yahoo.com
Lycos: http://www.lycos.com

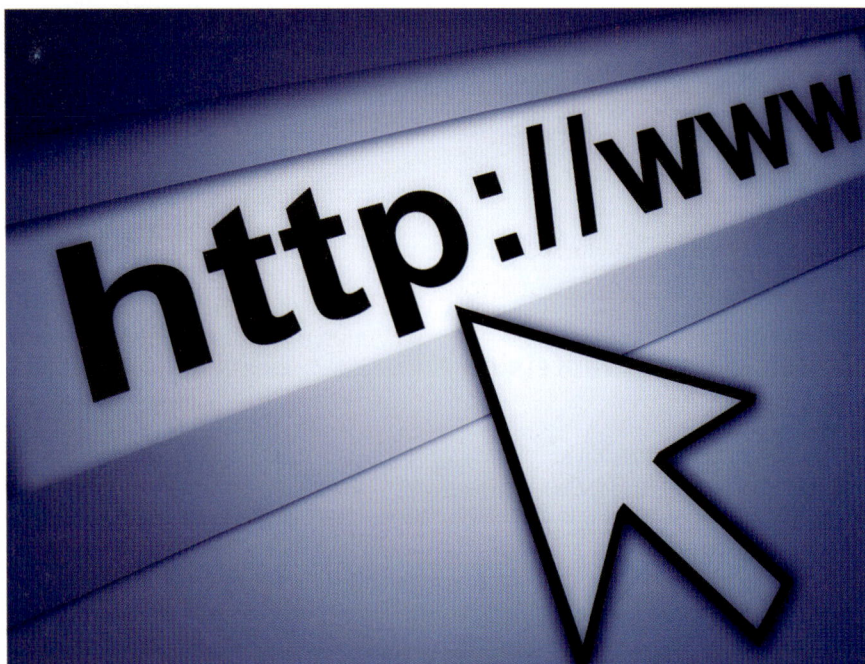

Enw unigryw ar dudalen we ar y Rhyngrwyd yw Lleolydd Adnoddau Unffurf (*URL*)

Sut mae peiriant chwilio'n gweithio

Cyn i beiriant chwilio allu dod o hyd i wybodaeth, rhaid iddo storio gwybodaeth am yr holl dudalennau gwe sydd ar y Rhyngrwyd. Mae cannoedd o filiynau o dudalennau gwe wedi'u storio, felly mae hon yn dasg anodd iawn, ond nid yw'n cael ei gwneud â llaw. Yn lle hynny, mae porwr gwe o'r enw gwe-ymlusgwr (neu we-gopyn) yn dilyn yr holl gysylltau y gall ddod o hyd iddynt yn awtomatig, ac wrth ddod ar draws pob tudalen we unigol mae'n dadansoddi ei chynnwys – o ran penawdau, is-benawdau, geiriau allweddol, ac ati – i benderfynu sut y dylai gael ei mynegeio. Caiff y mynegai hwn ei storio mewn cronfa ddata. Mae'r data y mae'r gwe-ymlusgwr yn eu darganfod am bob tudalen yn cael eu hychwanegu at y mynegai hwn.

Pan ddefnyddir peiriant chwilio i wneud chwiliad drwy ddefnyddio nifer o eiriau allweddol, bydd yn chwilio'r mynegai am y geiriau hyn ac yn cynhyrchu rhestr o'r tudalennau gwe sy'n cyfateb orau.

Fel rheol, mae'r rhestr yn rhoi teitl y dudalen a rhai manylion cryno am ei chynnwys. Mae llawer o beiriannau chwilio yn graddio'r tudalennau gwe yn ôl eu perthnasedd i'r amod chwilio, gan restru'r rhai mwyaf perthnasol yn gyntaf.

Gwe-ymlusgwr

Mae gwe-ymlusgwr, neu we-gopyn, yn rhaglen sy'n pori pob tudalen we yn awtomatig mewn ffordd systematig. Mae'n gwneud hyn i ddarparu data cyfoes am y tudalennau gwe er mwyn cynhyrchu mynegai y gall peiriant chwilio ei ddefnyddio i wneud chwiliadau cyflym yn bosibl.

Sut mae tudalennau gwe'n cael eu hychwanegu at restri peiriant chwilio

Caiff tudalennau gwe eu hychwanegu at restr peiriant chwilio mewn sawl ffordd, gan ddibynnu ar y peiriant chwilio sy'n cael ei ddefnyddio. Fel arfer bydd peiriant chwilio yn graddio'r tudalennau sy'n cael eu cynhyrchu gan chwiliad yn ôl y meini prawf canlynol:

- y gyfatebiaeth fwyaf perthnasol i'r amod chwilio
- y wefan fwyaf poblogaidd
- y wefan fwyaf awdurdodol
- gallu cwmni i dalu am safle uwch yn y rhestr (nid yw pob peiriant chwilio yn caniatáu hyn).

Gallwch chwilio am ddelweddau yn ogystal â thestun

Chwiliadau Boole

Mae chwiliadau Boole yn eich helpu i arbed amser wrth chwilio am wybodaeth, felly maen nhw'n ddefnyddiol iawn. Pan wnewch chwiliad syml mae'n bosibl y cewch eich llethu gan yr holl wybodaeth. Mae chwiliadau Boole yn eich helpu i wneud chwiliad mwy penodol.

AND (AC)

Os teipiwch y chwiliad **USA AND flag**, fe gewch yr holl ddogfennau sy'n cynnwys y ddau air hyn.

Os teipiech **USA Flag**, byddwch yn dal i gael yr holl ddogfennau sy'n cynnwys y ddau air.

Gyda'r mwyafrif o beiriannau chwilio nid oes angen i chi deipio'r gair 'AND' rhwng y geiriau.

Mae AND yn golygu 'Rydw i **dim ond** eisiau dogfennau sy'n cynnwys y **ddau** air.'

OR (NEU)

Os teipiwch y chwiliad **USA OR flag**, fe gewch yr holl ddogfennau sy'n cynnwys y gair **USA** a'r holl ddogfennau sy'n cynnwys y gair **flag** a'r holl ddogfennau sy'n cynnwys y ddau air.

Mae OR yn golygu 'Rydw i eisiau dogfennau sy'n cynnwys y naill air neu'r llall. Does dim gwahaniaeth pa air.'

NOT (NID)

Tybiwch eich bod chi'n chwilio am wybodaeth am wahanol anifeiliaid anwes ond eich bod chi'n casáu cathod. Gallwch beidio â chynnwys cathod fel hyn: **Pets NOT cats**.

Felly os ydych am chwilio am wybodaeth am anifeiliaid anwes yn gyffredinol ond nid am gathod, teipiwch **Pets NOT cats**.

Chwilio am union gyfatebiaeth

Os ydych chi'n chwilio am eiriau sy'n cyfateb yn union (h.y. mae'r geiriau ochr-yn-ochr ac yn yr un drefn), yna rhowch y geiriau mewn dyfynodau fel hyn: '*Cynulliad Cenedlaethol Cymru*'.

Chwilio am ddyfyniad

Os ydych chi'n gwybod beth yw union eiriad y dyfyniad, gallwch ei deipio yn y peiriant chwilio. I gael union gyfatebiaeth mae angen rhoi'r geiriau mewn dyfynodau fel hyn: '*Yma o hyd*'.

Dewis y wybodaeth i'w defnyddio

Pan chwiliwch am wybodaeth, rydych chi'n cael gormod ohoni fel rheol. Wrth benderfynu pa wybodaeth i'w defnyddio, mae angen i chi ofyn y cwestiynau canlynol:

- A yw'r wybodaeth yn berthnasol i'r pwrpas?
- A yw ffynhonnell y wybodaeth yn ddibynadwy?
- A yw'r wybodaeth yn gywir (h.y. a ellir ei gwirio)?
- A yw'r wybodaeth yn ddiduedd?
- A yw'r wybodaeth yn fanwl?
- A yw'r wybodaeth yn briodol i'r gynulleidfa dan sylw?

> ### ➤ GEIRIAU ALLWEDDOL
>
> Lleolydd Adnoddau Unffurf (*URL: Uniform Resource Locator*) – y cyfeiriad gwe a ddefnyddir i leoli tudalen we neu wefan
>
> Peiriant chwilio – meddalwedd y gellir ei ddefnyddio i chwilio am wybodaeth ar y Rhyngrwyd

Effaith y Rhyngrwyd ar fusnes *(parhad)*

Cyfrifiadura gwasgaredig drwy'r Rhyngrwyd

Un broblem fawr wrth gasglu data yw bod cymaint o wybodaeth i'w dadansoddi. Mae angen llawer o rym cyfrifiadurol i ddadansoddi swm mawr o ddata ac mae'n bosibl na fydd y grym hwn ar gael oherwydd y gost.

Yn aml iawn gellir datrys y broblem hon drwy ddefnyddio cyfrifiadura gwasgaredig drwy'r Rhyngrwyd. Yn lle defnyddio un uwch-gyfrifiadur enfawr a drudfawr i wneud y gwaith, gellir defnyddio ychydig o gyfrifiaduron llai grymus sydd wedi'u cysylltu â'i gilydd drwy'r Rhyngrwyd ond sydd i gyd yn gweithio ar yr un broblem. Mae llawer o gyfrifiaduron yn segur am gyfnodau hir, felly pam na ddylem ddefnyddio eu grym prosesu?

Mae rhai systemau gwasgaredig yn gofyn i ddefnyddwyr cartref gyfrannu peth o'u hadnoddau cyfrifiadurol. Er enghraifft, mae ymchwil i newid yn yr hinsawdd yn cael ei wneud ar hyn o bryd sy'n gofyn am brosesu data ar raddfa enfawr. Gall defnyddwyr cartref gyfrannu peth o amser segur eu cyfrifiaduron i helpu'r project hwn.

Manteision cyfrifiadura gwasgaredig

- mae'n lleihau costau oherwydd nad oes angen uwch-gyfrifiadur grymus a drud
- gall ddefnyddio'r Rhyngrwyd i drosglwyddo gwaith i gyfrifiaduron rhywle yn y byd
- perfformiad gwell oherwydd y gall pob cyfrifiadur weithio ar ran o'r data.

Anfantais cyfrifiadura gwasgaredig

- problemau gyda diogelwch data sydd wedi'u gwasgaru ar draws cymaint o wahanol gyfrifiaduron.

Y project i Chwilio am Ddeallusrwydd Allfydol (SETI: Search for Extraterrestrial Intelligence)

Pwrpas y project hwn yw chwilio am fywyd deallus y tu hwnt i'r Ddaear, ac i wneud hyn fe ddefnyddir telesgop radio fel yr un yn y ffotograff. Mae telesgopau radio yn codi signalau radio o'r gofod. Pe bai'r signalau wedi'u cyfyngu i amrediad cul o amleddau, byddai gennym dystiolaeth o fywyd allfydol.

Y broblem sydd gan y project yw bod llawer iawn o sŵn cefndir, gan gynnwys signalau radio o orsafoedd teledu, radar, lloerennau a ffynonellau wybrennol. Mae'n anodd iawn dadansoddi'r data o delesgopau radio a chwilio am signalau eraill a allai gadarnhau bywyd allfydol.

Er mwyn chwilio am signalau mewn lled band cul, mae angen llawer o rym cyfrifiadura. Ar y dechrau câi uwch-gyfrifiaduron a oedd yn cynnwys prosesyddion paralel eu defnyddio

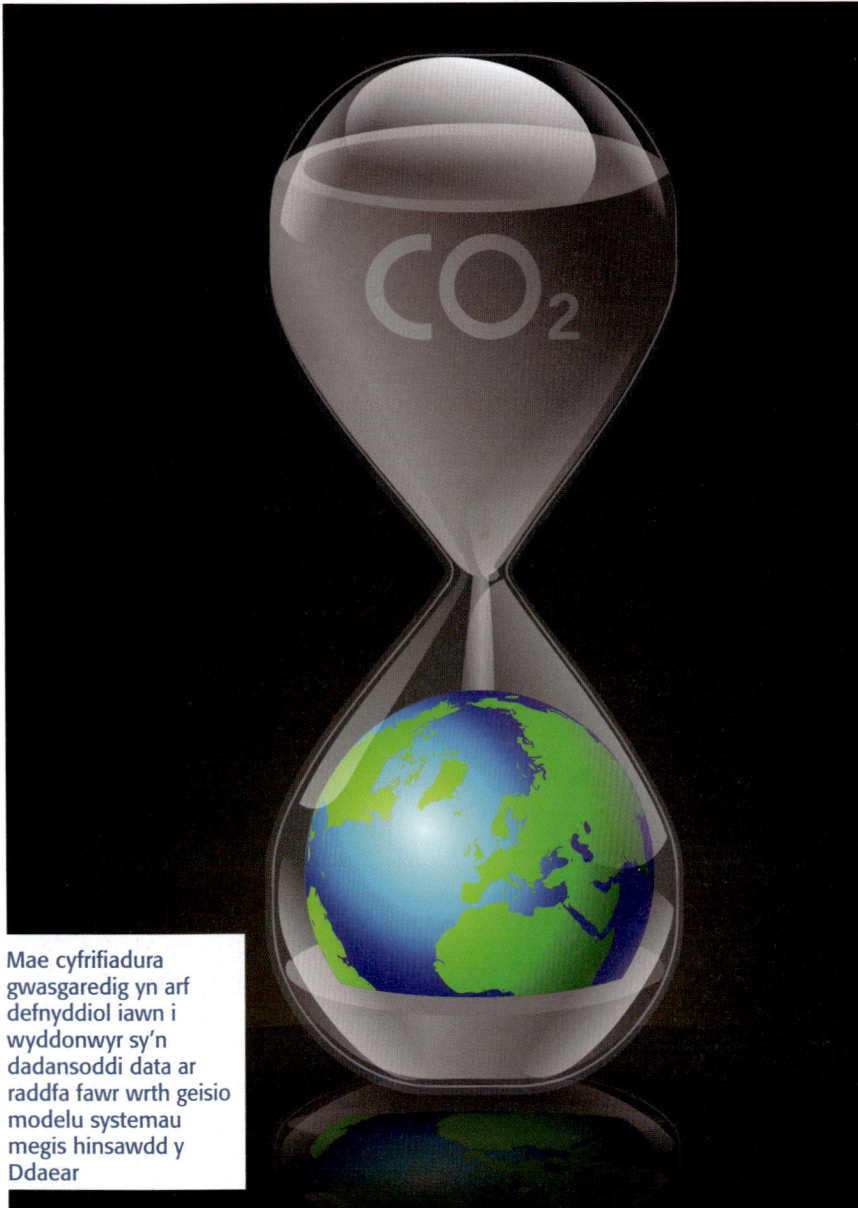

Mae cyfrifiadura gwasgaredig yn arf defnyddiol iawn i wyddonwyr sy'n dadansoddi data ar raddfa fawr wrth geisio modelu systemau megis hinsawdd y Ddaear

i brosesu'r swm enfawr o ddata o'r telesgopau. Yna meddyliodd rhywun am y syniad o ddefnyddio cyfrifiadur rhith yn cynnwys nifer mawr o gyfrifiaduron cartref wedi'u cysylltu â'r Rhyngrwyd. Cafodd y project ei enwi'n *SETI@home* a bu'n rhedeg er 1999.

Ar hyn o bryd mae 170,000 o wirfoddolwyr ar hyd a lled y byd yn cymryd rhan yn y project *SETI@home* ac mae'n defnyddio 320,000 o gyfrifiaduron, ond mae angen llawer mwy!

Y project Popular Power: *helpu i ddatblygu brechlynnau ffliw*

Project ymchwil nad yw er elw yw *Popular Power* a'i nod yw helpu i ddatblygu brechlynnau ffliw. Mae'r system yn defnyddio model cyfrifiadurol o'r system imiwn ddynol ac mae efelychiadau'n cael eu cynnal i ddarganfod effeithiolrwydd gwahanol frechlynnau. Y drafferth yw bod firysau ffliw yn mwtanu, felly mae'n bosibl na fydd brechlyn sy'n gweithio un flwyddyn yn gweithio'r flwyddyn nesaf.

Ers dechrau'r project, mae miloedd o gyfrifiaduron cartref wedi cwblhau miliynau o dasgau, ac mae hyn yn debyg i un cyfrifiadur pwerus iawn yn treulio cannoedd o flynyddoedd ar y broblem.

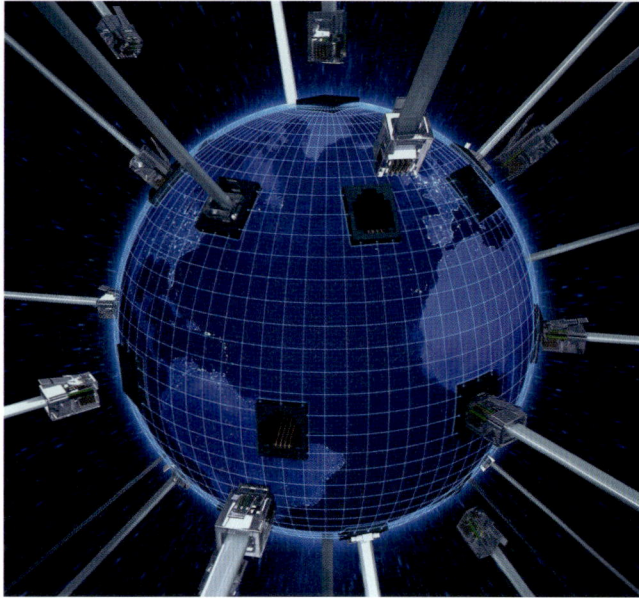

Mae'r project *SETI@home* wedi datrys y broblem o brosesu llawer iawn o ddata drwy rannu'r prosesu gyda chyfrifiaduron cartref drwy'r Rhyngrwyd

Mae Telesgop Allen yn gyfres o 42 o delesgopau radio sy'n casglu signalau o alaethau pell ac yn bwydo'r data i'r project *SETI*

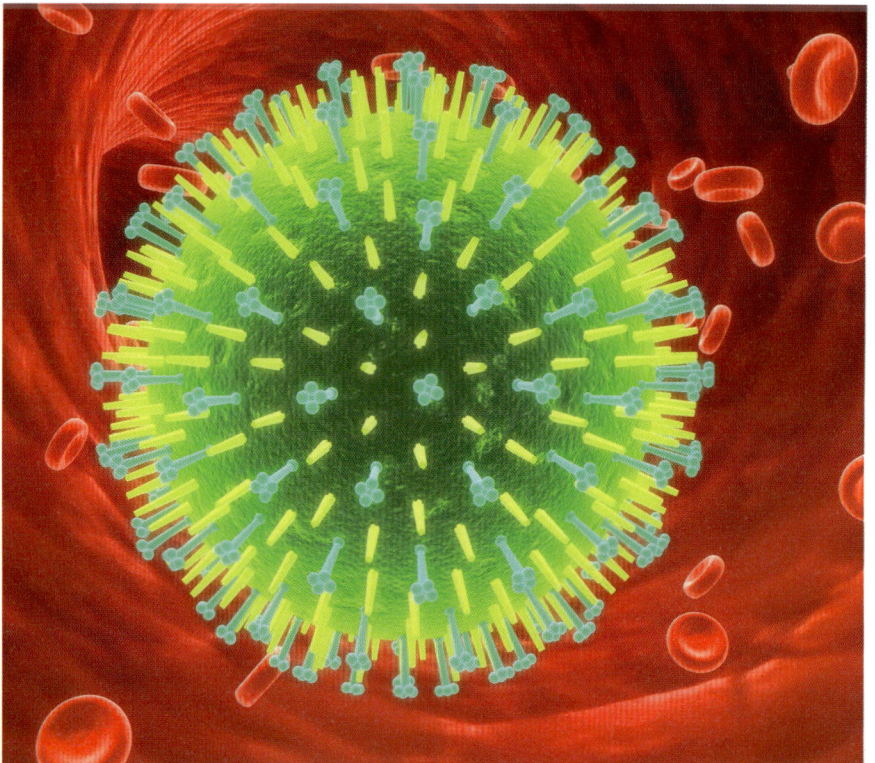

Mae cynnal efelychiadau i ddarganfod brechlyn rhag y ffliw yn un o gymwysiadau pwysig prosesu gwasgaredig gan ddefnyddio'r Rhyngrwyd

YN YR ARHOLIAD

Peidiwch â drysu rhwng cyfrifiadura gwasgaredig a chronfeydd data gwasgaredig. Mewn cyfrifiadura gwasgaredig mae'r prosesu'n cael ei rannu ar draws llawer o wahanol gyfrifiaduron, ond mewn cronfeydd data gwasgaredig, mae'r data sydd wedi'u storio yn cael eu gwasgaru ar draws sawl gweinydd, er mwyn cyflymu mynediad a gwella diogelwch yn bennaf.

Cysylltu â'r Rhyngrwyd

Cyflwyniad

Gallwn gysylltu â'r Rhyngrwyd drwy gebl neu'n ddiwifr. Bydd y dull a ddefnyddiwch yn effeithio ar ba mor gyflym y mae tudalennau'n ymddangos, pa mor gyflym y mae ffeiliau'n cael eu llwytho i lawr, ac ati.

Cysylltiad cebl â'r Rhyngrwyd

Mae dwy ffordd o gysylltu â'r Rhyngrwyd: deialu neu fand llydan. Yn y DU rydym ar y blaen i'r rhan fwyaf o wledydd y byd gan fod y mwyafrif o bobl wedi'u cysylltu â'r Rhyngrwyd drwy gysylltiad band llydan cyflym. I rai pobl yn y wlad hon ac mewn llawer o wledydd eraill, mae cysylltiad araf â'r Rhyngrwyd drwy fodem deialu yn arferol.

Modem deialu

Mae defnyddio modem deialu i gysylltu â'r Rhyngrwyd yn araf iawn ac mae hyn yn cyfyngu ar ei ddefnydd. Os ydych wedi arfer â band llydan ac yna'n defnyddio cysylltiad deialu, byddwch yn sylwi ei fod yn rhwystredig o araf. Yn wir, mae'n effeithio ar eich defnydd o'r Rhyngrwyd. Gall gymryd awr neu ragor i lwytho data i lawr yn lle ychydig funudau gyda band llydan.

Mae deialu'n defnyddio modem sy'n trawsnewid y signalau digidol yn gyfres o seiniau sydd wedyn yn cael eu hanfon ar hyd llinell ffôn. Caiff y signal sain ei droi'n ôl yn signal digidol, y gall y cyfrifiadur ei ddeall, ym mhen arall y wifren.

Gyda chysylltiad deialu, caiff cebl estyniad ffôn ei ddefnyddio i gysylltu'r cyfrifiadur â'r Rhyngrwyd

Mae defnyddio modem deialu i drosglwyddo data yn debyg i drosglwyddo dŵr ar hyd pibell gul gyda thap arni. Gan fod y bibell yn gul, mae'n anodd i lawer o ddŵr fynd drwyddi. Rhaid troi'r tap i gael y dŵr i lifo.

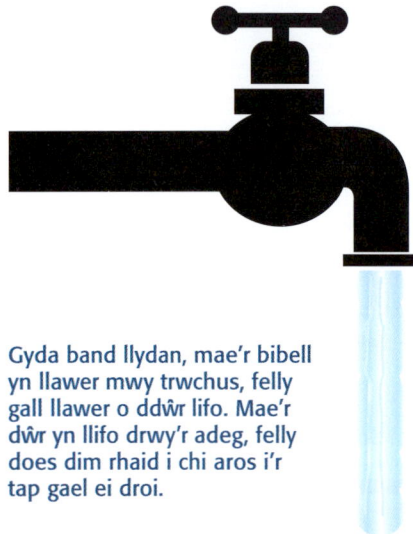

Gyda band llydan, mae'r bibell yn llawer mwy trwchus, felly gall llawer o ddŵr lifo. Mae'r dŵr yn llifo drwy'r adeg, felly does dim rhaid i chi aros i'r tap gael ei droi.

Band llydan

Mae band llydan yn cynnig mynediad a chyflymderau trosglwyddo llawer cyflymach i'r Rhyngrwyd ac yn caniatáu i chi wylio fideos ar-lein, llwytho ffeiliau i lawr yn gyflym, gwrando ar radio ar-lein, gwylio rhaglenni teledu yr ydych wedi'u colli, ac ati. Byddai'r rhain i gyd yn amhosibl heb gysylltiad band llydan cyflym.

Manteision band llydan

Mae gan fand llydan lawer o fanteision, e.e.:

- Mae amserau llwytho i lawr cyflym yn golygu y gall cerddoriaeth a ffilmiau a ffeiliau amlgyfrwng mawr gael eu llwytho i lawr yn gyflym.
- Nid yw'n peri bod eich llinell ffôn yn rhy brysur, felly gallwch ffonio a bod ar-lein yr un pryd.
- Gallwch wrando ar radio, gwylio

▼ **Byddwch yn dysgu**

▶ Am fynediad i'r Rhyngrwyd drwy gebl

▶ Am ddefnyddio modem deialu i gysylltu

▶ Am fand llydan

▶ Am fynediad symudol i'r Rhyngrwyd

fideo sgrin lawn, gwylio rhaglenni teledu a chwarae gemau ar-lein mewn amser real bron.
- Mae peiriannau chwilio'n gweithio'n llawer cyflymach, gan arbed amser pan fyddwch chi'n chwilio am wybodaeth.
- Nid ydych yn gwastraffu amser yn ceisio cysylltu â'r Rhyngrwyd.
- Rydych chi ar-lein drwy'r adeg, felly gall rhaglenni fel gwirwyr firysau ddiweddaru eu ffeiliau'n awtomatig.
- Gallwch ddefnyddio gwe-gam a fideo-gynadledda.
- Gallwch ddefnyddio'r Rhyngrwyd i wneud galwadau ffôn rhad.

Anfanteision band llydan

Nid oes llawer o anfanteision, ond dyma rai:

- Mae'n ddrud gan fod angen tanysgrifiad misol drutach i'r gwasanaeth.
- Mae'n bosibl nad yw ar gael yn yr ardal neu wlad lle'r ydych chi'n byw.

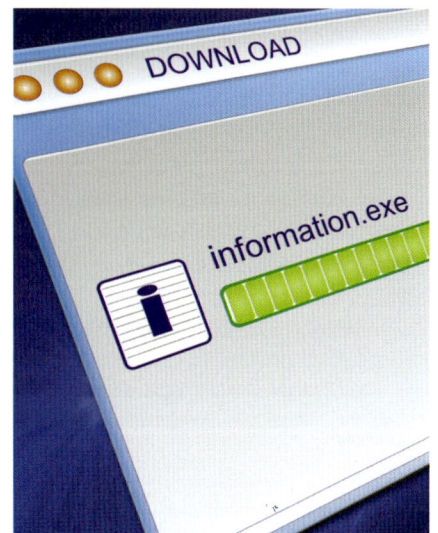

Mae band llydan yn cynnig amserau llwytho i lawr cyflym sy'n hanfodol wrth lwytho ffeiliau mawr i lawr

Rhaid cael cysylltiad band llydan i ddefnyddio gwe-gam yn effeithiol.

Mynediad symudol i'r Rhyngrwyd

Rhaid i lawer o bobl deithio fel rhan o'u gwaith, felly mae angen iddynt gael mynediad symudol i'w negeseuon e-bost a'r Rhyngrwyd. Mae llawer o ddyfeisiau'n cynnig mynediad symudol i'r Rhyngrwyd, gan gynnwys:

- gliniaduron
- cynorthwywyr digidol personol *(PDAs: Personal Digital Assistants)*
- ffonau symudol.

Mae gan fynediad symudol i'r Rhyngrwyd nifer o fanteision ac anfanteision, ac edrychwn ar y rhain yn yr adrannau sy'n dilyn.

→ GEIRIAU ALLWEDDOL

Lled band – mesur o ba mor gyflym y gall data gael eu trosglwyddo ar hyd sianel gyfathrebu, a allai fod yn gebl neu'n ddiwifr. Mae gan fand llydan led band llawer uwch na deialu.

Deialu – dull o gysylltu â'r Rhyngrwyd lle defnyddir modem a llinell ffôn. Mae'n cynnig cysylltiad araf.

Band llydan – cysylltiad cyflym â'r Rhyngrwyd nad yw'n defnyddio modem.

Manteision mynediad symudol i'r Rhyngrwyd

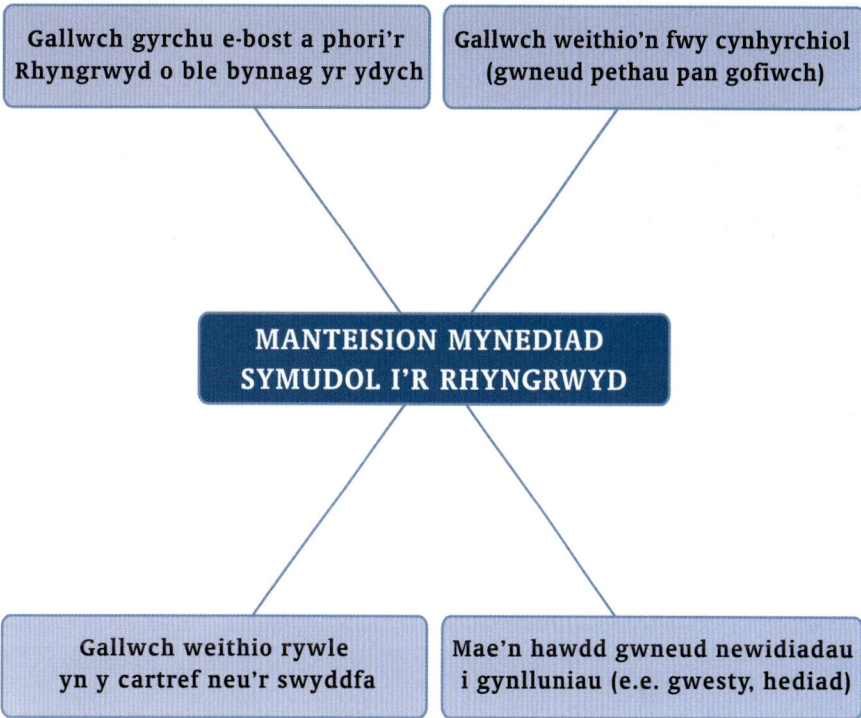

Gallwch gyrchu e-bost a phori'r Rhyngrwyd o ble bynnag yr ydych

Gallwch weithio'n fwy cynhyrchiol (gwneud pethau pan gofiwch)

MANTEISION MYNEDIAD SYMUDOL I'R RHYNGRWYD

Gallwch weithio rywle yn y cartref neu'r swyddfa

Mae'n hawdd gwneud newidiadau i gynlluniau (e.e. gwesty, hediad)

Anfanteision mynediad symudol i'r Rhyngrwyd

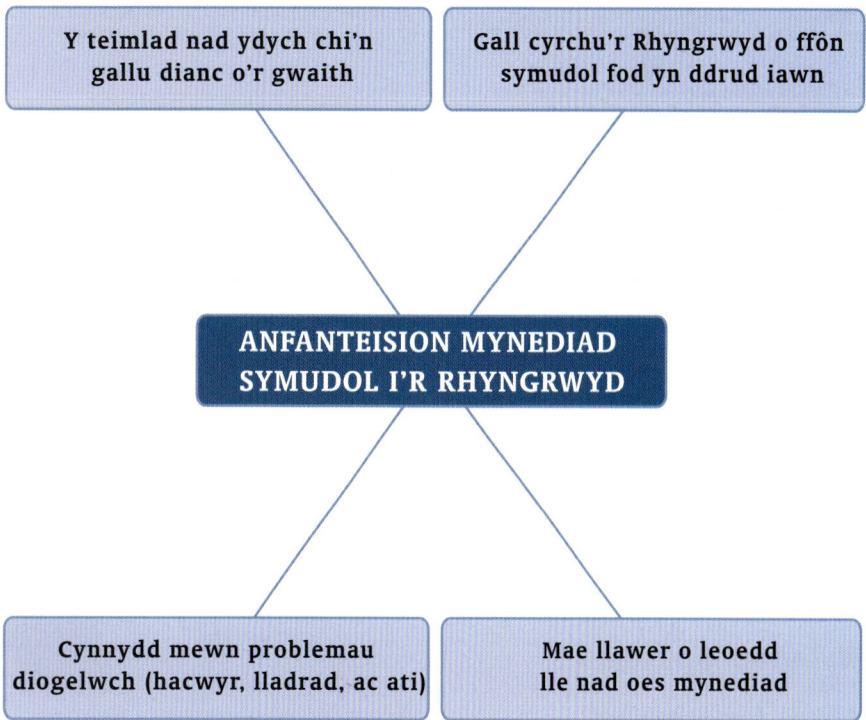

Y teimlad nad ydych chi'n gallu dianc o'r gwaith

Gall cyrchu'r Rhyngrwyd o ffôn symudol fod yn ddrud iawn

ANFANTEISION MYNEDIAD SYMUDOL I'R RHYNGRWYD

Cynnydd mewn problemau diogelwch (hacwyr, lladrad, ac ati)

Mae llawer o leoedd lle nad oes mynediad

Materion moesol, cymdeithasol a moesegol sy'n gysylltiedig â'r Rhyngrwyd

Cyflwyniad

Mae defnyddio'r Rhyngrwyd gan unigolion a sefydliadau wedi codi nifer o faterion y mae angen eu hystyried, e.e. materion moesol, cymdeithasol a moesegol.

Materion moesol

Mae nifer o faterion moesol yn gysylltiedig â defnyddio'r Rhyngrwyd, gan gynnwys:

- Sefydlu gwefannau sy'n cynnwys gwybodaeth anghywir yn fwriadol – gall pobl ddefnyddio'r wybodaeth hon a dibynnu arni gan feddwl ei bod yn gywir.
- Bwlio – mewn ystafelloedd sgwrsio, drwy e-bost, mewn blogiau, mewn negeseuon testun, yn enwedig ymhlith pobl ifanc.
- Gwefannau amhriodol – gall pobl weld deunydd amhriodol fel pornograffi, hiliaeth, fideos treisgar, sut i wneud ffrwydron, ac ati.
- Defnyddio e-bost i roi newyddion drwg (e.e. colli gwaith, israddio, diswyddo) pan fuasai egluro wyneb yn wyneb yn well.
- Lledu sïon – mae'n hawdd defnyddio'r Rhyngrwyd i ledu sïon. Os byddwch yn dweud wrth ychydig o bobl mewn ystafell sgwrsio, bydd y stori'n lledu'n gyflym. Fel rheol, os bydd rhywun yn dechrau si sy'n gelwydd ac sy'n achosi trallod i rywun arall, gallai'r person sy'n dechrau'r si gael ei siwio. Pan gaiff sïon eu dechrau dros y Rhyngrwyd, mae'n anodd dod o hyd i'r sawl sy'n gyfrifol.

Materion moesegol

Materion amheus nad ydynt o anghenraid yn anghyfreithlon yw materion moesegol, megis:

- Llên-ladrad – copïo deunydd heb ei briodoli neu heb gyfeirio at ffynhonnell y wybodaeth. Gallai hyn hefyd gynnwys gwefannau sy'n gwerthu traethodau neu waith cwrs.
- Anfon sbam (h.y. yr un e-bost hysbysebu i filiynau o bobl) – mae pobl yn gwastraffu amser yn dileu sbam os yw'n osgoi'r hidlydd sbam.
- Cwmnïau'n monitro defnydd y staff o'r Rhyngrwyd ac e-bost. Bydd rhai sefydliadau yn darllen e-byst personol hyd yn oed.
- Defnyddio cysylltiad Rhyngrwyd diwifr heb ganiatâd. Weithiau gellir defnyddio rhwydwaith agored i gysylltu â'r Rhyngrwyd. Y canlyniad yw bod y rhwydwaith yn arafach ar gyfer defnyddwyr cyfreithlon.
- Defnyddio meddalwedd ffoto-olygu i aflunio realiti – gellir defnyddio meddalwedd ffoto/fideo-olygu i aflunio realiti fel nad oes modd credu'r hyn a welwch mewn fideos, rhaglenni teledu a phapurau newydd ac ar wefannau.

Materion cymdeithasol

Mae materion cymdeithasol yn effeithio ar y ffordd mae cymdeithas yn gweithredu, a gellir dweud bod y Rhyngrwyd yn achosi nifer o broblemau, e.e.:

- Materion preifatrwydd – mae safleoedd rhwydweithio cymdeithasol, safleoedd e-fasnach, cofnodion darparwyr gwasanaethau Rhyngrwyd, monitro e-bost yn y gwaith, ac ati, i gyd yn tanseilio preifatrwydd defnyddwyr.
- Mynd yn gaeth i hapchwarae – gall hapchwarae achosi llawer o broblemau cymdeithasol, ac mae ar gynnydd gan ei bod hi mor hawdd gosod betiau dros y Rhyngrwyd.
- Gordewdra – roedd llawer o weithgareddau yn y gorffennol yn gofyn am ymarfer corff a gadwai pobl yn iach. Nid yw defnyddio'r Rhyngrwyd am gyfnodau hir yn iachus iawn.
- Mynd yn gaeth i gemau cyfrifiadur – bydd llawer o blant yn treulio oriau'n chwarae gemau cyfrifiadur, a gall eu sgiliau cymdeithasol a'u gwaith ysgol ddioddef.

Hawlfraint 2005 gan Randy Glasbergen
www.glasbergen.com

"Rydyn ni wedi dod o hyd i rywun mewn gwlad dramor sy'n gallu yfed coffi a siarad am chwaraeon drwy'r dydd am ffracsiwn o'ch cyflog chi."

- Mae mynediad i'r Rhyngrwyd yn cynyddu'r bwlch rhwng gwledydd ac unigolion tlawd a chyfoethog.
- Mae cwmnïau'n symud canolfannau galw i wledydd tramor – oherwydd bod y bil cyflogau'n is ac y gellir darparu'r un gwasanaeth yn rhad drwy ddefnyddio'r Rhyngrwyd a chysylltiadau ffôn y Rhyngrwyd.
- Gall twf e-fasnach olygu bod siopau'n gorfod cau, gyda'r canlyniad bod canol rhai dinasoedd yn edrych yn ddiffaith.

Perchenogaeth a rheolaeth ar y Rhyngrwyd

Mae'r Rhyngrwyd i bawb ac nid oes neb yn berchen arni. Nid oes fawr ddim rheolaeth ar gynnwys y deunydd ar y Rhyngrwyd, er bod rhai llywodraethau wedi dechrau rheoli'r hyn y gellir ei weld. Hefyd nid oes unrhyw reolaeth dros y bobl sy'n gallu cyrchu'r deunydd ar y Rhyngrwyd. Mae hyn yn golygu y gall plant gyrchu deunydd pornograffig neu dreisgar yn hawdd os na ddefnyddir meddalwedd arbennig i'w rhwystro.

Gan nad yw'r Rhyngrwyd yn cael ei 'phlismona', mae'n golygu nad yw'r wybodaeth yn cael ei gwirio am gywirdeb. Felly cyfrifoldeb y defnyddwyr yw sicrhau ei bod yn gywir. Wrth ddefnyddio gwybodaeth o'r Rhyngrwyd, mae angen i chi allu gwirio bod y deunydd yn addas a chywir.

Deunydd ffiaidd, anghyfreithlon neu anfoesegol ar y Rhyngrwyd

Mae llawer o ddelweddau a fideos pornograffig ar y Rhyngrwyd. Mae yna ddeddfau sy'n ymdrin â chynhyrchu a dosbarthu'r deunydd hwn ond gan fod llawer ohono'n dod o wledydd eraill, lle mae'n hollol gyfreithlon, nid oes llawer y gellir ei wneud i'w atal. Prif bryder oedolion yw y gallai plant ifanc gyrchu'r deunydd hwn yn ddamweiniol. Mae meddalwedd arbennig ar gael sy'n hidlo'r deunydd hwn allan, ond ni allwch byth fod yn hollol sicr.

Y pryder mawr yw bod pedoffilyddion yn defnyddio'r Rhyngrwyd i ddosbarthu lluniau pornograffig o blant ifanc a'u bod yn hudo plant i gyfarfod â nhw ar ôl siarad â nhw mewn ystafelloedd sgwrsio. Felly rhaid bod yn eithriadol o ofalus cyn trefnu i gyfarfod â rhywun yr ydych wedi siarad ag ef/hi ar-lein.

Nid oes angen i ddeunydd fod yn bornograffig i fod yn ffiaidd. Gallai delwedd o helgwn yn ymosod ar lwynog beri gofid i bobl sy'n hoffi anifeiliaid. Mae gwahanol bethau yn ffiaidd i wahanol bobl. Ond, wrth gwrs, mae yna ddeunydd y byddai bron pawb yn ystyried ei fod yn ffiaidd.

Sensoriaeth

Mae llawer o wledydd yn y byd sydd heb fod yn ddemocrataidd ac nid yw'r bobl sy'n eu rheoli yn caniatáu i wybodaeth symud yn rhydd i mewn ac allan o'r gwledydd hynny. Maen nhw'n aml yn rheoli'r cyfryngau (papurau newydd, teledu a radio) a'r unig newyddion a ddangosant yw newyddion sy'n ffafriol i'r llywodraeth. Mae sensro'r Rhyngrwyd gan lywodraeth yn rhoi mwy o reolaeth iddi dros yr hyn y gall pobl ei weld.

Yr effeithiau ar gymunedau

Mae'r Rhyngwyd yn effeithio ar bob un ohonom ac mae mynediad i'r Rhyngrwyd o fudd i gymunedau. Dyma rai o'r effeithiau cadarnhaol ar gymunedau:

- Gall blogiau a sgyrsiau gael eu sefydlu i ganiatáu i gymunedau drafod materion lleol fel ceisiadau cynllunio, gangiau a materion plismona eraill, digwyddiadau cymunedol, ac ati.
- Mae pobl sy'n gaeth i'w cartrefi yn llai ynysig gan fod pobl yn cysylltu â nhw i sicrhau eu bod yn iawn.
- Gellir hysbysebu cyfleoedd ar gyfer cyflogaeth yn y gymuned ar wefannau lleol.
- Gellir sefydlu gwefannau cynghori lleol i ymdrin â phroblemau trigolion.

Rhai effeithiau negyddol ar gymunedau yw:

- Llai o ryngweithio cymdeithasol – mae pobl yn treulio amser yn pori'r Rhyngrwyd, chwarae gemau cyfrifiadur a chyfathrebu â'u ffrindiau 'seiber' yn hytrach na threulio amser gyda phobl yn y gymuned.
- Siopau lleol yn cau – mae mwy o archebion am nwyddau'n cael eu gwneud dros y Rhyngrwyd, felly mae siopau lleol yn gorfod cau.

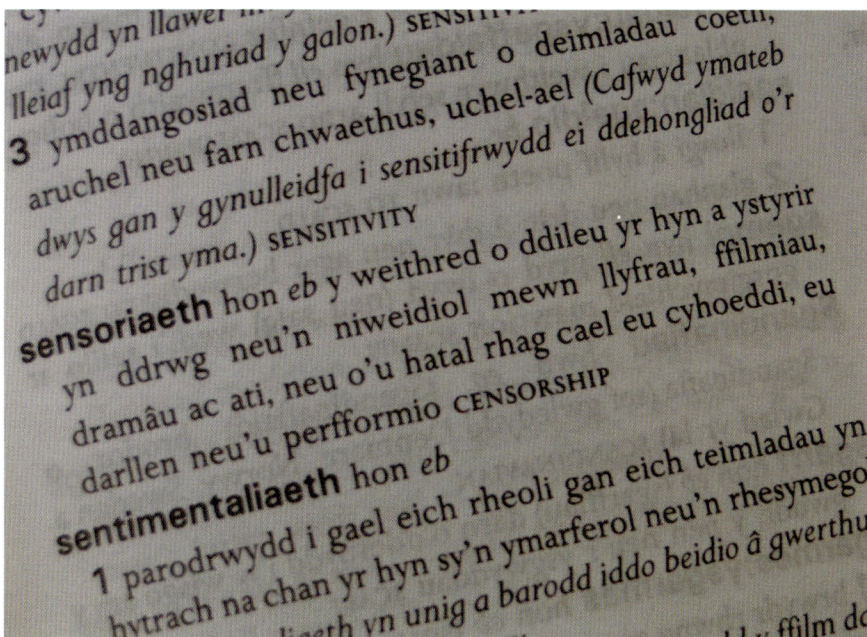

newydd yn llawer ... y ...
lleiaf yng nghyhuriad y galon.) SENSITIV...

3 ymddangosiad neu fynegiant o deimladau coeth, aruchel neu farn chwaethus, uchel-ael (Cafwyd ymateb dwys gan y gynulleidfa i sensitifrwydd ei ddehongliad o'r darn trist yma.) SENSITIVITY

sensoriaeth hon eb y weithred o ddileu yr hyn a ystyrir yn ddrwg neu'n niweidiol mewn llyfrau, ffilmiau, dramâu ac ati, neu o'u hatal rhag cael eu cyhoeddi, eu darllen neu'u perfformio CENSORSHIP

sentimentaliaeth hon eb
1 parodrwydd i gael eich rheoli gan eich teimladau yn hytrach na chan yr hyn sy'n ymarferol neu'n rhesymegol
...iaeth yn unig a barodd iddo beidio â gwerthu
... y ffilm dd...

Ystyr sensoriaeth yw'r arfer o sensro. Mae hyn yn cynnwys gwirio cynnwys a chael gwared â deunydd annymunol (e.e. deunydd mae'r llywodraeth am ei gadw'n gyfrinachol, deunydd treisgar neu rywiol amlwg)

Astudiaethau achos a Gweithgareddau

▶ Astudiaeth achos 1 | tt. 24–25

Sut y gall cyfrifiaduron cartref helpu gwyddonwyr i ddeall dirgelion y bydysawd yn well

Mae project ym Mhrifysgol Illinois yn America yn defnyddio cyfrifiadura gwasgaredig ar draws y Rhyngrwyd i ddeall y bydysawd yn well. Oherwydd y grym cyfrifiadura a lled band sydd eu hangen ar gyfer y project enfawr hwn, mae cynllunwyr y project wedi gofyn i ddefnyddwyr cyfrifiaduron cartref am eu help gyda'r ymchwil.

Enw'r project yw *Cosmology@Home* ac mae'n debyg i'r project *SETI@home*. Pan fydd defnyddiwr cartref yn rhedeg *Cosmology@Home* ar ei gyfrifiadur cartref, bydd yn defnyddio rhan o rym prosesu, cof a lled band y cyfrifiadur i helpu gyda'r project.

Yn syml, mae gan y brifysgol fodelau cosmolegol sy'n disgrifio sut mae'r bydysawd yn gweithio, a'r dasg yw darganfod pa un o'r modelau hyn sy'n disgrifio orau'r swm enfawr o ddata arbrofol o arbrofion ffiseg ronynnol a mesuriadau seryddol. Caiff y paramedrau yn y modelau eu newid i gymharu'r rhagfynegiadau damcaniaethol â beth sy'n digwydd mewn gwirionedd.

1 Eglurwch ystyr bob un o'r termau canlynol:
   (a) Cyfrifiadura gwasgaredig (2 farc)
   (b) Lled band (1 marc)
   (c) Model cyfrifiadurol. (2 farc)
2 Yn y project hwn, gofynnwyd i ddefnyddwyr cyfrifiaduron cartref am eu cymorth.
   (a) Rhowch un rheswm pam nad oedd yn bosibl defnyddio cyfleusterau cyfrifiadura'r brifysgol ei hun i redeg y model. (1 marc)
   (b) Nodwch ddau adnodd cyfrifiadura y mae cyfrifiadur cartref yn eu rhoi i'r project hwn. (1 marc)
   (c) Eglurwch pam y mae angen cymaint o rym cyfrifiadura ar gyfer y model cosmolegol. (2 farc)
3 Cymharwch a chyferbynnwch fanteision ac anfanteision defnyddio naill ai cysylltiad deialu neu gysylltiad band llydan i gyrchu'r Rhyngrwyd. (4 marc)
4 Mae llawer o arbrofion gwyddonol ar y gweill sy'n dibynnu ar gyfrifiadura gwasgaredig.
   (a) Eglurwch beth yw cyfrifiadura gwasgaredig a rhowch ddau reswm pam mae'n cael ei ddefnyddio wrth ddadansoddi canlyniadau o arbrofion gwyddonol enfawr. (4 marc)
   (b) Disgrifiwch un cymhwysiad ar gyfer cyfrifiadura gwasgaredig. Yn eich disgrifiad, dylech egluro'r rhesymau dros ddefnyddio cyfrifiadura gwasgaredig a'i fanteision. (4 marc)

▶ Astudiaeth achos 2 | tt. 28–29

Problem wedi'i hachosi gan negesfwrdd ar y Rhyngrwyd

Difethodd tua 500 o labystiaid Rhyngrwyd barti pen-blwydd merch ifanc 14 oed mewn neuadd eglwys.

Roedd hi ar fin dathlu ei phen-blwydd yn 14 oed gyda'i theulu a'i chyfeillion pan wthiodd tua 500 o bobl ifanc eu ffordd i mewn i'r neuadd a meddiannu'r parti. Heb yn wybod i'r ferch, roedd neges am y parti wedi ymddangos ar negesfwrdd partïon ar y Rhyngrwyd.

Cafodd y ferch a'i chyfeillion a'i theulu fraw wrth i'r llabystiaid dorri poteli, chwydu, a malu drysau. Galwyd ar yr heddlu ac roedd angen nifer o blismyn, a ddaeth mewn pum car, i dawelu'r bobl ifanc feddw. Cafodd pedwar ohonynt eu harestio am fod yn feddw ac afreolus.

1 Mae'r enghraifft yn yr astudiaeth achos yn dangos y problemau sy'n cael eu hachosi gan symudiad rhydd gwybodaeth ar y Rhyngrwyd:
   (a) Eglurwch beth yw ystyr negesfwrdd Rhyngrwyd. (2 farc)
   (b) Caiff rhai negesfyrddau eu monitro gan ddarparwr y gwasanaeth Rhyngrwyd. Sut y byddai hyn wedi helpu i osgoi'r broblem hon? (1 marc)
2 Gall symudiad rhydd gwybodaeth greu problemau i gymdeithas. Disgrifiwch sefyllfa wahanol lle mae symudiad rhydd gwybodaeth, gan ddefnyddio systemau TGCh fel y Rhyngrwyd, wedi creu problemau i gymdeithas. (3 marc)

Uchelgais Rwanda yw dod yn ganolfan TGCh Affrica

Mae Rwanda, gwlad yn Affrica, yn ceisio dod yn ganolfan TG Affrica, er gwaethaf y ffaith ei bod yn wlad dlawd iawn o'i chymharu â'i chymdogion mwy cyfoethog sy'n meddu ar adnoddau naturiol megis olew, diemyntau a chopr.

Mae'r llywodraeth yn Rwanda yn awyddus i newid o economi amaethyddol i economi'n seiliedig ar wybodaeth. Mae dau gylch ffibr-optig yn amgylchynu'r brifddinas ac mae cebl arall wedi cael ei osod ar draws y wlad.

Rwanda yw un o'r gwledydd lleiaf datblygedig yn y byd ond mae ei harweinwyr yn ystyried mai technoleg yw'r ffordd orau o geisio cystadlu ar raddfa fyd-eang. Ond mae pobl eraill yn anghytuno, gan gredu y dylai mwy o arian gael ei wario ar gynlluniau hanfodol fel trydan i bawb.

Mae'r dyfodol yn edrych yn ddisglair gan fod bron 70% o'r boblogaeth yn gallu darllen ac maent i gyd yn siarad yr un iaith, sy'n ei gwneud yn haws sefydlu presenoldeb ar y we.

Mae Rwanda'n gobeithio mewnforio cyfrifiaduron rhad wedi'u hailwampio. Gan fod trydan yn brin mewn rhannau o'r wlad, y bwriad yw defnyddio cyfrifiaduron pŵer isel gyda batrïau y gellir eu hailwefru drwy ddefnyddio pŵer solar.

Y bwriad yw cael llawer o'r cyfrifiaduron mewn teleganolfannau sydd â chysylltiadau band llydan â'r Rhyngrwyd. Byddai hyn yn galluogi rhai Rwandiaid i gael cyfleoedd busnes, yn gymorth i eraill ddod o hyd i swyddi, ac yn gymorth i ffermwyr ddod o hyd i ffyrdd o dyfu cnydau gwell. Mae pobl na allant ddarllen nac ysgrifennu yn gallu defnyddio TGCh i'w helpu i ddatblygu'r sgiliau hynny.

Mae rhai pobl o fewn y wlad yn dweud bod yr holl syniad yn anghywir. Mae nhw'n credu na allwch adeiladu ysgol i fyfyrwyr astudio ynddi cyn i chi roi bwyd iddynt i'w fwyta. Mae pobl eraill yn credu os bydd y wlad yn cael isadeiledd y bydd y sector preifat yn buddsoddi ac y bydd pawb yn cael budd o hynny.

1   Cyrchwch wefan y *CIA, World Factbook*, gan ddefnyddio'r cyfeiriad gwe canlynol:
    https://www.cia.gov/library/publications/the-world-factbook/index.html
    Teipiwch neu dewiswch y wlad Rwanda. Defnyddiwch y wybodaeth ar y tudalennau gwe i lenwi'r manylion yn y golofn ar gyfer Rwanda.
    Nawr dewch o hyd i'r ffigurau ar gyfer y DU a llenwch y golofn arall.
    Ysgrifennwch baragraff byr am y gwahaniaethau rhwng gwlad lai datblygedig fel Rwanda a gwlad lawer mwy datblygedig fel y DU. (5 marc)

Cwestiwn	Rwanda	Y DU
Poblogaeth		
Disgwyliad oes (gwryw)		
Disgwyliad oes (benyw)		
Llythrennedd y boblogaeth gyfan		
Cynnyrch Mewnwladol Crynswth (CMC) y pen (h.y. y person)		
Ffôn (prif linellau)		
Ffôn (symudol)		
Gwesteiwyr rhyngrwyd (h.y. Darparwyr Gwasanaeth Rhyngrwyd)		
Nifer o ddefnyddwyr Rhyngrwyd		

2   (a)   Disgrifiwch **dair** ffordd y byddai dyfodiad y Rhyngrwyd yn helpu rhai o bobl Rwanda. (3 marc)

   (b)   Rhowch **un** rheswm pam y mae'n bosibl na fyddai pawb yn Rwanda yn elwa drwy gael mynediad i'r Rhyngrwyd. (1 marc)

3   A ydych chi'n meddwl bod y buddsoddiad hwn mewn TGCh yn dda i Rwanda? Rhowch reswm cryno dros eich ateb. (2 farc)

Gwe-hidlo yn China

Mewn rhai gwledydd, fel China, mae'r llywodraeth yn defnyddio mur gwarchod sy'n hidlo mynediad i safleoedd nad ydyn nhw'n eu caniatáu, hynny yw, safleoedd sy'n beirniadu'r llywodraeth neu'n portreadu'r wlad mewn ffordd negyddol. Mae'r mur gwarchod yn rhwystro pynciau sydd wedi'u gwahardd, cyfeiriadau gwe penodol a geiriau penodol.

Mae llawer o bobl yn poeni y bydd y llywodraeth yn gallu eu hadnabod a chymryd camau yn eu herbyn os ceisiant ddefnyddio'r geiriau neu wefannau hyn.

1   Bydd gwe-hidlo'n cael ei wneud yn aml gan ysgolion neu rieni plant ifanc. Rhowch **ddau** reswm pam mae hyn yn angenrheidiol. (2 farc)

2   (a)   Rhowch **un** rheswm pam mae'r awdurdodau yn China yn defnyddio gwe-hidlo. (2 farc)

   (b)   Rhowch enw'r ddyfais neu'r meddalwedd a ddefnyddir i rwystro mynediad at gynnwys penodol ar y Rhyngrwyd. (1 marc)

Astudiaethau achos a Gweithgareddau (parhad)

▶ Astudiaeth achos 5 | tt. 28–29

Cronfeydd traethodau ar-lein yn siomi myfyrwyr

Bydd rhai myfyrwyr sy'n rhy brysur neu'n rhy ddiog i ysgrifennu eu traethodau, aseiniadau neu waith cwrs eu hunain yn troi at y Rhyngrwyd am gymorth.

Maen nhw'n hen gyfarwydd â thorri a phastio deunydd o wefannau i greu traethodau sy'n cael eu cyflwyno fel eu gwaith eu hun. Yn ffodus mae gan lawer o awdurdodau addysg feddalwedd arbennig sy'n gallu darganfod llên-ladrad drwy chwilio'r Rhyngrwyd am frawddegau sydd yr un fath.

Bydd rhai myfyrwyr yn datrys y broblem hon drwy dalu i rywun gynhyrchu darn gwreiddiol o waith iddyn nhw. Twyllo amlwg yw hyn wrth gwrs. Heblaw am y materion moesegol, gall ansawdd y gwaith fod yn wael iawn a methu ag ennill y radd sydd wedi cael ei haddo.

Meddai un myfyriwr a brynodd draethawd a gafodd radd 'F', 'Roedd yn wastraff arian llwyr – fydda i byth yn gwneud hynny eto'.

1 Rhowch ddau reswm pam y gellid ystyried bod cronfeydd traethodau'n anfoesegol. (2 farc)

2 Disgrifiwch un ffordd y gallai defnyddio cronfeydd traethodau effeithio ar gymdeithas yn gyffredinol. (1 marc)

3 A ydych chi'n meddwl y dylid gwneud cronfeydd traethodau yn anghyfreithlon? Rhowch reswm dros eich ateb. (2 farc)

▶ Gweithgaredd 1: Y Rhyngrwyd a hawliau dynol

Mae gwefan wedi cael ei sefydlu gan gorff o'r enw 'Human Rights Watch' sy'n rhoi manylion achosion o sathru hawliau dynol ledled y byd.

Y cyfeiriad gwe yw www.hrw.org. Ewch i'r wefan i gael amcan o'r broblem mewn gwledydd eraill.

Mae safle arall, www.cyber-rights.org/, yn ymdrin â'r rhyddid a'r hawliau y dylai pobl sy'n defnyddio'r Rhyngrwyd eu mwynhau.

Defnyddiwch amodau chwilio gyda gwahanol beiriannau chwilio i ddarganfod gwefannau eraill sy'n ymdrin â hawliau dynol a'r Rhyngrwyd. Gallech ddefnyddio amodau chwilio amryfal fel hyn: 'human rights' AND 'Internet'. Os dewch chi o hyd i wefannau da, ysgrifennwch eu cyfeiriadau gwe a defnyddiwch nhw yn eich gwaith.

Ysgrifennwch grynodeb byr o sut y mae rhai llywodraethau mewn gwledydd ar hyd a lled y byd yn rhwystro eu dinasyddion rhag cael mynediad llawn i wasanaethau rhyngrwyd.

▶ Gweithgaredd 2: Y project SETI

Ar gyfer y gweithgaredd hwn mae gofyn i chi gynhyrchu crynodeb byr o'r defnydd diweddaraf o gyfrifiadura gwasgaredig ar gyfer y project SETI. Gallwch ddod o hyd i'r wybodaeth ar y wefan ganlynol:
http://setiathome.berkeley.edu/

▶ Gweithgaredd 3: Y model hinsawdd

Ar gyfer y gweithgaredd hwn mae gofyn i chi gynhyrchu esboniad byr o sut y defnyddir cyfrifiadura gwasgaredig i helpu gydag ymchwil i'r hinsawdd. Gallwch ddod o hyd i'r wybodaeth ar y wefan ganlynol:
http://www.bbc.co.uk/sn/climateexperiment/theexperiment/distributedcomputing.shtml

Cwestiynau

1 Mae person yn defnyddio modem ar hyn o bryd ond gan ei bod hi'n treulio llawer o amser ar y Rhyngrwyd hoffai allu pori'r Rhyngrwyd yn gyflymach.
   (a) Eglurwch ystyr y term 'pori'r Rhyngrwyd'. (1 marc)
   (b) Eglurwch pam y byddai band llydan yn well na defnyddio modem. (2 farc)
   (c) Mae band llydan yn cynnig 'fideo ar alw'. Eglurwch ystyr hyn. (2 farc)
   (ch) Rhowch enwau **dau** beth y gallech eu gwneud gyda chysylltiad band llydan y byddai'n anodd i chi eu gwneud gyda chysylltiad modem. (4 marc)
   (d) Pam mae lled band mor bwysig pan fyddwch chi'n trosglwyddo ffeiliau mawr o le i le? (2 farc)

2 Mae cwmni'n sefydlu gwefan i hysbysebu ei gynhyrchion ac i ganiatáu i gwsmeriaid archebu nwyddau ar-lein.
   (a) Disgrifiwch **ddwy** ffordd y gallai cwsmer ddod o hyd i'r wefan drwy ddefnyddio'r Rhyngrwyd. (4 marc)
   (b) Ar ôl i gwsmeriaid ddod o hyd i wefan y cwmni, rhaid iddynt ddod o hyd i'r cynhyrchion sydd o ddiddordeb iddynt. Amlinellwch **ddwy** ffordd y gallent gael y wybodaeth hon yn gyflym. (4 marc)

3 Gall y Rhyngrwyd gael ei defnyddio gan gyrff at lawer o wahanol ddibenion.
   Eglurwch, drwy roi enghraifft, sut y gallai pob un o'r canlynol gael ei ddefnyddio gan gorff:
   (a) Protocol trosglwyddo ffeiliau (*FTP*) (2 farc)
   (b) Cronfeydd data ar-lein (2 farc)
   (c) E-fasnach. (2 farc)

4 Trafodwch y gwahanol ddulliau o gysylltu â'r Rhyngrwyd. Yn eich trafodaeth, dylech ystyried y dulliau a'r manteision ac anfanteision cysylltiedig. (4 marc)

5 Gall y wybodaeth mewn cronfeydd data ar-lein gael ei chyrchu mewn sawl ffordd.
   Disgrifiwch yn fyr sut y gall pob un o'r canlynol gael ei ddefnyddio i gyrchu gwybodaeth:
   (a) Lleolwyr Adnoddau Unffurf (*URLs*) (2 farc)
   (b) Chwiliadau Boole (2 farc)

(c) Hypergysylltiadau. (2 farc)

6 (a) Er mwyn i gwmni roi gwasanaeth ar-lein rhyngweithiol ar waith, rhaid bodloni nifer o ofynion. Nodwch **dri** gofyniad ac, ar gyfer pob un, disgrifiwch pam y mae ei angen. (3 marc)
   (b) Trafodwch fanteision ac anfanteision e-fasnach i
      (i) y cwsmer
      (ii) y busnes. (4 marc)

7 (a) Eglurwch bwrpas peiriant chwilio. (2 farc)
   (b) Eglurwch sut y mae peiriant chwilio yn gweithio a sut y mae ffeiliau'n cael eu hychwanegu at restri peiriant chwilio. (4 marc)

8 Trafodwch fanteision ac anfanteision mynediad symudol i'r Rhyngrwyd. (4 marc)

9 Mae llawer o brojectau ymchwil mawr yn defnyddio cyfrifiadura gwasgaredig drwy gyfrwng y Rhyngrwyd.
   (a) Eglurwch ystyr cyfrifiadura gwasgaredig. (3 marc)
   (b) Disgrifiwch gymhwysiad yr ydych chi'n gwybod amdano lle mae cyfrifiadura gwasgaredig yn cael ei ddefnyddio. Nodwch yn glir pam y cafodd prosesu gwasgaredig ei ddewis ar gyfer y cymhwysiad yr ydych chi'n ei ddisgrifio. (6 marc)

10 Mae erthygl mewn papur newydd yn dweud 'Does neb yn berchen ar y Rhyngrwyd nac yn ei rheoli'. Gan roi enghreifftiau addas, cyflwynwch ddadl argyhoeddiadol o blaid neu yn erbyn y gosodiad hwn. (D.S. Cewch gytuno neu anghytuno – fe gewch eich marcio ar sail cryfder eich dadleuon.) (6 marc)

Cymorth gyda'r arholiad

Enghraifft 1

1 Mae defnyddio'r Rhyngrwyd yn codi materion moesol, cymdeithasol a moesegol mawr. Heblaw am droseddu, amlinellwch **ddau** o'r materion uchod, gan roi **dwy** enghraifft briodol i'ch helpu i'w hegluro. (8 marc)

Ateb myfyriwr 1

1 Un o effeithiau cymdeithasol defnyddio'r Rhyngrwyd yw bod siopau'n cau ar y stryd fawr gan fod pobl yn dewis prynu nwyddau dros y Rhyngrwyd. Bydd llawer o ddinasoedd yn gweld nifer o siopau fel gwerthwyr gwyliau, siopau llyfrau ac ati yn cau. Bydd pobl yn colli eu swyddi a bydd yr economi leol yn dioddef. Hefyd ni fydd dim i'w wneud ar ddydd Sadwrn gan na fyddwch chi'n gallu mynd i'r siopau.

Mae yna lawer o wefannau sy'n cynnwys gwybodaeth anghywir, felly bydd plant ifanc yn cael gwybodaeth ffug wrth wneud projectau. Ni fydd plant ifanc yn gallu barnu a yw'r wybodaeth yn gywir ai peidio ac rydw i'n meddwl y dylai'r gwefannau hyn gael eu gwahardd.

Sylwadau'r arholwr

1 Dylai'r myfyriwr fod wedi edrych ar y marciau ar gyfer y cwestiwn hwn. Mae 8 marc ar gael, felly dylai'r myfyriwr fod wedi darllen y cwestiwn yn ofalus ac wedi rhoi ateb mwy cynhwysfawr.

Rhaid egluro dau fater o blith: cymdeithasol, moesegol neu foesol. Mae'r myfyriwr wedi camddehongli'r cwestiwn ac wedi ymdrin â dwy broblem yn hytrach na sawl problem o dan ddau fater.

Mae'n hollbwysig darllen a deall y cwestiwn er mwyn ennill y marciau uchaf sy'n bosibl, hyd yn oed os na allwch ateb ond rhan o'r cwestiwn.

Er bod ateb y myfyriwr yn iawn, roedd angen ysgrifennu am broblemau eraill hefyd, felly mae diffyg ymdriniaeth wedi cyfyngu ar y marciau.

(4 marc allan o 8)

Ateb yr arholwr

1 Dylai'r myfyrwyr ymdrin â nifer o'r canlynol:
Y bwlch economaidd rhwng pobl dlawd a phobl sy'n fwy cyfoethog yn mynd yn fwy
Effeithiau cau siopau ar gymunedau yn sgil e-fasnach/Rhyngrwyd
Sensro'r Rhyngrwyd gan lywodraethau tramor
Materion preifatrwydd (camerâu, cofnodi manylion, ymyrraeth y llywodraeth â bywyd pob dydd, ac ati)
Llên-ladrad (prynu gwaith cwrs, cronfeydd traethodau, copïo'n syth o wefannau)
Mynd yn gaeth i ystafelloedd sgwrsio, chwarae gemau a gamblo

Ateb myfyriwr 2

1 Mae llawer o faterion moesegol. Er enghraifft, mae bron yn amhosibl i rywun gael preifatrwydd. Mae'r wladwriaeth yn ymyrryd fwyfwy â bywyd pob dydd. Mae cardiau adnabod, cwmnïau ffôn yn cadw manylion galwadau, darparwyr gwasanaeth rhyngrwyd yn cadw manylion e-byst, camerâu'n eich gwylio wrth i chi gerdded i lawr y stryd neu ddefnyddio eich car i gyd yn tanseilio preifatrwydd. Gall systemau adnabod wynebau eich adnabod ac mae hyd yn oed ysgolion yn defnyddio olion bysedd i adnabod disgyblion.
Mater moesegol arall yw lledaeniad canolfannau galwadau, lle y mae pobl yn aml yn gorfod gweithio am gyflog isel, ac weithiau caiff y canolfannau hyn eu symud i wledydd tramor gan eu bod yn rhatach i'w rhedeg. A ddylid caniatáu hyn pan ddylid cadw'r swyddi yn y wlad hon?
Mae rhai myfyrwyr yn defnyddio cronfeydd traethodau i gopïo darnau o draethodau a'u cyfuno â'u gwaith eu hunain er mwyn cael marciau da, ac nid yw hyn yn deg. Gall plant brynu gwaith cwrs TGAU ar-lein o safleoedd gwerthu ar y Rhyngrwyd.
Un mater cymdeithasol yw bod defnyddio'r Rhyngrwyd yn lledu'r bwlch rhwng pobl dlawd a phobl fwy cyfoethog. Ni all pobl siopa ar-lein heb gerdyn debyd neu gredyd, ac ni all pobl ar incwm isel neu sydd â hanes credyd gwael gael y rhain. Hefyd gall pobl fwy cyfoethog fanteisio ar gynigion arbennig ar y Rhyngrwyd ond nid yw'r rhain ar gael i bobl nad ydynt yn gallu ei chyrchu.

Sylwadau'r arholwr

1 Yn yr ateb hwn mae'r myfyriwr wedi egluro'r problemau canlynol: preifatrwydd, colli swyddi i wledydd tramor, llên-ladrad a'r bwlch cynyddol rhwng dosbarthiadau cymdeithasol.

Mae'r myfyriwr wedi rhoi enghreifftiau lle mae hynny'n berthnasol ac wedi llunio'r atebion yn dda, ac mae ei sillafu, atalnodi a gramadeg yn gywir.

Mae wedi defnyddio'r termau cywir yn yr atebion.

(8 marc allan o 8)

Diffyg rhyngweithio cymdeithasol (e.e. cael ffrindiau rhith yn lle ffrindiau go iawn)

Gwefannau sy'n mynd ati i dwyllo neu roi gwybodaeth anghywir

Problemau gyda materion hawlfraint/eiddo deallusol

Gwefannau amhriodol sy'n hyrwyddo terfysgaeth, hunanladdiad, casineb hiliol

Defnyddio'r Rhyngrwyd a safleoedd rhwydweithio i fwlio pobl eraill

Lledu sïon mewn blogiau neu ystafelloedd sgwrsio

6-8 marc Mae'r ymgeiswyr yn rhoi ateb eglur a rhesymegol sy'n disgrifio ac yn egluro tri mater o leiaf yn llawn a manwl ac yn rhoi enghreifftiau perthnasol. Defnyddiant dermau priodol a sillafu, atalnodi a gramadeg cywir.

3-5 marc Mae'r ymgeiswyr yn esbonio amrywiaeth o faterion ond nid yw'r atebion yn ddigon eglur nac yn cynnwys enghreifftiau perthnasol. Mae ychydig o gamgymeriadau sillafu, atalnodi a gramadeg.

0-2 farc Nid yw'r ymgeiswyr ond yn rhestru tri mater neu'n rhoi disgrifiad byr o un neu ddau o faterion. Nid yw'r ateb yn glir iawn ac mae llawer o gamgymeriadau sillafu, atalnodi a gramadeg.

Enghraifft 2

2 Mae siop ar-lein newydd yn cael ei chreu sy'n cynnig nwyddau dylunydd yn rhatach na siopau'r stryd fawr.

(a) Mae'r siop ar-lein yn cynnig profiad siopa rhyngweithiol a lefel uchel o wasanaeth i gwsmeriaid. Manylwch ar bedwar peth sydd ei angen, heblaw am galedwedd, ar gyfer gweithredu'r system hon yn llwyddiannus. **(4 marc)**

(b) Rhowch **ddwy** o fanteision siopa ar-lein i'r siop. **(2 farc)**

(c) Rhowch **ddwy** o fanteision siopa ar-lein i'r cwsmer. **(2 farc)**

(ch) Disgrifiwch **ddwy** broblem bosibl y gallai cwsmer eu hwynebu wrth siopa ar-lein. **(2 farc)**

Ateb myfyriwr 1

2 (a) Bydd angen meddalwedd dylunio gwefan i greu'r wefan a chadw'r cynnwys yn gyfoes, ychwanegu tudalennau newydd, ac ati. Bydd angen staff dylunio gwefan sydd â'r wybodaeth dechnegol a'r sgiliau dylunio i greu gwefan dda a fydd yn denu cwsmeriaid, ac a fydd yn diweddaru'r wefan.
Rhaid cael cysylltiad â'r Rhyngrwyd.

(b) Nid oes angen adeilad.
Mae'r siop yn gwneud mwy o arian.

(c) Gallwch siopa o'ch cartref, sy'n golygu nad oes rhaid i chi wario ar betrol neu barcio ac nad oes angen i chi wastraffu amser yn chwilio am nwyddau yr ydych chi eu heisiau ond nad ydynt ar gael yn y siopau.
Gall y cwsmer arbed arian yn aml gan ei bod hi'n haws pori am yr eitem rataf neu ddefnyddio un o'r gwefannau cymharu prisiau. Hefyd, mae gan siopau ar-lein gynigion arbennig nad ydynt ar gael ond i gwsmeriaid ar-lein.

(ch) Mae'n dipyn o strach prynu dillad dros y Rhyngrwyd gan na allwch weld a ydyn nhw'n ffitio cyn eu prynu, ac os nad ydyn nhw'n ffitio rhaid i chi wastraffu amser yn y Swyddfa Bost i'w hanfon yn ôl. Weithiau pan brynwch nwyddau o wledydd tramor, rydych chi'n meddwl eich bod chi'n cael bargen, ond pan ddeuant i mewn i'r wlad mae tollau'n cael eu codi arnynt a rhaid i chi dalu'r rhain cyn gallu eu derbyn. Felly gallant gostio cymaint ag yn y wlad hon – neu fwy.

Sylwadau'r arholwr

2 (a) Ar gyfer pob marc mae angen i'r myfyriwr ddisgrifio gofyniad yn fanwl. Mae'r ddau ofyniad cyntaf wedi cael eu hegluro'n ddigon manwl i ennill marc yr un. Gosodiad syml yw'r trydydd heb unrhyw fanylion, felly ni roddir unrhyw farciau. Nid oes pedwerydd gofyniad.

(b) Nid yw 'Nid oes angen adeilad' yn hollol wir. Mae angen adeilad ond nid oes rhaid iddo fod mor fawr nac mewn lleoliad drud yng nghanol y ddinas. Dim marciau am hyn.
Nid yw 'Mae'r siop yn gwneud mwy o arian' yn ddigon manwl – gallai'r siop wneud mwy o arian mewn llawer o wahanol ffyrdd. Pe bai'r myfyriwr wedi ychwanegu 'drwy gyflogi llai o staff na siop draddodiadol' neu 'drwy fod â gorbenion llai', byddai wedi cael y marc.

(c) Mae'r ddwy fantais wedi cael eu disgrifio'n dda ac yn haeddu dau farc.

(ch) Dyma ddau ateb da iawn, felly marciau llawn eto.
(6 marc allan o 10)

Cymorth gyda'r arholiad *(parhad)*

Ateb myfyriwr 2

2 (a) Bydd angen meddalwedd i greu'r wefan ar gyfer y siop.
Bydd angen pobl gymwysedig addas sydd â gwybodaeth arbenigol ym maes sefydlu gwefannau.
Bydd angen cysylltiad â'r Rhyngrwyd a allai gael ei ddarparu drwy weinydd Rhyngrwyd arbennig neu gan gwmni arall megis BT neu Virgin.

(b) Mae pobl yn dod atoch chi, felly mae'n haws.
Gallant wneud mwy o arian na siop.

(c) Nid oes angen iddynt adael y cartref, felly mae'n cymryd llawer llai o amser na siopa traddodiadol.
Gallwch arbed arian gan ei bod hi'n llawer haws siopa ar y Rhyngrwyd i gael y pris gorau.

(ch) Gall fod yn anodd cael eich arian yn ôl os yw'r nwyddau'n ddiffygiol.
Mae'n bosibl na fydd y gwasanaeth cwsmeriaid cystal ag mewn siop draddodiadol.

Sylwadau'r arholwr

2 (a) Dyma dri phwynt addas, gyda disgrifiad addas, felly rhoddir marc am bob un.

(b) Nid yw'r myfyriwr wedi egluro at beth mae'r pwynt cyntaf yn cyfeirio, felly dim marciau am y rhan hon o'r cwestiwn. Nid yw'r ail bwynt yn wir o anghenraid – mae llawer o siopau traddodiadol yn fwy proffidiol na siopau ar-lein, felly dim marciau am hyn.

(c) Mae dau bwynt da wedi'u gwneud, felly rhoddir dau farc.

(ch) Nid yw'r pwynt cyntaf yn wir gan fod y gyfraith yn amddiffyn cwsmeriaid siopau ar-lein a siopau traddodiadol fel ei gilydd.
Rhoddir marc am yr ail bwynt er nad yw'n fanwl iawn.

(6 marc allan o 10)

Ateb yr arholwr

2 (a) Pedwar gofyniad (un marc yr un), e.e.:
Darparwr/cyflenwr gwasanaeth rhyngrwyd (*ISP*)
Cynnal gwefan cwmni (staff profiadol gyda gwybodaeth am wefannau)
Catalog o stoc ar ffurf cronfa ddata
Dulliau o dalu'n ddiogel (troli/basged siopa, desg dalu, ac ati)
Cronfa ddata o archebion cwsmeriaid

(b) Dwy fantais (un marc yr un), e.e.:
Gorbenion is na siopau'r stryd fawr – dim adeiladau drud yng nghanol y ddinas, ditectifs siop, ac ati
Gall gyrraedd cwsmeriaid mewn unrhyw ran o'r wlad neu'r byd am gost weddol fach
Gellir cyflogi llai o staff gan fod llawer o systemau TGCh yn awtomatig
Gellir cysylltu'r system e-fasnach â systemau eraill fel cyfrifon, rheolaeth stoc, ac ati, i arbed costau gweinyddol

(c) Dwy fantais (un marc yr un), e.e.:
Bydd pobl sy'n byw mewn ardaloedd gwledig anghysbell yn gallu cyrchu'r cynhyrchion hynny sydd ond ar gael mewn dinasoedd mawr
Gall pobl anabl neu oedrannus siopa am nwyddau heb adael eu cartrefi
Bydd cwmnïau danfon nwyddau'n gwneud yn dda oherwydd y cynnydd mewn nwyddau sy'n cael eu hanfon yn uniongyrchol i gartrefi pobl
Bydd llai o deithiau'n cael eu gwneud i siopau gan leihau tagfeydd a llygredd mewn trefi a dinasoedd
Bydd siopa'n cymryd llai o amser a bydd yn rhyddhau pobl i wneud pethau eraill

(ch) Dwy broblem (un marc yr un) megis:
Gall gwefannau ffug gael eu sefydlu i gael arian gan gwsmeriaid, felly mae'n anodd i gwsmeriaid wybod a yw gwefan nad ydynt wedi clywed amdani'n ddilys
Mae gan rai gwefannau siopa wasanaeth cwsmeriaid gwael ac nid yw'n hawdd dychwelyd neu newid nwyddau weithiau
Gall siopau yng nghanol dinasoedd fynd yn wag oherwydd y cynnydd mewn cystadleuaeth, ac felly gall staff golli swyddi
Mae pobl yn hoffi pori a chyfarfod â phobl eraill wrth siopa

Mapiau meddwl cryno

Manteision e-fasnach i gwsmeriaid a busnesau

MANTEISION E-FASNACH

I'R CWSMER
- Gellir archebu 24/7
- Mwy o ddewis
- Arbed ar gostau
- Gall pobl anabl wneud eu siopa eu hunain
- Dim costau teithio

I'R SIOP/BUSNES
- Costau sefydlu a rhedeg is
- Llai o staff
- Hawdd diweddaru gwybodaeth
- Prisio hyblyg
- Gall cwsmeriaid ddod o rywle
- Archebu 24/7

Anfanteision e-fasnach i gwsmeriaid a busnesau

ANFANTEISION E-FASNACH

I'R CWSMER
- Pryderon diogelwch
- Gwefannau twyllodrus
- Strach os oes angen dychwelyd nwyddau
- Colli pleser cymdeithasu
- Costau cudd

I'R SIOP/BUSNES
- Dibyniaeth ar y Rhyngrwyd
- Cystadleuaeth gynyddol o dramor
- Gall rhai cwsmeriaid weld cost danfon fel problem
- Dibyniaeth ar gwmnïau danfon

Ffyrdd o gyrchu'r Rhyngrwyd

FFYRDD O GYRCHU'R RHYNGRWYD

DEIALU
- Araf
- Mae'n cyfyngu ar ddefnydd o'r Rhyngrwyd
- Yr unig gysylltiad sydd ar gael weithiau

CEBL
- Mynediad i fand llydan
- Amserau llwytho i lawr cyflym
- Gellir gwylio fideos, gwrando ar y radio, ac ati
- Does dim angen gwastraffu amser yn cysylltu
- Gellir defnyddio gwe-gamerâu a fideo-gynadledda
- Nid yw ar gael ym mhob gwlad

DIWIFR
- Gellir gweithio i ffwrdd o'r swyddfa
- Gellir gwneud trefniadau teithio funud olaf
- Gellir cyrchu dyfeisiau cludadwy lle mae rhwydwaith
- Dim angen ceblau
- Mae angen llwybrydd diwifr yn y cartref

TOPIG 3: Rhyngwyneb cyfrifiadur-dyn

Mae'r topig hwn yn ymhelaethu ar Dopig 9 yn Uned 1 UG ac yn edrych yn fanylach ar ryngwynebau cyfrifiadur-dyn (*HCI: Human-Computer Interface*) a'r ffactorau y dylid eu cymryd i ystyriaeth wrth eu dylunio.

Er mwyn defnyddio cyfrifiaduron, rhaid bod rhyngwyneb rhwng y cyfrifiadur a'r defnyddiwr dynol ac mae nifer o wahanol fathau o ryngwyneb. Yn nes ymlaen pan fyddwch yn cynhyrchu eich gwaith project, bydd yn rhaid i chi ddylunio eich rhyngwynebau eich hun ar gyfer pobl eraill. Cyn dylunio rhyngwyneb cyfrifiadur-dyn, mae'n hanfodol i chi ddeall anghenion y bobl fydd yn ei ddefnyddio. Bydd yr anghenion hyn yn eang yn aml, a rhaid i'r rhyngwyneb ddarparu ar gyfer pobl ag amrywiaeth fawr o sgiliau a galluoedd TGCh. Weithiau bydd y rhyngwyneb yn cael ei ddefnyddio gan grŵp penodol o bobl fel plant, peirianwyr, meddygon neu bobl ag anabledd penodol.

▼ Y cysyniadau allweddol sy'n cael sylw yn y topig hwn yw:

▶ Y ffactorau i'w hystyried wrth ddylunio rhyngwyneb defnyddiwr da

CYNNWYS

Y ffactorau i'w hystyried wrth ddylunio rhyngwyneb defnyddiwr da

- edrych yn daclus
- gofyn am gyn lleied o ddefnydd o'r bysellfwrdd â phosibl
- defnyddio synthesis lleferydd fel y gallant glywed geiriau.

Dylai'r rhyngwyneb ar gyfer *CAD*:

- allu cael ei addasu neu ei gwsmereiddio – dylai ganiatáu i'r defnyddiwr addasu'r sgrin (e.e. cynnwys yr eitemau maent yn eu defnyddio a dileu'r eitemau nad ydynt yn eu defnyddio)
- lleihau symudiadau'r llygoden – dylai fod gan ryngwyneb ddyluniad sy'n gwneud hyn
- cadw'r adran gweithio ar y sgrin mor fawr â phosibl – dylai gynnwys cwymplenni i hwyluso hyn
- defnyddio sgrin fawr os yw'r rhyngwyneb yn gymhleth – fel y gellir gweld mwy o'r diagram ar y sgrin ar yr un pryd
- defnyddio dyfeisiau mewnbynnu eraill – megis padiau neu lechi graffeg, fel y gellir gwneud dewisiadau o'r llechen yn hytrach na'r sgrin, gan adael y sgrin yn gliriach ar gyfer y cynllun sy'n cael ei ddangos arni
- defnyddio cwymplenni sy'n dangos y dewisiadau a ddefnyddir amlaf ar ben y ddewislen, fel nad oes angen symud i lawr drwy'r ddewislen yn fwy nag sydd raid.

Cyflwyniad

Mae rhai mathau o galedwedd a meddalwedd wedi cael eu dylunio'n arbennig ar gyfer grwpiau penodol o ddefnyddwyr. Ond defnyddir y mwyafrif llethol o systemau TGCh gan amrywiaeth eang o ddefnyddwyr. Yn yr adran hon byddwch yn ystyried beth sy'n gwneud rhyngwyneb defnyddiwr da.

Wrth greu rhyngwyneb defnyddiwr da, rhaid ystyried nifer o ffactorau, gan gynnwys:

- cysondeb arwyddbostio a gwybodaeth naidlen
- cymorth ar y sgrin
- dyluniad sy'n briodol i'r dasg
- gwahaniaethu rhwng arbenigedd defnyddwyr
- strwythur llywio clir
- defnydd gan bobl anabl.

Cysondeb arwyddbostio a gwybodaeth naidlen

Bydd rhyngwyneb defnyddiwr da yn cynnwys arwyddbostio a gwybodaeth naidlen sydd yn gyson.

Yr hyn y mae hynny'n ei olygu yw bod yr holl gymhorthion llywio megis y botymau Nesaf a Blaenorol, naidlenni, eiconau, ac ati yn edrych yr un fath o sgrin i sgrin ac yn ymddangos yn yr un lle ar y sgrin.

Cymorth ar y sgrin

Mae pawb sy'n defnyddio TGCh wedi bod yn ddysgwr ar un adeg. Gall fod yn rhwystredig i wybod beth yr ydych am ei wneud ond heb fod yn siŵr sut i'w wneud. Dyma lle mae cymorth ar y sgrin yn ddefnyddiol.

Dylai fod gan bob pecyn meddalwedd gyfleuster cymorth ar-lein. Drwy ddefnyddio'r cyfleuster hwn gall defnyddwyr gael cymorth sy'n cael ei gyflenwi gan y pecyn yn hytrach na gorfod edrych drwy lawlyfrau mawr.

Darparu cymorth ar-lein i ddefnyddwyr

Cymorth sgrin i ddefnyddwyr newydd

Gall rhai sgriniau cymorth edrych yn anodd a gallant ddefnyddio termau anghyfarwydd. Dylai sgriniau cymorth egluro pethau'n syml a gellir gwneud hyn orau drwy roi enghreifftiau i'r defnyddiwr.

Bydd y mwyafrif o ddefnyddwyr yn defnyddio'r meddalwedd i wneud amrywiaeth o dasgau. Byddant yn arbenigwyr ar rai tasgau ac yn ddibrofiad gyda thasgau eraill. Heblaw am ddarparu cymorth os oes ei angen, dylai'r meddalwedd adnabod a rhagweld gofynion y defnyddiwr a chynnig cymorth i wneud y dasg yn haws. Mae Microsoft Office yn defnyddio dewiniaid sy'n eich helpu drwy rai o'r tasgau anoddaf. Mae'r cymorth hwn yn caniatáu i ddefnyddwyr gyflawni tasgau mewn cyn lleied o amser â phosibl.

Dyluniad sy'n briodol i'r dasg

Ni fyddai'r dyluniad neu gynllun ar gyfer darn o feddalwedd sy'n dysgu plant sut i sillafu yr un fath â hwnnw ar gyfer darn o feddalwedd Cynllunio drwy Gymorth Cyfrifiadur (*CAD: computer-aided design*) a ddefnyddir gan bensaer i ddylunio adeiladau cymhleth. Gan fod y defnyddwyr hyn yn cyflawni tasgau cwbl wahanol, mae angen dyluniad arnynt sy'n briodol i'r dasg.

Dylai'r rhyngwyneb ar gyfer dysgu sillafu i blant:

- gynnwys cyn lleied o destun â phosibl ar y sgrin
- defnyddio lliwiau disglair i ddenu plant ifanc at y pecyn

Gwahaniaethu rhwng arbenigedd defnyddwyr

Gall defnyddwyr gael eu rhannu'n:

- ddysgwyr – pobl nad ydynt wedi defnyddio'r meddalwedd o'r blaen
- profiadol – pobl sydd wedi defnyddio'r meddalwedd, ond dim ond rhai o'i nodweddion
- arbenigwyr – pobl â blynyddoedd lawer o brofiad o'r meddalwedd sy'n defnyddio ei holl nodweddion.

Bydd gan bob un o'r defnyddwyr hyn anghenion gwahanol mewn perthynas â'r rhyngwyneb. Blaenoriaeth y dysgwr fydd rhyngwyneb hawdd ei ddysgu a chymorth hawdd ei gyrchu. Bydd y defnyddiwr arbenigol am wneud y dasg yn yr amser lleiaf posibl. Dyma rai ffyrdd y gellir defnyddio rhyngwyneb cyfrifiadur-dyn i wahaniaethu rhwng defnyddwyr:

Darparu llwybrau byr i arbenigwyr – bydd arbenigwyr yn teipio'n gyflym iawn a byddant yn arbed amser drwy ddysgu gorchmynion sy'n defnyddio cyfuniad o fysellau ar y cof yn hytrach na defnyddio'r llygoden a chlicio ar eiconau a chwymplenni. Gall dysgwr ddefnyddio'r llygoden, felly mae llawer o becynnau'n darparu rhyngwyneb arall.

Mwy o ffyrdd o wneud yr un gweithrediad – mae rhyngwynebau defnyddiwr yn cynnig mwy nag un ffordd o wneud yr un peth, a'r defnyddiwr sy'n cael dewis. Er enghraifft, efallai ei bod yn well gan ddysgwr ddefnyddio cwymplen neu glicio ar eicon er mwyn argraffu ffeil ond gallai defnyddiwr profiadol ddefnyddio dull cyflymach megis rhoi gorchymyn ar ffurf dilyniant o fysellau, er enghraifft, Ctrl+P.

Bydd gan bob math o ddefnyddiwr wahanol anghenion

Strwythur llywio clir

Dylai fod yn glir i'r defnyddwyr sut i wneud tasgau penodol. Felly os yw'r dasg yn gymhleth, gall y defnyddiwr gael ei dywys drwyddi mewn cyfres o gamau bach. Er enghraifft, mae'r dewin postgyfuno yn y pecyn prosesu geiriau Word yn mynd â'r defnyddiwr un cam ar y tro drwy'r broses o greu dogfen bostgyfuno gan ddefnyddio ei ddata ei hun.

Bydd llawer o wefannau'n defnyddio mannau poeth heb ddweud wrth y defnyddiwr am glicio arnynt. Ni ddylai'r defnyddiwr byth orfod dyfalu beth i'w wneud nesaf.

Dylai botymau Ymlaen ac Yn ôl bob amser gael eu gosod yn yr un lle ar y sgrin.

Dylai botymau Nesaf fod yr un fath a chael eu rhoi yn yr un lle. Os yw defnyddiwr am symud yn gyflym drwy gyfres o dudalennau, mae'n ddiflas gorfod lleoli'r llygoden cyn clicio. Mae'n llawer haws cadw'r botwm Nesaf yn yr un lle. Y cyfan mae'n rhaid i'r defnyddiwr ei wneud wedyn yw clicio i fynd i'r dudalen nesaf.

Defnydd gan bobl anabl

Mae angen i bobl ag anghenion arbennig neu benodol a phobl anabl ddefnyddio cyfrifiaduron. Gall systemau TGCh gynnig llawer o gyfleoedd newydd i'r bobl hyn a gwella eu bywydau.

Bydd defnyddwyr sydd â nam ar eu synhwyrau – efallai na allant ddarllen y llythrennau ar sgrin yn iawn – yn defnyddio ffont sy'n hawdd ei ddarllen ac yn sicrhau bod y ffont yn ddigon mawr iddynt ei ddarllen.

Os yw rhywun yn ddall, yna gall TGCh ei helpu drwy ddefnyddio cyfrifiadur 'siarad' lle mae'r geiriau'n cael eu siarad wrth gael eu teipio gan y defnyddiwr, neu eu hallbynnu ar y cyfrifiadur. Hefyd gall defnyddwyr dall ddefnyddio bysellfyrddau Braille arbennig i fewnbynnu data a gallant ddefnyddio argraffyddion Braille i gynhyrchu allbwn mewn Braille.

Egluro geiriau neu ymadroddion anodd

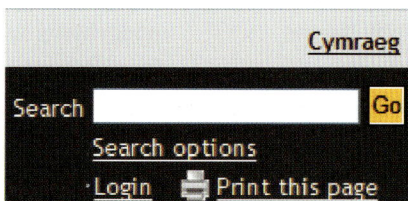

Cynnig ieithoedd gwahanol. Mae'r wefan hon ar gael yn y Gymraeg

Gall y dudalen we gael ei darllen i'r defnyddiwr – delfrydol i bobl â nam golwg

Mae defnyddwyr ag anabledd corfforol fel arfer yn cael trafferth symud mewn rhyw ffordd neu'i gilydd. Er enghraifft, efallai nad ydynt yn gallu defnyddio eu breichiau neu eu dwylo neu efallai na allant gerdded. Mewn rhai achosion, mae'r anabledd mor ddifrifol fel bod y person bron wedi'i barlysu.

Mae pobl na allant ysgrifennu oherwydd eu hanabledd yn gallu defnyddio systemau wedi'u hysgogi gan y llais i roi data i mewn i'r cyfrifiadur.

Ffyrdd o wneud gwefan neu system TGCh arall yn fwy hygyrch

Caiff gwefannau a systemau TGCh eraill eu dylunio'n aml i gael eu defnyddio gan y cyhoedd. Mae hyn yn cynnwys pobl o bob oed ac o bob cenedl, mewn gwahanol sefyllfaoedd (yn y gwaith, ar gyfer hamdden, ac ati), ag amrywiaeth o anableddau, sy'n dod o wahanol gefndiroedd addysgol, ac sydd â gwahanol sgiliau TGCh. Hynny yw, bron pawb.

Dyma rai ffyrdd o wneud gwefan yn fwy hygyrch i ddefnyddwyr:

Dylid defnyddio iaith eglur a syml.

Gall y defnyddiwr newid maint y testun

Gall y defnyddiwr newid lliw y testun a'r cefndir

Astudiaeth achos a Chwestiynau

▶ Astudiaeth achos   tt. 40–41

Defnyddio rhyngwyneb cyfrifiadur-dyn i sicrhau bod unigolyn yn cyrraedd ei botensial

Mae'n debyg mai Stephen Hawking yw'r ffisegwr enwocaf sy'n fyw heddiw gan ei fod wedi ymddangos ar lawer o raglenni teledu ac wedi ysgrifennu llawer o lyfrau ar wyddoniaeth boblogaidd gan gynnwys *A Brief History of Time*. Yr hyn sy'n fwy rhyfeddol fyth yw ei fod wedi goresgyn anabledd corfforol enfawr i wneud hyn. Mae gan Stephen ffurf ar glefyd niwron echddygol (*motor neurone disease*), clefyd cynyddol sy'n gwneud i chi golli'ch rheolaeth niwrogyhyrol yn gyfan gwbl bron. Mae hyn yn golygu eich bod chi'n cael eich parlysu'n llwyr.

Mae meddwl Stephen Hawking ymhlith y gorau yn y byd, ac eto ni all gyfathrebu â'r byd y tu allan heb ddefnyddio cyfarpar arbennig. Pe na bai'r holl gyfarpar hwn ar gael, ni fyddai wedi gallu cyfathrebu â neb. Ond yn lle hynny mae wedi ysgrifennu llyfrau, wedi teithio i bedwar ban byd, ac mae'n athro ym Mhrifysgol Caergrawnt.

Ar y dechrau bu Stephen yn defnyddio cyfrifiadur a system o'r enw Equaliser. Gallai weithredu'r system hon drwy symud switsh ag un o'i fysedd. Mae cyfres o linellau ar y sgrin sy'n cynnwys sawl gair yr un, ac mae system sganio yn symud bar amlygu dros bob gair yn ei dro. Mae Stephen yn aros nes bod y bar yn symud dros y gair sydd ei angen arno ac yna'n pwyso'r switsh â'i fys.

Collodd Stephen ei lais oherwydd ei afiechyd ond gall siarad yn electronig drwy ddefnyddio syntheseiddydd llais/lleferydd, er bod gan y llais acen Americanaidd!

Wrth i'r clefyd waethygu mae wedi gorfod addasu rhyngwyneb cyfrifiadur-dyn i allu ymdopi â'i barlys cynyddol. Yr unig ffordd y gall gyfathrebu bellach yw drwy grychu ei foch dde i weithredu switsh isgoch sydd ynghlwm wrth ei sbectol. Gall ddefnyddio'r switsh hwn i siarad, ysgrifennu nodiadau a llyfrau, pori'r Rhyngrwyd ac anfon e-byst. Mae'r cyfrifiadur ynghlwm wrth ei gadair olwyn ac mae'n caniatáu mynediad diwifr i rwydweithiau gan gynnwys y Rhyngrwyd.

Mae'r system hefyd yn ei alluogi i reoli'r drysau yn ei gartref.

1   Eglurwch ystyr y term 'rhyngwyneb cyfrifiadur-dyn'. (2 farc)
2   Disgrifiwch **ddau** o ofynion rhyngwyneb cyfrifiadur-dyn a ddefnyddir gan Stephen Hawking. (2 farc)
3   Disgrifiwch ryngwyneb cyfrifiadur-dyn gwahanol a all gael ei ddefnyddio gyda math arbennig o anabledd. Dylech egluro beth yw'r anabledd a sut mae'r rhyngwyneb yn gweithio. (4 marc)

▶ Cwestiynau 1   tt. 40–41

1   Wrth ddylunio unrhyw system TGCh, mae'r rhyngwyneb cyfrifiadur-dyn yn rhan bwysig ohoni. Enwch **bedwar** ffactor y dylid eu cymryd i ystyriaeth wrth ddylunio rhyngwyneb cyfrifiadur-dyn ac, ar gyfer pob ffactor, disgrifiwch pam y mae'n bwysig. (8 marc)
2   Dyma rai o'r pethau a allai fod yn ddefnyddiol i'r sawl sy'n defnyddio rhyngwyneb defnyddiwr. Eglurwch yn fyr pam mae pob un o'r nodweddion hyn yn ddefnyddiol.
    (a) Gellir dadwneud gweithredoedd yn hawdd. (1 marc)
    (b) Gellir mynd yn ôl i sgriniau blaenorol. (1 marc)
    (c) Defnyddio lliwiau/testun yn fflachio i roi negeseuon rhybuddio. (1 marc)
    (ch) Cynnig peth adborth i'r defnyddiwr megis sŵn clic, neu lun o amserydd wyau. (1 marc)
    (d) Cynnig dewis i'r defnyddiwr o ran y ffordd y rhoddir gorchmynion. (1 marc)
3   Rhaid ystyried nifer o ffactorau wrth ddylunio rhyngwyneb defnyddiwr da. Rydych chi'n dylunio rhyngwyneb defnyddiwr ar gyfer cyflwyniad amlgyfrwng sy'n rhedeg yn barhaus.
    Disgrifiwch sut y byddech chi'n dylunio pob un o'r ffactorau canlynol i sicrhau bod y cyflwyniad yn hawdd ei ddefnyddio.
    (a) Strwythur llywio clir. (2 farc)
    (b) Dyluniad sy'n briodol i'r dasg. (2 farc)
    (c) Cymorth ar y sgrin. (2 farc)
4   (a) Mewn perthynas â meddalwedd, eglurwch, gan roi enghraifft, sut y bydd anghenion defnyddiwr arbenigol yn wahanol i anghenion defnyddiwr newydd. (3 marc)
    (b) Eglurwch **un** ffordd y gellir darparu ar gyfer anghenion defnyddwyr arbenigol a newydd wrth ddylunio rhyngwyneb cyfrifiadur-dyn ar gyfer darn o feddalwedd. (2 farc)

Cymorth gyda'r arholiad

Enghraifft 1

1 Wrth ddylunio unrhyw system TGCh, mae rhyngwyneb cyfrifiadur-dyn yn rhan bwysig ohoni. Enwch **bedwar** ffactor y dylid eu cymryd i ystyriaeth wrth ddylunio rhyngwyneb cyfrifiadur-dyn ac, ar gyfer pob ffactor, disgrifiwch pam mae'n bwysig. **(8 marc)**

Ateb myfyriwr 1

1 Y gallu i newid maint y testun. Er enghraifft, mewn ffeil pdf gallwch addasu maint y ddogfen drwy ei chwyddo fel y gall pobl â golwg gwael ei gweld.

Dylai'r botwm Nesaf neu'r botymau eraill a ddefnyddir i lywio fod yn yr un lle ar y sgrin fel bod defnyddwyr yn gwybod ble maen nhw ar unwaith ar bob sgrin.

Dylid darparu cymorth sgrin fel bod defnyddwyr yn gallu teipio ymadrodd neu frawddeg a gadael i'r meddalwedd wneud y dasg iddynt.

Dylai'r defnyddiwr allu addasu'r cyfuniadau lliw o destun a chefndir ar y sgrin.

Sylwadau'r arholwr

1 Mae'r ffactor a'r rheswm ar gyfer y pwynt cyntaf yn gywir. Mae'r ail ffactor yn cael un marc yn unig gan nad yw'r myfyriwr wedi dweud pam mae'n bwysig. Dylai fod wedi nodi pam mae'n bwysig i Nesaf fod yn yr un lle ar bob sgrin.

Mae'r trydydd ffactor yn cael un marc am grybwyll cymorth sgrin ond nid yw'r rheswm yn gwneud synnwyr. Mae'r ffactor olaf yn gywir ond ni roddir rheswm dros ei phwysigrwydd. Efallai bod y myfyriwr yn meddwl bod y rheswm yn amlwg, ond er hynny rhaid ei gynnwys i ennill y marc. **(5 marc allan o 8)**

Ateb myfyriwr 2

1 Mae maint y ffont yn bwysig oherwydd bod angen ffont mawr ar bobl oedrannus a phlant ifanc iawn. Ond mae ffont llai yn addas ar gyfer defnyddwyr eraill er mwyn iddynt allu gweld mwy o wybodaeth ar y sgrin.

Dylai fod strwythur llywio clir fel y gall y defnyddiwr symud yn ôl ac ymlaen o sgrin i sgrin heb wastraffu amser.

Dylai'r rhyngwyneb fod yn reddfol fel ei fod yn hawdd ei ddefnyddio. Bydd dewisiadau lliw da ar gyfer y sgrin a'r cefndir yn gwneud y sgrin yn hawdd ei gweld, sy'n bwysig i ddefnyddwyr â golwg gwael neu sy'n ddall i liwiau, ac mewn rhai pecynnau neu wefannau gallwch newid y cyfuniadau lliw yn hawdd.

Sylwadau'r arholwr

1 Mae angen esboniad llawn o bob ffactor a pham mae'n bwysig i ennill y ddau farc ar gyfer pob ffactor. Mae'r ffactor gyntaf, maint y ffont, yn cael ei hesbonio'n ddigon da i ennill dau farc.

Rhoddir dau esboniad ar gyfer yr ail ffactor, felly dau farc am hyn.

Er bod y trydydd ffactor yn gywir, nid oes esboniad digon manwl, felly ni roddir unrhyw farciau.

Mae'r ffactor olaf am liw wedi cael ei hesbonio'n dda ac mae'n werth dau farc. **(6 marc allan o 8)**

Ateb yr arholwr

1 Nid yw nodi'r ffactor yn unig yn ennill marciau, felly rhaid i'r myfyrwyr roi disgrifiad pellach o'r ffactor a/neu fanylu ar beth sy'n gwneud y ffactor yn berthnasol i ryngwyneb defnyddiwr da i ennill y ddau farc.

Dylai unrhyw bedwar o'r ffactorau canlynol gael eu disgrifio'n fanwl:

Maint y ffont – bydd rhai defnyddwyr am weld mwy ar y sgrin, felly dylai maint y ffont fod yn fach (1) neu bydd angen ffont mwy o faint ar ddefnyddwyr ifanc a defnyddwyr hŷn sydd â golwg gwael (1).

Cymorth ar y sgrin – mae'n bwysig os nad yw cymorth ar gael o ffynonellau eraill (1) gan ei fod bob amser ar gael o fewn y rhaglen (1).

Dyluniad sy'n briodol i'r dasg. Gall fod yn well gan arbenigwyr deipio gorchmynion i mewn yn hytrach na defnyddio llygoden (1). Bydd yn well gan ddefnyddwyr llai profiadol ddefnyddio'r llygoden a rhyngwyneb defnyddiwr graffigol (1) oherwydd nad oes angen iddynt ddysgu gorchmynion.

Strwythur llywio clir. Dylai fod yn amlwg i ddefnyddwyr sut i fynd i'r cam neu sgrin nesaf (1) a dylai nodweddion llywio (e.e. Nesaf, saethau Ymlaen ac Yn ôl) gael eu gosod yn yr un lle (1).

Dewis o liw. Dylid defnyddio cyfuniadau sy'n darparu cyferbyniad rhwng y testun a'r cefndir (1) a dylai defnyddwyr allu newid y lliwiau (1) fel y gallant osgoi'r cyfuniad coch/gwyrdd os ydynt yn ddall i liwiau (1).

Cysondeb arwyddbostio a gwybodaeth naidlen. Fel bod popeth ar y rhyngwyneb i'w weld yn syth (1), sy'n gwneud y meddalwedd yn haws ei ddysgu (1).

Cymorth gyda'r arholiad (parhad)

Enghraifft 2

2 Trafodwch, gan roi enghreifftiau, sut y gall rhyngwyneb cyfrifiadur-dyn gael ei ddylunio sy'n addas ar gyfer defnyddwyr anabl. **(4 marc)**

Ateb myfyriwr 1

2 Gallai defnyddwyr na allant ddefnyddio eu dwylo fanteisio ar adnabod lleferydd fel ffordd o roi gorchmynion a mewnbynnu gwybodaeth.

Gallai defnyddwyr â nam ar eu golwg ddefnyddio rhyngwynebau lle gallwch addasu maint eitemau fel ffontiau ar y sgrin a bydd angen iddynt weld y testun ar y cefndir.

Sylwadau'r arholwr

2 Dim ond dau bwynt da sy'n cael eu gwneud yma, felly rhoddir dau farc yn unig. **(2 farc allan o 4)**

Ateb myfyriwr 2

2 Bydd defnyddwyr na allant ddefnyddio bysellfwrdd yn defnyddio dyfeisiau mewnbynnu eraill fel adnabod lleferydd, dyfeisiau mewnbynnu arbennig fel pibau yr ydych chi'n chwythu iddynt neu synwyryddion arbennig sy'n canfod symudiad y llygaid.

Dylai dyluniad y rhyngwyneb ddefnyddio digon o gyferbyniad rhwng y testun a'r cefndir i helpu defnyddwyr â golwg gwael.

Dylid ystyried cyfuniadau o liwiau fel y gall defnyddwyr sy'n ddall i liwiau ddefnyddio'r rhyngwyneb.

Mewn rhai gwefannau mae gennych ddewis o gyfuniadau lliw. Gall pobl â phroblemau cydsymud ddefnyddio llygod mwy o faint gyda botymau mawr.

Sylwadau'r arholwr

2 Un marc yr un am y ddwy ddyfais fewnbynnu arbenigol ar gyfer defnyddwyr anabl.

Mae'r wybodaeth am gyferbyniad a defnyddio gwahanol gyfuniadau o liwiau yn ennill dau farc. Rhoddir un marc am y wybodaeth am bobl â phroblemau cydsymud.

Sylwer: er bod pum marc posibl yma, ni ellir rhoi ond pedwar ar y mwyaf. **(4 marc allan o 4)**

Ateb yr arholwr

2 Un marc am bob pwynt hyd at uchafswm o 4 marc.

Gallai pobl na allant ddefnyddio bysellfwrdd neu lygoden ddefnyddio adnabod lleferydd (1).

Defnyddio dyfeisiau mewnbynnu arbenigol fel y rheiny sy'n defnyddio pibau chwythu neu symudiadau'r llygaid (1).

Y gallu i chwyddo rhannau o'r sgrin i helpu defnyddwyr â golwg gwael (1).

Y gallu i gynyddu maint y ffont i helpu defnyddwyr â golwg gwael (1).

Defnyddio cynlluniau lliw cywir i helpu pobl sy'n ddall i liwiau (1).

Defnyddio negeseuon gweledol yn hytrach na seiniau rhybuddio ar gyfer defnyddwyr byddar (1).

Defnyddio llygoden fwy o faint i helpu pobl â phroblemau cydsymud (1).

Defnyddio digon o gyferbyniad rhwng y testun a'r cefndir i helpu pobl â golwg gwael (1).

Mapiau meddwl cryno

Y ffactorau i'w hystyried wrth ddylunio rhyngwyneb defnyddiwr da

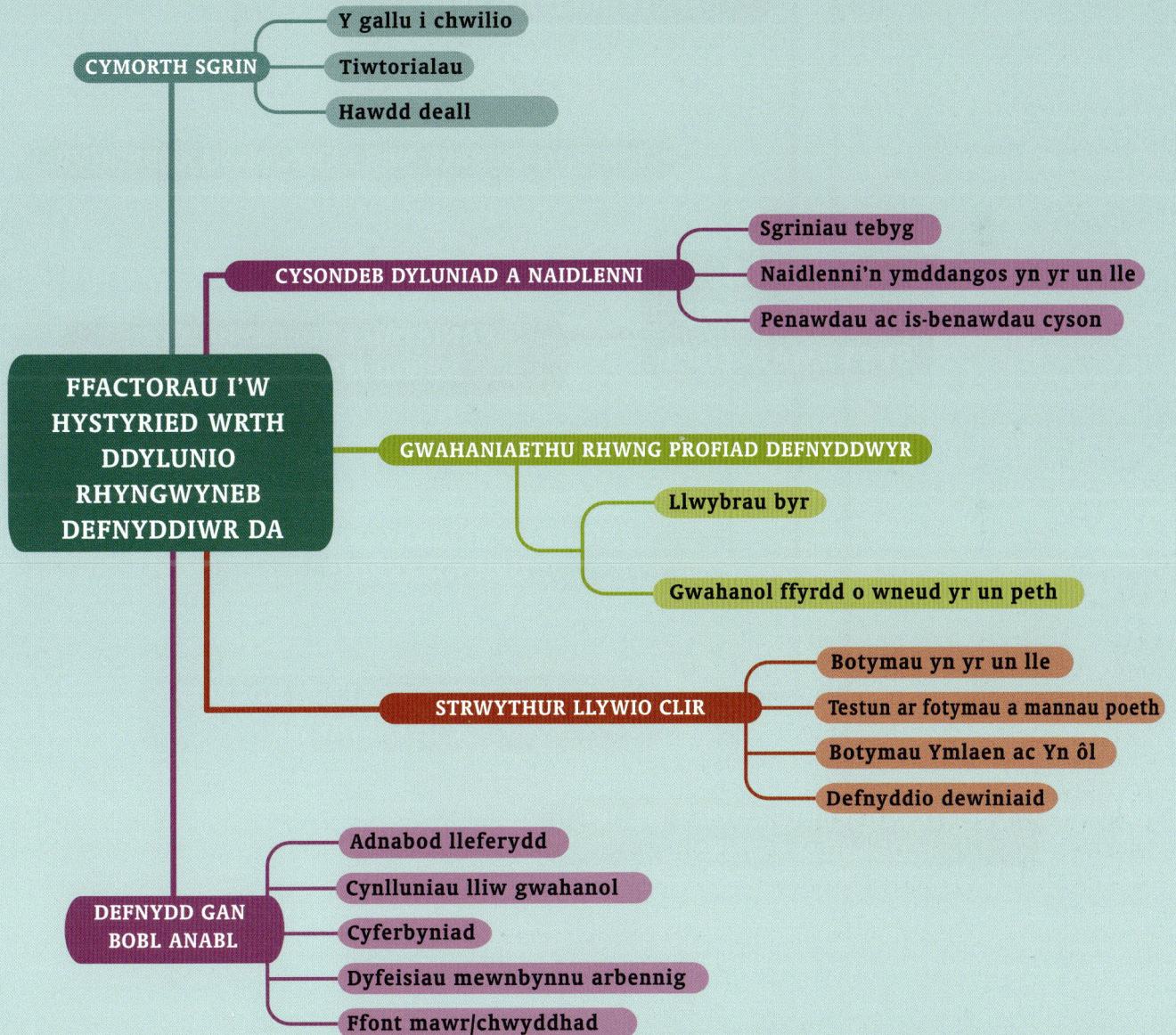

CYMORTH SGRIN
- Y gallu i chwilio
- Tiwtorialau
- Hawdd deall

FFACTORAU I'W HYSTYRIED WRTH DDYLUNIO RHYNGWYNEB DEFNYDDIWR DA

CYSONDEB DYLUNIAD A NAIDLENNI
- Sgriniau tebyg
- Naidlenni'n ymddangos yn yr un lle
- Penawdau ac is-benawdau cyson

GWAHANIAETHU RHWNG PROFIAD DEFNYDDWYR
- Llwybrau byr
- Gwahanol ffyrdd o wneud yr un peth

STRWYTHUR LLYWIO CLIR
- Botymau yn yr un lle
- Testun ar fotymau a mannau poeth
- Botymau Ymlaen ac Yn ôl
- Defnyddio dewiniaid

DEFNYDD GAN BOBL ANABL
- Adnabod lleferydd
- Cynlluniau lliw gwahanol
- Cyferbyniad
- Dyfeisiau mewnbynnu arbennig
- Ffont mawr/chwyddhad

Ffyrdd o wneud gwefan neu system TGCh arall yn fwy hygyrch

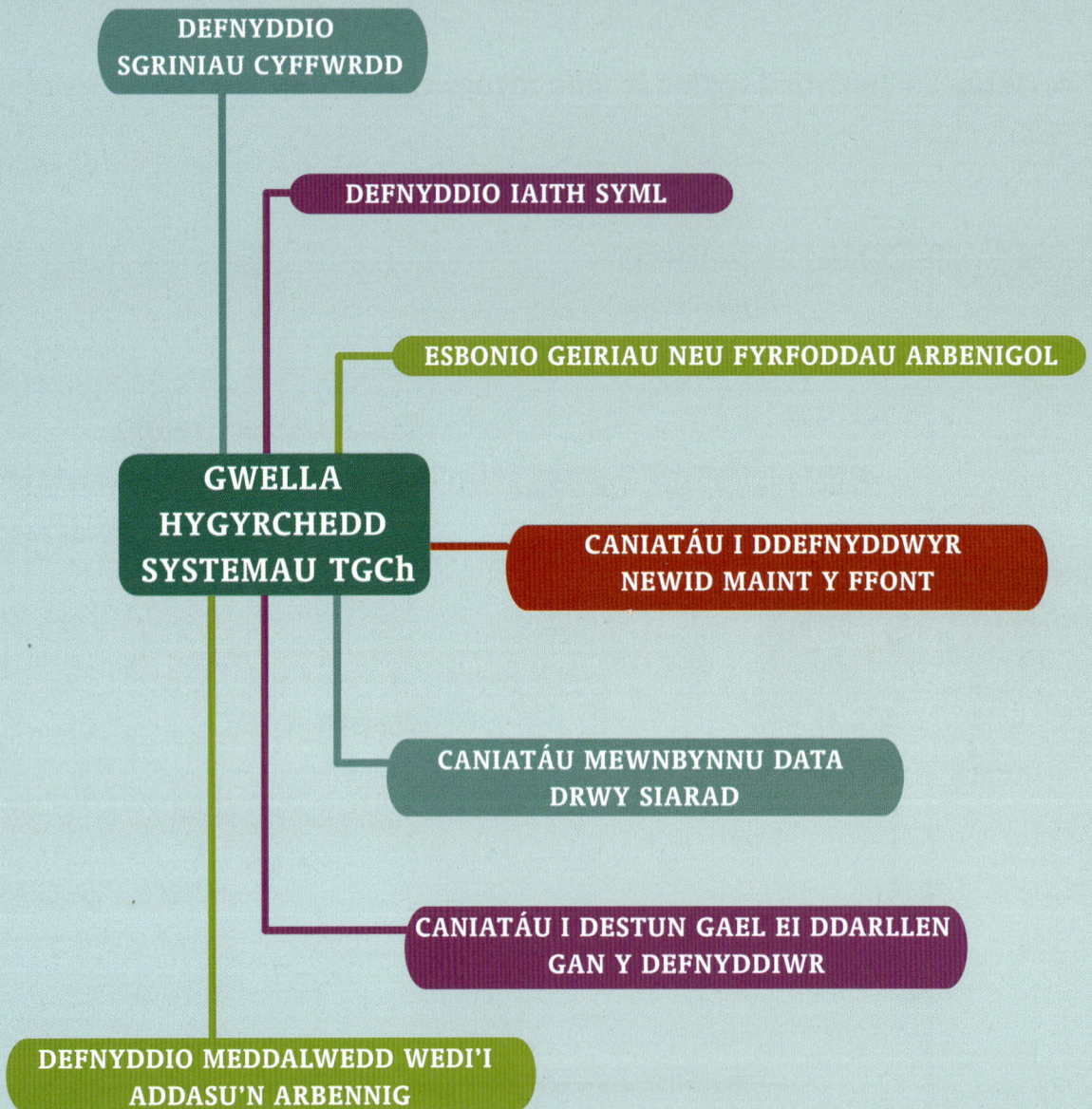

DEFNYDDIO SGRINIAU CYFFWRDD

DEFNYDDIO IAITH SYML

ESBONIO GEIRIAU NEU FYRFODDAU ARBENIGOL

GWELLA HYGYRCHEDD SYSTEMAU TGCh

CANIATÁU I DDEFNYDDWYR NEWID MAINT Y FFONT

CANIATÁU MEWNBYNNU DATA DRWY SIARAD

CANIATÁU I DESTUN GAEL EI DDARLLEN GAN Y DEFNYDDIWR

DEFNYDDIO MEDDALWEDD WEDI'I ADDASU'N ARBENNIG

TOPIG 4: Gweithio gyda TGCh

I'r mwyafrif o weithwyr, yr unig ddewis sydd ar gael iddynt yw'r swyddfa neu'r ffatri. Byddai'n well gan lawer o bobl fyw mewn rhan wahanol o'r wlad, felly mae'n gwneud synnwyr, os oes modd, i gyflogwyr fodloni'r dymuniad hwn. Pe baech chi'n siarad â gweithwyr cwmni, mae'n ddigon posibl y byddech chi'n darganfod bod llawer ohonynt yn teithio pellter mawr i'r gwaith bob dydd, yn enwedig os ydyn nhw'n cymudo i ganol dinas fawr. Wrth reswm, mae'n wastraff mawr i bobl dreulio'r holl amser hwn yn teithio rhwng y gwaith a'r cartref. Hefyd rhaid ystyried yr agwedd amgylcheddol. Pan fydd ceir yn llonydd mewn tagfeydd, cynhyrchant fygdarthau sy'n gwenwyno'r atmosffer. Mae adnoddau gwerthfawr fel petrol yn cael eu defnyddio.

Nawr, oherwydd y defnydd cynyddol o dechnoleg, mae'n bosibl i lawer o weithwyr, yn enwedig y rheiny sy'n cael eu cyflogi i wneud tasgau gweinyddol, weithio yn eu cartrefi, gan osgoi'r problemau hyn. Mae'r datblygiadau sydd wedi gwneud hyn yn bosibl yn cynnwys y gostyngiad mewn costau cyfathrebu a'r defnydd cynyddol o rwydweithiau, gan gynnwys y Rhyngrwyd, i rannu gwaith.

▼ Y cysyniadau allweddol sy'n cael sylw yn y topig hwn yw:

▶ Telegymudo

▶ Codau ymddygiad

CYNNWYS

Telegymudo/teleweithio

Cyflwyniad

Yn y topig hwn byddwch yn dysgu sut mae llawer o weithwyr bellach yn defnyddio telegymudo/teleweithio yn eu swyddi ac am fanteision ac anfanteision gwneud hyn iddyn nhw a'u cyflogwyr.

Byddwch hefyd yn edrych ar agwedd ar weithio gyda TGCh sy'n ceisio lleihau'r problemau sy'n wynebu gweithwyr wrth ddefnyddio cyfarpar TGCh. Mae'r agwedd hon yn cael ei galw'n god ymarfer, sef dogfen a roddir i weithwyr sy'n egluro'r hyn y gallant ei wneud ac na allant ei wneud wrth ddefnyddio cyfarpar TGCh.

Telegymudo

Mae telegymudo'n golygu cyflawni tasgau'n ymwneud â'ch gwaith drwy ddefnyddio telathrebu i anfon a derbyn data rhyngoch chi a swyddfa ganolog heb orfod bod yno'n gorfforol. Mae hefyd yn golygu y gallwch weithio o'ch cartref yn hytrach na gorfod cymudo i swyddfa bob dydd.

Mae gweithwyr sy'n telegymudo yn mwynhau peth hyblygrwydd o ran lleoliad ac oriau gwaith. Mae hyn yn golygu bod telegymudwyr yn gallu gweithio yn eu ffordd eu hunain ond bod rhaid iddynt wneud y gwaith.

Mae'r datblygiadau mewn TGCh sydd wedi galluogi mwy o bobl i weithio o'u cartrefi'n cynnwys:

- mynediad i'r Rhyngrwyd
- mynediad i e-bost
- cyfrifiadura symudol
- ffonau symudol
- warysau data (lle caiff holl ddata'r corff eu cadw mewn cronfa ddata enfawr)
- fideo-gynadledda
- cysylltiadau band llydan cyflym iawn.

Mathau o swyddi a phobl y mae telegymudo/teleweithio yn addas iddynt

Nid yw telegymudo/teleweithio yn addas i bob math o swydd a pherson.

- Mae rhai swyddi'n ddelfrydol ar gyfer gweithio o'r cartref. Rhai enghreifftiau yw rhaglenwyr, dylunwyr gwefannau, cyfrifwyr, clercod mewnbynnu data, clercod sy'n prosesu hawliadau yswiriant, ac ati.
- Rhaid gallu ymddiried yn y gweithwyr. Os oes angen anfon data personol neu sensitif, rhaid i'r gweithiwr gadw'r data hyn yn ddiogel a phreifat.
- Rhaid bod gan y gweithiwr hunangymhelliant. Mae llawer o bethau i dynnu ei sylw yn y cartref.
- Rhaid bod gan y gweithiwr le i weithio. Bydd angen lle yn y tŷ ar gyfer y cyfarpar sydd ei angen.

Telegymudo a theleweithio, a oes gwahaniaeth?

Oes, mae gwahaniaeth bach, ond ni chewch eich profi ar hyn gan fod y ddau air yn aml yn cael eu defnyddio i olygu'r un peth. Ond, a bod yn fanwl, mae teleweithio yn golygu defnyddio cyfathrebiadau i arbed taith. Er enghraifft, gallech ddefnyddio fideo-gynadledda mewn swyddfa i gymryd rhan mewn cyfarfod gyda chynrychiolwyr o bedwar ban byd. Nid ydych wedi teithio ac yn lle hynny rydych wedi defnyddio telathrebu, felly teleweithio yw hyn.

Gallwch weld nawr y byddai defnyddio TGCh gartref i wneud eich gwaith hefyd yn cael ei ystyried yn deleweithio gan eich bod chi'n defnyddio telathrebu yn hytrach na theithio i'r gwaith.

O hyn ymlaen, ac yn y cwestiynau arholiad, cewch gymryd bod gan deleweithio a thelegymudo yr un ystyr.

Caledwedd ar gyfer teleweithio

Gallai'r caledwedd ar gyfer teleweithio gynnwys y canlynol:

- cyfrifiadur
- sgrin
- cysylltiad Rhyngrwyd (llwybrydd diwifr, llwybrydd cebl, ac ati)
- microffon (ar gyfer fideo-gynadledda)
- gwe-gamera (ar gyfer fideo-gynadledda)
- ffôn Rhyngrwyd (ar gyfer gwneud galwadau rhad dros y Rhyngrwyd)
- argraffydd.

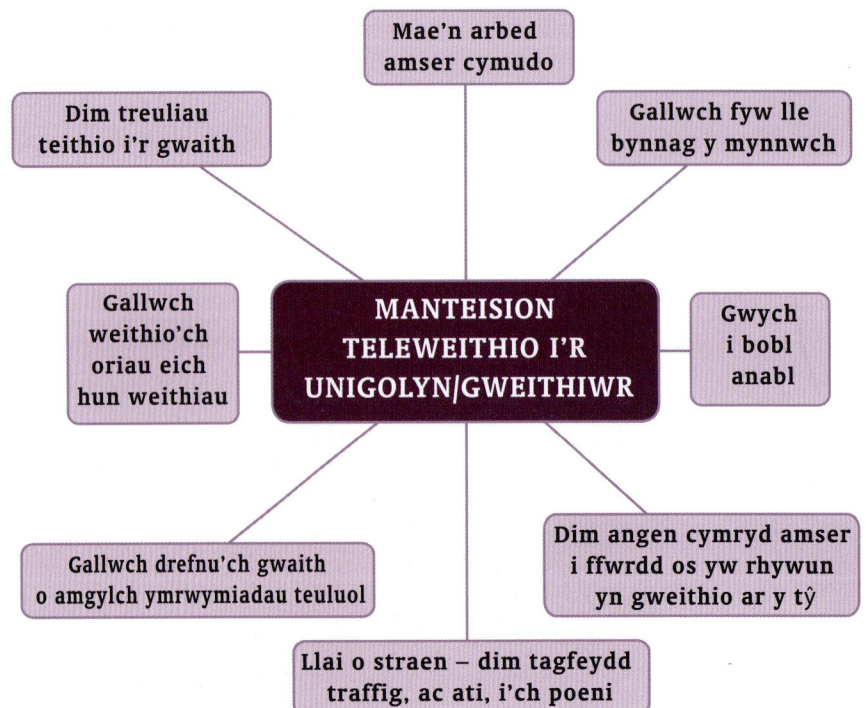

Mae gweithio o'r cartref, neu deleweithio, yn boblogaidd iawn gyda gweithwyr

Manteision teleweithio i'r gweithiwr

Mae teleweithio yn ei gwneud hi'n haws i bobl fyw a gweithio lle bynnag y mynnant, gan fod rhai pobl yn gallu gweithio gartref. Mae hyn yn lleihau tagfeydd traffig ac allyriadau carbon deuocsid ac felly mae'n 'fwy gwyrdd'. Mae lleihau cymudo yn llesol iawn i'r amgylchedd. Hefyd mae o gymorth i ardaloedd gwledig lle mae pobl yn gorfod symud i ffwrdd i gael gwaith fel rheol.

Manteision teleweithio i'r cyflogwr

Mae gan deleweithio fanteision i'r cyflogwr hefyd sy'n cael eu dangos yn y diagram isod:

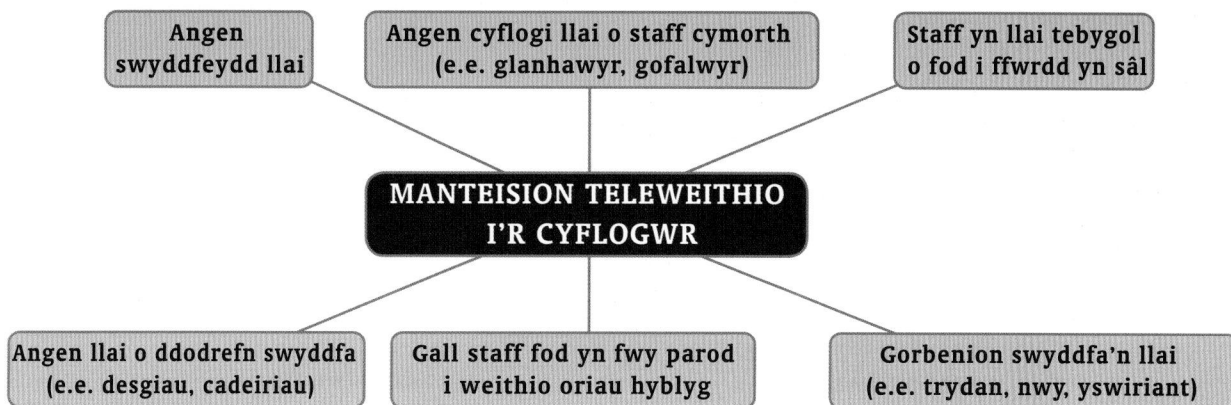

Angen swyddfeydd llai	Angen cyflogi llai o staff cymorth (e.e. glanhawyr, gofalwyr)	Staff yn llai tebygol o fod i ffwrdd yn sâl

MANTEISION TELEWEITHIO I'R CYFLOGWR

Angen llai o ddodrefn swyddfa (e.e. desgiau, cadeiriau)	Gall staff fod yn fwy parod i weithio oriau hyblyg	Gorbenion swyddfa'n llai (e.e. trydan, nwy, yswiriant)

Anfanteision teleweithio i'r gweithiwr

Dyma rai o anfanteision teleweithio i'r gweithiwr:

Cynnydd mewn costau megis gwresogi a goleuo'r cartref	Gall gweithwyr deimlo'n unig	Gall rhai cyflogwyr dalu llai i deleweithwyr gan fod mwy o gystadleuaeth am swyddi
Dim cydweithwyr i fynd allan gyda nhw		Collir y ffin rhwng y cartref a gwaith
Rhai staff yn colli statws – dim swyddfeydd moethus, ac ati	Efallai nad oes lle tawel yn y tŷ i weithio	Gall pobl eraill yn y tŷ darfu arnoch

ANFANTEISION TELEWEITHIO I'R GWEITHIWR

"Na, dydw i ddim yn unig yn gweithio o adre. Ddim ers i mi gyflogi fy nghynorthwywr, Mr Clustia Mawr!"

GEIRIAU ALLWEDDOL

Telegymudo – cyflawni tasgau'n ymwneud â'ch gwaith drwy ddefnyddio telathrebu i anfon a derbyn data rhyngoch chi a'r swyddfa ganolog heb orfod bod yno'n gorfforol

Teleweithio – defnyddio cyfathrebiadau i arbed taith, er enghraifft, gallech ddefnyddio fideo-gynadledda

Telegymudo/teleweithio *(parhad)*

Anfanteision teleweithio i'r cyflogwr

Mae gan gyflogwyr lai o reolaeth dros eu teleweithwyr. Mae nifer o anfanteision sy'n cael eu crynhoi yn y diagram canlynol:

Efallai y bydd angen newid strwythur trefniadaeth

Anodd gwybod pa mor galed mae staff yn gweithio

Anoddach i reolwyr reoli gwaith

ANFANTEISION TELEWEITHIO I'R CYFLOGWR

Anoddach cynnal cyfarfodydd

Y cyflogwr sy'n talu am gyfarpar TGCh y gweithwyr fel rheol

Gall cael cyfarpar TGCh ar fwy o safleoedd gynyddu risgiau diogelwch

Manteision teleweithio i gymdeithas

Mae teleweithio o fudd i gymdeithas yn y ffyrdd canlynol:

Llai o dagfeydd: llai o bobl yn teithio i'r gwaith

Llai o nwyon tŷ gwydr: yn lleihau cynhesu byd-eang

Gall helpu i gadw gwaith mewn cymunedau gwledig

MANTEISION TELEWEITHIO I GYMDEITHAS

Llai o lygredd aer

Perthnasoedd teuluol yn gwella: pobl yn treulio mwy o amser gartref

Pob rhan o'r wlad yn ffynnu, yn hytrach na dinasoedd yn bennaf

Fideo-gynadledda

Mae fideo-gynadledda'n caniatáu i gyfarfodydd wyneb yn wyneb gael eu cynnal heb i'r rheiny sy'n cymryd rhan orfod bod yn yr un ystafell neu hyd yn oed yr un ardal ddaearyddol. Yn ogystal â galluogi pobl i weld a chlywed ei gilydd, gellir defnyddio fideo-gynadledda i rannu dogfennau a chyflwyniadau.

Oherwydd bod angen lled band uchel ar gyfer y fideo a'r awdio, rhaid cywasgu a datgywasgu'r signalau mewn amser real, a defnyddir dyfais neu feddalwedd o'r enw *codec* (*coder/ decoder*) i wneud hyn.

Mae gan systemau fideo-gynadledda y cydrannau canlynol:

Caledwedd

- Cyfrifiadur amlgyfrwng safon uchel
- Camerâu fideo neu we-gamerâu – weithiau gall y camera gael ei weithredu o bell fel y gellir chwyddo mewn ar siaradwr neillltuol
- Microffonau
- Sgrin (gall hon fod yn sgrin gyfrifiadur neu'n sgrin blasma fawr sy'n caniatáu i lawer o bobl weld y llun)
- Uchelseinyddion
- Weithiau mae'r *codec* (i godio/ dadgodio'r signalau) ar ffurf caledwedd.

Meddalwedd

- Weithiau mae'r *codec* (i godio/ dadgodio'r signalau) ar ffurf meddalwedd.

Cyfathrebiadau data

Cysylltiad â rhwydwaith digidol cyflym iawn.

Cyfarfod wyneb yn wyneb yn cael ei gynnal dros ardal ddaearyddol fawr

Manteision defnyddio fideo-gynadledda

Mae gan fideo-gynadledda lawer o fanteision a gall y rhain gael eu dosbarthu'n fanteision i wahanol grwpiau o bobl.

Manteision i'r gweithiwr

- Llai o straen gan nad yw gweithwyr yn gorfod dygymod ag oedi mewn meysydd awyr, damweiniau, gwaith ar y ffordd, ac ati.
- Gwell bywyd teuluol gan fod llai o amser yn cael ei dreulio i ffwrdd o'r cartref mewn gwestai.
- Nid oes angen iddynt dreulio llawer o amser yn teithio i gyfarfodydd ac adref.

Manteision i'r corff

- Llawer rhatach gan nad yw'n gwario arian ar dreuliau teithio, gwestai, prydau, ac ati.
- Staff yn fwy cynhyrchiol gan nad ydynt yn gwastraffu amser yn teithio.
- Gellir cynnal cyfarfodydd ar fyr rybudd heb lawer o gynllunio.
- Gellir cynnal cyfarfodydd byr mewn achosion lle na fyddai'n ymarferol i staff deithio pellter mawr i fynychu cyfarfod byr.

Manteision i gymdeithas

- Bydd llai o bobl yn hedfan i gyfarfodydd, gan leihau hediadau a'r carbon deuocsid sy'n cael ei allyrru. Bydd hyn yn helpu i leihau cynhesu byd-eang.
- Ni fydd ffyrdd wedi'u tagu â thraffig ac felly fe fydd llai o straen ar weithwyr a llai o lygredd.

Anfanteision/cyfyngiadau fideo-gynadledda

- Cost y cyfarpar – mae cyfarpar fideo-gynadledda arbenigol yn ddrud.
- Ansawdd delweddau a sain yn wael – nid yw ansawdd y delweddau cystal ag y byddech chi'n ei gael ar deledu oherwydd yr angen i gywasgu a datgywasgu signalau a anfonir dros gysylltau cyfathrebu.
- Gall pobl deimlo'n hunanymwybodol iawn wrth ddefnyddio fideo-gynadledda a methu gwneud argraff dda.
- Er y gall dogfennau a diagramau ar ffurf ddigidol gael eu rhannu, nid yw hyn yn bosibl gyda chynnyrch neu gydran go iawn.
- Gall diffyg cysylltiad wyneb yn wyneb olygu nad yw trafodaeth mor effeithiol.

Defnyddio fideo-gynadledda

- **Addysg** – bydd rhai ysgolion yn defnyddio fideo-gynadledda ar gyfer dysgu pynciau Safon Uwch os nad oes digon o fyfyrwyr i gyfiawnhau cael athro arbenigol. Felly gellir defnyddio un athro i ddysgu myfyrwyr o sawl ysgol.
- **Meddygaeth** – gall meddygon prysur ddefnyddio fideo-gynadledda i gynnal cyfarfodydd am gleifion, a gall lluniau pelydr X, delweddau uwchsain a dadansoddiadau labordy gael eu trosglwyddo'n ddigidol i'r fideo-gynadleddwyr.
- **Busnes** – defnyddir fideo-gynadledda'n helaeth mewn busnes i gynnal cyfarfodydd o bell.

Fideo-gynadledda mewn ysbytai

Mae fideo-gynadledda yn dod yn boblogaidd mewn ysbytai ac mae ganddo sawl defnydd fel:

- Darparu cysylltiad rhwng y claf a'r meddyg, a chyfieithydd os na all y claf siarad Saesneg.
- Helpu cleifion sy'n byw yng nghefn gwlad sy'n cael trafferth i ddod i'r ysbyty. Mae hyn yn arbennig o bwysig mewn ardaloedd gwledig lle gall yr ysbyty agosaf fod gannoedd o filltiroedd i ffwrdd.
- Hyfforddi staff meddygol. Gall staff ddysgu oddi wrth arbenigwyr yn gyflym heb dreulio amser yn mynychu cyrsiau mewn lleoliad anghysbell.
- Ymgynghori ag arbenigwr sy'n byw'n bell oddi wrth y claf a'r meddygon. Mae gan rai cleifion glefydau cymhleth ac mae'n bosibl nad oes gan y meddygon yn yr ysbyty brofiad o ymdrin â nhw ac felly mae angen cymorth o'r tu allan arnynt.
- Ar gyfer cyfarfodydd rhanbarthol y byddai staff yn gorfod teithio iddynt fel rheol.

Mae fideo-gynadledda wedi datblygu cymaint erbyn hyn fel ei bod hi'n anodd gwybod pwy sydd yn yr ystafell a phwy sydd ddim!

> **GEIRIAU ALLWEDDOL**
>
> **Lled band** – mesur o faint o ddata y gellir eu trosglwyddo gan ddefnyddio cyfrwng trosglwyddo data

51

Codau ymddygiad

Cyflwyniad

Mae llawer o ddeddfau'n ymwneud â defnyddio TGCh, gan gynnwys y Ddeddf Gwarchod Data a'r Ddeddf Camddefnyddio Cyfrifiaduron, a rhaid i bob gweithiwr weithio o fewn y gyfraith. Yn ogystal ag ufuddhau i'r deddfau, rhaid i staff sy'n defnyddio TGCh weithredu mewn ffordd nad yw'n peri niwed na phoen i ddefnyddwyr eraill. Os bydd pawb yn cytuno i ddefnyddio'r cyfleusterau TGCh mewn ffordd benodol, gallant fynd ymlaen â'u gwaith yn ddi-rwystr.

Yn yr adran hon byddwch yn dysgu pam mae gan gyrff god ymddygiad a beth y byddai cod ymddygiad nodweddiadol yn ei gynnwys.

Beth yw cod ymddygiad?

Cytundeb rhwng gweithiwr a chorff yw cod ymddygiad lle mae'r gweithiwr yn ufuddhau i reolau'r corff ac yn gweithio o fewn canllawiau penodedig mewn perthynas â defnyddio TGCh a'r Rhyngrwyd.

Pam cael cod ymddygiad?

Mae cyflogwyr yn wynebu llawer o broblemau wrth adael i'w staff ddefnyddio eu cyfarpar TGCh. Heb ganllawiau priodol ar ffurf dogfen o'r enw cod ymddygiad, byddai'r staff yn gallu gwneud fel y mynnant a gall hyn achosi problemau i'r cyflogwr, gan arwain at ddirwyon i'r corff mewn rhai achosion.

Yr ateb yw cael cod ymddygiad sy'n nodi'n glir i'r staff y canllawiau sydd wedi cael eu pennu gan y corff ynghylch defnyddio TGCh a'r Rhyngrwyd. Mae ufuddhau i'r cod ymddygiad yn cael ei gynnwys fel amod mewn cytundebau cyflogaeth. Pe bai aelod o staff yn penderfynu peidio ag ufuddhau i'r cod ymddygiad, byddai'n wynebu cyfres o fesurau disgyblu a allai arwain at ddiswyddiad.

Dylid parchu hawliau pobl eraill

Y problemau posibl

Mae llawer o broblemau'n codi wrth i staff ddefnyddio TGCh yn y gweithle, gan gynnwys:

- Cyflwyno firysau – drwy lwytho gemau i lawr, peidio â sganio cyfryngau cludadwy, peidio â diweddaru sganwyr firysau, ac ati.
- Camddefnyddio cyfleusterau TGCh gan y staff, e.e. defnyddio telathrebiadau at eu dibenion eu hun (e.e. galwadau ffôn, e-bost, fideo-gynadledda, ac ati) a defnyddio argraffyddion at ddefnydd personol.
- Dosbarthu deunydd hiliol neu rywiol – e.e. anfon jôcs cas drwy e-bost neu negeseuon testun, dosbarthu delweddau atgas dros rwydwaith y corff, ac ati.
- Camddefnyddio data at ddibenion anghyfreithlon – e.e. defnyddio e-bost a negeseuon testun i fwlio rhywun yn y gwaith neu'r ysgol/coleg.
- Defnydd amhriodol o ffonau symudol – mewn tai bwyta ac ysgolion ac ar gludiant cyhoeddus.
- Blacmel, twyll cyfrifiadurol neu werthu i gyrff eraill.
- Torri telerau hawlfraint neu gytundebau meddalwedd fel bod y cwmni'n wynebu camau cyfreithiol gan gyflenwyr meddalwedd neu gyrff eraill.

Cod ymddygiad i'r staff

Mae cod ymddygiad i'r staff yn cynnwys rheolau wedi'u llunio gan yr uwch reolwyr neu eu cynghorwyr sy'n nodi'r hyn y gall aelodau staff eu gwneud a'r hyn na allant eu gwneud yn ystod eu cyflogaeth. Mae hefyd yn nodi'r cosbau a gaiff eu gweithredu os na fydd yr aelod staff yn ufuddhau i'r rheolau. Fel rheol, bydd y cod ymddygiad yn rhan o bolisi diogelwch TG corfforaethol y cwmni neu sefydliad, ond weithiau bydd yn cael ei gynnwys mewn dogfen ar wahân y mae'n rhaid i'r staff ei darllen, cytuno ar ei chynnwys, ac yna ei llofnodi.

Beth mae cod ymddygiad yn ei gynnwys?

Bydd cod ymddygiad fel rheol yn cynnwys y canlynol.

Cyfrifoldebau

Rhaid i holl aelodau cyrff sy'n defnyddio'r cyfleusterau TG weithredu'n gyfrifol a dylai pob defnyddiwr fod yn gyfrifol am ddiogelwch a chyflwr da yr adnoddau dan ei reolaeth.

Rhaid i ddefnyddwyr barchu hawliau pobl eraill, parchu'r cyfleusterau a'r dulliau rheoli corfforol, a chydymffurfio

Rhaid i chi wybod am y deddfau sy'n ymwneud â TGCh

â'r trwyddedau a chytundebau cytundebol perthnasol.

Gan fod cyfrifiaduron a rhwydweithiau'n rhoi mynediad i adnoddau y tu mewn ac y tu allan i'r corff, dylai mynediad agored o'r fath gael ei drin fel braint a rhaid i bob defnyddiwr unigol ymddwyn yn gyfrifol.

Rhaid i ddefnyddwyr ymddwyn yn gyfrifol drwy sicrhau nad yw firysau'n cael eu cyflwyno drwy fethu â sganio cyfryngau, ac ati. Hefyd rhaid iddynt beidio â bod yn esgeulus a gadael eu cyfrifiaduron wedi'u mewngofnodi fel bod pobl eraill yn gallu cyrchu'r data, neu adael gliniaduron lle gallant gael eu dwyn yn hawdd.

Parchu hawliau pobl eraill

Mae gan bobl sy'n defnyddio systemau TGCh hawliau y mae angen eu parchu. Gall systemau TGCh fel e-bost, blogiau ac ystafelloedd sgwrsio gael eu camddefnyddio gan bobl eraill. Er enghraifft, gallent ledu stori ffug am rywun neu ddefnyddio TGCh i fwlio unigolyn. Mae angen i staff deimlo'n ddiogel yn y gweithle, felly rhaid iddynt i gyd barchu ei gilydd.

Cydymffurfio â'r ddeddfwriaeth

Mae llawer o ddeddfau'n ymdrin â'r ffordd y gall TGCh gael ei defnyddio. Mae'r rhai presennol yn cynnwys Deddf Gwarchod Data 1998, Deddf Camddefnyddio Cyfrifiaduron 1990 a Deddf Hawlfraint, Dyluniadau a Phatentau 1988. Rhaid i staff lynu wrth y deddfau hyn ac os na wnânt mae'n bosibl y cânt eu diswyddo a'u herlyn gan yr heddlu.

Rhaid i hysbysebion ar gyfer nwyddau ar y Rhyngrwyd gydymffurfio â'r Ddeddf Disgrifiadau Masnach. Rhaid i wasanaethau ariannol fel benthyciadau, morgeisi ac yswiriant ufuddhau i'r rheolau yn y Ddeddf Gwasanaethau Ariannol.

Cydymffurfio â chytundebau trwyddedu

Pan brynwch ddarn o feddalwedd nid ydych chi'n berchen arno ac nid ydych yn hollol rydd i'w ddefnyddio fel y mynnwch. Pan brynwch feddalwedd, rydych chi'n prynu trwydded i'w ddefnyddio. Pan fydd cyrff yn prynu meddalwedd i'w ddefnyddio dros rwydwaith, maen nhw'n prynu trwydded i ganiatáu i nifer penodol o gyfrifiaduron ddefnyddio'r meddalwedd ar yr un pryd. Mae'n bwysig nad yw staff yn defnyddio mwy o fersiynau o'r meddalwedd nag y mae'r drwydded yn ei ganiatáu.

Awdurdodi

Bydd cyrchu adnoddau gwybodaeth y corff heb gael ganiatâd priodol gan y rheolwr diogelwch, defnyddio cyfleusterau'r corff heb awdurdod, a llygru neu gamddefnyddio adnoddau gwybodaeth yn fwriadol yn cael ei ystyried yn drosedd yn erbyn y cod ymddygiad.

Polisi diogelwch

Gallai'r polisi diogelwch gynnwys y canlynol:

- peidio â datgelu cyfrineiriau
- peidio â datgelu data'r cwmni i drydydd parti
- sicrhau nad yw allbrintiau o ddata'r cwmni yn cael eu gweld heb ganiatâd
- sicrhau bod unrhyw ddogfennau nad oes eu hangen yn cael eu gwaredu'n ofalus
- rhaid i ddefnyddwyr a gweinyddwyr

systemau fod ar eu gwyliadwriaeth rhag camddefnydd sy'n tarfu ar y system neu'n bygwth ei hyfywedd.

Caniatâd i gyrchu data

Ni ddylai gweithwyr gyrchu data neu ffeiliau oni bai bod ganddynt ganiatâd i wneud hynny. Mae hyn yn golygu, er enghraifft, na ddylai pobl gyrchu unrhyw ffeiliau heblaw am y rheiny sy'n hanfodol ar gyfer gwneud eu gwaith.

Diogelwch

Rhaid i staff gymryd diogelwch caledwedd, meddalwedd a data o ddifrif ac ufuddhau i reolau a fydd yn lleihau'r tebygolrwydd y caiff yr adnoddau hyn eu colli. Dyma rai rheolau y gellir eu cynnwys mewn cod ymddygiad i wella diogelwch:

- Rheolau am beidio â datgelu cyfrineiriau i bobl eraill ac am newid cyfrineiriau'n rheolaidd.
- Rheolau'n ymwneud â'r defnydd personol o e-bost.
- Rheolau'n ymwneud â defnyddio'r Rhyngrwyd (e.e. ni all staff gyrchu gwefannau rhwydweithio cymdeithasol, ystafelloedd sgwrsio na gwefannau arwerthu yn ystod oriau gwaith).
- Rheolau'n ymwneud â throsglwyddo data. Ni fydd llawer o gyrff yn caniatáu i unrhyw ddeunydd gael ei lwytho i lawr o'r Rhyngrwyd i gyfrifiaduron y cwmni. Gall lawrlwytho ffeiliau mawr megis ffeiliau cerddoriaeth neu fideos ddefnyddio lled band gwerthfawr a gwneud y Rhyngrwyd yn arafach i staff eraill.

"Mae diogelwch gwybodaeth yn flaenoriaeth yn y cwmni hwn. Rydyn ni wedi gwneud llawer o bethau gwirion rydyn ni eisiau eu cadw'n gyfrinachol."

Codau ymddygiad *(parhad)*

Y cosbau am gamddefnydd

Ni fydd y cod ymddygiad i staff yn cael ei gymryd o ddifrif oni bai bod cosbau am gamddefnyddio cyfleusterau TGCh. Bydd yn rhaid i'r cosbau hyn fod yn gymesur â difrifoldeb y drosedd. Mae'r cosbau'n cynnwys:

- rhybuddion anffurfiol – rhybuddion llafar na chânt eu cofnodi mewn ffeil
- rhybuddion ysgrifenedig – rhybuddion mwy ffurfiol sy'n cael eu cofnodi mewn ffeil
- diswyddiad
- erlyniad.

Bydd ymchwiliad yn cael ei wneud i droseddau honedig yn erbyn y cod ymddygiad ac os oes sail i'r honiad fe gymerir mesurau pellach. Mewn rhai achosion gallai hyn gynnwys achos sifil neu droseddol.

Rhybuddion anffurfiol

Gallai troseddau bach yn erbyn y cod fod yn anfwriadol. Er enghraifft, dewis cyfrineiriau gwael, defnyddio gormod o le ar y disg oherwydd cymhennu ffeiliau gwael. Bydd rheolwr llinell y defnyddiwr yn ymdrin yn anffurfiol â throseddau o'r fath.

Rhybuddion ysgrifenedig

Rhaid ymdrin â throseddau mwy difrifol mewn ffordd fwy ffurfiol rhag ofn bod angen cymryd camau pellach yn y dyfodol. Byddai troseddau o'r fath yn cynnwys rhannu cyfrifon neu gyfrineiriau, gadael dogfennau heb eu goruchwylio ar eich desg, ac ati. Bydd troseddau difrifol yn cael eu trin yn ffurfiol fel arfer drwy roi rhybudd ysgrifenedig am y drosedd gyntaf, a diswyddo efallai am ail drosedd. Rhai troseddau difrifol yw:

- defnyddio terfynellau heb awdurdod
- ymdrechion i ddwyn data neu gyfrineiriau
- defnyddio neu gopïo meddalwedd heb awdurdod
- cyflawni mân droseddau'n gyson.

Diswyddo

Gall troseddau difrifol iawn arwain at ddiswyddo gweithiwr. Gall hyn ddigwydd hefyd os yw gweithiwr wedi anwybyddu nifer o rybuddion ysgrifenedig. Bydd gweithiwr yn aml yn cael ei ddiswyddo os yw wedi gwneud rhywbeth sy'n peri embaras i'w gyflogwr neu a allai arwain at erlyn y cwmni.

Erlyn

Os yw'r drosedd yn ddifrifol iawn gall y gweithiwr gael ei ddiswyddo'n syth, ac os yw'r gweithiwr wedi torri'r gyfraith gall yr heddlu gael eu galw i mewn a gall erlyniad troseddol ddilyn. Er y bydd llawer o gyrff yn ceisio osgoi galw'r heddlu i mewn oherwydd y cyhoeddusrwydd gwael, efallai na fydd unrhyw ddewis arall os yw'r gweithiwr wedi cyflawni trosedd fawr, yn enwedig os yw wedi peri colledion mawr i'r corff.

Byddai troseddau o'r fath yn cynnwys dwyn meddalwedd neu ddata, peidio dilyn y Ddeddf Gwarchod Dadta, llwytho pornograffi i lawr o'r Rhyngrwyd ayb.

Y gwahaniaethau rhwng materion cyfreithiol a moesol mewn perthynas â chodau ymddygiad

Mater sy'n ddifrifol ac yn erbyn y gyfraith yw mater cyfreithiol. Os yw'r weithred yn anghyfreithlon gall yr heddlu neu awdurdod arall erlyn yr aelod staff a gall hyn arwain at ddirwy neu garchar hyd yn oed.

Mater y byddai'r mwyafrif o bobl yn meddwl ei fod yn anghywir, ond nad yw'n anghyfreithlon, yw mater moesol.

Gwybodaeth ffug

Bwriad gwybodaeth ffug yw twyllo neu gamarwain. Gall fod yn fater moesol neu gyfreithiol, fel y dengys yr enghreifftiau hyn.

Gwybodaeth ffug fel mater moesol:

- Contractwr (e.e. dylunydd gwefannau, rhaglennwr, peiriannydd rhwydwaith) yn gwneud cynnig am gontract er nad yw'n gwybod digon am y maes.
- Dadansoddydd systemau yn cynghori cleient i brynu math arbennig o galedwedd gan y bydd yn cael gwyliau rhad os caiff y contract ei lofnodi.
- Gwerthwr cyfrifiaduron mewn siop yn cynghori cleient i brynu cyfrifiadur gan wneuthurwr penodol gan ei fod yn cael comisiwn uwch ac nid oherwydd mai hwnnw yw'r cyfrifiadur gorau i gwrdd â'i anghenion.
- Peidio â rhoi gwybod i ddarpar gwsmeriaid neu gleientiaid am yr holl ffeithiau sydd ar gael am gynhyrchion neu wasanaethau (e.e. y ffaith bod model newydd, gwasanaeth gwell neu gynnig arbennig ar fin cael ei gyflwyno).

Gwybodaeth ffug fel mater cyfreithiol:

- Gwerthwr eiddo yn rhoi gwybodaeth ffug am dŷ ar ei wefan yn groes i'r Ddeddf Eiddo sy'n nodi'r hyn na all gwerthwyr eiddo ei wneud.
- Gwerthwr mewn siop yn gwerthu meddalwedd gan honni y gall wneud pethau na all eu gwneud mewn gwirionedd, yn groes i'r Ddeddf Disgrifiadau Masnach.
- Ysbyty yn gwrthod rhoi gwybodaeth i glaf am ei afiechyd i guddio camgymeriad yn y diagnosis meddygol. Mae gwrthod rhoi'r wybodaeth bersonol gywir yn groes i'r Ddeddf Gwarchod Data.

Gall y troseddau mwyaf difrifol arwain at ddiswyddiad

RYDYM YN EICH DISWYDDO

Preifatrwydd

Mae preifatrwydd yn ymwneud â gallu pobl i gadw gwybodaeth am eu bywyd yn breifat. Mae'r gyfraith yn gwarchod data personol rhag cael eu camddefnyddio ond mae yna rai pethau nad ydynt yn anghyfreithlon ond sydd yn erbyn moesau'r mwyafrif o bobl.

Preifatrwydd fel mater moesol:

- Cwmnïau'n prynu rhestri o wybodaeth oddi wrth ei gilydd er mwyn gallu ffonio pobl i geisio gwerthu nwyddau neu wasanaethau.
- Cwmnïau'n cadw llygad ar ddefnydd eu staff o e-bost neu'r Rhyngrwyd gyda'r nod o ganfod camddefnydd neu fel mater o chwilfrydedd.

Preifatrwydd fel mater cyfreithiol:

- Gwerthu gwybodaeth bersonol am bobl nad ydynt wedi rhoi caniatâd i'r manylion gael eu trosglwyddo, yn groes i'r Ddeddf Gwarchod Data.
- Peidio â rhoi gwybod i'r Comisiynydd Gwybodaeth fod y corff yn prosesu data personol. Rhaid gwneud hyn o dan y Ddeddf Gwarchod Data.

A wyddoch chi fod yr holl chwiliadau a wnewch a'r holl e-byst a anfonwch yn cael eu cofnodi gan eich darparwr gwasanaeth rhyngrwyd?

Patrymau cyflogaeth

Mae cyflwyno TGCh wedi arwain at newid mewn patrymau cyflogaeth gan fod llai o bobl yn gweithio o 9 tan 5 o ddydd Llun i ddydd Gwener.

Materion cyfreithiol

Mae peth risg yn gysylltiedig â gweithio gyda chyfarpar TGCh, fel anaf straen ailadroddus (*RSI*), poen cefn a straen. Mae dyletswydd gyfreithiol ar gyflogwyr i ddiogelu'r gweithiwr rhag y risgiau hyn drwy ddefnyddio'r cyfarpar cywir (goleuadau priodol, dyfeisiau sy'n cynnal yr arddyrnau, cadeiriau cymwysadwy, ac ati) neu arferion gweithio cywir (newid tasgau, seibiau rheolaidd, ac ati).

Bydd cyrff yn ceisio sicrhau nad yw staff yn gallu mynd â nhw i'r llys drwy eu dysgu sut i ddefnyddio cyfarpar TGCh yn ddiogel a'u gorfodi i fabwysiadu arferion gweithio diogel yn unol â'r cod ymddygiad.

Materion moesol

Mae newidiadau mewn patrymau cyflogaeth wedi codi nifer o faterion moesol:

- Gweithwyr yn teimlo bod y cyflogwr yn cadw golwg arnynt. Mae defnyddio cyfrifiaduron i fonitro perfformiad, a rheolwyr yn edrych ar e-bost personol ac amser sy'n cael ei dreulio ar y Rhyngrwyd, yn gwneud i weithwyr deimlo dan fygythiad.
- Bydd rhai cyflogwyr yn manteisio ar amgylchiadau personol gweithwyr (e.e. mamau gyda phlant ifanc, gofalwyr, pobl anabl) i gynnig gwaith â chyflog is iddynt gartref, gan ddefnyddio TGCh.
- Mae dyfodiad TGCh yn golygu bod llawer mwy o bobl yn gweithio rhan-amser. Er bod hyn yn rhoi hyblygrwydd i'r gweithwyr, nid oes ganddynt yr un hawliau bob amser â gweithwyr amser-llawn.

Tegwch

Gall gwledydd cyfoethog fanteisio ar y datblygiadau diweddaraf ym maes TGCh, ond rhaid i wledydd tlotach fodloni ar gyfarpar hŷn.

Gellir rhannu'r boblogaeth o fewn gwledydd yn bobl sy'n dlawd yn nhermau gwybodaeth (gwybodaeth wael am gyfrifiaduron a'r Rhyngrwyd) a phobl sy'n gyfoethog yn nhermau gwybodaeth (mynediad band llydan cyflym a gwybodaeth dda am TGCh). Gall pobl sy'n hyddysg mewn TGCh fanteisio ar fenthyciadau a gwyliau rhad, ac ati, ond nid yw'r wybodaeth hon ar gael i bobl nad ydynt yn gwybod llawer am TGCh.

Hawliau eiddo deallusol

Mae'n iawn bod pobl sy'n datblygu meddalwedd, caledwedd a dulliau cyfathrebu newydd yn cael eu gwobrwyo am eu gwaith. Nid yw'n iawn bod y gwaith hwn yn cael ei gopïo gan bobl eraill. Mae llawer o gyrff yn nodi yn eu cod ymddygiad fod unrhyw waith a gynhyrchir yn ystod oriau gwaith yn perthyn iddyn nhw.

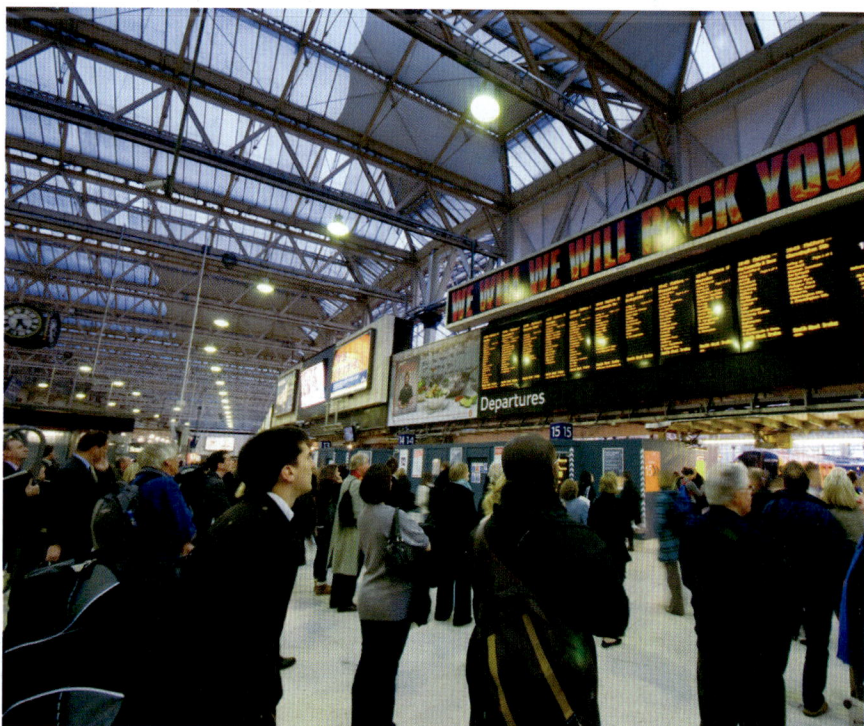

Gyda dyfodiad TGCh, mae llai o bobl yn gorfod cymudo i'r gwaith

Cwestiynau a Gweithgareddau

▶ Cwestiynau 1 tt. 48–55

1 (a) Eglurwch ystyr y term teleweithio. (2 farc)
   (b) Rhowch **ddwy** o fanteision teleweithio i'r gweithiwr. (2 farc)
   (c) Enwch swydd sy'n addas ar gyfer teleweithio a rhowch reswm pam mae'r swydd hon yn arbennig o addas ar gyfer teleweithio. (2 farc)

2 (a) Enwch **bedwar** datblygiad TGCh sydd wedi galluogi staff fyddai'n gweithio mewn swyddfa fel arfer i weithio gartref. (2 farc)
   (b) Nid yw gweithio o'r cartref gan ddefnyddio cyfarpar TGCh yn briodol ar gyfer pob math o swydd. Enwch **ddwy** swydd a fyddai'n briodol ar gyfer gweithio o'r cartref, gan roi rhesymau dros bob dewis. (4 marc)

3 (a) Eglurwch y gwahaniaeth rhwng y termau telegymudo a theleweithio. (2 farc)
   (b) Mae fideo-gynadledda yn gymhwysiad poblogaidd o TGCh mewn ysbytai. Disgrifiwch **ddau** ddefnydd mewn ysbytai ar gyfer fideo-gynadledda.

4 Mae cwmni'n ystyried gadael i rai o'i weithwyr weithio o'u cartrefi.
   Mae caniatáu telegymudo yn fanteisiol i'r gweithiwr yn ogystal â'r cyflogwr.
   (a) Trafodwch y caledwedd TGCh sydd ei angen ar weithwyr i delegymudo. Ar gyfer pob darn o galedwedd a ddisgrifiwch dylech egluro pam y mae ei angen. (4 marc)
   (b) Eglurwch **ddwy** o fanteision telegymudo i'r gweithiwr. (2 farc)
   (c) Eglurwch **ddwy** o fanteision telegymudo i'r cyflogwr. (2 farc)
   (ch) Colli bywyd cymdeithasol yw un o anfanteision telegymudo i'r gweithiwr. Eglurwch **ddwy** o anfanteision eraill telegymudo i'r gweithiwr. (2 farc)

5 Bydd rhai gweithwyr yn anffodus yn camddefnyddio cyfleusterau TGCh corff.
   (a) Disgrifiwch **dair** ffordd y gall gweithiwr gamddefnyddio cyfleusterau TGCh. (3 marc)
   (b) Mae gan lawer o gyrff god ymddygiad i rwystro eu gweithwyr rhag camddefnyddio'r cyfleusterau TGCh.
      (i) Disgrifiwch beth yw cod ymddygiad a disgrifiwch **dri** pheth y byddai'n eu cynnwys. (5 marc)
      (ii) Esboniwch sut mae cyflogwyr yn gorfodi gweithwyr i gydymffurfio â'r cod ymddygiad. (2 farc)

▶ Gweithgaredd 1: Cod ymddygiad

Dyma rai enghreifftiau o gamddefnyddio cyfleusterau TGCh sy'n cael eu cynnwys gan gorff yn ei god ymddygiad ar gyfer ei staff. Eich tasg chi yw penderfynu ar y lefel o drosedd – o fach i ddifrifol, i ddifrifol iawn – ac yna awgrymu'r camau disgyblu y dylai'r corff eu cymryd.

1   Defnyddio rhwydwaith y corff i gael mynediad heb ei awdurdodi i systemau cyfrifiadurol eraill.
2   Gwneud rhywbeth yn fwriadol neu'n ddiofal a fydd yn tarfu ar weithrediad arferol cyfrifiaduron, terfynellau, perifferolion neu rwydweithiau.
3   Ceisio trechu cynlluniau gwarchod data neu ddarganfod gwendidau diogelwch.
4   Torri telerau cytundebau trwyddedu meddalwedd neu ddeddfau hawlfraint.
5   Defnyddio e-bost i aflonyddu ar bobl eraill.
6   Symud ffeiliau mawr ar draws y rhwydwaith ar adegau brig gan arafu'r rhwydwaith yn sylweddol i ddefnyddwyr eraill.
7   Postio gwybodaeth ar y Rhyngrwyd a all fod yn enllibus neu'n ddifrïol ei natur.
8   Dangos delweddau ar systemau cyfrifiadurol y sefydliad sy'n ddelweddau rhywiol, yn annymunol eu cynnwys graffigol, neu'n aflonyddu'n rhywiol.
9   Dod â gemau cyfrifiadurol i mewn i'r swyddfa a'u rhedeg ar y peiriannau yno.
10 Defnyddio cyfleusterau TG y corff i wneud gwaith preifat a chael tâl amdano.
11 Cael data o gronfa ddata sy'n torri'r deddfau gwarchod data.
12 Peidio â newid eu cyfrinair yn ddigon rheolaidd.
13 Defnyddiwr yn gadael terfynell heb ei goruchwylio pan fo data personol i'w gweld ar y sgrin.
14 Peidio â sganio disgiau data am firysau cyn eu defnyddio ar gyfrifiaduron y corff.
15 Copïo meddalwedd sy'n eiddo i'r corff ac wedi'i drwyddedu iddo i'w werthu mewn gwerthiannau cist car.

Cymorth gyda'r arholiad

Enghraifft 1

1 Mae gan lawer o gwmnïau (cwmnïau amlwladol) swyddfeydd mewn gwahanol rannau o'r byd. Mae hyn yn ei gwneud yn anodd i staff deimlo eu bod yn perthyn i'r un corff.

    (a) Eglurwch sut y gall y corff ddefnyddio TGCh i alluogi staff sy'n gweithio mewn gwahanol leoedd ar hyd a lled y byd i gyfathrebu'n effeithiol. (8 marc)

    (b) Mae nifer o broblemau'n gysylltiedig â defnyddio TGCh ar gyfer cyfathrebu.

        Disgrifiwch **dair** o anfanteision defnyddio TGCh ar gyfer cyfathrebu. (6 marc)

Ateb myfyriwr 1

1 (a) Gallant ddefnyddio negeseuon testun gan eu bod yn rhatach na galwadau ffôn i wledydd tramor. Mae hefyd yn golygu y gallant eu darllen pan fydd yn gyfleus. Gallant anfon e-bost. Mae e-byst yn dda gan eu bod yn cael eu storio'n awtomatig ac yn gallu cyrraedd bron ar unwaith. Gellir ysgrifennu atebion yn gyflym a does dim angen i chi wastraffu amser yn egluro i beth yr ydych chi'n ymateb.

        Gallwch ddefnyddio ystafelloedd sgwrsio lle gallwch gyfarfod a siarad â'ch cydweithwyr os ydych chi'n teimlo'n unig.

    (b) Os bydd y cyfarpar yn torri, ni allwch gyfathrebu. Mae pobl yn hoffi siarad wyneb yn wyneb. Mae'n anoddach dweud a yw pobl yn dweud celwydd neu beth yw eu barn am yr hyn yr ydych chi'n ei ddweud dim ond drwy edrych ar eu hwynebau.

        Nid yw e-bost yn addas ar gyfer rhoi newyddion drwg – dydy pobl ddim eisiau cael eu diswyddo drwy e-bost a dylai hyn gael ei wneud yn bersonol.

Sylwadau'r arholwr

1 Mae wyth marc ar gael am ran (a), felly mae angen wyth ffaith berthnasol.

Er bod negeseuon testun yn ddull cyfathrebu pwysig, maen nhw'n addas ar gyfer negeseuon byr yn unig ac nid yw'r staff mewn cyrff yn cyfathrebu fel hyn, felly dim marciau am yr ateb hwn.

Mae'r ateb am e-bost yn gwneud tri phwynt perthnasol, felly tri marc am hyn.

Ni fyddai'r mwyafrif o gwmnïau am i'r staff drafod eu busnes mewn ystafell sgwrsio, yn enwedig pe bai ar agor i'r cyhoedd. Dim marciau yma.

Dylai atebion gyfeirio at ddulliau cyfathrebu ar lefel y busnes, nid at y ffordd y mae staff yn cyfathrebu ar lefel bersonol gyda chyfeillion. **(3 marc allan o 8)**

Mae chwe marc am ran (b), felly mae angen chwe phwynt dilys.

Mae dau ateb yma – pob un gyda gosodiad ac esboniad pellach sy'n gywir, felly rhoddir pedwar marc.

(4 marc allan o 6)

Ateb myfyriwr 2

1 (a) Gall y staff ddefnyddio e-bost i gyfathrebu â chydweithwyr. Mae hyn yn gyflymach na llythyrau cyffredin a all gymryd wythnosau i gyrraedd mewn rhai gwledydd.

        Mae'n hawdd ateb drwy glicio ar Ateb a theipio eich neges. Does dim angen i chi aros am y post na chiwio am stampiau. Gallai'r corff greu blog sy'n caniatáu i'r holl staff bostio eu sylwadau. Mae hyn yn rhoi cyfle i bobl wyntyllu problemau neu gwynion.

        Gallai rheolwyr y cwmni benderfynu gweithredu arnynt neu beidio.

        Gall y staff ddefnyddio fideo-gynadledda i drefnu cyfarfodydd rhith lle gall pawb weld ei gilydd a gallant drosglwyddo dogfennau electronig a gwylio cyflwyniadau. Nid yw amser yn cael ei wastraffu yn teithio i gyfarfodydd ac mae'r costau'n llawer is gan nad oes angen i'r rheolwyr dalu am gostau teithio, prydau bwyd, gwestai, ac ati.

    (b) Mae pobl weithiau'n mwynhau cael newid i ffwrdd o'r swyddfa ac maen nhw'n hoffi cymdeithasu. Mae fideo-gynadledda yn rhoi stop ar hyn.

        Mae'n anodd i bobl ymddwyn yn naturiol mewn sefyllfa fideo-gynadledda ac mae llawer o bobl yn hoffi trafod wyneb yn wyneb.

        Mae e-bost yn ddefnyddiol ar gyfer cyfathrebu cyflym ond gall ymddangos braidd yn swta.

        Efallai na fydd pobl yn meddwl bod postio deunydd ar flogiau a negesfyrddau yn syniad da gan y byddai'n hawdd olrhain eu sylwadau yn ôl iddynt.

Sylwadau'r arholwr

1 Mae rhan (a) wedi cael ei hateb yn dda. Mae'r holl bwyntiau'n berthnasol ac wedi'u mynegi'n glir. Mae wyth pwynt gwahanol sy'n gywir, felly rhoddir marciau llawn am yr adran hon. **(8 marc allan o 8)**

Mae'r ateb i ran (b) yn dda iawn hefyd, gyda phum pwynt dilys. **(5 marc allan o 6)**

Cymorth gyda'r arholiad *(parhad)*

Ateb yr arholwr

1 (a) Un marc am nodi dull cyfathrebu addas sy'n briodol i fusnes. Yna un marc am bob esboniad ynghylch sut y gellir defnyddio'r dull cyfathrebu.

Defnyddio e-bost i gyfathrebu â chydweithwyr (1). Mae'n darparu cyfathrebu cyflym gan nad oes angen aros am y post (1). Dull rhad gan fod e-byst bron yn ddi-dâl (1). Gall yr e-bost gwreiddiol a'r ateb gael eu cadw gyda'i gilydd, felly nid yw amser yn cael ei wastraffu yn chwilio am yr e-bost gwreiddiol (1). Gellir atodi ffeiliau a dogfennau eraill wrth e-byst i gael sylwadau arnynt (1). Gellir defnyddio fideo-gynadledda i gynnal cyfarfodydd rhith (1). Nid oes angen i staff deithio i gyfarfod go iawn (1). Mae'n arbed arian i'r cwmni gan nad oes talu am y man cyfarfod a chostau teithio, gwesty, prydau, ac ati (1). Gellir galw cyfarfodydd ar fyr rybudd neu yn ôl yr angen (1). Mae fforymau, blogiau a negesfyrddau yn caniatáu i staff gyfathrebu â'i gilydd (1). Gallant drefnu trafodaethau a gofyn i bobl eraill am sylwadau (1).

(b) Un marc am yr anfantais ac un marc am esbonio canlyniadau'r anfantais honno. Diffyg cysylltiad wyneb yn wyneb (1). Mae'n bwysig gweld wynebau pobl a darllen 'iaith y corff' (1). Mae'n dibynnu ar dechnoleg nad yw'n ddibynadwy bob amser (e.e. cysylltau lloeren) (1). Gall cyfarpar dorri i lawr a tharfu'n gyfan gwbl ar gyfathrebiadau (1). Weithiau mae'n fwy sensitif i roi newyddion drwg yn bersonol (1). Weithiau mae e-byst yn rhy fyr ac yn gallu achosi camddealltwriaeth (1).

Enghraifft 2

2 **O ganlyniad i'r twf mewn systemau cyfathrebu mae nifer cynyddol o bobl yn gweithio o'u cartrefi, neu'n 'telegymudo'.**

Trafodwch, gyda chymorth enghreifftiau addas, fanteision ac anfanteision y dull hwn o weithio i weithwyr a chyflogwyr. (8 marc)

Ateb myfyriwr 1

2 Gall telegymudo fod o fudd i famau gyda phlant bach neu ofalwyr sy'n gofalu am berthnasau oedrannus gan eu bod yn gallu gweithio oriau hyblyg. Does dim angen iddynt gymudo i'r gwaith, felly gallant fod yn y tŷ yn gofalu am bobl. Mae telegymudo'n ddelfrydol i bobl anabl gan nad oes angen iddynt fynd i'r gwaith ar gludiant cyhoeddus.

Ond mae'n anodd eich ysgogi eich hun i weithio yn eich cartref eich hun.

Mae yna fanteision i gyrff hefyd. Does dim angen swyddfeydd mor fawr gan fod pobl yn defnyddio rhan o'u cartrefi eu hunain. Does dim angen prynu cyfarpar cyfrifiadurol gan fod y gweithwyr yn gallu defnyddio eu cyfrifiaduron eu hunain. Hefyd does dim angen prynu meddalwedd gan y bydd gan y gweithwyr eu meddalwedd eu hunain.

Sylwadau'r arholwr

2 Mae'r manteision i'r gweithiwr wedi'u disgrifio'n dda a rhoddir enghreifftiau addas. Rhoddir tair mantais ddilys ac un anfantais i weithwyr.

Mae'r ateb am fanteision i'r corff yn wannach. Mae'r pwynt am swyddfeydd llai yn ddilys ond nid oes yr un rhan arall o'r ateb hwn sy'n haeddu marc. Bydd cyflogwyr bron bob amser yn darparu caledwedd a meddalwedd i sicrhau diogelwch eu data. **(5 marc allan o 8)**

Ateb myfyriwr 2

2 Mae manteision i'r gweithiwr yn cynnwys gallu trefnu eich bywyd preifat o amgylch eich gwaith, er enghraifft, gweithio gyda'r nos os oes gennych blant ifanc.

Hefyd, does dim angen i chi fyw mewn dinasoedd a gallwch gael bywyd gwell yng nghefn gwlad. Does dim cymudo drud yn y car neu ar y trên, felly mae mwy o arian i'w wario ar fwynhau.

Mae anfanteision i'r gweithiwr yn cynnwys teimlo'n unig gan nad oes gennych gydweithwyr i siarad â nhw a mynd allan gyda nhw. Efallai y bydd yn anodd gwahanu bywyd gwaith oddi wrth fywyd cartref, gan beri tyndra rhwng partneriaid. Bydd angen lle i wneud y gwaith a gall hyn fod yn amhosibl weithiau. Bydd angen cymhelliant da arnynt gan na fydd rheolwr yno drwy'r adeg i weld beth maen nhw'n ei wneud.

Mae manteision i'r cyflogwr yn cynnwys peidio â gorfod cael swyddfa mor fawr, sy'n arbed costau rhentu, gwresogi, goleuo a glanhau. Hefyd gallant recriwtio staff o ardal lawer ehangach, sy'n golygu eu bod nhw'n debygol o gael staff gwell gyda chymhelliant uwch. Mae gweithwyr sy'n telegymudo yn fwy teyrngar ac yn debygol o aros gyda'r cwmni'n hirach.

Mae anfanteision i'r cyflogwr yn cynnwys colli rheolaeth dros ddiogelwch ei ddata. Mae'n anodd gwybod pwy all fod yn edrych ar y data personol sy'n cael eu storio. Hefyd mae'n anoddach rheoli gweithwyr sy'n byw mewn lleoedd anghysbell.

> ## Sylwadau'r arholwr
>
> 2 Mae hwn yn ateb ardderchog. Sylwch ar y ffordd mae'r myfyriwr wedi trefnu'r manteision ac anfanteision mewn paragraffau ar wahân. Mae nifer mawr o bwyntiau dilys wedi cael eu gwneud yma a dim pwyntiau anghywir, ac mae'r myfyriwr wedi'i fynegi ei hun yn dda. Mae'r ateb hwn yn haeddu marciau llawn. **(8 marc allan o 8)**

Ateb yr arholwr

2 Dylai ymgeiswyr gynnwys pwyntiau tebyg i'r canlynol yn eu hatebion:

Manteision i'r gweithwyr:

Oriau gwaith hyblyg – ar yr amod bod y gwaith yn cael ei wneud

Gall fod yn gyfleus os oes plant ifanc, perthnasau anabl, pobl oedrannus i ofalu amdanynt

Gall y gweithiwr fyw yn unrhyw le yn y wlad

Dim amser yn cael ei dreulio ar deithio i'r gwaith, felly mae mwy o amser ar gyfer gweithgareddau hamdden a'r teulu

Dim costau teithio i'r gwaith

Anfanteision i'r gweithwyr:

Rhaid bod â chymhelliant uchel gan fod llawer o bethau i dynnu'ch sylw gartref

Mae angen lle i wneud y gwaith

Llai o gysylltiad cymdeithasol o'i gymharu â gweithio gyda phobl eraill

Cynnydd mewn biliau fel gwresogi, goleuo, yswiriant, ac ati

Manteision i'r corff

Mae angen swyddfa lai, felly mae costau'n is

Gellir dewis gweithwyr â sgiliau addas o ardal ehangach

Nid oes angen cyflogi cymaint o staff cymorth (e.e. glanhawyr, staff diogelwch, ac ati)

Mae staff yn fwy parod i weithio oriau anghymdeithasol os oes angen

Mae staff yn llai tebygol o fod i ffwrdd yn sâl

Anfanteision i'r corff

Mae'n anoddach i reolwyr reoli'r gwaith

Mae'n anodd dweud pa mor galed mae staff yn gweithio

Problemau diogelwch gan fod data'r cwmni ar gyfrifiaduron i ffwrdd o'r swyddfa

Efallai y bydd angen newid strwythur y corff

Yr angen i ddarparu cyfarpar i weithwyr cartref, sy'n gallu bod yn ddrud

6–8 marc Mae'r ymgeiswyr yn rhoi ateb clir a rhesymegol sy'n llawn a manwl ac yn disgrifio un fantais ac un anfantais i'r gweithwyr, ac un fantais ac un anfantais i'r corff. Defnyddir sillafu, atalnodi a gramadeg cywir.

3–5 marc Mae'r ymgeiswyr yn rhoi atebion ond mae diffyg eglurder neu brinder enghreifftiau perthnasol. Mae ychydig o wallau atalnodi a gramadeg.

0–2 farc Nid yw'r ymgeiswyr yn trafod, a'r cyfan a wnânt yw rhestru tair mantais/anfantais neu roi esboniad byr o un neu ddwy yn unig o fanteision/anfanteision. Mae diffyg eglurder yn yr ateb a llawer o wallau sillafu, atalnodi a gramadeg.

Map meddwl cryno

Cod ymddygiad

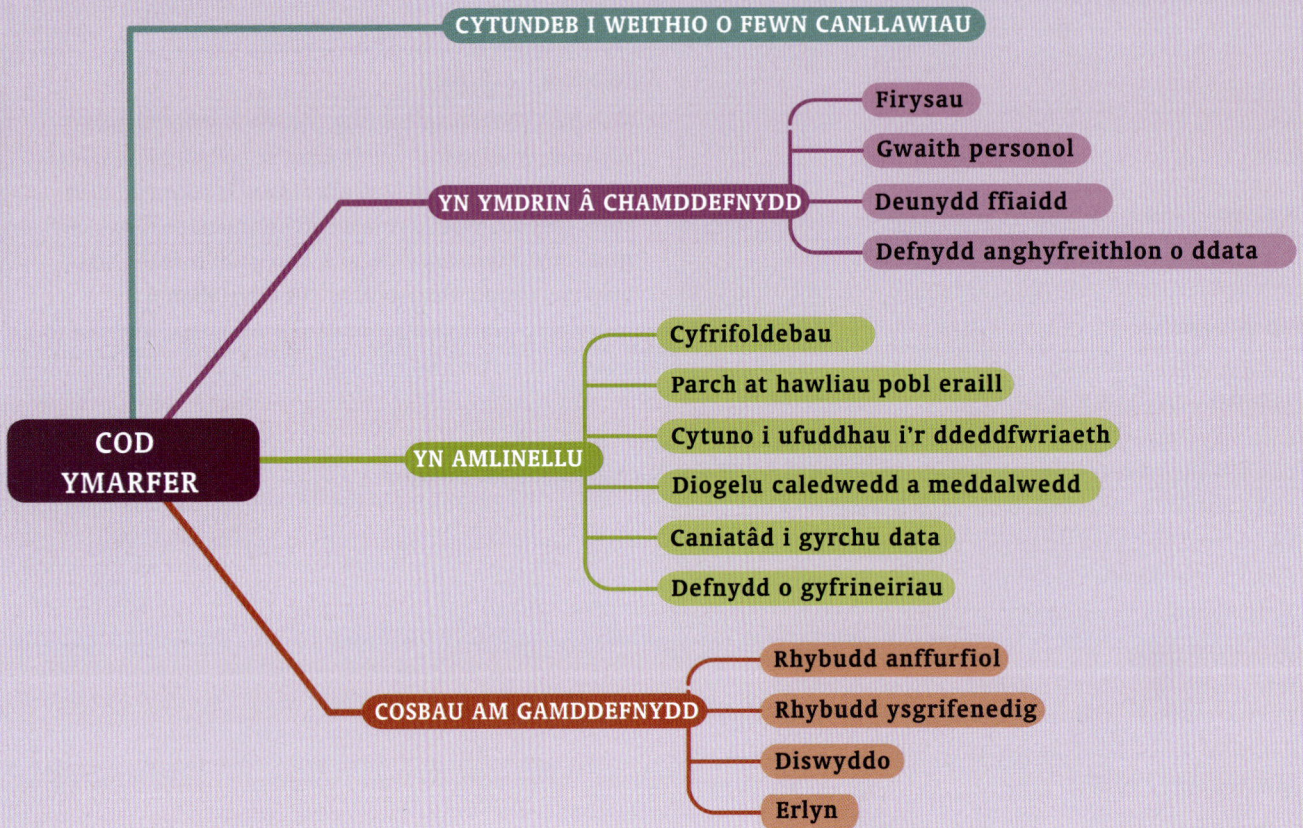

CYTUNDEB I WEITHIO O FEWN CANLLAWIAU

YN YMDRIN Â CHAMDDEFNYDD
- Firysau
- Gwaith personol
- Deunydd ffiaidd
- Defnydd anghyfreithlon o ddata

COD YMARFER

YN AMLINELLU
- Cyfrifoldebau
- Parch at hawliau pobl eraill
- Cytuno i ufuddhau i'r ddeddfwriaeth
- Diogelu caledwedd a meddalwedd
- Caniatâd i gyrchu data
- Defnydd o gyfrineiriau

COSBAU AM GAMDDEFNYDD
- Rhybudd anffurfiol
- Rhybudd ysgrifenedig
- Diswyddo
- Erlyn

TOPIG 5: Polisïau diogelwch

Mae gan bob cwmni da systemau diogelwch llym i amddiffyn ei adnoddau TGCh er mwyn cadw ei fantais gystadleuol, sicrhau nad yw ei ddelwedd yn dioddef oherwydd cyhoeddusrwydd gwael a diogelu data personol rhag cael eu datgelu. Mae'n hanfodol i gyrff amddiffyn eu systemau TGCh mewn ffordd cost effeithiol gan ddefnyddio lefelau priodol o ddiogelwch. Mae systemau TGCh yn agored i ddau ddosbarth o fygythiadau: damweiniol a bwriadol.

Mae bygythiadau damweiniol yn cynnwys pethau fel camgymeriadau dynol, tân, methiant cyfarpar, trychinebau naturiol (llifogydd, daeargrynfeydd, ac ati) ac mae bygythiadau bwriadol yn cynnwys twyll, difrod, fandaliaeth, llosgi bwriadol, ysbïo, ac ati. Gall y bygythiadau hyn ddod o'r tu allan neu o'r tu mewn i gorff.

I ymdrin â'r bygythiadau hyn, dylai fod gan bob corff bolisi diogelwch TGCh a ddylai gael ei gynhyrchu gan yr uwch reolwyr a'r cyfarwyddwyr a'i gefnogi ganddynt. Mae'r polisi diogelwch TGCh yn ddogfen sy'n cynnwys pob agwedd ar ddiogelwch o fewn y corff ac mae'n cynnwys amodau a rheolau y mae angen i'r holl staff ufuddhau iddynt.

Yn y topig hwn byddwch yn dysgu am y bygythiadau i systemau TGCh a sut y gall y bygythiadau hyn gael eu hatal neu eu lleihau. Byddwch hefyd yn dysgu sut y gall data gael eu hadfer os collir y data gwreiddiol, a byddwch yn edrych ar elfennau polisi diogelwch TGCh.

▼ Y cysyniadau allweddol sy'n cael sylw yn y topig hwn yw:

▶ Bygythiadau i systemau TGCh, canlyniadau a'r angen am weithdrefnau wrth gefn

▶ Y ffactorau i'w cymryd i ystyriaeth wrth lunio polisïau diogelwch

▶ Gweithdrefnau gweithredol ar gyfer atal camddefnydd

▶ Atal camddefnydd damweiniol

▶ Atal troseddau neu gamddefnydd bwriadol

▶ Y ffactorau sy'n penderfynu faint mae cwmni'n ei wario ar ddatblygu rheolaeth a lleihau risg

CYNNWYS

Bygythiadau a chanlyniadau posibl, a gweithdrefnau wrth gefn

▼ Byddwch yn dysgu

▶ Am y bygythiadau i systemau TGCh o ganlyniad i drychinebau naturiol

▶ Am fygythiadau caledwedd a meddalwedd diffygiol

▶ Am y bygythiadau oddi wrth dân

▶ Am fygythiadau colli pŵer

Cyflwyniad

Mae bygythiadau i system TGCh fel tân, lladrad, difrod bwriadol, colli pŵer, ac ati, yn gallu bod yn beryglus.

Mae llawer o fygythiadau posibl i systemau TGCh ac mae rhai yn fwy tebygol o ddigwydd nag eraill. Gall maint y difrod amrywio hefyd, o beri ychydig o anghyfleuster (e.e. gorfod dileu firws) i golli caledwedd, meddalwedd a data'n gyfan gwbl.

Rhaid i bawb sy'n defnyddio system TGCh fod yn ymwybodol o'r bygythiadau gan mai dim ond wedyn y gellir cymryd mesurau i leihau'r difrod os bydd y bygythiad yn troi'n realiti.

Gall bygythiadau i systemau TGCh arwain at golli data. Gall hyn arwain at golledion ariannol o ganlyniad i:

- gymryd camau cyfreithiol
- colli busnes
- colli hyder cwsmeriaid neu'r cyhoedd
- colli amser gweithio ar gyfrifiaduron
- yr angen i staff dreulio amser yn datrys problemau.

O ble mae'r bygythiadau'n dod?

Mae nifer enfawr o fygythiadau i systemau cyfrifiadurol. Dyma restr o'r prif fygythiadau ond mae'n debyg y gallwch feddwl am ragor:

- firysau
- mwydod
- hysbyswedd
- sbam
- difrod bwriadol gan staff
- tân
- trychineb naturiol
- terfysgaeth
- Trojanau
- ysbïwedd
- hacio
- difrod damweiniol gan staff
- lladrad
- caledwedd neu feddalwedd gwallus

Bygythiadau oddi wrth drychinebau naturiol

Ym Mhrydain nid yw'n arferol i ni ddioddef oddi wrth drychinebau naturiol fel llawer o wledydd eraill, ond gall hyn newid wrth i hinsawdd y byd newid. Dyma rai trychinebau naturiol a'u canlyniadau.

Daeargrynfeydd – colli pŵer, colli llinellau cyfathrebu, difrod i systemau TGCh wrth i adeiladau ddymchwel, ac ati.

Mellt – gall mellt beri i bŵer gael ei golli dros dro, gan arwain at golli data. Gall hefyd achosi difrod mwy difrifol, gyda chaledwedd, meddalwedd a data'n cael eu colli'n gyfan gwbl.

Llifogydd – difrod dŵr i galedwedd, meddalwedd a data, colli pŵer neu linellau cyfathrebu.

Tonnau llanw – maen nhw'n digwydd ar ôl daeargrynfeydd fel rheol, pan gânt eu galw'n tsunami.

Llosgfynyddoedd – difrod tân a mwg, dinistrio adeiladau.

Tymhestloedd – colli llinellau trydan, dinistrio cyfarpar cyfathrebu, ac ati.

Bygythiadau oddi wrth galedwedd neu feddalwedd gwallus

Fel unrhyw ddyfais drydanol, gall cyfrifiaduron fethu. Ond er y gall y caledwedd gael ei drwsio neu ei adnewyddu, nid yw'n hawdd adfer rhaglenni a data, yn enwedig os nad yw unrhyw gopïau wrth gefn wedi cael eu gwneud.

Caledwedd gwallus – mae caledwedd cyfrifiadurol yn weddol ddibynadwy ond mae'n torri weithiau, a rhaid i chi fod yn barod ar gyfer hyn. Y broblem yn aml yw bod y gyriant disg caled yn methu fel nad oes modd defnyddio'r data a'r rhaglenni.

Meddalwedd gwallus – gall meddalwedd, yn enwedig meddalwedd pwrpasol (*bespoke*), a meddalwedd pecyn weithiau, gynnwys namau sy'n gallu arwain at lygru neu golli data.

Bygythiadau oddi wrth dân

Mae tân yn fygythiad difrifol mewn unrhyw weithle a gall systemau TGCh gael eu colli, felly rhaid cymryd mesurau i leihau'r bygythiad, e.e.:

- dim ysmygu mewn ystafelloedd cyfrifiaduron
- peidio â gorlwytho socedi trydan
- archwilio gwifrau'n rheolaidd i sicrhau eu bod yn ddiogel
- gwagio biniau'n rheolaidd
- peidio â gadael llawer iawn o bapur ar hyd y gweithle
- larymau tân/canfodyddion mwg ym mhob ystafell
- gosod system ysgeintio
- defnyddio diogellau gwrthdan i storio cyfryngau sy'n cynnwys rhaglenni a data
- cadw copïau wrth gefn oddi ar y safle.

Dwyn cyfrifiaduron neu galedwedd

Gall cyfrifiaduron neu galedwedd arall gael eu dwyn. Os caiff cyfrifiadur ei ddwyn, yna bydd y caledwedd, meddalwedd a data'n cael eu colli. Mae'n bosibl y bydd y cwmni wedi gweithredu'n groes i Ddeddf Gwarchod Data 1998 os gellir profi nad oedd ganddo fesurau diogelwch digonol i atal data personol rhag cael eu colli.

Mae dwyn cyfrifiaduron, yn enwedig gliniaduron, yn ddigwyddiad cyffredin. Mae gliniaduron yn darged amlwg gan eu bod nhw:

- yn fach, ysgafn a hawdd eu cuddio
- yn cael eu defnyddio'n aml mewn lleoedd cyhoeddus (caffis, trenau, meysydd awyr, ac ati)
- yn cael eu rhoi mewn cistiau ceir
- yn boblogaidd iawn ac yn hawdd i ladron eu gwerthu.

Hacio

Mae hacio'n golygu ceisio torri i mewn i system gyfrifiadurol ddiogel. Fel rheol mae'r haciwr yn rhaglennwr medrus sydd â'r wybodaeth dechnegol i allu manteisio ar wendidau system ddiogelwch. Ar ôl i haciwr sicrhau mynediad i system TGCh, gall benderfynu:

- gwneud dim byd a bodloni ar y ffaith ei fod wedi llwyddo i sicrhau mynediad
- cyrchu data sensitif neu bersonol
- defnyddio data personol i flacmelio pobl
- difrodi'r data
- newid data'n fwriadol i gyflawni twyll.

Lledu firysau

Mae cysylltu cyfrifiaduron unrhyw gorff â'r Rhyngrwyd yn cynyddu'r perygl o ledu firysau i'r rhwydwaith mewnol. Dylai'r meddalwedd sganio a gwaredu firysau diweddaraf fod wedi'i osod ar bob cyfrifiadur.

Ymosodiadau rhwystro gwasanaeth

Ymosodiad ar system ddiogel fel bod y corff yn methu defnyddio rhai o'i adnoddau yw ymosodiad rhwystro gwasanaeth. Er enghraifft, gallai ymosodiad rhwystro gwasanaeth ar siop lyfrau ar-lein olygu bod rhwydwaith y siop lyfrau'n cael cymaint o geisiadau fel bod y rhwydwaith yn colli cysylltedd dros dro. Golygai hyn na fyddech chi'n gallu archebu llyfr tra oedd yr ymosodiad yn digwydd. Yn ogystal â pheri anghyfleustra i gwsmeriaid, gallai'r cwmni golli llawer o fusnes gan fod cwsmeriaid yn methu archebu.

Problemau gyda phŵer

Gall pŵer gael ei golli am lawer o resymau: er enghraifft, trychineb naturiol, tywydd garw, neu weithiwr yn torri drwy gebl â rhaw wrth drwsio'r ffordd.

Colli pŵer

Bydd system bŵer wrth gefn yn cynnal y cyflenwad ac yn cadw'r cyfrifiaduron i fynd nes bod y prif gyflenwad trydan yn cael ei adfer. Bydd pŵer wrth gefn fel rheol yn cynnwys naill ai pŵer wedi'i storio (ar gyfer colli pŵer am gyfnod byr) neu gyfuniad o bŵer wedi'i storio a generadur. Cyfresi o fatrïau yw pŵer wedi'i storio. Mae generaduron yn defnyddio diesel neu betrol i gynhyrchu trydan.

Gan fod oediad bach rhwng colli'r trydan a chychwyn y generadur, mae angen batrïau i ddarparu pŵer yn ystod y cyfnod hwn.

Mae pŵer wrth gefn weithiau'n cael ei alw'n gyflenwad trydan annhoradwy (*UPS: uninterruptible power supply*).

Newidiadau yn y cyflenwad pŵer

Mae anwadaliadau pŵer, neu sbigynnau ac ymchwyddiadau, yn digwydd yn amlach na cholli pŵer yn gyfan gwbl. Mae goleuadau'n fflicran yn arwydd o'r rhain. Gallant achosi problemau gyda chyfrifiaduron a pheri colli data.

Canlyniadau colli data

Prif ganlyniadau colli data yw:
- Colli busnes ac incwm – bydd colli manylion cwsmeriaid a'u harchebion yn golygu nad yw'r cwmni'n gwybod bod cwsmer wedi rhoi archeb, ac ni fydd yn gwybod a yw cwsmeriaid wedi talu.
- Colli enw da – ni fydd yn edrych yn dda os na all corff ofalu am ei ddata'n iawn. Bu llawer o hanesion yn y newyddion yn ddiweddar am adrannau'r llywodraeth yn colli data.
- Camau cyfreithiol – o dan Ddeddf Gwarchod Data 1998 mae'n ofynnol i gyrff gadw data personol yn ddiogel. Gall cyrff sy'n methu gwneud hyn gael eu herlyn.

Y ffactorau i'w cymryd i ystyriaeth wrth lunio polisïau diogelwch

▼ **Byddwch yn dysgu**

▶ Am ddiogelwch corfforol

▶ Am atal camddefnydd

▶ Am ymchwilio'n barhaus i afreoleidd-dra

▶ Am fynediad i'r system – sefydlu gweithdrefnau ar gyfer cyrchu data fel gweithdrefnau mewngofnodi, muriau gwarchod

▶ Am weinyddu personél

▶ Am weithdrefnau gweithredol

▶ Am god ymddygiad a chyfrifoldebau'r staff

▶ Am weithdrefnau disgyblu

Cyflwyniad

Mae llawer o bethau y gellir eu gwneud i ddileu neu leihau'r bygythiadau i systemau TGCh, ac yn yr adran hon byddwn yn edrych ar y ffactorau y mae angen eu cymryd i ystyriaeth wrth lunio polisi diogelwch.

Diogelwch corfforol

Mae diogelwch corfforol yn golygu defnyddio dulliau corfforol yn hytrach na dulliau meddalwedd i amddiffyn caledwedd a meddalwedd. Dau brif bwrpas darparu diogelwch corfforol yw:

- cyfyngu ar fynediad i'r cyfarpar cyfrifiadurol
- cyfyngu ar fynediad i'r cyfrwng storio.

Er mwyn cyfyngu ar fynediad i'r cyfarpar cyfrifiadurol, mae'n syniad da cyfyngu ar fynediad i'r adeilad neu ystafell fel cam cyntaf. Os bydd rhywun yn llwyddo i gael mynediad, rhaid cael ffordd o ddiogelu'r cyfrifiaduron rhag cael eu dwyn.

Atal camddefnydd

Defnyddir dau ddull i geisio atal camddefnyddio systemau TGCh:

- dulliau corfforol
- dulliau rhesymegol (meddalwedd).

Mae dulliau rhesymegol yn defnyddio meddalwedd i reoli mynediad i raglenni a data sydd wedi'u storio ar y system TGCh. Nid yw dulliau corfforol yn defnyddio meddalwedd ond pethau fel cloeon, dynion diogelwch, a cheblau i gysylltu cyfrifiaduron wrth ddesgiau, ac ati.

Dulliau corfforol

- Rheoli mynediad i'r ystafell – drwy ddefnyddio bysellbadiau ar y drysau fel bod angen teipio cod i gael mynediad, neu gardiau magnetig arbennig i agor drysau, neu ddulliau biometrig fel olion bysedd neu adnabod iris.
- Rheoli mynediad i'r adeilad – drwy ddefnyddio dynion diogelwch a fydd yn herio ymwelwyr, yn

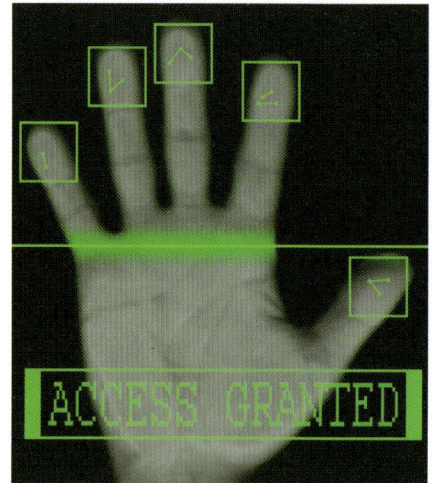

Rheoli mynediad i'r ystafell

Defnyddio cloeon ar gyfrifiaduron

Defnyddir dulliau corfforol a rhesymegol i amddiffyn systemau TGCh

Rhoi cyfrifiaduron dan glo gyda'r nos

Defnyddio dyfeisiau i rwystro gosod cyfryngau symudadwy mewn cyfrifiaduron

Defnyddio camerâu diogelwch

cadw cofnod o bobl sy'n dod i mewn i'r adeilad ac yn ei adael, ac yn sicrhau bod mynediad i ystafelloedd yn cael ei reoli, ac ati.

- Defnyddio cloeon ar gyfrifiaduron – fel na allant gael eu troi ymlaen.
- Rhoi'r cyfrifiaduron dan glo gyda'r nos neu eu diogelu o dan orchuddion dur – i atal mynediad neu ladrad.
- Defnyddio camerâu diogelwch mewn ystafelloedd cyfrifiaduron – mae'n llai tebygol y caiff cyfrifiaduron eu camddefnyddio neu eu dwyn.
- Defnyddio ceblau metel cryf i gysylltu cyfrifiaduron wrth ddesgiau – i atal lladrad.
- Defnyddio dyfeisiau i rwystro gosod cyfryngau symudadwy mewn cyfrifiaduron – gallwch gael dyfeisiau sy'n atal mynediad i yriannau *CD/ DVD* neu byrth *USB*, gan rwystro

Defnyddio diogell wrthdan

copïo data heb awdurdod.
- Defnyddio diogell wrthdan i storio data ar gyfryngau storio cludadwy.

Diogelwch meddalwedd (neu ddiogelwch rhesymegol)

Os bydd pobl yn llwyddo i gael mynediad heb awdurdod i'r ystafell lle mae'r cyfrifiaduron, bydd angen ail linell amddiffyn i'w hatal rhag cyrchu'r meddalwedd neu ddata. Defnyddir diogelwch meddalwedd i gyflawni hyn.

Mae hyn yn cynnwys:

- defnyddio cyfrineiriau a'r angen i fewngofnodi i gyfyngu ar fynediad i'r cyfrifiadur
- defnyddio lefelau o fynediad i rai

rhaglenni, ffeiliau a data.

Trywyddau archwilio

Wrth ddefnyddio systemau TGCh, bydd llawer o drafodion yn digwydd heb gynhyrchu unrhyw waith papur. Gallech dybio felly y byddai'n anodd iawn olrhain y trafodion hyn. Ond yn ffodus mae pob system TGCh yn darparu trywyddau archwilio sy'n ei gwneud hi'n bosibl i chi olrhain manylion trafodion penodol.

Er enghraifft, yn achos system prosesu gwerthiannau, gellid gwirio bod cwsmer wedi rhoi archeb, bod yr arian am yr archeb wedi'i dalu, a bod y nwyddau wedi cael eu danfon a'u derbyn. Pe bai unrhyw afreoleidd-dra yn y broses, gallai'r trywydd archwilio ei ddarganfod a gellid ymchwilio iddo. Mae trywyddau archwilio yn darparu cofnodion y gellir eu defnyddio i olrhain trafodion. Caiff archwiliadau rheolaidd eu gwneud sy'n gwneud i staff feddwl ddwywaith gan eu bod yn gwybod bod rhywun yn cadw golwg ar bethau.

Defnyddir trywyddau archwilio i gadw cofnod o:

- bwy sydd wedi newid data
- pryd y cafodd y data eu newid
- pa newidiadau gafodd eu gwneud.

Ymchwilio'n barhaus i afreoleidd-dra

Un ffordd o ddarganfod a yw cardiau credyd/debyd yn cael eu camddefnyddio yw cwestiynu trafodion sy'n anarferol ar gyfer y cwsmer hwnnw.

Afreoleidd-dra o ran lle – weithiau bydd taliadau cerdyn credyd am symiau mawr mewn gwlad dramor yn cael eu gwrthod gan y cwmni cerdyn credyd. Bydd y cwmni'n gofyn i'r cwsmer gysylltu fel y gall ofyn cyfres o gwestiynau i sicrhau mai ef/hi yw deiliad go iawn y cerdyn a bod y trafodion yn ddilys.

Afreoleidd-dra o ran swm – gall trafodion gwerth uchel ar gardiau credyd godi amheuon ar unwaith, yn enwedig os nad yw'r cwsmer erioed wedi prynu rhywbeth drud iawn o'r blaen. Cyn caniatáu i drafodion o'r fath fynd drwodd, bydd y banc yn gofyn i'r cwsmer gysylltu ag ef i sicrhau bod y trafodion yn ddilys ac nad oes neb yn defnyddio'u manylion personol yn anghyfreithlon.

Y ffactorau i'w cymryd i ystyriaeth wrth lunio polisïau diogelwch (parhad)

Mynediad i'r system – sefydlu gweithdrefnau ar gyfer cyrchu data fel gweithdrefnau mewngofnodi, muriau gwarchod

Dulliau rheoli mynediad

Mae dulliau rheoli mynediad yn sicrhau bod mynediad i systemau TGCh sy'n cynnwys data, rhaglenni a gwybodaeth y corff yn cael ei reoli mewn rhyw ffordd, fel bod mynediad awdurdodedig yn unig yn cael ei ganiatáu. Defnyddir y dulliau canlynol i reoli mynediad i'r system.

Gweithdrefnau mewngofnodi

Mae enw defnyddiwr neu rif adnabod defnyddiwr yn cael ei ddefnyddio i adnabod defnyddiwr rhwydwaith. Unwaith mae'r rhwydwaith yn gwybod pwy sy'n ei ddefnyddio, gall ddyrannu adnoddau fel lle storio a mynediad i rai ffeiliau.

Mae cyfrineiriau'n llinynnau o nodau sy'n hysbys i'r defnyddiwr yn unig ac a ddefnyddir i gyrchu'r system TGCh. Mae'r cyfrinair yn sicrhau mai'r person sy'n rhoi'r enw defnyddiwr neu rif adnabod defnyddiwr yw'r person hwnnw/ honno.

Ni ddylech byth ysgrifennu eich cyfrinair yn unman.

Hawliau mynediad

Mae hawliau mynediad yn cyfyngu ar fynediad defnyddwyr i'r ffeiliau hynny'n unig sydd eu hangen arnynt i wneud eu gwaith. Caiff eu hawliau eu dyrannu iddynt i ddechrau gan reolwr y rhwydwaith a phan fewngofnodant drwy roi eu henw defnyddiwr a chyfrinair, caiff yr hawliau hyn eu dyrannu gan y cyfrifiadur.

Gall fod gan ddefnyddiwr sawl gwahanol lefel o fynediad i ffeiliau, gan gynnwys:

- Darllen yn unig – ni all defnyddiwr ond darllen cynnwys y ffeil. Ni all newid na dileu'r data.
- Darllen/ysgrifennu – gall defnyddiwr ddarllen y data ar y ffeil a gall newid y data.
- Atodi – gall ychwanegu cofnodion newydd ond ni all newid na dileu'r cofnodion presennol.
- Dim mynediad – ni all agor y ffeil, felly ni all wneud dim iddi.

"Ddrwg gen i am yr arogl. Mae fy holl gyfrineiriau wedi'u tatŵo rhwng bysedd fy nhraed."

Muriau gwarchod

Mae mur gwarchod yn galedwedd a/ neu'n feddalwedd sy'n gweithio mewn rhwydwaith i rwystro cyfathrebu nad yw'n cael ei ganiatáu o un rhwydwaith i rwydwaith arall. Gan dybio bod mewnrwyd (sef rhwydwaith mewnol) yn gysylltiedig â'r Rhyngrwyd, bydd diogelwch llym yn gwarchod y fewnrwyd ond mae'r Rhyngrwyd yn rhwydwaith sydd â diogelwch isel iawn. Mae mur gwarchod yn rheoli'r traffig rhwng y ddau rwydwaith ac yn edrych ar bob pecyn o ddata i weld a oes unrhyw beth ynghylch y data sy'n torri'r polisi diogelwch. Os oes, ni fydd y data'n cael dod i mewn i'r fewnrwyd.

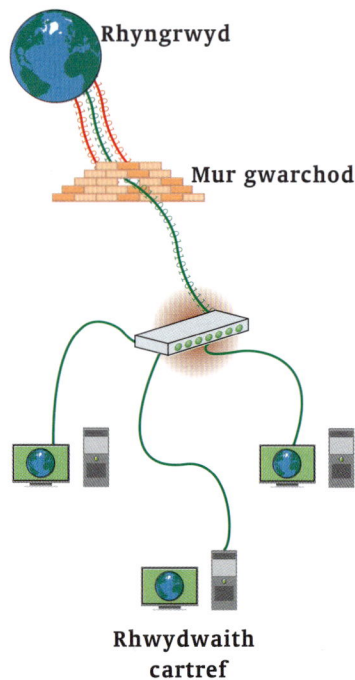

Rhyngrwyd

Mur gwarchod

Rhwydwaith cartref

Defnyddir muriau gwarchod i atal lledu firysau ac i rwystro mynediad heb ei awdurdodi (h.y. hacio)

Defnyddir muriau gwarchod hefyd i reoli'r adnoddau allanol y mae gan ddefnyddwyr fynediad iddynt ar y fewnrwyd.

Gall Norton Personal Firewall gael ei ddefnyddio i ddiogelu eich cyfrifiadur rhag mynediad heb ei awdurdodi pan fyddwch ar y Rhyngrwyd. Meddalwedd yw'r mur gwarchod hwn.

Gall muriau gwarchod fod yn:

- galedwedd
- meddalwedd
- cyfuniad o galedwedd a meddalwedd.

Gweinyddu personél

Mae rhwystro camddefnydd gan staff yn agwedd bwysig ar bolisi diogelwch. Dylai staff gael eu gweinyddu'n gywir fel eu bod yn deall, os camddefnyddiant gyfleusterau TGCh, y caiff hyn ei ddarganfod a byddant yn cael eu disgyblu. Byddai gweinyddu personél yn cynnwys:

- Hyfforddiant – bydd staff yn llai tebygol o dorri'r cod ymddygiad, yn llai tebygol o wneud camgymeriadau gyda data, yn llai tebygol o golli gwaith, ac ati.
- Trefnu i'r staff wneud tasgau addas – sicrhau bod gwybodaeth a sgiliau'r gweithiwr yn cyd-fynd â'r dasg y mae angen iddynt ei chyflawni.

- Sicrhau bod staff yn cael eu rheoli – dylai rheolwyr sicrhau bod staff yn gweithio'n ddiogel ac yn amddiffyn preifatrwydd a diogelwch y data y maen nhw'n eu defnyddio.

Gweithdrefnau gweithredol gan gynnwys cynlluniau adfer o drychineb ac ymdrin â bygythiadau oddi wrth firysau

Dylai cyrff roi gweithdrefnau gweithredol yn eu lle fel rhan o'u polisi diogelwch er mwyn lleihau y gwahanol fygythiadau i systemau TGCh. Dyma rai enghreifftiau o weithdrefnau gweithredol:

- Gellir rhoi cod ymarfer i bob defnyddiwr sy'n amlinellu pethau mae'n rhaid iddynt eu gwneud wrth gyrchu gwybodaeth.
- Mae mynediad i ddata a rhaglenni gweithredol gan staff datblygu, fel rhaglenwyr a dadansoddwyr systemau, yn syniad da fel nad yw staff yn cael cyfle i gyflawni twyll.
- Mae dyletswyddau staff yn cael eu cylchdroi hefyd i atal twyll, gan ei bod hi'n llai tebygol y byddai dau unigolyn yn meddwl yr un fath, a hefyd gallai'r naill ddarganfod unrhyw dwyll a gyflawnwyd gan y llall.

- Gall gweithdrefnau gael eu cyflwyno a fydd yn ymdrin ag adfer o drychinebau fel bod pawb yn gwybod beth i'w wneud os caiff peth neu'r cyfan o'r cyfleusterau TGCh eu colli.
- Dylai gweithdrefnau gael eu rhoi ar waith i leihau'r tebygolrwydd o gyflwyno firws, fel peidio â gadael i staff lwytho gemau, cerddoriaeth, ac ati, i lawr, a pheidio â rhoi cyfryngau symudadwy yn y cyfrifiadur.

Cod ymddygiad a chyfrifoldebau'r staff

Pan fydd gweithwyr yn defnyddio cyfleusterau TGCh, mae ganddynt nifer o gyfrifoldebau cyfrifiadura a chyfreithiol. Gan fod cyflogwr yn aml yn gyfrifol am weithredoedd ei staff pan fyddant yn y gwaith, mae'n bwysig i'r staff wybod beth yw'r cyfrifoldebau hyn. Hyd yn oed os nad yw corff yn gwybod am weithredoedd anghyfreithlon ei staff, mae'n bosibl ei fod yn gyfrifol amdanynt. Mae hyn yn digwydd os gellir profi nad oedd wedi cymryd camau digonol i atal y gweithgaredd anghyfreithlon.

Mae gan y mwyafrif o gyrff god ymddygiad sy'n nodi'n glir yr hyn y gall y staff ei wneud a'r hyn na allant ei wneud â'r cyfleusterau TGCh. Buoch yn astudio codau ymddygiad yn Nhopig 4.

Gweithdrefnau disgyblu

Ni fydd staff yn cymryd sylw oni bai eu bod yn gwybod y bydd canlyniadau difrifol os byddant yn camddefnyddio'r systemau TGCh. Mae angen hyfforddi'r staff i sicrhau eu bod yn gwybod am y problemau mae camddefnydd yn eu hachosi a'r canlyniadau os byddant yn cael eu dal.

"Fe dorrodd rhywun i mewn i dy gyfrifiadur, ond mae'n edrych fel gwaith haciwr dibrofiad."

Gweithdrefnau gweithredol ar gyfer atal camddefnydd

▼ Byddwch yn dysgu

▶ Am sgrinio darpar weithwyr

▶ Am drefniadau ar gyfer dosbarthu'r wybodaeth ddiweddaraf am firysau o'r Rhyngrwyd, defnyddio disgiau hyblyg a chyfryngau symudadwy eraill, a dulliau o wneud copïau wrth gefn

▶ Am sefydlu hawliau diogelwch ar gyfer diweddaru tudalennau gwe

▶ Am sefydlu rhaglen adfer o drychineb

▶ Am sefydlu gweithdrefnau archwilio (trywyddau archwilio) i ganfod camddefnydd

Cyflwyniad

Mae gweithdrefnau gweinyddol y gall cyrff eu rhoi yn eu lle i leihau neu atal bygythiadau oddi wrth staff. Mae'r gweithdrefnau hyn yn cynnwys mesurau fel sgrinio staff a sefydlu rhaglen i adfer o drychineb.

Sgrinio darpar weithwyr

Gellir atal problemau yn y dyfodol drwy ddewis staff yn ofalus. Rhaid gallu ymddiried mewn staff TGCh a gall nifer o fygythiadau godi os na chânt eu dewis yn ofalus.

Mae cael tystlythyrau a sgrinio staff TGCh yn drwyadl yn hanfodol.

Trefniadau ar gyfer dosbarthu'r wybodaeth ddiweddaraf am firysau o'r Rhyngrwyd a sganio am firysau

Firysau

Gall difrod maleisus godi o'r tu mewn neu y tu allan i gorff a gall amrywio o weithiwr anfodlon yn newid rhaglen fel nad yw'n gweithio'n iawn neu'n argraffu neges sarhaus ar bob anfoneb cyn gadael y cwmni, i gyflwyno firws yn fwriadol.

Firysau yw'r prif fygythiad yn y cyswllt hwn. Rhaglen sy'n ei dyblygu ei hun (ei chopïo ei hun) yn awtomatig yw firws ac fel rheol mae wedi cael ei greu i achosi difrod.

Unwaith y bydd cyfrifiadur neu gyfrwng wedi cael firws, dywedwn ei fod wedi'i heintio. Mae gan y mwyafrif o firysau bwrpas arall, heblaw am eu copïo eu hunain. Er enghraifft, gallant:

- arddangos negeseuon niwsans ar y sgrin
- dileu rhaglenni neu ddata
- defnyddio adnoddau, gan wneud i'ch cyfrifiadur redeg yn arafach.

Un o'r prif broblemau yw bod firysau newydd yn cael eu creu drwy'r adeg. Felly pan gewch heintiad firws, nid yw bob amser yn glir beth y bydd yn ei wneud i'r system TGCh.

Heblaw am y difrod y mae firysau'n ei wneud yn aml, rhaid treulio llawer iawn o amser yn datrys y problemau maen nhw'n eu creu.

Dylai pob cyfrifiadur gael gwirydd firysau, rhaglen sy'n gallu canfod a dileu'r firysau hyn. Rhaid diweddaru'r gwirwyr firysau hyn yn rheolaidd, gan fod firysau newydd yn cael eu datblygu o hyd ac na fyddai fersiynau hŷn o wirwyr firysau yn eu canfod bob amser.

Ffyrdd y gall firysau gael eu lledaenu

Gall firysau gael eu lledaenu drwy:

- e-bost allanol
- e-bost mewnol
- mewnrwyd y corff (rhwydwaith mewnol sy'n defnyddio technoleg y Rhyngrwyd yw mewnrwyd)
- rhannu disgiau
- clicio ar hysbysebion ar y Rhyngrwyd
- llwytho rhaglenni i lawr o wefannau gemau, ac ati.

E-bost

Lawrlwythiadau

Heintiad firws

Hysbysebion bras ar wefan

Cyfryngau optegol/magnetig

Sut i atal firysau

Y ffordd orau o osgoi firysau yw drwy ddefnyddio meddalwedd gwirio firysau a rhoi mesurau ar waith a fydd yn lleihau'r tebygolrwydd o gael heintiad gan firws.

Dyma rai camau y gellir eu cymryd i atal firysau rhag heintio system TGCh:

- gosod meddalwedd gwirio firysau
- peidio ag agor e-bost o ffynonellau anhysbys
- diweddaru'r meddalwedd gwirio firysau – ffurfweddu'r meddalwedd gwirio firysau fel bod diweddariadau'n cael eu gosod yn awtomatig
- cael polisi clir ar ddefnydd derbyniol ar gyfer yr holl staff sy'n defnyddio cyfrifiaduron
- hyfforddi staff i fod yn ymwybodol o'r problemau a beth y gallant ei wneud i helpu
- peidio â chaniatáu i raglenni fel gemau, fideos neu ffilmiau gael eu llwytho i lawr i systemau TGCh y corff
- peidio ag agor ffeiliau sydd wedi'u hatodi wrth e-byst os na wyddoch pwy sydd wedi'u hanfon
- os oes modd, peidio â gadael i staff ddefnyddio eu cyfryngau symudadwy eu hunain (disgiau hyblyg, gyriannau magnetig symudadwy).

Gweithdrefnau sganio am firysau

Rhaid i weithdrefnau sganio am firysau fod ar waith, ac fel arfer maent yn cynnwys:

- Amlder y sganio – pa mor aml y caiff sganiau llawn/rhannol eu gwneud
- Pa bryd y caiff gyriannau disg caled eu sganio am firysau – fel rheol pan nad yw'r cyfrifiaduron yn cael eu defnyddio at ddibenion eraill, gan fod sganio am firysau'n arafu'r cyfrifiaduron
- Sut y dylai cyfryngau cludadwy gael eu sganio cyn eu defnyddio
- Y dulliau a ddefnyddir i ddiweddaru meddalwedd sganio am firysau (e.e. defnyddio'r Rhyngrwyd i ddiweddaru'n awtomatig).

Diffinio gweithdrefnau ar gyfer llwytho i lawr o'r Rhyngrwyd, defnyddio cyfryngau storio cludadwy, a gwneud copïau wrth gefn

Bydd defnyddwyr am lwytho data i lawr o'r Rhyngrwyd ar ffurf data ystadegol, ffotograffau, diweddariadau i raglenni, cyflwyniadau, ac ati, felly rhaid cael gweithdrefnau i ganiatáu hyn. Y broblem yw atal defnyddwyr rhag lawrlwytho deunydd fel rhaglenni, cerddoriaeth a ffilmiau. Bydd hyfforddiant a defnyddio cod ymddygiad yn gweithio yn achos rhai defnyddwyr.

Mae gweithdrefnau ar gyfer llwytho ffeiliau i lawr o'r Rhyngrwyd yn cynnwys:

- Mynd bob amser am y dewis 'Cadwch y rhaglen hon ar ddisg', gan ei rhoi mewn ffolder dros dro i ffwrdd o'r prif rwydwaith. Rhedwch ac yna sganiwch y ffeil hon ac, os yw'n lân, ail-ffeiliwch hi yn y ffolder priodol ar eich system.
- Peidio â gosod unrhyw feddalwedd a lawrlwythwyd o'r Rhyngrwyd ar eich cyfrifiadur neu ar rwydweithiau'r cwmni. Mae'n bosibl eu bod yn cynnwys firysau a gallent fod yn anghyfreithlon.
- Bod yn ymwybodol o'r perygl o orlwytho eich cyfrifiadur drwy lwytho gormod o ffeiliau mawr i lawr a'u storio. Peidiwch â chadw'r ffeiliau hyn am fwy o amser nag sydd raid.

Bydd defnyddwyr yn aml yn dymuno storio eu data eu hun ar gyfryngau symudadwy neu mewn ffordd a fydd yn eu galluogi i weithio ar y ffeiliau gartref. Rhaid rhoi gweithdrefnau yn eu lle i ganiatáu i ddefnyddwyr wneud copïau o'u data eu hun.

Mae gweithdrefnau ar gyfer gwneud copïau wrth gefn yn cynnwys:

- Y mathau o gyfryngau storio a all gael eu defnyddio i wneud copïau wrth gefn.
- Pa mor aml y dylai copïau gael eu gwneud.
- Lle mae copïau i gael eu storio (e.e. diogell wrthdan, y tu allan i'r adeilad, ac ati).
- Gweithdrefnau ar gyfer defnyddio cyfryngau cludadwy sy'n cael eu cysylltu â'r cyfrifiadur drwy'r porth *USB* (e.e. a oes angen sganio'r cyfryngau'n gyntaf, pa ddata all gael eu trosglwyddo, ac ati).

- Gweithdrefnau ar gyfer trosglwyddo gwybodaeth bersonol. Ni fydd llawer o gyrff yn caniatáu i'r wybodaeth hon gael ei llwytho i lawr i liniaduron neu gyfryngau storio cludadwy na'i throsglwyddo fel atodiad ffeil drwy e-bost. Ni fydd rhai cyrff yn caniatáu iddi gael ei throsglwyddo y tu allan i'r adeilad.

Sefydlu hawliau diogelwch ar gyfer diweddaru tudalennau gwe

Os oes gennych wefan, byddwch yn ceisio atal pobl fel hacwyr a staff anfodlon rhag gwneud newidiadau amheus iddi.

Rhaid diogelu gwefannau fel mai dim ond staff sydd wedi'u hawdurdodi sy'n gallu gwneud newidiadau iddynt, a chaiff yr hawliau hyn eu pennu drwy system o rifau adnabod a chyfrineiriau. Mae'r hawliau diogelwch hyn yn caniatáu i rai staff newid tudalennau gwe.

Sefydlu rhaglen adfer o drychineb

Dylai pob corff ystyried beth y byddai'n ei wneud pe bai'n colli ei holl gyfleusterau TGCh (caledwedd a meddalwedd). Byddai'r data gan y corff o hyd gan y byddent yn cael eu cadw oddi ar y safle fel mesur diogelwch. Er nad yw trychinebau o'r fath yn digwydd yn aml, bydd y mwyafrif o gyrff yn colli rhan o'u cyfleusterau TGCh ar ryw adeg.

Cyfres o gamau i'w cymryd os caiff rhan neu'r cyfan o'r cyfleusterau eu colli yw rhaglen adfer o drychineb. Pwrpas rhaglen adfer o drychineb yw lleihau colli busnes a chael popeth yn ôl fel yr oedd mor fuan â phosibl.

Sefydlu gweithdrefnau archwilio (trywyddau archwilio) i ganfod camddefnydd

Pan ddefnyddir TGCh i gwblhau tasgau, nid oes unrhyw gofnodion papur fel arfer. Mae hyn yn golygu na allwch ddilyn darnau o bapur i ddarganfod beth sydd wedi digwydd.

Yn lle hynny, mae meddalwedd y cyfrifiadur yn cadw cofnod o'r newidiadau sy'n cael eu gwneud a phwy sydd wedi'u gwneud. Drwy edrych ar drywyddau archwilio mae'n bosibl darganfod twyll a phwy oedd yn gyfrifol amdano.

Atal camddefnydd damweiniol

▼ **Byddwch yn dysgu**

▶ Am weithdrefnau gwneud copïau wrth gefn ac adfer

▶ Am wneud copïau wrth gefn safonol ar ddisg hyblyg

▶ Am systemau *RAID*

▶ Am systemau taid-tad-mab

▶ Am wneud copïau wrth gefn o ffeiliau rhaglen

Cyflwyniad

Pa ddull bynnag a ddefnyddiwch i ddiogelu caledwedd, meddalwedd a data, mae posibilrwydd bob amser o golli cyfleusterau drwy gamddefnydd damweiniol. Gall caledwedd gael ei brynu eto, felly y peth pwysig yw diogelu'r meddalwedd a'r data. Rhaid cael dulliau o adfer rhaglenni a data os cânt eu llygru neu eu colli.

Gweithdrefnau gwneud copïau wrth gefn ac adfer

Pwysigrwydd gwneud copïau wrth gefn

Mae'n bwysig cadw copïau o feddalwedd a data fel bod modd adfer y data pe bai'r system TGCh yn cael ei cholli'n gyfan gwbl.

Mae'r data sy'n cael eu cadw gan gorff yn werthfawr iawn ac fel rheol mae'n werth mwy na'r holl galedwedd a meddalwedd gyda'i gilydd. Ystyriwch sut y gallai busnes weithredu heb y canlynol:

- cronfa ddata o gwsmeriaid
- cronfa ddata o gyflenwyr
- cronfa ddata o gynhyrchion
- cofnodion o'r holl ohebiaeth, rhestri prisiau, dyfynbrisiau, ac ati
- gwybodaeth am gyfrifon.

Cyplyswch golli'r holl ddata hyn â'r ffaith nad oes unrhyw galedwedd na meddalwedd ar gael ac mae gennych broblem enfawr.

Mae gwneud copïau o ddata yn gofyn am ymdrech, ond bydd yr offer diweddaraf ar gyfer gwneud copïau wrth gefn yn gwneud hynny'n awtomatig fel nad oes angen i neb feddwl amdano.

Mae astudiaeth ddiweddar am gwmnïau sy'n colli eu data mewn trychineb yn dangos:

- bod 29% yn rhoi'r gorau i fasnachu o fewn dwy flynedd
- bod bron 43% byth yn agor eu drysau eto ar ôl y drychineb.

Gweithdrefnau gwneud copïau wrth gefn

Mae gwneud copïau wrth gefn yn golygu creu copïau o ddata a rhaglenni fel ei fod yn bosibl eu hail-greu os cânt eu colli neu eu difrodi.

Gweithdrefnau gwneud copïau wrth gefn yw'r camau hynny y gall unigolyn neu gorff eu cymryd i sicrhau bod copïau rheolaidd yn cael eu gwneud. Dylid gwneud copïau'n rheolaidd a dylid eu cadw i ffwrdd o'r cyfrifiadur ac, os yw hynny'n bosibl, oddi ar y safle.

Ble i gadw copïau wrth gefn

Mae'n well cadw copïau wrth gefn oddi ar y safle ond mae hyn yn anghyfleus weithiau.

Cadw copïau wrth gefn mewn diogell wrthdan

Gallech gadw'r copïau wrth gefn mewn diogell wrthdan. Bydd hyn yn eu diogelu rhag lladrad a thanau bach fel rheol. Ond nid ydynt yn wrthdan ond am gyfnod penodol o amser – tua dwy awr fel rheol.

Defnyddio diogell wrthdan i gadw copïau wrth gefn

Cadw copïau wrth gefn oddi ar y safle

Mae angen cadw copïau wrth gefn oddi ar y safle os yw hynny'n bosibl, oherwydd pe bai adeilad yn cael ei ddinistrio'n gyfan gwbl, yna'r tebygolrwydd yw y byddai'r holl ddata, meddalwedd a chaledwedd yn cael eu dinistrio hefyd. Er ei bod hi'n anghyffredin i adeiladau cyfan gael eu dinistrio, mae hyn yn digwydd – mae'r ymosodiad gan derfysgwyr ar y Twin Towers, a ddinistriodd ddau adeilad yn gyfan gwbl, yn enghraifft.

Un dull o gadw copïau wrth gefn oddi ar y safle yw drwy ddefnyddio cyfryngau storio symudadwy. Mae gyriannau disg caled magnetig symudadwy, cofion pin, gyriannau sip a chardiau cof i gyd yn gludadwy ac yn gallu cael eu defnyddio i storio copïau wrth gefn.

Gwneud copïau wrth gefn

Dylai pawb fod o ddifrif ynghylch gwneud copïau wrth gefn. Dyma rai cynghorion hanfodol:

- defnyddiwch dâp neu ddisg gwahanol bob dydd a sefydlwch system ar gyfer eu cylchdroi
- gwnewch un person yn gyfrifol am wneud y copïau wrth gefn
- cadwch gopïau wrth gefn yn ddiogel (h.y. mewn diogell wrthdan) ac oddi ar y safle os oes modd
- ymarfer adfer data o gopïau wrth gefn – rhaid i chi fod yn sicr bod hyn yn bosibl a'ch bod chi'n gwybod sut i adfer data.

Defnyddio gwasanaethau ar-lein i wneud copïau wrth gefn

Mae nifer o gwmnïau'n darparu gwasanaeth sy'n eich galluogi i wneud copïau wrth gefn ar-lein o'ch ffeiliau hanfodol. Mae hwn yn wasanaeth da ac rydych yn talu amdano yn ôl faint o ddata a storiwch.

Mantais y gwasanaethau hyn yw eich bod chi'n gallu dweud wrth y system pa ddata yr ydych chi am eu copïo a bydd y cyfan yn cael ei wneud drosoch. Bydd eich data'n cael eu copïo yn y cefndir pan nad ydych chi'n defnyddio'r cyfrifiadur. Hefyd caiff y copïau eu storio ar weinydd y cwmni sy'n darparu'r gwasanaeth – felly nid yw'r copi'n cael ei gadw ar yr un safle â'r data gwreiddiol. I sicrhau diogelwch, caiff yr holl ddata eu hamgryptio cyn eu hanfon dros y Rhyngrwyd, a dim ond y sawl sy'n berchen ar y data sy'n gallu eu cyrchu.

Prif anfantais storio ar-lein yw eich bod chi'n ymddiried mewn corff arall, a allai fynd allan o fusnes, i ofalu am eich data.

I gael rhagor o wybodaeth am wneud copïau wrth gefn ar-lein, ewch i'r wefan ganlynol: http://www.datadepositbox. com

I gael gweld y meddalwedd a'i gyfleusterau, ewch i http://www. datadepositbox.com/demo-citytv.asp

Amserlennu gwneud copïau wrth gefn

Y prif ofyniad ar gyfer copïau wrth gefn yw y dylent fod yn hawdd i'w gwneud. Yn aml iawn gellir eu gwneud yn awtomatig drwy drefnu amser i'r copïau ddechrau cael eu gwneud. Bydd y system yn arafu fel rheol yn ystod y broses hon. Bydd defnyddwyr rhwydweithiau yn sylwi ar y lleihad mewn cyflymder, felly mae'n well gwneud y copïau wrth gefn pan nad yw'r cyfrifiaduron yn cael eu defnyddio. Gellir gwneud hyn dros nos ond mae llawer o systemau TGCh yn rhedeg 24 awr y dydd, felly rhaid amserlennu'r copïo i ddigwydd ar yr adegau lleiaf prysur.

Gall copïau wrth gefn gael:

- eu gwneud â llaw
- eu hamserlennu i gael eu gwneud yn awtomatig.

↪ GEIRIAU ALLWEDDOL

Copïau wrth gefn – copïau o feddalwedd a data sy'n cael eu cadw fel ei bod yn bosibl adfer y data os collir y system TGCh

Dyfeisiau a chyfryngau ar gyfer storio copïau wrth gefn

Mae yna lawer iawn o ddyfeisiau a chyfryngau a all gael eu defnyddio i storio copïau wrth gefn. Mae'r dewis o ddyfais/cyfrwng yn dibynnu ar:

- faint o le storio sydd ei angen
- pa mor gludadwy yw'r ddyfais/cyfrwng (pwysau a maint)
- cyflymder trosglwyddo data (pa mor gyflym y gall data gael eu hysgrifennu i neu eu darllen o'r ddyfais/cyfrwng)
- cyflymder cyrchu data (yr amser mae'n ei gymryd i'r ddyfais ddod o hyd i ddarn penodol o ddata sydd wedi'i storio)
- y gallu i gael ei chysylltu â gwahanol gyfrifiaduron neu ddyfeisiau eraill fel argraffwyr, camerâu, ac ati.

Tâp magnetig

Mae tâp magnetig yn gyfrwng storio delfrydol – mae'n rhad ac mae ganddo gynhwysedd storio mawr. Cyfryngau symudadwy yw tapiau magnetig, sy'n golygu y gellir eu defnyddio mewn unrhyw yriant cydwedd.

Bydd cyrff mawr yn defnyddio llyfrgelloedd tapiau i gadw eu copïau wrth gefn. Mae'r rhain yn defnyddio llawer o dapiau ac mae rhai'n defnyddio breichiau robotaidd i lwytho'r tapiau i mewn i'r gyriannau.

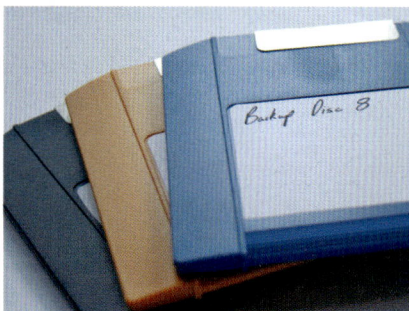

Cyfryngau a ddefnyddir i wneud copïau

Atal camddefnydd damweiniol *(parhad)*

Crynodeb o ddyfeisiau a chyfryngau ar gyfer gwneud copïau wrth gefn

Enw	Cost	Cynhwysedd storio	Cyflymder	Sylwadau
Disg hyblyg 3.5"	Rhad	1.44 MB	araf	Cynhwysedd storio bach iawn. Nid oes gan bob cyfrifiadur yriant i'w ddarllen. Hen-ffasiwn erbyn hyn.
Cof pin	Rhad	Hyd at 4 GB ac yn codi	gweddol gyflym	Cyfrwng cludadwy poblogaidd iawn. Mae wedi disodli'r disg hyblyg 3.5".
CD-RW	Di-dâl – yn dod gyda'r cyfrifiadur	Hyd at 700 MB	gweddol gyflym	Cyfrwng da ar gyfer cyfrifiaduron personol sy'n cynnwys ychydig bach o ddata.
DVD-RW	Di-dâl – rhan o'r cyfrifiadur	Hyd at 4.7 GB	gweddol gyflym	Cynhwysedd storio uwch, felly mae'n fwy addas ar gyfer copïo'r data ar gyfrifiadur personol.
Gyriant disg caled (sefydlog)	Di-dâl – rhan o'r cyfrifiadur	160 i 250 GB ac yn codi	cyflym	Cynhwysedd storio enfawr a gellid defnyddio peth ohono i wneud copïau, ond byddai ar yr un gyriant.
Gyriant disg caled (allanol neu symudadwy)	Gweddol rad	Hyd at 500 GB ac yn codi	cyflym	Gwych ar gyfer gwneud copïau'n gyflym. Mae angen ei ddatgysylltu a'i storio mewn diogell wrthdan neu oddi ar y safle.
Gyriant tâp magnetig	Drud	Enfawr, hyd at 8TB	cyflym	Dewis da os oes angen copïo llawer iawn o ddata. Mae gan rai systemau tâp dapiau symudadwy.
Gyriant sip	Rhad	Hyd at 750 MB ac yn codi	araf	Gellir tynnu'r disgiau allan a'u rhoi mewn lle diogel. Yn debyg i ddisg hyblyg symudadwy mwy trwchus.
Copïo i'r Rhyngrwyd	Rhad	Heb ei gyfyngu	araf	Y gost yn dibynnu ar faint o ddata sy'n cael eu storio. Caiff y data eu storio oddi ar y safle.

Disg magnetig

Mae cynnwys un disg magnetig yn cael ei gopïo i ddisg magnetig arall.

Cyfryngau optegol

Mae hyn yn cynnwys cyfryngau fel *CD-R, CD-RW, DVD-RW*, ac ati. Y brif broblem sy'n codi wrth ddefnyddio cyfryngau optegol ar gyfer storio data yw bod y gyfradd trosglwyddo (h.y. y gyfradd mae data'n cael eu darllen i'r cyfryngau neu eu copïo o'r cyfryngau) yn araf, a'i bod hi felly'n cymryd mwy o amser i wneud copïau wrth gefn.

Cofion pin

Mae'r rhain yn boblogaidd iawn gan eu bod yn fach a chludadwy ac yn ddelfrydol ar gyfer gwneud copïau o ffeiliau bach. Mae'n hawdd iddynt gael eu colli neu eu dwyn.

Disg magnetig

Cof pin

Systemau arae diangen o ddisgiau rhad (*RAID*) – disgiau drych

Mae llawer o systemau rhwydwaith yn defnyddio system arae diangen o ddisgiau rhad (*RAID: redundant array of inexpensive disks*) i wneud copïau wrth gefn. Mae'r system yn defnyddio cyfres o ddisgiau magnetig ar gyfer storio'r data. Mae yna wahanol systemau *RAID*, ac mewn un math fe fydd y system yn cymryd drosodd yn awtomatig os caiff y data gwreiddiol eu difrodi neu eu dinistrio, gan sicrhau bod y gwasanaeth yn parhau.

Clystyru

Defnyddir techneg o'r enw clystyru i wella diogelwch rhwydweithiau. Yma caiff y gweinyddion ffeiliau a'r dyfeisiau storio eu rhwydweithio â'i gilydd fel nad oes angen dibynnu ar un gweinydd ffeiliau ac un ddyfais storio.

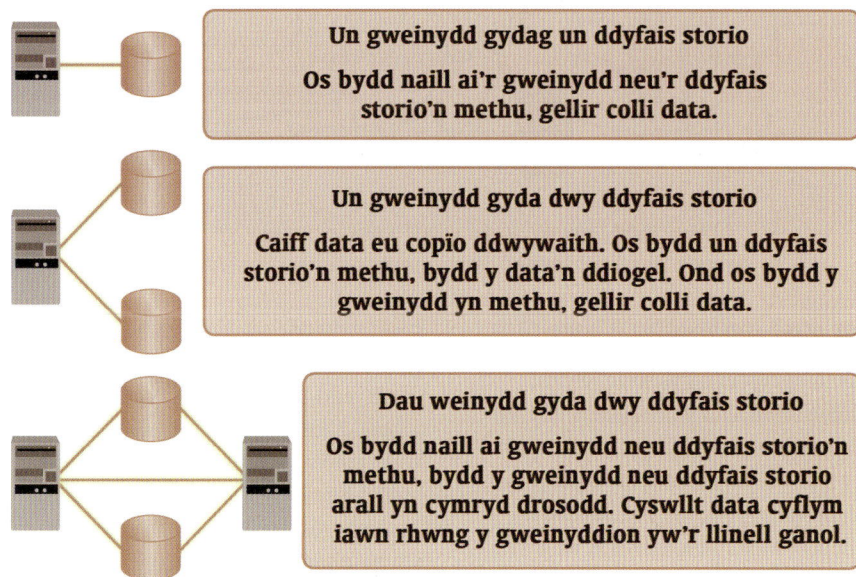

Un gweinydd gydag un ddyfais storio

Os bydd naill ai'r gweinydd neu'r ddyfais storio'n methu, gellir colli data.

Un gweinydd gyda dwy ddyfais storio

Caiff data eu copïo ddwywaith. Os bydd un ddyfais storio'n methu, bydd y data'n ddiogel. Ond os bydd y gweinydd yn methu, gellir colli data.

Dau weinydd gyda dwy ddyfais storio

Os bydd naill ai gweinydd neu ddyfais storio'n methu, bydd y gweinydd neu ddyfais storio arall yn cymryd drosodd. Cyswllt data cyflym iawn rhwng y gweinyddion yw'r llinell ganol.

Y system taid-tad-mab (y system ffeiliau hynafiadol)

Mewn rhai systemau TGCh mawr, lle mae angen storio llawer o ddata, mae dau fath o ffeil yn cael eu cadw: meistr-ffeiliau a ffeiliau trafod. Y feistr-ffeil yw'r fersiwn mwyaf cyflawn o ffeil, ac os byddai'n cael ei cholli neu ei difrodi fe fyddai'r system gyfan yn ddiwerth. Defnyddir ffeiliau trafod i gadw manylion yr holl drafodion (h.y. darnau o fusnes) sydd wedi digwydd ers i'r feistr-ffeil gael ei diweddaru ddiwethaf. Defnyddir y ffeil drafod i ddiweddaru'r feistr-ffeil o bryd i'w gilydd (e.e. bob nos, bob wythnos).

Mae bob amser yn bosibl y bydd y data sy'n cael eu cadw ar feistr-ffeil (disg neu dâp) yn cael eu dinistrio, er enghraifft gan ddefnyddiwr dibrofiad, toriad trydan, tân neu ladrad. I'r mwyafrif o gwmnïau, gallai colli data hanfodol fod yn drychinebus. Ond drwy ddefnyddio'r egwyddor taid-tad-mab, gellir ail-greu'r feistr-ffeil os caiff ei cholli.

Mae'r egwyddor yn gweithio fel hyn. Mae tair cenhedlaeth o ffeiliau'n cael eu cadw. Mae'r feistr-ffeil hynaf yn cael ei galw'n daid-ffeil ac yn cael ei chadw gyda'i ffeil drafod. Defnyddir y ddwy ffeil hyn i gynhyrchu meistr-ffeil newydd, y tad-ffeil, a ddefnyddir, gyda'i ffeil drafod, i gynhyrchu'r ffeil fwyaf cyfoes, y mab-ffeil. Caiff y broses ei hailadrodd a daw'r mab-ffeil yn dad-ffeil a'r tad-ffeil yn daid-ffeil, ac yn y blaen. Dim ond tair cenhedlaeth sydd eu hangen a gall y ffeiliau eraill gael eu hailddefnyddio. Fel rheol fe gaiff y system hon, y system ffeiliau hynafiadol, ei defnyddio ar gyfer tapiau, er y gellir ei defnyddio hefyd ar gyfer disgiau.

Gwneud copïau wrth gefn o ffeiliau rhaglen

Yn aml iawn gall rhaglenni sydd wedi eu colli, gael eu hadfer yn hawdd oherwydd gellir defnyddio'r disgiau gwreiddiol i lwytho'r ffeiliau yn ôl i'r system. Caiff niferoedd cynyddol o raglenni eu llwytho i lawr o'r Rhyngrwyd, felly os caiff y rhaglenni gwreiddiol eu difrodi neu eu colli, yna bydd cynhyrchwyr y meddalwedd yn caniatáu iddynt gael eu lawrlwytho eto.

Bydd llawer o gwmnïau'n datblygu eu meddalwedd eu hunain neu'n treulio llawer o amser yn addasu rhaglenni presennol fel eu bod yn gweithio'n well iddyn nhw. Yn yr achosion hyn, rhaid gwneud copïau wrth gefn ar y dechrau ac ar unrhyw adeg y caiff newidiadau eu gwneud.

Atal troseddau neu gamddefnydd bwriadol

▼ **Byddwch yn dysgu**

▶ Am ddulliau o reoli mynediad i ystafelloedd cyfrifiaduron

▶ Am ddulliau o ddiogelu data sy'n cael eu trawsyrru

▶ Am ddefnyddio gweinyddion

▶ Am ddulliau o ddiffinio statws diogelwch a hawliau mynediad

▶ Am ddulliau o ddiogelu caledwedd a meddalwedd yn gorfforol

▶ Am ddiogelwch systemau ffeilio dogfennau

Cyflwyniad

Rhaid diogelu systemau TGCh rhag pobl sy'n dymuno eu camddefnyddio neu eu difrodi'n fwriadol. Mae'r bobl hyn yn cynnwys hacwyr, ysgrifenwyr firysau, gweithwyr anfodlon, twyllwyr, blacmelwyr a lladron.

Hefyd rhaid i chi roi mesurau yn eu lle i atal camddefnydd gan weithwyr, fel edrych ar wybodaeth bersonol nad yw'n angenrheidiol ar gyfer eu swydd, gwneud castiau, a gadael cyfrifiaduron wedi'u mewngofnodi pan nad ydynt wrth eu desg.

Mae'r adran hon yn edrych ar y dulliau a ddefnyddir gan gyrff i helpu i atal troseddau neu gamddefnydd bwriadol.

Dulliau o reoli mynediad i ystafelloedd cyfrifiaduron

Cyfyngu ar fynediad

I atal dwyn a mynediad anghyfreithlon i systemau TGCh, gellir defnyddio bysellbadiau i gyfyngu ar fynediad i ystafelloedd cyfrifiaduron. Hefyd gellir defnyddio caledwedd sy'n darparu profi biometrig i gyfyngu ar fynediad i ystafelloedd a chyfrifiaduron. Mae profi biometrig yn cynnwys adnabod wyneb, adnabod llais, neu adnabod olion bysedd. Mantais y ffurf hon ar fynediad yw nad oes angen cofio cyfrineiriau na chodau.

Gall systemau adnabod wyneb gyfyngu ar fynediad i ystafelloedd a chyfrifiaduron. Gellir eu defnyddio hefyd i gofnodi faint o amser mae defnyddwyr yn ei dreulio wrth eu gweithfannau.

Dulliau o ddiogelu data sy'n cael eu trosglwyddo

Amgryptio

Mae amgryptio'n cymysgu data cyn eu hanfon i lawr llinell gyfathrebu neu'n ddiwifr. Felly, hyd yn oed os caiff y data eu rhyng-gipio, nid ydynt yn gwneud unrhyw synnwyr i'r rhyng-gipiwr. Mae amgryptio hefyd yn nodwedd o'r systemau gweithredu diweddaraf, lle caiff y data sydd wedi'u storio ar y gyriant disg caled eu hamgryptio'n awtomatig. Felly ni ellir deall y data os caiff y cyfrifiadur ei ddwyn neu'r data eu copïo.

"Mae meddalwedd amgryptio'n ddrud … felly yn lle hynny rydyn ni wedi aildrefnu'r llythrennau ar eich bysellfwrdd."

Sut mae amgryptio'n gweithio

Mae amgryptio'n gweithio fel hyn. Tybiwch fod Siân yng Nghaerdydd am anfon e-bost diogel at Jac yn Llundain. Ar ôl i Siân deipio'r e-bost mae hi'n pwyso'r dewis 'amgryptio' ar y meddalwedd e-bostio. Mae'r meddalwedd yn gofyn at bwy y mae hi am anfon yr e-bost. Mae hi'n dewis enw Jac o'r rhestr sy'n cael ei dangos. Mae'r holl bobl ar y rhestr hon yn bobl y mae gan Siân allwedd gyhoeddus ar eu cyfer ac y gall hi anfon negeseuon wedi'u hamgryptio atynt. Mae'r meddalwedd amgryptio yn cymysgu ac yn ail-gymysgu'n awtomatig bob did deuaidd o'r neges gyda phob did deuaidd yn allwedd gyhoeddus Jac. Y canlyniad yw cymysgedd deuaidd o ddata a all gael ei ddatgymysgu gan yr un meddalwedd yn unig, gan ddefnyddio allwedd breifat Jac. Pan fydd Jac yn derbyn y neges yn Llundain, bydd angen iddo ddewis 'dadgryptio' ac yna bydd y meddalwedd yn gofyn iddo am gyfrinair. Ar ôl teipio hwn i mewn fe gaiff ei allwedd breifat ei dadgryptio. Mae'r allwedd breifat yn rhif hir iawn a chaiff nifer mawr o gyfrifiaduau eu gwneud sy'n datgymysgu'r cymysgedd deuaidd i roi'r neges gan Siân. Pe câi'r neges ei rhyng-gipio, ni fyddai hacwyr yn gallu ei gweld gan nad yw'r allwedd breifat a ddefnyddir i wneud y cyfrifiadau sydd eu hangen i ddatgymysgu'r neges yn eu meddiant.

➡ GEIRIAU ALLWEDDOL

Amgryptio – y broses o godio ffeiliau cyn eu hanfon dros rwydwaith er mwyn eu diogelu rhag hacwyr. Hefyd y broses o godio ffeiliau sydd wedi'u storio ar gyfrifiadur fel na ellir eu darllen os caiff y cyfrifiadur ei ddwyn

Gweinyddion dirprwyol

Mae gweinydd dirprwyol, sy'n gallu bod yn galedwedd neu'n feddalwedd, yn derbyn ceisiadau gan ddefnyddwyr i gyrchu gweinyddion eraill ac mae naill ai'n eu hanfon ymlaen i'r gweinyddion eraill neu'n gwrthod mynediad iddynt. Gellir defnyddio gweinyddion dirprwyol i atal neu gyfyngu ar fynediad i gyfeiriadau gwe (Lleolwyr Adnoddau Unffurf) neu i wasanaethau gwe megis negeseua sydyn ac ystafelloedd sgwrsio.

Defnyddir gweinyddion dirprwyol gan ysgolion i sicrhau y gall disgyblion gyrchu peth gwybodaeth yn unig. Mae'n eu rhwystro rhag cyrchu gwefannau sy'n cynnwys deunydd amheus a rhag gwastraffu amser mewn ystafelloedd sgwrsio neu ar safleoedd rhwydweithio cymdeithasol pan ddylent fod yn gweithio.

Bydd llawer o gyrff yn defnyddio gweinyddion dirprwyol i hidlo cynnwys ac i sicrhau bod eu gweithwyr yn cydymffurfio â'u polisi defnydd derbyniol.

Dulliau o ddiffinio statws diogelwch a hawliau mynediad defnyddwyr

Defnyddir nifer o ddulliau i ddiffinio statws diogelwch a hawliau mynediad defnyddwyr, gan gynnwys:

- defnyddio hierarchaeth o gyfrineiriau
- dyrannu adnoddau rhwydwaith i ddefnyddwyr ar sail rhifau adnabod defnyddwyr a chyfrineiriau
- dyrannu hawliau mynediad i ddefnyddwyr ar sail eu swydd neu eu statws (e.e. darllen yn unig, atodi'n unig, creu cofnodion newydd yn unig).

Dulliau o ddiogelu caledwedd a meddalwedd yn gorfforol

Gall caledwedd a meddalwedd gael eu diogelu'n gorfforol mewn llawer o wahanol ffyrdd, gan gynnwys:

- cyfyngu ar fynediad i ystafelloedd cyfrifiaduron
- cloeon ar fysellfyrddau
- cloeon ar gyfrifiaduron sy'n defnyddio dulliau biometrig (sganio'r retina, olion bysedd, adnabod wyneb)
- cysylltu ceblau â phob cyfrifiadur fel na ellir eu symud
- cloeon ar yriannau i rwystro pobl rhag rhoi cyfryngau symudadwy ynddynt i gopïo data neu feddalwedd
- muriau gwarchod i ddiogelu rhwydweithiau rhag mynediad heb ei awdurdodi gan hacwyr
- cloi cyfrifiaduron yn yr ystafell ar ddiwedd y diwrnod
- storio copïau gwreiddiol o feddalwedd mewn diogell wrthdan
- gwneud copïau wrth gefn o feddalwedd yn rheolaidd a'u storio mewn diogell wrthdan neu oddi ar y safle.

Diogelwch systemau ffeilio dogfennau

Does dim pwynt cael diogelwch TG ardderchog os yw canlyniadau prosesu, ar ffurf allbrintiau, yn cael eu gadael ar ddesgiau i unrhyw un eu gweld ac, o bosibl, eu llungopïo. Yn ogystal â diogelwch corfforol a rhesymegol, mae angen i ni ystyried diogelwch dogfennau. Hefyd, rhaid i ddefnyddwyr ddiogelu dogfennau sydd wedi'u storio ar eu gliniaduron; mae'r rhain yn fwy tebygol o gael eu dwyn gan eu bod yn cael eu defnyddio mewn lleoedd cyhoeddus a'u rhoi mewn ceir. Mae'n bwysig i unrhyw allbrintiau neu adroddiadau gael eu rhoi dan glo pan nad ydynt yn cael eu defnyddio ac iddynt gael eu rhwygo'n fân cyn eu rhoi yn y bin sbwriel.

"O, roedd yr un fath ag unrhyw ddiwrnod arall yn y gwaith. Hynny ydy, nes i mi disian ger y peiriant rhwygo papur."

Rheoli risg: costau a rheolaeth

▼ **Byddwch yn dysgu**

▶ Am nodi risgiau posibl

▶ Am y tebygolrwydd y bydd risgiau'n cael eu gwireddu

▶ Am ganlyniadau tymor byr a thymor hir y bygythiad

▶ Am ba mor barod yw cwmni i ymdrin â'r bygythiad

Cyflwyniad

Mae llawer o fygythiadau i systemau TGCh, felly mae angen i gyrff benderfynu beth ydynt, pa mor debygol ydyw y byddant yn dod yn realiti a'r canlyniadau tebygol pe bai hyn yn digwydd, beth y gellir ei wneud i leihau'r risg, ac a ellir gwneud hyn am bris rhesymol.

Mae dadansoddi risg yn rhan bwysig o redeg systemau TGCh mewn cyrff gan fod llawer ohonynt yn dibynnu'n gyfan gwbl ar eu system TGCh. Yn yr adran hon byddwch yn dysgu am ddadansoddi risg a sut y gall corff asesu faint o arian y dylid ei wario ar leihau'r risg.

Dadansoddi risg

Prif bwrpas dadansoddi risg yw gwneud pawb yn y corff yn ymwybodol o'r bygythiadau diogelwch i'w galedwedd, meddalwedd a data. Rhaid iddynt wybod beth fyddai canlyniadau colled o'r fath am gyfnod byr neu hir, megis y golled ariannol sy'n digwydd ar unwaith a'r golled tymor hir sy'n deillio o golli ffydd cwsmeriaid, y cyhoeddusrwydd gwael yn y wasg, a'r anallu i ddarparu gwasanaeth cwsmeriaid.

Er mwyn gwneud dadansoddiad risg, bydd angen ystyried y canlynol:

- Rhoi gwerth ar bob cydran mewn system gwybodaeth lwyddiannus a fyddai'n cynnwys:
  - caledwedd
  - meddalwedd
  - dogfennaeth
  - pobl
  - sianeli cyfathrebu
  - data.
- Nodi'r bygythiadau i'r uchod a'r tebygolrwydd y byddant yn codi.

Bydd y mwyafrif o gyrff yn cynnal adolygiad o'u diogelwch technoleg gwybodaeth corfforaethol sy'n edrych ar y wybodaeth a brosesir gan eu cyfrifiaduron gyda'r bwriad o nodi risgiau megis colli'r wybodaeth, camgymeriadau a bylchau, camddefnydd, a datgelu heb awdurdod, a phenderfynu ar eu goblygiadau posibl. Bydd angen ystyried pob risg o safbwynt y bygythiad i ddiogelwch, y golled a achosai, a'r tebygolrwydd y bydd yn cael ei wireddu. Y nod yw darganfod y systemau hynny sy'n allweddol i'r corff ac edrych ar y goblygiadau tymor byr neu dymor hir os collir y systemau hyn.

Dyma ychydig o ganlyniadau niferus colli system:

- problemau llif arian wrth i anfonebau gael eu hanfon yn hwyr
- penderfyniadau busnes gwael o ganlyniad i ddiffyg gwybodaeth rheoli
- colli ewyllys da cwsmeriaid a chyflenwyr
- oediadau cynhyrchu gan nad yw'r stoc gywir ar gael
- archebion yn cyrraedd yn hwyr a chwsmeriaid yn mynd at gwmnïau eraill
- prinder stoc neu ormod o stoc oherwydd rheolaeth stoc anaddas.

Nodi risgiau posibl

Er mwyn diogelu systemau TGCh, rhaid nodi'r risgiau. Dyma restr o'r risgiau i'r mwyafrif o systemau TGCh:

- firysau
- tân
- difrod naturiol (llifogydd, daeargrynfeydd, mellt, llosgfynyddoedd, ac ati)
- hacio (torri i mewn i linellau cyfathrebu)
- methiant systemau o ganlyniad i beiriant yn torri i lawr
- twyll
- toriad trydan
- difrod bwriadol
- dwyn (caledwedd, meddalwedd a data)
- blacmel
- ysbïo
- bomiau terfysgwyr
- colli cemegion

- nwy'n gollwng
- fandaliaeth
- colli diod dros gyfarpar cyfrifiadurol
- methiant cysylltau telathrebu
- problemau gyda cheblau data mewn rhwydweithiau
- bothau a llwybryddion, ac ati, yn torri i lawr
- methiant meddalwedd
- meddalwedd systemau sy'n cynnwys namau ac yn peri i'r cyfrifiadur chwalu
- difrodi/colli'r gyriant disg caled
- streiciau.

Y tebygolrwydd y bydd risgiau'n cael eu gwireddu

Mae rhai pethau bron yn sicr o ddigwydd yn hwyr neu'n hwyrach, e.e. toriad trydan, ond mae pethau eraill, fel ffrwydradau, yn digwydd yn llawer llai aml. Serch hynny, rhaid cymryd yr holl risgiau i ystyriaeth. Rhaid i'r uwch reolwyr benderfynu pa mor debygol ydyw y bydd y risg yn cael ei wireddu a sut y gellir lleihau'r risg am bris rhesymol, a pha lefelau o risg sy'n dderbyniol i'r corff.

▶ **GEIRIAU ALLWEDDOL**

Dadansoddiad risg – y broses o asesu pa mor debygol ydyw y bydd rhai pethau'n digwydd ac amcangyfrif cost y difrod y gallent eu hachosi a beth y gellir ei wneud am gost resymol i ddileu neu leihau'r risg

Rhaglen adfer o drychineb – cynllun sy'n adfer cyfleusterau TGCh mewn cyn lleied o amser â phosibl er mwyn lleihau'r colledion a achosir os bydd rhan neu'r cyfan o gyfleusterau TGCh y corff yn cael eu colli

Canlyniadau tymor byr a thymor hir y bygythiad

Rhai o ganlyniadau tymor byr colli data yw:

- Mae angen defnyddio adnoddau (e.e. staff, cyfarpar, ac ati) i adfer y data.
- Mae'n bosibl y bydd angen talu iawndal i bobl sydd wedi dioddef o ganlyniad i golli'r data.
- Colledion ariannol o ganlyniad i golli busnes gan nad yw'r cwmni'n gallu cymryd archebion.
- Embaras os bydd y wasg yn rhoi sylw i'r digwyddiad.
- Y posibilrwydd y caiff y cwmni ei erlyn am golli data personol oherwydd diogelwch gwael. Gellid dwyn achos o dan y Ddeddf Gwarchod Data.

Rhai o ganlyniadau tymor hir colli data yw:

- Os bydd y cwmni'n colli ei enw da, bydd ei gwsmeriaid yn troi eu cefnau arno.
- Gall colledion ariannol o ganlyniad i golli archebion olygu bod y cwmni'n mynd i'r wal.
- Y gost uchel o adnewyddu caledwedd, meddalwedd a data a gollwyd.

Pa mor barod yw'r cwmni i ymdrin â'r bygythiad?

Rhaid i gyrff ofyn iddyn nhw eu hunain yn gyson pa mor dda y gallent ymdrin â bygythiad pe câi ei wireddu. Mae bygythiadau a chyrff yn newid dros y blynyddoedd, felly rhaid adolygu hyn o bryd i'w gilydd.

Y rhaglen adfer o drychineb

Pwrpas y rhaglen adfer o drychineb yw sicrhau bod adnoddau (staff, adeiladau, trydan, ac ati) a chyfarpar cyfrifiadurol hanfodol ar gael pe bai trychineb yn digwydd. Bydd y cynllun yn ymdrin â'r canlynol fel rheol:

- colli cyfarpar cyfrifiadurol yn rhannol neu'n gyfan gwbl
- colli gwasanaethau hanfodol fel trydan, gwres neu aerdymheru
- colli staff allweddol (e.e. colli'r holl

staff rhwydwaith cymwysedig i gyd ar unwaith gan eu bod yn penderfynu ffurfio eu cwmni cyfleusterau eu hunain)
- colli gwasanaethau cynnal neu wasanaethau cymorth
- colli data neu feddalwedd
- colli cyfarpar neu wasanaethau telathrebu'n rhannol neu'n gyfan gwbl
- colli'r adeilad sy'n gartref i'r cyfarpar TG yn rhannol neu'n gyfan gwbl.

Mae ffrwydradau'n llawer llai tebygol o ddigwydd

▶ Gweithgaredd: Creu rhaglen adfer o drychineb

Mae llawer o fygythiadau i systemau TGCh a gellir lleihau'r difrod y gallent ei achosi drwy lunio rhaglen adfer o drychineb.

Rydych wedi cael eich rhoi yng ngofal y cyfleusterau TGCh yn eich ysgol/coleg. Mae'r cyfleusterau hyn yn cynnwys rhwydweithiau addysgu'r ysgol/coleg a'r rhwydweithiau gweinyddu, sy'n storio gwybodaeth bwysig fel data am staff a myfyrwyr a manylion yr holl drafodion ariannol.

Mae gofyn i chi greu rhaglen adfer o drychineb ar gyfer eich ysgol neu goleg. Drwy ddarllen y nodiadau ar raglenni adfer o drychineb yn y topig hwn a thrwy wneud eich ymchwil eich hun ar y Rhyngrwyd, dylech ddod i ddeall y pwnc yn dda.

Cynhyrchwch y rhaglen adfer o drychineb ar gyfer eich ysgol neu goleg. Sicrhewch fod eich rhaglen yn cynnwys y canlynol:

- **Sut y gallai'r ysgol/coleg ymdopi pe byddai caledwedd a meddalwedd TGCh yn cael eu colli.**
- **Sut y gallai'r ysgol/coleg adfer unrhyw ddata a gollir.**
- **Sut y gallai'r ysgol/coleg ymdopi pe byddai rhai staff allweddol yn gadael.**
- **Sut y gallai'r ysgol/coleg ymdopi pe byddai cyfarpar a gwasanaethau telathrebu yn cael eu colli.**

Bydd tymhestloedd yn dinistrio llinellau trydan neu delathrebu'n aml

Astudiaethau achos

▶ **Astudiaeth achos 1** | t. 74 |

Microsoft Vista yn hwyluso gwneud copïau wrth gefn

Cyn rhyddhau Windows 7, Microsoft Windows Vista oedd system weithredu ddiweddaraf Microsoft.

Roedd angen system weithredu newydd oherwydd y defnydd cynyddol sy'n cael ei wneud gan ddefnyddwyr cartref a busnes o fideos, cerddoriaeth, gemau a meddalwedd amlgyfrwng arall.

Mae'r meddalwedd yn cynnwys llawer o nodweddion diogelwch newydd i gadw data a ffeiliau'n ddiogel. Yn arbennig, mae cyfyngiadau newydd yn sicrhau nad yw defnyddwyr yn gallu gwneud newidiadau sy'n rhoi eu system mewn perygl.

Hyd yn oed os caiff cyfrifiadur ei ddwyn, mae un nodwedd newydd o'r enw technoleg BitLocker yn amgryptio'r gyriant disg caled fel na all neb arall ei ddefnyddio. Ni allant gyrchu'r data na'r rhaglenni ar y gyriant. Mae hyn yn arbennig o ddefnyddiol gan fod mwy a mwy o bobl yn defnyddio gliniaduron mewn lleoedd cyhoeddus fel awyrennau, trenau, caffis, ac ati.

Mae gwneud copïau wrth gefn yn llawer haws gyda Windows Vista.

Un o nodweddion Vista yw Dewin hawdd ei ddefnyddio sy'n dangos i chi sut i wneud copi wrth gefn gam wrth gam. Mae'n hawdd i chi nodi pa ffeiliau yr ydych chi am eu copïo, y ddyfais i storio'r copi arni a'r amser i ddechrau'r broses.

Darllenwch yr erthygl uchod yn ofalus ac yna atebwch y cwestiynau sy'n dilyn. Mae'n bosibl y bydd angen i chi wneud peth ymchwil er mwyn ateb y cwestiynau hyn. Defnyddiwch lyfrau a geirfâu ar-lein i'ch helpu. Pob lwc!

1 System weithredu yw Vista.
   (a) Ysgrifennwch ddisgrifiad byr o bwrpas system weithredu. (4 marc)
   (b) Disgrifiwch **ddwy** dasg y byddai system weithredu yn eu cyflawni. (2 farc)
   (c) Enghraifft o system weithredu yw Microsoft Vista. Rhowch enwau brand **dwy** system weithredu arall. (2 farc)

2 Yn yr erthygl mae'n dweud 'mae cyfyngiadau newydd yn sicrhau nad yw defnyddwyr yn gallu gwneud newidiadau sy'n rhoi eu system mewn perygl'.

Drwy roi enghraifft, disgrifiwch newid y gellid ei wneud i system weithredu a fyddai'n rhoi'r system TGCh mewn mwy o berygl. (2 farc)

3 Mae Vista yn defnyddio Dewin i helpu defnyddwyr i wneud copïau wrth gefn o'u data.
   (a) Disgrifiwch beth yw ystyr y term Dewin. (1 marc)
   (b) Eglurwch pam mae Dewin yn nodwedd ddefnyddiol i ddefnyddiwr. (2 farc)

4 Pan ddefnyddir system TGCh mewn busnes, mae'r data sy'n cael eu storio yn aml yn fwy gwerthfawr na'r caledwedd a'r meddalwedd.
   (a) Nodwch, drwy roi enghraifft, pam mae'r gosodiad uchod yn aml yn wir. (2 farc)
   (b) Eglurwch **ddwy** ffordd y gall data gael eu colli o gyfrifiadur. (2 farc)

(c) Caiff copïau wrth gefn eu gwneud er mwyn gallu adfer ffeiliau os cânt eu colli neu eu difrodi. Rhowch enwau **dau** fath o gyfrwng storio sy'n briodol ar gyfer gwneud copïau. (2 farc)

5 (a) Eglurwch ystyr y term amgryptio. (3 marc)
   (b) Eglurwch sut mae amgryptio'r data ar yriant disg caled cyfrifiadur yn ei wneud yn ddiwerth i leidr sy'n dwyn y cyfrifiadur. (2 farc)

6 Gellir trefnu i gyfrifiadur wneud copïau wrth gefn ar unrhyw adeg o'r dydd. Nodwch, gan roi rheswm, yr amser gorau o'r dydd i wneud copïau wrth gefn. (2 farc)

Cynllunio da yn cadw busnesau a foddwyd ar dir sych

Yn ystod llifogydd difrifol Mehefin 2007 llifodd y dŵr i swyddfeydd llawer o fusnesau. Roedd rhai busnesau wedi paratoi ar gyfer hyn: roedd ganddynt gynlluniau i adleoli staff i ganolfan adfer o drychineb lle y gallent ddefnyddio'r systemau TGCh a'u data wrth gefn i roi'r busnes ar ei draed eto.

Cafodd cyfrifiaduron rhai cwmnïau eu dinistrio'n llwyr gan y dŵr. Cafodd cwmnïau eraill drafferth gyda'u cyflenwad trydan wrth i ddŵr fynd i mewn i'r ceblau.

Dywedodd un aelod o'r gymuned fusnes ei bod hi'n bosibl na fyddai cwmnïau bach llai trefnus yn goroesi, ond y byddai cwmnïau mawr yn dod drwyddi oherwydd bod ganddynt gynlluniau i gadw'r busnes i fynd

1 Disgrifiwch ddau gynllun y gallai corff eu rhoi ar waith i gadw busnes i redeg er gwaethaf y ffaith na ellir defnyddio'r cyfleusterau TGCh mewn swyddfa a bod y data ar y cyfrifiadur wedi'u colli. (6 marc)

2 Mae'n well cadw data wrth gefn oddi ar y safle.
   (a) Rhowch un rheswm pam y dylai data wrth gefn gael eu cadw oddi ar y safle. (1 marc)
   (b) Disgrifiwch un dull o gadw data wrth gefn oddi ar y safle. (2 farc)

Yr heddlu'n mynd i drafferth

Roedd gan gwmni olew ymgyrch hyrwyddo i annog gyrwyr i brynu eu brand nhw o betrol yn eu gorsafoedd petrol nhw yn hytrach na brandiau eraill. I'r perwyl hwn, rhoddodd y cwmni restr o rifau cofrestru ceir ym mhob gorsaf betrol a phe baech chi'n berchen ar un o'r rhifau hyn byddech chi'n ennill £1000. Daeth plismyn i'r orsaf betrol a sylwi ar y rhestr o rifau. Roeddynt yn sylweddoli y gallant deipio'r rhifau i mewn i Gyfrifiadur Cenedlaethol yr Heddlu (PNC) a chael cyfeiriadau'r perchenogion. Fe wnaethant hynny ac yna buont yn ffonio'r bobl ar y rhestr gan ddweud y gallent roi gwybod iddynt, am swm o arian, sut i gael eu dwylo ar £1000. Cytunodd y rhan fwyaf o bobl wrth gwrs. Ond cafodd y plismyn eu dal yn y diwedd a'u disgyblu'n llym. Yn dilyn ymchwiliad, darganfuwyd bod Cyfrifiadur Cenedlaethol yr Heddlu yn cael ei gamddefnyddio'n helaeth a bod llawer o gwmnïau preifat yn talu i gael y wybodaeth oedd wedi'i storio arno. Roedd yn rhaid i hyn ddod i ben, ac yn awr, er mwyn gallu holi'r gronfa ddata, rhaid i blismyn ddangos pwy ydynt a rhoi rheswm pam y maen nhw eisiau'r wybodaeth. Yn ogystal, mae'n bosibl ymgymryd ag archwiliad (audit) o unrhyw chwiliad penodol i sicrhau bod angen ei wneud a'i fod yn gyfreithlon.

1 Caiff rhwydweithiau mawr fel Cyfrifiadur Cenedlaethol yr Heddlu eu defnyddio gan filoedd o ddefnyddwyr bob dydd.
   (a) Eglurwch sut y gall rhifau adnabod defnyddwyr a chyfrineiriau gael eu defnyddio i gyfyngu mynediad i rai staff yn unig. (2 farc)
   (b) Mae gan Gyfrifiadur Cenedlaethol yr Heddlu gyfres o hawliau mynediad i'r bobl hynny sydd wedi'u hawdurdodi i'w ddefnyddio. Eglurwch ystyr 'hawliau mynediad' a rhowch enghraifft o sut y gallent gael eu defnyddio gyda'r system hon. (4 marc)
   (c) Defnyddir meddalwedd archwilio yn aml gyda Chyfrifiadur Cenedlaethol yr Heddlu. Eglurwch pam mae'r meddalwedd hwn yn ddefnyddiol a sut y gall rwystro'r math o fethiant diogelwch sy'n cael ei ddisgrifio yn yr astudiaeth achos. (2 farc)

2 Mae camddefnyddio adnoddau rhwydwaith yn broblem i bob corff sy'n defnyddio rhwydwaith.
   (a) Heblaw am fynediad heb ei awdurdodi i ddata personol, disgrifiwch bedair ffordd y gall rhwydwaith gael ei gamddefnyddio. (4 marc)
   (b) Eglurwch ddau beth y gallai'r heddlu eu gwneud i atal ei staff ei hun rhag camddefnyddio Cyfrifiadur Cenedlaethol yr Heddlu. (2 farc)

Astudiaethau achos *(parhad)*

▶ Astudiaeth achos 4   tt. 76–77

Ymosodiad gan derfysgwyr

Ar ddydd Sadwrn yn ystod mis Mehefin 1996 derbyniodd yr heddlu rybudd wedi'i godio fod bom wedi cael ei blannu yng Nghanolfan Arndale, prif ganolfan siopa Manceinion. Ychydig dros awr ar ôl i'r ganolfan a'r ardal o'i chwmpas gael ei chlirio fe ffrwydrodd y bom, gan anafu dros ddau gant o bobl a dinistrio rhan helaeth o'r ganolfan. Achosodd y bom, y mwyaf o'i fath ym Mhrydain ar adeg o heddwch, ddifrod helaeth i swyddfeydd cwmni yswiriant Royal and Sun Alliance lle y cafodd rhai o'r staff eu niweidio.

    Gan fod prif gyfrifiadur y cwmni yn yr adeilad hwn, yr oedd ofn y byddai gweithgareddau pob dydd y cwmni yn dioddef yn enbyd. Roedd gan y cwmni swyddfa yn Lerpwl, ac roedd terfynellau yma a oedd wedi'u cysylltu â'r prif gyfrifiadur. Darganfu'r staff fod peth bywyd o hyd yn y system er bod difrod helaeth i'r adeilad lle'r oedd y prif gyfrifiadur. Roedd peth gobaith y gellid adfer y system, ond oherwydd bod perygl o ffrwydradau gan fod y bibell nwy wedi'i thorri, fe dorrodd y dynion tân y cyflenwad trydan i ffwrdd. Roedd y rhan fwyaf o'r caledwedd wedi cael ei ddinistrio i bob pwrpas yn y ffrwydrad, ond, fel pob cwmni doeth, roedd gan hwn gynllun wrth gefn. Roedd ganddo gytundeb gyda chwmni adfer data arbenigol a oedd yn berchen ar galedwedd cyffelyb a chopïau o'r un math o feddalwedd ag a oedd yn cael ei ddefnyddio gan Royal and Sun Alliance. Oherwydd bod angen staff arnynt a oedd yn deall y busnes yswiriant i weithredu'r cyfrifiadur, cafodd staff y Royal eu cludo i swyddfeydd y cwmni adfer ac aethant ati i adfer y data o'r cyfryngau storio wrth gefn a oedd yn cael eu cadw oddi ar y safle. Erbyn bore Llun roedd yr holl ddata wedi cael eu hadfer ac roedd switsfwrdd dros dro wedi'i sefydlu, ac nid oedd yr un diwrnod o fasnachu wedi cael ei golli.

1   Mae gan gwmni yswiriant Royal and Sun Alliance bolisi diogelwch TGCh. Fel rhan o'r polisi byddent wedi gwneud dadansoddiad risg ac wedi sefydlu rhaglen adfer o drychineb.
   (a) Eglurwch beth sy'n cael ei wneud mewn dadansoddiad risg. (2 farc)
   (b) Disgrifiwch y technegau a ddefnyddir mewn rhaglen adfer o drychineb i adfer y data a ddefnyddir gan y systemau TGCh. (6 marc)

2   Mae gweithdrefnau gwneud copïau wrth gefn a gweithdrefnau adfer yn hanfodol i sicrhau diogelwch systemau TGCh.
   (a) Eglurwch y gwahaniaeth rhwng gweithdrefnau gwneud copïau wrth gefn a gweithdrefnau adfer. (2 farc)
   (b) Rhowch **dri** pheth y mae'n rhaid eu hystyried wrth ddewis gweithdrefnau gwneud copïau wrth gefn. (3 marc)
   (c) Eglurwch pam mae gweithdrefnau adfer yn hanfodol os yw corff fel y cwmni yswiriant hwn yn dymuno sicrhau diogelwch ei ddata. (2 farc)

Cwestiynau a Gweithgareddau

▶ **Cwestiynau 1** | tt. 62–73

1 Beth y gall cyrff ei wneud i leihau'r difrod sy'n cael ei achosi gan:
   (a) doriad trydan (2 farc)
   (b) ymosodiad gan firws (2 farc)
   (c) gweithiwr dibrofiad yn dileu'r data ar gyfryngau magnetig yn ddamweiniol (2 farc)
   (ch) hacwyr yn newid data pwysig yn fwriadol (2 farc)
   (d) aelodau staff allweddol yn gadael gyda holl arbenigedd TG a gwybodaeth y corff (2 farc)
   (dd) colli'r llinellau cyfathrebu mewn Rhwydwaith Ardal Eang (RhAE/*WAN – Wide Area Network*) (2 farc)
   (e) colli mynediad i swyddfeydd y corff o ganlyniad i dân mewn adeilad gerllaw (2 farc)
   (f) aelod pwdlyd o'r staff yn dinistrio rhaglenni'r cwmni yn fwriadol. (2 farc)

2 Mae rheolwr TGCh yn poeni bod rhai aelodau staff yn camddefnyddio'r cyfleusterau TGCh.
   (a) Rhowch enwau **dri** math o gamddefnyddio data. (3 marc)
   (b) Eglurwch sut y gallech ganfod y mathau o gamddefnydd a nodwyd gennych yn rhan (a). (3 marc)
   (c) Mae'r rhan fwyaf o gyrff yn defnyddio 'polisi diogelwch TGCh'. Disgrifiwch fanteision cael polisi o'r fath. (3 marc)

3 Mae rhai cymwysiadau TGCh yn defnyddio meddalwedd sy'n cynnal trywydd archwilio.
   Enwch **un** cymhwysiad o'r fath a nodwch pam mae'r cyfleuster hwn yn angenrheidiol. (4 marc)

4 'Mae systemau TGCh yn hanfodol i gorff; gallai canlyniadau methiant fod yn drychinebus.'
   Trafodwch y gosodiad hwn, gan gynnwys y canlynol yn eich trafodaeth:
   • y bygythiadau posibl i'r system
   • y cysyniad o ddadansoddi risg
   • canlyniadau corfforaethol methiant y system

   • y ffactorau y dylid eu hystyried wrth lunio'r 'rhaglen adfer o drychineb' i'w gwneud hi'n bosibl i adfer o drychineb. (8 marc)

5 Mae systemau TGCh yn wynebu llawer o fygythiadau.
   (a) Nodwch **bum** bygythiad i system TGCh ac ar gyfer pob un eglurwch sut y gellir lleihau'r risg y bydd yn cael ei wireddu. (5 marc)
   (b) Pe bai'r bygythiadau yr ydych wedi'u crybwyll yn rhan (a) yn cael eu gwireddu, gallai'r canlyniadau i'r corff fod yn niferus.
       Disgrifiwch **dri** chanlyniad o'r fath i gorff. (3 marc)

6 Bydd rhan o bolisi diogelwch TGCh yn ymwneud â gweithdrefnau gwneud copïau wrth gefn a gweithdrefnau adfer.
   (a) Trafodwch y dulliau a ddefnyddir gan gyrff i wneud copïau wrth gefn o'u rhaglenni a'u data. (6 marc)
   (b) Eglurwch sut y gall corff sicrhau bod modd adfer y data gwreiddiol os caiff caledwedd, meddalwedd a data eu colli. (4 marc)

7 Bydd y mwyafrif o gyrff yn defnyddio cod ymddygiad staff i leihau'r risg oddi wrth eu staff eu hunain.
   (a) Disgrifiwch **ddwy** broblem sy'n cael eu hachosi gan staff sy'n camddefnyddio systemau TGCh. (2 farc)
   (b) Disgrifiwch sut y gall cod ymddygiad helpu i leihau camddefnydd bwriadol a damweiniol gan staff. (3 marc)

8 Dylai pob corff TGCh feddu ar bolisi diogelwch TGCh.
   (a) Trafodwch y ffactorau y mae angen eu cymryd i ystyriaeth wrth lunio polisi diogelwch TGCh er mwyn ymdrin â'r canlynol:
       (i) camddefnydd damweiniol
       (ii) camddefnydd bwriadol. (4 marc)
   (b) Eglurwch ystyr dadansoddiad risg a rhowch enghraifft o sut y byddai corff yn mynd ati i wneud un. (4 marc)

▶ Cwestiynau 2 | tt. 74–77

1 Mae tân yn fygythiad mawr i unrhyw system TGCh. Nodwch dri pheth y gellir eu gwneud i leihau'r risg i system TGCh oddi wrth dân. (3 marc)

2 Gall y bobl sy'n gweithio i gorff fod yn gyfrifol am golli neu ddifrodi ei ddata. Gall hyn gael ei wneud yn faleisus neu'n ddamweiniol. Bydd cyrff yn gwneud copïau wrth gefn rheolaidd fel y gellir adfer data sy'n cael eu colli neu eu difrodi ond byddant hefyd yn ceisio atal gweithwyr rhag ymyrryd â'u data.

(a) Disgrifiwch **ddau** fesur y gellid eu hymgorffori yn y caledwedd i atal colli neu ddifrodi data. (4 marc)

(b) Disgrifiwch **ddau** fesur meddalwedd y gellid eu defnyddio i atal colli neu ddifrodi data. (4 marc)

(c) Disgrifiwch **ddwy** weithdrefn y gallai'r corff eu mabwysiadu i atal colli neu ddifrodi data. (4 marc)

3 Bydd mynediad i adnoddau rhwydwaith yn cael ei reoli'n aml gan gyfrineiriau a gwahanol lefelau o fynediad a ganiateir ar gyfer defnyddwyr.

(a) Eglurwch beth yw cyfrinair. (2 farc)

(b) Eglurwch ystyr y term 'lefelau o fynediad a ganiateir ar gyfer defnyddwyr'. (2 farc)

4 Mae llawer o fygythiadau posibl i systemau TGCh, felly rhoddir nifer o fesurau diogelwch ar waith. Disgrifiwch y mathau canlynol o ddiogelwch, gan roi enghraifft berthnasol ym mhob achos.

(a) Diogelwch corfforol (1 marc)

(b) Diogelwch personél (1 marc)

(c) Diogelwch meddalwedd/rhesymegol (1 marc)

(ch) Diogelwch cyfathrebiadau (1 marc)

(d) Diogelwch dogfennau. (1 marc)

▶ Gweithgaredd 1: Cynhyrchu cyflwyniad ar wneud copïau wrth gefn

Mae gofyn i chi ddefnyddio meddalwedd cyflwyno i wneud cyflwyniad ar bwysigrwydd gwneud copïau wrth gefn a'r gwahanol ddulliau sydd ar gael.

Mae gan y cwmni yr ydych chi'n cyflwyno'r wybodaeth iddo agwedd rywsut-rywsut at gadw copïau wrth gefn ar hyn o bryd, a bydd yn rhaid i chi ei argyhoeddi bod yr amser a'r ymdrech sydd ynghlwm wrth wneud copïau yn hollol angenrheidiol.

Bydd y cyflwyniad yn para am bum munud ac yn ystod yr amser hwn bydd angen i chi ymdrin â'r canlynol:

- Y mathau o fygythiad i'r data.
- Canlyniadau colli data i'r cwmni.
- Yr angen i wneud copïau rheolaidd.
- Yr angen i roi cyfrifoldeb i rywun dros wneud copïau.
- Y mathau o systemau gwneud copïau wrth gefn sydd ar gael.

Dyma'r wybodaeth sydd gennych am y ffordd y mae'r cwmni'n ymdrin â gwneud copïau wrth gefn ar hyn o bryd.

- Rhaid i rywun gofio gwneud y copi a mynd ag ef oddi ar y safle.
- Yn aml nid yw copïau'n cael eu cludo oddi ar y safle.
- Mae pawb yn meddwl mai gwaith rhywun arall yw gwneud y copïau.
- Dydy copïau byth yn cael eu profi, felly does neb yn gwybod a fyddai'n bosibl adfer y data sydd ynddynt.

▶ Gweithgaredd 2: Paratoi polisi diogelwch TGCh ar gyfer ysgol neu goleg

Gan ddefnyddio eich gwybodaeth am y topig hwn, cynhyrchwch bolisi diogelwch TGCh ar gyfer ysgol neu goleg. Dylai'r polisi hwn ymdrin â'r defnydd sy'n cael ei wneud gan y staff gweinyddol o'r systemau TGCh yn ogystal â'r defnydd sy'n cael ei wneud gan athrawon/darlithwyr a myfyrwyr/disgyblion o'r cyfleusterau TGCh.

Wrth gynhyrchu'r ddogfen hon dylech gofio'r canlynol:

- Mae colegau ac ysgolion yn cadw llawer o ddata personol (cofnodion myfyrwyr, cofnodion personél, manylion meddygol, canlyniadau arholiadau, tystlythyrau, ac ati).
- Mae gan bob ysgol a choleg gysylltiad â'r Rhyngrwyd ac mae hyn yn codi amrywiaeth o gwestiynau diogelwch y bydd angen i chi ymdrin â nhw.
- Bydd angen i chi benderfynu pwy fydd yn gweithredu'r rheolau y byddwch chi'n gorfod eu llunio mewn perthynas â defnyddio cyfarpar TGCh.
- Bydd llawer o fyfyrwyr yn chwarae gemau ar systemau'r ysgol/coleg pan ddylent fod yn gweithio.
- Gan fod cymaint o fyfyrwyr yn dod â gwaith ar gofion pin i'r ysgol/coleg mae firysau'n broblem ddifrifol.

Cofiwch at bwy yr ydych chi'n anelu'r ddogfen. Ni fydd pawb yn deall y termau technegol a ddefnyddir, ac mae'n bosibl na fyddant mor gyfarwydd â chyfrifiaduron ag yr ydych chi'n meddwl.

Cymorth gyda'r arholiad

Enghraifft 1

1 Dylai fod gan bob corff bolisi diogelwch TGCh. Mae gofyn i chi gynhyrchu polisi diogelwch TGCh ar gyfer busnes. Trafodwch y dulliau y gallai'r busnes eu mabwysiadu i osgoi camddefnyddio cyfarpar TGCh (a) yn ddamweiniol a (b) yn fwriadol.

Yn eich atebion i ran (a) dylech gyfeirio at weithdrefnau gwneud copïau wrth gefn a gweithdrefnau adfer, ac yn eich atebion i ran (b) dylech ystyried dulliau o ddiogelu'r data sy'n cael eu trawsyrru. (8 marc)

Ateb myfyriwr 1

1 (a) Er gwaethaf popeth, bydd damweiniau'n digwydd, felly mae'n well ystyried ffyrdd o adfer y systemau TGCh pe bai'r camddefnydd yn achosi difrod.

Gallech hyfforddi'r defnyddwyr fel nad ydynt yn dileu ffeiliau'n ddamweiniol neu'n copïo fersiwn hŷn o ffeil ar ben ffeil newydd.

Dylid gwneud copïau wrth gefn o ddata ar gyfryngau symudadwy (CD-RW, tapiau magnetig, gyriannau disg caled cludadwy) a dylid cadw'r rhain oddi ar y safle. Wedyn, os caiff y data eu dinistrio gellir eu hadfer.

(b) Dylai defnyddwyr allgofnodi os byddant i ffwrdd o'u desgiau i sicrhau na all pobl eraill ddefnyddio eu cyfrifiadur i gyrchu ffeiliau nad oes ganddynt hawl i'w gweld. Gellir defnyddio cloeon allwedd i rwystro mynediad heb ei awdurdodi i rwydweithiau.

Mae rhifau adnabod defnyddwyr a chyfrineiriau yn sicrhau rheolaeth dros fynediad i ffeiliau ac mae lefelau o fynediad yn pennu pa ffeiliau y gall defnyddiwr eu gweld a'r gweithrediadau ffeil y gallant eu gwneud (e.e. atodi, creu cofnodion newydd, darllen yn unig, ac ati).

Pan anfonir manylion ariannol dros rwydweithiau, gellir amgryptio'r data fel na fydd unrhyw un sy'n eu rhyng-gipio yn gallu eu deall.

Ateb myfyriwr 2

1 (a) Mae camddefnydd damweiniol yn golygu gwneud camgymeriadau drwy ddileu data, colli gwaith, difrodi cyfrifiaduron, anghofio gwneud copïau wrth gefn, colli coffi ar eich cyfrifiadur ac ati. Mae'n anodd atal y rhain rhag digwydd, felly rhaid i staff fod yn ofalus.

(b) Mae lawrlwytho rhaglenni yn anghyfreithlon yn gallu arwain at erlyn y cwmni. Felly dylai fod gan y busnes god ymddygiad i'r staff sy'n gwahardd hyn.

Mae gadael data personol ar eich cyfrifiadur pan ewch am ginio yn beth drwg, oherwydd gall pobl eraill eu gweld a gallent ddefnyddio'r cyfrifiadur i gyrchu gwybodaeth bersonol arall.

Pan gaiff data eu trawsyrru mae angen eu diogelu rhag hacwyr. Mae'n anodd gwarchod signalau radio mewn systemau diwifr. Yr unig ffordd y gallwch wneud hyn yw drwy amgodio.

Sylwadau'r arholwr

1 (a) Roedd angen i'r myfyriwr sylweddoli bod y cwestiwn hwn yn ymwneud ag eitemau a ddylai fod yn y polisi diogelwch TGCh.

Mae'n gwneud y pwynt pwysig ei bod yn anodd diogelu rhag camddefnydd damweiniol ac mai'r polisi gorau yw ystyried sut i adfer data a gollir.

Mae'r rhan am hyfforddi yn ennill marc a hefyd y ddau bwynt am wneud copïau wrth gefn (un am grybwyll cyfryngau nodweddiadol ac un am eu cadw oddi ar y safle).

Gallai'r myfyriwr fod wedi sôn am wneud copïau wrth gefn yn rheolaidd a sut y dylid eu profi i sicrhau y gall y data gwreiddiol gael eu hadfer. (3 marc allan o 4)

(b) Mae'r myfyriwr wedi gwneud rhai pwyntiau da yma ac wedi cael marciau llawn am yr adran hon.

(4 marc allan o 4)

(Cyfanswm o 7 marc allan o 8)

Sylwadau'r arholwr

1 (a) Nid yw'r myfyriwr wedi darllen y cwestiwn yn iawn gan ei fod wedi cynhyrchu ateb sy'n sôn am gamddefnydd damweiniol mewn ffordd gyffredinol. Roedd angen i'r ateb drafod yr hyn a ddylai gael ei gynnwys yn y polisi diogelwch TGCh i ymdrin â chamddefnydd damweiniol. Ni ellir rhoi unrhyw farciau am yr adran hon.

(0 marc allan o 4)

(b) Mae crybwyll cod ymddygiad i'r staff yn ennill 1 marc.

Nid yw'r darn am adael data personol ar y sgrin yn ennill marc gan nad oes unrhyw sôn am allgofnodi.

Mae'r ateb olaf yn sôn am 'amgodio' yn lle 'amgryptio', felly dim marciau am yr ateb hwn. (1 marc allan o 4)

(Cyfanswm o 1 marc allan o 8)

Cymorth gyda'r arholiad *(parhad)*

Ateb yr arholwr

1 (a) Un marc yr un hyd at uchafswm o bedwar am bedwar pwynt, e.e.:

Yr angen i wneud copïau wrth gefn ar gyfryngau symudadwy neu i drosglwyddo data i gyfrifiadur arall

Eu symud i leoliad arall oddi ar y safle

Cael amserlen ar gyfer gwneud copïau wrth gefn

Rhoi prawf ar y drefn gwneud copïau i sicrhau y gellir defnyddio'r copïau i adfer data.

  (b) Un marc yr un hyd at uchafswm o bedwar am bedwar pwynt, e.e.:

Peidio â gadael cyfrifiaduron wedi'u mewngofnodi os ydych chi'n gadael y gweithfan/allgofnodi'n awtomatig ar ôl i'r cyfrifiadur fod yn segur am gyfnod penodol

Defnyddio rhifau adnabod defnyddwyr a chyfrineiriau i ddilysu defnyddwyr

Defnyddio amgryptio i sicrhau na all hacwyr ryng-gipio data

Sefydlu gweithdrefnau archwilio i ganfod camddefnydd bwriadol.

Enghraifft 2

2 Mae myfyriwr yn gweithio ar broject TGCh yn yr ysgol gan ddefnyddio cyfrifiaduron yr ysgol. Mae'n awyddus i barhau i weithio ar y project gartref ar ei liniadur.

(a) Disgrifiwch weithdrefn gwneud copïau wrth gefn y gallai ef ei defnyddio. (4 marc)

(b) Mae'r bygythiadau i liniaduron yn aml yn fwy na'r bygythiadau i gyfrifiaduron bwrdd gwaith. Rhowch **un rheswm** pam. (1 marc)

Ateb myfyriwr 1

2 (a) Gellir storio'r gwaith ar ddisg a chadw'r disg mewn lle diogel.

  (b) Mae gliniaduron yn ysgafnach ac yn hawdd eu cuddio/dwyn.

Sylwadau'r arholwr

2 Yn rhan (a) nid oes unrhyw sôn am y math o ddisg na bod angen tynnu'r disg o'r cyfrifiadur, felly dim marc. Mae'n sôn am gadw'r disg mewn lle diogel ond nid i ffwrdd o'r cyfrifiadur, ond gellir rhoi un marc.

Roedd angen i'r myfyriwr edrych ar y marciau a oedd ar gael. Mae 4 marc fel rheol yn golygu bod angen 4 pwynt dilys, neu efallai 3 gyda marc ychwanegol am eglurder yr esboniad.

Rhan (b): mae'n wir bod gliniaduron yn ysgafnach, ond nid yw'r myfyriwr wedi dweud yn benodol pam y mae hyn yn berthnasol. **(1 marc allan o 5)**

Ateb myfyriwr 2

2 (a) Anfon e-bost i'w gyfeiriad e-bost cartref a phori am y ffeiliau mae'n gweithio arnynt a'u hatodi wrth yr e-bost. Mae hyn yn golygu y bydd y data'n cael eu hanfon oddi ar y safle sy'n dda am ddiogelwch, a gall y ffeiliau gael eu llwytho i lawr gartref. Mantais hyn yw nad oes unrhyw gyfryngau i'w colli, er enghraifft, cofion pin, disgiau *CD-RW* ac ati. Gall y myfyriwr hefyd gyrchu'r ffeil o unrhyw gyfrifiadur sydd wedi'i gysylltu â'r Rhyngrwyd.

(b) Mae gliniaduron yn gludadwy ac, os ydych chi'n cario un yn y stryd mewn cas, mae'n amlwg bod gennych chi un yn y cas a gallai rhywun ymosod arnoch chi a dwyn y gliniadur.

Sylwadau'r arholwr

2 (a) Yma mae'r myfyriwr wedi ysgrifennu pedair brawddeg ac mae pob un yn berthnasol i'r cwestiwn ac yn ychwanegu mwy o wybodaeth am y dull gwneud copïau dan sylw a'r gweithdrefnau i'w dilyn. Marciau llawn am y rhan hon o'r cwestiwn.

(b) Yma mae'r myfyriwr wedi nodi'r bygythiad, sef dwyn, ac wedi rhoi rheswm pam mae gliniadur yn fwy tebygol o gael ei ddwyn. Marciau llawn am y rhan hon o'r cwestiwn. **(5 marc allan o 5)**

Atebion yr arholwr

2 (a) Un marc hyd at gyfanswm o bedwar am bob un o'r pwyntiau canlynol:
- Gwneud copi o'r ffolder/ffeiliau.
- Storio'r copi ar gyfryngau symudadwy fel disg hyblyg magnetig, disg caled magnetig allanol, storio ar y gweinydd os gall y myfyrwyr gyrchu'r gweinydd o'r Rhyngrwyd.
- Atodi'r ffeil wrth e-bost ac anfon yr e-bost i'ch cyfrif e-bost cartref/ysgol.
- Storio'r data ar gof pin symudadwy.
- Storio'r copi ar ddisg optegol megis *DVD-RW, CD-RW*.
- Dylai ffeiliau gael eu copïo'n rheolaidd (e.e. ar ddiwedd sesiwn).
- Mae angen cadw'r ffeiliau mewn lle diogel i ffwrdd o'r cyfrifiadur.
- Mae rhoi prawf ar y copïau wrth gefn yn hanfodol i sicrhau y gellir adfer y ffeiliau.

(b) Un marc am un rheswm megis:
- Mae gliniaduron yn llai ac felly'n haws eu dwyn.
- Defnyddir gliniaduron mewn lleoedd cyhoeddus yn aml, felly maen nhw'n fwy tebygol o gael eu dwyn.
- Gall cysylltiadau diwifr â rhwydweithiau mewn lleoedd cyhoeddus gyflwyno firysau.
- Mae gliniaduron yn fwy tebygol o gael eu difrodi (e.e. cael eu gollwng, colli hylif arnynt).

Cymorth gyda'r arholiad *(parhad)*

Enghraifft 3

3 Dylai fod gan bob corff weithdrefnau gwneud copïau wrth gefn er mwyn iddo allu adfer ei ddata os cânt eu colli o ganlyniad i fethiant diogelwch. Rhowch **bum** eitem, ynghyd â rhesymau, y byddai angen eu hystyried wrth benderfynu ar weithdrefnau gwneud copïau wrth gefn. **(10 marc)**

Ateb myfyriwr 1

3 Muriau gwarchod i rwystro hacwyr rhag hacio i mewn i systemau TGCh y cwmni.

Hyfforddiant i sicrhau nad yw defnyddwyr yn gwneud camgymeriadau ac yn dileu data'n ddamweiniol.

Y math o ddyfais storio fel cof pin neu ddisg magnetig symudadwy y rhoddir y copïau wrth gefn arni.

Ble i roi'r copïau wrth gefn. Mae'n well eu storio i ffwrdd o'r data gwreiddiol, er enghraifft, oddi ar y safle.

Faint o ddata sydd i'w storio – os oes llawer iawn o ddata byddai'n well eu storio ar dâp magnetig oherwydd bod gan hwn gynhwysedd storio mawr iawn.

Sylwadau'r arholwr

3 Mae'r ddau ateb cyntaf yn dangos bod y myfyriwr wedi gweld y geiriau 'methiant diogelwch' yn y cwestiwn ac wedi dechrau ysgrifennu am hynny. Mae'n gyffredin iawn i fyfyrwyr ateb y cwestiwn yr hoffent ei gael. Dim marciau am yr atebion hyn.

Mae'r tri ateb nesaf yn dda. Mae'r myfyriwr wedi dechrau ateb y cwestiwn. **(6 marc allan o 10)**

Ateb myfyriwr 2

3 Pa mor aml i wneud copïau wrth gefn. Er enghraifft, gyda system ar-lein lle mae'r data'n newid fesul munud rhaid gwneud copïau drwy'r adeg. Gyda systemau eraill, gallai unwaith y dydd fod yn ddigon.

Ble i gadw'r copïau wrth gefn – oddi ar y safle sydd orau, rhag ofn bod y copïau'n cael eu dinistrio'r un pryd â'r data gwreiddiol.

Pa gyfrwng i'w ddefnyddio i storio'r copïau wrth gefn. Mae hyn yn dibynnu ar faint o ddata sydd a pha mor aml y mae'n newid. Mae cofion pin yn dda am gadw symiau bach ac mae tâp magnetig yn dda am gopïo llawer iawn o ddata.

Pa fath o gopïau wrth gefn i'w gwneud – er enghraifft, mae yna gopïau wrth gefn cynyddol sydd ond yn gwneud copïau o'r data sydd wedi newid ers gwneud y copi diwethaf.

Pwy sy'n gwneud y copïau wrth gefn? Mae'n well gwneud un person yn gyfrifol am wneud copïau wrth gefn.

Sylwadau'r arholwr

3 Ateb da iawn yw hwn ac mae'n haeddu marciau llawn. **(10 marc allan o 10)**

Atebion yr arholwr

3 Un marc am enw'r eitem ac un marc am ymhelaethu/egluro ymhellach.

Y math o gopi wrth gefn y dylid ei wneud (1) e.e. llawn, cynyddol neu wahaniaethol (1).

Pa mor aml y dylid gwneud y copïau wrth gefn (1), e.e. yn barhaus, bob awr, bob dydd, ac ati (1).

Y cyfrwng/dyfeisiau gwneud copïau wrth gefn a ddefnyddir (1) e.e. disg magnetig, tâp magnetig, ac ati (1).

Ble mae'r copïau wrth gefn i gael eu cadw/storio (1) – oddi ar y safle, mewn diogell wrthdan, eu trosglwyddo gan ddefnyddio'r Rhyngrwyd, ac ati (1).

Pwy a ddylai fod yn gyfrifol am wneud a storio copïau wrth gefn (1) – fel bod blaenoriaeth uchel yn cael ei rhoi i gopïau wrth gefn (1).

Wrth ateb y math hwn o gwestiwn mae'n bwysig ystyried sut y gall y marciau gael eu dyrannu. Bydd gan yr arholwr restr o bwyntiau y mae angen eu gwneud a rhaid i chi ragweld beth y gallant fod. Peidiwch â gwastraffu amser yn ysgrifennu brawddegau nad ydynt yn ychwanegu dim at yr ateb neu sy'n ailadrodd gwybodaeth a roddir yn y cwestiwn.

Mapiau meddwl cryno

Bygythiadau posibl i systemau TGCh

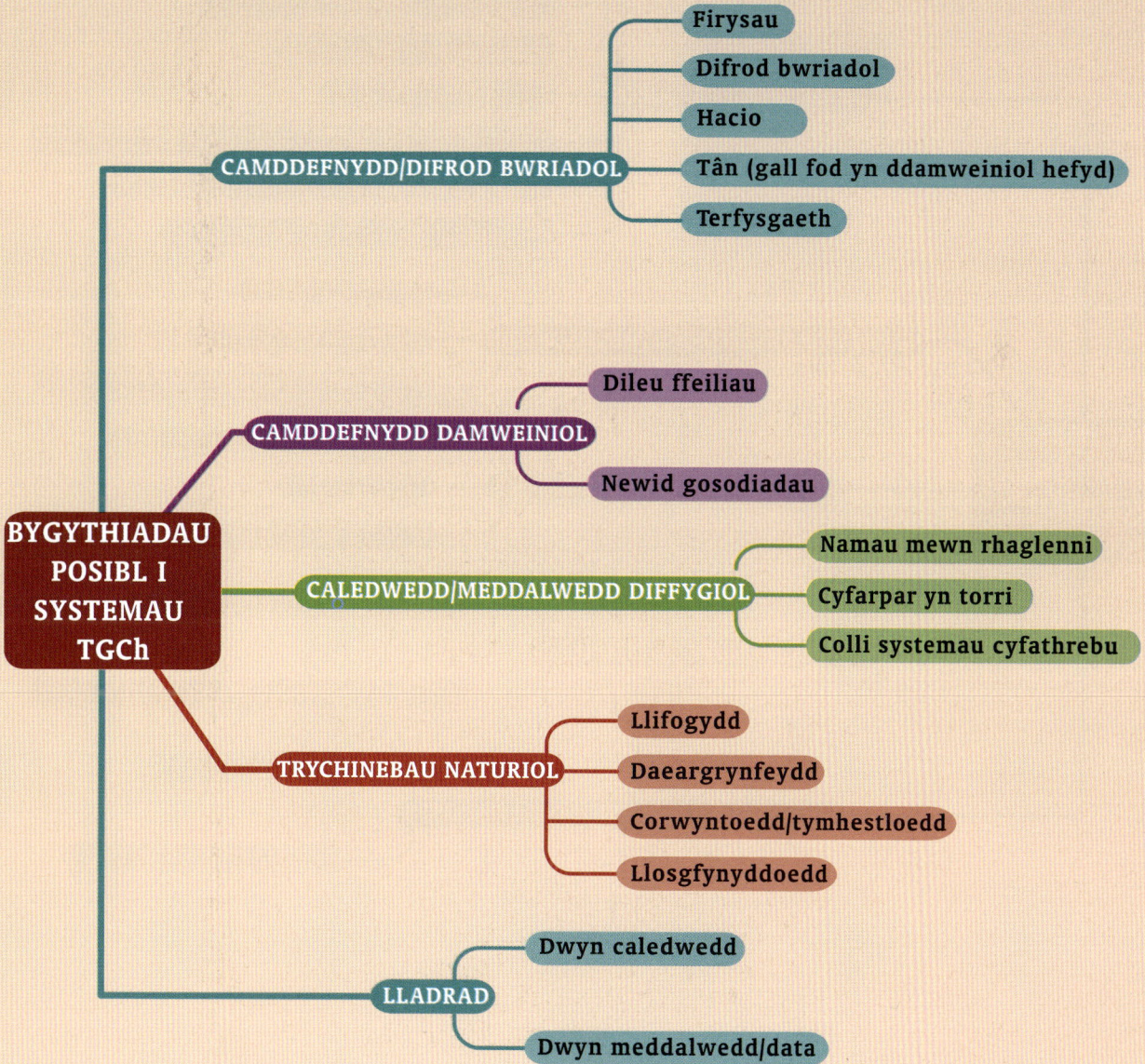

```
BYGYTHIADAU
POSIBL I
SYSTEMAU
TGCh
```

- CAMDDEFNYDD/DIFROD BWRIADOL
  - Firysau
  - Difrod bwriadol
  - Hacio
  - Tân (gall fod yn ddamweiniol hefyd)
  - Terfysgaeth

- CAMDDEFNYDD DAMWEINIOL
  - Dileu ffeiliau
  - Newid gosodiadau

- CALEDWEDD/MEDDALWEDD DIFFYGIOL
  - Namau mewn rhaglenni
  - Cyfarpar yn torri
  - Colli systemau cyfathrebu

- TRYCHINEBAU NATURIOL
  - Llifogydd
  - Daeargrynfeydd
  - Corwyntoedd/tymhestloedd
  - Llosgfynyddoedd

- LLADRAD
  - Dwyn caledwedd
  - Dwyn meddalwedd/data

Mapiau meddwl cryno *(parhad)*

Ffactorau wrth lunio polisïau diogelwch

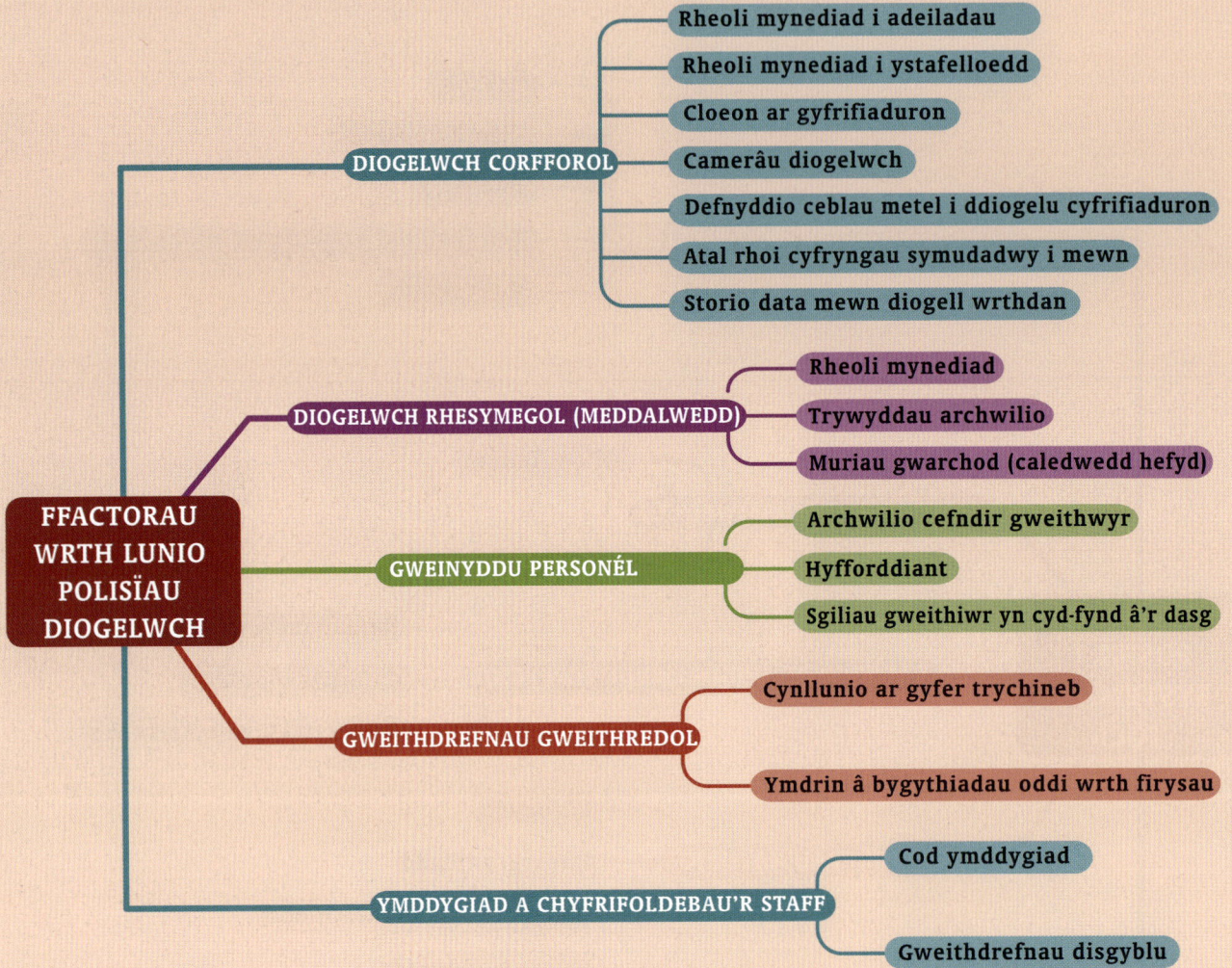

FFACTORAU WRTH LUNIO POLISÏAU DIOGELWCH

DIOGELWCH CORFFOROL
- Rheoli mynediad i adeiladau
- Rheoli mynediad i ystafelloedd
- Cloeon ar gyfrifiaduron
- Camerâu diogelwch
- Defnyddio ceblau metel i ddiogelu cyfrifiaduron
- Atal rhoi cyfryngau symudadwy i mewn
- Storio data mewn diogell wrthdan

DIOGELWCH RHESYMEGOL (MEDDALWEDD)
- Rheoli mynediad
- Trywyddau archwilio
- Muriau gwarchod (caledwedd hefyd)

GWEINYDDU PERSONÉL
- Archwilio cefndir gweithwyr
- Hyfforddiant
- Sgiliau gweithiwr yn cyd-fynd â'r dasg

GWEITHDREFNAU GWEITHREDOL
- Cynllunio ar gyfer trychineb
- Ymdrin â bygythiadau oddi wrth firysau

YMDDYGIAD A CHYFRIFOLDEBAU'R STAFF
- Cod ymddygiad
- Gweithdrefnau disgyblu

TOPIG 6: Systemau cronfa ddata

Rhaid i bob corff reoli'r data mae'n eu cadw a'r ffordd arferol o wneud hyn yw drwy ddefnyddio cronfa ddata sy'n cael ei chreu gyda meddalwedd cronfa ddata. Mae meddalwedd cronfa ddata yn caniatáu i ddata gael eu mewnbynnu a'u storio mewn ffordd strwythuredig sy'n hwyluso adalw'r data. Gall cronfeydd data fod yn ffeiliau fflat, sydd â defnydd cyfyngedig iawn, neu'n berthynol, sy'n fwy hyblyg ond sy'n gofyn am wybodaeth arbenigol i'w sefydlu.

Yn y topig hwn byddwch yn dysgu am nodweddion cronfeydd data perthynol sy'n eu gwneud yn ffordd nerthol o storio ac adalw data. Byddwch yn dysgu am gydrannau cronfeydd data perthynol a'r ffordd fwyaf effeithlon o roi data mewn cronfa ddata berthynol.

▼ Y cysyniadau allweddol sy'n cael sylw yn y topig hwn yw:

▶ Cronfeydd data

▶ Cronfeydd data gwasgaredig

CYNNWYS

Cronfeydd data

▼ Byddwch yn dysgu

▶ Am y cysyniad o ffeil fflat

▶ Am broblemau storio data mewn ffeiliau fflat

▶ Am y gwahaniaethau rhwng ffeiliau fflat a chronfeydd data perthynol

▶ Am drefniadaeth cronfeydd data perthynol

Cyflwyniad

Daethoch ar draws y termau ffeil fflat a chronfa ddata berthynol yn Nhopig 11 o'r cwrs UG. Byddwn yn mynd dros beth o'r deunydd yma eto er mwyn darparu ymdriniaeth lawn a chynhwysfawr o'r pwnc. Yn yr adran hon byddwch yn dysgu pam mae cronfeydd data perthynol yn llawer mwy defnyddiol na ffeiliau fflat a byddwch hefyd yn dysgu am drefniadaeth cronfeydd data perthynol.

Ffeiliau fflat

Y system storio symlaf yw'r ffeil fflat. Mae ffeiliau fflat yn cynnwys un tabl o ddata'n unig, felly nid ydynt yn addas ond ar gyfer systemau storio ac adalw data syml. Gyda ffeil fflat, byddai'r meddalwedd rhaglenni yn cyrchu'r set o ddata sy'n cael ei chadw yn y tabl. Byddai hyn yn golygu, e.e., y byddai system gyflogau yn cyrchu'r ffeil sy'n cynnwys yr holl weithwyr a'u manylion cyflog. Byddai system adnoddau dynol yn cyrchu ffeil wahanol sy'n cynnwys gwybodaeth am y gweithwyr: e.e. manylion personol, cymwysterau, swyddi blaenorol a chyflog. Fel y gallwch ddychmygu, mae cryn dipyn o ddyblygu ar draws y ddwy ffeil hyn gan nad yw'n bosibl rhannu ffeiliau mewn systemau ffeiliau fflat.

Problem y ffeil fflat

Mae'n bosibl storio'r holl wybodaeth yn yr un tabl. E.e. gallai cwmni llogi offer storio manylion cwsmeriaid ac offer sy'n cael eu llogi mewn un tabl fel hyn:

Mae pob darn o feddalwedd rhaglenni'n cyrchu ei storfa ei hun o ddata; nid yw data'n cael eu rhannu mewn system ffeiliau fflat a bydd hyn yn arwain at nifer o broblemau

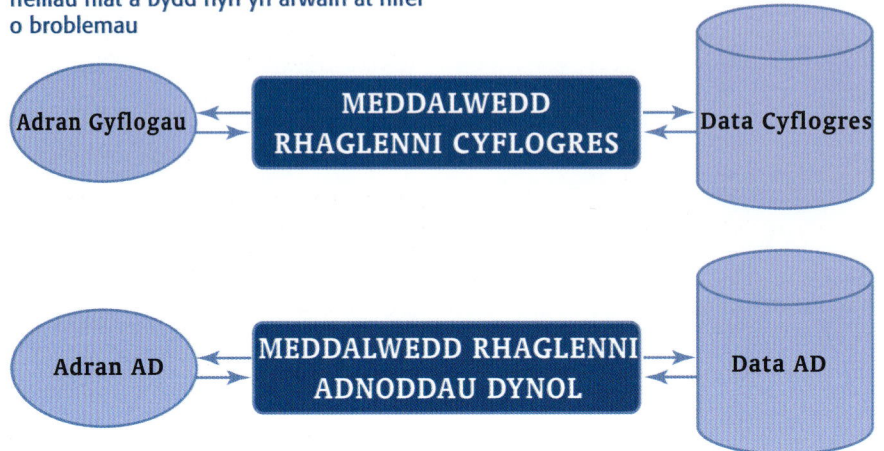

Mae hyn yn codi sawl problem:

- Nid oes cofnod o offer oni bai ei fod yn cael ei logi.
- Hyd yn oed os yw cwsmeriaid wedi llogi offer o'r blaen, mae angen teipio eu manylion (Rhif Cwsmer, Enw Cyntaf, ac ati) eto ac mae hyn yn gwastraffu amser. Os edrychwch ar ddwy res gyntaf y tabl fe welwch pa data sydd yr un fath.
- Rhaid teipio manylion yr offer wrth i bob cwsmer ei logi ac mae hyn yn gwastraffu amser hefyd.
- Mae problemau gyda chyfanrwydd

data. Gall anghysondebau godi, er enghraifft, os bydd cwsmer yn newid cyfeiriad, gan y bydd rhai o'r hen gofnodion (h.y. rhesi yn y tabl) yn cynnwys yr hen fanylion.

Mae ffordd lawer mwy effeithlon o drefnu'r data hyn, sef drwy storio'r data mewn tri thabl yn lle un a'u cysylltu â'i gilydd, gan ddefnyddio perthnasoedd i ffurfio cronfa ddata berthynol.

Rhif Cwsmer	Enw Cyntaf	Cyfenw	Llinell Gyntaf y Cyfeiriad	Cod Post	Rhif Ffôn	RhA Offer	Disgrifiad o'r Offer	Pris Llogi	Dyddiad Llogi	Dyddiad Dychwelyd
1200	Steve	Jones	1 Stryd Dafydd	L91 6TY	243 6782	0002	Stripiwr ager	£9.50	14/08/10	15/08/10
1200	Steve	Jones	1 Stryd Dafydd	L91 6TY	243 6782	0002	Stripiwr ager	£9.50	19/08/10	21/08/10
1201	Jenny	Chung	12 Heol Morris	L43 1WW	782 8722	0003	Peiriant glanhau carpedi	£43.00	14/08/10	19/08/10
1202	Ray	Thomas	8 Stryd Hir	L21 6TT	920 1111	0001	Golchwr gwasgedd	£15.00	13/08/10	14/08/10
1203	John	Rees	99 Ffordd Taf	L44 7TU	976 6121	0004	Sandiwr Llawr	£67.00	13/08/10	17/08/10
1203	John	Rees	99 Ffordd Taf	L44 7TU	976 6121	0001	Golchwr gwasgedd	£15.00	17/08/10	20/08/10

Enghraifft o dabl ffeil fflat

Ffeiliau fflat a chronfeydd data perthynol

Buom yn ystyried y gwahaniaethau rhwng ffeiliau fflat a chronfeydd data perthynol yn y gyfrol UG, ond dyma nodweddion y ddwy eto.

Nodweddion ffeiliau fflat:

- gellir defnyddio meddalwedd taenlen neu feddalwedd cronfa ddata i'w creu
- caiff yr holl ddata eu cadw mewn un tabl
- gallant ddioddef o afreidrwydd data (h.y. dyblygu data'n ddiangen)
- yn ddefnyddiol ar gyfer cymwysiadau syml yn unig
- hawdd iawn eu creu.

Nodweddion cronfeydd data perthynol:

- caiff y data eu cadw mewn dau neu ragor o dablau
- mae cysylltau rhwng y tablau
- gellir echdynnu data o unrhyw dabl
- mae angen mwy o arbenigedd i'w creu.

Cronfeydd data perthynol

Cydran sylfaenol unrhyw gronfa ddata berthynol yw'r tabl, sef casgliad o ddata wedi'u trefnu'n rhesi a cholofnau. Mae cronfeydd data perthynol yn cynnwys llawer o dablau ac mae pob tabl yn cynnwys data perthynol. Mae'r tablau wedi'u cysylltu â'i gilydd, sy'n golygu y gall data o sawl tabl gael eu cyfuno i roi gwybodaeth ystyrlon i'r defnyddiwr ar ffurf adroddiadau.

Er enghraifft, yn achos y cwmni llogi offer, gellir rhoi rhif unigryw (rhif adnabod) i bob offer (RhA_Offer), a manylion eraill (disgrifiad o'r offer, pris llogi), mewn un tabl fel hyn:

RhA Offer	Disgrifiad o'r Offer	Pris Llogi
0001	Golchwr gwasgedd	£15.00
0002	Stripiwr ager	£9.50
0003	Peiriant glanhau carpedi	£43.00
0004	Sandiwr llawr	£67.00

Tabl Offer

Gallai tabl arall restru'r cwsmeriaid, y rhoddir rhif cwsmer unigryw i bob un ohonynt, ynghyd â gwybodaeth arall fel enw, cyfeiriad a rhif ffôn, fel hyn:

Rhif Cwsmer	Enw Cyntaf	Cyfenw	Llinell Gyntaf y Cyfeiriad	Cod Post	Rhif Ffôn
1200	Steve	Jones	1 Stryd Dafydd	L91 6TY	243 6782
1201	Jenny	Chung	12 Heol Morris	L43 1WW	782 8722
1202	Ray	Thomas	8 Stryd Hir	L21 6TT	920 1111
1203	John	Rees	99 Ffordd Taf	L44 7TU	976 6121

Tabl Cwsmer

Gellir defnyddio tabl arall, o offer sy'n cael eu llogi (rhentol), i gofnodi pa offer sydd ym meddiant pwy, gan ddefnyddio'r rhif adnabod offer a rhif cwsmer priodol. Byddai'r tabl rhentol hefyd yn cynnwys y dyddiad llogi a'r dyddiad dychwelyd fel bod y siop yn gwybod os nad yw offer wedi cael ei ddychwelyd yn brydlon.

RhA Offer	Rhif Cwsmer	Dyddiad Llogi	Dyddiad Dychwelyd
0001	1202	13/08/10	14/08/10
0004	1203	13/08/10	17/08/10
0003	1201	14/08/10	19/08/10
0002	1200	14/08/10	15/08/10
0002	1200	19/08/10	21/08/10
0001	1203	17/08/10	20/08/10

Mae'r tabl Rhentol yn cynnwys meysydd (RhA Offer a Rhif Cwsmer) sy'n ymddangos yn y tablau eraill

Mewn cronfa ddata berthynol mae'r tablau'n cael eu cysylltu drwy gael maes cyfatebol yn y ddau dabl. Er enghraifft, mae RhA Offer i'w gael yn y tabl Offer a'r tabl Rhentol, a'r maes hwn sy'n darparu'r cyswllt. Hefyd mae cyswllt rhwng y tabl Rhentol a'r tabl Cwsmer, sef Rhif Cwsmer. Mae'r gallu i gysylltu tablau yn nodwedd bwysig o gronfeydd data perthynol.

Trefniadaeth cronfeydd data perthynol

Drwy ddefnyddio dull perthynol, gallwch osgoi rhai o'r problemau sy'n gysylltiedig â dulliau eraill o gadw data fel ffeiliau fflat. Mae trefniadaeth cronfeydd data perthynol yn ymdrin â'r materion canlynol:

- cysondeb data
- afreidrwydd data
- cyfanrwydd data
- annibyniaeth data.

> ### YN YR ARHOLIAD
>
> Sicrhewch eich bod chi'n cofio diffiniad o gronfa ddata berthynol. Fel arfer bydd yn rhaid i chi ddiffinio'r term cronfa ddata berthynol ac mae diffiniad o'r fath yn werth dau neu dri marc. Mae'n eithaf anodd rhoi diffiniad byr, felly mae'n well cofio'r diffiniad yn yr adran Geiriau Allweddol yn hytrach na cheisio meddwl am eich diffiniad eich hun.

Cysondeb data

Mewn rhai cyrff, mae'r un data i'w cael mewn gwahanol ffeiliau. Mae hyn yn wastraffus gan ei bod hi'n bosibl y bydd angen teipio'r data ddwywaith, a hefyd, os oes angen newid y data yn un o'r ffeiliau, yna rhaid eu newid hefyd yn y ffeil arall i sicrhau cysondeb. Os na chaiff hyn ei wneud, fe gewch sefyllfa lle mae'r data'n dibynnu ar y ffeil rydych chi'n ei defnyddio. Mae defnyddio mwy nag un ffeil i gadw'r data yn broblem pan nad oes cronfa ganolog o ddata, a bydd y data yn y cronfeydd data hyn yn colli cysondeb yn gyflym iawn.

Afreidrwydd data

Holl bwynt cynhyrchu cronfa ganolog o ddata yw y gellir rhannu'r data hyn ar ôl eu creu rhwng nifer o gymwysiadau. Prif fantais hyn yw nad oes angen dyblygu data. Os caiff y data eu cadw ar wahân, yna mae angen i unrhyw newid i un darn o ddata gael ei newid yn yr holl leoedd eraill lle mae'n cael ei gadw. Os nad yw hyn yn cael ei wneud, ni fydd y data'n gyson bellach. Er enghraifft, efallai fod yr adran farchnata a'r adran gyfrifon mewn cwmni yn cadw eu cronfeydd data eu hunain o gwsmeriaid. Os dywed y cwsmer wrth yr adran farchnata ei fod wedi newid ei gyfeiriad, bydd angen iddi newid ei chronfa ddata ei hun a dweud wrth yr adran gyfrifon am newid ei chofnodion hithau.

Cronfeydd data *(parhad)*

Y broblem, mewn cwmni mawr, yw ei bod hi'n bosibl nad yw'r bobl yn yr adran farchnata yn gwybod pa adrannau eraill sy'n cadw enwau a chyfeiriadau cwsmeriaid hefyd, felly mae'r system yn torri i lawr. Mae'n amlwg y byddai cael un gronfa ddata o gwsmeriaid a ddefnyddir gan yr holl adrannau yn golygu na fyddai'n rhaid teipio a storio'r un data yn ddiangen ac na fyddai angen gwneud newidiadau i'r un data mewn sawl lle gwahanol.

Mae'r mwyafrif o gyrff heddiw yn cadw'r data ar wahân i'r rhaglenni a ddefnyddir i brosesu'r data ac mae hyn yn golygu, os caiff y meddalwedd rhaglenni ei newid, y gall y data gael eu cadw er eu pen eu hun fel y gellir eu defnyddio gyda rhaglen arall.

Mae defnyddio storfa ganolog o ddata mewn cronfa ddata yn lleihau'r rhan fwyaf o'r dyblygu ond mae'n amhosibl dileu dyblygu'n gyfan gwbl. Mae data sy'n cael eu hailadrodd yn ddiangen yn cael eu galw'n ddata afraid.

Cyfanrwydd data

Mae cyfanrwydd data yn cyfeirio at gywirdeb data. Os bydd defnyddwyr yn darganfod bod peth o'r data mewn cronfa ddata yn anghywir, byddant yn dechrau colli ffydd yn yr holl ddata sydd yn y gronfa ddata. Mae sawl cam y gallwch eu cymryd i sicrhau cyfanrwydd neu gywirdeb data, gan gynnwys y canlynol.

Sicrhau nad yw gwallau'n digwydd wrth drawsgrifio data

Mewn rhai systemau, rhaid i rywun lenwi ffurflen, ac yna caiff y manylion o'r ffurflen eu teipio i gyfrifiadur. Mae llenwi'r ffurflen yn cael ei alw'n drawsgrifio, ac mae'r camgymeriadau sy'n digwydd yn ystod y broses hon yn cael eu galw'n wallau trawsysgrifol. Mae'r gwallau hyn yn anodd eu dileu ond mae gwirio gofalus gan reolwyr a hyfforddi staff yn drwyadl yn gallu eu lleihau.

Defnyddio dulliau gwireddu

Pan ddefnyddir dogfennau ffynhonnell fel anfonebau, ffurflenni cais, archebion, ac ati, caiff y data eu darllen oddi arnynt a'u teipio i'r cyfrifiadur. Dulliau gwireddu yw dulliau sy'n sicrhau bod y data sy'n cael eu teipio yr un fath yn union â'r ddogfen ffynhonnell. Bydd y gwiriad hwn yn cymharu'r hyn sydd wedi cael ei deipio â'r hyn sydd yn y ddogfen wreiddiol. Mewn geiriau eraill, bydd rhywun yn prawfddarllen yr hyn sydd wedi'i deipio i mewn.

Defnyddio dulliau dilysu

Er bod gwireddu data'n gwirio nad yw gwallau wedi cael eu cyflwyno wrth deipio, mae'n bosibl bod y data yn y ddogfen ffynhonnell yn anghywir i ddechrau. Caiff dilysu ei wneud gan y gronfa ddata ac mae'n gwirio cywirdeb y data drwy ddefnyddio gwiriadau math data, gwiriadau amrediad, ac ati.

Sicrhau bod gweithdrefnau yn eu lle ar gyfer diweddaru rheolaidd

Mae angen cynnal cronfeydd data'n rheolaidd. Bydd y gwaith cynnal hwn yn cynnwys diweddaru'r wybodaeth sydd ynddynt yn barhaus. Er enghraifft, mewn ysgol neu goleg, bydd angen llenwi ffurflen bob blwyddyn fel y gall y disgyblion/myfyrwyr hysbysu'r ysgol/coleg o unrhyw newidiadau yn eu manylion.

Sicrhau bod y gweithdrefnau gweithredol cywir yn cael eu dilyn

Gallai'r ffeil anghywir gael ei defnyddio i ddiweddaru'r feistr-ffeil, gyda'r canlyniad bod y manylion anghywir yn cael eu storio.

Annibyniaeth data

Mae annibyniaeth data yn golygu ei bod hi'n bosibl newid y data heb newid y rhaglen sy'n eu prosesu. Gan ddefnyddio system rheoli cronfeydd data (*DBMS: database management system*), gellir newid strwythur cronfa

ddata heb effeithio ar y rhaglenni presennol os nad yw meysydd newydd yn cael eu creu. Y rheswm am hyn yw bod y data, a'r rhaglenni sy'n defnyddio'r data, ar wahân i'w gilydd ac yn rhyngweithio â'i gilydd drwy'r *DBMS*.

Mae annibyniaeth data'n golygu y gall rhaglenni newydd gael eu datblygu i gyrchu'r data ac, os caiff systemau hollol newydd eu defnyddio, y gellir parhau i ddefnyddio'r data presennol.

Defnyddio prif allweddi, allweddi estron a chysylltau

Prif flociau adeiladu unrhyw gronfa ddata berthynol yw'r tablau. Mae pob tabl yn cynnwys meysydd ac mae yna nifer o feysydd arbennig o'r enw prif allweddi ac allweddi estron.

Prif allweddi

Maes a ddefnyddir i ddiffinio'n unigryw unrhyw gofnod neu linell/rhes benodol mewn tabl yw prif allwedd. Mae'n annhebyg y gallai unrhyw destun fod yn unigryw, felly mae prif feysydd bron bob amser yn feysydd rhifol.

Pe baech chi'n gwneud maes fel cyfenw yn brif allwedd, yna wrth i chi deipio data i mewn a cheisio teipio cyfenw sydd eisoes wedi cael ei ddefnyddio mewn cofnod gwahanol, ni fyddai'r gronfa ddata'n ei dderbyn. Rhaid sicrhau bod rhif sy'n diffinio'r cofnod yn unigryw. Mae prif allweddi o'r fath yn cynnwys rhif aelodaeth, rhif cynnyrch, rhif gweithiwr, rhif catalog, rhif rhan, rhif cyfrif, ac ati. Os nad yw prif allwedd wedi cael ei diffinio, bydd y mwyafrif o gronfeydd data yn creu un yn awtomatig. Er enghraifft, mae'r meddalwedd cronfa ddata berthynol Microsoft Access yn defnyddio prif allwedd o'r enw AutoNumber, ac mae hon yn rhoi rhif i bob cofnod ar sail y drefn y cafodd y cofnod ei deipio i'r gronfa ddata. Bydd y cofnod cyntaf i gael ei deipio yn cael y gwerth 1, a'r nesaf 2, ac yn y blaen.

Cysylltau

Os yw dau neu ragor o dablau'n cynnwys yr un maes, gallant gael eu cysylltu. Mae hyn yn golygu y gall y data gael eu cyfuno mewn gwahanol ffyrdd er eu bod wedi'u storio mewn tablau gwahanol. Mae cysylltau rhwng tablau'n aml yn cael eu galw'n berthnasoedd, a dyma un o brif nodweddion cronfeydd data perthynol.

Allweddi estron

Maes mewn un tabl sydd hefyd yn brif allwedd tabl arall yw allwedd estron. Defnyddir allweddi estron i sefydlu perthnasoedd rhwng tablau.

Normaleiddio cronfa ddata

Techneg fathemategol ar gyfer dadansoddi data yw normaleiddio. Proses mewn camau ydyw sy'n gwella dyluniad y gronfa ddata ym mhob cam. Ar y lefel hon mae angen i ni edrych ar normaleiddio fel proses dri cham o'r enw ffurf normal 1af, 2il ffurf normal a 3ydd ffurf normal.

Mae normaleiddio yn:

- Lleihau dyblygu data – nid oes angen i chi roi data i mewn fwy o weithiau nag sydd angen.
- Dileu data afraid – mae'n osgoi ailadrodd yr un data'n ddiangen.
- Sicrhau cyfanrwydd data – mae'n sicrhau nad oes gwallau yn y data ac nad oes gwahanol fersiynau o'r data, gan ddibynnu ar a yw'r data wedi cael eu diweddaru ai peidio.
- Caniatáu i wybodaeth gael ei hechdynnu o'r gronfa ddata mewn ffordd hyblyg.

Ailstrwythuro data i ffurf wedi'i normaleiddio

Wrth ddadansoddi system newydd ar gyfer ysbyty, cafwyd hyd i'r wybodaeth ganlynol.

Bydd gan bob ward yn yr ysbyty ei henw ei hun a chyfeirnod unigryw. Hefyd mae angen cofnodi nifer y gwelyau ym mhob ward ynghyd â'u henwau a'u cyfeirnodau. Mae staff nyrsio pob ward yn cael rhif staff sy'n unigryw i bob un ohonynt ac mae'r rhifau hyn yn cael eu cofnodi gyda'u henwau. Mae pob nyrs yn gweithio mewn un ward yn unig.

Rhoddir rhif claf i gleifion mewnol wrth iddynt gyrraedd ac mae hyn yn cael ei gofnodi gydag enw, cyfeiriad, rhif ffôn a dyddiad geni pob claf. Pan gânt eu derbyn i un o'r wardiau, mae pob claf yn cael ei neilltuo i un ymgynghorydd sy'n gyfrifol am ei ofal meddygol. Mae gan ymgynghorwyr eu rhifau staff unigryw eu hunain sydd wedi'u cofnodi gyda'u henwau a'u harbenigeddau.

Tybiwch ein bod ni'n penderfynu y gallwn roi'r holl fanylion mewn un tabl a bod y tabl hwn yn cael ei alw'n CLAF. Yna gallwn restru'r meysydd canlynol:

CLAF
Rhif_ward
Enw_ward
Nifer_o_welyau
Enw_nyrs
Rhif_staff_nyrs
Rhif_claf
Enw_claf
Cyfeiriad_claf
Rhif_ffôn_claf
Dyddiad_geni_claf
Rhif_ymgynghorydd
Enw_ymgynghorydd
Arbenigedd_ymgynghorydd

Dyma'r data yn eu ffurf heb ei normaleiddio.

Mynd o ffurf heb ei normaleiddio (UNF: un-normalised form) i ffurf normal gyntaf (1NF: first normal form)

Mae casglu data yn cael ei ystyried fel bod yn ei ffurf normal gyntaf os nad yw'n cynnwys unrhyw grwpiau ailadroddol o eitemau data. Felly mae angen i ni dynnu'r grwpiau ailadroddol o'r rhestr isod a'u rhoi yn eu rhestr eu hunain. Drwy archwilio'r rhestr gallwn weld nad oes un (Rhif-staff-nyrs, Enw-nyrs) yn union wedi'i gysylltu â'r maes rhif-claf gan y bydd llawer o nyrsys sy'n gweithio ar yr un ward yn gofalu am y claf. Mae hyn yn gwneud y grŵp hwn o

feysydd yn grŵp ailadroddol, ac felly i fynd i'r ffurf normal gyntaf mae angen i ni dynnu'r grŵp hwn a'i roi yn ei dabl ei hun a rhoi enw i'r tabl hwnnw sy'n adlewyrchu beth sydd ynddo. Mae CLAF-NYRSYS yn enw addas yn yr achos hwn. I gael manylion nyrsys claf penodol, rhaid gwybod beth yw'r Rhif-claf a'r Rhif-staff-nyrs. Mae angen i'r ddau faes hyn fod yn brif allwedd oherwydd bod angen y ddau at ddibenion adnabod.

Mae gennym y tablau canlynol bellach:

CLAF (Rhif-claf, Enw-claf, Cyfeiriad-claf, Rhif-ffôn-claf, Dyddiad-geni-claf, Rhif-ward, Enw-ward, Rhif-ymgynghorydd, Enw-ymgynghorydd, Arbenigedd-ymgynghorydd)

CLAF-NYRSYS (Rhif-claf, Rhif-staff-nyrs, Enw-nyrs)

Mynd o'r ffurf normal gyntaf i'r ail ffurf normal (2NF: second normal form)

Rydym nawr yn edrych ar y tablau sydd â dwy allwedd i weld a yw'r meysydd ynddynt yn dibynnu ar y ddwy allwedd neu un yn unig. Os yw maes yn dibynnu ar un yn unig o'r allweddi, yna dylai gael ei dynnu oddi yno gyda'i allwedd a'i grwpio mewn tabl newydd.

Yn ein henghraifft ni mae angen i ni edrych ar yr endid CLAF-NYRSYS gan fod y tabl hwn yn cynnwys dwy allwedd (cofiwch fod y prif allweddi wedi'u tanlinellu) a gwirio bod y meysydd eraill sydd heb eu tanlinellu yn dibynnu ar y ddwy allwedd (h.y. y rhai sydd wedi'u tanlinellu). Yr unig faes sydd gennym nad yw'n allwedd yw Enw-nyrs: er ei fod yn dibynnu ar Rhif-staff-nyrs nid yw'n dibynnu ar Rhif-claf. Mae angen i ni gymryd hwn a'i roi gyda chopi o'i allwedd i mewn i dabl newydd y byddwn yn ei alw'n NYRS. Ar ôl gwneud hyn, dywedwn fod y data bellach yn eu hail ffurf normal.

CLAF (Rhif-claf, Enw-claf, Cyfeiriad-claf, Rhif-ffôn-claf, Dyddiad-geni-claf, Rhif-ward, Enw-ward, Rhif-ymgynghorydd, Enw-ymgynghorydd, Arbenigedd-ymgynghorydd)

CLAF-NYRSYS (Rhif-claf, Rhif-staff-nyrs)

NYRS (Rhif-staff-nyrs, Enw-nyrs)

Cronfeydd data *(parhad)*

Mynd o'r ail ffurf normal i'r drydedd ffurf normal (3NF: third normal form)

I fynd i'r drydedd ffurf normal mae'n rhaid edrych ar y meysydd ym mhob un o'r tablau i weld a yw unrhyw feysydd yn dibynnu ar ei gilydd, ac os ydynt, mae angen eu symud i dabl ar wahân. Wrth symud y meysydd mae'n rhaid gadael un o'r meysydd yn y tabl gwreiddiol i'w ddefnyddio'n allwedd ar gyfer y tabl newydd.

Er enghraifft, wrth edrych ar y tablau a'r meysydd yn yr ail ffurf normal, gallwn weld fod Enw-ymgynghorydd ac Arbenigedd-ymgynghorydd yn dibynnu ar Rhif-ymgynghorydd ac fel arall. Felly rydym yn symud y grŵp hwn i dabl newydd o'r enw YMGYNGHORYDD ac yn gadael y maes Rhif-ymgynghorydd ar ôl i ddarparu cyswllt rhwng y tablau.

Hefyd, mae Enw-ward a Rhif-ward yn dibynnu ar ei gilydd, felly gellir symud y rhain a'u rhoi mewn tabl newydd o'r enw WARD. Caiff Rhif-ward ei adael yn y tabl gwreiddiol i ddarparu cyswllt.

Mae'r data bellach yn eu trydedd ffurf normal fel hyn:

CLAF (<u>Rhif-claf</u>, Enw-claf, Cyfeiriad-claf, Rhif-ffôn-claf, Dyddiad-geni-claf, Rhif-ward, Rhif-ymgynghorydd)

CLAF-NYRSYS (<u>Rhif-claf</u>, <u>Rhif-staff-nyrs</u>)

NYRS (<u>Rhif-staff-nyrs</u>, Enw-nyrs)

YMGYNGHORYDD (<u>Rhif-ymgynghorydd</u>, Enw-ymgynghorydd, Arbenigedd-ymgynghorydd)

WARD (<u>Rhif-ward</u>, Enw-ward)

Mae'r meysydd nawr yn eu trydedd ffurf normal a dywedwn eu bod wedi'u llwyr normaleiddio.

Manteision darparu data mewn gwahanol ffyrdd ar gyfer gwahanol ddefnyddwyr

Er mai'r un data yw'r data mewn cronfa ddata pwy bynnag sy'n eu defnyddio, gellir cyflwyno'r data mewn gwahanol ffyrdd i wahanol ddefnyddwyr. Er enghraifft, byddai angen darlun cyffredinol ar reolwyr yn hytrach na darlun manwl, ond byddai angen i staff sy'n ymdrin â chyfrifon cwsmeriaid wybod holl fanylion y trafodion.

Prif fantais hyn yw:

- Bydd gweinyddwyr yn cael darlun manwl o'r data.
- Gall rheolwyr gael darlun cyffredinol sydd heb ei gymylu gan lu o fanylion.
- Gellir cyflwyno data ar ffurf graff i'w gwneud hi'n haws gweld tueddiadau dros amser.

Diogelwch cronfeydd data

Mewn ffeil fflat mae'r holl ddata ar gyfer rhaglen benodol yn cael eu cadw gyda'i gilydd ac mae hyn yn golygu y gellir caniatáu mynediad i'r holl

ddata neu ddim data. Yn achos cronfa ddata berthynol, mae'n bosibl caniatáu mynediad i rai elfennau'n unig o'r data. Mae hyn yn golygu y gallwch roi gwahanol lefelau o fynediad i wahanol bobl gan ddibynnu ar ofynion eu swydd.

Sut i wella diogelwch mewn cronfa ddata berthynol

Mae diogelwch mewn cronfeydd data perthynol yn well na diogelwch mewn ffeiliau fflat mewn nifer o ffyrdd:

- Caiff y data eu storio ar wahân i'r rhaglenni a ddefnyddir i'w cyrchu.
- Mae hierarchaeth o gyfrineiriau sy'n caniatáu i bobl gyrchu'r wybodaeth sydd ei hangen arnynt i wneud eu gwaith a dim byd arall.
- Gellir cyfyngu mynediad i ran yn unig o raglen sy'n defnyddio peth o'r data'n unig.

Warysu data

A allwch chi ddychmygu system a fydd yn storio holl fanylion pob eitem yr ydych wedi'i phrynu mewn uwchfarchnad neilltuol yn ystod y deng mlynedd diwethaf? Os defnyddiwch

Mae angen nifer mawr o weinyddion a dyfeisiau storio ar gyfer warws data corff mawr

gerdyn teyrnged, gall yr holl eitemau a brynwch a'r ffordd y talwch amdanynt gael eu cysylltu â chi.

Gan fod swm y data yn enfawr, defnyddir cronfa ddata fawr iawn, o'r enw warws data, i storio'r data. Defnyddir warws data i storio holl ddata hanesyddol corff. Yna bydd systemau gwybodaeth rheoli'r corff yn defnyddio'r warws i gael y data a fydd yn helpu ei reolwyr i wneud penderfyniadau.

Adnodd corfforaethol yw'r warws data y gall pawb yn y corff ei ddefnyddio os oes ganddynt hawliau mynediad.

Enghreifftiau o sut y gall warws data gael ei ddefnyddio yw:

- Darganfod ar ba ddiwrnod o'r wythnos y gwerthodd siop benodol y nifer mwyaf o eitem benodol yn 2006.
- Sut yr oedd absenoldeb staff oherwydd salwch yn amrywio yn ystod y flwyddyn rhwng cangen Caerdydd a changen Abertawe.

Cloddio data

Ar ôl i'r data gael eu storio mewn warws data mae angen eu 'cloddio' i ddarganfod:

- patrymau yn y data
- cysylltiadau yn y data (e.e. mae pobl sy'n darllen y *Times* yn fwy tebygol o yfed gwin coch)
- tueddiadau dros amser (e.e. mae rhywun yn prynu bwyd iachach ac yn yfed llai o alcohol).

Drwy 'ddrilio i lawr' i mewn i'r corff o ddata, mae cloddio data yn galluogi defnyddwyr i ddeall y data'n well drwy ddarganfod patrymau ystyrlon. Gellir chwilio am batrymau ystyrlon yn y casgliad enfawr o ddata drwy ddefnyddio meddalwedd cloddio data sy'n cyflwyno'r canlyniadau ar ffurf tablau a graffiau.

Gall cloddio data gynhyrchu gwybodaeth fel:

- rhestri o gwsmeriaid sy'n debygol o brynu cynnyrch penodol (ar sail yr hyn y maen nhw wedi'i brynu o'r blaen)
- cymariaethau â chystadleuwyr
- canlyniadau 'beth os' defnyddiol o ymarferion modelu
- rhagfynegiadau o werthiant yn y dyfodol
- dadansoddiad o'r safleoedd gorau

ar gyfer siopau
- patrymau gwerthiant
- patrymau prynu cwsmeriaid
- pwy sy'n fwyaf tebygol o newid ei gerdyn credyd
- y cwsmeriaid hynny sy'n fwyaf teyrngar i'r cwmni/cynnyrch.

Cymwysiadau cloddio data

Defnyddir cloddio data mewn llawer ffordd. Dyma rai enghreifftiau:

- Helpu yn yr ymladd yn erbyn terfysgaeth – ers 9/11 bu llywodraeth UDA yn defnyddio cloddio data i ddadansoddi arferion teithio, gwario a chyfathrebu pobl er mwyn darganfod patrymau o ymddygiad anarferol a allai ei harwain at derfysgwyr.
- Ymladd dwyn o siopau Jaeger – gan ddefnyddio cloddio data a gwybodaeth am drafodion a lleoliad y dillad yn y siop, darganfu'r cwmni fod y rhan fwyaf o ddillad a oedd yn cael eu dwyn yn agos at y drysau, er bod tagiau diogelwch arnynt fel rheol. Drwy wario mwy ar deledu cylch-caeedig a chydweithredu â'r heddlu, fe gafodd y lladron eu dal a'r nwyddau eu dychwelyd.
- Darganfod anghenion cwsmeriaid – mae Virgin Media, sy'n cyflenwi pecynnau band llydan, ffôn a theledu cebl, yn defnyddio cloddio data i segmentu a thargedu cwsmeriaid sy'n fwyaf tebygol o brynu gwasanaethau newydd neu uwchraddio'r gwasanaethau sydd ganddynt eisoes.

Systemau rheoli cronfa ddata (*DBMS: database management systems*)

Pecynnau o raglenni yw systemau rheoli cronfa ddata (*DBMS*) sy'n galluogi cyrff i ddal casgliadau o ddata wedi'u canoli a'u strwythuro, a'u trin ymhellach mewn gwahanol ffyrdd. Mae pob system rheoli cronfa ddata yn caniatáu i'r defnyddiwr greu ei gronfeydd data ei hun ac mae'r mwyafrif o'r pecynnau'n weddol hyblyg yn hyn o beth. Mae'r systemau hyn yn cadw'r data ar wahân i'r rhaglenni ac mae hyn yn golygu bod rhaglenni, pan gânt eu datblygu, yn annibynnol ar y data sy'n cael eu storio. Mae systemau rheoli cronfa ddata yn:

- caniatáu i'r gronfa ddata gael ei diffinio
- caniatáu i'r defnyddwyr holi'r gronfa ddata
- caniatáu i'r data gael eu hatodi (ychwanegu), eu dileu a'u golygu
- caniatáu i'r defnyddiwr addasu strwythur y gronfa ddata
- darparu diogelwch digonol ar gyfer y data sy'n cael eu storio
- caniatáu i'r defnyddiwr fewnforio ac allforio data.

Mae'r ffigur isod yn dangos y dull hŷn o storio data lle caiff y data ar gyfer pob rhaglen eu storio ar wahân a lle gellir defnyddio'r rhaglenni i gyrchu'r data. Mae problemau'n codi pan mae angen i'r un data gael eu cyrchu gan wahanol raglenni.Mae hyn yn golygu

Mae'r dull ffeiliau fflat yn cadw'r data ar gyfer pob rhaglen ar wahân

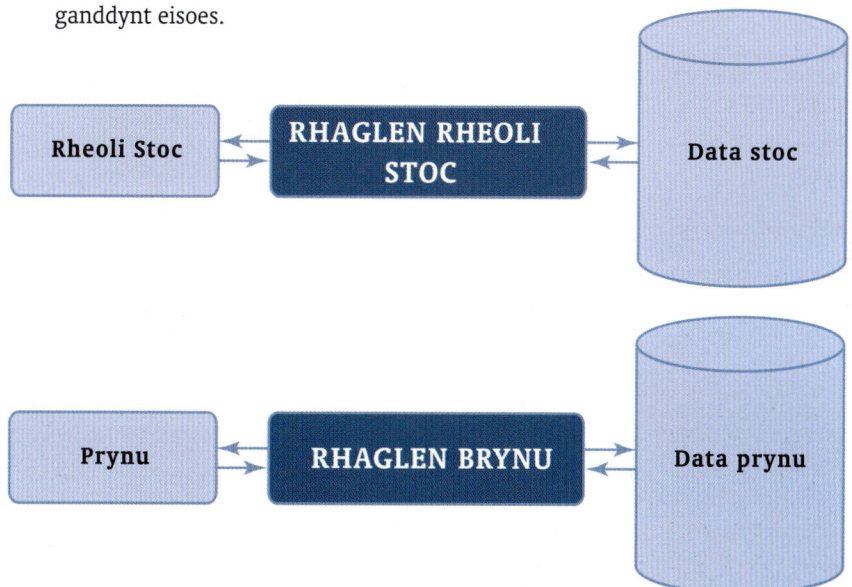

| Rheoli Stoc | ← | **RHAGLEN RHEOLI STOC** | → | Data stoc |

| Prynu | ← | **RHAGLEN BRYNU** | → | Data prynu |

Cronfeydd data *(parhad)*

bod yn rhaid i bob rhaglen gael ei diweddaru pan gaiff newidiadau eu gwneud, neu ni fydd data am yr un peth mewn gwahanol raglenni yn gyson.

Mae'r ffigur isod yn dangos dull y system rheoli cronfa ddata berthynol, lle caiff yr holl ddata eu cadw yn yr un lle a lle gall pob rhaglen eu cyrchu. Byddwn yn awr yn edrych ar fanteision y system rheoli cronfa ddata berthynol.

Manteision defnyddio Systemau Rheoli Cronfa Ddata

- Mae'n gwneud i bobl feddwl am y data sy'n cael eu storio, ac yn eu storio mewn ffordd resymegol a strwythuredig.
- Annibyniaeth data – mae'n bosibl cadw data ar wahân i'r rhaglenni sy'n eu defnyddio. Mae hyn yn ddefnyddiol os caiff y rhaglen gronfa ddata neu raglenni eraill eu newid. Nid oes angen ail-fewnbynnu'r data

er bod yn rhaid i chi ddefnyddio rhaglen arbennig i drawsnewid y data fel rheol.
- Mae'n osgoi afreidrwydd data. Caiff data eu mewnbynnu a'u storio unwaith yn unig, ni waeth faint o raglenni sy'n eu defnyddio.
- Oherwydd bod y data'n cael eu cadw'n ganolog, mae'n adnodd corfforaethol i'w ddefnyddio gan bob adran yn hytrach na pherthyn i un neu ragor o adrannau.
- Caiff cyfanrwydd y data ei gynnal. Mae diweddaru'r data mewn un man yn sicrhau bod y data yn gyfoes yn yr holl raglenni eraill sy'n defnyddio'r data.
- Mwy o ddiogelwch. Mae'n hawdd sefydlu mynediad wedi'i ganoli ac mae hyn yn golygu bod diogelwch yn well na phan oedd y data wedi'u gwasgaru drwy'r system.
- Caiff diffiniadau data eu safoni. Cyn

dyfodiad systemau rheoli cronfeydd data roedd yn beth cyffredin i wahanol raglenni ddefnyddio gwahanol enwau ar gyfer yr un eitem o ddata. Nid yw'r broblem hon yn codi gyda Systemau Rheoli Cronfeydd Data oherwydd bod gan y rhan fwyaf ohonynt eiriadur data sy'n sicrhau bod pawb yn defnyddio'r un enwau a diffiniadau.

Anfanteision defnyddio Systemau Rheoli Cronfeydd Data

- Gall fod yn anodd dysgu sut i ddefnyddio Sysytemau Rheoli Cronfeydd Data a gall gymryd llawer o amser. Mae'n system gymhleth iawn ac mae angen llawer o wybodaeth am ddadansoddi a dylunio cyn y gellir ei gweithredu'n llwyddiannus.
- Gall datblygu System Rheoli Cronfeydd Data fod yn ddrud iawn.

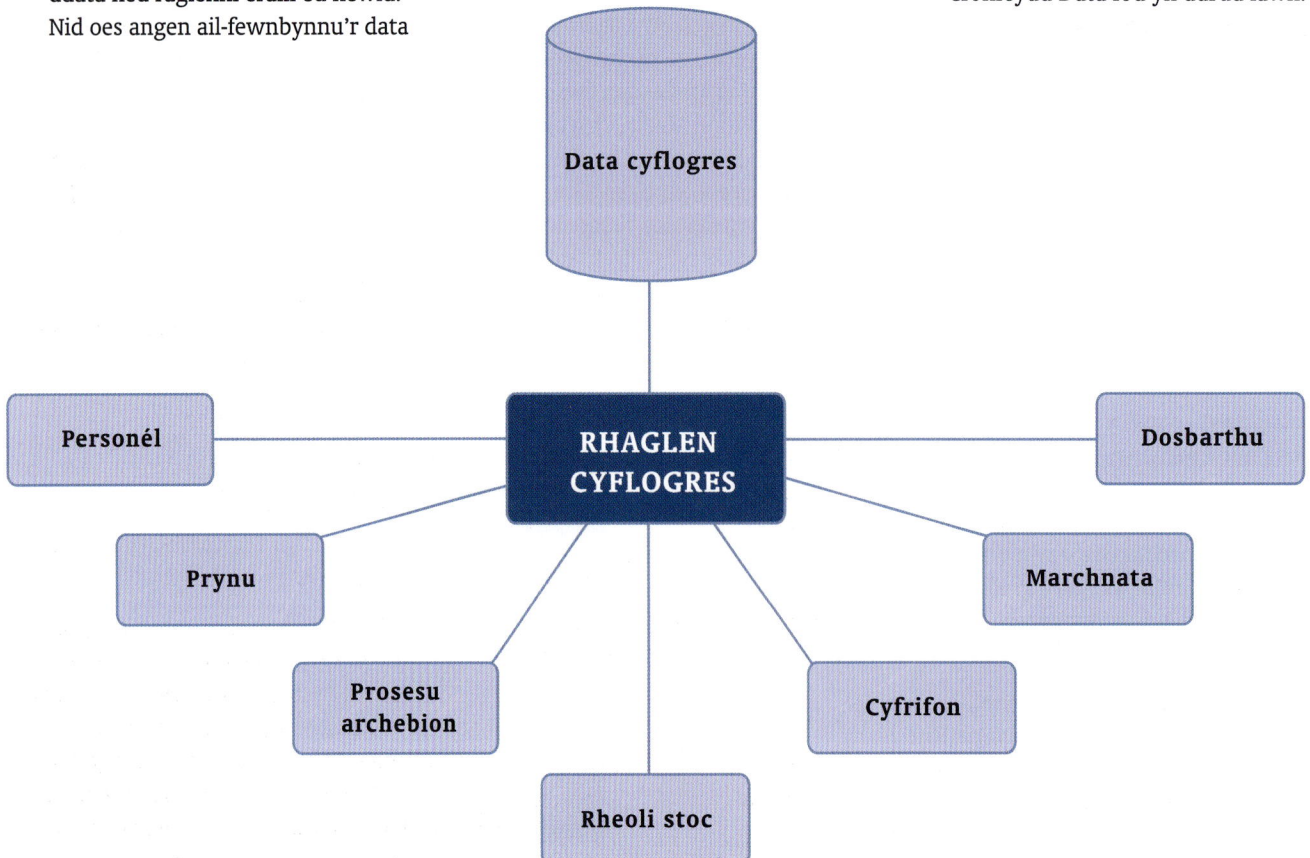

Data cyflogres

RHAGLEN CYFLOGRES

Personél

Dosbarthu

Prynu

Marchnata

Prosesu archebion

Cyfrifon

Rheoli stoc

Gall yr holl raglenni gyrchu'r storfa ddata drwy'r System Rheoli Cronfa Ddata

- Caiff yr holl ddata eu storio mewn lleoliad canolog ac mae hyn yn golygu bod mwy o berygl i ddiogelwch na phan mae'r data wedi'u gwasgaru drwy'r system. Mae angen diogelwch da ar system wedi'i chanoli, a chynllun adfer rhag trychineb, y dylid rhoi prawf arnynt yn rheolaidd.

Ymholiadau

Ceisiadau am wybodaeth benodol sy'n cael eu hanfon i gronfa ddata, ac sydd wedi'u hysgrifennu mewn iaith arbennig, yw ymholiadau. Er enghraifft, efallai eich bod chi eisiau rhestr o'r holl gofnodion mewn cronfa ddata sy'n cynnwys y cyfenw 'Jones', neu fanylion cyfrifon lle mae'r arian sy'n ddyledus yn fwy na'r credyd a ganiateir. Mae meddalwedd rheoli cronfeydd data yn caniatáu i ddefnyddwyr lunio ymholiadau penodol.

Ieithoedd ymholi

Mae gan wahanol gronfeydd data wahanol ddulliau o wneud ymholiadau i gyrchu gwybodaeth.

Mae nifer o ieithoedd ymholi ar gael ond yr un fwyaf cyffredin yw'r Iaith Ymholiadau Strwythuredig (SQL: Structured Query Language). Fel pob iaith gyfrifiadurol, mae angen i chi gofio rhestr o gyfarwyddiadau a sut i'w defnyddio er mwyn defnyddio SQL. Yn ffodus mae ffordd arall o'r enw Ymholi drwy Esiampl (QBE: Query by Example), sy'n galluogi'r defnyddiwr i lunio trefniadau, chwiliadau ac ymholiadau drwy glicio ar feysydd a meini prawf. Mae hyn yn newid yn y man i SQL ond nid oes angen i'r defnyddiwr wybod unrhyw orchmynion SQL i ddefnyddio QBE.

SQL

Mae SQL yn cynnwys nifer bach o orchmynion y gall y defnyddiwr eu cyfuno i gael manylion penodol o gronfa ddata. Hon yw'r iaith safonol bellach ar gyfer echdynnu gwybodaeth o gronfeydd data.

Mewn Iaith SQL, defnyddir y gorchymyn SELECT i holi'r gronfa ddata a chaiff y gorchymyn ei lunio fel hyn:

```
SELECT field list
FROM table
WHERE condition
```

Defnyddir rhestr o feysydd i restru'r meysydd yr ydym am eu hadalw. Mae'r rhestr o dablau yn cyfeirio at y tabl y gellir dod o hyd i'r meysydd hyn ynddo. Mae'r amod yn fynegiad Boole sy'n cael ei ddefnyddio i nodi'r cofnodion hynny sydd i gael eu hadalw.

Enghraifft 1

Rhaid echdynnu enwau'r holl weithwyr sy'n ennill dros £30,000 y flwyddyn yn adran gynhyrchu cwmni. Mae manylion y gweithwyr wedi'u storio mewn tabl o'r enw PERSONNEL.

Y gorchymyn Iaith SQL ar gyfer gwneud hyn yw:

```
SELECT Surname
FROM PERSONNEL
WHERE Department =
'Production' AND Salary >
30000
```

Enghraifft 2

Rydym ni eisiau rhestr o enwau a chyfeiriadau gweithwyr sy'n gweithio yn yr adrannau cynhyrchu neu farchnata o'r tabl PERSONNEL. Gallem ddefnyddio'r cyfarwyddyd SQL canlynol:

```
SELECT Surname, Street, Town,
Postcode
FROM PERSONNEL
WHERE Department =
'Production' OR 'Marketing'
```

Enghraifft 3

Rydych chi am roi pum diwrnod yn ychwanegol o wyliau i'r gweithwyr sy'n gweithio yn yr adrannau cynhyrchu neu farchnata. Gallech wneud beth sy'n cael ei alw'n 'ymholiad gweithredol'. Gallwch ddefnyddio ymholiad gweithredol i ddiweddaru rhai meysydd heb orfod mynd drwyddynt i gyd a newid rhai ohonynt â llaw.

```
UPDATE Personnel
SET No_of_days_holiday =
No_of_days_holiday + 5
WHERE Department =
'Production' OR 'Marketing'
```

Gan ddefnyddio SQL gallwch:

- gyfuno data o unrhyw dablau yn y gronfa ddata
- dewis pa feysydd i'w defnyddio
- pennu meini prawf chwilio
- dewis pa feysydd i'w rhoi yn yr adroddiad
- pennu trefn neu grwpiau
- rhoi enw i'r ymholiad a'i gadw fel y gellir ei ailddefnyddio
- cadw canlyniadau'r ymholiad.

Geiriadur data

Storfa ganolog o wybodaeth am ddata fel ystyron, perthnasoedd â data eraill, tarddiad, defnydd a fformat yw geiriadur data. Mae geiriadur data yn bwysig, yn enwedig i staff sy'n datblygu rhaglenni newydd a fydd yn defnyddio'r data mewn cronfa ddata. Mae'n caniatáu i'r gwahanol staff weithio'n gyson a deall sut mae'r data wedi cael eu sefydlu.

Pan gaiff geiriadur data ei gynhyrchu ar gyfer cronfa ddata bydd fel rheol yn cynnwys:

- manylion y tablau
- enwau'r meysydd
- mathau o feysydd
- hyd meysydd
- dulliau dilysu a ddefnyddir ar y meysydd.

Defnyddir geiriaduron data i ddisgrifio priodweddau'r meysydd mewn cronfa ddata. Mae'n cynnwys data am y data a gellir gwneud iddo ymddangos fel tabl naill ai ar bapur neu ar sgrin y cyfrifiadur.

Nid oes angen cwblhau'r geiriadur data â llaw. Mae llawer o systemau rheoli cronfeydd data yn gallu cynhyrchu geiriadur data'n awtomatig ar ôl i'r gronfa ddata gael ei sefydlu.

Cronfeydd data gwasgaredig

▼ **Byddwch yn dysgu**

▶ Am sut mae data'n cael eu storio mewn cronfa ddata wasgaredig

▶ Am fanteision defnyddio cronfa ddata wasgaredig

▶ Am anfanteision defnyddio cronfa ddata wasgaredig

Cyflwyniad

Mae llawer o gyrff nad ydynt yn storio eu holl ddata mewn un storfa. Yn lle hynny, mae'r data yn cael eu rhannu i wahanol leoliadau dros rwydwaith, ond o safbwynt y defnyddiwr mae'n ymddangos bod yr holl ddata yn yr un lle. Yn yr adran hon byddwch yn edrych ar fanteision defnyddio'r math hwn o gronfa ddata, sy'n cael ei galw'n gronfa ddata wasgaredig.

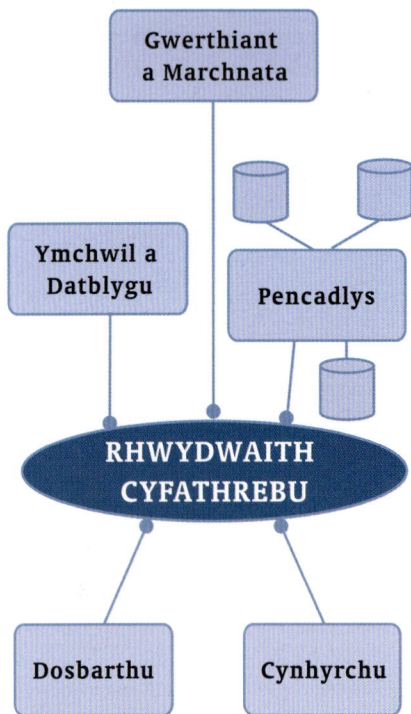

Rhwydwaith cyfathrebu ym mhencadlys cwmni

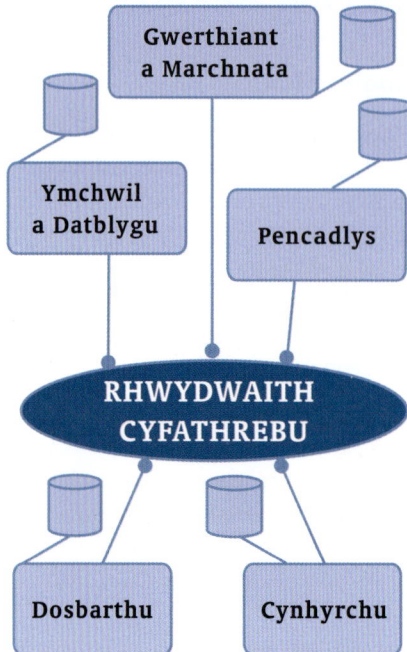

Rhwydwaith cyfathrebu wedi'i seilio ar gronfa ddata wasgaredig

Yn y rhwydwaith hwn mae'r holl ddata wedi'u storio ar y gweinydd ym mhencadlys y cwmni. Mae hyn yn golygu bod data sy'n cael eu creu yn y safleoedd yn cael eu trosglwyddo ar hyd y llinellau cyfathrebu i'r gronfa ddata ganolog yn y pencadlys. Os bydd llinell gyfathrebu'n cael ei thorri neu os bydd un o'r safleoedd yn methu, ni fydd yn bosibl prosesu'r data a gallent gael eu llygru. Hefyd, gan fod yn rhaid i holl draffig y rhwydwaith fynd drwodd i'r pencadlys, mae tagfa'n cael ei chreu a gall hyn arafu cyrchu'r data.

Mae'r diagram uchod yn dangos cronfa ddata wasgaredig. Yma mae'r data wedi'u dosbarthu ar draws sawl gweinydd ar wahanol safleoedd. Gellir storio'r data'n lleol, sy'n golygu bod y data sy'n cael eu creu ar y safle yn cael eu cadw ar y safle, sy'n golygu bod y cyflymder cyrchu'n uwch. Gellir cyrchu'r data ar y gweinyddion eraill o hyd, a bydd hyn yn gyflymach gan fod llai o draffig ar y rhwydwaith. I'r sawl sy'n defnyddio'r gronfa ddata mae'n ymddangos bod yr holl ddata yn yr un lle, ac mae'r gronfa ddata'n gweithredu'r un fath â phe bai'r holl ddata wedi'u lleoli gyda'i gilydd.

Manteision defnyddio cronfeydd data gwasgaredig

Rhai o fanteision defnyddio cronfeydd data gwasgaredig yw:

- Nid ydynt yn dibynnu ar storio data mewn un lleoliad, felly mae llai o berygl y bydd yr holl ddata'n mynd yn anhygyrch.
- Mae perfformiad y gronfa ddata'n gwella – mae ymholiadau'n cael eu prosesu'n gyflymach a chaiff cloddio data cymhleth ei gwblhau mewn llai o amser.
- Gall rhaglenni barhau i weithio os bydd y gweinydd lleol yn methu a bydd gweinyddion eraill sydd â data dyblygedig yn parhau'n hygyrch.
- Gall rhaglenni gyrchu cronfa ddata leol yn hytrach na gweinydd pell i leihau traffig ar y rhwydwaith a sicrhau'r perfformiad gorau posibl.

Anfanteision defnyddio cronfeydd data gwasgaredig

Rhai o anfanteision defnyddio cronfeydd data gwasgaredig yw:

- Maen nhw'n fwy cymhleth ac felly'n costio mwy i'w gosod a'u cynnal.
- Mae mwy o berygl i ddiogelwch gan fod ffeiliau'n cael eu trosglwyddo ar draws rhwydweithiau.
- Os bydd un o'r gweinyddion yn methu, gall gael effaith ar y gronfa ddata ac mae'n bosibl na fydd staff yn gallu cyrchu peth o'r data.
- Mae'r system yn dibynnu ar gyfathrebiadau data, felly os bydd llinell gyfathrebu'n methu mae'n bosibl na fydd modd cyrchu'r data.
- Gan fod llawer iawn o staff yn cyrchu'r gronfa ddata, mae'n bosibl y bydd anghysondebau yn y data.

Astudiaethau achos

▶ Astudiaeth achos 1 | tt. 94–95

Sut mae banciau'n defnyddio cloddio data

Er mwyn aros yn gystadleuol, mae angen i fanciau gynnig y cynnyrch iawn am y pris iawn ar yr adeg iawn i'w cwsmeriaid, a gwneud elw yn y broses wrth gwrs. Mae llawer o fanciau bellach yn defnyddio cloddio data i gyflawni hyn.

Gan ddefnyddio deallusrwydd artiffisial a chloddio data, gall banciau ddarganfod patrymau cymhleth neu fodelau mewn data sy'n eu helpu i ddatrys problemau busnes mewn meysydd fel marchnata uniongyrchol, gwerthuso risg credyd a chanfod twyll.

Dywedodd llefarydd ar ran un o'r banciau: 'Yn lle canolbwyntio ar gynhyrchion, rydym yn rhoi mwy o sylw i gwsmeriaid a'u teuluoedd a sut y gallwn eu helpu.'

Drwy ddefnyddio'r storfa enfawr o ddata trafodol sy'n cael eu cadw mewn warws data ar gyfer ei 6 miliwn o gwsmeriaid, mae un banc wedi cloddio'r data am yr holl gwsmeriaid hyn ac wedi cynhyrchu ffigur proffidioldeb ar gyfer pob un. Mae gwneud hyn yn rhoi gwell syniad i fanciau o sut y gallant wella eu gwasanaethau i gwsmeriaid a chynyddu elw'r banc o ganlyniad.

1 Defnyddir y termau canlynol yn yr astudiaeth achos. Rhowch ddiffiniad byr o bob un:
   (a) Cloddio data (1 marc)
   (b) Warws data (1 marc)
   (c) Data trafodol. (1 marc)

2 Gall cloddio data ddangos patrymau mewn data. Eglurwch, drwy roi enghraifft, beth yw ystyr y gosodiad hwn. (2 farc)

3 Disgrifiwch un fantais mae cloddio data yn ei chynnig i fanc.

   (2 farc)

4 Heblaw am ei ddefnydd mewn bancio, disgrifiwch yn fyr un enghraifft arall o gloddio data. (2 farc)

▶ Astudiaeth achos 2 | tt. 94–95

Y ffrwydrad data

Ym myd busnes heddiw, o ganlyniad i dwf, cyfuno a phrynu cwmnïau, a systemau TGCh newydd yn rhedeg ochr yn ochr â systemau hŷn, mae swm aruthrol o ddata'n cael ei storio ar wahanol systemau mewn gwahanol adrannau ac mewn gwahanol leoliadau daearyddol. Amcangyfrifwyd y bydd swm y data sy'n cael ei storio wedi cynyddu 30 o weithiau erbyn 2012 o'i gymharu â 2008.

Yr her i fusnesau yw gallu defnyddio'r data hyn i ddadansoddi proffidioldeb, cynhyrchu adroddiadau er mwyn cydymffurfio â deddfwriaeth, dadansoddi gwerthiant ac ymgyrchoedd marchnata, a rhagfynegi tueddiadau prynu.

Defnyddir warysau data i storio data trafodol a llawer o ddata eraill o systemau eraill, ac mae'r warws data fel rheol yn system ar wahân i sicrhau nad yw'n arafu systemau gweithredol. Mae hyn yn golygu nad yw'r systemau a ddefnyddir i gyflawni'r holl drafodion yn y busnes yr un fath â'r rheiny yn y warws data.

Rhaid i reolwyr gael gwybodaeth yn gyflym er mwyn gwneud penderfyniadau amserol, ac mae'r warws data, ynghyd ag offer cloddio data, yn caniatáu iddynt ddrilio i lawr i mewn i'r corff hwn o ddata i gasglu ffeithiau penodol am y busnes.

1 Eglurwch beth yw ystyr warws data a rhowch enw un system a ddefnyddir i gyflenwi data i'r warws data. (4 marc)

2 Cronfa ganolog o wybodaeth i'r cwmni cyfan yw warws data a gall unrhyw un ei gyrchu os oes ganddynt ganiatâd addas.
   (a) Nid oes gan yr holl staff ganiatâd i gyrchu'r holl wybodaeth yn y warws data. Disgrifiwch ddwy ffordd o atal mynediad i'r holl

   wybodaeth. (2 farc)
   (b) Defnyddir cloddio data i ddrilio i lawr i'r corff o ddata. Eglurwch ystyr cloddio data a disgrifiwch, drwy roi enghraifft, sut mae'n ddefnyddiol i fusnes. (4 marc)

Cwestiynau a Gweithgareddau

▶ Cwestiynau 1   tt. 90–94

1 Mae gan ysbyty gronfa ddata o gleifion sy'n cynnwys dau dabl. Mae strwythur y ddau dabl hyn i'w weld isod.

Cleifion
Rhif Claf
Cyfenw
Enw Cyntaf
Dyddiad Geni
Stryd
Tref
Cod Post
Rhif Ffôn Cysylltu
Rhif Ffôn Cartref
Rhif Ymgynghorydd

Ymgynghorwyr
Rhif Ymgynghorydd
Enw Ymgynghorydd

  (a) Pam mae'n well storio'r data mewn dau dabl ar wahân, yn hytrach na'u cadw mewn un tabl? (2 farc)

  (b) Pam mae Dyddiad Geni yn hytrach nag oedran yn cael ei storio yn y tabl Cleifion? (1 marc)

  (c) Pa faes fyddai'r brif allwedd:

    (i)   yn y tabl Cleifion?

    (ii) yn y tabl Ymgynghorwyr? (2 farc)

2 Bydd cyrff yn cadw llawer o ddata mewn cronfeydd data perthynol gyda System Rheoli Cronfeydd Data (*DBMS*).

  (a) Eglurwch beth yw ystyr cronfa ddata berthynol ac eglurwch ei phrif gydrannau. (2 farc)

  (b) Eglurwch beth yw ystyr System Rheoli Cronfeydd Data. (4 marc)

3 Mae rheolwr cwmni llogi offer yn dymuno defnyddio system rheoli cronfeydd data perthynol i'w helpu i reoli'r busnes. Mae'r gronfa ddata yn storio'r data mewn tri thabl, sef: OFFER, CWSMERIAID a RHENTOLION.

  (a) Beth yw'r prif fanteision i'r rheolwr hwn o storio'r data mewn cronfa ddata berthynol yn hytrach na chronfa ddata ffeiliau fflat? (3 marc)

  (b) Ar gyfer pob un o'r tablau, nodwch y maes sy'n brif allwedd a rhestrwch y meysydd eraill ar gyfer pob tabl a fyddai'n galluogi'r rheolwr i storio'r data gyda chyn lleied o afreidrwydd â phosibl. (4 marc)

4 Bydd llawer o gyrff yn defnyddio cronfa ddata wasgaredig i storio eu data.

  (a) Beth yw ystyr cronfa ddata wasgaredig? (3 marc)

  (b) Beth yw'r manteision ac anfanteision i gorff o ddefnyddio cronfa ddata wasgaredig? (6 marc)

5 Mae corff yn defnyddio system ffeiliau fflat gyfrifiadurol i storio ei ddata. Mae'r system ffeiliau fflat hon yn achosi problemau i'r cwmni.

  (a) Disgrifiwch **dair** mantais i'r cwmni o ddefnyddio cronfa ddata wasgaredig yn lle system ffeiliau fflat. (6 marc)

  (b) Mae'r corff yn defnyddio tair ffeil ar hyn o bryd: Cwsmer, Stoc ac Archebion. Wrth newid i system cronfa ddata wasgaredig byddai'n rhaid normaleiddio'r ffeiliau hyn. Eglurwch yn glir beth yw ystyr y term normaleiddio. (2 farc)

  (c) Mae rhannau o'r tair ffeil yn cael eu dangos isod. Normaleiddiwch y ffeiliau hyn, gan egluro unrhyw dybiaethau neu ychwanegiadau a wnewch at y ffeiliau. (5 marc)

Ffeil Cwsmer

Cyfenw	Enw Cyntaf	Stryd	Tref	Dinas	Cod Post
Charles	Ruby	21 Stryd y Dwyrain	Treforys	Abertawe	SA12 5DN
Hughes	Amy	188 Heol Morys	Treforys	Abertawe	SA9 5BZ

Ffeil Archebion

Cyfenw	Enw Cyntaf	Cod Post	Dyddiad Archebu	Eitem a Archebwyd	Nifer a Brynwyd	Pris	Cyfanswm Cost	Talwyd
Hughes	Amy	SA9 5BZ	02/01/11	Pren mesur	10	£1.50	£15.00	Do
Hughes	Amy	SA9 5BZ	02/01/11	Rîm o bapur llinellog A4	20	£3.50	£70.00	Do
Charles	Ruby	SA12 5DN	03/01/11	Styffylwr mawr	4	£5.50	£22.00	Naddo
Charles	Ruby	SA12 5DN	03/01/11	Styffylau	5	£1.20	£6.00	Naddo
Hughes	Amy	SA9 5BZ	05/01/11	Styffylwr mawr	10	£5.50	£55.00	Do

Ffeil Stoc

Enw'r Eitem	Pris	Nifer mewn Stoc
Pren mesur	£1.50	120
Rîm o bapur llinellog A4	£3.50	1298
Styffylwr mawr	£5.50	438
Styffylau	£1.20	654
Papur llungopïo	£2.85	1452

6 Bydd y mwyafrif o gyrff yn defnyddio cronfeydd data perthynol gyda Systemau Rheoli Cronfeydd Data.
(a) Eglurwch fanteision ac anfanteision cymharol defnyddio cronfa ddata berthynol yn hytrach na dull ffeiliau fflat ar gyfer storio data'r corff. (4 marc)
(b) Eglurwch ystyr System Rheoli Cronfeydd Data a chyferbynnwch ei manteision a'i hanfanteision. (6 marc)
7 Mae banc yn defnyddio warws data i gadw manylion ei gwsmeriaid a'u trafodion bancio. Mae'r storfa ganolog fawr hon o ddata yn galluogi staff y banc i ddefnyddio cloddio data i gael gwybodaeth am gwsmeriaid a'u harferion bancio.
(a) Disgrifiwch beth yw warws data ac eglurwch y manteision posibl i'r banc hwn. (6 marc)
(b) Eglurwch ystyr y term cloddio data a rhowch enghraifft o sut y gallai'r banc ei ddefnyddio. (3 marc)
8 Normaleiddio yw'r broses a ddefnyddir i sicrhau nad oes gan gronfa ddata unrhyw ddata afraid neu anghyson.
(a) Eglurwch ystyr:
(i) Data afraid (2 farc)
(ii) Data anghyson. (2 farc)
(b) Eglurwch berthnasedd prif allweddi, allweddi estron a pherthnasoedd mewn cronfa ddata berthynol. (6 marc)
(c) Mae'r data mewn cronfa ddata sydd heb gael ei normaleiddio yn achosi nifer o broblemau. Disgrifiwch **ddwy** broblem sy'n cael eu hachosi pan gaiff data heb eu normaleiddio eu storio mewn cronfa ddata. (4 marc)
9 Mae cwmni'n datblygu cronfa ddata fawr at ddefnydd pawb yn y cwmni. Bydd y staff sy'n datblygu'r gronfa ddata yn defnyddio geiriadur data.
(a) Eglurwch beth yw ystyr y term geiriadur data a rhowch **dri** pheth y gallai eu cynnwys. (4 marc)
(b) Disgrifiwch **un** rheswm pam y dylid defnyddio geiriadur data wrth ddatblygu'r gronfa ddata hon. (3 marc)

Cwestiynau a Gweithgareddau *(parhad)*

▶ Gweithgaredd 1: Enwi'r brif allwedd

Prif allwedd yw'r maes unigryw mewn grŵp o feysydd. Dyma rai meysydd mewn tabl. Rhaid i chi enwi'r brif allwedd. Dangosir enw'r tabl ynghyd â rhestr o'r meysydd y bydd y tabl yn eu cynnwys. Mae sampl o'r data a fydd yn cael eu rhoi yn y maes yn cael ei ddangos hefyd.

Ar gyfer pob tabl, nodwch enw'r maes y dylid ei ddewis yn brif allwedd, gan roi rheswm. Cofiwch feddwl am bwrpas y tabl wrth benderfynu ar y brif allwedd.

Tabl Myfyriwr	
Cyfenw	Jones
Enw cyntaf	Peter
Teitl	Mr
Rhyw	Gwryw
RhA myfyriwr	980006

Lleoedd Parcio	
Rhif lle parcio	190
Rhif cofrestru'r car	DR08TGH
Rhif gweithiwr	180041
Cod adran	104

Tabl Claf	
Cyfenw'r claf	Graham
Enw cyntaf y claf	Julie
Teitl	Miss
Rhif GIG	09-09809-8
Dyddiad geni	16/12/93

Tabl Gweithiwr	
Cod treth	416L
Cyfenw	Jones
Dyddiad geni	15/06/90
Rhif yswiriant gwladol	AB100136Y
Cod adran	029

Tabl Cynnyrch	
Disgrifiad o'r cynnyrch	Papur argraffu A4
Pris yr uned	£3.50
Nifer mewn stoc	127
Cod cynnyrch	1381
Rhif cyflenwr	2015

▶ Gweithgaredd 2: Normaleiddio data ar gyfer cronfa ddata coleg

Yn y gweithgaredd hwn mae gofyn i chi lwyr normaleiddio'r data ar gyfer cronfa ddata o fyfyrwyr mewn coleg. Mae prifathro'r coleg wedi nodi'r meysydd i'w storio ac wedi rhoi'r rhestr ganlynol i chi:

Rhif myfyriwr	Rhif cwrs
Cyfenw	Dyddiad geni
Enwau cyntaf	Teitl y cwrs
Cyfeiriad	Cost y cwrs
Rhif ffôn	Dyddiad cofrestru
Enw darlithydd	Rhif darlithydd

Mae'r prifathro hefyd wedi rhoi'r wybodaeth ganlynol i chi:
- Mae rhif myfyriwr unigryw yn cael ei roi i bob myfyriwr pan fydd yn ymuno â'r coleg.
- Gall myfyriwr ddilyn mwy nag un cwrs.
- Gall yr un cwrs (h.y. cyrsiau â'r un rhif cwrs) gael ei ddysgu gan un neu ragor o ddarlithwyr.

Gan ddefnyddio'r meysydd yn y rhestr uchod yn unig, ewch drwy'r broses o normaleiddio'r data.
Dylech ddisgrifio pob cam o'r broses normaleiddio (ffurf normal 1af, 2il a 3ydd) a nodi'n glir y prif allweddi ym mhob tabl.

▶ Gweithgaredd 3: Geiriadur data go iawn

Mae'r Gwasanaeth Iechyd Gwladol (GIG) yn defnyddio un o'r systemau TGCh mwyaf ym Mhrydain ac mae llawer o bobl yn gweithio ar systemau sy'n defnyddio'r swm enfawr o ddata sydd wedi'u storio.

I sicrhau cysondeb, ac i ddeall y system bresennol, mae angen iddynt gyfeirio at y geiriadur data. Gan fod cymaint o bobl yn defnyddio'r adnodd hwn, mae'r GIG wedi'i gynhyrchu ar-lein, ac ar gyfer y gweithgaredd hwn mae gofyn i chi ymchwilio iddo ac ysgrifennu paragraff byr sy'n amlinellu cynnwys y geiriadur data hwn.

Y cyfeiriad gwe ar gyfer geiriadur data'r GIG yw: http://www.datadictionary.nhs.uk/

▶ Gweithgaredd 4: Normaleiddio data ar gyfer cronfa ddata coleg

Mae'r eitemau data canlynol heb eu normaleiddio ac mae angen eu llwyr normaleiddio (h.y. eu trawsnewid yn drydedd ffurf normal) i greu tablau sy'n lleihau dyblygu data ar draws y tablau, gan ddatrys llawer o'r problemau sy'n gysylltiedig ag afreidrwydd data.

Mae enw'r tabl mewn teip trwm ac mae'r holl feysydd mewn cromfachau.

ARCHEB CWSMER (<u>Rhif-archeb-cwsmer</u>, Rhif-cwsmer, Enw-cwsmer, Cyfeiriad-cwsmer, Rhif-ffôn-cwsmer, Rhif-depo, Enw-depo, Rhif-cynnyrch, Enw-cynnyrch, Nifer-y-cynhyrchion, Pris-cynnyrch)

Cofiwch: Mae prif allweddi wedi'u tanlinellu.

Ewch drwy'r broses o normaleiddio, gan ddangos y gwahanol gamau (ffurf normal gyntaf, ail ffurf normal ac yn olaf trydydd ffurf normal).

I'ch helpu drwy'r broses, cofiwch y canlynol:

Ffurf normal gyntaf

Mae tabl yn y ffurf normal gyntaf os nad yw'n cynnwys unrhyw grwpiau ailadroddol.

Ail ffurf normal

Rhaid i'r tabl fod yn y ffurf normal gyntaf a rhaid iddo beidio â chynnwys unrhyw feysydd nad ydynt yn allweddi sy'n dibynnu ar ran yn unig o'r brif allwedd.

Trydydd ffurf normal

Rhaid i'r tabl fod yn yr ail ffurf normal a rhaid iddo beidio â chynnwys unrhyw feysydd nad ydynt yn allweddi sy'n dibynnu ar feysydd eraill nad ydynt yn allweddi.

Trawsnewidiwch y data yn drydedd ffurf normal, gan ddangos yr holl gamau.

▶ Gweithgaredd 5: Normaleiddio data ar gyfer cwmni llogi ceir

Mae cwmni llogi ceir yn defnyddio system draddodiadol ar hyn o bryd ond oherwydd cynnydd mewn cerbydau a busnes mae wedi penderfynu defnyddio cronfa ddata gyfrifiadurol i storio'r data.

Mae rhywun yn y cwmni wedi bod ar gwrs yn y coleg lleol ond mae wedi arfer â chronfeydd data ffeiliau fflat yn unig ac felly mae wedi penderfynu storio'r holl ddata mewn un ffeil o'r enw CERBYD. Mae'r meysydd y mae angen eu storio yn cael eu dangos o dan yr enw ffeil CERBYD.

CERBYD
Rhif-cofrestru
Gwneuthuriad
Model
Blwyddyn
Rhif-cwsmer
Cyfenw
Blaenlythyren
Cyfeiriad
Dyddiad-llogi
Dyddiad-dychwelyd

Weithiau mae'n bosibl y bydd un cwsmer yn llogi llawer o geir ar yr un adeg, felly bydd angen i chi gofio hyn wrth fynd drwy'r broses normaleiddio.

(a) Mae'r person sydd wedi bod ar gwrs yn y coleg wedi awgrymu y gellid defnyddio cronfa ddata ffeiliau fflat i storio'r data ond rydych chi'n anghytuno.

Cyflwynwch ddadl ysgrifenedig, gan gynnwys enghreifftiau, sy'n dangos y problemau sy'n debygol o godi a pham y byddai cronfa ddata ffeiliau fflat yn anaddas ar gyfer y cymhwysiad hwn. Hefyd, eglurwch fanteision storio'r data mewn cronfa ddata berthynol.

(b) Nawr mae'n rhaid i chi fynd drwy'r broses normaleiddio hyd nes bod y meysydd wedi'u gosod yn y drydedd ffurf normal. Dylech egluro sut yr ydych chi'n dod i'ch trefniant terfynol.

Cymorth gyda'r arholiad

Enghraifft 1

1 Mae manylion cleifion mewnol mewn ysbyty yn cael eu storio mewn cronfa ddata berthynol. Pan fydd cleifion yn cael eu derbyn i'r ysbyty mae ward yn cael ei neilltuo iddynt. Mae wardiau yn cael eu neilltuo i'r staff hefyd.

 (a) Eglurwch ystyr y term cronfa ddata berthynol. (2 farc)

 (b) Mae cronfeydd data perthynol yn cadw'r data mewn nifer o dablau. Yn system cleifion mewnol yr ysbyty mae tabl ar gyfer Ward ac mae hwn yn cynnwys y meysydd canlynol:

 WARD (RhAWard, NiferyGwelyau, RhAStaff)

 Yn y tabl hwn, RhAWard yw'r brif allwedd a RhAStaff yw'r allwedd estron.

 (i) Eglurwch y gwahaniaeth rhwng prif allwedd ac allwedd estron. (2 farc)

 (ii) Rhowch **ddau** dabl addas arall y byddech yn disgwyl eu gweld yn y gronfa ddata hon o gleifion mewnol, gan nodi unrhyw brif allweddi neu allweddi estron. (6 marc)

 (iii) Mae cronfeydd data perthynol yn fwy diogel na chronfeydd data ffeiliau fflat. Eglurwch pam. (3 marc)

 (c) Mae'r mwyafrif o ysbytai'n defnyddio cronfeydd data gwasgaredig. Disgrifiwch **ddwy** o fanteision a **dwy** o anfanteision defnyddio cronfeydd data gwasgaredig mewn ysbytai. (4 marc)

Ateb myfyriwr 1

1 (a) Cronfa ddata gyda pherthnasoedd rhwng y tablau yw cronfa ddata berthynol.

 (b) (i) Mae prif allwedd yn allwedd sy'n unigryw mewn tabl ond mae allwedd estron yn allwedd nad yw'n unigryw.

 (ii) CLAF, MEDDYG a NYRS

 CLAF prif allwedd RhAClaf, allwedd estron RhAWard

 MEDDYG prif allwedd RhAMeddyg, allwedd estron RhAClaf

 NYRS prif allwedd RhANyrs, allwedd estron RhAWard

 (iii) Mae'r data'n cael eu storio ar wahân i'r rhaglenni a ddefnyddir i'w cyrchu.

 Mae hierarchaeth o gyfrineiriau sy'n caniatáu i bobl gyrchu'r wybodaeth am gleifion sy'n angenrheidiol iddynt wneud eu gwaith a dim gwybodaeth arall.

 Ni allwch gopïo cronfa ddata berthynol.

 (c) Manteision yw ei bod hi'n llawer cyflymach i gyrchu'r data dros y rhwydwaith oherwydd y gall y data gael eu gwasgaru dros sawl gweinydd a'i bod hi'n haws gwneud copïau wrth gefn o'r data gan ei bod hi'n bosibl eu copïo ar unrhyw weinydd. Anfanteision yw bod trosglwyddo data ar hyd llinellau cyfathrebu yn risg i ddiogelwch. A phe bai un o'r cysylltau i weinydd yn methu, yna ni fyddai'n bosibl cael data o'r gweinydd hwnnw.

Sylwadau'r arholwr

1 (a) Dim marciau am yr ateb arwynebol iawn hwn. Mae termau fel hwn yn codi'n aml iawn a dylai myfyrwyr gofio'r diffiniad manwl air am air.

 (b) (i) Mae'r diffiniad o brif allwedd yn ddigon da i ennill 1 marc ond nid yw'n glir beth yw allwedd estron.

 (ii) Rhoddwyd enw'r tabl (h.y. WARD) a'r meysydd i'r myfyrwyr i'w helpu gyda'r rhan hon o'r cwestiwn. Dylai'r myfyriwr fod wedi sylwi y bydd RhAStaff yn cyfeirio at yr holl staff sy'n gysylltiedig â gofal y claf dan sylw ac y byddai hyn yn cynnwys meddygon, nyrsys a staff eraill. Dyma'r math o ateb sy'n eithaf anodd ei farcio. Gallai MEDDYG a NYRS fod yn dablau priodol ac mae'r myfyriwr wedi ysgrifennu prif feysydd a meysydd estron priodol.

 Rhoddwyd tri allan o chwe marc yma.

 (iii) Mae dau o'r atebion hyn yn gywir ond nid yw'n gywir i ddweud na ellir copïo cronfa ddata berthynol.

 (c) Mae'r holl atebion hyn yn gywir, felly marciau llawn.

 (10 marc allan o 17)

Ateb myfyriwr 2

1 (a) Mae'n gasgliad o ddata sy'n cael eu storio mewn dau neu ragor o dablau gyda chysylltau o'r enw perthnasoedd rhwng y tablau sy'n golygu y gellir echdynnu'r data sydd wedi'u cynnwys yn y tabl. Mae'r meddalwedd rhaglenni'n gallu defnyddio unrhyw ddata sydd wedi'u storio yn y tablau..

   (b) (i) Mae prif allwedd yn faes sy'n unigryw mewn rhes mewn tabl. Er enghraifft, byddai RhAStaff yn unigryw yn y tabl Staff gan na fyddai unrhyw ddau aelod o'r staff yn cael yr un RhAStaff.

   Mae allwedd estron yn brif allwedd mewn un tabl sy'n allwedd yn unig mewn tabl arall ac mae'n cael ei defnyddio i ddarparu'r cyswllt rhwng y tablau. Er enghraifft, mae RhAStaff yn faes prif allwedd yn y tabl Staff ond yn allwedd estron yn y tabl WARD.

   (ii) Dau dabl arall yw STAFF a CLAF.

   Yn y tabl STAFF, y brif allwedd yw RhAStaff a'r allwedd estron yw RhAWard. Yn y tabl CLAF, y brif allwedd yw RhAClaf a'r allwedd estron yw RhAWard.

   (iii) Gall hierarchaeth o gyfrineiriau gael ei defnyddio i sicrhau mynediad i rai tablau. Mae hyn yn golygu defnyddio enw defnyddiwr a chyfrinair i benderfynu pa dablau/ffeiliau y gall defnyddiwr eu cyrchu. Gallwch hefyd benderfynu a yw defnyddiwr yn gallu ychwanegu cofnod newydd, dim ond newid meysydd o fewn cofnod, neu dim ond edrych ar ac nid newid cofnodion. Mae hyn yn ei gwneud hi'n anos i ddefnyddwyr dibrofiad newid data'n ddamweiniol yn y ffeiliau.

   Mae data'n cael eu storio ar wahân i'r rhaglenni a ddefnyddir i'w cyrchu ac mae hyn yn golygu nad yw newid y rhaglen yn gallu llygru'r data.

   (c) Mae prosesu'r data'n cael ei rannu rhwng y cyfrifiaduron mewn gwahanol leoliadau, felly gall yr holl ddata gael eu prosesu ar yr un pryd, felly mae'n gyflymach.

Sylwadau'r arholwr

1 (a) Yma mae'r myfyriwr wedi nodi prif gydrannau cronfa ac er bod y diffiniad ychydig yn wahanol i'r diffiniad yn y fanyleb, mae'n dal yn dderbyniol ac yn ennill marciau llawn.

   (b) (i) Diffiniad da iawn yw hwn o'r ddau derm ac mae'r myfyriwr wedi cysylltu'r diffiniad â'r gronfa ddata sy'n dangos yn glir ei fod yn deall prif allweddi ac allweddi estron.

   (ii) Enwyd y ddau dabl yn gywir, a hefyd y prif allweddi a'r allweddi estron ar gyfer y ddau dabl. Marciau llawn.

   (iii) Rhoddwyd tri esboniad cywir yma, felly marciau llawn.

   (c) Yn anffodus mae'r myfyriwr wedi rhoi disgrifiad eithaf da o brosesu gwasgaredig yn lle cronfeydd data gwasgaredig, sy'n golygu na ellir rhoi unrhyw farciau.

   (13 marc allan o 17)

Ateb yr arholwr

1 (a) Dau farc ar y mwyaf gydag un marc am bob pwynt. Cronfa ddata berthynol – casgliad mawr o eitemau data wedi'u storio mewn tablau (1) sy'n cynnwys cysylltau rhwng y tablau (1) fel y gellir cyrchu'r data mewn llawer gwahanol ffordd (1) a gan amrywiaeth o wahanol raglenni (1).

   (b) (i) Un marc am bob diffiniad cywir sy'n debyg i'r canlynol.

   Maes sy'n cael ei ddefnyddio i ddiffinio'n unigryw gofnod neu linell/rhes benodol mewn tabl yw prif allwedd.

   Maes mewn un tabl sydd hefyd yn brif allwedd tabl arall yw allwedd estron. Defnyddir allweddi estron i sefydlu perthnasoedd rhwng y prif dabl a'r is-dablau eraill.

   (ii) Un marc yr un am y ddau dabl, un marc yr un am brif allweddi cywir ac un marc yr un am allweddi estron cywir.

   Tabl STAFF (prif allwedd RhAStaff ac allwedd estron RhAWard)

   Tabl CLAF (prif allwedd RhAClaf ac allwedd estron RhAWard)

   (iii) Un marc yr un am dri rheswm megis:

   Hierarchaeth o gyfrineiriau

   Storio data ar wahân i raglenni

Hawliau mynediad i rannau o'r rhaglen.

   (c) Un marc yr un am ddwy fantais ac un marc yr un am ddwy anfantais, e.e.:

Manteision:

Ymateb cyflymach i ymholiadau gan ddefnyddwyr y gronfa ddata

Nid oes dibyniaeth ar un storfa ganolog enfawr o ddata

Hawdd gwneud copïau wrth gefn a chopïo data rhwng gweinyddion

Os bydd un gweinydd yn methu bydd hi'n bosibl defnyddio'r lleill

Llai o draffig ar y rhwydwaith gan ei bod yn bosibl defnyddio'r data ar y gweinydd lleol i wneud ymholiadau lleol.

Anfanteision:

Maen nhw'n dibynnu'n drwm ar rwydweithiau a chyfathrebiadau nad ydynt bob amser yn ddibynadwy

Problemau diogelwch, yn enwedig os yw data personol sensitif yn cael eu trosglwyddo

Os bydd un o'r cysylltau i weinydd yn methu, yna ni fydd yn bosibl cael y data o'r gweinydd hwnnw

Mwy costus oherwydd bod angen defnyddio llinellau cyfathrebu drud

Mae'n fwy tebygol y bydd y data'n anghyson

Anoddach sicrhau diogelwch data sydd mewn llawer o wahanol leoliadau.

Cymorth gyda'r arholiad *(parhad)*

Enghraifft 2

2 Mae manylion staff, myfyrwyr a chyrsiau'n cael eu storio mewn un ffeil ac mae enghraifft o gynnwys y ffeil hon i'w gweld isod.

RhA Myfyriwr	Enw	Dyddiad Geni	Rhyw	Cod Cwrs	Enw Cwrs	RhA Darlithydd	Enw Darlithydd
0022	G Williams	12/12/95	B	FFIS1	Ffiseg Safon Uwch	211	D Preston
0012	T Ash	09/08/95	G	MATH2	Mathemateg Safon Uwch	310	H Pearce
0022	G Williams	12/12/95	B	MATH1	Mathemateg Safon Uwch	211	D Preston

Nid yw'r data yn y ffeil uchod wedi cael eu normaleiddio.

(a) Drwy ddefnyddio'r data yn y ffeil uchod i egluro eich ateb, disgrifiwch **ddwy** wahanol broblem sy'n codi pan gaiff data eu storio yn y ffeil yn y ffordd hon, hynny yw, heb eu normaleiddio. (4 marc)

(b) Gall y data uchod gael eu normaleiddio a'u rhoi mewn tablau gyda phrif allweddi ac allweddi estron a pherthnasoedd rhwng y tablau.
Gan ddefnyddio **dau dabl**, normaleiddiwch y data yn y tabl hwn a nodwch yn glir y prif allweddi a'r allweddi estron sydd eu hangen. (5 marc)

Ateb myfyriwr 1

2 (a) Rhaid i chi deipio mwy o ddata i mewn nag sydd eu hangen ac mae hyn yn cymryd amser ac yn costio arian.
Mae posibilrwydd y gallai'r data ar gyfer yr un person gael eu teipio i mewn mewn camgymeriad. Er enghraifft, gallai RhAMyfyriwr G Williams gael ei deipio fel 0022 a 0020, gan roi'r argraff bod dwy fyfyrwraig wahanol. Pe câi'r data eu normaleiddio, ni fyddai hyn yn digwydd gan y byddai RhAMyfyriwr yn ymddangos unwaith yn unig yn y tabl Myfyriwr.

(b) MYFYRIWR (RhAMyfyriwr, Enw, Dyddiad Geni, Rhyw, Cod Cwrs)

Sylwadau'r arholwr

2 (a) Mae'r frawddeg gyntaf am arbed amser ac arian yn wir ond dylai'r myfyriwr fod wedi bod yn fwy manwl ac wedi cyfeirio at y data dyblygedig yn y tabl. Hefyd mae'n well defnyddio'r termau cywir fel dyblygu data, anghysondeb data, ac ati.
Mae ail ran yr ateb yn llawer gwell gan fod y myfyriwr wedi dechrau cyfeirio at y data yn y ffeil.

(b) Mae'r myfyriwr wedi rhoi manylion ar gyfer un o'r tablau'n unig ac nid y ddau y mae'r cwestiwn yn gofyn amdanynt. Mae'r tabl a'i feysydd yn gywir ac mae prif allwedd wedi cael ei nodi drwy ei thanlinellu ond ni thynnwyd sylw at allwedd estron. Rhoddir dau farc, un am y tabl cywir ac un am y brif allwedd gywir.
(5 marc allan o 9)

Ateb myfyriwr 2

2 (a) Dyblygu data - rhaid i fanylion gael eu teipio i mewn fwy nag unwaith. Er enghraifft, mae manylion y darlithydd D Preston wedi'u storio fwy nag unwaith. Mae mewnbynnu data diangen yn gwastraffu amser ac mae mwy o siawns o wneud gwallau trawsysgrifol.

Anghysondeb data – mae dyblygu data yn golygu bod rhai o'r data yn yr un ffeil yr un fath. Os caiff y data eu newid yn un o'r ffeiliau, yna rhaid eu newid hefyd yn y ffeil arall i sicrhau cysondeb. Er enghraifft, mae'r manylion ar gyfer G Williams yn ymddangos ddwywaith ac os bydd camgymeriad yn cael ei wneud gyda rhai manylion – fel camgymeriad teipio – bydd yn anodd gwybod pa un sy'n gywir.

(b) Defnyddir dau dabl, MYFYRIWR a CWRS.

MYFYRIWR (<u>RhAMyfyriwr</u>, Enw, Dyddiad Geni, Rhyw, CodCwrs*)

CWRS (<u>CodCwrs</u>, EnwCwrs, RhADarlithydd*, EnwDarlithydd)

Yn y tabl Myfyriwr, RhAMyfyriwr yw'r brif allwedd a CodCwrs* yw'r allwedd estron.

Yn y tabl Cwrs, CodCwrs yw'r brif allwedd a RhADarlithydd* yw'r allwedd estron.

Sylwadau'r arholwr

2 (a) Mae hwn yn ateb da gan fod y myfyriwr wedi defnyddio'r termau cywir ac wedi rhoi esboniad da. Gallai fod wedi egluro anghysondeb data'n well, gan ddefnyddio data yn y tabl. Er enghraifft, gallai fod wedi rhoi enghraifft o sut y gallai'r data fynd yn anghyson, e.e. teipio'r Dyddiad Geni anghywir ar gyfer G Williams.

(b) Ateb da iawn yw hwn. Mae'r myfyriwr wedi nodi'r tablau cywir a'r meysydd cywir ym mhob tabl. Hefyd enwyd y prif allweddi cywir ar gyfer pob tabl drwy eu tanlinellu a defnyddiwyd yr arwydd clwyd i ddynodi'r allweddi estron. I fod yn hollol eglur, mae'r prif allweddi a'r allweddi estron wedi cael eu henwi mewn darn o destun. Mae hyn yn syniad da ac o gryn gymorth i'r arholwr. Mae hwn yn ateb perffaith ac mae'n haeddu marciau llawn.

(8 marc allan o 9)

Ateb yr arholwr

2 (a) Un marc am bob mantais ac enghraifft x 2.

Dyblygu data – caiff data eu dyblygu yn y ffeil sy'n golygu bod angen mewnbynnu mwy o ddata nag sydd eu hangen, gan wastraffu adnoddau. Er enghraifft, mae manylion y fyfyrwraig â'r RhAMyfyriwr 0022 wedi'u dyblygu yn y tabl.

Anghysondeb data – os yw data wedi'u dyblygu, ac os gwnewch gamgymeriad mewn un rhes o ddata, ni fyddwch yn gwybod pa fersiwn sy'n gywir. Er enghraifft, bydd gwneud camgymeriad gyda'r enw yn rhoi dwy set o ddata, a bydd yn anodd gwybod pa un sy'n gywir.

(b) Un marc am bob tabl cywir x 2, un marc am bob prif allwedd x 2 ac un marc am bob allwedd estron x 2 (hyd at uchafswm o 5 marc).

Enghreifftiau'n debyg i:

MYFYRIWR (<u>RhAMyfyriwr</u>, Enw, Dyddiad Geni, Rhyw, CodCwrs#)

CWRS (<u>CodCwrs</u>, EnwCwrs, RhADarlithydd#, EnwDarlithydd)

Mae prif allweddi wedi'u tanlinellu ac allweddi estron wedi'u dynodi ag arwydd clwyd.

I gael marciau llawn rhaid nodi'r prif allweddi a'r allweddi estron yn glir.

Map meddwl cryno

Cronfeydd data

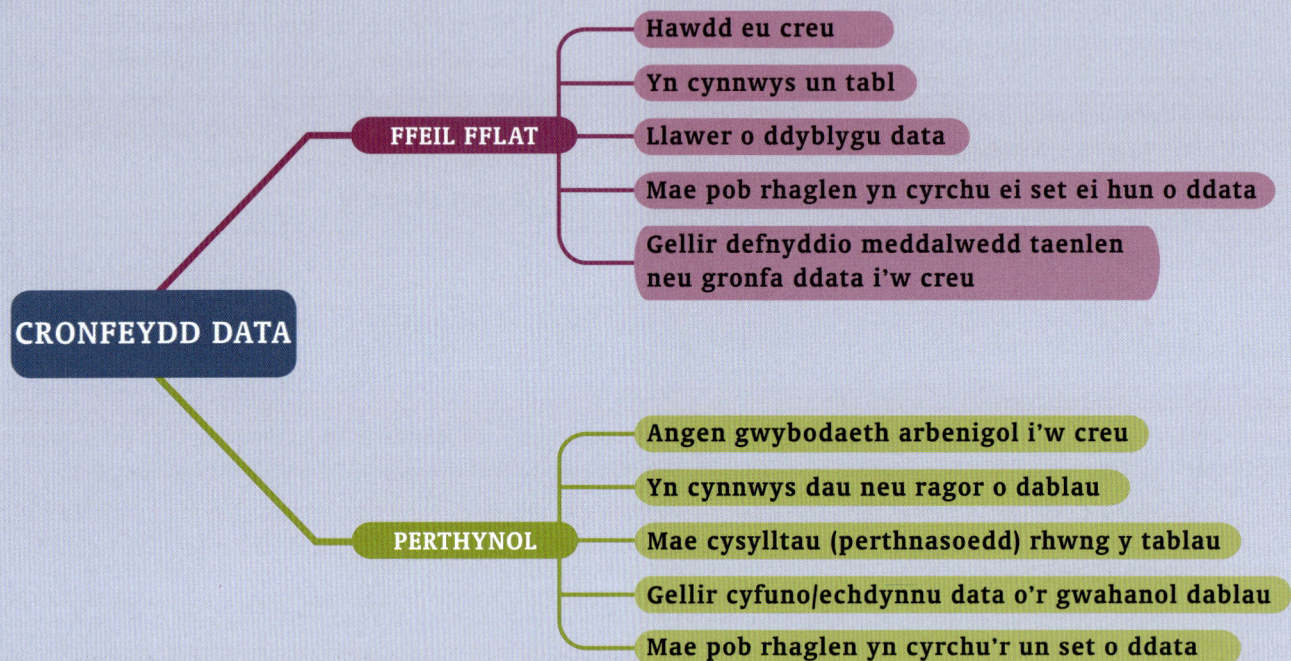

CRONFEYDD DATA

FFEIL FFLAT
- Hawdd eu creu
- Yn cynnwys un tabl
- Llawer o ddyblygu data
- Mae pob rhaglen yn cyrchu ei set ei hun o ddata
- Gellir defnyddio meddalwedd taenlen neu gronfa ddata i'w creu

PERTHYNOL
- Angen gwybodaeth arbenigol i'w creu
- Yn cynnwys dau neu ragor o dablau
- Mae cysylltau (perthnasoedd) rhwng y tablau
- Gellir cyfuno/echdynnu data o'r gwahanol dablau
- Mae pob rhaglen yn cyrchu'r un set o ddata

TOPIG 7: Rheoli newid

Bydd cyrff yn wynebu sawl her pan gaiff systemau TGCh newydd eu cyflwyno neu systemau presennol eu newid yn sylweddol. Mae'n bosibl y bydd angen newid y strwythurau trefniadaethol, y bydd angen i staff symud o'r adran neu'r lleoliad lle maen nhw'n gweithio, ac y bydd angen iddynt weithio mewn gwahanol grwpiau. A bydd angen i'r mwyafrif ohonynt gael hyfforddiant ychwanegol.

Yn y topig hwn byddwn yn edrych ar y materion sy'n gysylltiedig â rheoli newid a sut y gellir cyflwyno'r newid yn y ffordd leiaf trafferthus.

▼ Y cysyniad allweddol sy'n cael sylw yn y topig hwn yw:

▶ Canlyniadau newid

CYNNWYS

Canlyniadau newid

Cyflwyniad

Pan fydd corff yn cyflwyno system gwybodaeth newydd mae'n debyg y bydd yn arwain at newid yn arferion gweithio'r corff ac y bydd y newid yn effeithio ar y staff mewn rhyw ffordd. Bydd yn rhaid i'r rheolwyr a fydd yn goruchwylio datblygiad y system newydd allu rheoli newid yn fedrus er mwyn sicrhau cydweithrediad llawn y staff.

Yn sicr, bydd llawer o bobl yn gwrthwynebu newid hyd nes iddynt weld y gall fod o fantais iddynt.

Canlyniadau newid

Mae nifer o ganlyniadau newid y mae angen eu rheoli pan gaiff systemau TGCh newydd eu cyflwyno i gorff. Yn y topig hwn byddwn yn edrych ar wahanol ganlyniadau newid a sut y gellir eu rheoli.

Y sgiliau sydd eu hangen ac nad oes eu hangen

Mae angen sgiliau newydd i drin systemau TGCh newydd a rhaid i'r staff fod yn barod i ddysgu'r sgiliau hyn. Mae'n bosibl na fydd angen hen sgiliau ac y bydd yn rhaid i staff fynd ar gyrsiau er mwyn cadw eu swyddi. Pan gaiff systemau newydd eu cyflwyno, bydd angen hyfforddi'r staff i'w defnyddio.

Mae'r defnydd cynyddol o TGCh fel arfer yn golygu bod cynnydd yn nifer y swyddi medrus sydd ar gael (rheolwyr rhwydwaith, rheolwyr project, rhaglenwyr, peirianwyr cyfrifiadurol, ac ati), fel arfer ar draul swyddi llai medrus.

Mae nifer o swyddi a sgiliau wedi diflannu bellach, er enghraifft:

- Teipyddion – bydd y rhan fwyaf o bobl yn gairbrosesu eu dogfennau eu hunain.
- Clercod ffeilio – caiff data eu storio mewn cronfeydd data ar gyfrifiaduron bellach.
- Clercod post mewnol – byddai'r rhain yn dosbarthu post mewnol. Defnyddir e-bost gan mwyaf i wneud hyn heddiw.

Newidiadau i'r strwythur trefniadaethol

Pan gaiff systemau hollol newydd eu cyflwyno i gyrff, bydd llawer ohonynt yn manteisio ar y cyfle i newid strwythur y corff fel ei fod yn cyd-fynd yn well â'r system newydd. Mae cyflwyno systemau newydd yn golygu'n aml fod y ffiniau rhwng meysydd swyddogaethol yn cael eu pylu ac y bydd gofyn i'r staff wneud amrywiaeth fwy eang o dasgau. Er enghraifft, efallai y bydd cwsmer yn rhoi archeb i'r staff gwerthiant ac ar yr un pryd yn gofyn cwestiwn am ei gyfrif, ymholiad y byddai'r adran gyfrifon wedi ymdrin ag ef ar un adeg. Bydd y staff gwerthiant yn aml yn gwneud y ddau beth heddiw.

Mae'n bosibl y bydd y staff yn anfodlon wrth i grwpiau gwaith gael eu had-drefnu ac y bydd yn rhaid iddynt addasu i weithio mewn ffordd hollol wahanol gyda phobl wahanol. Weithiau, pan fydd cyrff yn ailstrwythuro, bydd aelod staff sydd wedi gweithio mewn swydd uwch am flynyddoedd yn gweld rhywun llawer iau yn cael ei ddyrchafu uwch ei ben.

Newidiadau i batrymau gwaith

O ganlyniad i'r defnydd cynyddol o gyfrifiaduron a chyfathrebiadau, mae llawer o gyrff yn gweithredu mewn marchnadoedd byd-eang bellach a rhaid iddynt allu ymateb i ofynion cwsmeriaid. Mae hyn yn golygu bod angen iddynt weithredu bedair awr ar hugain y dydd. Hefyd gall hyblygrwydd olygu bod mwy o waith rhan-amser a mwy o waith y tu allan i oriau swyddfa arferol ar gael. Mae hyn i'r dim i rai pobl fel mamau gyda phlant bach ond mae'n bosibl na fydd pobl eraill yn cael amser i ffwrdd yr un pryd â'u partneriaid.

Newidiadau i weithdrefnau mewnol

Bydd gofyn i staff ymgymryd â mwy o gyfrifoldeb ac amrywiaeth ehangach o dasgau'n aml pan gaiff systemau TGCh eu cyflwyno. Er enghraifft, gallai staff yr oedd ganddynt swyddi oedd yn gyfan gwbl weinyddol gynt orfod bod mewn cysylltiad uniongyrchol â'r cwsmeriaid a gwerthu nwyddau a gwasanaethau iddynt. Gyda'r system TG newydd mae'n bosibl y byddant yn gallu cwblhau eu gwaith traddodiadol yn gyflymach ac y bydd ganddynt amser i ymgymryd â gweithgareddau mwy proffidiol.

Bydd angen adolygu ffyrdd o wneud pethau a'u newid os oes rhaid, gan na fydd y dadansoddydd systemau fel rheol yn gweithio o amgylch y system bresennol wrth lunio'r system newydd ond, yn hytrach, yn ystyried y dull gorau o gyflawni'r dasg. Bydd dulliau gweithio'n newid yn aml a gall hyn beri straen i rai staff os na thrafodir y newidiadau gyda nhw ac os na chânt eu hyfforddi i ymdrin yn llwyddiannus â'r drefn newydd.

Rhaid i lawer o staff sy'n gweithio gyda systemau TGCh ufuddhau i god ymddygiad sy'n pennu'r hyn y gallant ei wneud ac na allant ei wneud gyda'r systemau TGCh.

Effeithiau ar y gweithlu (ofn newid)

Dyma rai enghreifftiau o adwaith anffafriol staff i gyflwyno system newydd:

- Ofn colli swyddi. Caiff rhai systemau eu cyflwyno i leihau'r nifer o bobl sydd eu hangen i wneud tasg benodol, gan mai cyflogau staff yw'r gost fwyaf yn aml.
- Ofn lleihad mewn statws a

bodlonrwydd swydd. Gyda chymorth y wybodaeth i reolwyr sy'n cael ei darparu gan systemau gwybodaeth rheoli cyfrifiadurol, gall rheolwyr wneud llawer mwy o waith ac nid oes angen cymaint o swyddi ar y lefel ganol. Gall bodlonrwydd swydd gael ei wanhau gan fod y cyfrifiadur yn gwneud cymaint o'r tasgau. Bydd rhai rheolwyr yn colli grym os caiff yr holl ddata ar gyfer y corff cyfan eu cadw'n ganolog, neu os bydd y corff yn defnyddio warws data. Mae hyn yn golygu nad yw'r data yn y system a ddarparwyd yn wreiddiol gan eu hadran nhw bellach yn perthyn iddyn nhw. Adnodd corfforaethol yw'r data a gall unrhyw un yn y corff eu defnyddio os oes ganddynt hawl.

- Ofn edrych yn wirion. Gallai rhai aelodau staff, yn enwedig staff hŷn, deimlo y byddai staff iau sy'n fwy hyddysg mewn TGCh yn gwneud hwyl am eu pennau oherwydd eu diffyg gwybodaeth.
- Newid lleoliad. Nid oes angen swyddfa mor fawr, felly mae'n bosibl y bydd y corff yn symud i adeilad llai i leihau costau (rhent, gwres, golau, ac ati) ac nid yn yr un lleoliad o anghenraid. Bydd unrhyw symudiad yn effeithio ar y staff mewn rhyw ffordd.

Goresgyn y gwrthwynebiad i systemau TGCh newydd

Un o'r problemau seicolegol mawr sy'n codi wrth gyflwyno systemau gwybodaeth newydd yw ofn. Yr ofn mwyaf yw colli swydd. Ofn arall gan lawer o staff yw y bydd eu hamodau gwaith yn gwaethygu. Oherwydd y gall yr ofnau hyn arwain at ysbryd isel a diffyg cydweithrediad wrth gyflwyno'r systemau hyn, dylai rheolwyr nodi'r ofnau a bod yn onest ynghylch canlyniadau'r system newydd. Mae llawer o gamau y gall rheolwyr eu cymryd, er enghraifft:

- Dylid egluro wrth yr holl staff pam mae angen y system newydd a dylid rhoi rhan iddynt mewn cyflwyno'r system, hyd yn oed os yw'n rhan fach. Os bydd y staff yn gweld bod gan y system fanteision amlwg ac os gallant gyfrannu at ei chynllunio, maen nhw'n llai tebygol o'i gwrthwynebu. Dylai rheolwyr annog y staff i roi cyngor ar y system newydd a dylid cydnabod eu cyfraniad.
- I wneud y broses ddysgu yn hawdd i'r staff, dylid cyflwyno hyfforddiant ac ailhyfforddiant cynhwysfawr. Gellir trefnu llawer o'r sesiynau hyfforddi hyn oddi ar y safle fod bod y staff yn gallu aros mewn gwesty braf ac fel bod yr hyfforddiant yn ddigwyddiad cymdeithasol. Bydd rhai cwmnïau'n cynnal hyfforddiant wedi'i seilio ar gyfrifiadur sy'n defnyddio dulliau amlgyfrwng. Gall hyn gael ei wneud ar eiddo'r cwmni fel nad yw'n tarfu cymaint ar ei weithrediadau busnes pob dydd.
- Dylai'r rheolwyr egluro manteision y system newydd o'i chymharu â'r hen un. Gall staff fod yn falch os bydd y system newydd yn datrys problemau a oedd yn fwrn yn yr hen system. Bydd hyn yn gwella

bodlonrwydd swydd.

- Dylai'r uwch reolwyr egluro goblygiadau'r system newydd cyn i sïon ddechrau lledu, yn enwedig mewn perthynas â diogelwch swyddi, cyflog, newidiadau mewn cytundebau cyflogaeth, gobaith o ddyrchafiad, newidiadau mewn amodau gwaith, ac ati. Yn anad dim, rhaid i wybodaeth am y system newydd gael ei chyfleu'n llawn ac yn agored.
- Bydd llawer o'r staff sy'n defnyddio'r system newydd yn dysgu sgiliau newydd a fydd yn gwella eu gobaith o ddyrchafiad ac yn rhoi cyfleoedd iddynt symud i swyddi gwell y tu allan i'r corff os dymunant.
- Dylai grwpiau cymdeithasol, fel cyfeillion sydd wedi gweithio gyda'i gilydd ers blynyddoedd lawer, gael eu cadw gyda'i gilydd os yw'n bosibl gan fod pobl sy'n mwynhau cwmni ei gilydd yn gweithio'n well fel tîm fel rheol.
- Dylai'r rheolwyr fod yn barod i dderbyn beirniadaeth o'r system newydd a gweithredu arni.

—GLASBERGEN—

"Mae angen mwy byth o fireinio ar yr ad-drefnu. Nes clywi di fel arall, rwyt ti'n blanhigyn pot yn y dderbynfa."

Cwestiynau a Gweithgareddau

▶ Cwestiynau 1   tt. 110–111

1 Pan gaiff systemau TGCh newydd eu cyflwyno, rhaid i'r staff ymdopi â'r newidiadau a fydd yn digwydd o ganlyniad i sefydlu'r system newydd.

Disgrifiwch sut mae pob un o'r canlynol yn newid ac yn effeithio ar y staff:

(a) Patrymau gwaith (2 farc)

(b) Y sgiliau sydd eu hangen/nad oes eu hangen (2 farc)

(c) Gweithdrefnau mewnol. (2 farc)

2 Mae corff yn cyflwyno llawer o systemau TGCh newydd er mwyn cystadlu â chwmnïau eraill. Mae hyn yn gofyn am lawer o newidiadau i'r corff a sut mae'n cael ei redeg.

(a) Eglurwch yr effaith y gallai'r datblygiadau TGCh newydd hyn eu cael ar swyddi a phatrymau gwaith. Dylech gynnwys **tair** enghraifft briodol a hollol wahanol i egluro eich ateb. (6 marc)

(b) Mae'r rheolwyr yn awyddus i sicrhau bod y newidiadau sy'n cael eu creu gan y system TGCh newydd yn achosi cyn lleied o straen â phosibl i'r gweithwyr. Amlinellwch yr hyn y gall y rheolwyr ei wneud i sicrhau hyn. (4 marc)

3 Cafodd system TGCh ei chyflwyno i gorff ac roedd yn cael ei hystyried yn fethiant. Nid problemau technegol a oedd yn gyfrifol am hyn ond anallu'r corff i reoli newid. Disgrifiwch **dri** ffactor sy'n dylanwadu ar reoli newid o fewn corff, gan roi enghreifftiau addas i egluro eich ateb. (6 marc)

4 Mae cwmni mawr yn ystyried cyflwyno system TGCh i gofnodi ymwelwyr. Mae'r system bresennol wedi'i seilio ar gofnod papur yn y dderbynfa. Bydd y system newydd yn cofnodi ymwelwyr a manylion eu hymweliad. Bydd cyflwyno'r system hon yn peri cryn newid i staff ac ymwelwyr. Yng nghyd-destun yr enghraifft hon, disgrifiwch **bedwar** ffactor y dylai'r rheolwyr eu hystyried wrth gyflwyno'r newid hwn. (8 marc)

▶ Gweithgaredd 1: Ofn newid

Dychmygwch eich bod chi'n gweithio i gwmni mawr sy'n cyfuno â chwmni mawr arall, gan arwain at ad-drefnu staff, gweithdrefnau a systemau TGCh.

Fel y rhan fwyaf o'r gweithwyr, byddech yn poeni am y newidiadau. Ysgrifennwch restr o'r pethau y byddech yn poeni amdanynt.

Ysgrifennwch rai pethau y gall rheolwyr y cwmni eu gwneud fel bod y newid yn rhoi'r gweithwyr dan lai o straen.

▶ Gweithgaredd 2: Cyflwyno meddalwedd newydd

Wrth i bobl fynd yn hŷn maen nhw'n ei chael hi'n anoddach ymdopi â newid. Mae corff yn cyflwyno system TGCh hollol newydd i gymryd lle'r hen un sydd wedi bod yn rhedeg, gyda mân newidiadau, ers y 15 mlynedd diwethaf. Wrth gwrs, mae rhai o aelodau hŷn y staff sydd wedi bod yn defnyddio'r hen system ers blynyddoedd yn pryderu am y system newydd gan ei bod, yn ôl y sôn, yn hynod o gymhleth.

Ysgrifennwch restr o'r pethau y gall rheolwyr eu gwneud i helpu staff hŷn i ddysgu'r system newydd.

Cymorth gyda'r arholiad

Enghraifft 1

1 Mae cwmni sydd wedi'i hen sefydlu wedi cael ei brynu, ac mae'r rheolwyr newydd yn awyddus i wneud y busnes yn fwy proffidiol drwy ddefnyddio'r systemau TGCh diweddaraf. Bydd y systemau newydd hyn yn cael effaith fawr ar y ffordd mae'r cwmni'n gweithio a hefyd ar y staff a gyflogant.

    (a) Eglurwch ystyr y term 'rheoli newid'. (2 farc)

    (b) Disgrifiwch yr effaith y gallai'r system TGCh newydd ei chael ar swyddi a phatrymau gwaith. Defnyddiwch **dair** enghraifft hollol wahanol i egluro eich ateb. (6 marc)

Ateb myfyriwr 1

1 (a) Mae rheoli newid yn golygu sut mae rheolwyr y corff yn rheoli'r newid a achosir gan y system TGCh newydd.

  (b) Bydd pobl yn colli eu swyddi o ganlyniad i'r system newydd. Mae'n bosibl y bydd pobl yn gorfod symud i wahanol safleoedd.

     Mae'n bosibl y bydd angen chwalu'r timau presennol o weithwyr, gan orfodi staff i weithio gyda phobl nad ydyn nhw'n dod ymlaen gyda nhw.

     Bydd yn rhaid i bobl weithio'n fwy hyblyg. Er enghraifft, mewn system e-fasnach, os bydd y wefan yn methu bydd angen ei thrwsio'n gyflym iawn neu fe fydd y cwmni'n colli archebion.

Sylwadau'r arholwr

1 (a) Ateb arwynebol yw hwn. Mae'n nodweddiadol o fyfyrwyr sy'n aildrefnu'r geiriau mewn cwestiwn i roi ateb. Dim marciau.

  (b) Mae'r rhan fwyaf o'r pwyntiau hyn yn ddilys ond ni roddir enghreifftiau ac eithrio yn rhan olaf yr ateb am e-fasnach.

    Nid yw 'pobl yn colli eu swyddi' yn rheswm addas oherwydd bod angen bod yn fwy penodol ynghylch y math o staff sydd mewn perygl.

    Mae'r atebion eraill yn dderbyniol ac yn cael marc yr un, ac mae'r ateb olaf yn cynnwys enghraifft sy'n werth dau farc. **(4 marc allan o 8)**

Ateb myfyriwr 2

1 (a) Mae hyn yn golygu sut mae rheolwyr yn ymdrin â'r newid fydd yn digwydd wrth gyflwyno'r system newydd. Byddai hyn yn cynnwys rheoli staff a all fod yn poeni am batrymau gwaith, gweithdrefnau mewnol, strwythurau trefniadaethol a'r sgiliau newydd y bydd eu hangen.

  (b) Gall ad-drefnu'r corff olygu bod gweithwyr yn gorfod symud i leoliadau gwahanol a gall hyn fod yn anghyfleus i rai staff.

     Mae'n bosibl na fydd staff hŷn am ddysgu sgiliau newydd a gallant deimlo dan fygythiad wrth weld staff iau yn dysgu sut i ddefnyddio'r system newydd yn gyflymach na nhw.

     Mae'n bosibl y bydd staff a oedd yn arfer gwneud gwaith gweinyddol papur, fel clercod ffeilio, pobl sy'n prosesu archebion, ac ati, yn colli eu swyddi.

     Mae'n bosibl y bydd yn rhaid i staff newid eu horiau gwaith a bod yn fwy hyblyg er mwyn cwrdd ag anghenion y busnes.

Sylwadau'r arholwr

1 (a) Nid yw'r ateb hwn yn manylu ar beth sy'n cael ei reoli er ei fod yn enwi'r holl newidiadau sy'n effeithio ar staff. Un marc am hyn.

  (b) Mae'r myfyriwr wedi rhoi ateb da yma ac wedi ychwanegu enghreifftiau at y rhan fwyaf o'r problemau. Marciau llawn am yr adran hon. **(7 marc allan o 8)**

Cymorth gyda'r arholiad *(parhad)*

Ateb yr arholwr

1 (a) Dim marciau am atebion sy'n cyfeirio'n syml at reolwyr yn rheoli newid.
        Un marc am grybwyll y canlynol gyda disgrifiad byr:

   Sgiliau newydd sydd eu hangen.   Newidiadau i'r strwythur trefniadaethol.
   Newid patrymau gwaith.   Newidiadau i weithdrefnau mewnol.
   Pryderon y gweithlu.

   (b) Un marc am bob pwynt hyd at uchafswm o chwech. Dylid rhoi 4 marc ar y mwyaf os na roddir unrhyw enghreifftiau.
        Efallai y bydd yn rhaid gweithio mewn lleoliad gwahanol (1) gan ei bod hi'n bosibl na fydd angen swyddfa mor fawr oherwydd bod angen llai o waith papur/staff/lle i gyfrifiaduron (1).
        Telegymudo/teleweithio – efallai y bydd hi'n bosibl gweithio o'r cartref gan ddefnyddio cyfrifiaduron a thelathrebu (1). Bydd hyn yn arbed amser a chostau teithio staff/yn gwneud bywyd personol yn haws/yn gwneud lles i'r amgylchedd (1).
        Ailhyfforddi – efallai y bydd angen dysgu sgiliau newydd (1) er mwyn defnyddio'r system TGCh newydd fel sgiliau cronfa ddata, defnyddio e-bost, sgiliau cloddio data, sgiliau diweddaru gwefan (1).
        Colli swyddi – mae systemau newydd yn disodli swyddi (1) a oedd yn cael eu gwneud o'r blaen gan bobl fel clercod ffeilio, clercod mewnbynnu data, clercod post (1).
        Oriau gwaith gwahanol (1) – gall systemau newydd weithredu 24 awr y dydd, felly mae'n bosibl y bydd angen i staff weithio'n fwy hyblyg (1).
        Caiff swyddi newydd eu creu (1) fel dylunwyr gwe, dadansoddwyr systemau, staff desg gymorth, gweinyddwyr rhwydwaith (1), ac ati.
        Bydd rhai swyddi'n cael eu symud dramor i ganolfannau galw (1) gan fod costau llafur yn rhatach a staff â chymwysterau da ar gael (1).

Enghraifft 2

2 (a) **Pan gaiff system TGCh newydd ei chyflwyno i'r gweithle, mae angen i'r gweithwyr ymdopi â nifer o newidiadau. Amlinellwch ddau newid o'r fath sy'n debygol o achosi straen i'r staff a disgrifiwch y pryderon a all fod gan y staff hyn. (4 marc)**

   (b) **Disgrifiwch a rhowch enghreifftiau o dri pheth y gall rheolwyr eu gwneud i leddfu unrhyw bryderon a all fod gan y staff am gyflwyno'r system TGCh newydd. (6 marc)**

Ateb myfyriwr 1

2 (a) Ofn diweithdra. Gall y system TGCh newydd olygu nad oes angen rhai swyddi a gâi eu gwneud o'r blaen gan bobl fel clercod mewnbynnu data a fyddai'n teipio data i mewn. Nawr mae dulliau cipio data uniongyrchol yn cael eu cynllunio, felly does dim angen y bobl hyn bellach. Ar ôl i rai pobl gael eu diswyddo, gall staff eraill boeni mai nhw fydd nesaf ac mae hyn yn rhoi'r holl weithwyr dan bwysau.

   (b) Dylai'r staff gael eu hailhyfforddi fel y gallant wneud y swyddi newydd sy'n cael eu creu gan y system TGCh newydd. Er enghraifft, gall staff ddysgu sgiliau cronfa ddata er mwyn gallu cael gwybodaeth am gwsmeriaid a'u targedu ar gyfer hyrwyddiadau gwerthu arbennig.
        Gall y rheolwyr egluro y bydd y newidiadau'n golygu y bydd llawer o'r tasgau pob dydd diflas yn cael eu disodli gan dasgau tra medrus mwy diddorol ac y bydd y staff yn gallu gwneud y rhain ar ôl cael hyfforddiant addas.

Sylwadau'r arholwr

2 (a) Nid yw'r myfyriwr wedi ysgrifennu am ddau newid ond mae'r newid a ddisgrifir yn ateb da, felly dau allan o bedwar marc am y rhan hon o'r cwestiwn.
        Ar ôl ateb cwestiwn ewch yn ôl iddo bob amser a sicrhau bod eich ateb yn cyd-fynd yn union â'r hyn sy'n cael ei ofyn.

   (b) Mae'r myfyriwr wedi rhoi dau ateb yn lle'r tri mae'r cwestiwn yn gofyn amdanynt. Mae'r ddau ateb yn dda.
        (6 marc allan o 10)

Ateb myfyriwr 2

2 (a) Colli swyddi. Gall y system newydd ddisodli rhai o'r tasgau yr arferai pobl eu gwneud a gall hyn arwain at ddiswyddiadau neu ymddeoliadau cynnar. Swyddi staff gweinyddol a rheolwyr canol fydd fwyaf mewn perygl.

Mwy o bwysau ar y staff. Weithiau rhaid i staff wneud mwy o waith mewn llai o amser ac weithiau gall y cyfrifiadur gael ei ddefnyddio gan y rheolwyr i fonitro faint o waith sy'n cael ei wneud.

Newidiadau mewn patrymau gwaith. Mae'n bosibl y bydd yn rhaid i'r staff weithio'n fwy hyblyg gan fod cwsmeriaid eisiau mynediad 24 awr y dydd i nwyddau a gwasanaethau. Gall y newidiadau hyn fod yn fuddiol i rai staff ond ni fydd staff eraill yn hoffi gweithio shifftiau.

(b) Dylai'r rheolwyr amlinellu manteision y system newydd i'r corff a sut y gallai hyn eu helpu i ehangu a chreu cyfleoedd newydd am ddyrchafiad i'r staff.

Darparu hyfforddiant priodol fel bod y staff yn gwybod beth maen nhw'n ei wneud.

Cynnwys yr holl staff mewn datblygu'r system newydd fel eu bod nhw'n teimlo eu bod nhw wedi helpu i'w chynllunio. Bydd hyn yn gwneud y staff yn hapusach gan mai'r system hon fydd yr un fydd yn eu helpu fwyaf.

Sylwadau'r arholwr

2 (a) Mae'r cwestiwn yn gofyn i'r myfyrwyr amlinellu dau newid ac mae'r ateb hwn yn cynnwys tri. Mae'r holl atebion yn gywir ac wedi'u disgrifio'n dda, felly marciau llawn am y rhan hon.

(b) Ateb da yw hwn a disgrifir dau bwynt yn llawn. Nid yw'r ail ateb yn ddisgrifiad llawn ac mae'n cael un marc yn unig.

(9 marc allan o 10)

Ateb yr arholwr

2 (a) Un marc am nodi'r newid ac un marc am egluro pam mae'n achosi straen.

Colli swyddi neu'r ofn y caiff swyddi eu colli (1) – gall y system newydd olygu bod staff sy'n cyflawni prosesau llaw megis teipio data/ffeilio/post mewnol/gwaith clerigol golli eu swyddi (1).

Gorfod dysgu sgiliau newydd (1). Gall staff hŷn fod dan bwysau gan eu bod yn teimlo eu bod yn edrych yn wirion oherwydd nad oes ganddynt y sgiliau TGCh sydd gan bobl iau (1).

Ofn newid i'r strwythur trefniadaethol (1). Oherwydd y newidiadau, efallai y bydd angen symud y swyddfa gan wneud y daith i'r gwaith yn llawer anoddach (1).

Newidiadau i batrymau gwaith (1). Efallai y bydd yn rhaid i staff weithio shifftiau, systemau 24 awr y dydd, ac ati (1).

Newidiadau i weithdrefnau mewnol (1). Efallai y bydd yn rhaid i'r staff dderbyn mwy o gyfrifoldeb am yr un cyflog (1).

Ofnau iechyd (1). Mae'n bosibl y bydd y staff yn poeni am y materion iechyd sy'n gysylltiedig â defnyddio cyfarpar TGCh am amser hir (1).

(b) Un marc am ddisgrifiad byr o'r ffactor ac un marc am esboniad pellach neu enghraifft.

Hyfforddiant/ailhyfforddiant priodol (1) i sicrhau bod yr holl staff yn deall y system newydd ac nad ydynt dan straen gan na wyddant beth i'w wneud (1).

Egluro'r manteision (1) fel bod y staff yn deall sut y byddant ar eu hennill gan y bydd eu swyddi'n haws/llai rhwystredig/mwy diddorol (1).

Egluro goblygiadau'r system newydd yn llawn (1) i atal sïon a all achosi pryder a straen (1).

Cyfle i ddysgu sgiliau newydd (1) a fydd yn galluogi staff i wella eu rhagolygon gwaith (1).

Cymryd rhan mewn datblygu'r system newydd (1) fel bod gan y staff system y gallant ei defnyddio'n hawdd (1).

Map meddwl cryno

Canlyniadau newid wrth gyflwyno systemau TGCh newydd

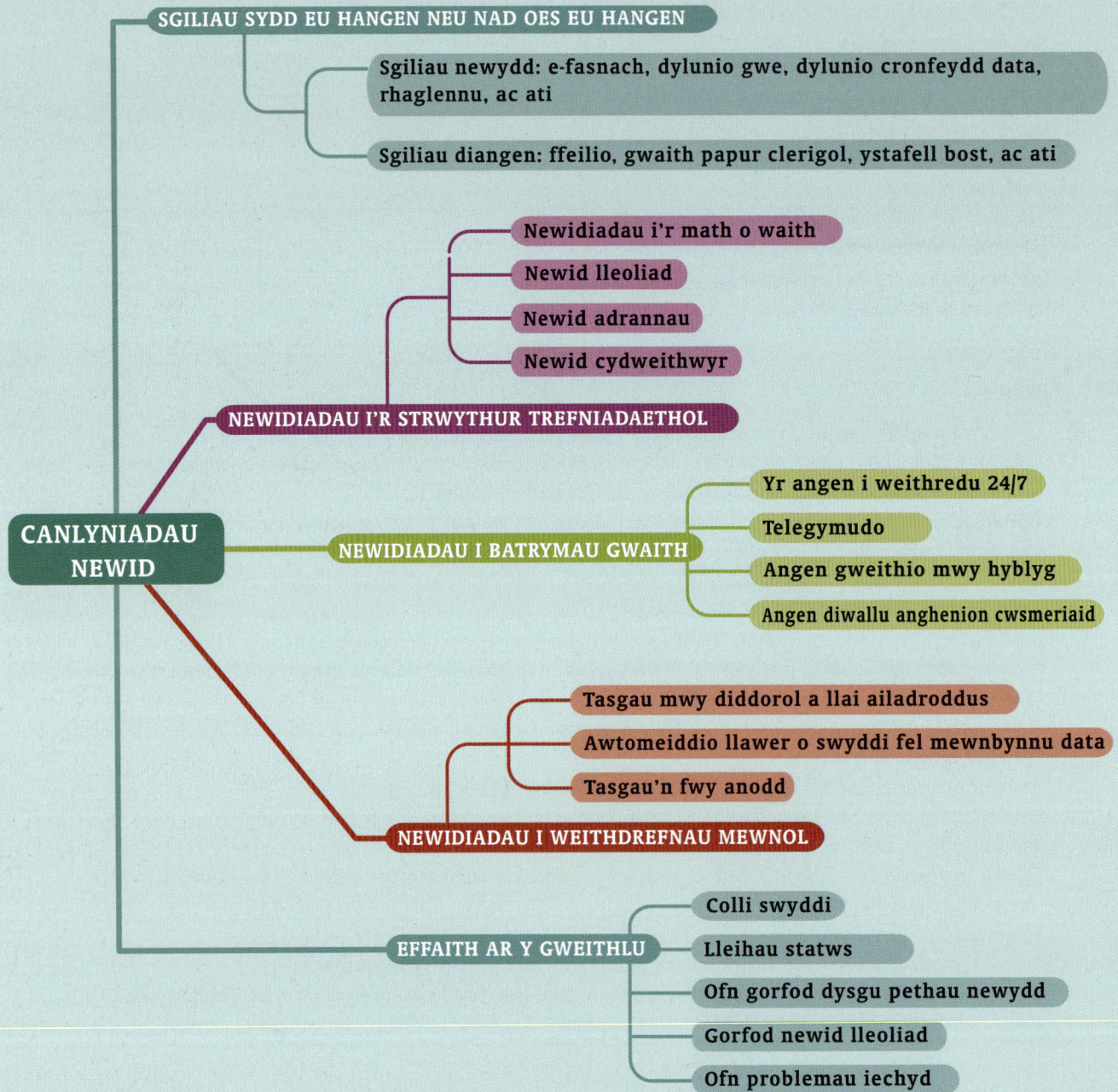

SGILIAU SYDD EU HANGEN NEU NAD OES EU HANGEN

Sgiliau newydd: e-fasnach, dylunio gwe, dylunio cronfeydd data, rhaglennu, ac ati

Sgiliau diangen: ffeilio, gwaith papur clerigol, ystafell bost, ac ati

Newidiadau i'r math o waith

Newid lleoliad

Newid adrannau

Newid cydweithwyr

NEWIDIADAU I'R STRWYTHUR TREFNIADAETHOL

Yr angen i weithredu 24/7

Telegymudo

Angen gweithio mwy hyblyg

Angen diwallu anghenion cwsmeriaid

NEWIDIADAU I BATRYMAU GWAITH

CANLYNIADAU NEWID

Tasgau mwy diddorol a llai ailadroddus

Awtomeiddio llawer o swyddi fel mewnbynnu data

Tasgau'n fwy anodd

NEWIDIADAU I WEITHDREFNAU MEWNOL

Colli swyddi

Lleihau statws

Ofn gorfod dysgu pethau newydd

Gorfod newid lleoliad

Ofn problemau iechyd

EFFAITH AR Y GWEITHLU

TOPIG 8: Systemau gwybodaeth rheoli

Mae gwybodaeth rheoli yn hollol allweddol i unrhyw fusnes neu gorff a gall amrywio o adroddiad unigol sy'n amlinellu pa mor broffidiol yw dewis o nwyddau i ddogfen gyfan sy'n amlinellu proffidioldeb yr holl nwyddau yn ôl rhanbarth. Mae llawer ffordd o gael y deunydd ar gyfer yr adroddiadau hyn, ond fel arfer mae'n dod o'r wybodaeth weithredol sy'n cael ei chyflenwi gan y system gyfrifiadurol. Fel arfer rhaid i'r data a'r wybodaeth sy'n dod o brosesu trafodion pob dydd gael eu prosesu ymhellach er mwyn iddynt fod yn ddefnyddiol i'r rheolwyr. Mae'n bwysig bod unrhyw wybodaeth rheoli y gofynnir amdani yn darparu'r cynnwys cywir, ar ffurf sy'n briodol i'r sawl fydd yn ei derbyn, ac yn cael ei chyflwyno ar yr adeg iawn. Nid yw pob system TGCh yn darparu gwybodaeth rheoli. Er enghraifft, mae system gyflogau sy'n cyfrifo'r holl ffigurau angenrheidiol ar gyfer talu'r gweithwyr ynghyd â manylion didyniadau fel treth, Yswiriant Gwladol, pensiynau, ac ati, yn gallu cyflawni'r holl weithrediadau angenrheidiol heb orfod cyflenwi gwybodaeth i reolwyr.

▼ Y cysyniadau allweddol sy'n cael sylw yn y topig hwn yw:

▶ Nodweddion system gwybodaeth rheoli (*MIS: management information system*) effeithiol

▶ Y llif o wybodaeth rhwng cydrannau allanol a mewnol system gwybodaeth rheoli

▶ Llwyddiant neu fethiant system gwybodaeth rheoli

▶ Nodweddion system gwybodaeth rheoli dda

▶ Ffactorau sy'n gallu arwain at System Gwybodaeth Rheoli wael

CYNNWYS

Nodweddion system gwybodaeth rheoli (*MIS: management information system*) effeithiol

Cyflwyniad

Rhaid i reolwyr wneud penderfyniadau am eu meysydd swyddogaethol yn y corff sy'n eu cyflogi, ac er mwyn sicrhau bod y penderfyniadau hyn yn rhai da, rhaid iddynt gael eu seilio ar wybodaeth. Pwrpas y system gwybodaeth rheoli yw darparu'r wybodaeth hon.

Diffiniad o *MIS*

Mae systemau sy'n trosi data o ffynonellau mewnol neu allanol i wybodaeth i'w defnyddio gan reolwyr yn cael eu galw'n systemau gwybodaeth rheoli. Casgliadau trefnedig o bobl, gweithdrefnau ac adnoddau sy'n helpu rheolwyr i wneud penderfyniadau yw'r rhain.

Rhaid i wybodaeth rheoli fod ar ffurf briodol i alluogi rheolwyr ar wahanol lefelau i wneud penderfyniadau effeithiol wrth gynllunio, cyfarwyddo a rheoli'r gweithgareddau y maen nhw'n gyfrifol amdanynt. Nid yw *MIS* yn bodoli ar wahân i systemau gwybodaeth, maen nhw'n rhan ohonynt. Defnyddir systemau gwybodaeth yn rheolaidd gan yr holl staff. Caiff *MIS* ei chynllunio i gwrdd â gofynion gwybodaeth y rheolwyr yn bennaf.

Un o broblemau llawer o reolwyr yw nad ydynt yn deall rôl gwybodaeth. Ond mae cael llawer o wybodaeth gywir yn lleihau eu hansicrwydd pan fyddant yn gwneud penderfyniadau. Gellir dod i benderfyniadau yn gyflym a gyda llai o bryder gan ei bod yn llawer mwy tebygol y byddant yn gywir. Pe na bai unrhyw ansicrwydd, ni fyddai angen y wybodaeth, gan y byddai'n bosibl i'r rheolwr ragweld beth fyddai'n digwydd. Yn anffodus, nid yw pethau mor syml â hyn fel arfer, felly rhaid cael cymaint o wybodaeth rheoli â phosibl

sy'n berthnasol i'r penderfyniad sy'n cael ei wneud.

Mae'n bwysig nodi bod gan lawer o reolwyr swyddi a chyfrifoldebau penodol a bod angen i unrhyw wybodaeth rheoli gael ei theilwra i anghenion y rheolwr unigol. Yn hanesyddol fe fyddai'r rheolwr wedi cysylltu â'r adran prosesu data neu reolwr cronfa ddata i geisio'r wybodaeth hon, a byddai'n aml yn cael yr ateb na allai'r system gynhyrchu'r wybodaeth yn y ffurf yr oedd wedi gofyn amdani. Roedd llawer o systemau cronfa ddata mawr yn anhyblyg iawn wrth gynhyrchu'r adroddiadau *ad hoc* a'r trefniadau hyn er y gallent gynhyrchu gwybodaeth reolaidd yn deillio o weithgareddau prosesu data.

Mae cronfeydd data modern yn llawer mwy hyblyg a chaiff llawer o reolwyr eu hyfforddi i ddefnyddio cyfrifiadur wedi'i gysylltu â'r rhwydwaith i gael y wybodaeth sydd ei hangen arnynt.

Enghraifft o MIS yn cael ei defnyddio

Er mwyn deall pwysigrwydd gwybodaeth rheoli, gadewch i ni edrych ar enghraifft wedi'i seilio ar gwmni sy'n cynhyrchu barbeciws. Mae'r cwmni hwn yn cyflogi wyth gwerthwr ac mae pob un yn cael ardal i weithio ynddi. Eu gwaith yw gwneud ymweliadau, hyrwyddo a gwerthu yn eu hardal nhw. Tybiwch fod y rheolwr gwerthu sy'n gyfrifol am y cynrychiolwyr yn dweud wrth y rheolwr cyfarwyddwr fod gwerthwr X, sy'n gyfrifol am ranbarth y gogledd-orllewin, wedi gwerthu nwyddau gwerth £100,000 y mis diwethaf. Mae hyn yn swnio'n dda i'r rheolwr gwerthu, ond ydy e mewn gwirionedd? Sut y gall y rheolwr benderfynu? Y ffordd orau fyddai iddo edrych ar werthiant y gwerthwyr eraill yn ystod yr un cyfnod er mwyn

gwneud cymhariaeth. Gallai hefyd gymharu'r gwerthiant â'r gwerthiant yn yr un rhanbarth yn ystod y blynyddoedd diwethaf. Gallai hyn ddatgelu tuedd, er enghraifft, bod barbeciws yn mynd yn fwy poblogaidd. Mae systemau gwybodaeth rheoli yn boblogaidd gan eu bod yn ei gwneud hi'n hawdd cael gwybodaeth o'r math hwn ac nad oes angen cyflogi staff arbenigol i echdynnu'r wybodaeth o'r system gyfrifiadurol. Mae'n gwneud synnwyr i adael i'r person sy'n deall y busnes wneud y penderfyniad am y wybodaeth y mae angen ei hechdynnu. Ond mae pethau'n fwy cymhleth na hyn, oherwydd bod angen cryn dipyn o hyfforddiant cyn y gall staff echdynnu gwybodaeth yn y ffordd yma.

Enghreifftiau eraill o MIS

Gall prif weithredwr cadwyn o uwchfarchnadoedd echdynnu gwybodaeth ariannol am bob uwchfarchnad yn y gadwyn er mwyn darganfod y rheiny sy'n gwneud y lleiaf o elw. Yna gall y rhain gael eu gwerthu a gellir defnyddio'r arian i agor uwchfarchnadoedd newydd mewn ardaloedd sy'n debygol o fod yn fwy proffidiol.

Mae rheolwr cwmni danfon parseli cenedlaethol yn defnyddio *MIS* i edrych ar y pellter mae pob cerbyd yn ei deithio er mwyn penderfynu a oes angen agor depo arall.

Beth sy'n gwneud *MIS* yn wahanol i system prosesu data?

Mae systemau gwybodaeth rheoli yn:

- cynhyrchu mwy o wybodaeth nag sydd ei hangen fel rheol ar gyfer prosesu data rheolaidd
- cynhyrchu data lle mae amseru'n hollbwysig

- cael eu defnyddio i roi gwybodaeth i reolwyr i'w helpu i wneud penderfyniadau
- wedi'u seilio o amgylch cronfeydd data.

Nodweddion MIS effeithiol

Bydd *MIS* effeithiol yn:

- Cynnwys gwybodaeth sy'n berthnasol a chywir. Mae gormod o fanylion bron cynddrwg â rhy ychydig oherwydd bod y rheolwyr yn gorfod treulio amser yn dewis y wybodaeth sydd ei hangen arnynt o'r llwyth o wybodaeth amherthnasol. Mae cywirdeb y wybodaeth yn dibynnu ar gywirdeb y data sy'n cael eu mewnbynnu i'r system. Rhaid dilysu a gwireddu'r data i sicrhau nad yw data anghywir yn cael eu prosesu.
- Rhoi'r wybodaeth yn ôl y gofyn. Mae rhai systemau gwybodaeth ond yn cynhyrchu allbwn penodol ar adeg benodol. Er enghraifft, gall adroddiadau gael eu cynhyrchu ar ddiwedd rhediad misol ar gyfer prosesu taliadau. Rhaid i reolwyr wneud penderfyniadau'n gyflym ac felly ni ddylent orfod aros am wybodaeth. Gan fod y *MIS* yn cael ei ddefnyddio gan y rheolwyr eu hunain, gallant echdynnu'r wybodaeth pryd bynnag y mae ei hangen.
- Hygyrch i amrywiaeth eang o ddefnyddwyr. Bydd angen gwybodaeth sy'n berthnasol i'w maes ar reolwyr mewn gwahanol rannau o'r busnes. Felly rhaid i'r *MIS* allu defnyddio'r holl ddata sydd yn y gronfa ddata gorfforaethol a'u cyflenwi i'r amrywiaeth lawn o ddefnyddwyr. Er mwyn i reolwyr allu manteisio ar y *MIS*, rhaid iddynt gael hyfforddiant ar sut i'w defnyddio.
- Cyflwyno'r data yn y fformat mwyaf priodol. Dylai'r rheolwyr allu dewis y ffordd y caiff y data eu cyflwyno. Er enghraifft, mewn tablau neu ar ffurf gwahanol fathau o graffiau. Gallant ddewis cadw'r wybodaeth mewn fformatau ffeil y gellir eu defnyddio i gynhyrchu taenlenni neu gyflwyniadau.
- Hyblyg. Cwyn gyffredin ymhlith rheolwyr yw nad yw'r *MIS* yn rhoi'r wybodaeth sydd ei hangen arnynt yn y fformat a ddewisant.

Felly bydd *MIS* dda yn caniatáu hyblygrwydd llwyr o ran y wybodaeth y gall ei chynhyrchu a'r ffordd y caiff ei chyflwyno. Dylai'r system fod yno i'r rheolwyr, nid fel arall.

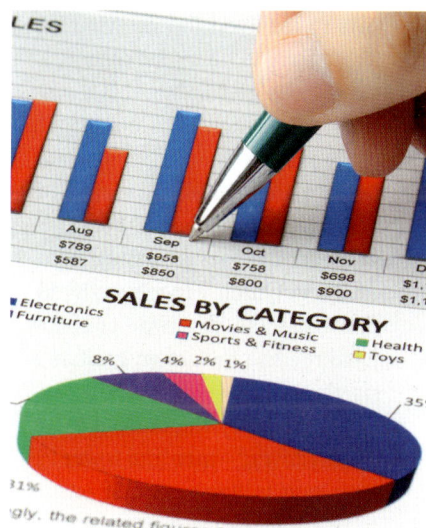

Dylai *MIS* gyflwyno gwybodaeth yn y fformat cywir: yma mae siartiau a thablau'n cael eu defnyddio

Y llif o wybodaeth rhwng cydrannau allanol a mewnol MIS

Bydd *MIS* yn defnyddio data o'r tu mewn ac o'r tu allan i'r sefydliad. Er enghraifft, gallai *MIS* ar gyfer cynhyrchu gwybodaeth am effeithiolrwydd ymgyrch farchnata gynnwys gwybodaeth am gystadleuwyr yn ogystal â gwybodaeth am gostau a gwerthiant o'u systemau TGCh mewnol.

Bydd ysgolion yn defnyddio *MIS* mewn sawl ffordd. Er enghraifft, i gynllunio ar gyfer y tymor hir, byddai'n rhaid sicrhau bod ganddynt yr adnoddau i ymdopi â chynnydd yn y boblogaeth.

Gallai gwybodaeth fewnol gynnwys ffigurau derbyn o flynyddoedd blaenorol, manylion brodyr a chwiorydd y disgyblion presennol a fydd yn dod i'r ysgol yn y man, manylion adnoddau fel staff, ystafelloedd, desgiau, ac ati.

Rhai ffynonellau gwybodaeth allanol posibl yw ffigurau o ysgolion cynradd, manylion cyfrifiad (i gael amcan o'r boblogaeth), manylion mewnfudo, data'r awdurdod lleol, ac ati.

Mae'r diagram isod yn dangos y ffynonellau gwybodaeth a'r llifoedd gwybodaeth. Dangosir yr endidau (h.y. cyflenwyr neu dderbynwyr gwybodaeth) mewn blychau hirgrwn a'r prosesau yn y blwch petryalog mewnol. Mae'r blwch allanol yn nodi ffin y system gyfan.

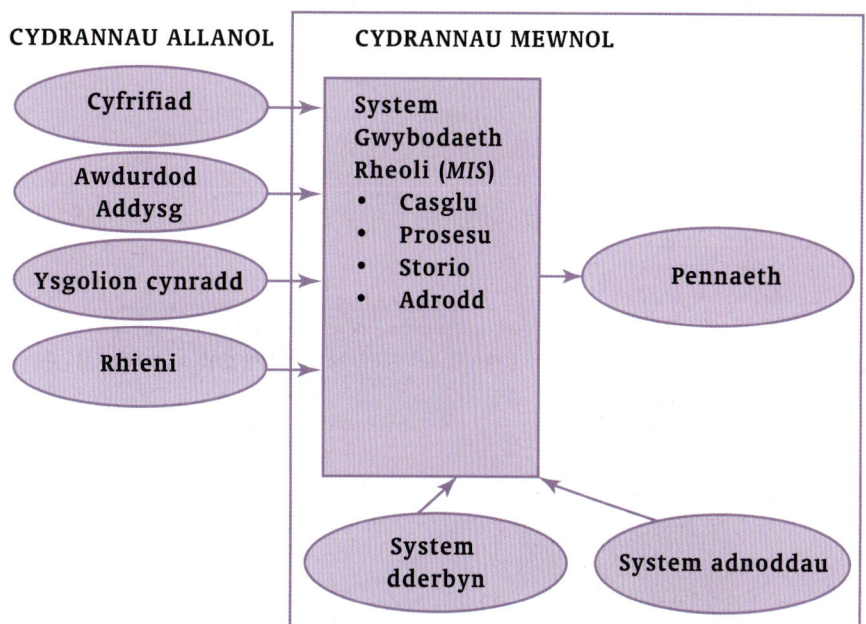

Y llif o wybodaeth o gydrannau allanol a mewnol *MIS*

Llwyddiant neu fethiant system gwybodaeth rheoli

▼ Byddwch yn dysgu

▶ Am y rhesymau dros fethiant *MIS*

▶ Am y cwynion y gall rheolwyr eu gwneud am eu *MIS*

Cyflwyniad

Yn yr adran hon byddwn yn edrych ar nodweddion *MIS* dda a'r ffactorau a all arwain at system gwybodaeth wael

Y rhesymau dros fethiant *MIS*

Mae llawer o systemau gwybodaeth rheoli yn fethiant. Ni lwyddant i roi'r wybodaeth y mae rheolwyr ei heisiau nac i gynhyrchu system sy'n gweithio. Weithiau fe gollir rheolaeth ar y project, mae costau'n codi ac mae cynnydd yn simsanu. Gellir colli rheolaeth ar broject am sawl rheswm: gall staff allweddol adael; gall rheolwyr fethu â chadw gafael ar amserlenni a chostau, ac yn y blaen. Nid yw rhai systemau yn cwrdd â'r disgwyliadau ac mae rhai o'r problemau sy'n codi wrth ddefnyddio systemau gwael yn cael eu crynhoi isod.

Nodweddion **MIS** *dda*

Mae'r nodweddion a fydd yn arwain at *MIS* dda yn cynnwys:

- Cywirdeb y data – rhaid i'r data o'r systemau trafodion sy'n cyflenwi data i'r rheolwyr fod yn gywir.
- Hyblygrwydd o ran dadansoddi data – mae gan wahanol reolwyr wahanol ofynion a rhaid i *MIS* allu ymdopi â hyn.
- Darparu data/gwybodaeth ar ffurf briodol – bydd angen i'r data/gwybodaeth gael eu cyflwyno yn unol â gofynion y rheolwyr. Bydd rhai eisiau tablau a bydd eraill am weld tueddiadau mewn graffiau a siartiau.
- Hygyrch i amrywiaeth eang o ddefnyddwyr – mae gan lawer o reolwyr sgiliau a gwybodaeth eang ym maes TGCh ond rhaid i'r *MIS* gael ei ddefnyddio gan bob rheolwr. Felly dylai'r *MIS* ei gwneud hi'n hawdd i'r holl reolwyr gael y wybodaeth sydd ei hangen arnynt.
- Gwella cyfathrebiadau rhyngbersonol – ymhlith rheolwyr a staff: mae'r rheolwyr yn cael gwybodaeth fanwl y gallant ei rhannu â rheolwyr mewn meysydd eraill. Hefyd mae'n haws iddynt gyfathrebu â staff is gan fod eu penderfyniadau wedi'u seilio ar wybodaeth gadarn.
- Mae'n caniatáu cynllunio projectau unigol – rhan o waith unrhyw reolwr yw cynllunio at y dyfodol. Gellir defnyddio'r wybodaeth o'r *MIS* i helpu i gynllunio datblygiadau newydd fel agor canghennau newydd, darparu rhwydwaith dosbarthu newydd, lleoli siopau newydd, ac ati.
- Mae'n osgoi gorlwytho gwybodaeth – rhaid i'r *MIS* beidio â chynhyrchu gwybodaeth ormodol gan y gall hyn wastraffu amser a'i gwneud hi'n anoddach defnyddio'r wybodaeth hanfodol.

'Fydda i byth yn deall y system gwybodaeth rheoli hon'

'Bob mis mae'n rhaid i mi frwydro drwy'r adroddiadau hyn i gael y wybodaeth sydd ei hangen arna i'

'Mae'r wybodaeth hon yn dweud pob math o bethau wrthyf, ond nid beth ydw i wir eisiau ei wybod'

CWYNION Y GALL RHEOLWYR EU GWNEUD AM EU *MIS*

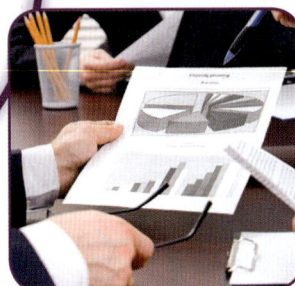

'Fe fyddai'r wybodaeth hon wedi bod yn ddefnyddiol iawn pe bai hi gen i yn y cyfarfod gwerthiant ddoe'

Ffactorau a all arwain at *MIS* wael

Mae'r ffactorau a all arwain at *MIS* wael yn cynnwys:

- Cymhlethdod y system – mae angen i systemau fod yn ddigon cymhleth i ddarparu'r manylion y bydd rheolwyr eu hangen ond ni ddylent fod mor gymhleth fel eu bod yn anodd eu defnyddio. Mae angen i system gwybodaeth rheoli fod yn syml i'w defnyddio fel bod yr holl reolwyr yn gallu ei defnyddio yn eu gwaith pob dydd.

- Diffyg dulliau ffurfiol – mae llawer o wahanol ffyrdd o ddatblygu system ac mae pob dull yn cael ei alw'n fethodoleg. Yr hyn sydd gan yr holl fethodolegau hyn yn gyffredin yw bod y system yn cael ei datblygu mewn ffordd ffurfiol. Mae hyn yn sicrhau nad yw dadansoddwyr systemau yn torri corneli ac yn datblygu system lai na pherffaith. Mae gwneud dadansoddiad systemau iawn yn gofyn am amser ac ymdrech ond o ganlyniad mae'r *MIS* a gynhyrchir yn llawer gwell na *MIS* sy'n cael ei datblygu mewn ffordd lai ffurfiol.

- Dadansoddiad cychwynnol annigonol – os caiff y dadansoddiad cychwynnol ei wneud gan rywun dibrofiad, mae'n bosibl na fydd yn gwerthfawrogi'r angen i ddadansoddi gofynion gwybodaeth y rheolwyr yn llawn. Caiff dadansoddiad o'r fath ei seilio ar wybodaeth anghyflawn ac ni fydd byth yn cynhyrchu *MIS* hyblyg sy'n gweithio'n dda. Bydd dadansoddiad annigonol yn ei amlygu ei hun fel rheol pan nad yw'r system yn gallu cyflawni tasg allweddol, neu pan nad yw'r system yn gweithredu mewn ffordd sy'n gyson â'r amcanion gwreiddiol.

- Peidio â chynnwys y rheolwyr wrth lunio'r system – arbenigwyr TGCh sy'n datblygu'r system gwybodaeth fel arfer gan mai nhw sy'n gwybod sut i wneud dadansoddiad systemau trwyadl, ond nid oes ganddynt gymaint o wybodaeth fusnes am y maes dan sylw â'r rheolwyr. Felly mae'n bwysig i'r rheolwyr fod yn rhan o'r tîm datblygu a bod ymgynghori cyson rhyngddynt a datblygwyr y system. Y gobaith yw y bydd hyn yn arwain at greu system dechnegol dda sy'n cwrdd ag anghenion y rheolwyr.

- Caledwedd a meddalwedd amhriodol – mae cyflymder y gweinydd ffeiliau a chyflymder trosglwyddo data yn effeithio ar y cyflymder y gellir echdynnu gwybodaeth rheoli o'r data sy'n cael eu cadw. Os yw'r system yn rhy araf, ni fydd y rheolwyr yn ei defnyddio ond pan mae'n hollol angenrheidiol. Dylai'r meddalwedd system gwybodaeth rheoli ei gwneud hi'n haws i reolwyr lunio amodau ar gyfer echdynnu'r wybodaeth sydd ei hangen arnynt a dylent allu dewis sut y cyflwynir y wybodaeth.

- Diffyg gwybodaeth ymhlith rheolwyr am eu systemau TGCh a'r hyn y gallant ei wneud – mae'n bosibl na fydd y rheolwyr yn gwybod am y datblygiadau diweddaraf mewn TGCh a rhoddant eu ffydd fel arfer mewn arbenigwyr TGCh gan gredu y bydd yr holl atebion ganddyn nhw. Ond ni all arbenigwyr TGCh fod yn arbenigwyr ar bob rhan o'r busnes, felly bydd disgwyl i'r rheolwyr wybod am y datblygiadau diweddaraf yn eu maes rheoli nhw hefyd. Er enghraifft, mae'n debyg y bydd rheolwyr personél yn tanysgrifio i gylchgronau proffesiynol ac y bydd erthyglau yn y rhain ar ddatblygiadau TGCh yn eu maes nhw. Dylai'r rheolwyr fynd ar gyrsiau hyfforddi i feithrin y sgiliau sydd eu hangen i echdynnu gwybodaeth rheoli ac i ddysgu gwerth y wybodaeth hon.

- Cyfathrebu gwael rhwng staff proffesiynol – dylai rheolwyr ac arbenigwyr TGCh weithio gyda'i gilydd i gynhyrchu system gwybodaeth rheoli sy'n cwrdd â gofynion y rheolwyr a fydd yn defnyddio'r system. Os yw cyfathrebu'n wael, mae'n bosibl na fydd y system yn bodloni anghenion y rheolwyr. Efallai na fydd y system yn echdynnu'r wybodaeth sydd ei hangen, neu efallai y bydd mor llafurus i'w defnyddio fel bod y rheolwyr yn ceisio ei hosgoi.

- Diffyg safonau proffesiynol – mae Cymdeithas Cyfrifiadurwyr Prydain (*BCS: British Computer Society*) yn gosod safonau ymarfer lleiaf y mae'n rhaid i'w holl aelodau lynu wrthynt. Mae llawer o staff TGCh yn aelodau o'r gymdeithas, ac mae aelodaeth yn dangos i'r cyflogwr neu'r sawl sy'n talu am eu gwasanaeth y bydd y gwaith yn cael ei wneud mewn ffordd broffesiynol. Nid yw pob arbenigwr cyfrifiadurol yn aelod o'r gymdeithas ac felly nid oes rheidrwydd arnynt i gydymffurfio â'i safonau. Prif gymhelliant rhai arbenigwyr cyfrifiadurol yw hyrwyddo eu gyrfa ac mae popeth a wnânt yn seiliedig ar hyn. Er enghraifft, os oes angen profiad arnynt mewn maes newydd fel defnyddio mewnrwydi, byddant yn ceisio perswadio'u cyflogwr i gael mewnrwyd, pa un a oes angen un ai peidio, dim ond er mwyn gallu dweud ar eu CV eu bod wedi cael profiad o ddefnyddio un. Bydd rhai 'arbenigwyr', yn enwedig y rhai sy'n gwneud gwaith contract, yn ymgymryd â gwaith nad oes ganddynt lawer o brofiad ohono yn y gobaith y byddant yn dod i arfer â'r gwaith wrth fynd yn eu blaen.

Astudiaeth Achos, Cwestiynau a Gweithgareddau

▶ Astudiaeth achos `tt. 118–119`

System Rheoli Gwybodaeth i Ysgolion (SIMS: Schools Information Management System)

Mae rhedeg ysgol neu goleg yn gymhleth iawn oherwydd bod cymaint o bobl a digwyddiadau i'w rheoli. Nid yw'n syndod bod systemau gwybodaeth rheoli ar gael sy'n helpu i wneud hyn a'r un fwyaf poblogaidd yw'r System Rheoli Gwybodaeth i Ysgolion (*SIMS*).

Gall y system hon gynhyrchu gwybodaeth ar wahanol lefelau, o'r wybodaeth sydd ei hangen ar y pennaeth ar lefel strategol i'r wybodaeth dactegol sydd ei hangen ar reolwyr a'r wybodaeth weithredol sydd ei hangen ar athrawon a staff gweinyddol yr ysgol.

Mae'r systemau hyn yn helpu ysgolion mewn llawer ffordd:

- Maen nhw'n lleihau'r baich gwaith ym mhob maes gwaith yn yr ysgol drwy awtomeiddio llawer o'r tasgau sy'n cael eu gwneud gan athrawon a staff cymorth.
- Gall rheolwyr tactegol fel penaethiaid adrannau ddefnyddio'r system i gynllunio beth i'w ddysgu yn eu hadran.
- Rhaid i'r pennaeth wneud penderfyniadau strategol am staffio. Gall y system gwybodaeth rheoli (*MIS*) ei helpu i wneud hyn drwy ddadansoddi'r effaith y gallai penodi cynorthwyydd dysgu ychwanegol ei chael ar y gyllideb, maint dosbarthiadau, a

chyflawniadau disgyblion, ar sail yr hyn sydd wedi digwydd yn y gorffennol. Yna mae'n haws gwneud y penderfyniad gan ei fod wedi'i seilio ar ffeithiau.

- Gall athrawon ddarparu gwybodaeth weithredol am ymddygiad da neu ddrwg mewn gwersi a gall y staff ddefnyddio hyn i ddarganfod patrymau ymddygiad.
- Gall staff strategol fel y pennaeth a'r is-benaethiaid ddefnyddio'r system gwybodaeth rheoli i gynllunio amserlenni, gan wneud y defnydd gorau o staff ac adnoddau.
- Ar lefel weithredol, gall athrawon ystafell ddosbarth leihau'r amser mae'n ei gymryd i ysgrifennu adroddiadau disgyblion drwy ddefnyddio cronfa sylwadau, sy'n cynnwys detholiad eang o sylwadau, i'w helpu i ysgrifennu'r adroddiadau.
- Gall y pennaeth gynllunio cyllid tymor hir a thymor byr a gwneud penderfyniadau gwybodus am y gyllideb.
- Gall penaethiaid blwyddyn gryfhau cysylltiadau rhwng yr ysgol a'r cartref drwy ddefnyddio'r offer adrodd i rannu gwybodaeth am ddisgyblion a thrwy ddefnyddio negeseuon testun ac e-bost i rannu'r wybodaeth hon â rhieni.

1 Mae tair lefel o staff mewn ysgol:
   Gweithredol
   Tactegol
   Strategol
   Ar gyfer pob un o'r staff canlynol, eglurwch pa un o'r tair lefel sy'n fwyaf priodol iddynt. (8 marc)
   (a) Pennaeth
   (b) Athrawes ddosbarth
   (c) Cynorthwyydd dysgu
   (ch) Clerc yn swyddfa'r ysgol
   (d) Pennaeth adran
   (dd) Pennaeth blwyddyn

   (e) Dirprwy bennaeth neu is-bennaeth
   (f) Gofalwr
2 Bydd ysgolion a cholegau'n defnyddio systemau gwybodaeth rheoli (*MIS*). Mae gwahanol lefelau o staff yn gofyn am wahanol bethau o'r system.
   (a) Eglurwch beth yw ystyr system gwybodaeth rheoli (*MIS*). (3 marc)
   (b) Disgrifiwch, gan roi enghraifft, dasg y byddai pennaeth ysgol yn defnyddio *MIS* i'w chyflawni ar y lefel strategol. (2 farc)
   (c) Disgrifiwch, gan roi enghraifft, dasg y byddai pennaeth adran/rheolwr yn defnyddio *MIS* i'w chyflawni ar y lefel dactegol. (2 farc)

▶ Gweithgaredd 1: Anghenion gwybodaeth rheoli ar wahanol lefelau

Mae cwmni'n gweithgynhyrchu barbeciws sy'n cael eu gwerthu drwy siopau *DIY* mawr a chanolfannau garddio ar hyd a lled Prydain. Mae deg model ar hyn o bryd, sy'n amrywio mewn pris o £55 am y rhataf i £1,400 am y model mwyaf soffistigedig.

Eich tasg chi yw edrych ar y wybodaeth rheoli y gall fod ei hangen ar y rheolwr cynhyrchu, a chymharu a chyferbynnu ei anghenion gwybodaeth ag anghenion y rheolwr gyfarwyddwr sydd â chyfrifoldeb cyffredinol dros y busnes.

1 Rheolwyr yw'r bobl mewn corff sy'n gyfrifol am wneud y penderfyniadau. Mae systemau gwybodaeth a ddefnyddir i roi gwybodaeth i reolwyr i seilio eu penderfyniadau arnynt yn cael eu galw'n systemau gwybodaeth rheoli (*MIS*). Mae systemau gwybodaeth rheoli yn mynd law yn llaw â systemau prosesu data ond maent yn cael eu defnyddio at wahanol ddibenion.

(a) Drwy roi enghraifft addas ar gyfer pob un, eglurwch y gwahaniaeth rhwng system prosesu data/trafodion a system gwybodaeth rheoli (*MIS*). (4 marc)

(b) Defnyddir systemau gwybodaeth rheoli (*MIS*) yn y mwyafrif o gyrff fel ysbytai ac ysgolion. Disgrifiwch, drwy roi enghraifft berthnasol ym mhob achos, sut y gall *MIS* gael ei defnyddio yn y ddau gorff hyn. (4 marc)

2 (a) Mae'r mwyafrif o gyrff yn defnyddio system gwybodaeth rheoli (*MIS*). Eglurwch bwrpas *MIS*. (3 marc)

(b) Eglurwch pam mae *MIS* yn arbennig o ddefnyddiol i reolwyr. (2 farc)

(c) Rhowch enghraifft o *un MIS* yr ydych wedi dod ar ei thraws ac eglurwch yn gryno sut y cafodd ei defnyddio. (3 marc)

3 Mae system gwybodaeth rheoli (*MIS*) yn troi'r data o ffynonellau mewnol ac allanol yn wybodaeth.

(a) Eglurwch beth yw ystyr y term 'ffynonellau mewnol ac allanol'. (2 farc)

(b) Disgrifiwch yn gryno *MIS* yr ydych wedi'i gweld neu wedi darllen amdani a disgrifiwch y data sy'n cael eu mewnbynnu i'r system a sut mae'r data'n cael eu hallbynnu i roi'r wybodaeth rheoli. (4 marc)

4 Defnyddir systemau gwybodaeth rheoli (*MIS*) gan lawer o gyrff.

(a) Bydd rheolwyr weithiau'n cwyno am eu *MIS* am nifer o resymau.
Disgrifiwch **bedwar** ffactor a all arwain at *MIS* wael. (8 marc)

(b) Disgrifiwch **bedair** o nodweddion *MIS* dda. (8 marc)

5 Un ffactor a all arwain at gynhyrchu system gwybodaeth rheoli (*MIS*) dda yw dealltwriaeth ac ymwneud y rheolwyr.
Disgrifiwch **dri** pheth y gallai'r rheolwyr eu gwneud i'w gwneud hi'n fwy tebygol y bydd *MIS* yn llwyddo. (6 marc)

► **Gweithgaredd 2: A yw'n system prosesu data/trafodion neu'n system gwybodaeth rheoli?**

Mae angen i chi allu gwahaniaethu rhwng system prosesu data a system gwybodaeth rheoli. Yn y bôn, os yw'r system yn prosesu data crai i gynhyrchu gwybodaeth reolaidd, mae'n debygol o fod yn system prosesu data. Ond os yw'r data a gynhyrchir yn cael ei chyfleu i reolwyr ar wahanol lefelau, ar ffurf briodol, i'w galluogi i wneud penderfyniadau effeithiol am gynllunio, cyfarwyddo a rheoli'r gweithgareddau y maen nhw'n gyfrifol amdanynt, yna mae'n system gwybodaeth rheoli.

Dyma rai systemau. Eich tasg chi yw dweud i ba gategori y maen nhw'n perthyn: systemau prosesu data neu systemau gwybodaeth rheoli.

- System gyflogau ar gyfer prosesu taflenni amser ac argraffu slipiau cyflog.
- System sy'n cymharu gwerthiant car o'r un gwneuthuriad yn ystod yr un mis ar gyfer y pum mlynedd ddiwethaf.
- Cynhyrchu rhestr, i'w rhoi i gwsmer, o'r holl brif werthwyr ceir yn y wlad sydd â char o liw a model penodol mewn stoc.
- Rhestr o'r eitemau i'w harchebu gan gyflenwyr sydd wedi cael ei chynhyrchu gan system pwynt talu electronig (*EPOS: electronic point of sale*).
- System dadansoddi gwerthiant sy'n ymchwilio i dueddiadau mewn gwerthiant dros gyfnod penodol o amser.
- Cynhyrchu rhestr o ddyledwyr (cwsmeriaid y mae arnynt arian i'r cwmni) i'w hanfon i'r rheolwyr i gael eu penderfyniad ar gamau pellach i'w cymryd.
- Ffigurau gwerthiant o gyfnodau cyffelyb blaenorol ar gyfer cynllunio lefelau cynhyrchu.
- System sy'n dadansoddi prisiau cystadleuwyr ar gyfer cynhyrchion cyffelyb.
- System sy'n cynhyrchu adroddiad sy'n nodi nifer y galwadau i ddesg gymorth a'r amser a gymerodd ar gyfartaledd i ddatrys pob problem.
- System ar gyfer cynhyrchu rhestr o fyfyrwyr sydd wedi talu eu ffioedd cwrs, i'w rhoi i diwtor y cwrs.

Cymorth gyda'r arholiad

Enghraifft 1

1 Mae'r mwyafrif o gyrff yn defnyddio systemau gwybodaeth rheoli.

Eglurwch beth yw system gwybodaeth rheoli ac ategwch eich esboniad drwy roi dwy enghraifft o sut y gall hi gael ei defnyddio.

Disgrifiwch y nodweddion/ffactorau hynny sy'n gwneud system gwybodaeth rheoli naill ai'n dda neu'n wael. Dylai eich ateb gynnwys enghreifftiau i ategu eich disgrifiad. **(13 marc)**

Ateb myfyriwr 1

1 Mae system gwybodaeth rheoli yn system a ddefnyddir gan reolwyr i'w helpu i reoli'r busnes neu'r corff. Gall y rheolwyr ddefnyddio'r system i'w helpu i reoli'r busnes. Gallant gael llawer o wybodaeth ddefnyddiol o'r system.

Mae system gwybodaeth rheoli dda yn gyflym gan nad yw'r rheolwyr eisiau aros yn rhy hir i'r wybodaeth gael ei harddangos. Rhaid i'r wybodaeth o'r system gwybodaeth rheoli (*MIS*) fod yn berthnasol fel nad oes rhaid iddynt fynd drwy lwyth o wybodaeth amherthnasol i fynd at y wybodaeth sydd ei hangen arnynt. Mae'n bosibl y bydd yn rhaid iddynt brosesu'r wybodaeth ymhellach, felly rhaid gallu ei hallforio mewn fformatau ffeil y gall meddalwedd arall, fel taenlenni, eu defnyddio.

Rhaid i'r wybodaeth sy'n cael ei chyflenwi gan y *MIS* fod yn gywir neu fe fydd y rheolwyr yn cymryd mwy o risg wrth wneud penderfyniadau.

Mae *MIS* wael yn llafurus i'w defnyddio. Mae angen i *MIS* fod yn hawdd ei defnyddio neu ni fydd y rheolwyr yn defnyddio'r system.

Sylwadau'r arholwr

1 Mae'r disgrifiad o sysytem gwybodaeth rheoli yn nodweddiadol o'r rheiny a gynhyrchir gan fyfyrwyr gwan nad ydynt wedi deall neu gofio'r prif bwyntiau. Yn lle hynny maen nhw'n malu awyr gan ddefnyddio geiriau yn y cwestiwn. Ni roddir unrhyw enghreifftiau o sut y gellir defnyddio *MIS*. Dim marciau am yr ateb hwn.

Mae'r ateb am ffactorau a nodweddion yn well ond braidd yn gyffredinol. **(5 marc allan o 13)**

Ateb myfyriwr 2

1 Mae system gwybodaeth rheoli yn system gyfrifiadurol sy'n darparu gwybodaeth ar ffurf y gall rheolwyr ei deall a'i defnyddio'n hawdd fel y gallant wneud penderfyniadau busnes da. Gall gwybodaeth o'r fath leihau'r risg wrth wneud penderfyniadau. Bydd system gwybodaeth rheoli (*MIS*) dda yn darparu gwybodaeth berthnasol fel nad yw rheolwyr yn treulio amser yn echdynnu'r data sydd eu hangen arnynt. Er enghraifft, pe bai angen ffigurau gwerthiant ar y rheolwyr ar gyfer cynnyrch penodol yn ystod y flwyddyn flaenorol, ni ddylent gael rhestr o ffigurau ar gyfer yr holl gynhyrchion.

Rhaid i'r wybodaeth fod yn gywir. Rhaid bod y data sy'n cael eu cipio yn gywir a rhaid iddynt gael eu dilysu a'u gwireddu. Bydd hyn yn gwneud penderfyniadau'r rheolwyr yn fwy cywir.

Bydd *MIS* dda yn hawdd ei defnyddio gan yr holl reolwyr, hyd yn oed y rheiny â sgiliau TGCh gwael, fel y gallant i gyd wneud defnydd o'r wybodaeth mae'r system yn ei darparu. Dylai'r corff helpu'r rheolwyr drwy ddarparu sesiynau hyfforddi a desgiau cymorth.

Mae *MIS* wael yn un sydd wedi cael ei chynhyrchu heb drafod â'r rheolwyr a fydd yn defnyddio'r wybodaeth. Os bydd y dadansoddiad cychwynnol o anghenion y defnyddwyr yn annigonol, ni fydd y system yn bodloni eu gofynion.

Bydd systemau gwybodaeth rheoli gwael yn codi'n aml o ganlyniad i ddefnyddio caledwedd a meddalwedd annigonol. Ni all y meddalwedd sy'n cael ei ddewis neu ei ddatblygu roi'r wybodaeth sydd ei hangen ar y rheolwyr.

Sylwadau'r arholwr

1 Mae'r rhan gyntaf, y disgrifiad o'r system gwybodaeth rheoli, yn dda ond nid yw'n crybwyll bod y *MIS* yn cynnwys pobl a gweithdrefnau yn ogystal â systemau TGCh. Mae gweddill y diffiniad yn dda ond ni roddir y ddwy enghraifft mae'r cwestiwn yn gofyn amdanynt. Mae tri disgrifiad da o *MIS* dda ynghyd ag enghreifftiau priodol.

Mae dau ateb da yn yr adran ar system gwybodaeth rheoli wael.

Mae hwn yn ateb da ar y cyfan, wedi'i strwythuro'n ofalus, ac mae'r myfyriwr yn amlwg yn deall beth yw *MIS* a sut mae'n cael ei ddefnyddio. Drwy edrych ar y dyraniad marciau yn ateb yr arholwr, fe welwch fod yr ateb hwn yn bodloni meini prawf ar gyfer y band uchaf o farciau. **(10 marc allan o 13)**

Ateb yr arholwr

1 Casgliad trefnedig o bobl, gweithdrefnau ac adnoddau sy'n helpu rheolwyr i wneud penderfyniadau yw system gwybodaeth rheoli (*MIS*).

Rhai enghreifftiau o'i defnydd yw:

Pennaeth ysgol yn dadansoddi'r disgyblion hynny sy'n mynd ar ei hôl hi gyda'u gwaith, yn ôl tystiolaeth profion, ac y mae eu presenoldeb yn wael, fel y gellir trefnu cyfweliadau â rhieni.

Rheolwr cynhyrchu cwmni yn defnyddio system gwybodaeth rheoli i wneud rhagfynegiadau ynghylch faint o gynnyrch penodol i'w wneud ar sail gwerthiant yn yr un chwarter mewn blynyddoedd blaenorol.

Mae nodweddion *MIS* dda yn cynnwys y canlynol:

Cywirdeb y wybodaeth a gynhyrchir (mae'n dibynnu ar gywirdeb y data sy'n cael eu mewnbynnu fel rheol).

Y gallu i alluogi rheolwyr i greu eu hymholiadau eu hunain yn hyblyg.

Cyflwyno'r data ar ffurf briodol i'w gwneud yn hawdd eu deall.

Mae'n osgoi rhoi gwybodaeth nad oes ei hangen.

Yn gallu cael ei defnyddio gan reolwyr y mae eu profiad a'u sgiliau TGCh yn amrywio.

Y gallu i drosglwyddo data i becynnau eraill i'w prosesu/dadansoddi ymhellach, e.e. pecyn taenlen.

Ffactorau a all arwain at *MIS* wael:

Ymgynghori annigonol gyda'r rheolwyr wrth ddadansoddi'r system i ddarganfod beth sydd ei hangen arnynt o'r system.

Mae diffyg hyfforddiant yn golygu nad yw llawer o reolwyr yn defnyddio'r system fel y dylent.

Defnyddio caledwedd neu feddalwedd amhriodol. Er enghraifft, mae'n bosibl y bydd y rhwydwaith yn rhedeg yn araf wrth brosesu'r wybodaeth sydd ei hangen wrth gynhyrchu adroddiadau *MIS*.

Cyfathrebu gwael rhwng staff proffesiynol. Mae staff wedi methu â chydweithredu wrth sefydlu'r system gwybodaeth rheoli.

Dadansoddiad cychwynnol annigonol. Nid yw'r system yn gwneud yn union beth y dylai ei wneud.

10–13 marc Mae'r ymgeiswyr wedi rhoi diffiniad cywir o *MIS* ac wedi rhoi enghreifftiau clir o ddau ddefnydd. Mae'r ymgeiswyr wedi rhoi atebion clir a rhesymegol sy'n cymharu ac yn cyferbynnu pedair nodwedd/pedwar ffactor yn fanwl gan roi enghreifftiau perthnasol. Defnyddiant dermau priodol a sillafu, atalnodi a gramadeg cywir.

5–9 marc Mae'r ymgeiswyr wedi rhoi diffiniad rhannol gywir o *MIS* ac wedi rhoi un enghraifft gywir. Mae'r ymgeiswyr wedi rhoi ychydig o nodweddion/ffactorau ond nid oes eglurder neu enghreifftiau. Mae ychydig o gamgymeriadau sillafu, atalnodi a gramadeg.

0–4 marc Mae'r ymgeiswyr wedi rhoi diffiniad arwynebol o *MIS* a dim enghreifftiau. Mae'r ymgeiswyr wedi rhestru pedair nodwedd ond nid ydynt yn rhoi eglurhad neu'r cyfan a wnânt yw disgrifio dwy nodwedd/dau ffactor yn llawnach. Mae diffyg eglurder yn yr ymateb ac mae llawer o gamgymeriadau sillafu, atalnodi a gramadeg.

Enghraifft 2

2 **Caiff systemau gwybodaeth rheoli eu defnyddio gan gyrff i droi data o ffynonellau mewnol ac allanol yn wybodaeth. Un o nodweddion system gwybodaeth rheoli effeithiol yw bod y wybodaeth yn cael ei chyfleu i'r rheolwr ar ffurf briodol.**

Heblaw am y nodweddion a roddir yn yr enghraifft, rhowch dair nodwedd hollol wahanol sydd gan system gwybodaeth rheoli effeithiol, a rhowch enghraifft addas i ategu pob ateb. (6 marc)

Ateb myfyriwr 1

2 Dyma'r tair nodwedd:

Mae'n gwella cyfathrebu personol rhwng y rheolwyr a'r staff.

Cywirdeb y data.

Gall rheolwyr sydd â sgiliau a gwybodaeth eang ym meysydd TGCh a *MIS* ddefnyddio'r system yn hawdd.

Sylwadau'r arholwr

2 Mae'r myfyriwr wedi rhestru tair o nodweddion system gwybodaeth rheoli ond heb gynnwys unrhyw enghreifftiau nac esboniad pellach.

Dylai'r myfyriwr fod wedi darllen y cwestiwn yn fwy gofalus ac wedi sylweddoli bod nifer y marciau'n dangos bod angen ateb manylach. **(3 marc allan o 6)**

Cymorth gyda'r arholiad *(parhad)*

Ateb myfyriwr 2

2 Mae'r system gwybodaeth rheoli yn rhoi'r wybodaeth/ data mewn ffurf briodol. Er enghraifft, mae'n haws gweld tueddiadau gwerthiant os caiff graffiau eu cynhyrchu yn hytrach na thablau o rifau. Dylai'r *MIS* adael i'r rheolwr ddewis sut mae'r wybodaeth yn cael ei chyflwyno. Dylai'r rheolwyr gael yr union wybodaeth sydd ei hangen arnynt a dim byd arall; wedyn ni fydd angen iddynt chwilio drwy lwyth o wybodaeth am y wybodaeth berthnasol. Ni ddylai'r rheolwyr orfod gwneud prosesu pellach i gael y wybodaeth sydd ei hangen. Er enghraifft, dylai rhestr o ffigurau gwerthiant ar gyfer un cynnyrch dros y pum mlynedd diwethaf gael ei chynhyrchu ar ei phen ei hun yn hytrach na gyda manylion yr holl gynhyrchion eraill. Dylai'r *MIS* fod yn hawdd ei defnyddio. Dylai ei gwneud hi'n hawdd i'r rheolwyr ddewis y wybodaeth sydd ei hangen arnynt o'r system. Er enghraifft, dylai'r rheolwr allu dewis y meysydd sydd i gael eu harddangos a gwneud unrhyw grwpio neu drefnu drwy glicio ar restr. Dylai rheolwyr â sgiliau a gwybodaeth sy'n amrywio'n eang allu defnyddio'r *MIS*.

Sylwadau'r arholwr

2 Mae hwn yn ateb da iawn, ym mhob rhan. Mae'r myfyriwr wedi rhoi'r nodweddion yn glir, wedi ychwanegu esboniad pellach, ac wedi ategu'r atebion ag enghraifft gywir.
Rhoddir marciau llawn yma. **(6 marc allan o 6)**

Ateb yr arholwr

2 Rhoddir un marc am y nodwedd, ac un marc am esboniad pellach ac am enghraifft.
Cywirdeb y data (1) sy'n cael eu prosesu gan y *MIS* (1) i roi gwybodaeth y gellir dibynnu arni fel y gellir gwneud penderfyniadau cywir (1).
Hyblygrwydd dadansoddi data (1) fel y gallu i ddarparu cymariaethau rhwng dwy set o ffigurau (1) drwy ddefnyddio gwahaniaethau, canrannau, cynnydd/gostyngiad canrannol (1), ac ati.
Darparu'r data ar ffurf briodol (1) fel cynhyrchu graffiau yn hytrach na thablau o rifau (1) er mwyn gweld tueddiadau'n haws (1).
Yn hygyrch i amrywiaeth eang o ddefnyddwyr ac yn cefnogi amrywiaeth eang o sgiliau a gwybodaeth (1) fel bod yr holl reolwyr yn gallu cael y wybodaeth (1) sydd ei hangen arnynt i wneud penderfyniadau yn eu maes gan ddefnyddio'r *MIS* (1).
Mae'n gwella cyfathrebu rhyngbersonol ymhlith rheolwyr a gweithwyr (1) drwy roi gwybodaeth fanwl gywir iddynt i'w thrafod mewn cyfarfodydd (1), ac i sicrhau bod y penderfyniadau sy'n cael eu gwneud mor gadarn â phosibl (1).
Mae'n caniatáu cynllunio projectau unigol (1). Mae hyn yn golygu y gall rheolwyr gynllunio ymgyrchoedd gwerthu, staffio, cyllidebau (1), ac ati, gan ddefnyddio gwybodaeth a gânt o'r *MIS* fel tueddiadau gwerthiant, costau staff (1), ac ati.
Mae'n osgoi gorlwytho gwybodaeth (1). Bydd y *MIS* yn caniatáu i'r rheolwyr echdynnu gwybodaeth benodol iawn (1) nad yw'n cynnwys unrhyw wybodaeth amherthnasol sy'n gwastraffu eu hamser wrth ei darllen (1).

Mapiau meddwl cryno

Nodweddion system gwybodaeth rheoli (*MIS*) effeithiol

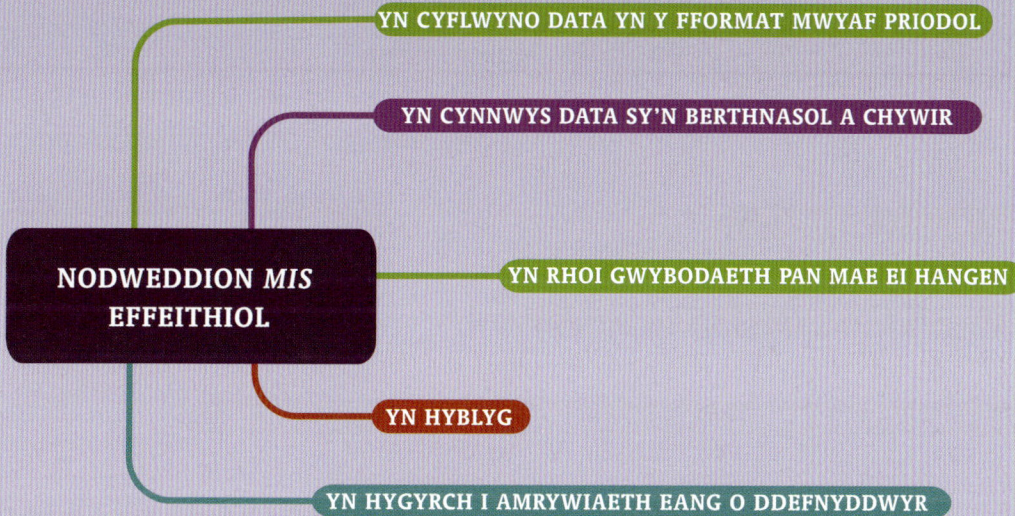

YN CYFLWYNO DATA YN Y FFORMAT MWYAF PRIODOL

YN CYNNWYS DATA SY'N BERTHNASOL A CHYWIR

NODWEDDION *MIS* EFFEITHIOL

YN RHOI GWYBODAETH PAN MAE EI HANGEN

YN HYBLYG

YN HYGYRCH I AMRYWIAETH EANG O DDEFNYDDWYR

Nodweddion system gwybodaeth rheoli (*MIS*) dda

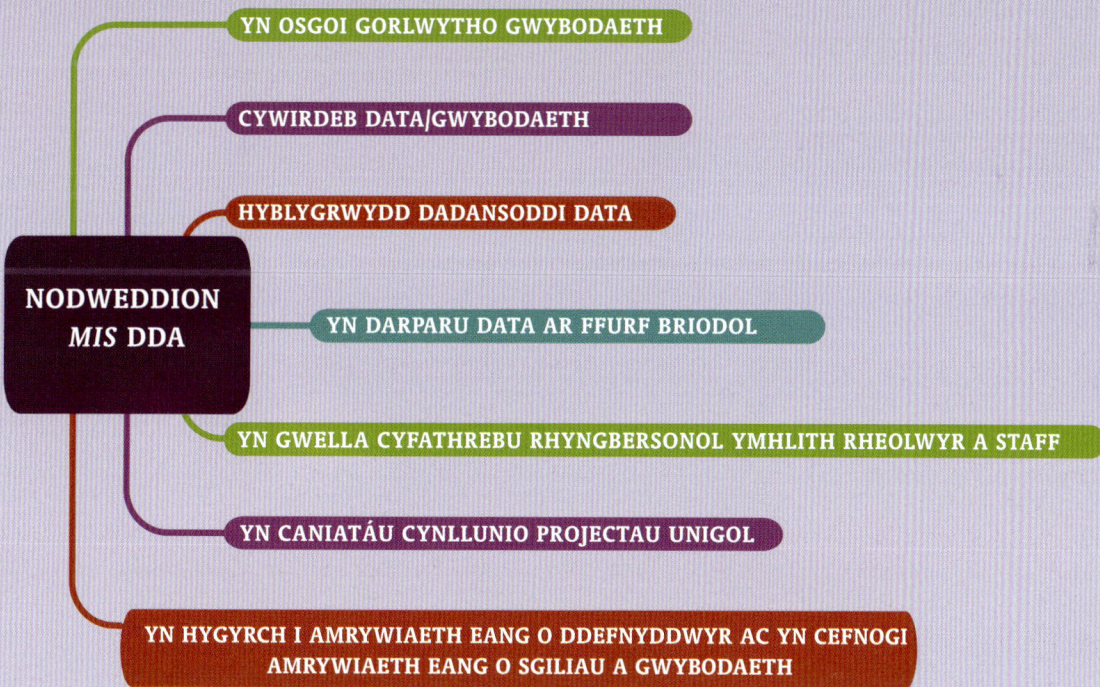

YN OSGOI GORLWYTHO GWYBODAETH

CYWIRDEB DATA/GWYBODAETH

HYBLYGRWYDD DADANSODDI DATA

NODWEDDION *MIS* DDA

YN DARPARU DATA AR FFURF BRIODOL

YN GWELLA CYFATHREBU RHYNGBERSONOL YMHLITH RHEOLWYR A STAFF

YN CANIATÁU CYNLLUNIO PROJECTAU UNIGOL

YN HYGYRCH I AMRYWIAETH EANG O DDEFNYDDWYR AC YN CEFNOGI AMRYWIAETH EANG O SGILIAU A GWYBODAETH

Ffactorau a all arwain at system gwybodaeth rheoli (*MIS*) wael

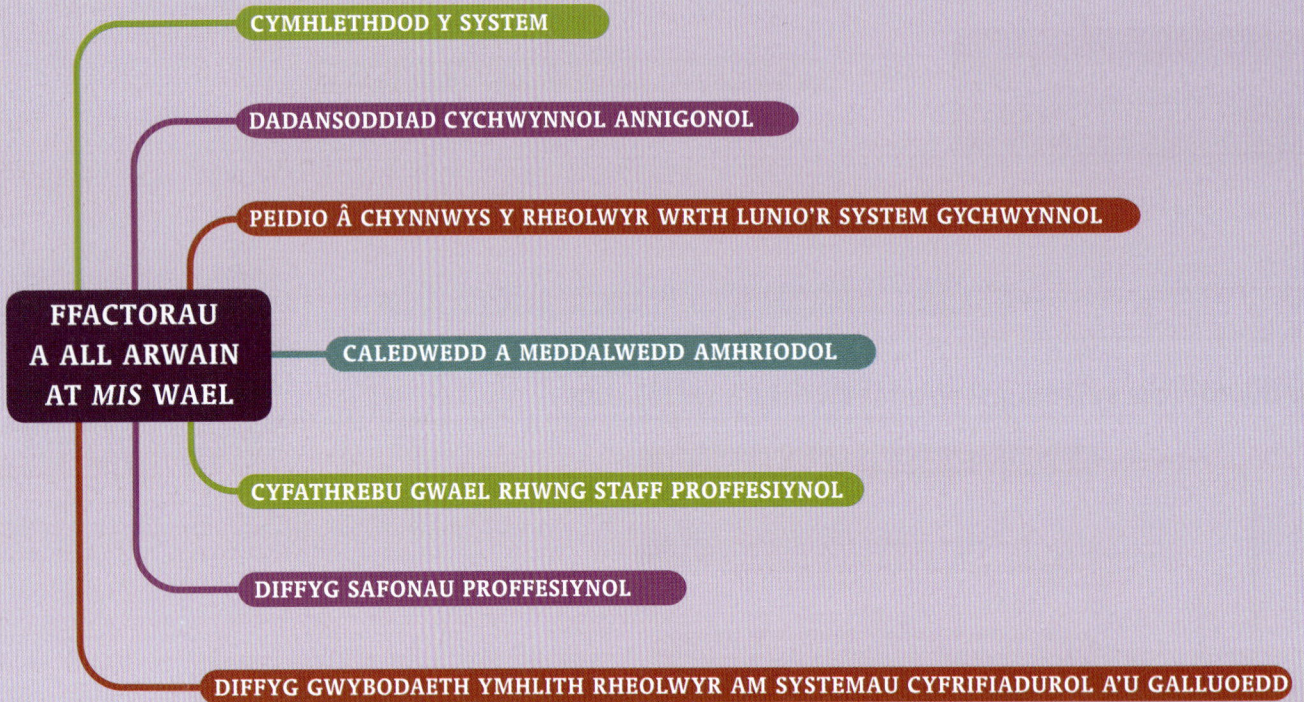

CYMHLETHDOD Y SYSTEM

DADANSODDIAD CYCHWYNNOL ANNIGONOL

PEIDIO Â CHYNNWYS Y RHEOLWYR WRTH LUNIO'R SYSTEM GYCHWYNNOL

FFACTORAU A ALL ARWAIN AT *MIS* WAEL

CALEDWEDD A MEDDALWEDD AMHRIODOL

CYFATHREBU GWAEL RHWNG STAFF PROFFESIYNOL

DIFFYG SAFONAU PROFFESIYNOL

DIFFYG GWYBODAETH YMHLITH RHEOLWYR AM SYSTEMAU CYFRIFIADUROL A'U GALLUOEDD

TOPIG 9: Cylchred oes datblygu system

Er mwyn i systemau gwrdd â'r disgwyliadau, mae angen eu datblygu mewn ffordd ffurfiol. Cyfres o gamau yw'r gylchred oes datblygu system (*SDLC: system development life cycle*) sy'n cael eu cymryd wrth ddatblygu system newydd neu newid system bresennol.

Yn y topig hwn, byddwch yn dysgu am y camau hyn a sut y cânt eu cymhwyso wrth ddatblygu system gyfrifiadurol newydd.

▼ Y cysyniadau allweddol sy'n cael sylw yn y topig hwn yw:

▶ Prif elfennau'r gylchred oes datblygu system a sut y cânt eu cymhwyso wrth ddatblygu datrysiad cyfrifiadurol

▶ Ymchwilio i'r system

▶ Dadansoddi'r system

▶ Dylunio'r system

▶ Gweithredu'r system

▶ Cynnal y system

▶ Gwerthuso'r system

CYNNWYS

Ymchwilio i'r system

Cyflwyniad

Yn yr adran hon byddwn yn ystyried beth yw system, y gylchred oes datblygu system (*SDLC: system development life cycle*) a pham mae angen systemau newydd, a'r cam cychwynnol, sef ymchwilio i'r system.

Beth yw system?

Ffordd o wneud pethau yw system. Disgrifiad arall o system yw cyfanwaith cymhleth y mae ei ddarnau cydrannol wedi'u trefnu at ddiben cyffredin.

Mae gan gyrff wahanol systemau i ymdrin â'r gwahanol feysydd swyddogaethol. Mae angen systemau ar gyfer talu staff, prynu stoc neu ddefnyddiau crai, rheoli faint o stoc sy'n cael ei gadw, cadw cyfrifon, sicrhau cydymffurfiad â'r ddeddfwriaeth, cadw cofnodion o'r staff sy'n gweithio i'r corff, ac yn y blaen.

Y gylchred oes datblygu system

Dilyniant o weithgareddau sy'n cael eu cyflawni wrth ddadansoddi system yw'r gylchred oes datblygu system. Edrychwch ar y diagram ac fe sylwch ei bod yn gylchol. Y rheswm am hyn yw ei bod hi'n arferol, pan fydd y system yn weithredol, i edrych arni o bryd i'w gilydd i ystyried pa welliannau y gellir eu gwneud iddi. O ganlyniad caiff y gweithgareddau yn y diagram eu hailadrodd.

Y gylchred oes datblygu system

Y tasgau sy'n gysylltiedig ag ymchwilio i'r system

Cyn gallu dechrau dadansoddi'r system, rhaid darganfod llawer o ffeithiau am y corff a'r tasgau y bydd disgwyl i'r system newydd eu cyflawni. Felly mae angen i chi ymchwilio i'r system bresennol a hefyd i ofynion y system newydd er mwyn gallu gwneud y dadansoddiad.

Er mwyn gwella system sy'n bodoli mae'n rhaid deall sut mae'n gweithio. Mae'r broses o ddarganfod sut yn cael ei galw'n ddarganfod ffeithiau. Mae sawl ffordd wahanol o gael gwybodaeth am y system bresennol:

- cyfweliadau
- arsylwi
- archwilio cofnodion
- holiaduron.

Cyfweliadau – bydd cyfweliadau â rheolwyr fel rheol yn dangos sut mae eu hadran nhw yn gweithio ac yn dangos unrhyw broblemau maen nhw'n eu cael gyda'r system bresennol. Byddant yn gallu cynnig gwybodaeth am y ffordd yr hoffent weld y system newydd yn gweithio a byddant yn gallu dweud wrthych am y wybodaeth yr hoffent i'r system newydd ei darparu.

Staff gweithredol yw'r aelodau hynny o'r staff sy'n gwneud y rhan fwyaf o waith pob dydd y corff ac mae eu gwybodaeth am y corff wedi'i chyfyngu i'w maes eu hunain. Byddant yn gallu rhoi manylion y tasgau bach a sut maen nhw'n cael eu cyflawni.

Mae casglu gwybodaeth mewn cyfweliadau yn cymryd llawer o amser.

Arsylwi – pe baech chi am ddysgu swydd rhywun arall, y ffordd orau o wneud hyn fyddai drwy eistedd am ychydig o ddyddiau gyda'r person y byddwch chi'n gwneud ei swydd neu eistedd gyda rhywun sy'n gwneud swydd debyg. Mae arsylwi'n golygu eistedd gyda rhywun i weld beth maen nhw'n ei wneud er mwyn deall y llifoedd gwybodaeth a phrosesau sy'n rhan o'u swydd.

Archwilio cofnodion – mae llawer o gyrff yn parhau i ddefnyddio a chynhyrchu dogfennau papur ar gyfer eu busnes. Drwy archwilio'r dogfennau hyn gallwch ddeall pa wybodaeth sy'n cael ei chadw a sut mae'n cael ei throsglwyddo rhwng y gwahanol adrannau neu rhwng y corff a'i gyflenwyr a'i gwsmeriaid.

Byddai dogfennau sy'n rhoi gwybodaeth gyffredinol yn cynnwys:

- siartiau trefniadaeth (siart yn dangos yr hierarchaeth yn y corff; gellir ei ddefnyddio i weld pwy sy'n adrodd i bwy)
- CVs y staff – defnyddiol ar gyfer asesu anghenion sgiliau neu hyfforddiant y staff
- disgrifiadau swydd – rhoddant fanylion y tasgau sy'n cael eu gwneud gan wahanol bobl
- llawlyfrau polisi/gweithdrefnau – defnyddiol ar gyfer deall y ffordd mae'r corff yn gweithio
- dogfennaeth flaenorol am systemau – dogfennau papur sy'n disgrifio systemau blaenorol.

Byddai dogfennau sy'n rhoi gwybodaeth benodol yn cynnwys:

- catalogau o gynhyrchion
- ffurflenni archebu
- anfonebau
- nodion anfon
- rhestri dewis (i staff warws)

Drwy edrych ar CV gallwch ddarganfod pa sgiliau sydd gan y staff a'r hyfforddiant tebygol y bydd ei angen arnynt

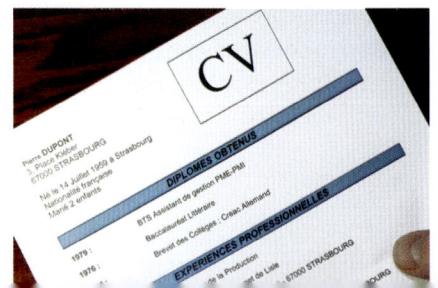

Mae siartiau trefniadaeth yn eich helpu i ddeall strwythur y busnes

Holiaduron – ar yr olwg gyntaf mae holiaduron yn edrych fel ffordd ddelfrydol o gasglu gwybodaeth am gwmni. Does dim angen treulio amser yn holi pobl ac mae holiadur yn cadw at y pwyntiau pwysig heb grwydro, sy'n gallu digwydd mewn cyfweliad.

Ond mae gan holiaduron anfanteision. Mae llawer o bobl yn anghofio eu llenwi, a gall hyn roi darlun anghyflawn o'r system. Gall pobl eraill gamddeall rhai o'r cwestiynau os caiff y ffurflenni eu postio iddynt ac os nad oes cymorth personol ar gael.

Er hynny, mae holiaduron yn ddefnyddiol pan mae angen casglu gwybodaeth gan nifer mawr o unigolion, gan eu bod yn llyncu llawer llai o amser staff na chyfweliadau.

Wrth lunio holiadur, dylid cofio'r canlynol:

- Sicrhewch fod y cwestiynau wedi'u geirio'n fanwl fel nad oes angen i'r defnyddwyr ddehongli'r cwestiynau.
- Mae'n well os nad oes rhaid i'r staff roi eu henwau ar yr holiadur, neu mae'n bosibl na chewch atebion gonest.
- Strwythurwch yr holiadur fel bod y cwestiynau cyffredinol yn dod gyntaf, ac yna'r rhai mwy penodol. Hefyd mae'n werth rhannu'r holiadur yn feysydd swyddogaethol: er enghraifft, gallai un rhan ymdrin â phrosesu archebion ac un arall â rheolaeth stoc. Bydd hyn yn amrywio yn ôl y math o gorff.
- Osgowch gwestiynau arweiniol (cwestiynau sy'n awgrymu'r ateb yr hoffech ei gael).
- Ar ddiwedd yr holiadur, gofynnwch y cwestiwn hwn bob amser: 'A oes unrhyw beth arall y dylwn i wybod amdano yn eich barn chi?'

Dichonoldeb a'r adroddiad dichonoldeb

Ymchwiliad cychwynnol i edrych ar y tebygolrwydd o allu creu system newydd am gost resymol yn unol â nodau ac amcanion penodol yw ymarfer dichonoldeb. Caiff y canlyniadau eu crynhoi mewn dogfen o'r enw adroddiad dichonoldeb, dogfen a ddefnyddir gan uwch reolwyr a chyfarwyddwyr i asesu dichonoldeb y project. Ar ôl gwneud yr asesiad, byddant yn penderfynu a ddylid mynd ymlaen â'r project neu roi'r gorau iddo.

Mae cynnal ymarfer dichonoldeb yn weithgaredd pwysig gan ei fod yn sicrhau nad yw systemau sy'n debygol o fethu yn cael eu datblygu. Mae llawer o enghreifftiau o roi'r gorau i systemau newydd ar ôl gwneud cryn dipyn o waith arnynt.

Bydd ymarfer dichonoldeb fel rheol yn cynnwys:

- Ymarfer darganfod ffeithiau cychwynnol i gael gwybodaeth am yr hyn sydd ei angen o'r project.
- Ymchwiliad i'r goblygiadau technegol, cyfreithiol, economaidd, gweithredol ac amserlennu.
- Nodi costau a manteision y system newydd a'u pwyso a'u mesur yn erbyn ei gilydd.
- Gwneud argymhellion ynghylch dichonoldeb y project.
- Cynllun drafft ar gyfer gweithredu'r project.

Beth y dylid ei gynnwys yn yr adroddiad dichonoldeb?

Ar ôl y dadansoddiad cychwynnol o'r system bresennol, caiff adroddiad dichonoldeb ei gynhyrchu. Defnyddir hwn i benderfynu a ddylid mynd ymlaen â'r project neu anghofio amdano.

Bydd adroddiad dichonoldeb fel arfer yn cynnwys rhai neu'r cyfan o'r canlynol:

- Gofynion y defnyddiwr – dylid nodi gofynion y system o safbwynt y defnyddwyr, cytuno arnynt gyda'r dadansoddydd, a'u hysgrifennu fel rhestr. Bydd y system newydd yn cael ei barnu yn erbyn pa mor dda mae'n cwrdd â'r gofynion hyn.

- Manylion y caledwedd a'r meddalwedd presennol – efallai y gellir eu hailddefnyddio neu efallai bod angen caledwedd neu feddalwedd newydd neu ychwanegol.
- Diffiniad o gwmpas (hyd a lled) y system bresennol – fel bod pawb yn deall beth mae'r system yn ei wneud a ddim yn ei wneud. Bydd maint a chymhlethdod y project arfaethedig yn dibynnu ar gwmpas y system. Mae cwmpas y system yn cynnwys manylion fel siartiau trefniadaeth, ffynonellau data a dulliau o gipio data.
- Y prif weithrediadau a phrosesau prosesu data – mae'n well dangos hyn ar ffurf diagram cyd-destun, sef diagram llif data arbennig sy'n cynrychioli'r system gyfan fel petryal ac yn dangos sut mae'r wybodaeth yn llifo o'r systemau i gyrff a systemau allanol ac yn ôl. Fe ddysgwch sut i lunio diagram cyd-destun yn nes ymlaen yn y topig hwn pan edrychwn ar yr ystod lawn o ddiagramau llif data.
- Darganfod y problemau gyda'r system bresennol – yma fe fyddech chi'n rhestru'r problemau hyn gyda'r bwriad o adeiladu system newydd a all eu datrys.
- Dadansoddi costau a manteision y system newydd – rhaid talu am gostau staff, cyfarpar, caledwedd, meddalwedd, trwyddedau, systemau cyfathrebu, hyfforddiant, ac ati. Rhaid pwyso a mesur y rhain yn erbyn manteision y system newydd: e.e. llai o gamgymeriadau, mwy o wybodaeth i reolwyr, y gallu i gwblhau trafodion yn gyflymach, cynnydd mewn busnes gan fod cwsmeriaid yn fwy bodlon.

Dadansoddi'r system

▼ **Byddwch yn dysgu**

▶ Am y tasgau sy'n gysylltiedig â dadansoddi'r system bresennol

▶ Am ddiagramau llif data lefel uchel (golwg cyd-destunol)

▶ Am lefelu diagramau llif data

▶ Am dablau penderfyniad

▶ Am ddiagramau perthynas endidau

▶ Am y geiriadur data

Cyflwyniad

Ar ôl cwblhau'r ymchwiliad i'r system, a phenderfynu mynd ati i gyflwyno'r system newydd ar sail yr adroddiad dichonoldeb, bydd angen dealltwriaeth well o'r system sy'n cael ei datblygu. Mae'r broses o ddarganfod ffeithiau am y system bresennol a'u dogfennu'n cael ei galw'n ddadansoddiad systemau ac mae'n cael ei chyflawni gan ddadansoddydd systemau. Defnyddir nifer o offer a thechnegau i gynhyrchu diagramau a siartiau sy'n dogfennu'r system bresennol neu system newydd a byddwn yn ymdrin â'r rhain yn yr adran hon.

Dadansoddi'r system bresennol

Mae'r dadansoddiad yn edrych yn fanwl ar y system bresennol neu'r gofynion ar gyfer tasg sydd erioed wedi cael ei chyflawni o'r blaen. Mae'r dadansoddydd systemau yn ymchwilio i'r gofynion ar gyfer y system newydd.

Mae peth dadansoddi cychwynnol yn cael ei wneud wrth ymchwilio i'r system, gan fod yn rhaid cael gwybodaeth am y system sy'n cael ei datblygu (nodau ac amcanion, maint, cwmpas, ailddefnyddio caledwedd a meddalwedd, ac ati) gan y bydd angen hyn i gynhyrchu'r adroddiad dichonoldeb.

Bydd y dadansoddiad yn cynnwys y canlynol fel rheol:

- deall y system bresennol
- deall y system arfaethedig os nad oes system ar hyn o bryd
- casglu a dadansoddi gofynion gwahanol ddefnyddwyr
- cyflwyno'r datrysiad mewn ffordd resymegol gan ddefnyddio offer a thechnegau fel diagramau llif data, modelau data, manylebau prosesau a diagramau systemau
- cynhyrchu manyleb.

Adnabod a deall yr offer a'r technegau a ddefnyddir i ddeall system

Defnyddir nifer o offer a thechnegau i helpu i ddeall a dadansoddi systemau a rhoddwn sylw i'r rhain yma.

Defnyddiwn ddogfennaeth yn rheolaidd wrth ddadansoddi systemau a hynny ar wahanol lefelau yn y system. Fel arfer fe weithiwn o orolwg o'r system gyfan, ond wrth ddatblygu'r system caiff rhagor o fanylion eu hychwanegu. Er mwyn disgrifio a deall y system, defnyddiwn nifer o dechnegau a diagramau.

Diagramau llif data

Mae'n bwysig, wrth egluro'r system bresennol, i ni edrych ar y llifoedd data/gwybodaeth o fewn y corff. Mae hyn yn cael ei wneud ar ffurf diagramau llif data yn aml. Defnyddir y rhain ar y dechrau i ddisgrifio'r system ac maent yn edrych ar fewnbynnau, y prosesau sy'n cael eu cyflawni ar y mewnbynnau, a'r allbynnau a gynhyrchir.

Yn anffodus, mae llawer ffordd wahanol o lunio diagram llif data. Mae'r dull a welwch yma yn cael ei ddefnyddio gyda'r fethodoleg dadansoddi a dylunio systemau strwythuredig (*SSADM: Structured Systems Analysis and Design Methodology*).

Mae ymchwiliad cychwynnol o'r system yn cael ei wneud yn yr astudiaeth dichonoldeb. Mae hyn yn edrych ar y mewnbynnau i'r system, y prosesau sy'n cael eu cyflawni arnynt, a'r allbynnau o'r system. Hefyd mae cwmpas y system yn cael ei bennu yn y cam hwn. I hwyluso dadansoddi pellach, mae diagramau llif data yn cael eu llunio. Defnyddir diagramau llif data i ystyried y data, ond maent yn anwybyddu'r cyfarpar sy'n storio'r data. Diagramau llif data yw'r cam cyntaf a gymerir wrth ddisgrifio system.

Defnyddir cyfres o symbolau yn y diagramau hyn. Yn anffodus mae awduron gwahanol yn defnyddio symbolau gwahanol, sy'n gallu peri dryswch.

Proses neu weithred – blwch petryalog yw hwn ac mae'n cynrychioli proses sy'n gwneud rhywbeth gyda'r data (gallai drin y data mewn rhyw ffordd neu wneud cyfrifiadau arnynt, ac ati). Mae'r blwch wedi'i rannu'n dair rhan: mae rhif yn y blwch uchaf ar y chwith, sef cyfeirnod y blwch; defnyddir prif gorff y blwch i ddisgrifio'r broses; a defnyddir y blwch uchaf ar y dde i gofnodi'r person neu adran sy'n gyfrifol am y broses.

1	Clerc cyflogau
Cyfrifo cyflogau	

Proses

Ffynhonnell allanol data (o ble mae'r data'n dod) neu suddfan allanol data (i ble mae'r data'n mynd)

Blwch hirgrwn yw hwn a ddefnyddir i ddisgrifio o ble, y tu allan i'r system, y mae'r data'n dod, ac i ble mae'r data'n mynd. Nid ydym yn poeni am beth sy'n digwydd i'r data cyn iddynt gyrraedd y blwch (os yw'n ffynhonnell) nac am beth sy'n digwydd i'r data ar ôl iddynt fynd heibio i suddfan.

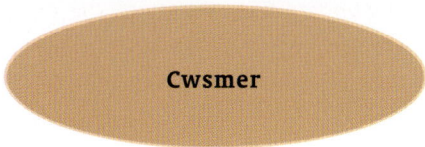

Gwrthrych allanol

YN YR ARHOLIAD

Dylai fod un blwch proses yn unig mewn diagram cyd-destun, a dim storfeydd

GEIRIAU ALLWEDDOL

Diagram cyd-destun – diagram sy'n cynrychioli system fel un broses diagram llif data

Diagram llif data – mae'n dangos y llifoedd, storfeydd a phrosesu data mewn system

Llif data

Defnyddir saeth sy'n pwyntio i gyfeiriad y llif i ddangos llif data. Fel rheol mae'n syniad da i chi roi disgrifiad o'r llif data ar y saeth. Yn ôl confensiwn, nid ydym byth yn defnyddio berf ar lif data.

Manylion cerdyn credyd

Llif data

Storfa ddata

Mae'r diagram yn dangos y symbol ar gyfer hon. Lleoliad lle mae data'n cael eu storio yw storfa ddata, er enghraifft: drôr lle cadwch lythyrau, blychau ffeil, ffolderi, llyfrau, cwpwrdd ffeilio (neu ddrôr penodol mewn cwpwrdd ffeilio), disg magnetig, CD, ac ati. Yma eto, mae rhif ar y symbol sy'n cael ei ddefnyddio'n gyfeirnod, ond rhoddir llythyren o flaen y rhif hefyd. Defnyddir M i ddynodi storfa law (*manual*) ac C i ddynodi storfa gyfrifiadurol (*computer store*).

C1	Cyfrifon cwsmeriaid

Mae'r storfa ddata hon yn cynnwys manylion cyfrifon cwsmeriaid ac mae ar gyfrifiadur

Lefelau o ddiagramau llif data

Wrth ddadansoddi systemau, yr arfer yw llunio'r diagram llif data ar wahanol lefelau. Mae'r lefel a ddefnyddir yn adlewyrchu dyfnder yr ymchwiliad i'r system drwy gyfrwng y diagram llif data.

Y diagram cyd-destun

Diagram cyd-destun yw'r enw a roddir ar y diagram llif data cyntaf sy'n cael ei lunio, gan ei fod yn rhoi'r system dan sylw yn ei chyd-destun. Diagram llif data lefel uchel ydyw sy'n dangos y system gyfan fel un blwch a'r llifoedd gwybodaeth rhwng endidau allanol fel cyflenwyr a chwsmeriaid. Gallai'r diagram ddangos y llif o archebion gan gwsmeriaid (endid allanol) i'r adran werthiant a nwyddau'n cael eu hanfon o'r warws i'r cwsmeriaid.

Nod y diagram cyd-destun yw dangos cwmpas neu hyd a lled y system.

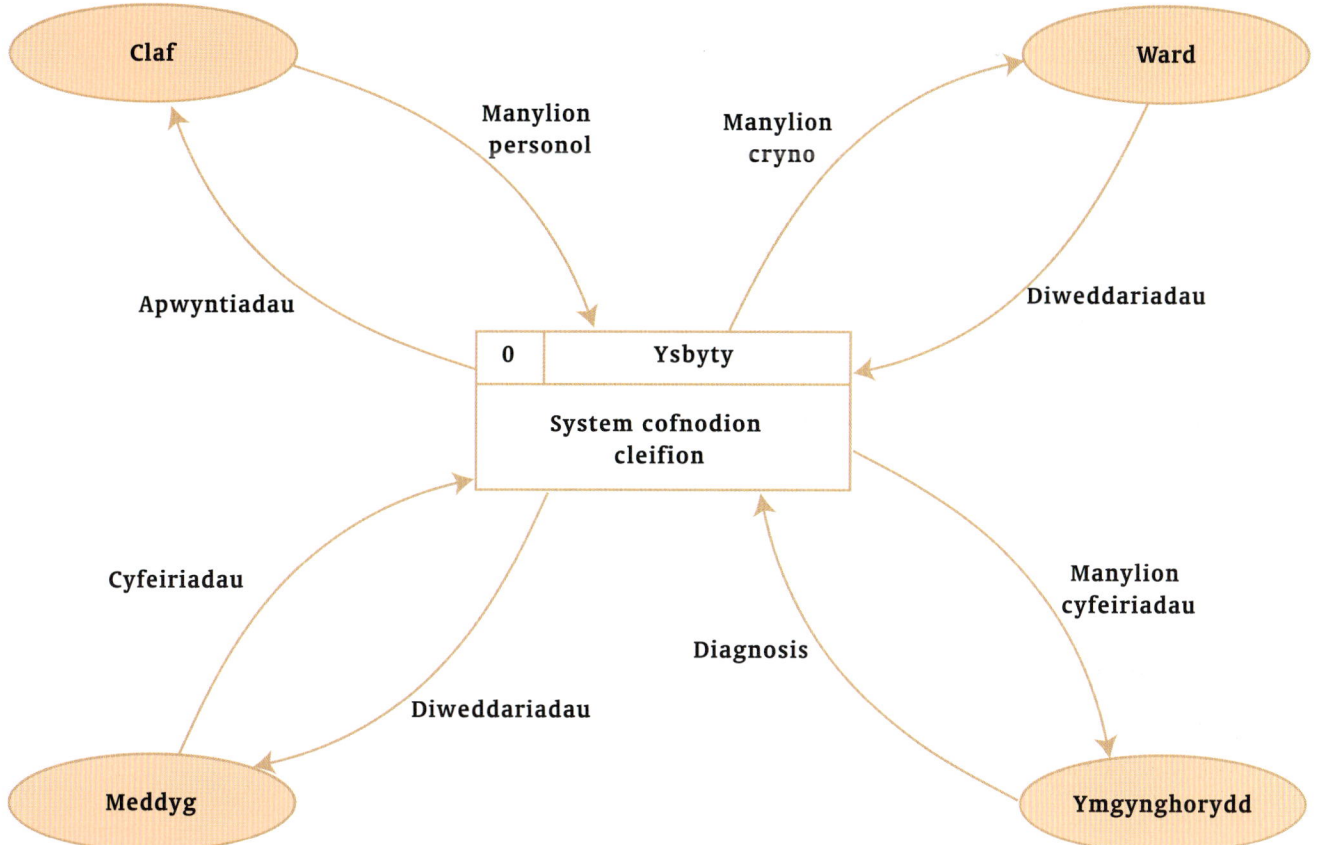

Diagram cyd-destun ar gyfer system cofnodion cleifion

Dadansoddi'r system (*parhad*)

Diagram llif data lefel 1

Mae'r diagram llif data lefel 1 yn dangos y prif brosesau yn y system ac yn rhoi darlun mwy manwl o'r llifoedd data. Dylech gynnwys tua chwe blwch proses ar y mwyaf yn y diagram hwn.

Diagram llif data lefel 2

Mae diagram llif data lefel 2 yn torri'r blychau proses yn y diagram llif data lefel 1 i lawr ac felly'n rhoi darlun manylach byth o'r prosesau. I wneud hyn, rhaid cymryd un o'r blychau proses (h.y. y blychau petryalog) o'r diagram llif

data lefel 1, a bydd y broses y mae'n ei chynrychioli yn cael ei thorri i lawr yn gyfres o is-brosesau mwy manwl.

Mantais y dull manylu hwn yw y gallwn, drwy edrych ar bob lefel gan ddechrau gyda'r diagram llif data amlinell, adeiladu darlun cymhleth o'r llifoedd gwybodaeth drwy'r system gam wrth gam. Yn lle cael un diagram llif data gyda nifer mawr o brosesau i'w gweld, mae gennym deulu cyfan o ddiagramau llif data sy'n dangos gwahanol lefelau o fanylder.

Gall diagramau llif data lefel uchel gael eu torri i lawr yn ddiagramau

pellach sy'n dangos mwy o fanylder.

Defnyddio diagramau llif data

Gall diagramau llif data gael eu defnyddio:

- wrth ymchwilio i systemau i gofnodi'r hyn sy'n cael ei ddarganfod
- wrth ddylunio system i egluro sut y bydd system arfaethedig yn gweithio
- wrth amlinellu manylebau systemau newydd.

Dadelfeniad (h.y. lefelu) y diagramau llif data

Lefel 0 Y diagram cyd-destun

Diagram llif data lefel 1

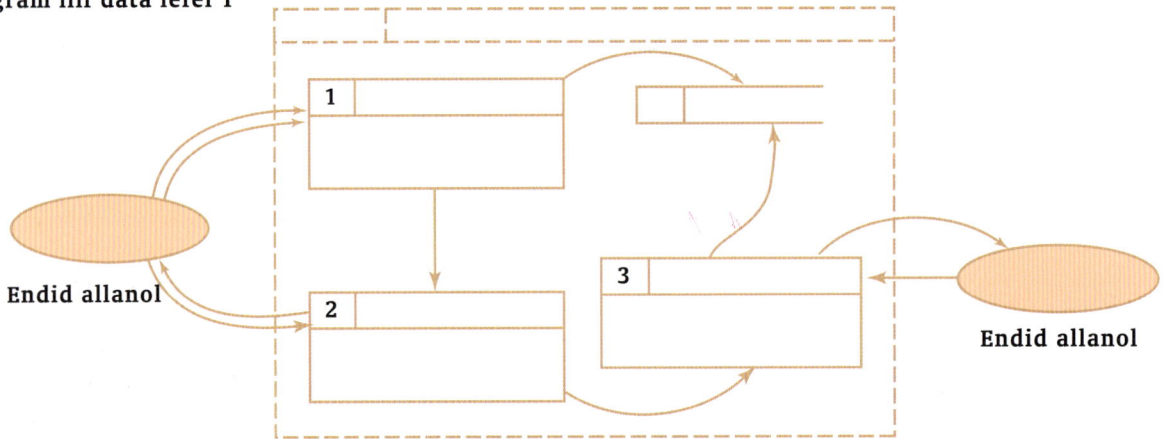

Diagram llif data lefel 2 (mae Blwch 2 yn y diagram uchod wedi'i ehangu)

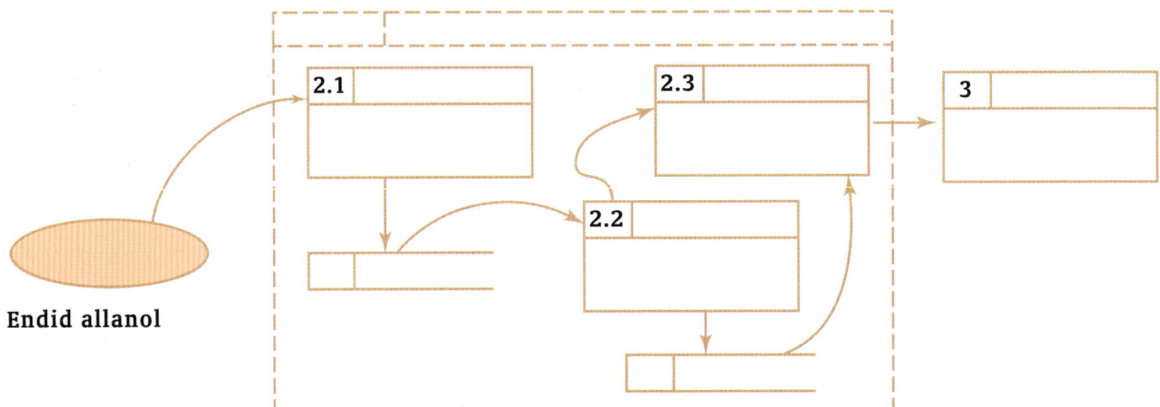

Tablau penderfyniad

Mae tabl penderfyniad yn cynnig ffordd syml o arddangos y gweithredoedd i'w rhoi ar waith o dan amodau penodol.

Gan edrych ar y diagram (ochr dde), byddwch yn gweld yr adrannau canlynol:

- Bonyn amodau: dyma'r sefyllfaoedd neu ddigwyddiadau y mae angen rhoi prawf arnynt. Y rhain yw achos y gweithredoedd y mae angen eu gweithredu yn adran y bonyn gweithredoedd. Amod nodweddiadol yw 'ydy oedran person dros 18?'.
- Bonyn gweithredoedd: y rhain yw'r gweithredoedd i'w rhoi ar waith gan ddibynnu ar y cyfuniad o amodau cyffredinol yn y bonyn amodau perthnasol. Er enghraifft, gallai 'gellir gweini diod feddwol iddo/iddi' fod yn weithred.
- Cofnodion amodau: mae'r rhain yn rhoi amcan o ba amodau sy'n berthnasol. Mae hyn yn cael ei wneud drwy roi Y (Ydy) neu N (Na) ger pob amod gan ddibynnu ar a yw'r amod yn berthnasol ai peidio.
- Cofnodion gweithredoedd: mae'r rhain yn rhoi'r weithred i'w gweithredu gan ddibynnu ar yr amodau sy'n berthnasol. Rhoddir croes ar y tabl penderfyniad i ddangos pa weithred neu weithredoedd a ddylai gael ei/eu rhoi ar waith.

Mae hyn yn swnio'n gymhleth! Ond maen nhw'n llawer mwy anodd eu disgrifio na'u gwneud, felly edrychwn ar enghraifft syml y dylech fod yn gyfarwydd â hi.

Enghraifft: Tabl penderfyniad ar gyfer set o oleuadau traffig

Er mwyn deall sut mae creu tablau penderfyniad, mae'n well edrych ar

Dyma'r ffordd safonol o gyflwyno tabl penderfyniad

enghraifft nad yw'n gysylltiedig â TGCh, e.e. gweithrediad goleuadau traffig.

Yr amodau cyffredinol yw lliwiau'r goleuadau a'r gweithredoedd cyffredinol yw a ddylai'r gyrrwr stopio, gyrru ymlaen, ac ati.

Y cam cyntaf yw ysgrifennu mewn iaith syml yr amodau cyffredinol, sef, yn yr achos hwn, liwiau'r tri golau yn y fan hyn. Hefyd, byddwn yn ysgrifennu'r gweithredoedd cyffredinol i'w gweithredu.

Ar gyfer yr amodau yn y bonyn amodau mae gennym:

    COCH
    AMBR
    GWYRDD

ac ar gyfer y gweithredoedd yn y bonyn gweithredoedd mae gennym:

    STOPIO
    MYND
    GALW'R HEDDLU (gan nad yw'r goleuadau'n gweithio)

Rhaid i ni greu'r tabl nesaf, felly mae angen i ni wybod faint o reolau fydd. Yn ddamcaniaethol, gallwn gyfrifo hyn drwy ddefnyddio'r fformiwla 2 i bŵer nifer yr amodau yn y bonyn amodau, sef 3 yn yr achos hwn. Felly fe fydd 2^3 yn rhoi 8 rheol. Er mwyn symlrwydd, byddwn yn cymryd bod yr holl reolau hyn yn bosibl (mewn gwirionedd mae systemau methu diogel (*fail-safe*) mewn systemau

goleuadau traffig go iawn, felly fe fydd rhai o'r rheolau hyn yn amhosibl).

Gallwn nawr lunio'r grid a llenwi'r bonyn amodau, y bonyn gweithredoedd a'r cyfuniadau o Y ac N sy'n ffurfio'r cofnodion amodau. Wrth ysgrifennu'r llythrennau, mae'n well cael system lle'r ydych chi'n rhoi pob Y i mewn yn gyntaf ac yna'r tri chyfuniad gyda dwy Y ac un N, yna un Y a dwy N, ac yn olaf y tair N.

I lenwi'r cofnodion gweithredoedd, edrychwn ar y cyfuniadau i weld pa weithred neu weithredoedd i'w gweithredu. Rhoddwn groes yn y lleoedd perthnasol i ddangos y weithred neu weithredoedd i'w gweithredu.

Manteision tablau penderfyniad

Prif fanteision tablau penderfyniad yw:

- Gallwch sicrhau bod yr holl gyfuniadau o amodau'n cael eu hystyried.
- Maen nhw'n hawdd eu deall, gan fod yr holl wybodaeth sydd ei hangen mewn un tabl.
- Mae gan y tabl gynllun safonol, felly mae pawb yn defnyddio'r un fformat.
- Gall rhaglenwyr eu defnyddio i ysgrifennu rhaglenni ac maen nhw'n ddefnyddiol ar gyfer darganfod amodau rhesymeg mewn taenlenni a chronfeydd data.
- Dangosant achos ac effaith ac felly maen nhw'n ddealladwy i'r mwyafrif o bobl.

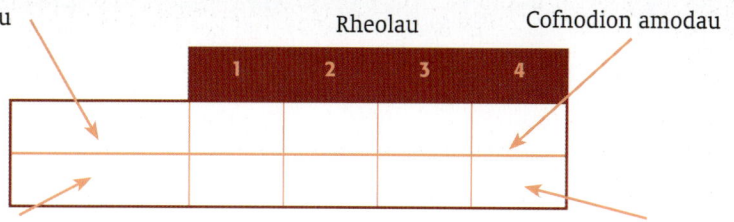

Gall y rheolau ar gyfer goleuadau traffig gael eu cyflwyno mewn tabl

	Rheolau							
	1	2	3	4	5	6	7	8
COCH	Y	Y	Y	N	N	N	Y	N
AMBR	Y	Y	N	Y	N	Y	N	N
GWYRDD	Y	N	Y	Y	Y	N	N	N
STOP	X	X	X	X		X	X	X
MYND					X			
GALW'R HEDDLU	X		X	X				X

Tabl penderfyniad ar gyfer goleuadau traffig

135

Dadansoddi'r system *(parhad)*

Creu tabl penderfyniad ar gyfer costau cludiant

Mewn rhai sefyllfaoedd, does dim pwynt ysgrifennu ac ystyried yr holl reolau, gan y bydd rhai ohonynt yn amhosibl. Dyma enghraifft:

Tybiwch fod y cludiant i'w dalu wrth archebu cryno ddisg o glwb fel a ganlyn:

    1–3 *CD*, cludiant £2.50
    4–6 *CD*, cludiant £3.00
    7+ *CD*, cludiant £4.00

Gan fod 3 amod cyffredinol, mae yna 8 rheol. Os edrychwn ar y rheolau hyn, gwelwn fod rhai ohonynt yn amhosibl. Er enghraifft fe fyddai YYY yn amhosibl gan na all nifer y cryno ddisgiau sy'n cael ei archebu gael ond un Y. Rydym yn cael rheolau amhosibl pan mae'r cwestiynau yn y tabl penderfyniad yn gysylltiedig â'i gilydd.

Rydym yn ymdrin â'r rheolau amhosibl hyn drwy eu gadael allan o'r tabl penderfyniad.

Drwy ddileu'r rheolau amhosibl, rydym yn cael y tabl penderfyniad canlynol:

	Rheolau		
	1	**2**	**3**
1–3 *CD* wedi'u harchebu	Y	N	N
4–6 *CD* wedi'u harchebu	N	Y	N
dros 7 *CD* wedi'u harchebu	N	N	Y
Cludiant £2.50	X		
Cludiant £3.00		X	
Cludiant £4.00			X

Tabl penderfyniad ar gyfer costau cludiant

Diagramau systemau

Mae diagramau systemau neu siartiau llif systemau yn darlunio sut mae system TGCh yn gweithio drwy ddangos y mewnbynnau i ran benodol o'r system, yr hyn sy'n cael ei wneud i brosesu'r data, a beth sy'n symud allan o'r rhan hon i rywle arall. Maen nhw hefyd yn dangos yn glir lif cyffredinol y gweithrediadau mewn system a'r caledwedd a chyfryngau sydd ynghlwm wrth y gweithrediadau hyn. Mae'r siart llif systemau hefyd yn dangos tarddiad y mewnbynnau a'r dull prosesu (llaw neu gyfrifiadurol).

Terfynwr (dechrau neu ddiwedd)

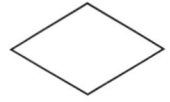

Penderfyniad

Mewnbwn â llaw (e.e. bysellfwrdd)

Uned arddangos weledol/sgrin

Mewnbwn neu allbwn

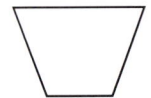

Gweithrediad â llaw (wedi'i gyflawni â llaw)

Llinell gyfathrebu

Tâp magnetig

Proses (wedi'i chyflawni gan gyfrifiadur)

Cysylltydd

Storfa ddisg magnetig

Dogfen neu allbrint

Y symbolau a ddefnyddir mewn siartiau llif systemau

Flowchart

CYCHWYN

Mewnbynnu manylion cyngerdd

Cofnodion cyngerdd

Cyrchu cofnod cyngerdd

Arddangos seddau sydd ar gael → Seddau sydd ar gael

Ydy'n bosibl gwneud bwciad? — Nac ydy → STOPIO

Ydy

Lleihau'r seddau sydd ar gael ar y cofnod cyngerdd

Cadarnhau bwciad ← Cadarnhau bod bwciad wedi'i wneud ac argraffu tocynnau → Argraffu tocynnau

Casglu arian

STOPIO

Siart llif systemau ar gyfer system bwcio tocynnau cyngerdd

Modelau perthynas endidau (*ERMs*)

Techneg ar gyfer diffinio anghenion gwybodaeth corff er mwyn creu sylfaen gadarn i adeiladu system briodol arni yw modelu perthynas endidau (*ERMs: Entity relationship models*). Yn syml, mae modelu perthynas endidau yn nodi'r ffactorau pwysicaf yn y corff dan sylw. Endidau yw'r enw a roddir ar y ffactorau hyn.

Mae priodweddau'r ffactorau hyn (**priodoleddau**) a'u perthynas â'i gilydd (**perthnasoedd**) yn cael eu hystyried hefyd. Mae modelau perthynas endidau yn rhesymegol, sy'n golygu nad ydynt yn dibynnu ar y dull gweithredu. Os yw dwy adran mewn corff yn gwneud yr un tasgau'n union ond mewn gwahanol ffyrdd, byddai eu *ERMs* yr un fath gan y byddent yn defnyddio'r un endidau a pherthnasoedd. Ond gallai eu diagramau llif ddata fod yn wahanol, oherwydd y gallai eu llifoedd gwybodaeth fod yn wahanol. Mae'r modelau hyn yn arbennig o ddefnyddiol gan eu bod yn annibynnol ar unrhyw storfa neu unrhyw ffyrdd o gyrchu'r data. Felly nid ydynt yn dibynnu ar unrhyw galedwedd neu feddalwedd eto.

Cynrychioliad haniaethol o'r data mewn corff yw model perthynas endidau, a nod modelu perthynas endidau yw cynhyrchu model manwl o anghenion gwybodaeth y corff a fydd naill ai o gymorth i ddatblygu system newydd neu i wella system sy'n bod eisoes.

Mae model perthynas endidau yn disgrifio system fel set o endidau data gyda pherthnasoedd rhyngddynt (un-i-un, un-i-lawer, llawer-i-lawer).

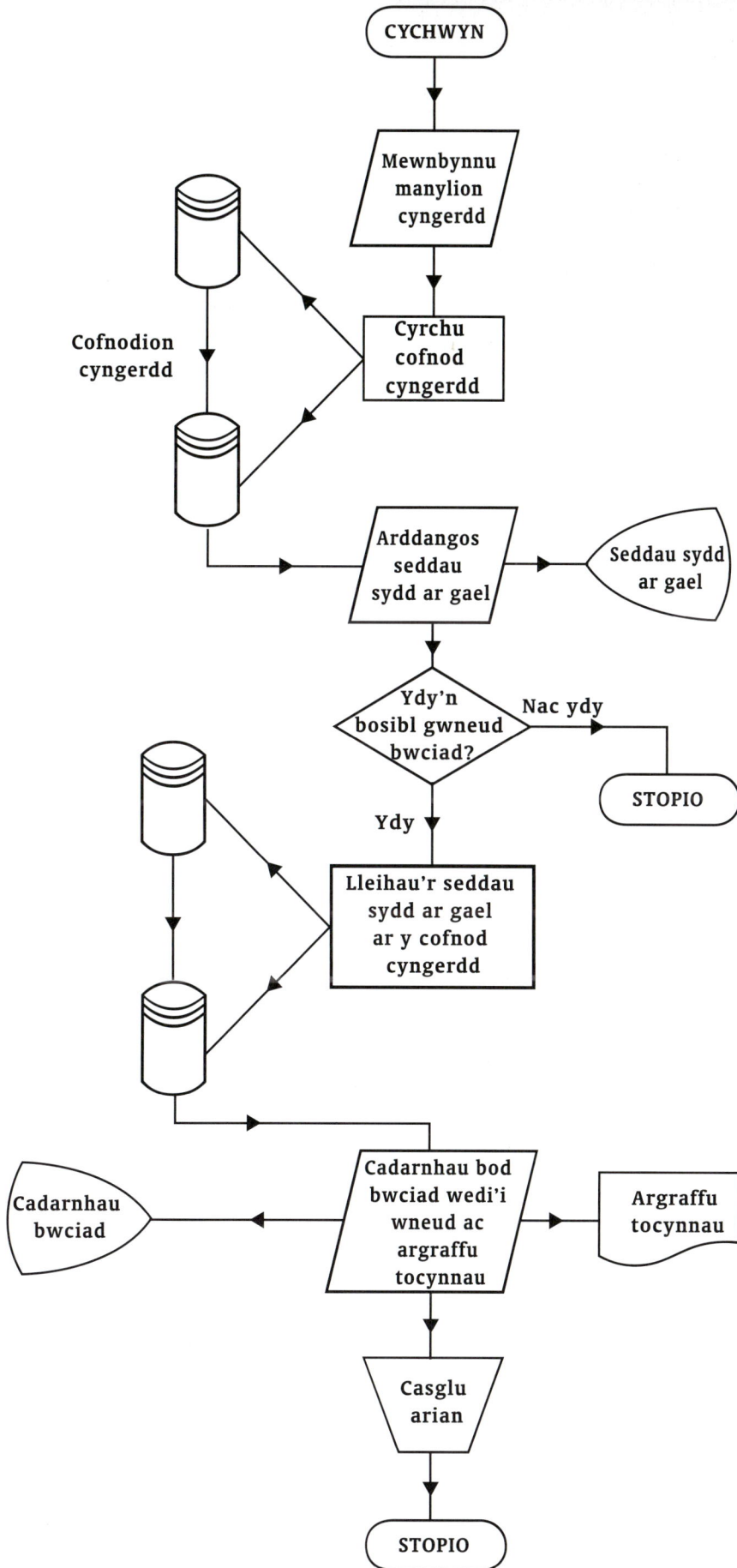

Dadansoddi'r system (parhad)

Diagramau perthynas endidau

Mae diagramau perthynas endidau'n edrych ar unrhyw gydrannau sy'n bwysig i'r system a'r perthnasoedd rhyngddynt.

Felly beth yw **endid**? Gall endid fod yn unrhyw beth mae data'n cael eu cofnodi amdano. Gall fod yn bobl, lleoedd, gwrthrychau, cwsmeriaid, gwerthiant, taliadau neu staff. Mae gan bob endid **briodoleddau** cysylltiedig. Mae priodoleddau'n manylu ar yr endid. Gadewch i ni edrych ar enghraifft. Yn y tabl canlynol mae'r priodoleddau ar gyfer yr endid 'cwsmer' yn rhoi manylion pellach.

Endid	Priodoleddau
CWSMER	Rhif cwsmer
	Enw'r cwmni
	Cod post
	Rhif ffôn
	Terfyn credyd
	Swm sy'n ddyledus

Mewn diagram endidau, mae pob endid wedi'i gynrychioli gan betryal gydag enw'r endid wedi'i ysgrifennu ynddo. Mae'r berthynas rhwng yr endidau yn cael ei dangos fel llinellau rhwng y blychau hyn. Caiff yr endid ei ysgrifennu mewn priflythrennau a defnyddir y ffurf unigol bob amser gan y byddai defnyddio'r lluosog yn awgrymu math o berthynas. Felly yn lle CWSMERIAID, defnyddiwn y ffurf CWSMER.

Gallai'r endid MYFYRIWR fod â'r priodoleddau canlynol: rhif myfyriwr, cyfeiriad myfyriwr, rhif ffôn, dyddiad geni, tiwtor, rhif cwrs. Wrth ddatblygu cronfeydd data, mae'n bwysig ystyried a yw'n bosibl torri pob priodoledd i lawr ymhellach yn rhagor o briodoleddau. Er enghraifft, gallai cyfeiriad myfyriwr gael ei dorri i lawr yn stryd, tref a chod post. Pan nad oes angen torri priodoleddau i lawr ymhellach, dywedwn eu bod yn atomig. Felly mae stryd, tref a chod post yn atomig.

Mae torri priodoleddau i lawr i gynhyrchu priodoleddau atomig yn caniatáu i ni drin y data mewn ffordd hyblyg. Er enghraifft, gallem chwilio am fyfyrwyr sydd â chod post penodol neu drefnu myfyrwyr yn ôl y dref maen nhw'n dod ohoni.

Perthnasoedd

Perthynas yw'r ffordd y mae'r endidau mewn system wedi'u cysylltu â'i gilydd. Gall perthynas fod yn un-i-un (1:1), yn un-i-lawer (1:m), neu'n llawer-i-lawer (m:m).

Mae'r mathau posibl hyn o berthynas rhwng dau endid A a B yn cael eu dangos isod.

Perthynas un-i-un, perthynas un-i-lawer, a pherthynas llawer-i-lawer

Modelu perthynas endidau ar gyfer siop e-fasnach sy'n gwerthu cryno ddisgiau

Dyma sut y gall model perthynas endidau gael ei lunio ar gyfer siop sy'n gwerthu cryno ddisgiau drwy'r post gan ddefnyddio'r Rhyngrwyd. Y peth cyntaf i'w wneud yw nodi'r endidau hynny sy'n hanfodol i'r system hon. Defnyddir pedwar endid yma:

> CWSMER
> CD
> ARCHEB
> CLUDIAD

Nawr mae angen i chi feddwl am y berthynas rhwng yr endidau hyn. Mae'n ddefnyddiol i'w hysgrifennu fel rhestr:

- mae cwsmer yn rhoi archeb
- archeb yw un neu ragor o gryno ddisgiau
- cludiad (*delivery*) yw'r cryno ddisgiau yn yr archeb
- mae cludiad yn cael ei wneud i'r cwsmer.

GEIRIAU ALLWEDDOL

Endid – gwrthrych yn y byd real sy'n berthnasol i system TGCh, e.e. lle, gwrthrych, unigolyn, cwsmer, anfoneb, cynnyrch, cwrs, ac ati

Priodoledd – un eitem o ddata sy'n cynrychioli ffaith am endid. Gellir meddwl am briodoledd fel rhywbeth sy'n ychwanegu manylion neu wybodaeth bellach at yr endid

Gellir yn awr lunio'r perthnasoedd rhwng yr endidau fel hyn:

Model perthnasoedd syml ar gyfer y system. Nid yw'r archebion (un-i-un, un-i-lawer, ac ati) wedi cael eu hychwanegu

Nid yw'r diagram yn berffaith o bell ffordd. Mae'n amhosibl gwybod a yw'r archeb yn cynnwys mwy nag un *CD*. Yn ogystal, os bydd y siop yn rhedeg allan o *CD* poblogaidd, bydd y rhan o'r archeb sydd mewn stoc yn cael ei hanfon, a bydd unrhyw gryno ddisgiau nad ydynt mewn stoc yn dilyn.

Gadewch i ni edrych nawr ar y berthynas rhwng dau o'r endidau: ARCHEB a *CD*. Mae angen i ni ystyried y berthynas o'r ddau ben. Gan edrych o ARCHEB i gyfeiriad *CD*, gallwn weld ei bod yn bosibl i un archeb fod am lawer o gryno ddisgiau. Gan edrych i'r cyfeiriad arall, gallwn weld ei bod yn bosibl i CD penodol (hynny yw, teitl penodol) fod mewn llawer o wahanol archebion. Mewn geiriau eraill, mae'r berthynas rhwng ARCHEB a *CD* yn un llawer-i-lawer.

Mae'r diagram uchod yn dangos y berthynas llawer-i-lawer rhwng ARCHEB a *CD*

Gallwn weld o'r diagram fod llawer o gwsmeriaid yn archebu llawer o gryno ddisgiau. Byddai hyn yn awgrymu y gallai archeb fod am lawer o gryno ddisgiau, ond mae'n amhosibl dweud am ba gryno ddisgiau y mae'r archeb. Mae angen ffordd o gysylltu'r cryno ddisgiau wrth ei gilydd fel ei bod yn bosibl eu croesgyfeirio. Byddwn yn gwneud hyn drwy greu endid newydd o'r enw LLINELL ARCHEBU. Mae'r endid hwn yn nodi'r *CD* sydd ar linell benodol mewn trefn benodol.

Rydym yn osgoi cael perthynas llawer-i-lawer drwy gyflwyno'r endid LLINELL ARCHEBU

Pan fyddwch yn creu eich modelau perthynas endidau eich hun, bydd yn rhaid i chi gynhyrchu endid newydd bob tro y dewch ar draws perthynas llawer-i-lawer.

Ar ôl creu'r endid newydd LLINELL ARCHEBU mae'n rhaid edrych ar y perthnasoedd rhwng ARCHEB a LLINELL ARCHEBU. Perthynas un-i-lawer yw hon oherwydd y gall un archeb gynnwys llawer o linellau archebu. Gan edrych ar y berthynas rhwng *CD* a LLINELL ARCHEBU, gallwn weld fod hon hefyd yn berthynas un-i-lawer oherwydd y gall *CD* penodol fod mewn llawer o wahanol linellau archebu.

Dyma'r model perthynas endidau ar gyfer y system:

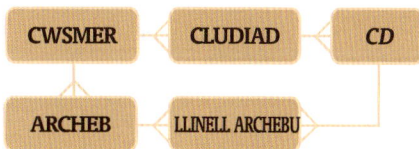

Y model perthynas endidau cyflawn ar gyfer y system

Penderfynu ar y math o berthynas

Wrth lunio modelau perthynas endidau, mae'n rhaid penderfynu ar y math o berthynas rhwng yr endidau. Mae'n well gwneud hyn drwy edrych ar yr endid o'r ddau ben. Cymerwch y berthynas rhwng myfyrwyr a chyrsiau mewn coleg. Gan edrych ar y berthynas o gyfeiriad y myfyriwr yn gyntaf, gallwn weld y gallai myfyriwr gymryd mwy nag un cwrs. Gan edrych ar y berthynas o gyfeiriad y coleg, gwelwn y gallai un cwrs gael ei ddilyn gan lawer o fyfyrwyr. Felly perthynas llawer-i-lawer sydd rhwng CWRS a MYFYRIWR.

Sylwch ar y defnydd o eiriau ar y perthnasoedd

Mae'r diagram perthynas endidau'n dweud wrthym:

- fod un cwrs yn cynnwys llawer o fyfyrwyr
- y gall un myfyriwr fod wedi'i gofrestru ar un neu ragor o gyrsiau.

Fel rydym wedi ei weld yn barod, ni all perthnasoedd llawer-i-lawer gael eu gweithredu, felly yn yr achos hwn rhaid i ni greu endid newydd a fydd yn cysylltu myfyriwr â chwrs. Mewn cwrs mae hyn yn cael ei alw'n COFRESTRIAD. Os bydd myfyrwyr yn dilyn sawl cwrs, bydd cofrestriad ar gyfer pob cwrs a bydd priodoleddau'r cofnod cofrestriad yn cynnwys rhif myfyriwr (i adnabod y myfyriwr) a rhif cwrs (i adnabod y cwrs) yn brif allweddi.

Dyma'r endidau sydd gennym bellach:

Y diagram perthynas endidau terfynol yn dangos y perthnasoedd rhwng yr endidau: CWRS, MYFYRIWR a CHOFRESTRIAD

Defnyddio allweddi

Rydym wedi gweld eisoes y gellir defnyddio nifer o briodoleddau i ddisgrifio pob endid. Priodoleddau sydd ag arwyddocâd arbennig yw allweddi. Mae dau brif fath o allwedd: prif allweddi ac allweddi estron.

Prif allwedd

Mae prif allwedd fel rheol yn un neu'n sawl priodoledd a all ddiffinio endid neilltuol yn unigryw. Er enghraifft, gallai'r endid CWSMER gael ei ddiffinio'n unigryw gan briodoledd unigryw fel rhif cwsmer. Gallai'r endid RHESTR GYFLOGAU gael ei ddiffinio'n unigryw gan yr endid rhif gweithiwr.

Allwedd estron

Pan fydd un o'r priodoleddau sy'n brif allwedd mewn un endid hefyd yn ymddangos mewn endid arall, mae'n golygu bod perthynas rhwng yr endidau. Yn yr endid arall, nid yw'r priodoledd yn brif allwedd gan ei bod

hi'n bosibl nad yw'n unigryw. Felly mae'n cael ei alw'n allwedd estron.

Tybiwch eich bod chi'n cyflenwi nwyddau i gwmnïau eraill a'ch bod chi am sefydlu system i gadw manylion y cysylltiadau hyn. Mae dau endid, sef CWMNI a CYSYLLTIAD. Dyma briodoleddau'r endidau hyn:

CYSYLLTIAD	CWMNI
RhA_Cysylltiad	RhA_Cwmni
EnwCyntaf	Enw
EnwOlaf	Cyfeiriad
RhA_Cwmni	RhifFfôn

Rhoddir rhif unigryw i bob cysylltiad, felly RhA_Cysylltiad yw'r brif allwedd ar gyfer yr endid CYSYLLTIAD. Yn yr endid CWMNI, mae'r priodoledd RhA_Cwmni yn unigryw ac felly hon yw'r brif allwedd. Sylwch fod y priodoledd RhA_Cwmni i'w weld yn y ddau endid ac mai hwn yw'r brif allwedd yn yr endid CWMNI. Yn yr endid CYSYLLTIAD, mae RhA_Cwmni yn allwedd estron a gan fod y priodoleddau yn ymddangos yn y ddau endid, gallwn eu cysylltu. Felly mae gan yr endid CYSYLLTIAD berthynas â'r endid CWMNI.

Efallai eich bod chi'n meddwl tybed pam mae gennym ddau endid, yn hytrach na'u cyfuno a chael un yn unig sy'n cynnwys y cysylltiadau a'r cwmni maen nhw'n gweithio iddo. Pe bai un cysylltiad yn unig ar gyfer pob cwmni, byddai'n briodol rhoi'r cwmnïau a'r manylion cysylltu gyda'i gilydd. Ond mewn gwirionedd fe fydd mwy nag un cysylltiad ym mhob cwmni; os storiwn y manylion cysylltu, bydd enw'r cwmni a'i gyfeiriad i'w gweld mewn llawer o wahanol fanylion cysylltu. Felly fe fyddai enw a chyfeiriad y cwmni yn grŵp o briodoleddau sy'n cael ei ailadrodd. Gall hyn achosi problemau; er enghraifft, os bydd y cwmni'n newid cyfeiriad, bydd yn rhaid newid y cyfeiriad ar gyfer pob cysylltiad sy'n gweithio i'r cwmni hwnnw. Ond os defnyddir dau endid ac os bydd cyfeiriad y cwmni'n newid, unwaith yn unig bydd angen ei newid yn yr endid CWMNI.

Dadansoddi'r system *(parhad)*

Sicrhau bod priodoleddau'n atomig

Yn ogystal â sicrhau y rhoddir priodoleddau sy'n cael eu hailadrodd yn eu hendid eu hunain, rhaid i chi sicrhau bod yr holl briodoleddau'n atomig. Priodoleddau nad oes angen eu torri i lawr ymhellach yw priodoleddau atomig.

Er enghraifft, gallai rhywun sy'n newydd i ddadansoddi systemau a chynllunio cronfeydd data ddefnyddio priodoledd o'r enw Enw i gadw'r eitemau data canlynol:

Mr Stephen Doyle

Y drafferth gyda hyn yw os caiff trefniad ei wneud yn awr fe gaiff ei wneud yn nhrefn teitl, felly bydd enwau sy'n dechrau gyda Miss yn ymddangos o flaen enwau sy'n dechrau gyda Mr. Hefyd, os ydych am chwilio am y cyfenw Doyle, ni allwch wneud hyn oni wyddoch beth yw'r teitl a'r enw cyntaf. Mae defnyddio un priodoledd o'r enw Enw yn achosi problemau gan nad yw'r priodoledd yn atomig. Rhaid ei dorri i lawr yn dri phriodoledd megis Teitl, Enw Cyntaf a Chyfenw.

Creu strwythur y gronfa ddata

Ar ôl y broses normaleiddio, mae gennym sawl endid lle'r oedd gennym un o'r blaen, ac mae gan bob un restr o briodoleddau cysylltiedig. Ar ôl gwneud hyn rydym mewn sefyllfa i ddechrau meddwl am y tablau a ddefnyddir i storio'r data. Gall pob tabl gael yr un enw â'r endidau a gall yr enwau maes fod yr un fath â'r enwau a roddir i'r priodoleddau. Yna mae'n rhaid cynllunio strwythur pob tabl ar wahân.

Nodyn pwysig

Mewn perthynas llawer-i-lawer sydd wedi'i dadelfennu i berthynas un-i-lawer a llawer-i-un gan ddefnyddio tabl cyswllt, dylai fod gan y tabl cyswllt allwedd gyfansawdd (e.e. mewn tabl rhentol gallai fod angen allwedd gyfansawdd fel RhA_Cwsmer, RhA_Offer, a DyddiadAllan).

Sicrhewch bob amser bod yr allweddi a ddefnyddir yn diffinio'n unigryw res yn y tabl.

Geiriaduron data

Offer a ddefnyddir yn ystod dadansoddi systemau ac yn arbennig wrth gynllunio cronfeydd data yw geiriadur data. Pwrpas geiriadur data yw darparu gwybodaeth am y gronfa ddata, ei defnydd, a defnyddwyr y system. Gellid dweud bod geiriadur data yn darparu 'data am ddata'.

Cynnwys y geiriadur data

Mae geiriaduron data'n cynnwys rhai neu'r cyfan o'r nodweddion canlynol:

Enwau endidau

Mae angen rhoi enw i bob endid a rhaid cofnodi'r enwau hyn a'u disgrifio yn y geiriadur data.

Perthnasoedd rhwng yr endidau

Gall y perthnasoedd rhwng yr endidau gael eu disgrifio a'u dangos mewn diagram perthynas endidau *(ERD: entity relationship diagram)*.

Enwau priodoleddau

Dylid rhoi enw i bob priodoledd, a dylid dewis yr enwau hyn fel eu bod yn disgrifio'r data mor llawn â phosibl heb fod yn rhy hir. Mae'n well osgoi gadael bylchau mewn enwau maes: yn lle hynny defnyddiwch - neu _ i wahanu'r geiriau.

Cyfystyron

Gwahanol enwau am yr un peth yw cyfystyron. Mewn llawer o gyrff mawr, mae cronfa ddata yn ganolog i'r system gyfrifiadurol a chaiff ei defnyddio gan wahanol adrannau ar gyfer gwahanol weithgareddau. Felly mae'n gyffredin i weld defnyddwyr mewn gwahanol adrannau'n defnyddio enw gwahanol am yr un cysyniad; gall hyn fod yn ddryslyd iawn i'r dadansoddwyr systemau sy'n ceisio dylunio ac adeiladu system newydd. I osgoi dryswch, dylai defnyddwyr restru unrhyw enwau eraill am yr un peth yn adran cyfystyron y geiriadur data.

Math data

Mae angen pennu math data ar gyfer pob priodoledd. Ar ôl pennu math data y priodoledd, mae'n bwysig iawn rhoi'r un math data iddo ble bynnag mae'n digwydd. Fel rheol, i greu perthnasoedd rhwng priodoleddau, rhaid bod ganddynt yn union yr un enwau a mathau data.

Fformat

Dylai manylion fformatau gael eu cynnwys yn y geiriadur data ar gyfer y priodoleddau hynny y mae eu fformatau'n cael eu gosod gan y defnyddiwr. Er enghraifft, gall maes rhifol fod yn gyfanrif byr, yn gyfanrif hir, arian cyfred, ac ati, a gellir rhoi dyddiadau fel 12 Mehefin 2009, 12/06/09, ac ati.

Disgrifiad o'r priodoledd

Dylai fod gan bob priodoledd ddisgrifiad, a gellir rhoi'r disgrifiadau hyn yn yr adran hon o'r geiriadur data.

Hyd y priodoledd

Yma fe bennwn hyd y priodoledd ar gyfer y priodoleddau hynny y gellir gosod eu hyd.

⮕ GEIRIAU ALLWEDDOL

Geiriadur data − casgliad canolog o wybodaeth am ddata fel ystyr, perthnasoedd â data eraill, tarddiad, defnydd, ac ati, a ddefnyddir yn bennaf wrth gynllunio cronfeydd data i sicrhau cysondeb

Enwau endidau eraill lle mae'r priodoledd yn ymddangos

Gall priodoledd ymddangos mewn mwy nag un endid. Er enghraifft, gallai'r priodoledd Rhif_Archeb ymddangos yn yr endidau ARCHEB a LLINELL_ARCHEBU. Pe baem ni am newid enw priodoledd, byddai'n ddefnyddiol gwybod ym mha endidau mae'n cael ei ailadrodd.

Dyma gofnod mewn geiriadur data ar gyfer priodoledd o'r enw NIFER_AILARCHEBU:

Enw'r priodoledd	NIFER_AILARCHEBU
Cyfystyron	Dim
Math data	Rhifol
Fformat	0 lle degol (h.y. cyfanrif)
Disgrifiad o'r priodoledd	Nifer yr unedau o eitem stoc a all gael eu harchebu ar unrhyw un adeg
Hyd y priodoledd	4 digid
Enwau endidau	RHESTR_STOC
	EITHRIAD_RHESTR_STOC
	AILARCHEBU

Mae geiriaduron data'n sicrhau bod yr holl staff sy'n datblygu systemau TGCh yn mynd ati mewn ffordd gyson

Caiff geiriaduron data eu creu wrth adeiladu cronfeydd data masnachol mawr, yn enwedig os bydd y cronfeydd data yn cael eu defnyddio at lawer o wahanol gymwysiadau. Pwrpas y geiriadur data yw sicrhau bod y data ar draws y system gyfan yn gyson. Caiff systemau mawr eu hadeiladu gan dimau project mawr, a bydd pob tîm yn gweithio ar ran o'r system gyfan. Yn y fath sefyllfa, mae termau cyson yn hollbwysig. Yn aml, cyfeirir at y data mewn geiriaduron data fel metaddata.

Gall geiriaduron data gael eu cynhyrchu fel systemau papur ond mae gan eiriaduron data cyfrifiadurol fanteision amlwg. Mae gan y mwyafrif o systemau rheoli cronfeydd data perthynol feddalwedd sy'n creu ac yn cynnal geiriadur data. Mae'n cael ei gadw fel cronfa ddata ar wahân a chaiff ei ddiweddaru'n awtomatig wrth i newidiadau gael eu gwneud i strwythur y brif gronfa ddata.

Nodweddion dewisol y geiriadur data

Yn ogystal â'r nodweddion hanfodol uchod, mae rhai geiriaduron data yn darparu'r wybodaeth ychwanegol hon:

- gwiriadau dilysu
- manylion y gwiriadau dilysu a gyflawnir ar y data sy'n cael eu mewnbynnu ar gyfer pob priodoledd
- allwedd
- y math o allwedd.

Caiff y mwyafrif o eiriaduron data eu storio ar gyfrifiadur ac ar-lein

Dylunio'r system a gweithredu'r system

Cyflwyniad

Ar ôl ymchwilio i'r hen system neu system arfaethedig a'i dadansoddi, bydd y system newydd yn cael ei chynllunio gan ddefnyddio dogfennau sydd wedi cael eu casglu a'u creu yn ystod y camau blaenorol. Yn ystod y cam dylunio mae'r system yn cael ei chynllunio – nid yw'n cael ei rhoi ar waith ar yr adeg hon.

Ar ôl cwblhau'r gwaith dylunio, gall cam nesaf yr *SDLC* ddechrau, sef ei gweithredu. Yn ystod y cam gweithredu rydych chi'n cymryd y cynlluniau ac yn dechrau creu'r system gan ddefnyddio'r caledwedd, meddalwedd a data.

Yn yr adran hon byddwch hefyd yn edrych ar briodoldeb y gwahanol ddulliau newid drosodd o'r hen system TGCh i'r system TGCh newydd.

Dylunio'r system

I gwblhau'r cam dylunio fe fydd y dadansoddydd yn dechrau meddwl am y system y mae wedi'i dadansoddi ac yn dechrau dylunio'r system ei hun. Defnyddir yr holl ddogfennau sydd wedi cael eu cynhyrchu yn ystod y cam dadansoddi i ddylunio'r system yn unol â gofynion y defnyddwyr.

Bydd dylunio'r system fel rheol yn cynnwys rhai neu'r cyfan o'r canlynol:

- Creu'r fanyleb ddylunio ar gyfer y caledwedd a'r meddalwedd.
- Dylunio'r strwythur data a'r strwythur ffeiliau a fydd yn ei gwneud hi'n bosibl i adeiladu'r system. Byddai hyn yn cynnwys dylunio'r meysydd a strwythur y tablau ar gyfer y gronfa ddata berthynol.
- Dylunio'r dulliau mewnbynnu ac allbynnu gwybodaeth. Byddai hyn yn cynnwys dylunio'r ffurflenni a ddefnyddir i fewnbynnu data a'r adroddiadau ac ymholiadau a ddefnyddir i allbynnu gwybodaeth. Byddai dogfennau arbenigol megis anfonebau, slipiau cyflog, nodion trosglwyddo, ac ati, yn cael eu dylunio hefyd.

- Dylunio'r systemau gwybodaeth a fydd yn caniatáu i ddefnyddwyr echdynnu'r wybodaeth sydd ei hangen arnynt i wneud penderfyniadau.
- Dylunio rhwydweithiau a materion trawsyrru fel topoleg, cyfryngau trawsyrru data, protocolau, ac ati.
- Materion personél. Mae creu systemau hollol newydd yn golygu bod angen hyfforddi staff ac ad-drefnu adrannau'n aml.
- Prosesau a gweithdrefnau diogelwch. Os yw data personol yn cael eu storio, rhaid cofrestru hyn gyda'r Comisiynydd Gwybodaeth, felly rhaid cael gweithdrefnau yn eu lle i sicrhau bod hyn yn digwydd. Mae angen ystyried diogelwch y data a gwneud cynlluniau i warchod yr holl ddata sy'n cael eu cadw drwy ddefnyddio lefelau mynediad, rhifau adnabod a chyfrineiriau.

Mae'n bwysig nodi mai'r unig beth sy'n digwydd yn ystod y cam hwn yw cynhyrchu cynlluniau a dyluniadau. Mae'r fersiwn sy'n gweithio yn cael ei gynhyrchu yn y cam nesaf, y cam gweithredu.

YN YR ARHOLIAD

Byddwch chi'n creu cronfa ddata berthynol ar gyfer eich gwaith project, felly byddwch chi'n cael profiad o lawer o'r sgiliau cynllunio a dylunio uchod wrth gynhyrchu eich ateb.

Yn yr arholiad dylech ddefnyddio'r wybodaeth sydd gennych ar ôl gweithio ar eich project i'ch helpu i ateb cwestiynau a rhoi enghreifftiau.

Gweithredu'r system

Dyma'r cam lle caiff y system ei hadeiladu yn unol â'r cynlluniau a gynhyrchwyd yn y cam blaenorol. Bydd yr holl wahanol staff yn y tîm project yn dod â'u harbenigedd neilltuol i'r project ac yn gweithio gyda'i gilydd i roi'r system ar waith.

Bydd gweithredu'n cynnwys y canlynol fel rheol:

- prynu a gosod caledwedd a meddalwedd ac ailhyfforddi'r staff i ddefnyddio'r meddalwedd newydd
- addasu'r meddalwedd presennol
- cynhyrchu unrhyw god rhaglennu sydd ei angen ar gyfer y datrysiad gan raglenwyr
- gosod y meddalwedd ar y caledwedd gweithredu (sylwer: datblygir systemau'n aml ar gyfarpar nad yw'n cael ei ddefnyddio ar gyfer gweithrediadau pob dydd am resymau diogelwch)
- cynhyrchu'r fframwaith sydd ei angen ar gyfer y cronfeydd data
- hyfforddi defnyddwyr.

Yn y mwyafrif o brojectau TGCh bydd llawer o bobl yn gweithio ar wahanol agweddau ar y system. Bydd pob aelod o'r tîm yn gyfrifol am ei ran ei hun o'r project a bydd yn rhoi prawf ar ei waith i sicrhau bob popeth yn gweithio.

Bydd profi'n cynnwys y canlynol:

- profi ar lefelau manwl iawn (h.y. bydd unigolion yn profi eu rhan nhw yn drwyadl)
- profi ar lefel uwch (h.y. pan gaiff gwahanol rannau'r system eu huno)

- profi ar lefel systemau lle caiff yr holl rannau eu profi fel system gyfan
- profi pob maes gan ddefnyddio data cywir, anghywir ac eithafol
- profi'r system gyda data go iawn.

Priodoldeb gwahanol ddulliau newid drosodd

Er mwyn newid o un system i system arall mae'n rhaid cael strategaeth newid drosodd. Mae'r gwahanol ddulliau'n cael eu hamlinellu yma:

- **Newid drosodd uniongyrchol** – gyda'r strategaeth hon rydych chi'n rhoi'r gorau i ddefnyddio'r hen system un diwrnod ac yn dechrau defnyddio'r system newydd drannoeth. Anfantais y system hon yw bod elfen o risg, yn enwedig os yw'r caledwedd a'r meddalwedd diweddaraf yn cael eu defnyddio. Os bydd y system yn methu, gall y canlyniadau fod yn drychinebus i'r busnes. Mantais y dull hwn yw ei fod yn gofyn am lai o adnoddau (pobl, arian, cyfarpar) a'i fod yn syml os aiff popeth yn iawn.
- **Newid drosodd ochr yn ochr** – defnyddir y dull hwn i leihau'r risg wrth gyflwyno system TGCh newydd. Mae'r hen system yn rhedeg ochr yn ochr â'r system newydd am gyfnod o amser nes bob pawb yn fodlon bod y system newydd yn gweithio'n iawn. Yna rhoddir y gorau i'r hen system ac mae'r holl waith yn cael ei wneud ar y system newydd. Anfantais y dull hwn yw ei fod yn gofyn am lawer o waith diangen (gan fod y gwaith yn cael ei wneud ddwywaith) ac mae felly'n ddrud o ran amser y staff. Hefyd mae'n gofyn am fwy o gynllunio.
- **Trawsnewid gam wrth gam** – gall un modiwl ar y tro gael ei newid i'r system newydd nes bod y system gyfan wedi'i throsglwyddo. Y fantais yw y gall y staff TG ymdrin ag unrhyw broblemau sy'n cael eu hachosi gan fodiwl cyn symud ymlaen i fodiwlau newydd. Er hynny, dim ond systemau sy'n cynnwys modiwlau ar wahân a all gael eu trin yn y modd hwn.
- **Trawsnewid peilot** – mae'r dull hwn yn ddelfrydol ar gyfer cyrff mawr sydd â llawer o leoliadau neu ganghennau. Gellir cyflwyno'r system i un gangen yn gyntaf ac yna ei throsglwyddo i'r lleill dros gyfnod o amser. Mantais y dull hwn yw bod y gweithredu'n cael ei wneud ar raddfa lawer llai a mwy hydrin. Yr anfantais yw ei bod hi'n cymryd mwy o amser i roi'r system ar waith ym mhob cangen.

Dewis dull newid drosodd

Mae'r dewis o ddull newid drosodd yn dibynnu ar sawl ffactor:

- A oes system yn barod? Os nad oes system (papur neu TGCh), ni ellir defnyddio newid drosodd ochr yn ochr.
- Ai'r system ddiweddaraf sy'n cael ei chyflwyno (technoleg newydd, meddalwedd newydd, ac ati)? Bydd newid drosodd ochr yn ochr yn lleihau'r risg neu bydd trawsnewid peilot yn lleihau'r problemau a all godi.
- A yw'r system i gael ei gweithredu mewn llawer o wahanol leoliadau? Os felly, byddai trawsnewid peilot yn ddelfrydol. Gallech gael y system i weithio mewn un gangen cyn ei gweithredu mewn canghennau eraill.

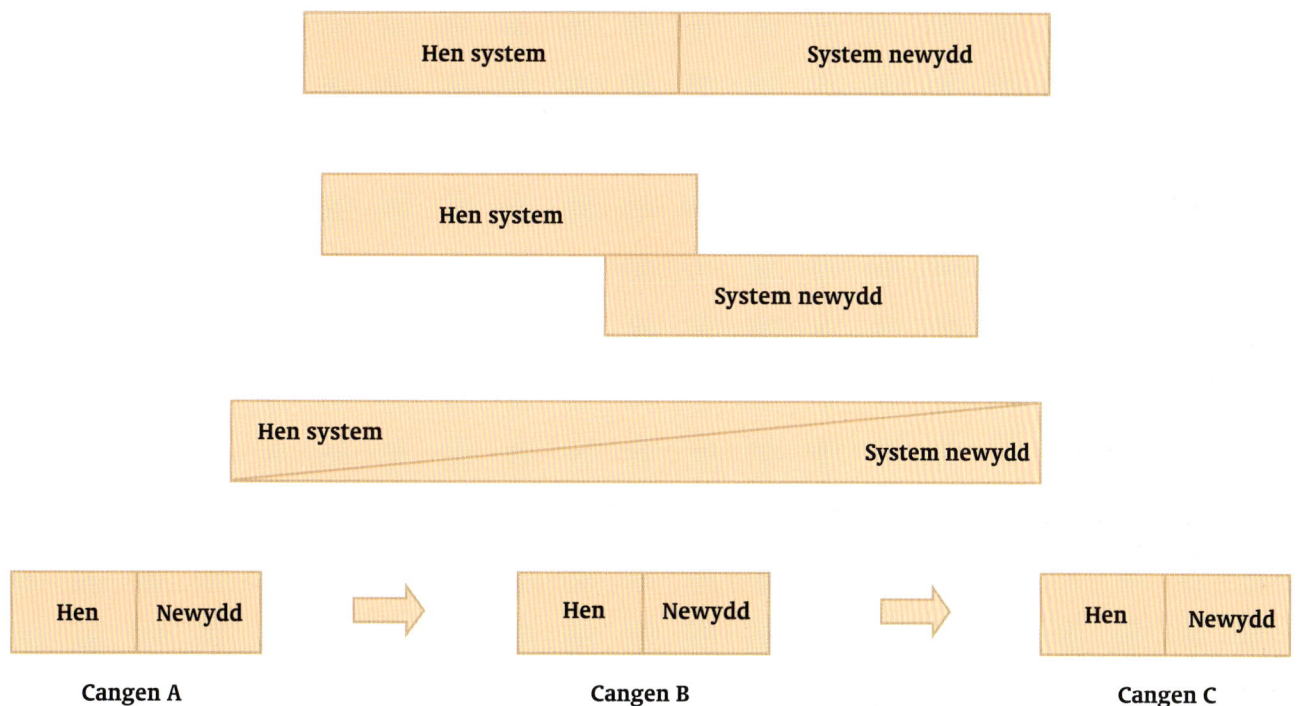

Y pedwar dull newid drosodd: uniongyrchol, ochr yn ochr, cam wrth gam a pheilot

Cynnal y system a gwerthuso'r system

Cyflwyniad

Ar ôl gweithredu'r system bydd angen ei chynnal. Mae'n bwysig nodi y gall y bobl a oedd ynghlwm wrth y project gwreiddiol adael y corff, felly bydd angen i staff newydd ymgyfarwyddo â'r system. Bydd angen dogfennau arnynt i allu gwneud hyn, felly y cam cyntaf yn y broses o gynnal system yw cynhyrchu dogfennaeth i'r defnyddiwr a dogfennaeth dechnegol.

Yn yr adran hon byddwch yn dysgu am y gwaith cynnal system parhaus mae angen ei wneud wrth i anghenion busnes a gofynion defnyddwyr newid. Yn y man fe gyrhaeddir pwynt pan nad yw'r system bellach yn cwrdd ag anghenion y busnes ac felly bydd angen creu system newydd a bydd angen ailadrodd yr holl dasgau sy'n gysylltiedig â'r gylchred oes datblygu system. Byddwn hefyd yn edrych ar werthuso'r system, sef ystyried pa mor dda mae'r project wedi mynd ac i ba raddau mae'n bodloni anghenion y defnyddwyr.

Cynnal y system

Bydd cynnal y system newydd yn cynnwys y canlynol:

- hyfforddi'r holl ddefnyddwyr a fydd yn defnyddio'r system
- cynhyrchu dogfennaeth dechnegol a dogfennaeth i'r defnyddiwr i helpu pobl i ddeall y system
- sefydlu cyfleusterau desg gymorth i helpu defnyddwyr sy'n cael trafferth gyda'r system newydd.

Ni ellir gadael system fel y mae nes bod system newydd yn cael ei datblygu i gymryd ei lle. Bydd angen newid y system wrth i'r busnes neu gorff newid. Efallai y bydd angen ysgrifennu neu newid rhaglenni. Er enghraifft, gallai'r gyfradd TAW newid, neu gallai cyfraddau treth incwm newydd olygu bod angen gwneud newidiadau. Bydd busnesau'n newid cyfeiriad neu'n ymgymryd â mentrau newydd. Er enghraifft, mae Nwy

Prydain, sy'n cyflenwi nwy i gartrefi, bellach yn cyflenwi trydan ac mae gan Anglia Water, sy'n cyflenwi dŵr i dai a busnesau, is-gwmni sy'n adeiladu cartrefi.

Fel arfer bydd cynnal system yn cynnwys y canlynol:

- nodi swyddogaethau newydd y mae angen eu hychwanegu at y system bresennol mewn cyfarfodydd adolygu
- timau cynnal yn newid y rhaglenni presennol neu'n creu rhai ychwanegol
- nodi unrhyw broblemau gweithredol fel perfformiad gwael neu namau meddalwedd yn y cyfarfod adolygu, a'u cywiro gan y staff priodol
- ymchwilio i unrhyw chwalfeydd system i ddarganfod y rhesymau drostynt
- rheoli rhyngwynebau gyda systemau eraill, fel y Rhyngrwyd, e-bost a mewnrwydi.

Y tri math o gynnal system: perffeithiol, addasol a chywirol

Gall cynnal system TGCh (h.y. caledwedd, meddalwedd a chyfathrebiadau) gael ei ddosbarthu'n un o'r canlynol:

- **Perffeithiol** – gwaith cynnal fydd yn gwella perfformiad y system TGCh yw hwn. Fel rheol bydd hyn yn cynnwys ychwanegu nodweddion newydd at y meddalwedd i wneud iddo gynhyrchu'r wybodaeth o gronfa ddata yn gyflymach neu i wella cyflymder rhwydwaith.
- **Addasol** – mae'r ffordd o wneud pethau'n newid mewn corff. Er enghraifft, gall deddfau newydd olygu bod angen newid y system. Gall newidiadau yn y ffordd mae corff yn gweithredu olygu bod angen iddo newid ei system.
- **Cywirol** – mae hyn yn golygu cywiro'r namau na ddaethant i'r amlwg yn

ystod y profi. Weithiau mae'r cyfuniad o feddalwedd a chaledwedd newydd yn peri i'r hen feddalwedd chwalu. Bydd gwneuthurwyr meddalwedd yn cynhyrchu diweddariadau i ddatrys y problemau hyn a rhaid gosod y rhain ar y system.

I sicrhau bod system TGCh yn gweithio'n iawn mae'n rhaid rhoi sylw i nifer o faterion cynnal, gan gynnwys:

- Darganfod gwallau – gall gwallau godi ni waeth pa mor drwyadl y caiff system ei phrofi, a rhaid ymdrin â'r rhain. Maent yn cael eu hachosi gan wrthdrawiadau gyda meddalwedd arall fel arfer, yn enwedig gwirwyr firysau a muriau gwarchod. Rhaid darganfod a chywiro'r gwallau hyn.
- Materion diogelwch – gall rhai rhaglenni (yn enwedig systemau gweithredu) achosi gwendidau diogelwch na chawsant eu rhagweld pan gafodd y meddalwedd ei greu'n wreiddiol. Mae problemau o'r fath yn cael eu datrys drwy ysgrifennu cod meddalwedd i newid y meddalwedd wreiddiol a chau'r bwlch yn yr amddiffynfeydd.
- Newidiadau yn yr amgylchedd busnes – mae anghenion busnes yn newid gydag amser a bydd angen newid y meddalwedd i ymdopi â'r newidiadau hyn. Mae enghreifftiau'n cynnwys newidiadau i ddeddfwriaeth a threthi, ac ati.
- Anfodlonrwydd o ran y caledwedd a'r meddalwedd – efallai y bydd y defnyddwyr yn diflasu ar y caledwedd a'r meddalwedd. Mae'n bosibl na fydd yn gweithio'n ddigon cyflym neu'n methu rhoi wybodaeth sydd ei hangen i gadw'r busnes yn gystadleuol.
- Diweddaru'r system – mae angen diweddaru'r meddalwedd yn aml i

ymdopi â newidiadau i'r systemau gweithredu neu drwy ychwanegu rhagor o swyddogaethau.

Dogfennaeth dechnegol a dogfennaeth defnyddiwr

Mae dau fath o ddogfennaeth: dogfennaeth dechnegol a dogfennaeth defnyddiwr.

Dogfennaeth defnyddiwr (neu ganllaw defnyddiwr) ar gyfer y system

Dogfen y gall defnyddiwr droi ati i ddysgu tasgau newydd neu i ddatrys problemau yw canllaw neu lawlyfr y defnyddiwr. Dylai'r canllaw ymdrin â'r canlynol:

- y gofynion caledwedd a meddalwedd lleiaf
- sut i lwytho'r meddalwedd
- sut i wneud tasgau penodol
- sut i gadw ffeiliau
- sut i argraffu
- cwestiynau cyffredin
- sut i ymdrin â negeseuon gwall a datrys problemau
- sut i wneud copïau wrth gefn o ddata.

Mae'n syniad da cynnwys enghreifftiau ac ymarferion i helpu'r defnyddiwr i ddeall y system. Nid oes gan ddefnyddwyr wybodaeth dechnegol fel rheol, felly dylid osgoi iaith dechnegol.

Dylai'r canllaw fanylu ar yr hyn i'w wneud o dan amgylchiadau eithriadol. Er enghraifft, os bydd y system yn methu â darllen disg neu os bydd data yn cael eu hanfon at argraffydd sydd heb ei droi ymlaen, ac os caiff y peiriant ei gloi, bydd angen i'r defnyddiwr wybod beth i'w wneud. Dylai cyfarwyddiadau ar wneud copïau wrth gefn a chau i lawr gael eu cynnwys hefyd.

Y defnyddwyr sy'n gwybod orau sut mae system yn gweithio wrth gwrs, a dylid gofyn iddynt werthuso unrhyw ganllaw arfaethedig. Yna dylid cynnwys eu sylwadau yn y canllaw. Mae'n siŵr eich bod chi wedi chwilio am gyngor mewn llawlyfrau eich hun, felly fe wyddoch pa mor bwysig ydynt.

Dogfennaeth dechnegol

Mae'n bwysig i systemau gael eu dogfennu'n llawn gan ei bod hi'n bosibl na fydd y sawl a ddyluniodd ac a weithredodd y system ar gael bellach i helpu gyda

problemau. Mae'n bosibl y bydd rhywun arall nad yw mor gyfarwydd â'r system yn gorfod ei chynnal. Er mwyn gwneud hyn bydd angen i'r person hwn ddeall sut y cafodd y system ei dylunio a'i gweithredu. Caiff gwybodaeth dechnegol ei chynhyrchu i helpu dadansoddwyr systemau a rhaglenwyr i ddeall manylion technegol y system.

Byddech yn disgwyl gweld y canlynol yn y ddogfennaeth dechnegol:

- copi o fanyleb ddylunio'r system
- yr holl ddiagramau a ddefnyddiwyd i gynrychioli'r system (siartiau llif, siartiau llif systemau, diagramau llif data, diagramau perthynas endidau, ac ati)
- y geiriadur data
- dyluniadau macros, fformiwlâu ar gyfer taenlenni, rhestri rhaglen
- dyluniadau ar gyfer cynlluniau sgrin
- dyluniadau rhyngwyneb defnyddiwr
- cynllun profi.

Gwerthuso'r system

Mae'r cam gwerthuso ac adolygu'n digwydd fel arfer ychydig o wythnosau ar ôl y gweithredu gan mai dim ond bryd hynny y bydd y defnyddwyr a'r bobl eraill sydd ynghlwm wrth ddatblygu'r system yn darganfod unrhyw broblemau neu ddiffygion.

Bydd y gwerthuso a'r adolygu'n cynnwys y canlynol fel arfer:

- gwirio bod y system newydd yn cwrdd yn llawn â gofynion gwreiddiol y defnyddwyr
- asesu pa mor fodlon yw'r cleientiaid ar ddatblygiad y system newydd
- sefydlu cylch adolygu fel bod y system yn cael ei gwirio o bryd i'w gilydd i sicrhau ei bod hi'n dal i gwrdd â'r gofynion.

Meini prawf ar gyfer gwerthuso system

Cyn gwerthuso system dylid rhestru rhai meini prawf gwerthuso fel:

- pa mor agos mae'r datrysiad yn cyd-fynd â gofynion y defnyddwyr
- pa mor hawdd ei defnyddio yw'r system newydd
- pa mor fodlon yw'r defnyddwyr ar y wybodaeth a roddir gan y system
- pa mor dda mae'r system wedi cael ei phrofi, gan gyfeirio at y dystiolaeth

mewn logiau o broblemau defnyddwyr fel chwalfeydd system
- pa mor ddibynadwy yw'r system
- pa mor dda mae'r system yn sicrhau diogelwch y data a'r rhaglenni.

Offer ar gyfer casglu gwybodaeth i'r adroddiad gwerthuso

Mae nifer o offer ar gyfer casglu gwybodaeth i'r adroddiad gwerthuso, gan gynnwys:

- **Prawf meintiol** – mae bob amser yn haws gwerthuso bodlonrwydd ar system neu berfformiad system os yw'n bosibl ei fynegi ar ffurf rifiadol. Er enghraifft, gellid gofyn i'r defnyddwyr asesu bodlonrwydd cyffredinol ar y datrysiad ar raddfa o 1 i 5 lle mae 5 yn cyfleu'r bodlonrwydd mwyaf. Yna gallwch gynhyrchu ystadegau o'r rhifau (e.e. y bodlonrwydd cyfartalog ymhlith defnyddwyr oedd 3.8).
- **Cyfweliadau logio gwallau** – caiff yr holl alwadau i ddesgiau cymorth eu logio ac mae'r logiau hyn yn darparu gwybodaeth bwysig am bethau fel:
  - pa mor hawdd ei ddefnyddio oedd y meddalwedd
  - i ba raddau yr oedd yr hyfforddiant a gafodd y defnyddwyr yn cwrdd â'u hanghenion
  - pa mor aml y methodd y meddalwedd.
- Cyfweliadau gyda rheolwyr y ddesg gymorth i asesu pa mor dda mae'r system TGCh newydd yn gweithio.
- **Holiaduron** – gellir rhoi holiaduron i ddefnyddwyr i gasglu gwybodaeth am eu bodlonrwydd neu fel arall ar y system newydd. Mae defnyddwyr yn fwy tebygol o roi atebion gonest i'r cwestiynau os gallant aros yn ddienw, ac er mwyn symleiddio prosesu'r holiaduron gellir defnyddio dulliau mewnbynnu awtomatig fel adnabod marciau gweledol (AMG/OMR: *optical mark recognition*) neu adnabod nodau gweledol (ANG/OCR: *optical character recognition*) i'w hasesu.

Dylai'r defnyddwyr gael cyfle i roi awgrymiadau ynghylch cyfyngiadau'r system newydd a gwelliannau y gellid eu gwneud.

Cynnal y system a gwerthuso'r system *(parhad)*

Dulliau o osgoi costau ôl-weithredu

Nid yw costau system yn gorffen ar ôl iddi gael ei hadeiladu a dechrau cael ei defnyddio. Rhai o'r costau sy'n codi wedyn yw:

- Costau hyfforddi – mae'n bosibl nad oedd yr hyfforddiant cychwynnol yn cwmpasu'r holl feddalwedd, neu fod angen addasu'r meddalwedd a hyfforddi'r staff o'r newydd.

- Costau addasu – mae'n bosibl na fydd y meddalwedd yn cyflawni'r holl dasgau angenrheidiol, felly bydd angen addasu'r system. Gellir lleihau'r costau ôl-weithredu hyn drwy roi mwy o sylw i ofynion defnyddwyr.
- Costau desg gymorth a chefnogi – gellir lleihau'r costau hyn os yw'r hyfforddiant yn gynhwysfawr a'r meddalwedd yn hawdd ei ddefnyddio.
- Yr angen i brynu caledwedd

ychwanegol – bydd cyrff yn ehangu'n aml felly bydd angen gwneud y systemau'n fwy drwy gynyddu cynhwysedd storio, cael prosesyddion cyflymach, ac ati. Wrth adeiladu systemau, dylid cynnwys digonedd o gynhwysedd ychwanegol i ddechrau i leihau'r costau hyn.
- Cywiro namau – os caiff y system ei phrofi'n drwyadl, ni ddylai namau ddigwydd neu dylent gael eu lleihau'n sylweddol.

```
        ┌─────────────────────────────────────┐
        │ Costau desg gymorth a chefnogi       │
        └─────────────────────────────────────┘

┌──────────────────┐                      ┌────────────────────────────┐
│ Costau hyfforddi │                      │ Cost caledwedd ychwanegol  │
└──────────────────┘                      └────────────────────────────┘

              ┌──────────────────────────────┐
              │   COSTAU ÔL-WEITHREDU         │
              └──────────────────────────────┘

┌──────────────────┐                      ┌────────────────────────────┐
│ Costau addasu    │                      │ Cywiro namau               │
└──────────────────┘                      └────────────────────────────┘
```

Costau ar ôl gweithredu

Hawlfraint 2001 gan Randy Glasbergen.
www.glasbergen.com

"Byddwch cystal â gwrando'n ofalus gan fod rhai o'r dewisiadau yn ein dewislen wedi newid. I gael gwasanaeth cwsmeriaid, ewch i ganu. I gael cymorth technegol, ewch i gyfri defaid. I gael y gwasanaeth trwsio, arhoswch tan Sul y pys…"

Os yw'r system yn un anodd ei defnyddio, gall defnyddwyr fod ar y ffôn yn hir yn ceisio mynd drwodd at y ddesg gymorth

Cwestiynau a Gweithgareddau

▶ **Cwestiynau 1**   tt. 128–146

1 (a) Eglurwch ystyr y termau canlynol, gan roi enghraifft addas ar gyfer pob un:
    (i) Endid (2 farc)
    (ii) Priodoledd. (2 farc)
   (b) Dylai priodoleddau fod yn atomig. Eglurwch ystyr hyn. (2 farc)

2 (a) Mae system TGCh newydd gael ei datblygu ar gyfer cwmni. Cymharwch a chyferbynnwch y gwahanol ddulliau y gallai'r cwmni eu defnyddio i newid drosodd i'r system newydd. (6 marc)
   (b) Ar ôl newid drosodd, bydd angen cynnal y system.
    (i) Eglurwch beth yw ystyr cynnal y system a rhowch **ddwy** enghraifft o dasgau sy'n cael eu gwneud fel rheol fel rhan o'r gwaith cynnal hwn. (4 marc)
    (ii) Disgrifiwch **ddau** wahanol ddull o gynnal system, gan roi enghraifft ym mhob achos i egluro eich ateb. (4 marc)

3 Caiff systemau TGCh eu datblygu am amrywiaeth o resymau.
   (a) Eglurwch **dri** rheswm dros ddatblygu system TGCh newydd. (3 marc)
   (b) Cyn datblygu system newydd, bydd ymchwiliad i'r system yn cael ei wneud a bydd hyn yn arwain at gynhyrchu'r adroddiad dichonoldeb. Amlinellwch y tasgau sy'n cael eu gwneud wrth ymchwilio i'r system ac eglurwch bwrpas yr adroddiad dichonoldeb a'r hyn mae'n ei gynnwys fel arfer. (6 marc)

4 Prif elfennau'r gylchred oes datblygu system yw:
   Ymchwilio i'r system
   Dadansoddi'r system
   Dylunio'r system
   Gweithredu'r system
   Cynnal y system
   Gwerthuso'r system
   (a) Eglurwch pam mae datblygu system yn broses gylchol. (2 farc)
   (b) Disgrifiwch **dair** techneg wahanol a ddefnyddir i ddadansoddi system. (3 marc)
   (c) Disgrifiwch **dair** tasg sy'n cael eu cwblhau wrth ddylunio'r system. (3 marc)

5 Gall nifer o wahanol ddulliau trawsnewid neu newid drosodd gael eu defnyddio wrth i un system TGCh gael ei disodli gan system newydd.
   (a) Rhowch enwau **tri** dull gwahanol o drawsnewid o un system TGCh i un arall a thrafodwch fanteision cymharol pob dull. (6 marc)
   (b) Ar ôl gweithredu'r system, bydd angen darparu dogfennaeth dechnegol a dogfennaeth defnyddiwr. Eglurwch y gwahaniaeth rhwng y ddau fath hyn o ddogfennaeth a rhowch **dair** enghraifft o eitemau a gâi eu cynnwys yn y ddau fath. (8 marc)

6 Mae llyfrgell coleg yn defnyddio system rheoli cronfeydd data perthynol i weithredu system aelodaeth a benthyciadau. Gall staff a myfyrwyr fenthyca cymaint o lyfrau ag sydd eu heisiau arnynt ar unrhyw adeg benodol.
   (a) Enwch **dri** endid y byddech chi'n disgwyl eu cael yn y system hon ac, ar gyfer pob un, rhowch restr o'r priodoleddau ar gyfer pob tabl, gan nodi ym mhob achos y prif allweddi a'r allweddi estron. (9 marc)
   (b) Lluniwch ddiagram perthynas endidau i ddangos y cysylltau rhwng y tablau cronfa ddata a enwyd gennych yn (a). (3 marc)

Cwestiynau a Gweithgareddau *(parhad)*

▶ Gweithgaredd 1: Creu tabl penderfyniad

I lwyddo mewn cwrs astudiaethau cyfrifiadurol mewn coleg, rhaid i fyfyriwr fodloni'r amodau canlynol:

- rhaid iddynt lwyddo yn yr holl unedau cyfrifiadura
- rhaid iddynt lwyddo mewn Cymraeg a Mathemateg
- rhaid i'w presenoldeb fod yn 80% o leiaf (os bydd eu presenoldeb yn is na hyn fe fyddant yn methu, hyd yn oed os llwyddant yn yr holl unedau eraill).

Y gweithredoedd a all ddigwydd yw:

- pasio'r cwrs
- ail-wneud Cymraeg neu Fathemateg neu'r ddau bwnc
- ail-wneud modiwlau cyfrifiadura a fethwyd
- methu'r cwrs.

Lluniwch dabl penderfyniad i ddangos y rheolau hyn.

▶ Gweithgaredd 2: Diagramau perthynas endidau.

Lluniwch ddiagram perthynas endidau i ddangos pob un o'r perthnasoedd canlynol:

1. Mae dosbarthiadau'n cynnwys llawer o fyfyrwyr.
2. Mae gan un cwsmer lawer o archebion.
3. Mae un tiwtor yn darlithio ar lawer o gyrsiau.
4. Mae pob modiwl yn cael ei ddysgu gan un tiwtor.
5. Mae llawer o fyfyrwyr yn cofrestru ar lawer o gyrsiau.
6. Mae llawer o gwsmeriaid yn archebu llawer o gynhyrchion.

▶ Gweithgaredd 3: Pa rai o'r rhain sy'n atomig?

Pa rai o'r priodoleddau canlynol y gellid ystyried eu bod yn atomig?

- Manylion fideo
- Dyddiad Geni
- Rhif Aelodaeth
- Enw Aelod
- Cofrestriad
- Nifer mewn stoc
- Rhif Cynnyrch.

Ar gyfer pob un o'r priodoleddau uchod nad ydynt yn atomig, awgrymwch briodoleddau addas a all gymryd eu lle sydd yn atomig.

▶ Gweithgaredd 4: Creu model perthynas endidau ar gyfer system TGCh

Mae cwmni ymchwil marchnata yn cyflogi staff pan mae eu hangen ar gyfer contractau, ond wrth i un contract orffen mae un arall yn dechrau. Mae'r cwmni wedi gofyn i ymgynghorydd adeiladu system gronfa ddata i storio manylion y staff, contractau, cyfraddau, a'r oriau mae'r staff yn eu gweithio. Mae'r dadansoddiad cychwynnol wedi darganfod yr endidau canlynol:

> GWEITHIWR
> DYDDIAD
> AWR
> CONTRACT
> CYFRADD

Mae ymchwilio pellach wedi darganfod y canlynol:

Ar unrhyw un adeg gall un gweithiwr weithio ar un contract yn unig ac mae gan y contract hwnnw un gyfradd yn unig. Gall fod llawer o oriau gan yr un contract a gall gael ei wneud dros lawer o ddyddiadau. Hefyd gall yr un gyfradd gael ei thalu dros lawer o oriau a chael ei defnyddio dros lawer o ddyddiadau. Yn olaf, gall un gweithiwr weithio llawer o oriau dros lawer o ddyddiadau.

Cynhyrchwch fodel perthynas endidau i ddangos y system hon.

▶ Gweithgaredd 5: Penderfynu ar briodoleddau ar gyfer endidau

Mae ymchwiliad wedi cael ei wneud i'r system fenthyca yn llyfrgell coleg. Mae ganddi'r endidau canlynol:

- AELOD (rhywun sy'n aelod o'r llyfrgell ac yn gymwys i fenthyca llyfrau)
- LLYFR (llyfr a all gael ei fenthyca)
- BENTHYCIAD (cysylltiad rhwng llyfr penodol a'r sawl sy'n ei fenthyca)
- CAIS CADW (gall aelodau ofyn i lyfrau gael eu rhoi ar gadw, fel y gallant eu benthyca pan gânt eu dychwelyd gan aelod arall)

Rhan yn unig o system gyfan y llyfrgell yw'r system uchod.

Ar gyfer y gweithgaredd hwn, rhaid i chi adnabod a rhestru'r priodoleddau ar gyfer pob un o'r endidau uchod. Bydd angen rhai priodoleddau sy'n diffinio'n unigryw yr endidau AELOD a LLYFR. Er bod yr *ISBN* (Rhif Llyfr Safonol Rhyngwladol) yn cael ei ddefnyddio gan siopau llyfrau i adnabod teitlau llyfrau, mae'n bosibl y bydd gan lyfrgell lawer o gopïau o'r un teitl. Os felly, ni ellid defnyddio'r *ISBN* i wahaniaethu rhwng pob copi.

Cynhyrchwch eich rhestr a dangoswch hi i'ch tiwtor.

Cymorth gyda'r arholiad

Enghraifft 1

1 Ni waeth pa mor dda mae meddalwedd yn cael ei brofi, bydd bob amser angen ei gynnal wedyn. Eglurwch y mathau o gynnal y gall fod angen eu gwneud a pham mae angen gwneud y gwaith cynnal hwn. Defnyddiwch enghreifftiau addas i egluro eich atebion. (8 marc)

Ateb myfyriwr 1

1 Efallai y bydd angen uwchraddio'r meddalwedd. Bydd hyn yn cael gwared ag unrhyw namau neu broblemau na chawsant eu darganfod yn ystod y profi.

Bydd angen newid meddalwedd wrth i anghenion y busnes newid. Efallai y bydd angen i'r meddalwedd allu trosglwyddo ffeiliau i ddarn arall o feddalwedd sydd wedi cael ei brynu'n ddiweddar. Bydd hyn yn ei gwneud hi'n bosibl i'r data o'r meddalwedd gael eu hanfon i a'u defnyddio gan ddarn arall o feddalwedd.

Mae'n bosibl y caiff deddfau newydd eu pasio sy'n golygu y bydd angen newid y meddalwedd. Er enghraifft, gall deddfau gwarchod data newydd olygu bod y meddalwedd yn gweithio mewn ffordd sy'n anghyfreithlon o dan y deddfau newydd.

Sylwadau'r arholwr

1 Mae rhai pwyntiau dilys yma ond byddai wedi bod yn well i enwi'r tri math o gynnal: perffeithiol, addasol a chywirol a strwythuro'r ateb yn unol â hyn.

Mae pum pwynt yn yr ateb sy'n haeddu marciau.

(5 marc allan o 8)

Ateb myfyriwr 2

1 Mae cynnal perffeithiol yn golygu bod perfformiad y meddalwedd neu'r system gyfan yn cael ei wella. Un enghraifft yw lle mae fersiwn newydd yn cael ei gynhyrchu sy'n rhedeg yn gyflymach na'r fersiwn blaenorol fel bod gwybodaeth i'r rheolwyr yn cael ei chynhyrchu mewn llawer llai o amser.

Cynnal cywirol yw lle mae problemau gyda'r meddalwedd sy'n achosi chwalfeydd, canlyniadau anghywir, problemau gyda gosodiadau tudalen, ac ati. Mae'r rhaglenwyr neu wneuthurwyr y rhaglen yn cywiro'r meddalwedd.

Gall fod angen cynnal addasol os yw anghenion y corff wedi newid ers datblygu'r meddalwedd. Efallai y bydd y rheolwyr yn penderfynu y dylai'r meddalwedd weithio mewn ffordd wahanol a chynhyrchu mwy o wybodaeth rheoli. Bydd cyrff yn cyfuno weithiau, felly bydd angen newid y systemau fel eu bod i gyd yn cynhyrchu allbwn mewn ffordd debyg.

Sylwadau'r arholwr

1 Dyma ateb llawer gwell na'r un blaenorol. Mae'r ateb wedi'i strwythuro ac felly bu'n bosibl rhoi esboniadau ac enghreifftiau da.

Mae hwn yn ateb gwych sy'n gwneud llawer mwy o bwyntiau na'r wyth sydd eu hangen i ennill marciau llawn.

(8 marc allan o 8)

Ateb yr arholwr

1 Un marc am bob pwynt hyd at uchafswm o wyth. Uchafswm o chwe marc yn unig os nad yw'r myfyrwyr wedi rhoi unrhyw enghreifftiau.

Cynnal perffeithiol (1) – gwella perfformiad y meddalwedd (1).

Enghreifftiau: Ffurfweddu'r meddalwedd rheoli rhwydwaith (1) i wella perfformiad fel gwella amserau cyrchu data, cynhyrchu adroddiadau'n gyflymach, ac ati (1).

Efallai y bydd angen addasu'r meddalwedd i wella'r rhyngwyneb defnyddiwr (1) ar ôl cael adborth gan ddefnyddwyr sy'n cael mwy o drafferth i'w ddefnyddio nag y dylent (1).

Datblygu tiwtorialau ar-lein a mwy o sgriniau cymorth (1) i helpu staff newydd i ddysgu defnyddio'r meddalwedd (1).

Mae cyflenwr y meddalwedd yn darparu uwchraddiadau (1) a fydd yn gwella perfformiad y meddalwedd (1).

Cynnal cywirol (1) – mae'n bosibl y bydd angen cywiro namau yn y meddalwedd nad oeddynt yn amlwg adeg y profi (1).

Enghraifft: Gall darn o feddalwedd chwalu (1) wrth gael ei ddefnyddio gyda darn arall o feddalwedd (1).

Gall darn o feddalwedd chwalu (1) wrth gael ei ddefnyddio gydag eitem arbennig o galedwedd (1).

Gall meddalwedd fod yn risg i ddiogelwch a gall fod angen cywiro hyn (1).

Efallai nad yw adroddiadau'n cael eu hargraffu'n gywir (1).

parhad drosodd

Cymorth gyda'r arholiad *(parhad)*

Cynnal addasol (1) – gall fod angen newid meddalwedd wrth i anghenion y busnes neu gorff newid (1).

Enghraifft: Efallai y bydd angen newid y meddalwedd fel ei fod yn gallu darparu gwybodaeth ar gyfer rheolwyr (1) na ragwelwyd y byddai ei hangen pan gafodd ei ddatblygu (1).

Bydd angen newid y meddalwedd wrth i werthoedd fel y gyfradd TAW neu gyfraddau treth incwm newid (1).

Mae'r corff yn ehangu (1) felly mae angen newid y meddalwedd er mwyn iddo allu ymdopi â chynnydd yn nifer y defnyddwyr (1).

Addasu'r meddalwedd i weithio gyda meddalwedd systemau gweithredu sydd newydd ei ddatblygu neu galedwedd newydd(1)

Mae'n bosibl y bydd angen addasu'r meddalwedd i'w ddiogelu rhag bygythiadau firws a hacwyr (1).

Enghraifft 2

2 Ar ôl ymchwilio i system a'i dadansoddi, fel ei bod yn barod i'w gweithredu, mae angen rhoi ystyriaeth i'r strategaeth newid drosodd.

  (a) Cymharwch a chyferbynnwch **ddwy** strategaeth wahanol ar gyfer newid drosodd o hen system TGCh i system TGCh newydd. (6 marc)

  (b) Trafodwch y rhesymau pam y gall defnyddwyr fynd yn fwyfwy anfodlon ar ddatrysiad TGCh newydd dros gyfnod o amser. (4 marc)

Ateb myfyriwr 1

2 (a) Gallwch redeg y ddwy system ochr yn ochr â'i gilydd nes bod y defnyddwyr yn hapus gyda pherfformiad y system newydd. Gallan nhw wneud yn siŵr nad oes unrhyw wallau ac yna gallan nhw ddechrau defnyddio'r system newydd. Ffordd arall yw dechrau defnyddio'r system newydd yn unig. Mae'r ffordd hon yn hawdd gan nad oes angen gwastraffu amser.

  (b) Mae'n bosibl na fydd y system newydd mor hawdd ei defnyddio ag yr oedden nhw'n meddwl ac nad ydyn nhw wedi cael digon o hyfforddiant, felly maen nhw'n gwastraffu amser yn aros am rywun i'w helpu. Efallai nad ydyn nhw'n hoffi cyflymder y rhwydwaith gan ei bod hi'n cymryd gormod o amser i wneud tasgau. Efallai y byddan nhw'n darganfod nad yw'r system yn gwneud popeth yr oedd i fod i'w wneud, felly mae'n wastraff ar arian.

Sylwadau'r arholwr

2 (a) Er bod y myfyriwr wedi sôn am drawsnewid ochr yn ochr nid yw wedi egluro ei fanteision cymharol yn ddigon clir. Rhaid nodi bod hyn yn cynyddu baich gwaith y staff gan fod popeth yn cael ei wneud ddwywaith yn ystod y cyfnod o weithredu ochr yn ochr. Mae ail ran yr ateb yn arwynebol. Dylech roi enw'r dull trawsnewid bob amser (h.y. newid drosodd uniongyrchol yn yr achos hwn). Mae angen llawer mwy o fanylder yma, a chymhariaeth. Rhoddwyd dau allan o chwe marc am ran (a).

  (b) Mae'r ateb yn dda ond yn haeddu dau o'r pedwar marc yn unig.
Dylai gyfeirio at fwy o bethau sy'n newid dros amser fel newid yn natur y busnes.
(4 marc allan o 10)

Ateb myfyriwr 2

2 (a) Trawsnewid ochr yn ochr lle mae'r system newydd yn rhedeg ochr yn ochr â'r hen system am ychydig o ddyddiau neu wythnosau nes bod y defnyddwyr yn sicr y bydd y system newydd yn gweithio yn ôl y disgwyl. Mae'r dull hwn yn cynyddu'r gwaith mae'r defnyddwyr yn gorfod ei wneud gan eu bod yn gwneud y gwaith ddwywaith am gyfnod. Y fantais yw eich bod chi'n gallu cadw'r hen system nes bod yr un newydd yn gweithio'n berffaith, felly dull risg isel yw hwn o'i gymharu â'r dull nesaf.

Enw'r dull nesaf yw newid drosodd uniongyrchol, sy'n golygu bod yr hen system yn cael ei stopio a bod y system newydd yn cael ei rhoi ar waith yn syth. Mae tipyn o risg yma o'i gymharu â thrawsnewid ochr yn ochr gan nad oes hen system i droi ati os bydd problemau'n codi.

(b) Gall fod anfodlonrwydd gan nad yw'r system TGCh yn cwrdd yn llawn â gofynion y defnyddwyr sy'n golygu bod pethau na allant eu gwneud gyda'r system newydd. Gall y busnes ehangu ac mae'n bosibl na fydd y system newydd yn gallu ymdopi â'r gofynion perfformiad sy'n cael eu rhoi arni. Er enghraifft, gall fod mwy o ddefnyddwyr, gyda'r canlyniad bod y system gyfan yn arafu.

Gall fod yn gostus iawn i gynnal y system.

Sylwadau'r arholwr

2 (a) Mae'r ddau ddull trawsnewid wedi cael eu disgrifio'n dda ac mae'r ateb yn crybwyll manteision ac anfanteision. Marciau llawn am yr ateb ardderchog hwn.

(b) Tri phwynt yn unig sydd wedi cael eu hegluro'n ddigonol. Ni roddir marc am y pwynt olaf oherwydd nid yw'n ymhelaethu drwy roi enghraifft o gynnal.

(9 marc allan o 10)

Ateb yr arholwr

2 (a) Un marc am yr enw cywir ac esboniad byr cywir ar gyfer pob strategaeth x 2.

Un marc am fantais neu anfantais y dull x 2.

Un marc am gymharu'r dull x 2.

Newid drosodd uniongyrchol – rhoi'r gorau i ddefnyddio'r hen system un diwrnod a dechrau defnyddio'r system newydd y diwrnod nesaf (1). Elfen o risg, yn enwedig os yw'r caledwedd a'r meddalwedd ddiweddaraf wedi cael eu defnyddio (1). Os bydd y system yn methu gall fod yn drychinebus i'r busnes (1).

Angen llai o adnoddau (pobl, arian, cyfarpar), ac mae'n syml os nad aiff dim o'i le (1).

Newid drosodd ochr yn ochr – yn cael ei ddefnyddio i leihau'r risg wrth gyflwyno system TGCh newydd (1). Mae'r hen system TGCh yn cael ei gweithredu ochr yn ochr â'r system TGCh newydd am gyfnod o amser nes bod pawb sy'n defnyddio'r system newydd yn fodlon ei bod yn gweithio'n iawn (1). Yna rhoddir y gorau i'r hen system ac mae'r holl waith yn cael ei wneud ar y system newydd (1). Anfanteision: llawer o waith diangen (gan fod y gwaith yn cael ei wneud ddwywaith) ac felly'n ddrud o ran amser staff (1). Hefyd mae'n golygu bod angen gwneud mwy o waith cynllunio cyn gweithredu'r system (1).

Trawsnewid gam wrth gam – gall modiwl ar y tro gael ei drawsnewid i'r system newydd mewn camau nes bod y system gyfan wedi'i throsglwyddo (1). Un fantais yw bod staff TG yn gallu datrys problemau sy'n cael eu hachosi gan fodiwl cyn symud ymlaen i fodiwlau newydd (1). Un anfantais yw nad yw'n addas ond ar gyfer systemau sy'n cynnwys modiwlau ar wahân (1).

Trawsnewid peilot – mae'r dull hwn yn ddelfrydol ar gyfer cyrff mawr gyda llawer o leoliadau neu ganghennau lle gall y system newydd gael ei defnyddio gan un gangen ac yna ei throsglwyddo i'r lleill dros gyfnod o amser (1). Mantais: mae'r gweithredu ar raddfa lawer llai ac yn haws ei drin (1). Anfantais: mae'n cymryd mwy o amser i weithredu'r system yn yr holl ganghennau (1).

(b) Un marc am bob pwynt hyd at uchafswm o dri. Marc arall am bwynt sy'n cyfeirio at anfodlonrwydd cynyddol dros gyfnod o amser. Rhai atebion posibl yw:

Nid yw holl ofynion y defnyddwyr wedi cael eu bodloni, felly nid yw'r system yn cwrdd â disgwyliadau'r defnyddwyr.

Mae newid yn anghenion y busnes yn golygu na all y system ymdopi â'r galwadau newydd arni.

Methiant i ddarparu'r wybodaeth sydd ei hangen ar ddefnyddwyr.

Mae'r rhyngwyneb defnyddiwr yn peri llawer o broblemau i ddefnyddwyr a mwy o ddefnydd o'r ddesg gymorth.

Mae'r meddalwedd neu'r system yn methu gan nad oedd y profi'n ddigon trwyadl.

Mae perfformiad y rhwydwaith neu'r cyflymder cyrchu data yn mynd yn annerbyniol wrth i ragor o ddefnyddwyr gael eu hychwanegu at y system.

Mae angen addasu'r system yn rheolaidd a'i newid am system newydd.

Mae gormod o amser yn cael ei dreulio'n diweddaru'r system newydd.

Mae cost cymorth i'r defnyddwyr yn rhy uchel.

Mae yna wendidau diogelwch na chawsant eu rhagweld pan gafodd y system ei datblygu'n wreiddiol.

Map meddwl cryno

Cylchred oes datblygu system (*SDLC*)

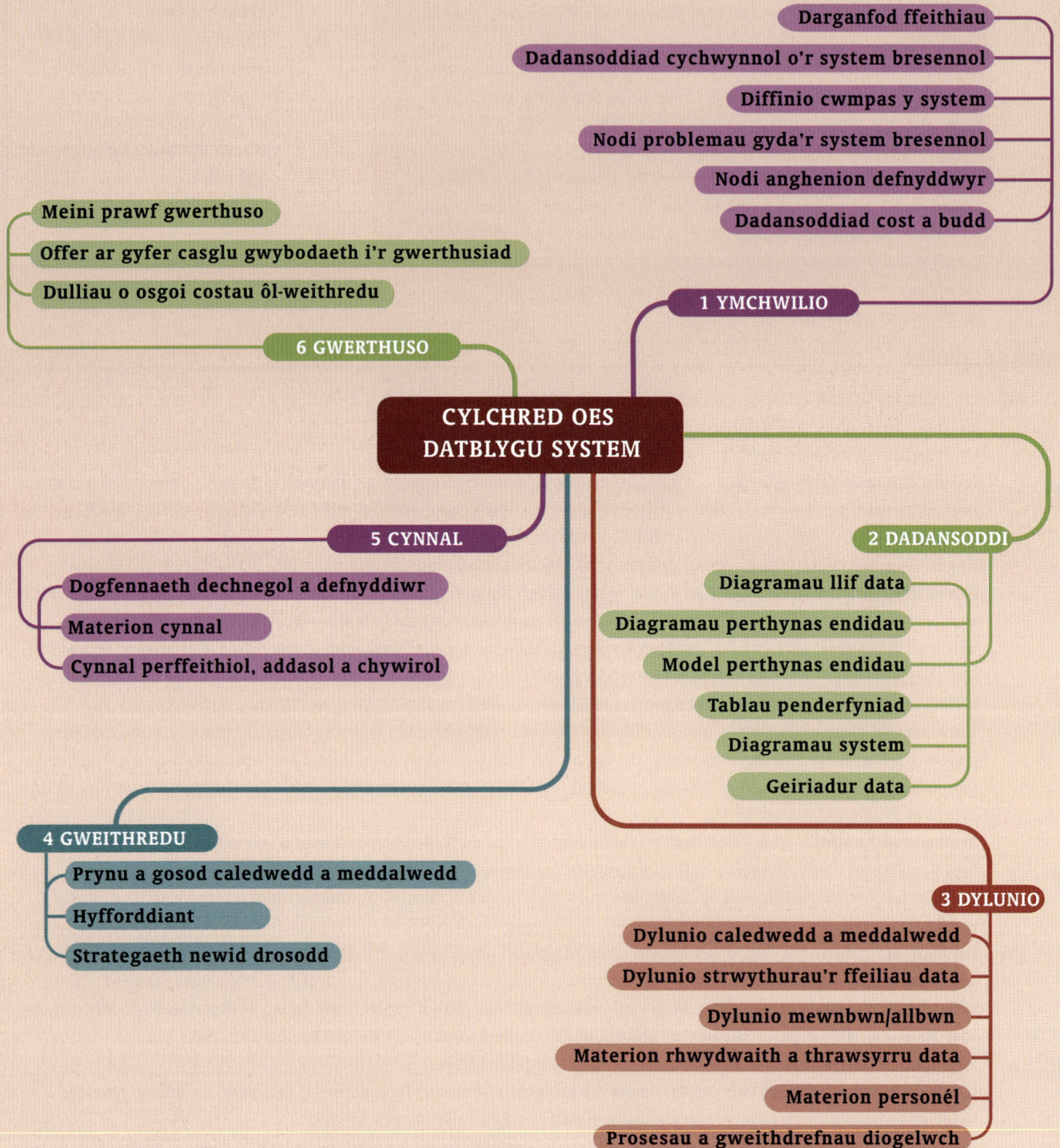

Darganfod ffeithiau

Dadansoddiad cychwynnol o'r system bresennol

Diffinio cwmpas y system

Nodi problemau gyda'r system bresennol

Nodi anghenion defnyddwyr

Dadansoddiad cost a budd

1 YMCHWILIO

Meini prawf gwerthuso

Offer ar gyfer casglu gwybodaeth i'r gwerthusiad

Dulliau o osgoi costau ôl-weithredu

6 GWERTHUSO

CYLCHRED OES DATBLYGU SYSTEM

5 CYNNAL

Dogfennaeth dechnegol a defnyddiwr

Materion cynnal

Cynnal perffeithiol, addasol a chywirol

2 DADANSODDI

Diagramau llif data

Diagramau perthynas endidau

Model perthynas endidau

Tablau penderfyniad

Diagramau system

Geiriadur data

4 GWEITHREDU

Prynu a gosod caledwedd a meddalwedd

Hyfforddiant

Strategaeth newid drosodd

3 DYLUNIO

Dylunio caledwedd a meddalwedd

Dylunio strwythurau'r ffeiliau data

Dylunio mewnbwn/allbwn

Materion rhwydwaith a thrawsyrru data

Materion personél

Prosesau a gweithdrefnau diogelwch

UNED IT4: Cronfeydd Data Perthynol

Yn Uned IT4, sy'n cyfrif am 20% o gyfanswm y marciau am yr arholiad Safon Uwch, rhaid i chi gynhyrchu datrysiad i broblem o'ch dewis sy'n gofyn am ddefnyddio cronfa ddata berthynol. Bydd y datrysiad hwn yn ddarn sylweddol o waith a fydd yn cynnwys prosesau dadansoddi, dylunio, gweithredu, profi a gwerthuso. Rhaid i chi hefyd ddangos eich bod chi wedi rheoli'ch gwaith yn effeithiol.

Byddwch yn gwneud y gwaith dros gyfnod estynedig o amser a chaiff ei asesu'n fewnol (h.y. gan eich athro/athrawes neu ddarlithydd) a'i safoni'n allanol gan CBAC.

▼ Elfennau allweddol yr asesiad ar gyfer IT4 yw:

▶ Dewis problem

▶ Gofynion y defnyddiwr

▶ Manyleb ddylunio

▶ Gweithredu

▶ Profi

▶ Dogfennaeth defnyddiwr

▶ Gwerthuso

▶ Cynllunio'r project

CYNNWYS

Dewis problem

▼ **Byddwch yn dysgu**

► Am ba broblem i'w dewis

► Am gynghorion ar gael syniadau ar gyfer projectau

► Am syniadau ar gyfer projectau cronfa ddata

► Am gwestiynau ac atebion am y project

► Am y marciau a roddir am brif elfennau'r project

Cyflwyniad

Mae dewis y project cywir ar gyfer creu cronfa ddata yn hanfodol i sicrhau llwyddiant, felly mae angen rhoi ystyriaeth ofalus iddo. Os dewiswch yr un iawn, fe fyddwch chi'n cynhyrchu project y gellir ei gwblhau mewn pryd a byddwch chi'n gallu darparu tystiolaeth ar gyfer yr holl elfennau sydd i gael eu hasesu. Os dewiswch yr un anghywir, gallech wynebu pob math o broblemau technegol gan fod y project yn rhy gymhleth, a gall yr amser a dreuliwch yn datrys y problemau hyn olygu nad oes gennych ddigon o amser wedyn i gasglu tystiolaeth ddigonol ar gyfer yr elfennau y mae angen eu hasesu.

Yn yr adran hon cynigiwn gyngor pwysig ar ddewis eich project.

Pa broblem i'w dewis?

Cyn i chi hyd yn oed feddwl am broblem, mae angen i chi ddarllen drwy'r topig hwn yn ofalus. Dyma'r unig ffordd o ddeall yr hyn sydd ei angen. Rhaid i chi ddewis problem a fydd yn caniatáu i chi ymdrin â'r holl elfennau. Rhaid osgoi dewis problem rhy gymhleth, ond rhaid iddi fod yn ddigon cymhleth i fodloni'r holl feini prawf.

Mae dewis problem addas yn hanfodol i lwyddiant y project. Rhaid i chi beidio â bod yn rhy uchelgeisiol a rhaid i chi wybod beth yw eich cyfyngiadau. Peidiwch â chael eich sugno i mewn i gymhlethdodau'r meddalwedd cronfa ddata a chofiwch nad oes gennych lawer o amser i gwblhau'r project.

Cynghorion ar gael syniadau ar gyfer projectau
Defnyddiwch eich cysylltiadau

Gofynnwch i'ch rhieni, perthnasau, cymdogion, ffrindiau, ac ati, beth maen nhw'n ei wneud yn eu swyddi. Efallai y byddan nhw'n gallu awgrymu syniad i chi. Os gallwch gysylltu â nhw i gael y wybodaeth sydd ei hangen arnoch, bydd o gymorth mawr.

Y projectau gorau yw'r rheiny sydd â defnyddiwr terfynol go iawn ac sy'n cyfeirio'n ôl yn gyson at y defnyddiwr terfynol i gael ei ymateb a'i sylwadau ar ddatblygiadau. Bydd angen ymgynghori â'r defnyddiwr terfynol ym mhob cam o'r project.

Mae projectau go iawn yn aml yn well na rhai ffug

Does dim byd o'i le mewn creu system newydd sbon yn hytrach na gwella system bresennol. Bydd llawer o bobl yn dechrau o'r dechrau, felly mae'n briodol i chi ddylunio system i ymdopi â'r tasgau y byddai'n rhaid iddynt eu cyflawni. Ond bydd angen tipyn o wybodaeth flaenorol arnoch am y math o fusnes yr ydych chi'n creu system ar ei gyfer, ac ni fydd system bresennol i chi ei harchwilio.

Defnyddiwch ddeunydd cyfeiriol i'ch helpu i gael syniadau

Bydd bwrw golwg dros gylchgronau neu bapurau newydd yn rhoi syniadau i chi am fusnesau, er enghraifft, y busnesau archebu drwy'r post sy'n hysbysebu ynddynt. Mae gan y rhain i gyd yr un math o broblemau, felly ni ddylai fod yn rhy anodd i chi feddwl am system ar eu cyfer.

Peidiwch â dewis system sydd y tu allan i'ch profiad neu bydd yn anodd i chi gael gwybodaeth amdani

Er enghraifft, efallai y byddwch chi'n dewis creu system i gadw cofnodion deintyddion, ond os nad ydych yn gwybod sut mae deintyddion yn ymdrin â'u tasgau gweinyddol, neu os na allwch gael gwybodaeth am hyn, gallech wneud pethau'n anodd i chi eich hun a chreu system afrealistig iawn yn y diwedd.

Byddwch yn realistig

Mae'n bwysig bod yn realistig am yr hyn y gallwch ei gyflawni yn yr amser sydd ar gael. Fel rheol mae gan y projectau gorau lai o dasgau, maen nhw wedi'u dylunio a'u dogfennu'n well, wedi'u profi a'u gwerthuso'n llawn, ac yn canolbwyntio ar y broblem dan sylw.

Syniadau ar gyfer projectau

Os nad oes gennych gysylltiadau i'ch helpu gyda system go iawn, gallech greu eich system eich hun ar sail un o'r syniadau sy'n dilyn. Bydd angen i chi ymgynghori â'ch athro/athrawes neu ddarlithydd neu gysylltu â rhywun sy'n barod i'ch helpu yn y maes o'ch dewis.

Dyma rai syniadau y gallech eu datblygu:

- System sy'n cofnodi cyfarpar a roddir ar fenthyg mewn coleg
- System bwcio ar gyfer campfa
- System syml ar gyfer storio cofnodion disgyblion mewn ysgol
- Cynelau cŵn neu lety cathod
- Deintyddfa
- Asiantaeth cariadon
- Asiantaeth swyddi ar gyfer cyfrifiadurwyr contract
- Asiantaeth tiwtoriaid
- Ysgol yrru
- Meddygfa
- Asiantaeth trefnu teithiau annibynnol fach
- Cwmni llogi ceir
- Cwmni llogi offer a chyfarpar
- Cwmni rhentu fflatiau/tai
- Cwmni llogi gwisgoedd priodas
- Cwmni archebu gwin drwy'r post
- Tŷ bwyta
- Siop lyfrau archebu drwy'r post
- Cyfanwerthwr
- System rheoli ysgol
- Cwmni llogi bysiau
- Cwmni archebu cydrannau cyfrifiadurol drwy'r post
- System aelodaeth ar gyfer clwb chwaraeon

- Gwerthwr eiddo
- System rheoli stoc ar gyfer cyflenwr defnyddiau adeiladu
- Rhentu tai gwyliau tramor
- Llogi ystafelloedd digwyddiadau
- Cronfa ddata gwerthwr eiddo
- Cronfa ddata o droseddau i'r heddlu
- Cronfa ddata o farciau disgyblion ar gyfer athro
- Cwmni archebu drwy'r post (cryno ddisgiau, llyfrau, dillad, gwin, nwyddau tŷ, cosmetigau, garddio, ac ati)
- Canolfan iechyd
- System apwyntiadau ar gyfer busnes trin gwallt/harddwch
- Cynhyrchu system rheoli stoc
- Asiantaeth staff dros dro
- Asiantaeth doniau/modelu
- Cwmni llogi dillad
- Cronfa ddata ar gyfer gweinyddu ysgol farchogaeth
- Llogi gwisg ffansi.

Cwestiynau ac atebion am y project

A oes rhaid cael defnyddiwr go iawn?

Nac oes, ond gall fod yn well cael un gan y bydd yn haws i chi ddeall ei ofynion drwy ei holi, a gallwch ofyn am ei sylwadau wrth weithio ar y datrysiad. Hefyd gall fod o gymorth wrth i chi werthuso'ch datrysiad gan y bydd yn gallu rhoi sylwadau ar eich gwaith.

Faint o amser sydd ar gael?

Bydd gennych 50 awr, dan oruchwyliaeth, i wneud y project.

Beth yw'r ffordd orau o ddechrau?

Cyn dechrau'ch project cronfa ddata, bydd angen i chi ddysgu sgiliau a gwybodaeth uwch yn y maes. Mae tuedd wedi bod i brojectau gael eu datblygu heb ymgynghori â neb ac i'r athro beidio â gweld y canlyniadau tan ar ôl i'r project gael ei orffen. Dylai eich athro gael gwybod am gynnydd eich project er mwyn gallu eich cywiro os dechreuwch fynd i'r cyfeiriad anghywir. Mae athrawon o hyd yn derbyn gormod o brojectau gorffenedig sy'n amhriodol i ofynion Safon Uwch, ac wedyn does dim

amser i wneud dim byd yn eu cylch. Cyn dechrau ar eich gwaith project, mae'n bosibl y cewch lyfrynnau gydag ymarferion gan eich athro i ddatblygu eich gwybodaeth o feddalwedd cronfa ddata. Dylech sicrhau eich bod chi'n gyfarwydd â nodweddion uwch cronfeydd data. Un ffordd o ddysgu

Mae'n well dysgu'r sgiliau a gwybodaeth angenrheidiol cyn dechrau datblygu'ch cronfa ddata

amdanynt yw drwy gael gafael ar ganllawiau uwch i'r meddalwedd, yn enwedig y rheiny sy'n canolbwyntio mwy ar ddatblygu rhaglenni. Ond cofiwch nad yw llên-ladrad yn cael ei ganiatáu a'i fod yn hawdd ei ddarganfod!

Peth cyngor defnyddiol cyn dechrau'r project

Mae'r cyngor canlynol yn berthnasol i bob rhan o'r project:

- Peidiwch â gwastraffu amser yn darparu tystiolaeth nad oes ei hangen.
- Peidiwch â thocio'ch gwaith yn ormodol gan fod hyn weithiau'n dinistrio'r dystiolaeth.
- Peidiwch â darparu sgrinluniau sy'n rhy fach – bydd yn rhaid i'r marcwyr allu eu darllen.
- Rhaid cynhyrchu gwaith dylunio priodol – nid yw anodi (h.y. ychwanegu sylwadau at) y datrysiadau a roddwyd ar waith yn

ddylunio priodol. Rhaid i ddylunio beidio â bod yn ôl-ystyriaeth.

Nodiadau pwysig am waith project

Dyma rai nodiadau pwysig am y gwaith project:

- Does dim rhaid i'ch project fod yn gymhleth i ennill marciau da.
- Mae dogfennu'ch project yn bwysig iawn. Mae'n rhoi tystiolaeth o'ch gwaith i'ch athro ac yn ei alluogi i farcio'ch project yn fanwl.
- Sicrhewch eich bod chi'n gwirio'ch gwaith am wallau gramadegol a sillafu. Peidiwch â dibynnu ar y cyfrifiadur i wneud hyn. Dylech ddarllen dros eich gwaith yn ofalus bob amser.

Y marciau a roddir am brif elfennau'r project

Dyma grynodeb o brif elfennau'r project a'r marciau sydd ar gael am bob un. Yn yr adrannau sy'n dilyn, byddwch chi'n edrych ar bob elfen yn ei thro ac ar y tasgau a'r dystiolaeth sydd eu hangen i gwblhau pob adran.

Elfennau	Marc mwyaf
Gofynion y defnyddiwr	12
Dylunio	24
Gweithredu	25
Profi	16
Dogfennaeth defnyddiwr	15
Gwerthuso	8
Cyfanswm	100

Tabl dyrannu marciau

Gwefannau defnyddiol ar gyfer dysgu am Microsoft Access

Dyma wefan Microsoft ar gyfer pob fersiwn o Access. Mae'n darparu tiwtorialau ac adrannau ar sut i wneud pethau: http://office.microsoft.com/en-us/access/FX100487571033.aspx

Crynodeb o wahanol gysylltau i diwtorialau Access: http://www.techtutorials.info/appaccess.html

Adrannau defnyddiol ar holl brif nodweddion y gronfa ddata Access: http://www.teach-ict.net/software/access/access.htm

Gofynion y defnyddiwr

Cyflwyniad

Ar ôl i chi ddewis eich project rhaid i chi nodi gofynion y defnyddiwr. Mae'n well os oes gennych ddefnyddiwr go iawn, ond nid yw hyn yn hollol angenrheidiol. Mae gofynion y defnyddiwr yn amlinellu beth sydd angen ei gynhyrchu. Yn yr adran hon byddwn yn edrych ar sut i gasglu gwybodaeth am ofynion y defnyddiwr a'r dogfennau y bydd angen i chi eu darparu ar gyfer y rhan hon o'r project.

Cynhyrchu gofynion y defnyddiwr

Cyn creu rhestr o ofynion y defnyddiwr, bydd yn rhaid i chi wneud ymchwiliad i'r system. Bydd angen casglu gwybodaeth am beth mae'n rhaid i'r system newydd ei wneud neu beth oedd yr hen system yn ei wneud a'r problemau a achosai.

I ddarparu tystiolaeth o ofynion y defnyddiwr, bydd gofyn i chi gynhyrchu dogfennau sy'n ymdrin â'r canlynol:

- Disgrifiad cefndir cyffredinol o'r corff yr ydych chi'n datblygu'r datrysiad ar ei gyfer
- Disgrifiad o ddelwedd ac ethos y corff
- Manylu ar nodau ac amcanion y system
- Amlinelliad o ofynion y rhyngwyneb defnyddiwr
- Amlinelliad o'r gofynion diogelwch
- Manylu ar ofynion caledwedd a meddalwedd y system newydd
- Diagram perthynas endidau ar gyfer y system bresennol neu system arfaethedig.

Disgrifiad cefndir cyffredinol o'r corff yr ydych chi'n datblygu'r datrysiad ar ei gyfer

Y peth cyntaf y bydd angen i chi ei wneud fydd disgrifio'r project. Gallai'r disgrifiad o'ch problem ddechrau fel hyn:

> 'Mae fy nhad yn gweithio mewn modurdy ac, er eu bod nhw'n gwerthu ac yn trwsio ceir ail-law yn bennaf, mae ganddynt fusnes llogi ceir hefyd. Mae ganddynt tua 20 o geir i'w llogi ac mae'r rhan hon o'r busnes yn ehangu'n gyflym. Hoffai fy nhad gael ffordd fwy proffesiynol o gofnodi a storio manylion y ceir, cwsmeriaid a rhentolion. Ar hyn o bryd caiff yr holl fanylion eu storio â llaw ar gardiau a gedwir mewn cypyrddau ffeilio, neu mewn dyddiaduron. Roedd hyn yn iawn pan oedd y modurdy'n llogi ychydig o geir bob wythnos, ond gyda 20 o geir a'r posibilrwydd o ddyblu eu nifer y flwyddyn nesaf, mae angen iddynt gael gwybodaeth yn gyflym a chyflwyno delwedd fwy proffesiynol i'w cwsmeriaid.'

Yn yr adran hon mae'n syniad da i chi grybwyll cwmpas y project. Mae hyn yn golygu pa mor bellgyrhaeddol y bydd y project. Er enghraifft, yn ogystal â dweud beth yn benodol y bydd y project yn ei wneud, gallwch nodi beth na fydd yn ei wneud. Os yw cwmpas y project yn rhy eang, mae'n bosibl na chewch ddigon o amser i'w gwblhau neu y bydd yn rhy gymhleth. Er enghraifft, gallai'r enghraifft uchod ymdrin â cheir, cwsmeriaid a rhentolion yn unig, ond gellid ehangu ei chwmpas drwy ychwanegu manylion fel gwasanaethau, profion MOT, yswiriant, ac ati.

Disgrifiad o ddelwedd ac ethos y corff

Yma mae angen i chi ystyried delwedd ac ethos y corff, gan y bydd hyn yn dylanwadu ar arddull tŷ y dogfennau a gynhyrchir gan y system gronfa ddata. Bydd angen i'r dogfennau neu ddyluniadau'r sgriniau adlewyrchu delwedd ac ethos y corff, felly rhaid ysgrifennu adran fer ar hyn.

Ethos

I ddeall beth yw ystyr y gair 'ethos', mae'n well edrych mewn geiriadur: ethos yw natur neu ysbryd hanfodol corff; y syniad sylfaenol sy'n sail i gredoau, defodau neu arferion y corff.

Delwedd

Mae gan y gair 'delwedd' ddiffiniad cyffelyb: delwedd yw'r canfyddiad cyffredinol neu gyhoeddus o gorff, cwmni, ac ati, sy'n cael ei gyflawni'n aml drwy ymdrechion gofalus i greu ewyllys da ar raddfa eang.

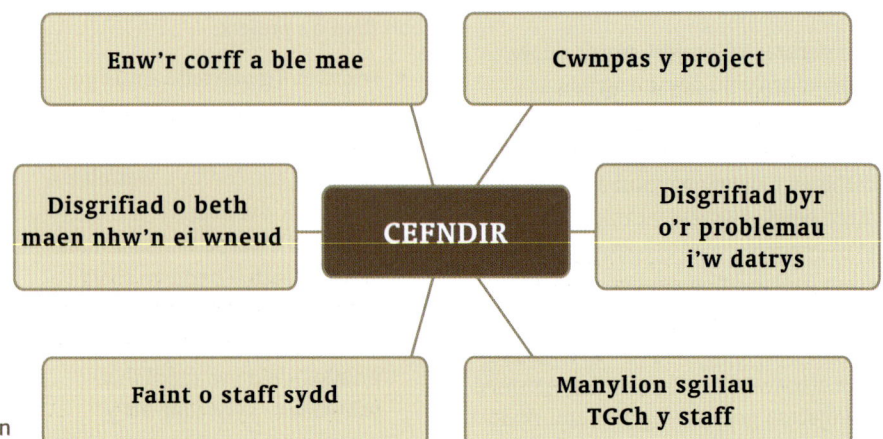

Pethau mae angen i chi gael gwybodaeth amdanynt i'w cynnwys yn eich disgrifiad cefndir

Enw'r corff a ble mae		Cwmpas y project
Disgrifiad o beth maen nhw'n ei wneud	CEFNDIR	Disgrifiad byr o'r problemau i'w datrys
Faint o staff sydd		Manylion sgiliau TGCh y staff

Arddull tŷ

Mae llawer o wahanol bobl mewn corff yn cynhyrchu dogfennau a rhaid i'r rhain edrych yn debyg – fel pe bai'r un person wedi'u cynhyrchu i gyd. Mae arddull tŷ yn ymwneud â:

- Defnyddio logos – defnyddir logos yn gyson (h.y. maint, lliw a safle).
- Defnyddio geiriau – mae gwahanol ffyrdd o sillafu'r un gair, felly rhaid cael cysondeb.
- Defnyddio lliwiau – defnyddir cynlluniau lliw yn gyson.
- Arddull ysgrifennu – rhaid i bobl eu mynegi eu hunain mewn ffordd gyson.
- Cywair – mae angen i rai dogfennau fabwysiadu cywair arbennig.

Manylu ar nodau ac amcanion y system

Nod system yw ei bwriad neu bwrpas. Er enghraifft, nod posibl system ar gyfer cwmni llogi ceir yw:

> 'Gwella cadw cofnodion o geir, cwsmeriaid a rhentolion fel bod cwsmeriaid yn mwynhau lefel broffesiynol o wasanaeth gan y busnes gyda'r canlyniad bod proffidioldeb cyffredinol y busnes yn gwella.'

Mae angen i nod gwmpasu'r system gyfan ac nid un rhan yn unig ohoni.

Gellir meddwl am amcanion fel cyfres o dargedau y dylai'r system anelu atynt. Mae'r amcanion ar gyfer system yn rhestr o'r pethau mae'n rhaid i'r system eu gwneud. Er enghraifft, amcanion posibl ar gyfer system yw:

- Y gallu i gael ei defnyddio gan bobl y mae eu sgiliau a'u gwybodaeth TG yn amrywio.
- Lleihau nifer y camgymeriadau sy'n cael eu gwneud gyda bwciadau.
- Lleihau costau postio gan fod nwyddau'n cael eu hanfon i'r cyfeiriad anghywir.
- Y gallu i adnabod y cwsmeriaid hynny sy'n defnyddio gwasanaeth yn fwyaf aml a'u targedu ar gyfer disgowntiau arbennig.
- Darparu gwybodaeth rheoli i seilio penderfyniadau arni.
- Galluogi'r holl staff i gyrchu adroddiadau ac ymholiadau sydd wedi'u storio yn ôl y gofyn.
- Rhoi gwybodaeth i gwsmeriaid rheolaidd am hyrwyddiadau arbennig a bargeinion drwy ddefnyddio'r gronfa ddata a chyfleuster postgyfuno meddalwedd prosesu geiriau. Bydd hyn yn helpu i gadw'r prif gwsmeriaid a sicrhau nad ydynt yn mynd at gystadleuwyr.

Amlinelliad o ofynion y rhyngwyneb defnyddiwr

Ni ddylai defnyddwyr orfod brwydro i ddeall y meddalwedd, felly rhaid i chi ei wneud mor hawdd â phosibl. Mae angen i chi restru gofynion y defnyddwyr yn nhermau arbenigedd defnyddwyr y system. Mae nifer o bethau y dylech eu hystyried wrth ddylunio'r rhyngwyneb defnyddiwr:

- Ni ddylai defnyddwyr orfod symud y llygoden neu ddefnyddio'r bysellfwrdd i deipio data fwy nag sy'n angenrheidiol.
- Dylai'r eitemau pwysicaf neu'r eitem a ddefnyddir amlaf ymddangos gyntaf.
- Ni ddylid gadael y defnyddwyr mewn gwactod, yn meddwl beth i'w wneud nesaf. Rhowch gyfarwyddiadau clir iddynt i'w helpu.
- Ni ddylid disgwyl i ddefnyddwyr fewnbynnu eu data'n syth i dablau.
- Rhaid i chi gadw eich defnyddwyr i ffwrdd o gymhlethdodau'r meddalwedd.
- Dylai fod cysondeb rhwng dewislenni. Dylent i gyd edrych yn debyg.

Dylech ystyried:

- Defnyddio switsfyrddau. Mae switsfyrddau'n symleiddio'r broses o gychwyn y gwahanol ffurflenni ac adroddiadau mewn cronfa ddata.
- Defnyddio ffurflenni mewnbynnu neu gofnodi data i ychwanegu data at eich cronfa ddata neu i edrych ar, golygu neu ddileu'r data presennol.
- Creu blychau deialog arbennig ar gyfer yr adegau y bydd angen i'r cyfrifiadur weithredu wrth dderbyn mewnbwn gan y defnyddiwr. Er enghraifft, gallwch greu blychau deialog chwilio sy'n gofyn am baramedrau chwilio gan y defnyddwyr, ac yna'n dychwelyd data sy'n cyd-fynd â'r paramedrau hyn.

Amlinelliad o'r gofynion diogelwch

Yma mae angen i chi ystyried gofynion diogelwch y system gyda'r defnyddiwr. Byddai'r rhain yn cynnwys:

- sut y gellir diogelu'r gronfa ddata drwy ddefnyddio cyfrineiriau
- sut y gellir cadw fersiynau archifol o'r data
- sut a pha bryd y caiff copïau wrth gefn eu gwneud o'r data.

Manylu ar ofynion caledwedd a meddalwedd y system newydd

Bydd y caledwedd sydd ei angen yn dibynnu ar y meddalwedd cronfa ddata a ddefnyddiwch. Y peth pwysicaf i'w wneud yw mynd i wefan gwneuthurwr y meddalwedd i weld beth yw ei ofynion lleiaf. Bydd angen i chi roi manylion y meddalwedd system y bydd y meddalwedd cronfa ddata yn rhedeg oddi tano.

Cofiwch gynnwys pethau fel cysylltiad a chyfarpar Rhyngrwyd os oes angen a gwirwyr firysau ac unrhyw wasanaethau (utilities) hanfodol eraill.

Diagram perthynas endidau

Dylech lunio diagram sy'n dangos yr endidau a'u perthynas â'i gilydd yn naill ai'r hen system neu'r system newydd arfaethedig os nad oes hen system.

Manyleb ddylunio

Cyflwyniad

Ar ôl penderfynu ar ofynion y defnyddiwr, y cam nesaf yw cynhyrchu manyleb ddylunio. Pwrpas y fanyleb ddylunio yw dylunio gwahanol elfennau'r datrysiad. Mae cynhyrchu'r elfennau hyn yn cael ei wneud yn y cam nesaf, y cam gweithredu.

Dylunio adroddiadau/ ymholiadau

Mae'n well dylunio'r allbwn yn gyntaf gan mai'r allbwn sy'n penderfynu pa brosesu a mewnbynnau sydd eu hangen. Os oes angen allbynnu maes, yna mae angen ei fewnbynnu neu ei gyfrifo o feysydd eraill sy'n cael eu mewnbynnu.

Nodyn pwysig

Peidiwch â chynnwys sgrinluniau sy'n dangos y gweithredu yn yr adran hon. Cynhwyswch ddyluniadau'n unig.

Wrth ddylunio adroddiadau, mae angen i chi ddarparu tystiolaeth o:

- adroddiadau o dablau neu o ymholiadau
- penynnau a throedynnau addas
- grwpio data trefnedig
- cyfrifiadau, cyfansymiau neu feysydd ystadegol sy'n dangos y meysydd a'r data'n glir – gellir gwneud cyfrifiadau ar unrhyw feysydd rhifol sydd wedi'u storio yn y tablau.

Wrth ddylunio ymholiadau, mae angen i chi ddarparu tystiolaeth o:

- Ymholiadau (un maes a sawl maes) am resymau/pwrpasau penodol – wrth echdynnu gwybodaeth o'ch cronfa ddata rhaid i chi bob amser ychwanegu cyd-destun drwy nodi'n glir y rheswm dros yr ymholiad.
- Ymholiadau sy'n defnyddio cysylltau perthynol a rhesymeg rhwng tablau am resymau/pwrpasau penodol – defnyddir AC, NEU a NID Boole yn yr ymholiadau a chaiff meysydd o sawl tabl eu cyfuno.

- Ymholiadau sy'n defnyddio paramedrau am resymau/pwrpasau penodol – er enghraifft, gallech ddod o hyd i ddisgyblion mewn dosbarth y mae eu dyddiad geni rhwng dau ddyddiad.
- Atodi, dileu neu ddiweddaru ymholiadau am resymau/ pwrpasau penodol – er enghraifft, gallech ddefnyddio ymholiad i gynyddu'r holl brisiau mewn tabl o gynhyrchion o 5%.

Dylunio mewnbynnau

Wrth ddylunio mewnbynnau, dylech gynhyrchu dogfennaeth ar gyfer y canlynol:

- geiriadur data
- normaleiddio
- dylunio technegau dilysu
- dylunio technegau diogelu cyfrinair
- dylunio ffurflenni mewnbynnu data ar-sgrin.

Geiriadur data

Pwrpas geiriadur data yw egluro wrth bobl eraill (e.e. cyd-ddatblygwyr neu bobl sy'n addasu'r gronfa ddata ar eich ôl) pa dablau, meysydd a phriodweddau maes sydd wedi cael eu defnyddio.

Gallwch ddechrau gyda rhestr o'r meysydd a gaiff eu cynnwys yn y gronfa ddata. Ni ddylid rhoi hyn yn y tablau ar hyn o bryd oherwydd bod normaleiddio (h.y. pan roddir y data yn y tablau) yn cael ei ddechrau ar ôl cwblhau'r geiriadur data.

Enwau maes – mae angen i'r rhain ddisgrifio'r data yn llawn ond ni

Enw'r Maes	Dyma enw'r maes
Math data	Gall data fod yn nod rhifol, dyddiad neu resymegol (Ie/ Na), memo, ac ati.
Fformat	Arian cyfred, nifer o leoedd degol, ffurf safonol (ffordd o storio rhifau mawr iawn neu fach iawn).
Disgrifiad o'r maes	Gwybodaeth am beth yw'r maes a sut mae'n cael ei ddefnyddio.
Hyd y maes	Nifer y nodau sydd ei angen ar gyfer y testun neu nifer y digidau a ddefnyddir ar gyfer rhifau.
Gwiriadau dilysu a gyflawnir ar y maes	Manylion masgiau mewnbwn, gwiriadau amrediad, cyfyngu defnyddwyr i restr o ddata, ac ati.
A oes angen y maes	Rhaid llenwi rhai meysydd â data, nid oes angen llenwi meysydd eraill.

Dyma'r hyn y dylid ei gynnwys yn y geiriadur data

Enw'r Maes	NIFER_AILARCHEBU
Math data	Rhifol
Fformat	0 lle degol (h.y. cyfanrif)
Disgrifiad o'r maes	Nifer yr unedau o eitem o stoc a all gael eu harchebu ar unrhyw un adeg
Hyd y maes	4 digid
Gwiriadau dilysu a gyflawnir ar y maes	>0
A oes angen y maes	Na

Cofnod mewn geiriadur data ar gyfer maes o'r enw NIFER_AILARCHEBU

ddylent fod yn rhy hir gan eu bod ar ben y colofnau ac felly'n pennu eu lled – a hyn fydd yn pennu faint o ddata y gellir eu cael ar y sgrin ar y tro.

Peidiwch â gadael bylchau yn enwau meysydd. Yn lle hynny, defnyddiwch - neu _ i wahanu'r geiriau, e.e. eitem-o-stoc, nifer_mewn_stoc, ac ati.

Math data – mae angen pennu math data ar gyfer pob maes. Mae hwn weithiau'n cael ei osod yn awtomatig gan y meddalwedd ac weithiau mae angen i'r defnyddiwr ei ddewis o restr. Ar ôl gosod math data ar gyfer maes, mae'n bwysig iawn bod gan y maes hwn yr un math data bob tro mae'n digwydd mewn gwahanol dablau. Fel arfer, rhaid ffurfio perthnasoedd rhwng y meysydd hynny sydd â'r un enwau maes a mathau data.

Fformat – dylech gynnwys manylion y fformatau a ddefnyddir ar gyfer y meysydd hynny y mae'r defnyddiwr yn gallu gosod eu fformatau. Er enghraifft, gall maes rhifol fod yn gyfanrif byr, cyfanrif hir, arian cyfred, ac ati; gellir rhoi dyddiadau mewn sawl fformat, e.e. 12 Mehefin 2011, 12/06/11, ac ati.

Disgrifiad o'r maes – gallwch gynnwys disgrifiad llawn o'r maes os nad yw enw'r maes yn hunaneglurhaol.

Hyd y maes – mae hyd y maes yn cael ei gynnwys ar gyfer y meysydd hynny y gall eu hydoedd gael eu pennu.

Gwiriadau dilysu a gyflawnir ar y maes – dylech gynnwys manylion dilysu, fel y defnydd o fasgiau mewnbwn, gwiriadau amrediad, rhestri o ddata y gellir eu caniatáu, ac ati.

A oes rhaid i'r defnyddiwr fewnbynnu data i'r maes bob amser – rhaid i rai meysydd gynnwys data bob amser, ond nid oes angen llenwi meysydd eraill, e.e. rhif_ffôn_symudol.

Dylunio gwiriadau dilysu

Gall gwiriadau dilysu gael eu dylunio yma ar sail y wybodaeth yn adran gwiriadau dilysu y geiriadur data.

Dylech wneud gwiriadau dilysu ar feysydd lle mae hyn yn briodol a dylech gymryd camau i gyfyngu ar y dewis o ddata sydd gan ddefnyddiwr lle mae hynny'n bosibl. Mae'n well defnyddio rhestri parod (rhestri am-edrych) os yw'r defnyddiwr yn gallu dewis nifer cyfyngedig o eitemau yn unig.

Dylunio prosesau

Wrth ddylunio prosesau, mae angen i chi ddarparu tystiolaeth o:

- ddylunio rheolweithiau awtomataidd gan ddefnyddio cod rhaglennu
- dylunio cyfrifiadau sy'n cael eu gwneud mewn adroddiadau neu ffurflenni.

Mae'n debyg y byddwch wedi cynnwys cyfrifiadau yn eich dyluniadau ar gyfer adroddiadau neu ffurflenni, ac felly y cyfan sydd ei angen yma yw disgrifiad o beth mae'r cyfrifiadau'n ei wneud a chyfeiriad at y rhif tudalen lle maent i'w gweld ar y dyluniadau o'r adroddiadau a ffurflenni.

Ar gyfer pob rheolwaith awtomataidd gallwch strwythuro'r gweithredoedd sydd eu hangen fel hyn:

GWEITHDREFN enw macro
Gweithred
Gweithred
Gweithred ac ati
DIWEDDGWEITHDREFN

Dylunio technegau diogelu cyfrinair

Dim ond ychydig o frawddegau sydd eu hangen i amlinellu'r dyluniadau ar gyfer system o gyfrineiriau i atal mynediad heb ei awdurdodi ac i rwystro defnyddwyr rhag newid strwythur y gronfa ddata drwy ddamwain neu fwriad.

Gall algorithmau ar gyfer rheolweithiau unigol sy'n gwella datrysiadau drwy ddefnyddio galluoedd rhaglennu'r pecyn meddalwedd (e.e. rheolwaith cyfrinair) gael eu cynnwys yma.

Dylunio ffurflenni mewnbynnu data ar-sgrin

Cofiwch mai dylunio ydych chi, felly'r nod yma yw darparu brasluniau wedi'u tynnu â llaw neu fersiynau bras wedi'u cynhyrchu â meddalwedd prosesu geiriau. Y cyfan mae'n rhaid ei wneud yw dangos

lleoliad logo'r corff, y pennawd ar gyfer y ffurflen, cyfarwyddiadau i'r defnyddiwr, a safleoedd y meysydd ynghyd ag enw'r tabl lle gellir dod o hyd i'r maes.

Dyma rai pethau y bydd angen eu hystyried:

- Defnyddio is-ffurflenni o fewn ffurflenni os bydd angen mewnbynnu data i fwy nag un tabl.
- Manylion y logo (blwch i nodi'r maint, enw'r ffeil, safle) a manylion graffigau eraill a ddefnyddir.
- Ffont, maint ffont a lliw ffont a ddefnyddir.
- Nodi safle penawdau.
- Penawdau/labeli ar gyfer meysydd (i hysbysu defnyddwyr pa ddata mae'n rhaid eu mewnbynnu).
- Blychau maes i ddangos safle pob maes.
- Botymau gorchymyn i ganiatáu i'r defnyddiwr symud i'r cofnod nesaf neu'r cofnod blaenorol, neu gau'r ffurflen, ac ati.
- Y meysydd sy'n cynnwys cwymprestri (*drop-down lists*) i ganiatáu i'r defnyddiwr ddewis o restr.

Dylunio'r rhyngwyneb defnyddiwr

Mae angen i chi ddarparu tystiolaeth o ddyluniad eich rhyngwyneb yma. Unwaith eto, rhaid cofio mai dyluniad yw hwn, felly ni ddylech ddefnyddio sgrinluniau o ddewislenni a ffurflenni a weithredwyd gennych at y pwrpas hwn. Mae brasluniau wedi'u tynnu â llaw i'r dim ar gyfer hyn.

Dyma rai pethau y dylech eu hystyried yn eich dyluniadau:

- Rhagsgriniau (*splash screens*) – sgriniau deniadol yr olwg yw'r rhain sy'n ymddangos wrth i raglen lwytho.
- Systemau dewislen – dyluniadau o'r holl ddewislenni a ddefnyddir a manylion unrhyw ddewislenni sydd wedi'u cysylltu â'i gilydd a sut.
- Lliwiau a ddefnyddir – ar gyfer y cefndir a'r testun.
- Ffontiau a meintiau ffont – mae angen nodi'r rhain ar eich dyluniad.
- Labeli/penawdau – cyfarwyddiadau i'r defnyddiwr ynghylch beth mae botwm gorchymyn yn ei wneud neu beth mae'n rhaid ei fewnbynnu.
- Botymau gorchymyn – i ganiatáu i'r defnyddiwr symud i'r cofnod nesaf neu'r cofnod blaenorol, neu gau'r ffurflen neu redeg macro.

Gweithredu, profi, dogfennaeth defnyddiwr, gwerthuso

▼ **Byddwch yn dysgu**

▶ Am y tasgau sy'n gysylltiedig â gweithredu'r datrysiad ar gyfer y gronfa ddata berthynol

▶ Am y tasgau sy'n gysylltiedig â rhoi prawf ar y datrysiad

▶ Am y tasgau sy'n gysylltiedig â darparu dogfennaeth defnyddiwr ar gyfer y datrysiad

▶ Am y tasgau sy'n gysylltiedig â gwerthuso'r datrysiad

▶ Am y cynllunio project y mae'n rhaid ei wneud wrth lunio'r datrysiad

Cyflwyniad

Gweithredu yw pan ydych chi'n cymryd eich dyluniad ac yn llunio'r datrysiad gan ddefnyddio'r meddalwedd cronfa ddata. Mae'n bwysig i'r gweithredu gyfateb i'r dyluniad cymaint â phosibl. Ond os ydych wedi gorfod newid y dyluniad (fel rheol oherwydd nad oes gennych y sgiliau i wneud yr hyn yr oeddech am ei wneud yn wreiddiol) mae angen i chi anodi eich dyluniad i ddangos y newidiadau.

Proses sy'n digwydd drwy gydol y gweithredu ac ar ôl llunio'r datrysiad yw profi. Mae'n gwirio pethau fel gwiriadau dilysu, ffurflenni ac adroddiadau, ac ymholiadau, ac yn sicrhau bod y datrysiad yn cwrdd â gofynion y defnyddiwr.

Dogfennaeth defnyddiwr yw'r dogfennau a ddatblygir i gynorthwyo'r defnyddiwr dibrofiad ac mae'n rhan bwysig o'r datrysiad.

Mae'r gwerthusiad yn adolygu'r datrysiad ac yn ystyried pa mor dda mae'r datrysiad wedi cwrdd â gofynion y defnyddiwr. Mae'n nodi'r problemau a gododd wrth ddatblygu'r gronfa ddata a pha mor dda y cawsant eu datrys.

Y tasgau sy'n gysylltiedig â gweithredu'r datrysiad ar gyfer y gronfa ddata berthynol

Dyma'r tasgau y byddai angen i chi ddarparu tystiolaeth ohonynt i ennill y marciau am yr adran hon. Peidiwch â phoeni os nad yw eich trefn chi yr un fath, gan fod y tasgau'n cael eu cwblhau ochr yn ochr weithiau.

- creu tablau a chysylltau
- cynnwys technegau dilysu data
- creu ffurflenni
- cynnwys cyfrifiadau
- creu ffurflenni gydag is-ffurflenni
- creu rhyngwyneb cyfeillgar
- creu macros
- cyflawni ymholiadau un tabl (cofiwch nodi'r cyd-destun, h.y. pam maen nhw'n cael eu cynhyrchu)
- cyflawni ymholiadau sawl tabl gan ddefnyddio cysylltau perthynol
- ymholiadau sy'n defnyddio paramedrau
- atodi, dileu neu ddiweddaru ymholiadau
- adroddiadau
- gwella rhannau unigol o'r datrysiad drwy ddefnyddio galluoedd rhaglennu'r pecyn meddalwedd.

Profi

Rhaid rhoi prawf trylwyr ar y gronfa ddata a gynhyrchwch drwy yn gyntaf baratoi cynllun profi sy'n rhoi prawf ar y canlyniadau disgwyliedig yn erbyn y canlyniadau gwirioneddol. Dylai fod tystiolaeth glir o'r holl ganlyniadau ar ffurf sgrinluniau neu allbrintiau ble bynnag y mae hynny'n briodol.

Yn eich cynllun profi mae angen i chi sicrhau eich bod chi'n profi:

- y rhyngwyneb defnyddiwr a'r holl lwybrau drwy'r system
- yr holl ffurflenni mewnbynnu data â data dilys, annilys ac eithafol
- pob gweithdrefn ddilysu

CYNGOR AR Y PROJECT

Os byddwch yn defnyddio sgrinluniau o'r meddalwedd cronfa ddata i ddarparu diagramau o'r tablau a'u perthnasoedd, sicrhewch fod y tablau'n cael eu hagor allan fel y gall eich athro/safonwr weld yr holl feysydd yn y tablau ac adnabod y prif allweddi a'r allweddi estron.

- pob adroddiad
- pob ymholiad
- systemau diogelwch
- pob rheolwaith unigol ac awtomataidd
- bod pob cyfrifiad yn gywir.

Gallwch drefnu eich cynllun profi fel a ganlyn, gan rifo pob prawf. Wrth i chi gynnwys tystiolaeth o ganlyniadau'r profion, gallwch gyfeirio at rif y prawf perthnasol.

Cynllun profi

Rhif y prawf	Disgrifiad o'r hyn sy'n cael ei brofi	Sut mae'r prawf yn cael ei gynnal	Prawf-ddata a ddefnyddir	Canlyniadau disgwyliedig	Canlyniadau	Camau cywiro a gymerwyd
1	Gwiriadau dilysu ar gyfer meysydd yn y tabl cwsmer	Mewnbynnir prawf-ddata i'r meysydd hynny sydd â gwiriadau dilysu	Set 1 tudalen 20	Dylai gwiriadau dilysu dderbyn neu wrthod y prawf-ddata yn unol â thabl Set 1	Ni weithiodd y maes ar gyfer cod post yn iawn – gweithiodd pob gwiriad dilysu arall	Cafodd y masg mewnbwn ei addasu i dderbyn yr amrediad llawn o fformatau derbyniol ar gyfer cod post
2						
3						
4						
ac ati						

Nodyn pwysig

Er bod yr adran ar brofi'n cael ei chynnwys yma ar ôl yr adran ar weithredu, mae'n well ei wneud yn ystod y gweithredu mewn gwirionedd er mwyn sicrhau bod pethau'n gweithio cyn symud ymlaen i rannau eraill o'r project.

Dogfennaeth defnyddiwr

Pwrpas y ddogfennaeth a gynhyrchir ar gyfer defnyddwyr yw galluogi defnyddwyr i ddeall a defnyddio'r system yn effeithiol. Dylai'r ddogfennaeth defnyddiwr a ddarparwch gynnwys y wybodaeth ganlynol:

- ble i ddod o hyd i'r gronfa ddata (h.y. y cyfeiriaduron lle mae'r data wedi'u storio)
- sut i gychwyn y gronfa ddata
- sut i deipio cyfrineiriau neu weithdrefnau diogelwch eraill
- sut i ddefnyddio'r rhyngwyneb defnyddiwr
- sut i ychwanegu, dileu, golygu, argraffu a chadw'r data mewn cofnodion drwy roi enghreifftiau ar ffurf sgrinluniau o ffurflenni mewnbynnu data
- enghreifftiau o destun dilysu i gefnogi gweithdrefnau dilysu
- cyfarwyddiadau ar ddefnyddio gwahanol fathau o ymholiadau a chynhyrchu adroddiad
- cyfarwyddiadau ar dechnegau adfer o drychineb.

© 2001 gan Randy Glasbergen
www.glasbergen.com

"Fe gymerodd bum diwrnod i ni feddwl sut i orffen ein project ddau ddiwrnod yn gynnar. Dyna pam yr ydyn ni dri diwrnod yn hwyr."

Defnyddiwch sgrinluniau i ddangos eich dogfennaeth defnyddiwr a sicrhewch ei bod hi'n cael ei hysgrifennu gyda sgiliau'r defnyddiwr cyffredin mewn golwg.

Gwerthuso

Mae gwerthuso'n golygu asesu'ch datrysiad yn feirniadol yn erbyn:

- Gofynion y defnyddiwr – yma mae angen i chi drafod pa mor agos y mae eich datrysiad yn cyfateb i ofynion y defnyddiwr. Gellir gwneud hyn orau mewn tabl ynghyd â disgrifiad byr.
- Problemau y daethoch ar eu traws a'r strategaethau a ddefnyddiwyd gennych i'w datrys – yma mae angen i chi nodi sut yr aethoch ati i ateb unrhyw broblemau. Er enghraifft, a wnaethoch chi eu datrys drwy ddarllen llyfrau, defnyddio cymorth ar-lein, ac ati, neu a fu'n rhaid i chi symleiddio'r system rhyw ychydig i oresgyn y broblem? Rhaid rhoi esboniad llawn yma.

Dylech osgoi gwneud sylwadau fel hyn:

> 'Roeddwn i'n meddwl ei fod yn gweithio'n dda iawn ac fe gefais i lawer o bleser o wneud y project.'

Os ydych wedi defnyddio defnyddiwr terfynol (eich athro/athrawes, ffrind neu berthynas, rhywun yr ydych chi'n ei adnabod gyda busnes neu gyda phroblem i'w datrys), gallwch roi holiadur i'r defnyddiwr ei lenwi. Os defnyddiwch holiadur, cynhwyswch ef gyda'ch tystiolaeth.

Mae angen i chi reoli amser yn ofalus yn ystod eich project

Cynllunio'r project

Nid yw'r adran hon wedi cael ei rhoi'n olaf oherwydd ei bod hi'n cael ei gwneud yn olaf. Dylech fod yn gwneud eich gwaith cynllunio drwy gydol y project. Drwy gydol y project bydd angen i chi ddangos eich bod chi wedi:

- defnyddio enwau addas ar gyfer y gronfa ddata, tablau, ffurflenni, ymholiadau, adroddiadau a rheolweithiau awtomataidd fel macros
- cadw eich gwaith yn rheolaidd
- cadw copïau wedi'u dyddio o ffeiliau ar gyfryngau symudadwy eraill a'u cadw mewn lleoliad arall neu mewn storfa ar-lein
- gweithio yn ôl cynllun amser – dylech hefyd gynnwys tystiolaeth o reoli amser yn llwyddiannus (siartiau Gantt, amserlenni, dyddiaduron, ac ati)
- diogelu cyfrinachedd ac wedi cydymffurfio â'r deddfau hawlfraint.

Siartiau Gantt

Siartiau bar llorweddol yw siartiau Gantt. Maent yn cael eu defnyddio i ddangos pryd mae'r tasgau sy'n rhan o waith cyfan yn dechrau ac yn gorffen. Mae ganddynt raddfa amser sy'n mynd ar draws y dudalen a rhestr o weithgareddau i'w gwneud sy'n rhedeg i lawr y dudalen. Mae'r blociau sy'n dangos hyd y gweithgareddau wedi'u tywyllu i ddangos yr amser a gymerir i wneud pob tasg. Maent yn cael eu defnyddio fel arf cynllunio, gan ei bod hi'n bosibl gosod pren mesur ar y siart i ddarganfod pa dasgau sy'n hwyr a pha rai sydd o flaen eu hamser.

Mae'n arferol llunio siartiau Gantt ar bapur sgwariau er mwyn gallu darllen yr amserau'n fanwl gywir. Mae pren mesur tryloyw wedi'i osod ar yr amser presennol yn help i weld pa weithgareddau sydd ar ei hôl hi. Mae'r rhan fwyaf o feddalwedd rheoli projectau yn defnyddio siartiau Gantt.

CYNGOR AR Y PROJECT

Mae'n syniad da i chi fod yn gyson wrth roi enwau i wrthrychau (e.e. tablau, ffurflenni, ymholiadau ac adroddiadau), felly gallai fod gennych enwau fel y rhain ymhAbsenoldebDisgyblionYnPeriPryder ffurfRhentolionCwsmeriaid adrGwerthiantMisolTrefnedig, ac ati.

Tasgau		WYTHNOSAU												
		1	2	3	4	5	6	7	8	9	10	11	12	13
1	Gofynion y defnyddiwr	■	■											
2	Dylunio			■	■									
3	Gweithredu					■	■	■	■					
4	Profi									■	■			
5	Dogfennaeth defnyddiwr											■	■	
6	Gwerthuso													■

161

Cyngor defnyddiol ar gyfer y project a sut y caiff y project ei asesu

Cyflwyniad

Mae'r gwaith project yn gydran bwysig yn TGCh Safon Uwch, felly mae'n werth rhoi amser o'r neilltu i ystyried rhai cynghorion buddiol. Cewch wybod hefyd sut mae'r project yn cael ei asesu.

Pethau y dylech fod yn ymwybodol ohonynt

Mae nifer o bethau y dylech fod yn ymwybodol ohonynt wrth weithio ar eich project:

Llên-ladrad (h.y. copïo) – Rhaid mai chi yn unig fydd wedi gwneud yr holl waith project y byddwch yn ei gyflwyno. Does dim pwynt copïo rhannau o enghreifftiau mewn gwerslyfrau neu rannau o ddeunyddiau project enghreifftiol y bwrdd arholi gan ei bod hi'n debyg y bydd eich athro/athrawes wedi eu gweld o'r blaen, ac mae'n sicr y bydd y safonwr yn gyfarwydd â nhw. Peidiwch â phrynu projectau Safon Uwch o safleoedd arwerthu ar y Rhyngrwyd fel e-Bay.

Mewn achosion lle gellir profi bod copïo wedi digwydd, gall y myfyriwr gael ei wahardd rhag sefyll yr arholiadau TGCh ac, o bosibl, pob pwnc arall mae'n eu cymryd. Bydd yn rhaid i chi a'ch athro lofnodi datganiad i gadarnhau mai eich gwaith chi yw'r gwaith yr ydych wedi'i gyflwyno a dyma pam mae angen i'ch athro/darlithydd eich gweld chi'n cwblhau'r rhan fwyaf o'ch gwaith yn y dosbarth.

Gwnewch gopïau wrth gefn yn rheolaidd – Fel myfyrwyr TGCh Safon Uwch, dylech fod yn ymwybodol o beryglon cadw un copi'n unig o'ch gwaith. Dylech wneud copïau wrth gefn yn rheolaidd ac ni ddylech gadw'r rhain yn agos at y gwreiddiol. Hefyd mae'n ddoeth argraffu'ch gwaith o bryd i'w gilydd a chadw'r allbrint.

Amserwch eich hun – Mae tuedd ymhlith myfyrwyr TGCh i fynd ati i ddatblygu'r system a gadael yr holl ddogfennaeth hyd y diwedd, pan fydd ganddynt system sy'n gweithio (gobeithio). Ond mae'n bwysig darparu'r ddogfennaeth wrth i chi fynd yn eich blaen. Os oes angen i chi wneud unrhyw newidiadau wrth i chi fynd yn eich blaen, mae hyn yn iawn os eglurwch pam yr ydych wedi'u gwneud. Mae rheoli amser yn sgìl pwysig drwy gydol eich bywyd a dylai fod gennych ryw syniad o'r hyn y gobeithiwch ei gyflawni yn y sesiynau ymarferol yn yr ysgol neu'r coleg.

Dogfennwch bopeth – Er y bydd yr athrawon/darlithwyr sy'n marcio'ch gwaith yn gyfarwydd â'r meddalwedd yr ydych wedi'i ddefnyddio, mae'n bosibl na fydd y safonwyr wedi'i ddefnyddio o'r blaen neu na fydd ganddynt gopi ohoni. Mae hyn yn golygu na fydd y safonwyr yn gallu marcio eich projectau ond ar sail y ddogfennaeth yr ydych chi wedi'i darparu. Os ydych, er enghraifft, wedi cynnwys system cymorth ar-lein arloesol fel rhan o broject cronfa ddata, ni fyddai'r safonwr yn gwybod amdani oni bai eich bod chi wedi darparu tystiolaeth ar ffurf dyluniadau sgrin ac argraffiadau sgrin. Os na fyddwch wedi cynnwys rhan/rhannau o'r ddogfennaeth, fe gollwch farciau, ni waeth pa mor glyfar yw'r datrysiad TGCh.

Cwestiynau cyffredin

Dyma rai atebion i gwestiynau a ofynnir yn aml gan fyfyrwyr am waith project.

Pwy fydd yn marcio fy ngwaith project?

Bydd eich gwaith project yn cael ei farcio gan eich athro/darlithydd ac yna bydd safonwyr yn cadarnhau bod y gwaith wedi cael ei farcio i safon sy'n cymharu ag ysgolion/colegau eraill. Unigolion sydd ag arbenigedd mewn TGCh fydd y safonwyr

fel rheol (athrawon neu ddarlithwyr eraill) ond ni fydd ganddynt arbenigedd o anghenraid yn y meddalwedd cronfa ddata yr ydych chi wedi'i ddefnyddio. Dylech nodi bod llawer o becynnau meddalwedd eraill y gellir eu defnyddio i greu cronfeydd data, heblaw am Microsoft Access. Felly bydd yn rhaid i chi egluro a dogfennu'ch gwaith yn ofalus fel y gallant weld yn glir beth yr ydych wedi'i wneud. Bydd hyn yn eich helpu i gael y marciau mwyaf posibl am eich ymdrechion. Er i chi wneud project penigamp, os na ddarparwch ddogfennaeth a thystiolaeth briodol, ni chewch ond marc isel amdano.

Pa feddalwedd y gallaf ei ddefnyddio?

Y peth gorau i'w wneud yw defnyddio'r meddalwedd cronfa ddata yr ydych chi'n fwyaf cyfarwydd ag ef. Cewch ddefnyddio unrhyw feddalwedd cronfa ddata, ond rhaid iddo fod yn feddalwedd a ddefnyddir gan fusnesau. Mae'r rhan fwyaf o becynnau integredig yn addas, ond ni fyddai'r rheiny sy'n cynnwys cronfa ddata nad yw'n berthynol (h.y. cronfa ddata ffeiliau fflat), megis Works, yn cael eu hystyried yn briodol. Dim ond meddalwedd cronfa ddata berthynol arbenigol y dylech ei ddefnyddio ar gyfer y project hwn.

Cewch ddefnyddio Microsoft Access ar gyfer y project ond mae llawer o becynnau cronfa ddata cystal os nad gwell eraill ar gael, a bydd eich athro/ darlithydd yn gallu dweud wrthych ba un i'w ddefnyddio.

Faint o gymorth y gall fy athro/ darlithydd ei roi i mi?

Dylai eich athro/darlithydd fod wedi rhoi digon o gyfle i chi ymarfer rhai o nodweddion uwch pecynnau cronfa ddata fel eich bod chi'n gwybod pa rai fydd yn eich helpu i ddatrys problem benodol. Caniateir i'ch athro ddefnyddio'r project yn gyfrwng ar gyfer addysgu, felly fe fyddwch chi'n dysgu sut i ddefnyddio'r meddalwedd, ond bydd yr athro'n cyflwyno'r meddalwedd yn nhermau cyffredinol, felly bydd yn rhaid i chi addasu'r hyn a ddysgwch i'ch datrysiad eich hun. Os oes rhaid i'ch athro roi cymorth uniongyrchol i chi oherwydd eich bod chi'n mynd ar gyfeiliorn, bydd yn gorfod cofnodi hyn ar y daflen asesu ac mae'n bosibl y cewch farciau is am eich project o ganlyniad.

A alla i wneud y gwaith gartref?

Rhaid i'ch athro allu goruchwylio'ch gwaith er mwyn sicrhau mai chi sydd wedi'i wneud. Er hynny fe gewch weithio ar y project gartref, ond rhaid i'r athro weld digon o'r gwaith yn yr ysgol neu goleg i fod yn fodlon mai eich gwaith chi ydyw.

A alla i wneud gwaith rhaglennu?

Cewch wneud peth rhaglennu fel rhan o'r project, ond yr unig raglennu y bydd y mwyafrif o fyfyrwyr yn ei wneud fydd cynhyrchu macro, neu gyfarwyddiadau Iaith Ymholiadau Strwythuredig (*SQL: Structured Query Language*) wrth echdynnu data.

Cewch wneud ychydig o waith rhaglennu os oes angen i chi

Cyflwyno'r projectau

Wrth gwrs mae'n rhaid gairbrosesu holl ddogfennaeth y project a dylai diagramau (argraffiadau sgrin fel rheol) gael eu cynnwys yn y dogfennau. Er y bydd llawer ohonoch yn gallu cynhyrchu gwaith gorffenedig 'sgleiniog' o ansawdd uchel, os nad oes llawer o dystiolaeth o ddadansoddi, dylunio a phrofi ni chewch farciau uchel. Mewn geiriau eraill, ni fydd yr arholwr yn edrych ar gyflwyniad y project yn unig.

Pa bwyntiau cyffredinol am ddogfennaeth y dylwn eu cadw mewn cof?

Dylai fod gan bob project ddalen glawr wedi'i llofnodi a dalen broject sy'n cynnwys enw'r Ganolfan, eich enw chi a theitl y project. Sicrhewch fod eich dogfennaeth yn cynnwys y canlynol: tudalen gynnwys, penynnau (sy'n addas i'ch project) a throedynnau yn nodi'r rhifau tudalen y mae'r cynnwys yn cyfeirio atynt.

Beth yw'r ffordd orau o gadw tudalennau fy mhroject?

Rhaid rhwymo projectau mewn ffordd sy'n lleihau eu maint ac yn ei gwneud hi'n hawdd i'r athro farcio'r deunydd. Wrth rwymo eich project, ni ddylech ddefnyddio'r dulliau canlynol:

- Ffeiliau modrwy neu ffeiliau bwa lifer, gan fod hyn yn gwneud y deunydd yn llawer rhy swmpus

ac yn achosi problemau storio a chludo i'r athro a'r safonwr.

- Rhwymwyr llithro (*slide binders*) sy'n ei gwneud hi'n anodd i bobl ddarllen y testun neu ddiagramau ac sy'n tueddu i ddod yn rhydd yn y post neu wrth gael eu darllen.

Peidiwch â rhoi eich deunydd mewn amlenni plastig na ffolderi cyflwyno plastig. Efallai y bydd eich athro am roi marciau neu sylwadau ar eich gwaith ac mae tynnu tudalennau o'r amlenni plastig yn cymryd amser. Hefyd mae defnyddio amlenni plastig yn gwneud y project yn drymach nag sydd angen. Mae'n well defnyddio un o'r dulliau canlynol:

- ffolder denau
- tagiau drwy dyllau wedi'u pwnsio.

A fydda i'n cael marciau am sillafu, atalnodi a gramadeg?

Er mwyn i chi gael y marciau am bob elfen o'r project, rhaid i chi gyfleu'ch ystyr yn glir. Felly rhaid i chi:

- ddefnyddio'r gwirydd sillafu
- cadw at yr holl reolau atalnodi
- prawfddarllen popeth yr ydych wedi'i ysgrifennu sawl gwaith cyn ei roi i'ch athro
- defnyddio'r gwirydd gramadeg
- gofyn i bobl eraill edrych dros eich gwaith i sicrhau ei fod yn gwneud synnwyr.

Peidiwch â rhoi eich gwaith project mewn ffeiliau bwa lifer fel hwn na ffeiliau modrwy gan eu bod nhw'n cymryd gormod o le. Cofiwch am yr athro/safonwr druan sy'n gorfod cario eich projectau i gyfarfodydd

Cyngor defnyddiol ar gyfer y project a sut y caiff y project ei asesu (parhad)

Sut y caiff y project ei asesu?

Er mwyn cael y marciau mwyaf, mae angen i chi ymdrin â chymaint â phosibl o'r meini prawf a roddir yn y tabl canlynol. Bydd eich athro a'r safonwr yn defnyddio'r meini prawf hyn i farcio gwahanol adrannau ac elfennau eich project. Dylech weithio drwy'r rhestr a chadarnhau bod gennych adrannau yn eich project sy'n cyfateb i'r adrannau ac elfennau yn y tabl. Hefyd gallwch ddefnyddio'r marciau sydd wedi'u dyrannu i'r elfennau i benderfynu faint o dystiolaeth y dylech ei chynhyrchu ar gyfer pob elfen.

Cynllun asesu manwl

Elfennau	Meini prawf	Marc
Gofynion y defnyddiwr		
Cefndir	Dangos dealltwriaeth glir o'r cefndir i'r broblem	2
Canlyniadau disgwyliedig / nodau ac amcanion	Datganiad clir o nodau ac amcanion y system, ynghyd â'r canlyniadau disgwyliedig ac arddull tŷ ac ethos y corff	6
Gofynion rhyngwyneb defnyddiwr	Manylion gofynion rhyngwyneb defnyddiwr penodol ar gyfer y system	1
Caledwedd	Manylion y gofynion caledwedd lleiaf ar gyfer y system	1
	Diagram Perthynas Endidau	2
Dylunio		
Dylunio mewnbynnau	Geiriadur data Normaleiddio Dylunio technegau dilysu Dylunio techneg diogelu cyfrineiriau Dylunio ffurflenni mewnbynnu data ar-sgrin	4 2 2 1 3
Dylunio rhyngwyneb defnyddiwr	Dylunio rhyngwyneb cyfeillgar, dewisiad, pen-blaen Dylunio ymholiadau (gan gynnwys pwrpas a strwythur)	1 6
Dylunio allbynnau	Dylunio adroddiad	2
Dylunio prosesau	Rheolweithiau awtomataidd gan ddefnyddio cod rhaglennu Dylunio cyfrifiadau mewn adroddiadau neu ffurflenni	2 1
Gweithredu	Creu tablau a chysylltau Technegau dilysu data Maes wedi'i gyfrifo Creu ffurflenni Creu ffurflenni gydag is-ffurflenni Creu rhyngwyneb cyfeillgar Macros Ymholiadau un tabl Ymholiadau sawl tabl Ymholiadau sawl tabl sy'n defnyddio cysylltau perthynol Ymholiadau paramedr Atodi, dileu neu ddiweddaru ymholiadau Adroddiadau Gwella rhannau unigol o'r datrysiad drwy ddefnyddio galluoedd rhaglennu'r pecyn meddalwedd	4 2 1 2 1 1 2 2 1 1 1 1 4 2
Profi ar sail cynllun profi	Rhoi prawf ar y rhyngwyneb defnyddiwr ac ar bob llwybr drwy'r system Profi gyda data dilys a data eithafol Rhoi prawf ar bob gweithdrefn ddilysu gyda data annilys Rhoi prawf ar adroddiad Rhoi prawf ar bob ymholiad Rhoi prawf ar systemau diogelwch Rhoi prawf ar bob rheolwaith unigol ac awtomataidd Profi bod pob cyfrifiad yn gywir	1 2 2 1 6 1 2 1
Dogfennaeth defnyddiwr	Manylion ble i ddod o hyd i'r gronfa ddata (cyfeiriaduron) a sut i gychwyn y gronfa ddata Manylion sut i deipio cyfrineiriau neu weithdrefnau diogelwch eraill Manylion sut i ddefnyddio'r rhyngwyneb defnyddiwr Manylion sut i ychwanegu, dileu, golygu, argraffu a chadw data mewn cofnodion drwy roi enghreifftiau ar ffurf sgrinluniau o ffurflenni mewnbynnu data Enghreifftiau o destun dilysu i gefnogi gweithdrefnau dilysu Cyfarwyddiadau ar ddefnyddio gwahanol fathau o ymholiadau a chynhyrchu adroddiad Cyfarwyddiadau ar dechnegau adfer o drychineb	1 1 1 5 2 4 1
Gwerthuso	Yn erbyn gofynion y defnyddiwr Problemau y daethpwyd ar eu traws a'r strategaethau a ddefnyddiwyd i'w datrys	4 4

Geirfa

Adnabod llais (*Voice recognition*) Mae systemau adnabod llais yn caniatáu i chi fewnbynnu data drwy ficroffon yn uniongyrchol i gyfrifiadur.

Adnabod marciau gweledol (AMG/*OMR: Optical mark recognition*) Dull mewnbynnu sy'n defnyddio ffurflenni neu gardiau papur gyda marciau arnynt sy'n cael eu darllen yn awtomatig gan ddyfais o'r enw darllenydd marciau gweledol.

Adnabod nodau gweledol (ANG/*OCR: Optical character recognition*) Dull mewnbynnu sy'n defnyddio sganiwr yn ddyfais fewnbynnu ynghyd â meddalwedd arbennig sy'n edrych ar siâp pob llythyren fel y gellir ei hadnabod ar wahân.

Adnabod nodau inc magnetig (ANIM/*MICR: Magnetic ink character recognition*) Dull mewnbynnu sy'n defnyddio rhifau wedi'u hargraffu ar ddogfen, fel siec gydag inc magnetig arbennig arni, a all gael eu darllen gan ddarllenydd nodau inc magnetig ar gyflymder uchel iawn.

Allbwn (*Output*) Y canlyniadau o brosesu data.

Allgofnodi (*Log-out*) Hysbysu'r rhwydwaith eich bod am gau mynediad i'r cyfleusterau rhwydwaith tan y mewngofnodi nesaf.

Allrwyd (*Extranet*) Rhwydwaith allanol a all gael ei ddefnyddio gan gwsmeriaid, cyflenwyr a phartneriaid corff yn ogystal â'r corff ei hun.

Am-edrychiadau ffeil/tabl (*File/Table lookups*) Defnyddir y rhain i sicrhau bod y codau sy'n cael eu defnyddio yr un fath â'r rheiny mewn ffeil neu dabl o godau.

Amgodio (*Encoding*) Y broses o roi data neu wybodaeth (e.e. testun, rhifau, symbolau, delweddau, sain a fideo) mewn fformat penodol fel y gall system TGCh eu trawsyrru neu eu storio'n effeithiol.

Amgryptio (*Encryption*) Codio data wrth iddynt gael eu hanfon dros rwydwaith fel mai'r unig berson sy'n gallu eu darllen yw'r sawl yr anfonwyd y data ato/ati. Pe byddai'r data yn cael eu rhyng-gipio gan haciwr, byddai mewn cod ac yn hollol ddiystyr.

Amlgyfrwng (*Multimedia*) Dull o gyfathrebu sy'n cyfuno mwy nag un cyfrwng at ddibenion cyflwyno, fel sain, graffeg a fideo.

Anaf straen ailadroddus (*RSI: Repetitive strain injury*) Cyflwr poenus sy'n effeithio ar y cyhyrau. Mae'n cael ei achosi pan mae rhai cyhyrau yn cael eu defnyddio'n rheolaidd yn yr un ffordd.

Arae ddiangen o ddisgiau rhad (*RAID: Redundant array of inexpensive disks*) System a ddefnyddir gan rwydweithiau i gadw copïau wrth gefn.

Archwilio (*Audit*) Cadw cofnod o bwy sydd wedi gwneud beth ar y rhwydwaith er mwyn darganfod achosion o gamddefnyddio'r system gan ddefnyddwyr awdurdodedig ac achosion o fynediad heb awdurdod gan hacwyr.

Argraffydd chwistrell (*Ink-jet printer*) Argraffydd sy'n gweithio drwy chwistrellu inc drwy dyllau ac ar y papur.

Argraffydd laser (*Laser printer*) Argraffydd sy'n defnyddio paladr laser i ffurfio nodau ar y papur.

Arlliwydd (*Toner*) Gronynnau plastig du a ddefnyddir gan argraffyddion laser fel 'inc'.

ASCII gweler Cod Safonol Americanaidd ar gyfer Ymgyfnewid Gwybodaeth.

Ateb (*Reply*) Mae'n caniatáu i chi ddarllen e-bost ac yna ysgrifennu'r ateb heb orfod mynd at gyfeiriad e-bost y sawl sydd wedi anfon yr e-bost gwreiddiol.

Atodi (*Append*) Gall defnyddwyr ychwanegu cofnodion newydd ond ni fyddant yn gallu newid na dileu'r cofnodion presennol.

Atodiadau ffeil (*File attachments*) Ffeiliau sy'n cael eu trosglwyddo gydag e-bost.

Band llydan (*Broadband*) Cysylltiad cyflym â'r Rhyngrwyd nad yw'n defnyddio modem.

Bar tasgau (*Taskbar*) Mae'n dangos y rhaglenni sydd ar agor. Mae'r cyfleuster hwn yn ddefnyddiol wrth weithio ar sawl rhaglen ar yr un pryd.

Bwrdd gwaith (*Desktop*) Man gwaith y rhyngwyneb defnyddiwr graffigol. Dyma lle mae'r holl eiconau.

Bws (*Bus*) Math o dopoleg rhwydwaith lle mae'r holl gyfrifiaduron wedi'u cysylltu â chebl cyffredin o'r enw bws.

Bygythiad allanol (*External threat*) Bygythiad i system TGCh sy'n dod o'r tu allan i'r corff.

Bygythiad mewnol (*Internal threat*) Bygythiad i system TGCh sy'n dod o'r tu mewn i'r corff.

CAD gweler Cynllunio drwy gymorth cyfrifiadur.

Caledwedd (*Hardware*) Cydrannau corfforol system gyfrifiadurol.

Camymarfer (*Malpractice*) Defnydd amhriodol neu ddiofal o rywbeth neu gamymddwyn.

Cerdyn cof (*Memory card*) Y cerdyn tenau a ddefnyddir mewn camerâu digidol i storio ffotograffau. Gellir defnyddio cardiau cof i storio data eraill hefyd.

Cilobeit neu 1024 beit (*Kilobyte or 1024 bytes*) Wedi'i dalfyrru'n KB weithiau. Mesur o gynhwysedd storio disgiau a chof.

Cipio data (*Data capture*) Term ar gyfer y gwahanol ddulliau o fewnbynnu data i'ch cyfrifiadur fel y gellir eu prosesu.

Cleient/gweinydd (*Client server*) Rhwydwaith lle mae nifer o gyfrifiaduron wedi'u cysylltu ag un neu ragor o weinyddion.

Cod deuaidd (*Binary code*) Cod wedi'i wneud o gyfres o ddigidau deuaidd – 0 neu 1.

Cod Safonol Americanaidd ar gyfer Ymgyfnewid Gwybodaeth (*ASCII: American Standard Code for Information Interchange*) Cod ar gyfer cynrychioli nodau'n ddeuaidd.

Codio (*Coding*) Cynhyrchu fersiynau byrrach o'r data i hwyluso teipio data i mewn a dilysu data.

Cof anghyfnewidiol (*Non-volatile memory*) Cof wedi'i storio ar sglodyn nad yw'n colli data pan gaiff y trydan ei ddiffodd.

Cof cyfnewidiol (*Volatile memory*) Cof sy'n colli data pan gaiff y trydan ei ddiffodd.

Cof darllen yn unig (*ROM: Read only memory*) Cof wedi'i storio ar sglodyn nad yw'n colli data pan ddiffoddwch y trydan.

Cof hapgyrch (*RAM: Random access memory*) Defnyddir *RAM* i gadw data dros dro tra bo'r cyfrifiadur yn gweithio arnynt. Caiff y cynnwys ei golli pan ddiffoddwch y cyfrifiadur.

Cof pin (*Pen drive*) Cyfrwng storio poblogaidd sy'n cynnig cynwyseddau storio rhad a mawr ac sy'n ddelfrydol ar gyfer storio ffotograffau, cerddoriaeth, a ffeiliau data eraill. Maen nhw ar ffurf bwrdd cylched brintiedig mewn cas plastig.

Comisiynydd Gwybodaeth (*Information Commissioner*) Y person sy'n gyfrifol am weithredu'r Ddeddf Gwarchod Data. Mae hefyd yn hybu arfer da ac yn sicrhau bod pawb yn gwybod beth yw goblygiadau'r Ddeddf.

Copi caled (*Hard copy*) Allbwn wedi'i argraffu o gyfrifiadur y gellir mynd ag ef i ffwrdd a'i astudio.

Copi wrth gefn (*Back up*) Rhaid gwneud copïau wrth gefn o feddalwedd a data fel y gellir adfer y data os bydd y system TGCh gyfan yn cael ei dinistrio.

CPU *gweler* Uned brosesu ganolog.

Cwci (*Cookie*) Ffeil destun fach a lwythir i lawr i'ch cyfrifiadur ac a ddefnyddir gan wefannau i gasglu gwybodaeth am y ffordd y defnyddiwch y wefan.

Cydraniad (*Resolution*) Eglurder delwedd.

Cyfanswm stwnsh (*Hash total*) Cyfanswm diystyr o rifau a ddefnyddir i wirio bod yr holl rifau wedi'u mewnbynnu i'r cyfrifiadur.

Cyflenwad trydan annhoradwy (*UPS: Uninterruptible power supply*) Cyflenwad trydan wrth gefn (generadur a batri) a fydd yn cadw'r cyfrifiadur i redeg os bydd y prif gyflenwad trydan yn methu.

Cyfradd trawsyrru (*Transmission rate*) Cyflymder llif data mewn didau yr eiliad drwy gyfrwng trawsyrru.

Cyfrifiadura gwasgaredig (Distributed computing) Rhwydweithio cyfres o gyfrifiaduron i weithio ar yr un broblem. Mae pob cyfrifiadur yn rhannu data, prosesu, lle storio a lled band i ddatrys y broblem.

Cyfrinair (*Password*) Cyfres o nodau y mae angen eu teipio cyn y bydd mynediad i system TGCh yn cael ei ganiatáu.

Cyfrwng mewnbwn (*Input media*) Y defnydd y caiff y data eu hamgodio arno fel y gellir eu darllen gan ddyfais fewnbynnu a'u digido er mwyn eu mewnbynnu, eu prosesu, a'u troi'n wybodaeth gan y system TGCh.

Cyfrwng trawsyrru (*Transmission medium*) Y defnydd sy'n ffurfio'r cysylltiad rhwng y cyfrifiaduron mewn rhwydwaith (h.y. aer yn achos diwifr, gwifren fetel, ffibr optegol).

Cyfryngau magnetig (*Magnetic media*) Cyfryngau fel tâp a disg lle mae'r data wedi'u storio fel patrwm magnetig.

Cylch Math o dopoleg rhwydwaith lle mae'r holl gyfrifiaduron wedi'u trefnu mewn cylch a lle mae ganddynt yr un statws.

Cymar wrth gymar (*Peer-to-peer*) Trefniant rhwydwaith lle mae gan bob cyfrifiadur statws cyfartal.

Cynhwysedd storio (*Storage capacity*) Faint o ddata y gall y ddyfais neu gyfrwng storio eu dal? Wedi'i fesur mewn MB neu GB fel rheol.

Cynllun profi (*Test plan*) Y dull a ddefnyddir i roi prawf ar system gyfan. Mae'n cynnwys cyfres o brofion.

Cynllunio drwy gymorth cyfrifiadur (*CAD: Computer-aided design*) Dull o ddefnyddio'r cyfrifiadur i gynhyrchu lluniadau technegol.

Cywasgu ffeiliau (*File compression*) Caiff ffeiliau eu cywasgu'n aml cyn eu storio neu eu hanfon dros rwydwaith.

Dadansoddi (*Analysis*) Torri problem i lawr fel ei bod yn haws ei deall a'i datrys.

Dadosodwr (*Uninstaller*) Meddalwedd sy'n cael ei ddefnyddio i gael gwared â'r holl ffeiliau a roddwyd ar gyfrifiadur pan gafodd darn o feddalwedd ei osod.

Darllen proflenni (*Proof reading*) Darllen yn ofalus yr hyn sydd wedi cael ei deipio a'i gymharu â'r hyn sydd ar y ffynhonnell data (ffurflenni archebu, ffurflenni cais, anfonebau, ac ati) i ddarganfod unrhyw gamgymeriadau, a all wedyn gael eu cywiro.

Darllen yn unig (*Read only*) Y cyfan y gall defnyddwyr ei wneud yw darllen cynnwys y ffeil. Ni allant newid na dileu'r data.

Darllen/ysgrifennu (*Read/write*) Gall defnyddwyr ddarllen y data sy'n cael eu cadw mewn ffeil a gallant newid y data.

Darllenydd marciau gweledol (*OMR: Optical mark recognition*) Darllenydd sy'n canfod marciau ar ddalen o bapur. Mae ardaloedd sydd wedi'u tywyllu yn cael eu canfod a gall y cyfrifiadur ddeall y wybodaeth sydd wedi'i chynnwys ynddynt.

Darllenydd stribed magnetig (*Magnetic strip reader*) Dyfais galedwedd sy'n darllen y data mewn stribed magnetig, fel y stribedi ar gefn cardiau credyd.

Darparwr Gwasanaeth Rhyngrwyd (*ISP: Internet service provider*) Y corff sy'n darparu eich cysylltiad Rhyngrwyd.

Data (*Data*) Ffeithiau a ffigurau crai neu set o werthoedd, mesuriadau neu gofnodion o drafodion.

Data gwallus (*Erroneous data*) Data sy'n chwerthinllyd neu'n hollol anaddas.

Data normal (*Normal data*) Data a ddylai fod yn dderbyniol.

Data personol (*Personal data*) Data am unigolyn byw sy'n benodol i'r unigolyn hwnnw.

Deddf Camddefnyddio Cyfrifiaduron 1990 (*Computer Misuse Act 1990*) Y Ddeddf sy'n gwneud nifer o weithgareddau'n anghyfreithlon, e.e. cyflwyno firysau'n fwriadol, hacio, defnyddio cyfarpar TGCh i gyflawni twyll, ac ati.

Deddf Gwarchod Data 1998 (*Data Protection Act 1998*) Deddf sy'n gwarchod yr unigolyn rhag camddefnyddio data.

Deddf Hawlfraint, Dyluniadau a Phatentau 1988 (*Copyright, Designs and Patents Act 1988*) Deddf sydd, ymhlith pethau eraill, yn ei gwneud hi'n drosedd i gopïo neu ddwyn meddalwedd.

Deddf Iechyd a Diogelwch yn y Gwaith 1974 (*Health and Safety at Work Act 1974*) Deddf sy'n sicrhau amodau gwaith a dulliau diogel i weithwyr.

Deddf Rhyddid Gwybodaeth 2000 (*Freedom of Information Act 2000*) Deddf sy'n rhoi'r hawl i bobl weld gwybodaeth sy'n cael ei chadw gan awdurdodau cyhoeddus.

Dewislenni (*Menus*) Maen nhw'n caniatáu i ddefnyddwyr wneud dewisiadau o restr.

Did (*Bit*) Digid deuaidd 0 neu 1.

Diogelwch (*Security*) Sicrhau bod y caledwedd, meddalwedd a data mewn system TGCh yn cael eu diogelu rhag difrod.

Dotiau y fodfedd (*Dpi: dots per inch*) Mesur o gydraniad delweddau. Y mwyaf o ddotiau y fodfedd sydd mewn delwedd, yr uchaf yw'r cydraniad.

Dwyn/twyll hunaniaeth (*Identity theft/fraud*) Dwyn eich manylion bancio/cerdyn credyd/personol i gyflawni twyll.

Dyfais fewnbynnu (*Input device*) Y ddyfais galedwedd a ddefnyddir i borthi'r data mewnbwn i system TGCh fel bysellfwrdd neu sganiwr.

Eiconau (*Icons*) Darluniau bach a ddefnyddir i gynrychioli gorchmynion, ffeiliau neu ffenestri.

Enw defnyddiwr (*Username*) Ffordd o adnabod pwy sy'n defnyddio'r system TGCh er mwyn dyrannu adnoddau rhwydwaith.

Ergonomeg (*Ergonomics*) Gwyddor gymhwysol yn ymwneud â dylunio a threfnu pethau mae pobl yn eu defnyddio fel bod y bobl a'r pethau yn rhyngweithio'n fwy effeithlon a diogel.

Fideo-gynadledda (*Videoconferencing*) System TGCh sy'n caniatáu i gyfarfodydd wyneb yn wyneb gael eu cynnal heb i'r rheiny sy'n cymryd rhan orfod bod yn yr un ystafell neu hyd yn oed yr un ardal ddaearyddol.

Firws (*Virus*) Rhaglen sy'n ei dyblygu ei hun (yn ei gopïo ei hun) yn awtomatig; fel rheol mae wedi cael ei chreu i achosi difrod.

Ffederasiwn yn erbyn Dwyn Meddalwedd (*Federation Against Software Theft*) Corff gwrth-ladrad sy'n gwarchod gwaith cyhoeddwyr meddalwedd.

Ffefrynnau (*Favourites*) Mannau storio lle gall Lleolydd Adnoddau Unffurf (*URL: Uniform resource locator*) gwefan (h.y. y cyfeiriad gwe) gael ei storio fel y gellir ei gyrchu yn nes ymlaen drwy gyswllt.

Ffeil wrth gefn (*Backup file*) Copi o ffeil sy'n cael ei ddefnyddio os caiff y ffeil wreiddiol ei llygru (ei difrodi).

Ffenestri, Eiconau, Dewislenni, Dyfeisiau Pwyntio (*WIMP: Windows Icons Menus Pointing devices*) Defnyddio rhyngwyneb defnyddiwr graffigol (RhDG/*GUI: Graphical user interface*) i reoli rhaglenni yn hytrach na theipio gorchmynion yn y llinell orchymyn.

GIGO gweler Sbwriel i mewn sbwriel allan.

Grwpiau (*Groups*) Rhestri o bobl a'u cyfeiriadau e-bost.

Gwall trawsosod (*Transposition error*) Camgymeriad sy'n cael ei wneud pan gaiff nodau eu cyfnewid fel eu bod yn y drefn anghywir.

Gwall trawsysgrifiol (*Transcription error*) Camgymeriad sy'n cael ei wneud wrth deipio data i mewn i gyfrifiadur gan ddefnyddio dogfen yn ffynhonnell data.

Gwasanaeth (*Utility*) Rhan o'r meddalwedd systemau sy'n cyflawni tasg benodol.

Gwe-gam (*Webcam*) Camera fideo bach a ddefnyddir fel dyfais fewnbynnu i anfon delwedd symudol dros fewnrwyd neu'r Rhyngrwyd.

Gweinydd dirprwyol (*Proxy server*) Gweinydd, a all fod yn galedwedd neu'n feddalwedd, sy'n derbyn ceisiadau gan ddefnyddwyr i gyrchu gweinyddion eraill ac sydd naill ai'n eu hanfon ymlaen i'r gweinyddion eraill neu'n gwahardd mynediad i'r gweinyddion.

Gweithredu (*Implementation*) Y broses o gynhyrchu fersiwn sy'n gweithio o ddatrysiad i broblem a osodwyd gan gleient.

Gwerthuso (*Evaluation*) Y weithred o adolygu'r hyn a gyflawnwyd, sut y cafodd ei gyflawni a pha mor dda mae'r datrysiad yn gweithio.

Gwe-rwydo (*Phishing*) Twyllo pobl i ddatgelu eu manylion bancio neu gerdyn credyd.

Gwireddu (*Verification*) Gwirio bod y data sy'n cael eu mewnbynnu i system TGCh yn cyfateb yn union i ffynhonnell y data.

Gwiriad amrediad (*Range check*) Techneg dilysu data sy'n gwirio bod y data a fewnbynnir i'r cyfrifiadur o fewn amrediad penodol.

Gwiriad fformat (*Format check*) Gwiriad sy'n cael ei wneud ar godau i sicrhau eu bod yn cydymffurfio â'r cyfuniadau cywir o nodau.

Gwiriad hyd (*Length check*) Gwiriad i sicrhau bod gan y data sy'n cael eu mewnbynnu y nifer cywir o nodau.

Gwiriad math data (*Data type check*) Gwiriad i sicrhau bod y data sy'n cael eu mewnbynnu o'r un math â'r math data a bennwyd ar gyfer y maes.

Gwiriad presenoldeb (*Presence check*) Gwiriad i sicrhau bod data wedi'u mewnbynnu i faes.

Gwiriadau dilysu (*Validation checks*) Gwiriadau y mae datblygwr datrysiad yn eu creu, gan ddefnyddio'r meddalwedd, er mwyn cyfyngu ar y data y gall defnyddiwr eu mewnbynnu gyda'r nod o leihau gwallau.

Gwirydd gramadeg (*Grammar checker*) Defnyddir hwn i wirio'r gramadeg mewn brawddegau ac i dynnu sylw at broblemau ac awgrymu ffurfiau eraill.

Gwirydd sillafu (*Spellchecker*) Cyfleuster a gynigir gan feddalwedd sy'n cynnwys geiriadur y gallwch wirio pob gair a deipiwch yn ei erbyn.

Gwybodaeth (*Information*) Allbwn o system TGCh neu ddata sydd wedi cael eu prosesu ac sy'n rhoi gwybodaeth i ni.

Gyrrwr (*Driver*) Rhaglen fer wedi'i hysgrifennu'n arbennig sy'n deall sut mae'r ddyfais mae'n ei rheoli/ei gweithredu yn gweithio. Mae angen gyrwyr i ganiatáu i'r meddalwedd systemau neu feddalwedd rhaglenni ddefnyddio'r ddyfais gysylltiedig yn briodol.

Gyrrwr argraffydd (*Printer driver*) Meddalwedd sy'n trawsnewid gorchmynion o'r meddalwedd systemau neu feddalwedd rhaglenni i ffurf y gall argraffydd penodol ei deall.

Hacio (*Hacking*) Y broses o geisio torri i mewn i system gyfrifiadurol ddiogel.

Haciwr (*Hacker*) Rhywun sy'n ceisio neu'n llwyddo i dorri i mewn i system TGCh.

Hawliau mynediad (*Access rights*) Cyfyngu mynediad defnyddwyr i'r ffeiliau sydd eu hangen arnynt i wneud eu gwaith.

Hysbysu (*Notification*) Y broses o roi gwybod i Swyddfa'r Comisiynydd Gwybodaeth fod corff yn storio ac yn prosesu data personol.

ISP gweler Darparwr Gwasanaeth Rhyngrwyd.

Log defnyddiwr (*User log*) Cofnod o fewngofnodi llwyddiannus ac aflwyddiannus a hefyd o'r adnoddau a ddefnyddir gan y defnyddwyr hynny sydd â hawl i gyrchu adnoddau rhwydwaith.

Lladrata (*Piracy*) Y broses o gopïo meddalwedd yn anghyfreithlon.

Llechen graffeg (*Graphics tablet*) Dyfais fewnbynnu sy'n defnyddio 'llechen' (tabled) fawr sy'n cynnwys llawer o siapiau a gorchmynion a all gael eu dewis gan y defnyddiwr drwy symud cyrchwr a chlicio. Mae'n rhoi'r barrau offer ar y llechen yn hytrach na gwneud y sgrin yn anniben wrth ddefnyddio meddalwedd *CAD* i wneud lluniadau technegol mawr.

Lled band (*Bandwidth*) Mesur o faint o ddata a all gael eu trosglwyddo gan ddefnyddio cyfrwng trosglwyddo data.

Lleolydd Adnoddau Unffurf (*URL: Uniform resource locator*) Y cyfeiriad gwe a ddefnyddir i leoli tudalen we.

Llusgo a gollwng (*Drag and drop*) Mae'n caniatáu i chi ddewis gwrthrychau (eiconau, ffolderi, ffeiliau, ac ati) a'u llusgo fel y gallwch eu trin mewn rhyw ffordd, fel eu llusgo i'r bin ailgylchu i gael gwared â nhw, ychwanegu ffeil at ffolder, copïo ffeiliau i ffolder, ac ati.

Llwybrydd (*Router*) Dyfais galedwedd sy'n gallu gwneud penderfyniadau am y llwybr y dylai pecyn unigol o ddata ei gymryd fel y bydd yn cyrraedd yn yr amser byrraf posibl.

Macros (*Macros*) Defnyddir macros i recordio cyfres o drawiadau bysell. Er enghraifft, gallech greu macro sy'n ychwanegu eich enw a'ch cyfeiriad at ben tudalen drwy bwyso un fysell yn unig neu glicio ar y llygoden.

Man poeth (*Hotspot*) Ardal lle gellir cyrchu'r Rhyngrwyd yn ddiwifr.

Map meddwl (*Mind map*) Diagram hierarchaidd gyda phrif syniad (neu ddelwedd) yng nghanol y map wedi'i amgylchynu gan ganghennau sy'n ymestyn o'r syniad canolog.

Meddalwedd (*Software*) Rhaglenni sy'n cyflenwi cyfarwyddiadau i'r caledwedd.

Meddalwedd generig (*Generic software*) Rhaglen sy'n briodol i amrywiaeth eang o dasgau ac y gellir ei ddefnyddio mewn llawer o feysydd gwaith.

Meddalwedd integredig (*Integrated software*) Pecyn rhaglen sy'n cynnwys meddalwedd ar gyfer sawl rhaglen wahanol. Bydd dau neu ragor o becynnau rhaglen mewn meddalwedd integredig bob amser.

Meddalwedd oddi ar y silff (*Off-the-shelf software*) Meddalwedd sydd heb gael ei ddatblygu at ddefnydd penodol.

Meddalwedd pecyn (*Package software*) Swp o ffeiliau sy'n angenrheidiol i gael rhaglen i weithio ynghyd â dogfennaeth i helpu'r defnyddiwr i ddechrau'r rhaglen.

Meddalwedd penodol (*Specific software*) Meddalwedd sy'n cyflawni un swyddogaeth yn unig.

Meddalwedd rheoli ffeiliau (*File management software*) Rhan o'r meddalwedd systemau a ddefnyddir i greu ffolderi, copïo ffolderi/ffeiliau, ailenwi ffolderi/ffeiliau, dileu ffolderi/ffeiliau, symud ffolderi/ffeiliau, ac ati.

Meddalwedd rhwydweithio (*Networking software*) Meddalwedd systemau sy'n caniatáu i gyfrifiaduron sydd wedi'u cysylltu â'i gilydd weithredu fel rhwydwaith.

Meddalwedd systemau (*Systems software*) Unrhyw feddalwedd cyfrifiadurol sy'n rheoli'r caledwedd ac felly'n caniatáu i'r meddalwedd rhaglenni wneud gwaith defnyddiol. Mae meddalwedd systemau'n cynnwys grŵp o raglenni.

Mewnbynnu (*Input*) Y weithred o roi data i mewn i system TGCh.

Mewngofnodi (*Log-in*) Dweud wrth y rhwydwaith pwy ydych chi er mwyn cael mynediad.

Mewnrwyd (*Intranet*) Rhwydwaith preifat o fewn corff sy'n defnyddio technoleg Rhyngrwyd.

MIDI *gweler* Rhyngwyneb Digidol Offeryn Cerdd.

Modem deialu (*Dialup modem*) Dyfais sy'n trawsnewid signalau digidol yn gyfres o seiniau sydd wedyn yn cael eu trosglwyddo ar hyd llinell ffôn. Caiff y signal sain ei drawsnewid yn ôl yn signal digidol ym mhen arall y wifren. Mae'n darparu cysylltiad araf â'r Rhyngrwyd.

MP3 (*MP3*) Fformat ar gyfer ffeiliau cerddoriaeth sy'n defnyddio cywasgu i leihau maint y ffeil yn sylweddol. Dyma pam mae'r fformat hwn mor boblogaidd ar gyfer dyfeisiau chwarae cerddoriaeth gludadwy fel yr iPod.

Mur gwarchod (*Firewall*) Caledwedd a/neu feddalwedd sy'n gweithio mewn rhwydwaith i atal cyfathrebu nad yw'n cael ei ganiatáu o un rhwydwaith i rwydwaith arall.

Mwydyn (*Worm*) Rhaglen sy'n parhau i'w dyblygu ei hun yn awtomatig ac wrth iddi wneud hyn mae'n cymryd mwy a mwy o le ar y disg a hefyd yn defnyddio cyfran fwy o adnoddau'r system ar gyfer pob copi.

Mynegiad dilysu/rheol ddilysu (*Validation expression/rule*) Gorchymyn y mae'n rhaid i ddatblygwr ei deipio er mwyn gosod y dilysiad ar gyfer maes neu gell benodol.

Nam (*Bug*) Camgymeriad neu wall mewn rhaglen.

Neges ddilysu (*Validation message*) Neges sy'n ymddangos os torrir y rheol ddilysu.

Neges fewnbynnu (*Input message*) Neges sy'n rhoi cyngor i'r defnyddiwr ar y math o ddata i'w mewnbynnu pan gaiff maes neu gell ei dewis.

Nod tudalen (*Bookmark*) Man storio lle gall Lleolydd Adnoddau Unffurf (*URL: Uniform resource locator*) gwefan (h.y. y cyfeiriad gwe) gael ei storio fel y gellir ei gyrchu yn nes ymlaen drwy gyswllt.

Peiriant chwilio (*Search engine*) Rhaglen sy'n chwilio am y wybodaeth sydd ei hangen arnoch ar y Rhyngrwyd.

Pennyn (*Header*) Testun ar ben tudalen.

Perifferolyn (*Peripheral*) Dyfais sydd wedi'i chysylltu â'r uned brosesu ganolog (*CPU*) ac o dan ei rheolaeth.

Perthynas (*Relationship*) Y ffordd y mae tablau wedi'u cysylltu â'i gilydd. Gall perthnasoedd fod yn un-i-un, un-i-lawer neu lawer-i-un.

Picsel (*Pixel*) Y dot lleiaf o olau ar sgrin cyfrifiadur all gael ei reoli'n unigol.

Plotydd graff (*Graph plotter*) Dyfais sy'n lluniadu drwy symud pen. Mae'n ddefnyddiol ar gyfer gwneud lluniadau wrth raddfa ac fe'i defnyddir yn bennaf gyda phecynnau cynllunio drwy gymorth cyfrifiadur (*CAD*).

Podledu (*Podcasting*) Creu a chyhoeddi darllediad radio digidol gan ddefnyddio microffon, cyfrifiadur a meddalwedd golygu awdio. Caiff y ffeil a gynhyrchir ei chadw mewn fformat MP3 ac yna ei llwytho i fyny i weinydd Rhyngrwyd. Yna gellir defnyddio cyfleuster o'r enw RSS i'w llwytho i lawr i chwaraeydd MP3 megis iPod.

Polisi ar ddefnydd derbyniol (*Acceptable use policy*) Dogfen sy'n egluro i'r holl weithwyr neu ddefnyddwyr beth sy'n dderbyniol ac yn annerbyniol wrth ddefnyddio systemau TGCh.

Porwr gwe (*Web browser*) Y rhaglen meddalwedd a ddefnyddiwch i gyrchu'r Rhyngrwyd. Un enghraifft yw Microsoft Internet Explorer.

Postgyfuno (*Mail merge*) Cyfuno rhestr o enwau a chyfeiriadau gyda llythyr safonol fel bod cyfres o lythyrau'n cael ei chynhyrchu a phob llythyr wedi'i gyfeirio at berson gwahanol.

Preifatrwydd (*Privacy*) Yr hawl i gadw agweddau ar eich bywyd yn breifat.

Proses (*Process*) Unrhyw weithrediad sy'n troi data yn wybodaeth.

Prosesu (*Processing*) Gwneud cyfrifiadau neu drefnu data yn drefn ystyrlon.

Prosesu amser real (*Real-time processing*) Caiff y data mewnbwn eu prosesu ar unwaith wrth iddynt gyrraedd. Caiff y canlyniadau effaith uniongyrchol ar y set nesaf o ddata sydd ar gael.

Prosesu trafodion (*Transaction processing*) Prosesu pob trafodyn wrth iddo godi.

Protocol (*Protocol*) Set o safonau sy'n ei gwneud hi'n bosibl i drosglwyddo data rhwng y cyfrifiaduron ar rwydwaith.

Protocol trosglwyddo ffeiliau (*FTP: File transfer protocol*) Protocol (ffordd o wneud pethau) safonol ar gyfer y Rhyngrwyd sy'n darparu ffordd syml o drosglwyddo ffeiliau rhwng cyfrifiaduron gan ddefnyddio'r Rhyngrwyd. Caiff ei ddefnyddio i drawsyrru unrhyw fath o ffeil (rhaglenni cyfrifiadurol, ffeiliau testun, graffigau, ac ati) drwy broses sy'n casglu'r data yn becynnau.

Pwyntydd (*Pointer*) Dyma'r saeth fach sy'n ymddangos wrth ddefnyddio Windows.

RAID *gweler* Arae ddiangen o ddisgiau rhad

RAM *gweler* Cof hapgyrch

ROM *gweler* Cof darllen yn unig

RSI *gweler* Anaf straen ailadroddus.

Rhaglen wasanaethu (*Utility program*) Meddalwedd sy'n helpu'r defnyddiwr i gyflawni tasgau fel gwirio am firysau, cywasgu ffeiliau, ac ati.

Rhaglennydd (*Programmer*) Rhywun sy'n ysgrifennu rhaglenni cyfrifiadurol.

Rhagolwg argraffu (*Print preview*) Nodwedd sydd gan y rhan fwyaf o feddalwedd a ddefnyddir i gynhyrchu dogfennau. Mae'n caniatáu i ddefnyddwyr gael rhagolwg o dudalennau dogfen i weld sut yn union y byddant yn cael eu hargraffu. Yna gellir cywiro'r dogfennau os oes angen.

Rheoliadau Iechyd a Diogelwch (*Cyfarpar Sgrin Arddangos*) 1992 (*Health & Safety [Display Screen Equipment] Regulations 1992*) Rheoliadau sy'n gorfodi cyflogwyr i gymryd mesurau i ddiogelu iechyd a diogelwch staff sy'n defnyddio cyfarpar TGCh.

Rheolydd data (*Data controller*) Y person mewn corff sy'n gyfrifol am reoli'r ffordd y caiff data personol eu prosesu.

Rhwydwaith (*Network*) Grŵp o ddyfeisiau TGCh

(cyfrifiaduron, argraffyddion, sganwyr, ac ati) sy'n gallu cyfathrebu â'i gilydd.

Rhwyll (*Mesh*) Math o dopoleg rhwydwaith lle mae llawer o lwybrau y gall data eu cymryd rhwng y cyfrifiaduron.

Rhyngrwyd (*Internet*) Grŵp enfawr o rwydweithiau sydd wedi'u huno â'i gilydd.

Rhyngweithiol (*Interactive*) Lle mae deialog cyson rhwng y defnyddiwr a'r cyfrifiadur.

Rhyngwyneb (*Interface*) Y man lle mae dau wrthrych yn cyfarfod. Mewn TGCh, mae hyn fel rheol rhwng dyfais fel cyfrifiadur, argraffydd, sganiwr, ac ati, a bod dynol.

Rhyngwyneb defnyddiwr graffigol (*RhDG/GUI: Graphical User Interface*) Rhyngwyneb sy'n caniatáu i ddefnyddwyr gyfathrebu â chyfarpar TGCh drwy ddefnyddio eiconau a chwymplenni.

Rhyngwyneb Digidol Offeryn Cerdd (*MIDI: Musical Instrument Digital Interface*) Defnyddir hwn yn bennaf ar gyfer cyfathrebu rhwng allweddellau electronig, syntheseiddwyr a chyfrifiaduron. Mae'r ffeiliau wedi'u cywasgu ac maen nhw'n weddol fach.

Rhyngwyneb iaith naturiol (*Natural language interface*) Rhyngwyneb sy'n caniatáu i'r defnyddiwr ddefnyddio iaith ysgrifenedig neu lafar (e.e. Cymraeg) i ryngweithio yn hytrach nag iaith a gorchmynion cyfrifiadurol.

Sbam (*Spam*) E-hebiaeth nad ydych wedi gofyn amdano (h.y. e-hebiaeth gan bobl nad ydych yn eu hadnabod a anfonir at bawb gan obeithio y bydd canran bach ohonynt yn prynu'r nwyddau neu wasanaethau sy'n cael eu cynnig).

Sbwriel i mewn sbwriel allan (*GIGO: Garbage in garbage out*) Os rhoddwch sbwriel i mewn i gyfrifiadur fe gewch sbwriel allan.

Seren (*Star*) Math o dopoleg rhwydwaith lle mae'r cyfrifiaduron wedi'u cysylltu ag un man cysylltu canolog.

Sgam (*Scam*) Sefydlu cwmni ffug gyda gwefan ffug ac yna dianc gyda'r arian sy'n cael ei dalu gan gwsmeriaid.

Sganiwr (*Scanner*) Dyfais fewnbynnu y gellir ei defnyddio i gipio delwedd. Mae'n ddefnyddiol ar gyfer digido hen ffotograffau, dogfennau papur neu luniau mewn llyfrau.

Sgiliau personol (*Personal skills*) Y sgiliau hynny sydd gan unigolyn a all gael eu trosglwyddo i unrhyw swydd neu dasg.

Sgiliau technegol (*Technical skills*) Y sgiliau hynny sy'n angenrheidiol er mwyn cwblhau tasg benodol mewn TGCh.

Sgrin gyffwrdd (*Touch screen*) Sgrin sy'n caniatáu i rywun wneud dewisiadau drwy gyffwrdd â'r sgrin.

Storfa eilaidd (neu wrth gefn) (*Secondary [or backup] storage*) Storfa y tu allan i'r cyfrifiadur.

Storfa gynradd (*Primary storage*) Y sglodion y tu mewn i'r cyfrifiadur sy'n storio data.

Stribed magnetig (*Magnetic strip*) Caiff data eu hamgodio yn y stribed magnetig a phan gaiff y cerdyn ei roi drwy ddarllenydd stribed magnetig, defnyddir y data o'r cerdyn i gofnodi'r trafodyn.

System rheoli cronfeydd data perthynol (*RDMS: Relational database management system*) System cronfa ddata lle caiff y data eu cadw mewn tablau y sefydlwyd perthnasoedd rhyngddynt. Defnyddir y meddalwedd i drefnu a dal y data a hefyd i'w hechdynnu a'u trin ar ôl eu storio.

System TGCh (*ICT system*) Caledwedd a meddalwedd yn gweithio gyda'i gilydd gyda phobl a gweithdrefnau i gyflawni tasgau.

System weithredu (*Operating system*) Meddalwedd sy'n rheoli'r caledwedd a hefyd yn rhedeg eich rhaglenni. Mae'r system weithredu'n rheoli gweithrediadau fel mewnbynnu, allbynnu, ymyriadau, storio, a rheoli ffeiliau.

Teipio data ddwywaith (*Double entry of data*) Mae dau berson yn defnyddio'r un ffynhonnell data i deipio manylion i system TGCh a dim ond os yw'r ddwy set o ddata'r un fath y cânt eu derbyn ar gyfer prosesu.

Telathrebu (*Telecommunications*) Y maes technolegol sy'n ymwneud â chyfathrebu o bell (e.e. ffôn, radio, cebl, ac ati).

Telegymudo (*Telecommuting*) Cyflawni tasgau'n ymwneud â swydd drwy ddefnyddio telathrebiadau i anfon data i a derbyn data gan swyddfa ganolog heb orfod fod yno'n gorfforol.

Teleweithio (*Teleworking*) Defnyddio cyfathrebiadau i arbed taith. Er enghraifft, gallech arbed taith dramor drwy ddefnyddio fideo-gynadledda.

Templedi (*Templates*) Defnyddir templedi i bennu strwythur dogfen megis ffontiau, gosodiad y dudalen, fformatio ac arddulliau.

Testun data (*Data subject*) Yr unigolyn byw y mae'r wybodaeth bersonol yn ymwneud ag ef/hi.

Topoleg (*Topology*) Y ffordd mae rhwydwaith penodol wedi'i drefnu, er enghraifft, cylch, seren, bws.

Trafodyn/trafodion (*Transaction/transactions*) Darn/darnau o fusnes, e.e. archeb, pryniant, dychweliad, danfoniad, trosglwyddiad arian, ac ati.

Trefnu (*Sorting*) Rhoi data mewn trefn esgynnol neu ddisgynnol.

Troedyn (*Footer*) Testun a roddir ar waelod tudalen.

Trojanau (*Trojans*) Llinellau o god cyfrifiadurol sy'n cael eu storio yn eich cyfrifiadur heb yn wybod i chi.

Trosedd (*Crime*) Gweithred anghyfreithlon.

Trosedd seiber (*Cybercrime*) Trosedd a gyflawnir drwy ddefnyddio systemau TGCh yn bennaf.

Trwydded meddalwedd (*Software licence*) Dogfen (digidol neu ar bapur) sy'n nodi o dan ba delerau y gellir defnyddio'r meddalwedd. Bydd yn cyfeirio at nifer y cyfrifiaduron y gellir defnyddio'r meddalwedd arnynt ar yr un pryd.

Tudalen we (*Webpage*) Un ddogfen ar y We Fyd-Eang.

Thesawrws (*Thesaurus*) Rhan o brosesydd geiriau sy'n caniatáu i chi ddewis gair a gweld y cyfystyron (h.y. geiriau ag ystyron tebyg).

Uned brosesu ganolog (*CPU: Central processing unit*) Ymennydd y cyfrifiadur. Mae'n storio ac yn prosesu data. Mae tair rhan iddi: yr uned rifyddeg-resymeg (*ALU*), yr uned reoli, a'r cof.

URL gweler Lleolydd Adnoddau Unffurf.

WAV Fformat ffeil a ddefnyddir gyda Windows ar gyfer storio seiniau. Nid yw ffeiliau yn y fformat hwn wedi'u cywasgu'n sylweddol.

Wi-Fi Nod masnach sy'n ardystio bod cynhyrchion yn bodloni safonau ar gyfer trawsyrru data dros rwydweithiau diwifr.

WIMP gweler Ffenestri, Eiconau, Dewislenni, Dyfeisiau Pwyntio.

Y We Fyd-Eang (*World Wide Web*) Ffordd o gyrchu'r wybodaeth sydd ar y Rhyngrwyd. Mae hi'n fodel rhannu gwybodaeth sydd wedi'i adeiladu ar ben y Rhyngrwyd.

Ymlaen (*Forward*) Os derbyniwch e-bost yr ydych chi'n meddwl y dylai pobl eraill ei weld, gallwch ei anfon ymlaen atynt.

Ysbïwedd (*Spyware*) Meddalwedd sy'n casglu gwybodaeth am ddefnyddwyr cyfrifiaduron sydd wedi'u cysylltu â'r Rhyngrwyd heb ganiatâd y defnyddwyr.

Mynegai

Cydnabyddiaeth

Hoffai Folens Limited ddiolch i'r canlynol am roi caniatâd i ddefnyddio deunydd sydd dan hawlfraint.

t.2, © petrafler/Fotolia; t.3, © drx/Fotolia; t.4, © Buttons Inc/Fotolia; t.5, © aphaspirit/Fotolia; t.7, © Nathalie Dulex/Fotolia; t.7, dechefbloke/Fotolia; t.7, © Dominique Luzy/Fotolia; t.7, © Com Evolution/Fotolia; t.7, © jeff gynane/Fotolia; t.7, © Galyna Andrushko/Fotolia; t.7, © 2005 Sean MacLeay/Fotolia; t.9, © Helder Almeida; t.10, Glasbergen; t.20, © Sean Gladwell/Fotolia; t.21, © treenabeena/Fotolia; t.22, © Guy Erwood/Fotolia; t.23, © Aycan Zivana/Fotolia; t.24, © Onidji/Fotolia; t.25, © ktsdesign/Fotolia; t.25, © cbpix/Fotolia; t.25, © Sebastian Kaulitzki/Fotolia; t.26, © Petr Kratochvil/Fotolia; t.26, © e-pyton/Fotolia; t.27, © Andres Rodriguez/Fotolia; t.28, Glasbergen; t.40, © dotshock/Fotolia; t.41, © Ramona Heim/Fotolia; t.49, Glasbergen; t.50, © tetrex/Fotolia; t.51, Sony; t.52, © godfer/Fotolia; t.53, © Stephen Coburn/Fotolia; t.53, Glasbergen; t.54, © Konstantin Shevtsov/Fotolia; t.55, © Stephen Finn/Fotolia; t.55, © SFL Travel/Alamy; t.62, © ewerest/Fotolia; t.62, © Robert Paul Van Beets/Fotolia; t.62, © 2006 Ron Hudson/Fotolia; t.62, © NorthShoreSurfPhotos/Fotolia; t.62, © 2006 James Steidl, James Group Studios, Inc/Fotolia; t.62, © pattie/Fotolia; t.64, © Christopher Dodge/Fotolia; t.64, © alphaspirit/Fotolia; t.64, © Sean Gladwell/Fotolia;

t.65, © Fotofolia VIII/Fotolia; t.65, Lindy Electronics; t.65, © Secret Side/Fotolia; t.65, © robynmac/Fotolia; t.66, Glasbergen; t.67, Glasbergen; t.68, Greengate Publishing Services; t.68, © Anette Linnea Rasmussen/Fotofolia; t.70, © Clivia/Fotolia; t.71, © Ana Vasileva/Fotolia; t.71, © Nikolai Sorokin/Fotolia; t.71, © Paul Lockwood/Fotolia; t.71, © Andy Mac/Fotolia; t.71, © Yong Hian Lim/Fotolia; t.72, © Dasha Kalashnikova/Fotolia; t.72, © James Blacklock/Fotolia; t.74, Glasbergen; t.75, Glasbergen; t.77, © Cyrus Cornell/Fotofolia; t.77, © Warren Rosenburg/Fotolia; t.94, © Amy Walters/Fotolia; t.111, Glasbergen; t.119, © Kenishirotie/Shutterstock; t.120, © Blend Images/SuperStock; t.120, © Emin Ozkan/Fotolia; t.120, © pressmaster/Fotolia; t.120, © Mark Aplet/Fotolia; t.130, © Roland/Fotolia; t.135, © Ronald Hudson/Fotolia; t.136, © Peter Galbraith/Fotolia; t.141, © endostock/Fotolia; t.141, © Microtechware/Fotolia; t.146, Glasbergen; t.146, © George Dolgikh/Fotolia; t.155, © Dara/Fotolia; t.161, Glasbergen; t.163, © demarco; t.163, © tetrex/Fotolia.

Atgynhyrchir sgrinluniau o gynhyrchion Microsoft gyda chaniatâd Corfforaeth Microsoft.

Gwnaed pob ymdrech i gysylltu â deiliaid hawlfraint y deunyddiau a ddefnyddiwyd yn y cyhoeddiad hwn. Os oes unrhyw ddeiliad hawlfraint nad ydym wedi'i gydnabod, byddem yn falch o wneud unrhyw drefniadau angenrheidiol.